Third International Conference on Sustainable Infrastructure and Built Environment (SIBE 2017)

MATEC Web of Conferences Volume 147 (2017)

Bandung, Indonesia
26 - 27 September 2017

Editors:

B. Sugeng
D. Roosmini
K. Pribadi
I.K. Hadihardaja

ISBN: 978-1-5108-5614-1

Printed from e-media with permission by:

Curran Associates, Inc.
57 Morehouse Lane
Red Hook, NY 12571

Curran Associates Inc. proceedings.com

Some format issues inherent in the e-media version may also appear in this print version.

Printed by Curran Associates, Inc. (2018)

For additional information, please contact EDP Sciences – Web of Conferences at the address below.

EDP Sciences – Web of Conferences
17, Avenue du Hoggar
Parc d'Activité de Courtabœuf
BP 112
F-91944 Les Ulis Cedex A
France

Phone: +33 (0) 1 69 18 75 75
Fax: +33 (0) 1 69 28 84 91

contact-edps@webofconferences.org

Additional copies of this publication are available from:

Curran Associates, Inc.
57 Morehouse Lane
Red Hook, NY 12571 USA
Phone: 845-758-0400
Fax: 845-758-2633
Email: curran@proceedings.com
Web: www.proceedings.com

TABLE OF CONTENTS

Greetings!

The Faculty of Civil and Environmental Engineering of ITB in collaboration with the National Taiwan University and the Hokkaido University will host The Third International Conference on Sustainable Infrasructure and Built Environment (SIBE-2017). With 'Sustainable Infrastucture and Built Environment: Past, Present, and Future' as its theme, this event will be held on September 26-27, 2017 in Bandung, Indonesia.

The objectives of SIBE-2017 are:

- To provide a platform for exchange of ideas and information among academics, researchers, consultants, engineers, manufacturers and post graduate scholars in civil, ocean, and environmental engineering;
- To discuss and evaluate the past and present approaches, innovative technologies, policies, and new directions in infrastructure development, pollution prevention, and eco-friendly technologies adapted to developing countries;
- To promote cooperation and networking amongst practitioners and researchers involved in addressing infrastructure and built environment issues.

The first SIBE was successfully accomplished its objectives on November 2-3, 2009 in Bandung, Indonesia. The Second SIBE was also a successful event accomplished on November 19-20, 2013.

We herewith invite industries, organizations, research institutions, and government agencies that are related to the development of sustainable infrastructure and built environment to participate in this conference. With your support, this event will significantly contribute to the acceleration of infrastructure provision, to facilitate economic activities, and to offer a better quality of life.

Bandung, January 2017
Chair of SIBE-2017 Organizing Committee,

Prof. Ir. Ricky Lukman Tawekal, MSE., Ph.D

PREFACE

International Conference on Sustainable Infrastructure and Built Environment (SIBE) is held in every 4 years by the Faculty of Civil and Environmental Engineering of ITB. The First SIBE conference was held in November 2009. The Second SIBE Conference brought the same spirit of the First SIBE with the topic of "Accelerating Sustainable Infrastructure Development – Challenges, Opportunities, and Policy Direction". This year, we brought the topic of "Past, Present, and Future", which encourages us to learn from the past and present experiences regarding infrastructure and built environment to achieve a better future. The Third International Conference on Sustainable Infrastructure and Built Environment (SIBE-2017) was held on September 26-27, 2017 in Bandung, Indonesia.

The objectives of SIBE conference are:
• To provide a platform for exchange of ideas and information among academics, researchers, consultants, engineers, manufacturers and postgraduate scholars in civil, ocean, and environmental engineering;
• To discuss and evaluate the past and present approaches, innovative technologies, policies, and new directions in infrastructure development, pollution prevention, and eco-friendly technologies adapted to developing countries;
• To promote cooperation and networking amongst practitioners and researchers involved in addressing infrastructure and built environment issues.

The conference discussed 8 sub-topics, which consists of Structures and Materials, Transportation System and Engineering, Water Resources Engineering and Management, Water and Waste Engineering and Management, Geotechnical Engineering, Ocean and Maritime Engineering, Construction Management, and Environmental Protection and Management. Each sub-topic was discussed in a parallel session during the conference.

SIBE2017
Past, Present, and Future
BIM Model

September 26 – 27, 2017
Bandung,

APPROACH OF RBUI METHODOLOGY
SIBE
2017

SIBE
Hydrodynamics Modeling of Kung Krabaen Lagoon, Chanthaburi Province, Thailand
26 September 2017
SIBE
2017

SIBE
2017
Past, Present,
September 26-27, 2017
Institut Teknologi Bandung
Indonesia
Small Scale Port Development

SIBE2017
The Third International Conference on
Sustainable Infrastructure and Built Environment
Past, Present, and Future
SIBE
TOYOTA
pgn
IPC
BNI

SIBE2017
Infrastructure Development Framework
2015-2019
Infrastructure
to Support Leading Sectors
Urban Transport
Intermodal
Avoid
Shift
Improve

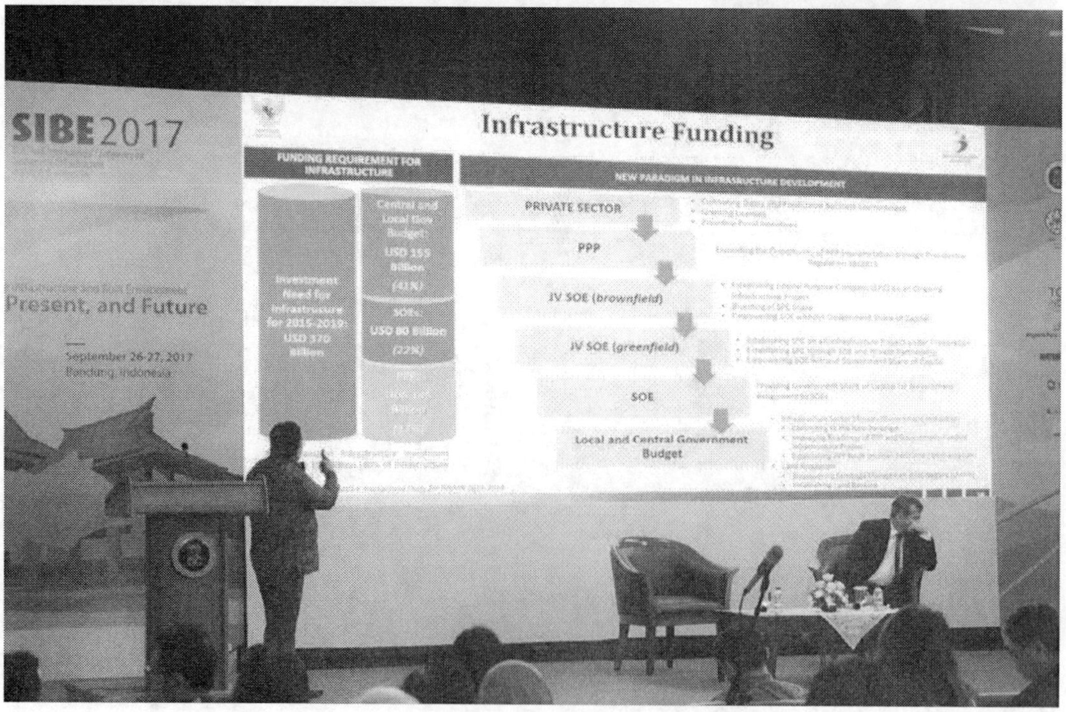
SIBE2017
Present, and Future
September 26-27, 2017
Bandung, Indonesia
Infrastructure Funding
FUNDING REQUIREMENT FOR INFRASTRUCTURE
Investment Need for Infrastructure for 2015-2019: USD 370 Billion
Central and Local Gov Budget: USD 155 Billion (41%)
SOEs: USD 80 Billion (22%)
PRIVATE SECTOR
PPP
JV SOE (brownfield)
JV SOE (greenfield)
SOE
Local and Central Government Budget

SIBE2017
Past, Present, and Future
Milestones of infrastructure development
Jatiluhur dam
Jagorawi toll
Semanggi bridge

structure and Built Environment
esent, and Future
September 26-27, 2
Bandung, Indonesia

SIBE2017
The Third International Conference on
Sustainable Infrastructure
and Built Environment
Long-term Management of Bridge S
-RC Bridge Deck Slab Fatigue
Repair with Advanced Compo
Professor Takashi Matsumoto, Ph.D
Laboratory of Bridge and Structural Design En
Hokkaido University
Japan
HOKKAIDO

SIBE2017
The Third International Conference on
Sustainable Infrastructure
and Built Environment
September 26 – 27, 2017
Bandung, Indonesia
Organize
Faculty of Civil and
Institut Teknologi
In collaborati

Statement of Peer review

In submitting conference proceedings to MATEC Web of Conferences, I certify to the Publisher that I adhere to the **Policy on Publishing Integrity** of the journal in order to safeguard good scientific practice in publishing.

1. All articles have been subjected to peer review administered by the proceedings editors.
2. Reviews have been conducted by expert referees, who have been requested to provide unbiased and constructive comments aimed, whenever possible, at improving the work.
3. Proceedings editors have taken all reasonable steps to ensure the quality of the materials they publish and their decision to accept or reject a paper for publication has been based only on the merits of the work and the relevance to the journal.

Title, date and place of the conference

Title : The Third International Conference on Sustainable Infrastructure and Built Environment
Date : September 26-27, 2017
Place : Institut Teknologi Bandung - Indonesia

Proceedings editor(s):

Prof. Dr. Ir. Bambang Sugeng, D.E.A.

Date and editor's signature

edp sciences

MATEC Web of Conferences

Statement of Peer review

In submitting conference proceedings to MATEC Web of Conferences, I certify to the Publisher that I adhere to the **Policy on Publishing Integrity** of the journal in order to safeguard good scientific practice in publishing.

1. All articles have been subjected to peer review administered by the proceedings editors.
2. Reviews have been conducted by expert referees, who have been requested to provide unbiased and constructive comments aimed, whenever possible, at improving the work.
3. Proceedings editors have taken all reasonable steps to ensure the quality of the materials they publish and their decision to accept or reject a paper for publication has been based only on the merits of the work and the relevance to the journal.

Title, date and place of the conference

Title : The Third International Conference on Sustainable Infrastructure and Built Environment
Date : September 26-27, 2017
Place : Institut Teknologi Bandung - Indonesia

Proceedings editor(s):

Prof. Ir. Iwan K. Hadihardaja, M.Sc., Ph.D.

Date and editor's signature

19/12/17

edp sciences

MATEC Web of Conferences

Statement of Peer review

In submitting conference proceedings to MATEC Web of Conferences, I certify to the Publisher that I adhere to the **Policy on Publishing Integrity** of the journal in order to safeguard good scientific practice in publishing.

1. All articles have been subjected to peer review administered by the proceedings editors.
2. Reviews have been conducted by expert referees, who have been requested to provide unbiased and constructive comments aimed, whenever possible, at improving the work.
3. Proceedings editors have taken all reasonable steps to ensure the quality of the materials they publish and their decision to accept or reject a paper for publication has been based only on the merits of the work and the relevance to the journal.

Title, date and place of the conference

Title : The Third International Conference on Sustainable Infrastructure and Built Environment
Date : September 26-27, 2017
Place : Institut Teknologi Bandung - Indonesia

Proceedings editor(s):

Dr. Ir. Dwina Roosmini, MS.

Date and editor's signature

12/12/17

edp sciences

MATEC Web of Conferences

Statement of Peer review

In submitting conference proceedings to MATEC Web of Conferences, I certify to the Publisher that I adhere to the **Policy on Publishing Integrity** of the journal in order to safeguard good scientific practice in publishing.

1. All articles have been subjected to peer review administered by the proceedings editors.
2. Reviews have been conducted by expert referees, who have been requested to provide unbiased and constructive comments aimed, whenever possible, at improving the work.
3. Proceedings editors have taken all reasonable steps to ensure the quality of the materials they publish and their decision to accept or reject a paper for publication has been based only on the merits of the work and the relevance to the journal.

Title, date and place of the conference

Title : The Third International Conference on Sustainable Infrastructure and Built Environment
Date : September 26-27, 2017
Place : Institut Teknologi Bandung - Indonesia

Proceedings editor(s):

Prof. Dr. Ir. Krishna Pribadi

Date and editor's signature

14/12/17

Numerical Analysis Of Viscous Wall Dampers On Steel Frame

Effy Hidayaty[1,*], *Herlien* D. Setio[2], *Adang* Surahman[2] and *Muslinang* Moestopo[2]

[1]Student of Doctoral Program at Faculty of Civil and Environmental Engineering, Bandung Institute of Technology
[2]Lecturer of Doctoral Program at Faculty of Civil and Environmental Engineering, Bandung Institute of Technology

Abstract. The structure is designed to minimize component damaged, one of them is using a damping system. Viscous Wall Damper is one of the damping systems, using high viscosity liquids as dampers. The viscous wall damper is represented by an Exponential Maxwell Damper model. Much of the literature shows the ability of viscous wall damper by experimental studies, but few of them had discussed numerically. This article will present an analytical model of the viscous wall damper, add viscous wall damper element to an existing frame and numerical analysis. The analysis was performed by a computer application on a steel space frame that excited a combination of three earthquake types. The influence of Viscous Wall Damper showed that a significant decrement of displacement u1 at the structure with Viscous Wall Damper in X & Y direction (50,78%) and with Viscous Wall Damper in Y direction (23,96%). The decrement displacement happened in all structure. At the end of the analysis shows the reduced of the structure periods, the structure response (displacement, velocity, and acceleration). All these results conclude that the structural components damaged due to loads can be greatly reduced.

1 Introduction

The basic principle of conventional earthquake resistant design that has been applied is intended to ensure an acceptable safety level while avoiding catastrophic failure and loss of life. Therefore, many types of research have been conducted into developing innovative earthquake – resistant systems in order to raise seismic performance level while keeping construction costs reasonable by using a damping system [1]. The damping system will dissipate most of the earthquake energy and less of them will transfer to the component [2]. Viscous Wall Damper (VWD) is one of the damping systems, using high viscosity liquid as dampers. The wall damper works on all of the frames, consisting of an inner plate/vane attached to the upper floor, the outer plates mounted on the lower floor, and the viscous fluid in the small gap in between. The outer plate is a kind of "house" for the inner plates. The viscous fluid is a non-toxic, odorless, transparent fluid with a viscosity of 90,000 poises [3].

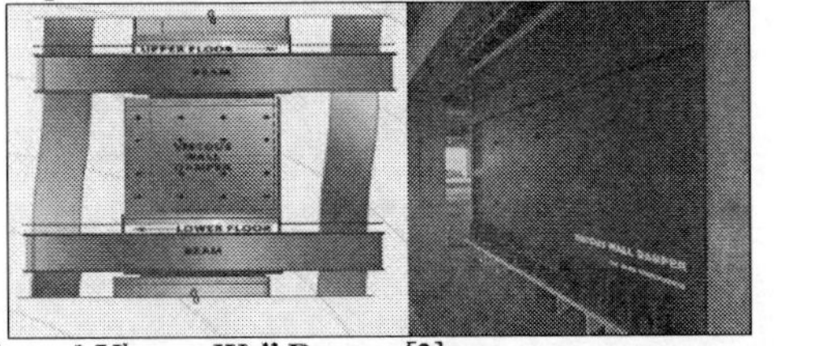

Figure 1 Viscous Wall Damper [3]

During seismic, or in strong winds excitation, the relative floor movement causes the vane move through the viscous fluid and produced a force against the inter-story drift. Thus, overall, structural deformation is reduced.

This viscous damper is a relatively new invention. Benefits of Viscous Wall Dampers are cost savings, which it reduces the weight of structural steel required and lower the total cost of the building; architectural flexibility, which the compact, rectangular shape of the damper is easier to incorporate and gives greater architectural freedom than diagonal braces or dampers; better performance, by reducing inter-story drift, the viscous wall damper provide superior seismic protection to the structure and its contents; maintenance free, the damper have no moving parts or seals, are under no internal pressure and do not require maintenance; retrofits, ideal for retrofits; easy to install and may require less structural strengthening than hydraulic damper retrofits [3].

Because of the added damping, displacements and stresses in the superstructure can be significantly reduced by up to 50%, resulting in less steel and lower structural cost. This savings is more than off-sets the cost of viscous wall damper and results in a better performing and more economical structure.

The concept of a viscous wall damper was first invented by Mitsuo Miyazaki [4] (Japan) in the 1980s, studied that effectiveness of the viscous wall damper was confirmed by 5-floor model. In 1988, Miyazaki, M; Arima, F. et al., [5] stated the viscous wall damper effectively absorbs energy, thus reducing the dynamic structural response due to earthquake; also effective for structures with moderate earthquakes; more effective for stronger earthquakes; the difference of damping effect on earthquake characteristic s is not known on the struc-

* Corresponding author: effy_hidayaty@yahoo.co.id

ture with this viscous wall damper; the acceleration and displacement of the structure may be reduced by 30% - 50%

Miyazaki's test of the viscous wall damper followed by Y. Matsusaka [6]; aims to produce effective dampers for weak to strong, reliable & durable inputs, maintenance free and economical, where the target damping constants 20% - 30% in the elastic range to reduce the dynamic response. They tested using the damper at SUT Building (2 basements and 14 floors) in Shizuoka, Japan. Time history analysis, the viscous wall damper matrix $[C_w]$ is added to the usual motion equation. Non-Newtonian fluid effect (α <1.0) is given in equation 1

$$[M]\{\ddot{x}\}+[C]\{\dot{x}\}+[C_w]\{\dot{x}^{\alpha}\}+[K]\{x\}=-[M]\{\ddot{z}\} \quad (1)$$

By comparing the dynamic response of different structures to the seismic input conditions, it was concluded that the viscous wall damper can reduce the structure response by about 70% to 80% with the maximum response still in the elastic range. All maximum responses can be reduced to the elastic range on the frame without any damage. [7, 8]

The reliability of the viscous wall damper was also tested by Ngai Yeung, et al. [9], to investigate the behavior and effectiveness of viscous wall damper in controlling the vibrations due to the effects of wind on high-rise buildings

Xilin Lu, et al [10] presented a thorough investigation of the viscous wall damper (VWD) for seismic response mitigation on reinforced concrete frames. The test results show a large number of additional dampers on the reinforced concrete frames of more than 20% and simultaneously drastically stiff the structure, decrease the displacement response by about 30% - 60%, the acceleration response and the reduced shear force on some cases, but increased in other cases.

The same case had been done by R Love et al [11], who had completed the construction of Hospitals in California, described an analysis of a 94,000 m^2 building. He concluded that it is possible to design a very important structure in areas with high seismicity. The efficiency of the designed structural system is evidenced by the reduced tonnage of building steel.

The same state is also written by Kazuhiko Sasaki, et al. [12]. Viscous dampers are able to absorb small and large vibrational energy such as wind or seismic movement efficiently. Characteristics of viscous wall damper were designed to effectively withstand earthquakes of magnitude that have been tested and predicted for the future. The earthquake that occurred in the Tohoku Region on March 11, 2011, is stronger than predicted.

All the research improve that the viscous wall damper is able to increase the damping ratio, reduced structural response due to earthquake and wind loads, and have been widely used in tall buildings in Japan and several other countries.

Much of the literature shows the ability of viscous wall damper in experimental studies (Reinhorn AM [13], Karunarathne, et al [14]), but few of them discusses numerically, especially about how to represented and modeling of the viscous wall damper in a software. Dynamic Isolation System, as a production the viscous wall damper, had released a guideline of viscous wall damper modeling in 2015.

This research presents the numerical analysis of viscous wall damper in the steel frame, including introducing the damper, modeling, specification, implementing in 3 story steel frames, the non-linear response history analysis in steel frame without and with the damper, and the conclusion.

2. Modeling of Viscous Wall Dampers

The seismic response of the viscous wall dampers can be readily modeled using existing nonlinear elements in SAP2000. This section of the modeling guide is written in the context of SAP2000 and assumes that nonlinear response history analysis is performed. Dynamic Isolation System Viscous Wall Dampers are best represented by an Exponential Maxwell Damper model, as shown in the figure on the below (from the CSI Analysis Reference Manual). The SAP2000 element type is NLLINK. The model consists of a linear spring, K, in series with an exponential damper characterized by C and α, such that the force in the damper is related to the velocity across the damper through the force-velocity relationship $F = C V^{\alpha}$ [3]. This formula is identical to apart of the formula (1).

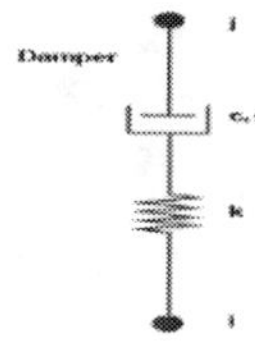

Figure 2 Modeling of Viscous Wall Dampers [3]

Unlike piston dampers, Dynamic Isolation System Viscous Wall Dampers connect to beams above and below. Therefore, to use the above 2-node element, a typical frame containing the dampers can be modeled as follows:

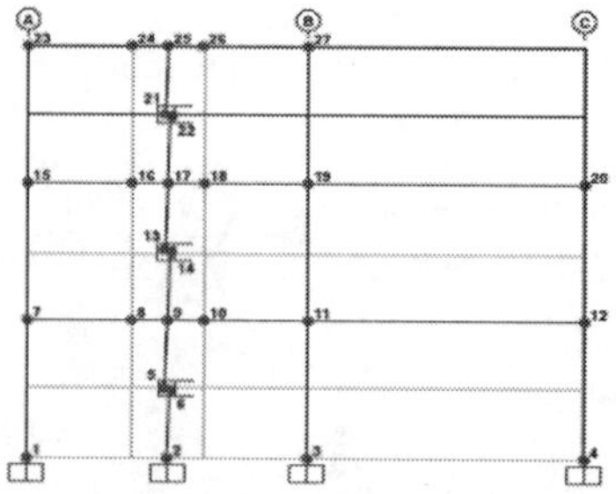

Figure 3 The Viscous Wall Damper in an existing frame [3]

3. Application The Viscous Wall Damper on Steel Frame

In order to verify the modeling viscous wall damper and to investigate its influence on the seismic response of a structure in a dynamic analysis, a 3D steel frame located in Yogyakarta (-7.7956; 110.3695), Indonesia, at the soft soil.

The geometrical and mechanical properties of the frame a given in table 1.

Table 1. Geometrical and mechanical properties of element structure

Geometrical Properties	
Element Structure	**Size**
Beam	WF 300 x 300 x 10 x 15
Beam which located the viscous wall damper	WF 350 x 350 x 12 x 19
Coloum	WF 350 x 350 x 12 x 19
Floor	120 mm
The viscous wall damper	DIS 7'x 12'
Mechanical Properties	
Steel Frame	A572 Gr 50
Modulus of Elasticity (Es)	200000 MPa
Yield Stress (fy)	345 MPa
Ultimate Stress (fu)	450 MPa

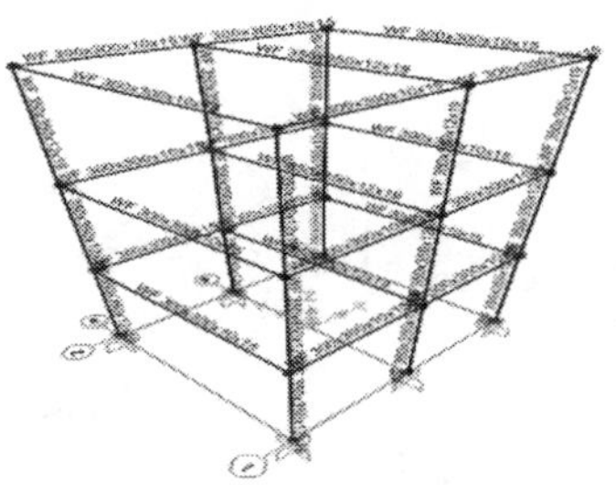

Figure 4 Numerical modeling of 3 D three-story steel frame building

The inherent damping of a structure is assumed to be equal to 5% and VWD is added as a supplementary damping to the system. Also, velocity exponential coefficient is assumed equal to 0,5.

Periods of fundamental approach structure Ta, was calculated as 0.195 seconds. The location of the building is in the city of Yogyakarta, which has historically experienced a major earthquake 6.2 Mw at May 27, 2016, with epicentre at coordinates (-7.977°; 110.318°) and 26 km from Jogyakarta city with a depth of 35 km due to a shift in the form of movement opposite (USGS), so the $R_{rupture}$ 43.6 km.

The seismic load had determined by matching the acceleration spectra at the surface from earthquake risk – targeted max. consider earthquake for Yogyakarta city.

A study is presented to evaluate the effect of the viscous wall dampers on building vibration control. Therefore, the influence of the damper in the considered structure's seismic response has been investigated using a nonlinear time history analysis of the frame model under the earthquake.

Table 2. The earthquake load used in structural analysis

No	Location	Year	Scale	Mag. (Mw)	R rupture
1	Imperial Valley Earthquake 1940	1940	0.841	6.95	6.09
2	Chi - chi Taiwan-06 Earthquake	1999	1.287	6.3	43.1
3	Iwate Japan Earthquake	2008	1.651	6.9	44. 75

3.1 Analysis of structure without the damper

Analysis of structure without the damper, as in figure 4.

3.2 Analysis of structure with the dampers in Y direction

The structure used The Dynamic Isolation System Viscous Wall Dampers (DIS VWD) size 7"x 12" double vane with a viscous fluid layer thickness of 5 mm. By using 3 dampers in the interior frame at Y direction, the structure had been analyzed.

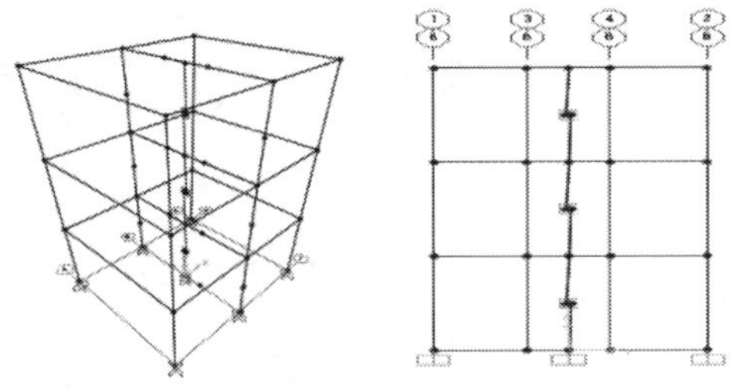

Figure 5. Structure & numerical model with damper in Y direction

3.3 Analysis of structure with the damper in x and y direction

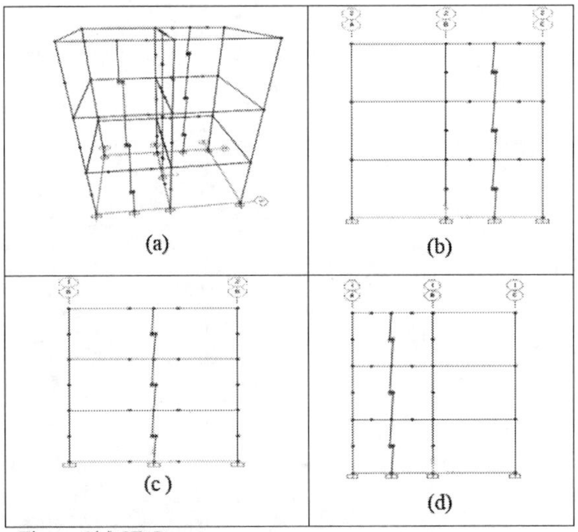

Figure 6(a) 3D Structure
(b) Frame in X direction (y=3.6 m) with VWD
(c.) Frame in Y direction (x=0 m) with VWD
(d) Frame in X direction (y=-3.6 m) with VWD

The structure used The Dynamic Isolation System Viscous Wall Dampers (DIS VWD) size 7"x 12" double vane with a viscous fluid layer thickness of 5 mm, the same size likes frame in Y direction.

By using 3 dampers in the interior frame at Y direction and 6 dampers in the exterior frame at X direction, the structure had been analyzed.

3.4 Analysis And Discussion

The analysis had been done by using SAP2000 version 19. The result of analysis given the following table.

Table 3. The modal periods and eigenvalue of 3 conditions structure

Mode	WITHOUT VWD		WITH VWD IN Y DIRECTION		WITH VWD IN X & Y DIRECTION	
	Periods (sec)	Eigen value (rad^2/s^2)	Periods (sec)	Eigen value (rad^2/s^2)	Periods (sec)	Eigenvalue (rad^2/s^2)
1	11.072	0.322	11,073	0,322	11,064	0,323
2	7.895	0.633	7,267	0,748	6,999	0,8057
3	6.845	0.843	6,313	0,991	4,562	1,897
4	3.424	3.367	3,424	3.367	2,854	4,846
5	2.387	6.926	0,835	56, 689	1,700	13,660
6	2.148	8.557	0,299	442,950	0,214	865,034
7	0.835	56.687	0,014	179467,6	0,180	1214,702
8	0.700	80.656	0,014	201249,1	0,120	2753,793
9	0.295	453.34	0,014	201265,7	0,023	72767,976

The table 3 presents the periods and eigenvalue of modal analysis of structure without VWD, with VWD on the interior frame in the Y direction and with VWD on the interior frame in Y direction & exterior frame in the X direction. In general, the periods had decreased in frame with VWD, otherwise frequency angular had increased. The periods didn't have the decrease at the first time, because its need velocity to move the vane through the viscous fluid.

The last statement was proven by figure 7, that shows the structure with dampers have similar periods, especially at the higher mode (mode 2, 3, and the others). The influence of VWD also showed at table 4, which there is a significant decrement of displacement u_1 at the structure with VWD in X & Y direction (50.78%) and with VWD in Y direction (23.96%). The decrement displacement happened in all structure, as table 4 presented.

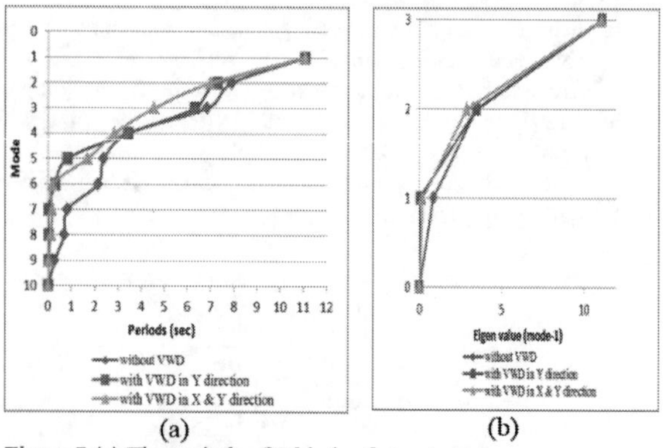

Figure 7 (a) The periods of 3 kinds of structure
(b) The Eigenvalue of them

The decrement was not happened only in the displacement, but also at the velocity, and at the acceleration, as figure 8.

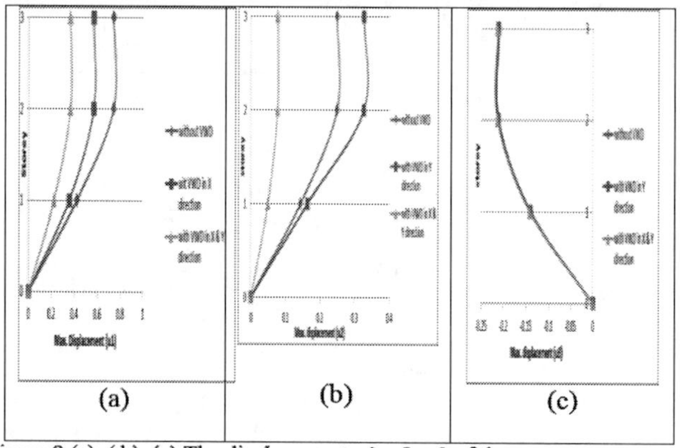

Figure 8 (a), (b), (c) The displacement u1, u2, u3 of three kinds structure respectively.

The time history analysis was conducted and the resulting peak displacements in the horizontal direction were obtained from the analysis.

Table 4. The decrement of displacement at structure with VWD

Structure	Story		Displacement					
			U1	U2	U3	R1	R2	R3
WITHOUT VWD	3	Max	0.7396	0.2483	-0.2105	0.0000	0.0024	0.0591
		Min	-0.7074	-0.2483	-1.0535	0.0000	-0.0025	-0.0591
	2	Max	0.7396	0.2483	-0.2104	0.0000	0.0024	0.0591
		Min	-0.7074	-0.2483	-1.0535	0.0000	-0.0025	-0.0591
	1	Max	0.4207	0.1426	-0.1392	0.0000	0.0018	0.0339
		Min	-0.4100	-0.1426	-0.6178	0.0000	-0.0019	-0.0339
WITH VWD IN Y FRAME	3	Max	0.5666	0.3258	-0.2122	0.0000	0.0019	0.0765
		Min	-0.6070	-0.3258	-1.0536	0.0000	-0.0021	-0.0776
	2	Max	0.5666	0.3258	-0.2122	0.0000	0.0019	0.0765
		Min	-0.6070	-0.325 8	-0.9411	0.0000	-0.0021	-0.0776
	1	Max	0.3553	0.1629	-0.1407	0.0000	0.0015	0.0382
		Min	-0.3478	-0.1629	-0.6179	0.0000	-0.0015	-0.0388
WITH VWD IN X & Y FRAME	3	Max	0.3640	0.0758	-0.2142	0.0000	0.0015	0.0174
		Min	-0.3846	-0.0758	-1.0537	0.0000	-0.0016	-0.0180
	2	Max	0.3640	0.0758	-0.2142	0.0000	0.0015	0.0174
		Min	-0.3846	-1.0537	-0.9412	-0.0000	-0.0016	-0.0180
	1	Max	0.2232	0.0487	-0.1418	0.0000	0.0012	0.0105
		Min	-0.2269	-0.0487	-0.6179	0.0000	-0.0013	-0.0116

https://doi.org/10.1051/matecconf/201814701001

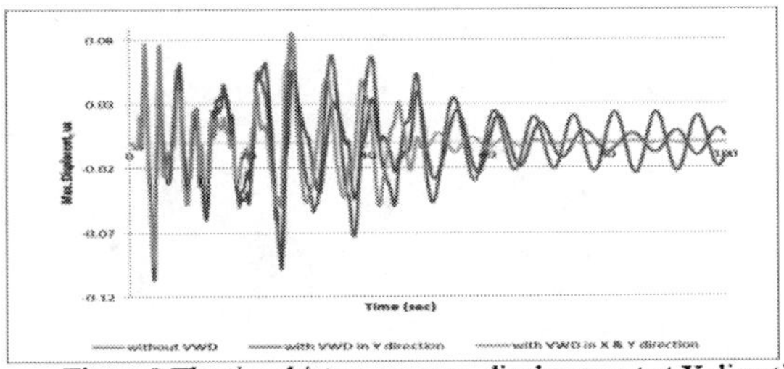

Figure 9 The time history vs max. displacement at X direction

This figure shows that structure with VWD in X & Y direction had most significant decrement than the others, especially at the steady state. At transient state, the effect of dampers was not too obvious in the first 30 seconds. After that, the structures begin to vibrate regularly until it reaches steady state. The similar case also happens at maximum velocity and maximum acceleration, as figure 10 and 11 respectively

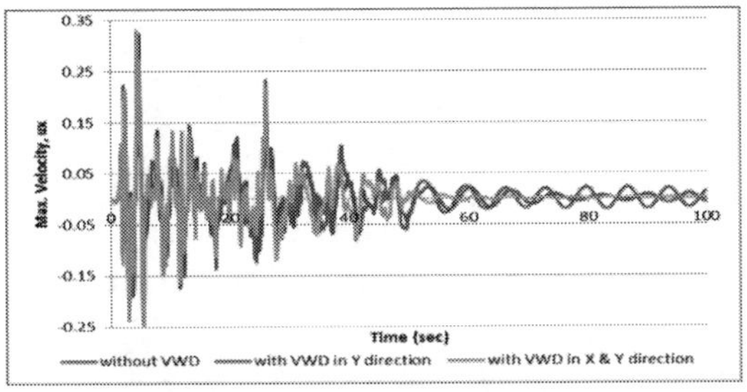

Figure 10 Time history versus max. velocity at X direction

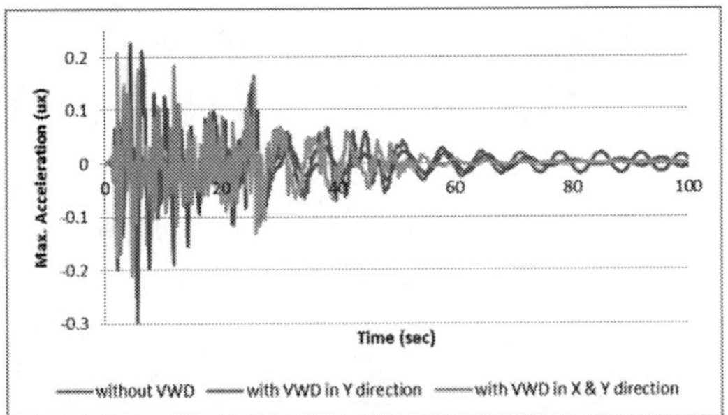

Figure 11 Time history versus max. acceleration at X direction

At the end, figures 9 - 11 make it obvious that the magnitude of the displacement decreased efficiently when the building was equipped with a VWD device.

4. Conclusion

Many efforts had been to develop the viscous wall damper as the equipment to controlled seismic response structure in experimental, but few of them had discussed numerically. A study to modeling the viscous wall damper by using existing nonlinear elements in SAP2000 had been done. It assumes that nonlinear response history analysis is performed. Dynamic Isolation System Viscous Wall Dampers are best represented by an Exponential Maxwell Damper model. The model consists of the a linear spring and in series with an exponential damper characterized by C and α, such that the force in the damper is related to the velocity across the damper through the force-velocity relationship

By adding the viscous wall damper at the un-control structure, the analysis was performed by a computer application on a steel space frame that excited a combination of three earthquake types at a three story steel frame with equipped viscous wall dampers at interior frame Y direction and with viscous wall damper in X & Y direction

From the analysis, it displays the ability of the damper due to a relative intense earthquake load. In general, the periods had decreased in frame with VWD, otherwise frequency angular had increased. The periods didn't have decrease at the first time, because its need velocity to move the vane through the viscous fluid and produced a force against the inter-story drift.

The influence of VWD showed that a significant decrement of displacement u_1 at the structure wwith VWD in X & Y direction (50,78%) and with VWD in Y direction (23,96%). The decrement displacement happened at all structure

At the end, all structure analysis shows the reduction of the structure periods and the structure response (displacement, velocity, and acceleration). All these results conclude that the structural components damaged due to loads can be greatly reduced.

References

1. E. Hidayaty, H. D. Setio, Prosiding KNPTS, *Peningkatan Kinerja Peredam Viskos Tipe Dinding Dengan Penambahan Kekasaran Plat Dalam Pada Struktur Baja Terhadap Beban Gempa,* pp 73-81, (2016)
2. E. Hidayaty, H. D. Setio, Prosiding KNPTS, *Analisis Penentuan Dimensi Plat Dalam Yang Telah Dikasarkan Pada Peredam Viskos Tipe Dinding Dalam Meningkatkan Kinerjanya Terhadap Beban Gempa,* pp 67-75, (2015)
3. Dynamic Isolation System, *Viscous Wall Dampers Guidelines for Modeling*, (2015)
4. F. Arima F, M. Miyazaki et al, SCTRRL, *Development of Viscous Damping Wall*, no. 13, (1986)
5. F. Arima, M. Miyazaki et al., Proceedings of the Ninth WCEE International Association for Earthquake Engineering, Tokyo, p. 821, (1988)
6. Y. Mitsusaka; WCEE, *Design of a Building with 20% or Greater Damping*, 1992 Balkema, Rotterdam, p 4143-4148, (1992)
7. F. Arima, H. Tanaka, N. Takase, H. Egasira, Nakamura, R., Miyazaki, M., and Sera, S., 1994a, SCTRRL, *Experimental Study on Seismic Respons Control Technologies for Building Structures : Vibration Test for High Rise Structures using Viscous Damping Walls*, (1994).

8. F. Arima, H. Tanaka, N. Takase, H. Egasira, R. Nakamura, M. Miyazaki, and S. Sera, S., 1994b, SCTRRL, *Experimental Study on Seismic Respons Control Technologies for Building Structures (Part II) : Vibration Test for High Rise Structures using Viscous Damping Walls*, (1994).
9. N. Yeung N; DE Pan Austin, AYT Leung; H C Chan, Structure in The Millenium , *Wind Tunnel Study Of Aeroelastic Building Model With Viscous-Damping Wall For Vibration Control*, , p 249 – 256, (1997).
10. L. Xilin, Z. Ying, Y. Feng, JSE, *Shaking Table Test and Numerical Analysis of RC Frames with Viscous Wall Damper*, Vol. 134, No. 1 January 1, 2008 © ASCE, ISSN 0733-9445/2008/1-64-76, (2008).
11. R. Love, S.E., J. Newell C. Poland, S.E., WCEE 13, *Perfomance-Based Design of an Essential Hospital with Supplemental Viscous Damping in a High Seismic Zone*,(2012)
12. K. Sasaki, et al, *WCEE13, Characteristics of Viscous Wall Damper of Intense Oscillation Test against Large Earthquakes ,(* 2012)
13. A. M. Reinhorn, and C. Li, NCEER , *Experimental and analytical investigation of seismic retrofit of structures with supplemental damping, Part III: viscous damping walls, Technical Report NCEER-95-0013*, (1995).
14. N. Karunarathne. N., Catholic University of America, *Damping of Frame Structure : An Educational Shake Table Test*, , 2006

On the accuracy and convergence of the hybrid FE-meshfree Q4-CNS element in surface fitting problems

Foek Tjong Wong[1,*], *Richo* Soetanto[1], and *Januar* Budiman[1]

[1]Master Program of Civil Engineering, Petra Christian University, Surabaya 60236, Indonesia

Abstract. In the last decade, several hybrid methods combining the finite element and meshfree methods have been proposed for solving elasticity problems. Among these methods, a novel quadrilateral four-node element with continuous nodal stress (Q4-CNS) is of our interest. In this method, the shape functions are constructed using the combination of the 'non-conforming' shape functions for the Kirchhoff's plate rectangular element and the shape functions obtained using an orthonormalized and constrained least-squares method. The key advantage of the Q4-CNS element is that it provides the continuity of the gradients at the element nodes so that the global gradient fields are smooth and highly accurate. This paper presents a numerical study on the accuracy and convergence of the Q4-CNS interpolation and its gradients in surface fitting problems. Several functions of two variables were employed to examine the accuracy and convergence. Furthermore, the consistency property of the Q4-CNS interpolation was also examined. The results show that the Q4-CNS interpolation possess a bi-linier order of consistency even in a distorted mesh. The Q4-CNS gives highly accurate surface fittings and possess excellent convergence characteristics. The accuracy and convergence rates are better than those of the standard Q4 element.

1 Introduction

The finite element method (FEM) is now a widely-used, well-establish numerical method for solving mathematical models of practical problems in engineering and science. In practice, FEM users often prefer to use simple, low order triangular or quadrilateral elements in 2D problems and tetrahedral elements in 3D problems since these elements can be automatically generated with ease for meshing a complicated geometry. Nevertheless, the standard low order elements produce discontinuous gradient fields on the element boundaries and their accuracy is sensitive to the quality of the mesh.

To overcome the FEM shortcomings, since the early 1990's up to present a vast amount of meshfree (or meshless) methods [1-2], which do not require a mesh in discretizing the problem domain, have been proposed. A recent review on meshfree methods presented by Liu [3]. While these newer methods are able to eliminate the FEM shortcomings, they also have their own, such as: (i) the computational cost is much more expensive than the FEM, and (ii) the computer implementation is quite different from that of the standard FEM.

To synergize the strengths of the finite element and meshfree methods while avoiding their weaknesses, in the last decade several hybrid methods combining the two classes of methods based on the concept of partition-of-unity have been developed [4-8]. Among several hybrid methods available in literature, the authors are interested in the four-node quadrilateral element with continuous nodal stress (Q4-CNS) proposed by Tang el al. [6] for the reason that this work is the pioneering hybrid method possessing the property of continuous nodal stress. The Q4-CNS can be regarded as an improved version of the FE-LSPIM Q4 [4-5]. In this novel method, the nonconforming shape functions for the Kirchhoff's plate rectangular element are combined with the shape functions obtained using an orthonormalized and constrained least-squares method. The benefits of the Q4-CNS are [6, 9, 10]: (1) the shape functions are C^1 continuous at nodes so that it naturally provides a globally smooth gradient fields. (2) The Q4-CNS can give higher accuracy and faster convergence rate than the standard quadrilateral element (Q4). (3) The Q4-CNS is more tolerant to mesh distortion.

The Q4-CNS has been developed and applied for the free and forced vibration analyses of 2D solids [9] and for 2D crack propagation analysis [10]. Recently the Q4-CNS has been further developed to its 3D counterpart, that is, the hybrid FE-meshfree eight-node hexahedral element with continuous nodal stress (Hexa8-CNS) [11]. However, examination of the Q4-CNS interpolation in fitting surfaces defined by functions of two variables has not been carried out. Thus, it is the purpose of this paper to present a numerical study on the on the accuracy and convergence of the Q4-CNS shape functions and their derivatives in surface fitting problems. Furthermore, the consistency (or completeness) property of the Q4-CNS shape functions is numerically examined in this study.

*Corresponding author: wftjong@petra.ac.id

2 The Q4-CNS interpolation

As in the standard finite element procedure, a 2D problem domain, $\bar{\Omega}$, is firstly divided into four-node quadrilateral elements to construct the Q4-CNS shape functions. Consider a typical element $\bar{\Omega}^e$ with the local node labels 1, 2, 3 and 4. The unknown function u on the interior and boundary of the element is approximated by

$$u^h(x,y)=\sum_{i=1}^{4} w_i(\xi,\eta)u_i(x,y) \quad (1)$$

where $w_i(\xi,\eta)$ and $u_i(x,y)$ are the weight functions and nodal approximations, respectively, associated with node i, i=1,…,4. Note that in the classical isoparametric four-node quadrilateral element (Q4), the weight functions are given as the shape functions and the nodal approximations are reduced to nodal values u_i. Here the weight functions are defined as the non-conforming shape functions for the Kirchhoff's plate rectangular element [6, 12], that is,

$$w_i(\xi,\eta)=\tfrac{1}{8}(1+\xi_0)(1+\eta_0)(2+\xi_0+\eta_0-\xi^2-\eta^2) \quad (2a)$$

$$\xi_0=\xi_i\xi,\quad \eta_0=\eta_i\eta,\quad i=1,2,3,4. \quad (2b)$$

where ξ and η are the natural coordinates of the classical Q4 with the values in the range of –1 to 1. The weight functions satisfy the partition of unity property, that is, $\sum_1^4 w_i(\xi,\eta)=1$. The nodal approximations $u_i(x,y)$ are constructed using the orthonormalized and constrained least-squares method (CO-LS) as presented by Tang et al. [6] and Yang et al. [9-10]. Here the CO-LS is briefly reviewed.

To construct the CO-LS approximation, nodal support domains of node i, $\bar{\Omega}_i$, i=1,…,4 of a typical quadrilateral element $\bar{\Omega}^e$ are firstly defined using the neighboring nodes of node i. For example, the nodal support domain of node 3 of element e is shown in Fig. 1(a). The element support domain $\hat{\Omega}^e$ is then defined as the union of the four nodal support domains, that is, $\hat{\Omega}^e=\cup_1^4\bar{\Omega}_i$, as shown in Fig. 1(b).

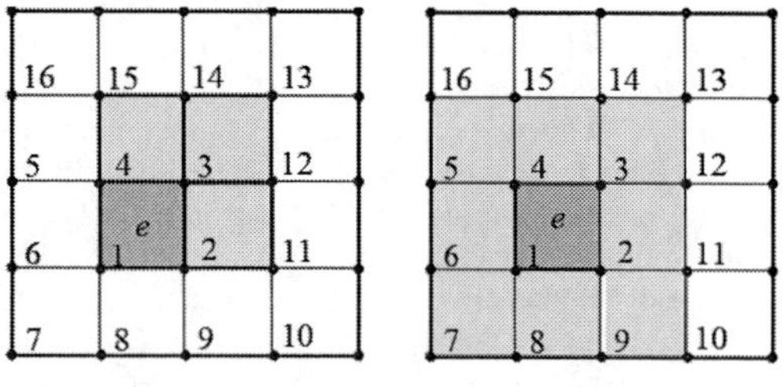

(a) $\bar{\Omega}_3$ (b) $\hat{\Omega}^e$

Fig. 1. Definitions of: (a) the nodal support domain of node 3 of element e and (b) the element support domain of element e.

Consider a nodal support domain of node i, $\bar{\Omega}_i$ with the total number of supporting nodes n. Let the labels for the nodes be j, j=1,…, n. Using the least-squares method, the nodal approximation $u_i(x,y)$ is given as

$$u_i(x,y)=\mathbf{p}^{\mathrm{T}}(x,y)\mathbf{A}^{-1}\mathbf{Ba} \quad (3)$$

where $\mathbf{p}(x, y)$ is a vector of polynomial basis functions, viz.

$$\mathbf{p}^{\mathrm{T}}(x,y)=\{1 \quad x \quad y \quad x^2 \quad \ldots \qquad (1\times m) \quad (4a)$$

In Eq. (4a) m is the number of monomial bases in $\mathbf{p}$. Following the original work [6], in this study the 'serendipity' basis function

$$\mathbf{p}^{\mathrm{T}}(x,y)=\{1 \quad x \quad y \quad x^2 \quad xy \quad y^2 \quad x^2y \quad xy^2\} \quad (4b)$$

is used if $n>8$ and the bi-linear basis function,

$$\mathbf{p}^{\mathrm{T}}(x,y)=\{1 \quad x \quad y \quad xy\} \quad (4c)$$

is used if $n\le 8$. Matrices $\mathbf{A}$ and $\mathbf{B}$ are the moment matrix and the basis matrix, respectively, given as

$$\mathbf{A}=\sum_{j=1}^{n}\mathbf{p}(x_j,y_j)\mathbf{p}^{\mathrm{T}}(x_j,y_j) \qquad (m\times m) \quad (5)$$

$$\mathbf{B}=[\mathbf{p}(x_1,y_1) \quad \ldots \qquad y_n)] \qquad (m\times n) \quad (6)$$

Vector $\mathbf{a}=\{a_1 \quad a_2 \quad \cdots$ is the vector of nodal parameters. Note that in general vector $\mathbf{a}$ is *not* a vector of nodal values because the approximation $u_i(x,y)$ does not necessarily pass through the nodal values.

Defining the inner product for any two basis functions $f(x,y)$ and $g(x,y)$ as

$$\left(f(x,y),g(x,y)\right)=\sum_{j=1}^{n} f(x_j,y_j)g(x_j,y_j) \quad (7)$$

and using the Gram-Schmidt orthonormalization algorithm [6], the basis vector $\mathbf{p}$ can be transformed into an orthonormal basis function vector $\mathbf{r}$ so that the moment matrix $\mathbf{A}$ becomes the identity matrix. Subsequently, the nodal approximation is constrained using the Lagrange multiplier method so that the nodal parameter $u_i(x,y)$ at node i is equal to the nodal value u_i. Going through the abovementioned process, the nodal approximation, Eqn. (3), turns into

$$u_i(x,y)=\mathbf{\Phi}(x,y)\mathbf{a}=\sum_{j=1}^{n}\phi_j^i(x,y)a_j \quad (8)$$

where

$$\begin{aligned}\mathbf{\Phi}(x,y)&=\left[\phi_1^i(x,y) \quad \phi_2^i(x,y) \quad \cdots \qquad y)\right]\\&=\mathbf{r}^{\mathrm{T}}(x,y)\mathbf{B}^i\end{aligned} \quad (9)$$

$$\mathbf{B}^i=\left[\mathbf{B}_1^i \quad \mathbf{B}_2^i \quad \cdots \right. \quad (10)$$

$$\mathbf{B}_j^i=\mathbf{r}(x_j,y_j)-f_j^i\mathbf{r}(x_i,y_i) \quad (11)$$

$$f_j^i = \begin{cases} \dfrac{\mathbf{r}^{\mathrm{T}}(x_i, y_i)\mathbf{r}(x_j, y_j)}{\mathbf{r}^{\mathrm{T}}(x_i, y_i)\mathbf{r}(x_i, y_i)} & \text{if } j \neq i \\ \dfrac{\mathbf{r}^{\mathrm{T}}(x_i, y_i)\mathbf{r}(x_j, y_j) - 1}{\mathbf{r}^{\mathrm{T}}(x_i, y_i)\mathbf{r}(x_i, y_i)} & \text{if } j = i \end{cases} \quad (12)$$

Note that n, the number of nodes in the nodal support domain of node i, in general varies with i.

Consider now the element support domain of element e, $\hat{\Omega}^e$, with the total number of nodes N. Let the node labels in $\hat{\Omega}^e$ be I=1, ..., N. Using this element level labelling system and substituting Eqn. (8) into Eqn. (1), the approximate function can be expressed as

$$\begin{aligned} u^h(x,y) &= \sum_{i=1}^{4} w_i(\xi,\eta) \sum_{I=1}^{N} \phi_I^i(x,y) a_I \\ &= \sum_{I=1}^{N} \psi_I(x,y) a_I \end{aligned} \quad (13)$$

in which $\psi_I(x,y)$ is the Q4-CNS shape function associated with node I in the element support domain. In this equation, if node I is not in the nodal support domain of node i, then $\phi_I^i(x,y)$ is defined to be zero. It is obvious that the shape function is the product of the nonconforming rectangular element shape functions $w_i(\xi,\eta)$ and the CO-LS shape functions $\phi_I^i(x,y)$, that is,

$$\psi_I(x,y) = \sum_{i=1}^{4} w_i(\xi,\eta) \phi_I^i(x,y) \quad (14)$$

3 Numerical tests

In this section, the accuracy and convergence of the Q4-CNS interpolation in fitting surfaces of $z = f(x,y)$ and their derivatives are examined. In this problem, both the stuctural system (the function domain) and the Q4-CNS element have one degrees of freedom at each node, that is, the surface level z. To measure the approximation errors, the following relative L_2 norm of error is used

$$r_z = \sqrt{\frac{\int_{\Omega^h} (z - z^h)^2 \, dA}{\int_{\Omega^h} z^2 \, dA}} \quad (15)$$

in which z is the function under consideration, z^h is the approximate function, and Ω^h is the approximate domain with the element characteristic size, h. This expression is also applicable to measure the relative error of the function partial derivatives (by replacing z and z^h with their derivatives). The integral in Eqn. (15) is evaluated numerically using Gaussian quadrature rule. The number of quadrature sampling points is taken to be 5×5. For the purpose of comparison, the accuracy and convergence of the standard Q4 interpolation and its partial derivatives are also presented.

3.1 Shape function consistency property

In a Rayleigh-Ritz based numerical method, a set of shape functions is required to be able to represent exactly all polynomial terms of order up to m in the Cartesian coordinates [13], where m is the variational index (that is, the highest order of the spatial derivatives that appears in the problem functional). A set of shape functions that satisfies this condition is called m-consistent [13]. This consistency property is a necessary condition for convergence (that is, as the mesh is refined, the solution approaches to the exact solution of the corresponding mathematical model).

To examine the consistency property of the Q4-CNS shape functions, consider a 10×10 square domain shown in Fig. 2. The domain is subdivided using 4×4 regular quadrilateral elements, Fig. 2(a), and irregular quadrilateral elements, Fig. 2(b). The functions under consideration are the polynomial bases up to the quadratic bases, that is, z = 1, $z = x$, $z = y$, $z = xy$, $z = x^2$ and $z = y^2$. The results of the relative errors for the Q4-CNS interpolation and its nonzero partial derivatives are presented in Tables 1 and 2, respectively, together with those of the standard Q4 interpolation.

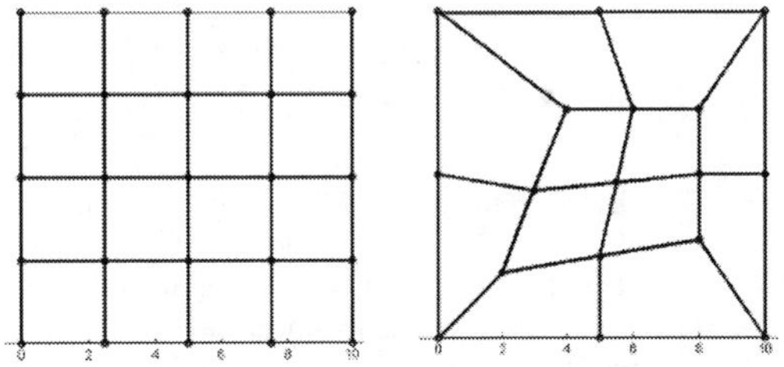

(a) Regular mesh (b) Irregular mesh

Fig. 2. Square function domain of size 10-by-10 subdivided into: (a) regular and (b) irregular quadrilateral elements.

Table 1. Relative L_2 norm of errors for the approximation of different polynomial basis functions using the regular and irregular meshes.

Function	Regular Mesh		Irregular Mesh	
	Q4-CNS	Q4	Q4-CNS	Q4
z=1	9.98E-16	1.32E-17	1.88E-15	1.35E-17
$z=x$	1.41E-15	0	2.82E-15	0
$z=y$	1.20E-15	0	1.45E-15	0
$z=xy$	1.39E-15	1.49E-16	4.59E-15	2.37%
$z=x^2$	1.22%	2.55%	2.65%	5.83%
$z=y^2$	1.22%	2.55%	2.33%	5.37%

The tables show that the Q4-CNS interpolation is capable to reproduce exact solutions up to the xy basis both for the domain with regular and irregular meshes. In other words, the Q4-CNS interpolation is consistent up to the xy basis. On the other hand, the Q4 interpolation is consistent up to the xy basis for the regular mesh only; for the irregular mesh it degrades to linear consistent. This finding may partly explain the reason why the Q4-CNS has a higher tolerance to mesh distortion [6]. For the x^2

and y^2 bases, both the Q4-CNS and Q4 interpolations are, as expected, not able to produce the exact solutions. For these bases, however, the Q4-CNS interpolation is consistently more accurate than the standard Q4.

Table 2. Relative L_2 norm of errors for the approximation of nonzero polynomial basis function derivatives using the regular and irregular meshes.

(a) Basis function derivatives with respect to x

Function Derivative w.r.t. x	Regular Mesh		Irregular Mesh	
	Q4-CNS	Q4	Q4-CNS	Q4
$z_{,x}=1$	9.11E-15	2.25E-16	2.15E-14	2.82E-16
$z_{,x}=y$	9.36E-15	2.55E-16	3.06E-14	11.32%
$z_{,x}=2x$	6.70%	12.50%	10.94%	16.58%

(b) Basis function derivatives with respect to y

Function Derivative w.r.t. y	Regular Mesh		Irregular Mesh	
	Q4-CNS	Q4	Q4-CNS	Q4
$z_{,y}=1$	8.71E-15	1.98E-16	9.61E-15	2.11E-16
$z_{,y}=x$	1.02E-14	2.93E-16	3.58E-14	12.53%
$z_{,y}=2y$	6.70%	12.50%	10.30%	15.90%

The tables clearly reveals that the Q4-CNS interpolation is not consistent up to all of the quadratic bases. As a consequence, the Q4-CNS is not applicable to variational problems possessing variational index $m = 2$, including the Love-Kirchhoff plate bending and shell models. This is in contradiction to the statement made in the original paper [6], which mentioned that the Q4-CNS "is potentially useful for the problems of bending plate and shell models". If the Reissner-Mindlin theory is adopted, however, the Q4-CNS is naturally applicable.

3.2 Accuracy and convergence

3.2.1 Quadratic function

The accuracy and convergence of the Q4-CNS interpolation in fitting functions in 2D domain are firstly examined using quadratic function (adapted from an example in Wong and Kanok-nukulchai [14]) given as

$$z = 1 - x^2 - y^2 \tag{16}$$

with two different domains, viz.

$$\overline{\Omega}_S = \{(x,y) | 0 \le x \le 1,\ 0 \le y \le 1\} \tag{17}$$

$$\overline{\Omega}_C = \{(x,y) | x^2 + y^2 \le 1, x \ge 0, y \ge 0\} \tag{18}$$

The first domain, Eqn. (17), is the unit square while the second one, Eqn. (18), is a quarter of the unit circle, both of which are located in the first quadrant of the Cartesian coordinate system. The unit square is subdivided using regular meshes of 2×2, 4×4, 8×8, and 16×16 square elements. The quarter of the unit circle is subdivided into 3, 12, 27, and 48 quadrilateral elements as shown in Fig. 3 (taken from an example in Katili [15]).

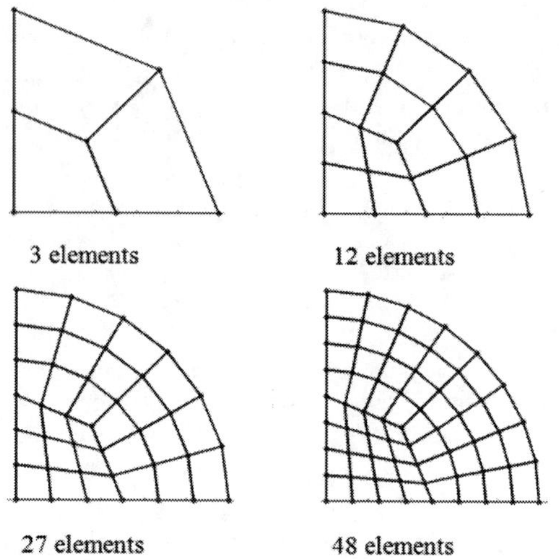

Fig. 3. A quarter of the unit circle subdivided into different number of quadrilateral elements (Katili [15], p.1899).

The relative error norms of the Q4-CNS and Q4 interpolations in approximating the quadratic function, Eqn. (16), and its partial derivatives, are presented in Table 3 for the square domain and in Table 4 for the quarter circle domain. The tables show that the Q4-CNS interpolation converges very well to the quadratic function z both for the regular mesh in the unit square domain and for the relatively irregular mesh in the quarter of the unit circle domain. The tables also confirm that the Q4-CNS interpolation is consistently more accurate than the Q4 interpolation. The finer the mesh the more accurate the Q4-CNS interpolation compared to the Q4.

The relative error norms are plotted against the number of elements on each edge, M, in log-log scale as shown in Fig. 4. The convergence graphs for the partial derivatives with respect to y are similar to Fig. 4(b) and have the same convergence rates. The graphs show that the average convergence rate of the Q4-CNS interpolation is about 25% faster than that of the Q4. It is worth mentioning here that the convergence rates of the Q4 interpolation, 2, and its partial derivatives, 1, are exactly the same as predicted by the interpolation theory [16].

https://doi.org/10.1051/matecconf/201814701002

Table 3. Relative L_2 norm of errors for the approximation of the quadratic function, r_z, and its partial derivatives, $r_{z,x}$ and $r_{z,y}$ over the unit square domain.

M	r_z		$r_{z,x}$		$r_{z,y}$	
	Q4-CNS	Q4	Q4-CNS	Q4	Q4-CNS	Q4
2	10.18%	16.26%	22.77%	25.00%	26.29%	28.87%
4	1.83%	4.07%	10.62%	12.50%	12.26%	14.43%
8	0.33%	1.02%	4.13%	6.25%	4.77%	7.22%
16	0.06%	0.25%	1.52%	3.13%	1.76%	3.61%
M: the number of elements on each edge						

Table 4. Relative L_2 norm of errors for the approximation of the quadratic function, r_z, and its partial derivatives, $r_{z,x}$ and $r_{z,y}$ over a quarter of the unit circle domain.

Number of elements	r_z		$r_{z,x}$		$r_{z,y}$	
	Q4-CNS	Q4	Q4-CNS	Q4	Q4-CNS	Q4
3	11.06%	16.59%	28.14%	33.92%	22.48%	27.10%
12	2.51%	4.52%	14.56%	16.16%	12.57%	13.96%
27	0.91%	2.04%	8.42%	10.68%	7.37%	9.36%
48	0.44%	1.15%	5.64%	7.99%	4.97%	7.03%

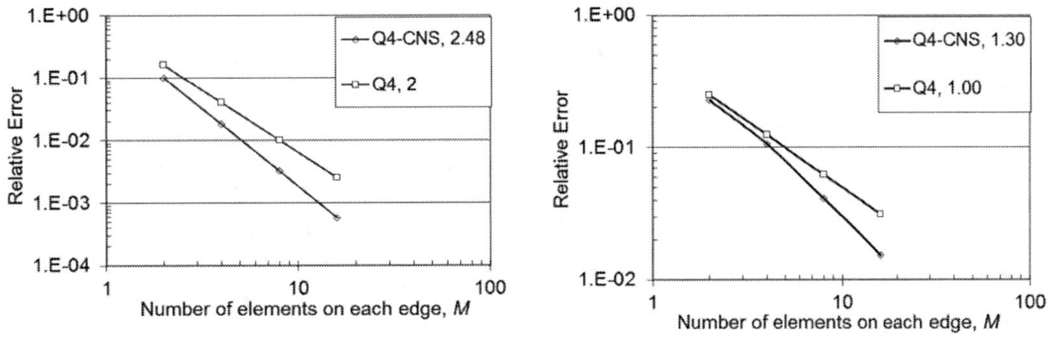

(a) Relative error norms of interpolations

(b) Relative error norms of interpolation x-partial derivative

Fig. 4. Convergence of the Q4-CNS and Q4 interpolations in approximating: (a) the quadratic function, (b) the partial derivatives of the function with respect to x, over the unit square. The number in the legend indicate the average convergence rate.

3.2.2 Bi-cosine function

The second function chosen to examine the accuracy and convergence of the Q4-CNS interpolation is

$$z = \cos(\frac{\pi}{2}x)\cos(\frac{\pi}{2}y) \tag{19}$$

defined over two different domains as in the previous example: (a) the square unit domain, Eqn. (17), and (b) a quarter of unit circular domain, Eqn. (18). The meshes used are the same as those in the previous example.

The relative error norms of the Q4-CNS and Q4 interpolations and their partial derivatives are presented in Table 5 for the unit square domain and in Table 6 for a quarter of circular domain. The convergence graphs of the relative error norms of the interpolations and their partial derivatives with respect to x for the bi-cosine function defined over the unit square are shown in Fig. 5. The tables and graphs confirm the findings in the previous example. The convergence rate for the Q4 interpolation, however, decreases a little bit from 2 to 1.9, whereas the convergence rate for the Q4-CNS remains the same as in the previous problem.

Table 5. Relative L_2 norm of errors for the approximation of the bi-cosine function, r_z, and its partial derivatives, $r_{z,x}$ and $r_{z,y}$ over the unit square domain.

M	r_z		$r_{z,x}$		$r_{z,y}$	
	Q4-CNS	Q4	Q4-CNS	Q4	Q4-CNS	Q4
2	6.76%	10.31%	20.58%	23.08%	16.16%	18.13%
4	1.22%	2.67%	9.50%	11.39%	7.46%	8.95%
8	0.22%	0.67%	3.74%	5.68%	2.94%	4.46%
16	0.04%	0.17%	1.38%	2.83%	1.09%	2.23%
M: the number of elements on each edge						

Table 6. Relative L_2 norm of errors for the approximation of the bi-cosine function, r_z, and its partial derivatives, $r_{z,x}$ and $r_{z,y}$ over a quarter of the unit circular domain.

Number of elements	r_z		$r_{z,x}$		$r_{z,y}$	
	Q4-CNS	Q4	Q4-CNS	Q4	Q4-CNS	Q4
3	4.89%	11.08%	19.35%	31.68%	13.20%	21.62%
12	1.12%	2.98%	10.25%	15.59%	7.36%	11.19%
27	0.41%	1.34%	5.99%	10.29%	4.33%	7.45%
48	0.20%	0.76%	3.99%	7.70%	2.90%	5.59%

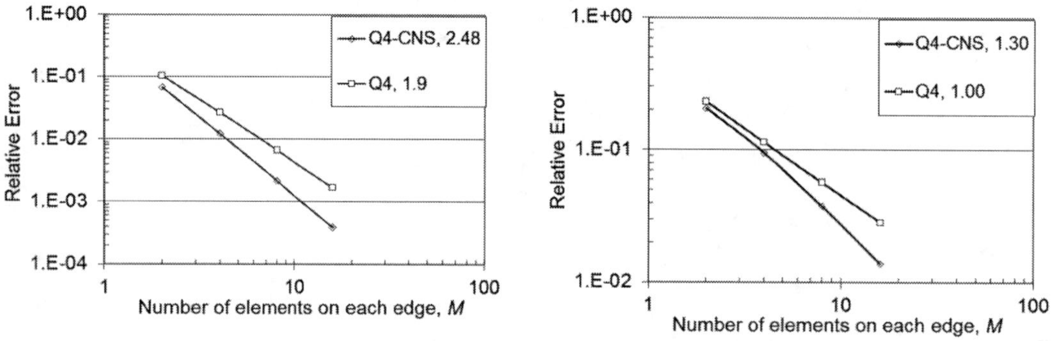

(a) Relative error norms of interpolations

(b) Relative error norms of interpolation *x*-partial derivative

Fig. 5. Convergence of the Q4-CNS and Q4 interpolations in approximating: (a) the bi-cosine function, (b) the partial derivatives of the function with respect to *x*, over the unit square. The number in the legend indicate the average convergence rate.

4 Conclusions

The consistency property, accuracy and convergence of the Q4-CNS interpolation in surface fitting problems have been numerically studied. The results show that the Q4-CNS interpolation is consistent up to the bilinear basis both for the regular and irregular meshes. It is more accurate than the Q4 in fitting the functions and their derivatives. In a sufficiently fine mesh, the error norm of the Q4-CNS interpolation is around 3 to 4 times smaller than that of the Q4, and the error norm of its derivatives is around 1.5 to 2 times smaller than that of the Q4. The Q4-CNS interpolation converge very well to the fitted function. Its convergence rate is approximately 25% faster than that of the Q4. Thus in terms of the accuracy and convergence, the Q4-CNS interpolation is a better choice to be empoyed as the trial and test functions in a Rayleigh-Ritz or Galerkin numerical method. The demerit of the present method is that the computational cost to construct the shape function is much higher than the Q4 shape function.

We gratefully acknowledge that this research is partially supported by the research grant of the Institute of Research and Community Service, Petra Christian University, Surabaya.

*Corresponding author: wftjong@petra.ac.id

References

1. G.R. Liu, *Mesh free methods: moving beyond the finite element method*, 1st ed. (CRC Press, Boca Raton, 2003, pp. 1-5)
2. Y.T. Gu, Int. J. Comput. Methods **2**, 477–515 (2005)
3. G.R. Liu, Int. J. Comput. Methods **13**, 1630001-1–1630001-42 (2016)
4. S. Rajendran, B.R. Zhang, Comput. Methods Appl. Mech. Eng. **197**, 128–147 (2007)
5. B.R. Zhang, S. Rajendran, Comput. Methods Appl. Mech. Eng. **197**, 3595–3604 (2008)
6. X.H. Tang, C. Zheng, S.C. Wu, J.H. Zhang, Appl. Math. Mech. **30**, 1519–1532 (2009)
7. Y. Yang, X.H. Tang, H. Zheng, Comput. Struct. **141**, 46–58 (2014)
8. Y. Yang, R. Bi, H. Zheng, Eng. Anal. Bound. Elem. **53**, 73–85 (2015)
9. Y. Yang, L. Chen, D. Xu, H. Zheng, Eng. Anal. Bound. Elem. **70**, 1–11 (2016)
10. Y. Yang, G. Sun, H. Zheng, X. Fu, Comput. Struct. **177**, 69–82 (2016)
11. Y. Yang, L. Chen, X.H. Tang, H. Zheng, Q.S. Liu, Comput. Struct. **178**, 17–28 (2017)
12. O.C. Zienkiewicz, R.L. Taylor, *The finite element method, Vol. 2: solid mechanics*, 5th ed. (Butterworth-Heinemann, 2000, p. 126)
13. C.A. Felippa, *Introduction to finite element methods (ASEN 5007)* (University of Colorado at Boulder, Fall 2016) http://www.colorado.edu/engineering/cas/courses.d/IFEM.d/ (14-Oct-2016)
14. F.T. Wong, W. Kanok-Nukulchai, Civ. Eng. Dimens. **11**, 15–22 (2009)
15. I. Katili, Int. J. Numer. Methods Eng. **36**, 1885–1908 (1993)
16. K.J. Bathe, *Finite element procedures* (Prentice-Hall, Upper Saddle River, 1996, pp. 244-250)

Implementation of Highly-Flowable Strain Hardening Fiber Reinforced Concrete in New RC Beam-Column Joints

Wen-Cheng Liao[1*], *Wei-Ru* Su[2]

[1]Associate Professor, Department of Civil Engineering, National Taiwan University, Taipei, Taiwan
[2]Graduate Student, Department of Civil Engineering, National Taiwan University, Taipei, Taiwan

Abstract. The purpose of New RC project was aimed to reduce the member sections and increase the available space of high rise buildings by using high strength concrete (f'_c > 70 MPa) and high strength rebars (f_y > 685 MPa). Material consumptions and member section sizes can be further reduced owing to the upgrade of strength. However, the nature of brittleness of high strength may also cause early cover spalling and other ductility issues. Addition of steel fibers is an alternative as transverse reinforcement. Highly flowable strain hardening fiber reinforced concrete (HF-SHFRC) has excellent workability in the fresh state and exhibits the strain-hardening and multiple cracking characteristics of high performance fiber reinforced cementitious composites (HPFRCC) in their hardened state. The objective of this study is to investigate the feasibility of implementing HF-SHFRC in New RC building systems, particularly for beam-column joints as an alternative of transverse reinforcements. Four full-scale exterior beam-column joints, including two specimens with intensive transverse reinforcements and two specimens made of HF-SHFRC without any stirrup, are tested. Test results show that the HF-SHFRC specimens perform as well as specimens with intensive transverse reinforcements regarding failure mode, ductility, energy dissipation and crack width control. Integration of New RC building systems and HF-SHFRC can assuring construction qualities and further diminish labor work and give infrastructure longer service life, and eventually lower the life-cycle cost.

1 Introduction

The purpose of New RC project was aimed to reduce the member sections and increase the available space of high rise buildings by using high strength concrete (f'_c > 70 MPa) and high strength rebars (f_y > 685 MPa). Material consumptions and member sections can be further reduced owing to the upgrade of strength. However, the nature of brittleness of high strength may also cause early cover spalling and other ductility issues. Addition of steel fibers is an alternative as transverse reinforcement. Highly flowable strain hardening fiber reinforced concrete (HF-SHFRC) has excellent workability in the fresh state and exhibits the strain-hardening and multiple cracking characteristics of high performance fiber reinforced cementitious composites (HPFRCC) in their hardened state.

A beam-column joint takes an important role in the force transferring mechanism in a special moment resisting frame. Thus, severe damage within a joint panel may cause deterioration in overall seismic performance of the frame. As a joint is subjected to high shear strength, code recommendations suggest arranging closely hoop transverse reinforcement to prevent joins from shear failure, which leads congested joints and results in difficult construction works. With high tensile strength and good toughness behavior, applying HF-SHFRC in joint panels is expected to reduce the amount of transverse reinforcement. The objective of this study is to investigate the feasibility of implementing HF-SHFRC in New RC building systems, such as beam-column joints and base columns as an alternative of transverse reinforcements. Test results show that the HF-SHFRC specimens perform as well as specimens with intensive transverse reinforcements regarding failure mode, ductility, energy dissipation and crack width control.

2 Experimental programs

Four full-scale exterior beam-column joints, including two specimens with intensive transverse reinforcements and two specimens made of HF-SHFRC without any stirrup, are tested.

2.1. Materials

The cementitious materials used in this study were ASTM Type I Portland cement; ground granulated blast furnace slag and class C fly ash. The coarse aggregate had a maximum size of 9.5 mm and consisted of solid crushed limestone from a local source, with a density of about 2.70 g/cm^3. A polycarboxylate-based

* Corresponding author: wcliao@ntu.edu.tw

superplasticizer was used in the mixture. Hooked steel fiber with circular cross-section was used, with normal tensile strength of 2300 MPa and aspect ratio of 79. The concrete compressive strengths were determined in accordance with ASTM C-39 standard compressive tests on at least six 100×200 mm cylinders. The concrete specified compressive strengths were 70 and 100 MPa. The slump flow of HF-SHFRC used in this study is 580mm so that the workability is not the issue during fabrication. Two different sizes of reinforcing bars were used: D25 (5.07 cm^2) with yield stress of 685 MPa for longitudinal reinforcement, and D13 (1.27 cm^2) with yield stress of 785 MPa for transverse reinforcement. The baseline tensile test for a single control reinforcing bar to determine yield strength, ultimate strength and bar strain at yield stress in accordance with ASTM E8/E8M and ASTM A370.

2.2 Specimen design

The seismic performances of exterior beam-column joints are assessed by applying a double curvature, cyclic lateral loading experiment. The columns had a clear height of 2800 mm with a square cross-section of 600×600 mm. Beam sectional size is 400×700mm and the net length is 3500mm. The four beam-column joints are labeled as LAMV, HAMV, LAMV_SF and HAMV_SF, respectively. LA represents as low axial load which is equal to 0.1 A_g/f'_c; HA represents as high axial load which is equal to 0.45 A_g/f'_c. MV represents medium shear ratio which is defined as the ratio of shear demand $V_{jh,u}$ and shear capacity V_n of joints, which is 0.7 in this experimental program, where $V_{jh,u}$ is the total horizontal joint shear force and V_n is the nominal shear strength of joint.

LAMV and HAMV are considered as control group (Chao [3]), and LAMV_SF and HAMV_SF are the experimental group. All these specimens have the identical flexural strength in column and beam section since their reinforcement layouts of beam and column sections are the same. The joints of LAMV and HAMV consist of dense transverse reinforcement to comply with ACI 318-11, while no transverse reinforcement is arranged in joints by using HF-SHFRC instead in LAMV_SF and HAMV_SF. The design parameters are listed in Table 1.

2.3 Test setup

The test of the specimens performed in National Center for Research on Earthquake Engineering (NCREE) using Multi-Axial Resting System (MATS) that built in 2007. The MATS conducted double-curvature cyclic deformation. The maximum axial and lateral load that can be applied by MATS is 60 MN and 7MN, respectively. The lateral force was set from the hydraulic actuator that placed in the bottom as shown in Fig. 1(a) and the loading protocol is shown in Fig. 1(b). For each cycle of each loading phase, once target drift was reached, loading was temporarily stopped and the crack patterns were marked and recorded. The experiment was continued until a significant loss in axial loading capacity of column was observed. Significant loss in axial loading capacity was interpreted using the following guidelines: rapid reduction in axial loading capacity; or sudden MATS loading stroke greater than or equal to 20 mm. Experiments in this study were conducted under high axial forces, resultantly the P-Δ effect was noticeable.

Table 1 Design parameters

Specimen ID Section	$\frac{A_{sh}}{sb_c}$ (%) V_f (%)	f'_c (MPa)	f_y (MPa)	f_{yt} (MPa)	$\frac{N_u}{A_g f'_c}$
LAMV 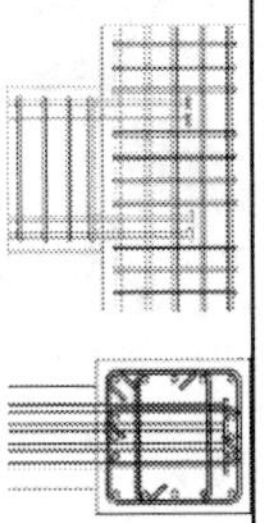	1.0 0	70	685 (D25)	785 (D13)	0.1
LAMV_SF 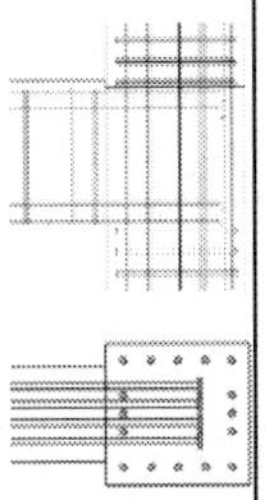	0 1.5	70	685 (D25)	-	0.1
HAMV 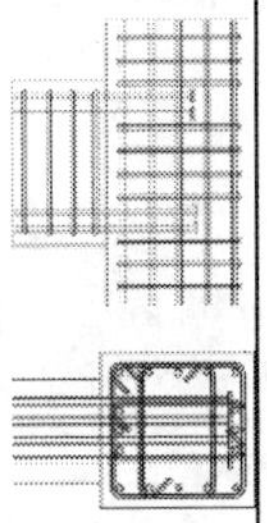	1.0 0	70	685 (D25)	785 (D13)	0.45
HAMV_SF 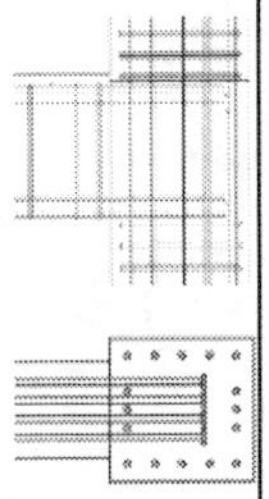	0 1.5	70	685 (D25)	-	0.45

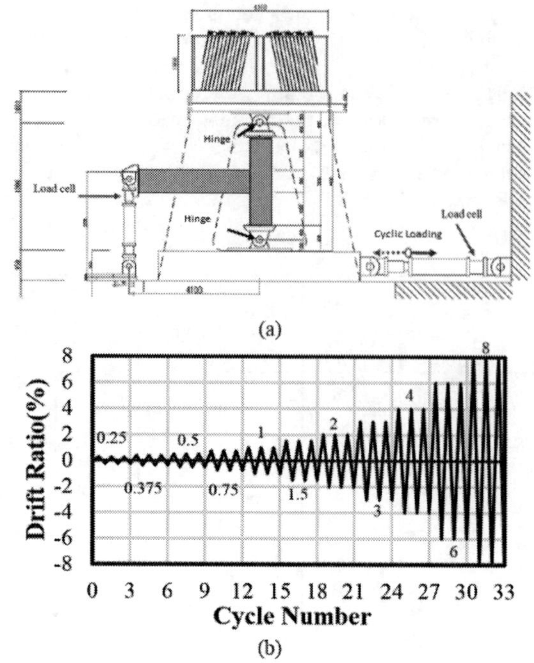

Fig. 1 (a) Test setup with MATS; (b) Loading protocol

3 Test results

3.1. LAMV and LAMV_SF

Figure 2 (a) shows the hysteresis loops of LAMV and LAMV_SF, which are modified with P-Delta effect. Their hysteresis loop envelopes are also shown in figure 2 (b). The apparent yielding plateau can be found in both LAMV and LAMV_SF, which stand for the B type failure mode. Their hysteresis loops and envelopes are almost identical since the yielding and failure of beam governs the cyclic behaviors. It is worth mentioning that there is no transverse reinforcement in the joint of LAMV_SF.

Their hysteresis loops present a linear status before the story drift ratio reaches 0.29%; as the story drift ratio exceeds 0.29%, lateral force declines slowly along with the increase of the story drift ratio. The lateral force declines obviously till the second and third cycle of the 8% drift ratio loop. The maximum lateral forces of LAMV and LAMV_SF are 487 kN and 488 kN as reaches 5.38% and 8%respectively. While the experiment progress to the third cycle of 6%, the longitudinal steel bar in the beam occurred local buckling, and failed in the plastic zone as 8%. The experiment ended up in the second cycle of 8%.

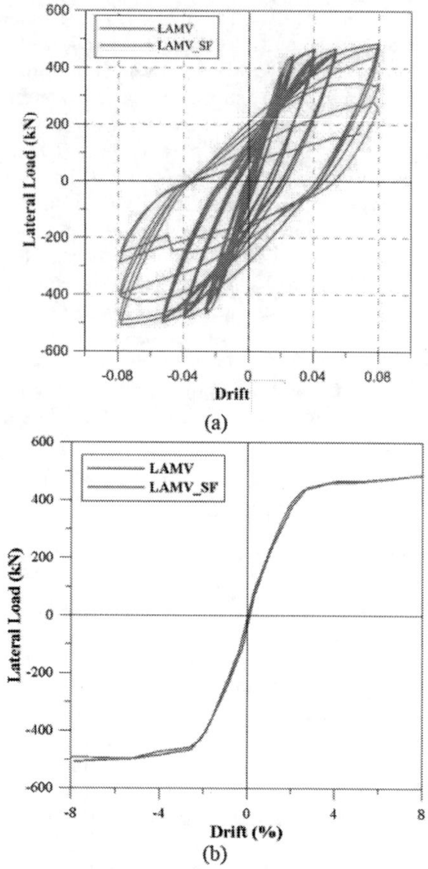

Fig. 2 (a) Hysteresis loops; (b) hysteresis loop envelopes of LAMV and LAMV_SF

3.2. HAMV and HAMV_SF

Figure 3(a) illustrates the hysteresis loops of HAMV and HAMV_SF, which are modified with P-Delta effect as well. Their hysteresis loop envelopes are compared in figure 3(b). It is noted that the inversion of the hysteresis loop can be observed before modified with P-Delta effect due to additional lateral force caused by high axial load. Thus, the lateral force of MATS needs to be reversed to get the expected displacement. It is found that there is also a clear yield plateau in HAMV with B type failure mode. The specimen performs as a linear status before the story drift ratio reaches 0.27% and its maximum lateral forces are 478kN as reaches ±7.93%.

Almost identical cyclic behavior of HAMV_SF can be observed before the second cycle of 4% since it suddenly fails. From the photo at specimen failed and the hysteresis loop (Fig. 3(a)), it can be conjectured that the failure of HAMV_SF caused by inadequacy of confinement, rather than insufficiency of shear capacity as BJ type failure. Even though HAMV_SF failed at 4.0% drift ratio, its performance is still qualified as acceptable joint in terms of stiffness, strength and energy dissipation required by ACI 374 [2].

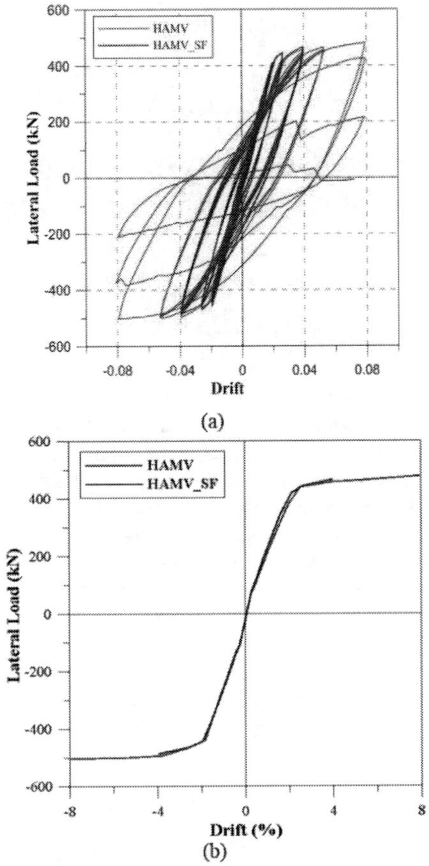

Fig. 3 (a) Hysteresis loops; (b) Hysteresis loop envelopes of HAMV and HAMV_SF

4 Conclusions

This study presents lateral cyclic loading tests on HF-SHFRC beam-column joints to examine its adequacy to substitute transverse reinforcement in the joint region under different axial loading levels. LAMV_SF and LAMV have similar or even better cyclic behaviors, whereas HAMV_SF does not perform as well as HAMV on cracking width control, this implies HF-SHFRC can fully replace transverse reinforcement under low axial load. Proper transverse reinforcement should be applied for confinement purpose in the case of high axial load even though shear capacity of HF-SHFRC is sufficient. HF-SHFRC specimens perform as well as specimens with intensive transverse reinforcements regarding failure mode, ductility, energy dissipation and crack width control. Application of HF-SHFRC to New RC building systems can assuring construction qualities and further diminish labor work and give infrastructure longer service life, and eventually lower the life-cycle cost.

References

1. ACI Committee 318, *Building Code Requirements for Structural Concrete* (ACI 318-14), American Concrete Institute, (2014)
2. ACI Committee 374, *Acceptance Criteria for Moment Frame Based on Structural Testing and Commentary*, American Concrete Institute, (2005)
3. Chao, W.-F., *Seismic Behavior of Headed-Bars in High-Strength RC Exterior Beam-Column Joints with P-Delta Effect*, Master Thesis, National Taiwan University of Science and Technology, Taipei, Taiwan, (2016)

Compressive Strength of Fly ash-based Geopolymer Concrete with a Variable of Sodium Hydroxide (NaOH) Solution Molarity

Herwani[1,*], *Ivindra* Pane[2], *Iswandi* Imran[2], and *Bambang* Budiono[2]

[1]Doctoral Program Student, Faculty of Civil and Environmental Engineering, ITB Bandung 40132, Indonesia
[2]Professor at Faculty of Civil and Environmental Engineering, ITB Bandung 40132, Indonesia

Abstract. Geopolymer concrete is a new material made by activating the raw materials which contain many elements of silica and alumina. Compressive strength of geopolymer concrete produced was influenced by the concentration of the activator solution. This paper presents an experimental investigation into fly ash-based geopolymer concrete. Research objective was to investigate the effects of alkaline activator solution (AAS) molarity on compressive strength of geopolymer concrete. Variable of the test were a solution to sodium hydroxide was chosen as the activator solution. Concentration of sodium hydroxide solution used was 10 M, 12 M and 14 M with ambient curing. The specimen is made of concrete cylinder with diameter 10 cm and height 20 cm as many as 9 pieces each variable. Compressive strength tests is performed when the concrete is 7, 14, and 28 days old. Results of the test are indicated that the increasing of sodium hydroxide (NaOH) solution concentration leads to improve the compressive strength of geopolymer concrete. The optimal compressive strength of geopolymer concrete was achieved at a concentration of sodium hydroxide solution (NaOH) of 12 M. Geopolymer concretes compressive strength only achieves around 50-60% of the planned.

1 Introduction

Geopolymer concrete is a kind of concrete that not uses Portland cement (PC) [1, 2]. Fly ash is one of the materials that can be used as a binder in its mixture [2, 3]. The geopolymer concrete is formed from a chemical reaction and not from hydration reaction [4]. The most impressive characteristic is that geopolymer concrete has high compressive strength and durability [5,6,7,8]. One of the conclusions of the research result was that constrained geopolymer concrete material is one of the materials suitable to develop for earthquake-disposed to areas [9]. To obtain a good geopolymer concrete, the type of activator ought to be adjusted to the compound contained in fly ash. A proper composition is highly expected for a chemical reaction to take place perfectly. The activator usually used is Sodium Hydroxide 8 M-14M and Sodium Silicate (N_2SiO_3) at a ratio between 0.4 and 2.5 [5].

Different areas have coals with different mineral contents, so that the characteristics of fly ash produced greatly vary [8]. The composition of fly ash mineral content affects the characteristics of the mechanic properties of geopolymer concrete. Compressive strength is one of the main characteristics of concrete.

Some researchers have developed mixed designs with diverse curing methods and variables to make good quality geopolymer concrete. Palomo et al. [10] researched geopolymer concrete by using class-F fly ash material. They used four different solutions at a ratio of mass between alkaline activator and fly ash from approximately 0.25 to 0.30. The ratio of the molars of SiO_2/K_2O or SiO_2/Na_2O solution was around 0.63-1.23. Compressive strength was found after carrying out a curing for 24 hours at 65°C, i.e., achieved more than 60 MPa for the mixture that used a combination of sodium hydroxide and sodium silicate activators. Xu, H and van Deventer [11] conducted a research and reported that the ratio of alkali solution mass and alumina-silicate was approximately 0.33 for geopolymer reaction to take place. The maximum compressive strength obtained achieved 19 MPa after curing for 72 hours at 35oC. Van Jaarsveld et al. [12] conducted a research by using a ratio of alkali mass of 0.39. The material used was 57% fly ash mixed with 15% kaolin. The alkaline solution was composed of 3.5% sodium silicate, 20% water, and 4% sodium or potassium hydroxide. The compressive strength obtained achieved 75 MPa. Hardjito and Rangan [5] conducted a research on a mixture of geopolymer concrete. The concentration of sodium hydroxide (NaOH) used was around 8M-16M. The ratio of mass between sodium silicate and sodium hydroxide was around 0.4-2.5. For the moment, the ratio of mass between alkaline activator and fly ash was around 35%. In conclusion that the higher molarity of sodium hydroxide causes the higher compressive strength of geopolymer concrete. The mass ratio between sodium silicate and sodium hydroxide is high then the compressive strength produced by the geopolymer concrete will increase. Compressive strength of

* Corresponding author: herwaniuntan@yahoo.co.id

geopolymer concrete by 7 days old achieved 67 MPa after curing for 24 hours at 600C. Januarti Jaya Ekaputri et al [13] conducted a research of the mechanical property of fly ash-based geopolymer concrete derived from Jawa Power Paiton. The variables used were activator solution molarity and ratio of sodium silicate (Na2SiO3) and sodium hydroxide (NOH). The optimum compressive strength was obtained at 10 M molarity with a ratio of sodium silicate to sodium hydroxide 1.5. The compressive strength produced achieved 48.59 Mpa. Tabassum R.K [14] conducted a research on the effect of the concentration of sodium hydroxide solution on geopolymer concrete mixture. A maximum compressive strength by 28 days old was 40.21 MPa. Sodium Hydroxide (NaOH) concentration gave a significant impact in enhancing the properties where the optimum concentration of NaOH of 12 M gave better strength properties of geopolymer Concrete [15].

2 Experimental Method

2.1 Materials

Geopolymer concrete material used is taken from local query. Fine aggregate used Cimalaka sands, and coarse aggregate used stones derived from Cimahi. Classification of coarse aggregate is passes through 20-mm and 10-mm sieve. Classification of sand is passes through a 4.75-mm sieve. The raw material fly ash from PLTU Suralaya waste was used. To activate silica and alumina elements and other contain in fly ash, alkaline solution was used. Sodium hydroxide and sodium sulphate were alkalis chosen activator in geopolymer concrete mixture. Sodium hydroxide was derived from a chemical store in form of flake. Sodium sulphate could be derived from chemical stores in bottle or drum container. Before being mixed in concrete, sodium hydroxide should be firstly solved according to the several of concentration. A XRF (X-ray fluorescence spectroscopy) test was done to determine the content of fly ash mineral. The content of elements in fly ash was as follows:

Table 1. The composition of chemical contents of Suralaya fly ash (in % of weight)

Fly Ash	SiO_2	Al_2O_3	Fe_2O_3	CaO	MgO	Na_2O	LoI
Specimen A	42.29	17.41	12.10	10.47	7.39	1.38	6.15
Specimen B	41.91	17.21	12.01	10.39	7.31	1.48	5.86
Specimen C	42.25	17.48	11.88	10.34	6.91	1.43	6.26

Super plasticizer was added on geopolymer concrete mixture to obtain good quality and workability. The quantity and kind of super plasticizer used as needed and recommended doses in Indonesian concrete standard (SNI).

2.2 Mixed Design and Manufacture of Specimens

The investigation of fine aggregate and coarse aggregate characteristics were done before making a mixed design. Sieve analysis is done on the coarse aggregate and fine aggregate. In addition to sieve analysis, there are also tests on specific gravity, water absorption, mud content, organic content and weight of volume. The results of the mud content on fine aggregate is 19.2%, it is necessary to wash before use on concrete. After washing, the fine aggregate was again tested for mud content. The magnitude only reaches 2%. The specific gravity of fine aggregate on SSD condition is 2,531 and water absorption is 2,147%. Whereas density of coarse aggregate diameter 9.5 - 19.5 mm is 2,618 and water absorption 3,609%.

The mixed design was created by using a nomogram proposed by Partogi. The nomogram is actually for paste phase geopolymer, but by a modification it can be used for geopolymer concrete mixed design. The nomogram shows the relation between the reactant molar ratio and weight of the aggregate to the paste ratio. The ratio of reactant molar was found from the comparison of water molarity and solid material molarity (H2O/Na2O+SiO2+Al203). The composition of mixed design used can be seen on table 2 below.

Table 2. The composition of geopolymer concrete mixed design (in Kg/m3)

Materials	Mix 1	Mix 2
Coarse Aggregate (SSD)	1200	1142.7
Dia. 4.5 – 9.5 mm	540	457.1
Dia. 9.5 – 19.5 mm	660	685.6
Fine Aggregate (SSD)	600	521.8
Fly Ash	450	410
NaOH Solutions	80	110
Na_2SiO_3 Solutions	120	120
Superplasticizer	9	8.2
Added water	-	2.3

The mixing of concrete using a pan mixer. Each mixture was prepared with the concentration of NaOH solution being 10, 12, and 14 M. The specimens were cylindrical diameter 10 cm and height 20 cm. Ambien curing performed on the specimens. The number of sample of 3 pieces for each variable. The tests were performed when geopolymer concrete was 7, 14, and 28 days old. Total sample is 27 pieces.

3. Results and discussion

3.1 Fresh Geopolymer Concrete

Characteristics of fresh geopolymer concrete is different from Portland cement-based normal concrete. The colour of geopolymer concrete is very dark (black), depending on the colour of the source of fly ash. The use of extra water in a very small amount of mixed design will have an effect on the characteristics of the geopolymer concrete. Fresh geopolymer concrete mixture behaves like a gel with a very thick consistency. In the first mixed design, there is no extra water. The mixture is rigid causing some porous specimens. Value of slump test was only 5 cm. The second Mix design showed different behaviours than the first mix design. The mixture consistence was relatively liquid, so the mortar of fresh concrete was more elastic. The value of its slump test was around 12-16 cm. Concrete casting into a mould was easier, and porous on the specimens became decreasing. Geopolymer concrete was more rapidly hardened as compared to Portland cement-based concrete. The result of observation during concrete casting in the mould, it needed around 150-250 minutes into hard. There was no bleeding on the surface during polymerization process. There were white stains on the surface of geopolymer concrete. The cause was sodium element existing in the concrete reacted with outside air to produce crystallization.

3.2 Concrete compressive strength

Fig. 1 and 2 showed the development of the compressive strength of geopolymer concrete at different concentration of NaOH solution. Maximum compressive strength of geopolymer concrete at 28 days was obtained on sodium hydroxide concentration of 14 M, i.e. 29.34 MPa. The biggest increase in compressive strength took place on a molarity of 12 M, i.e., 32.56%, while on molarity of 14 M it was only 8.51%. Other behaviour shown by geopolymer concrete was that it has a high initial compressive strength. By 7 days old, geopolymer concrete with NaOH molarity of 14 M has achieved 17.48 MPa. Then, the increase in concrete was not so large. The imperfect condition of sample, being porous, possibly caused less consistent behaviours. Fig. 3 showed the relation between the compressive strength of geopolymer concrete by 28 days old and the molarity of NaOH. From the graphic it could be seen that the trend of increases in compressive strength on molarities of 12 M and 14 M was not so significant. The indication that 12-M morality approached the optimal was not clearly seen yet. In the second mixture design, it could be seen that geopolymer concrete with the molarity of NaOH solution of 12 M was the optimum mixture (Fig. 4).

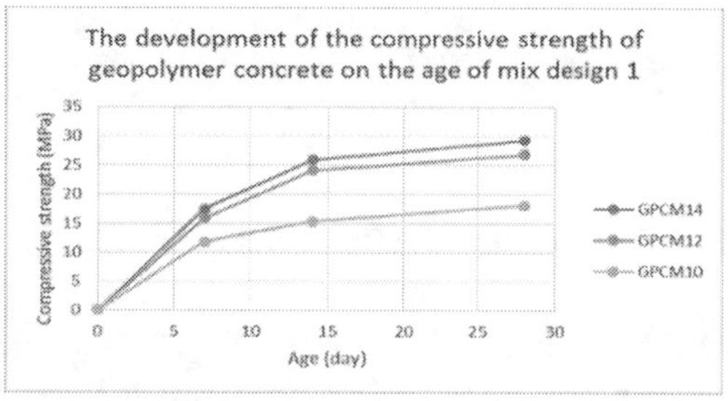

Fig. 1. The development of the compressive strength of geopolymer concrete on the age of the first mix design

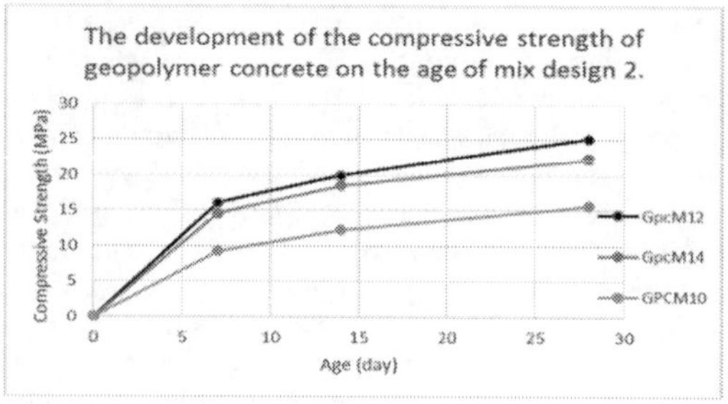

Fig. 2. The development of the compressive strength of geopolymer concrete on the age of the second mix design.

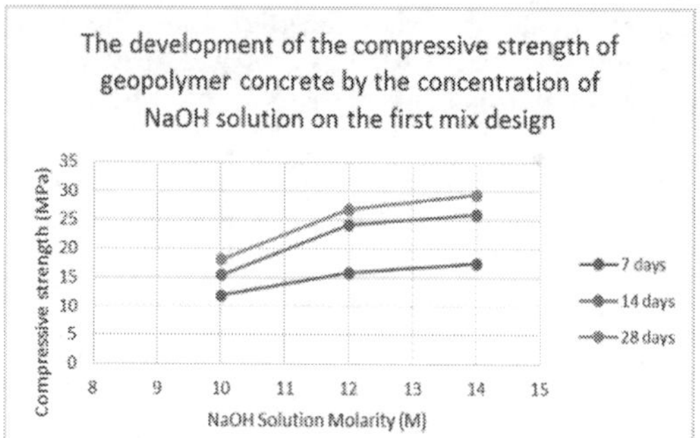

Fig. 3. The development of the compressive strength of geopolymer concrete by the concentration of NaOH solution on the first mix design.

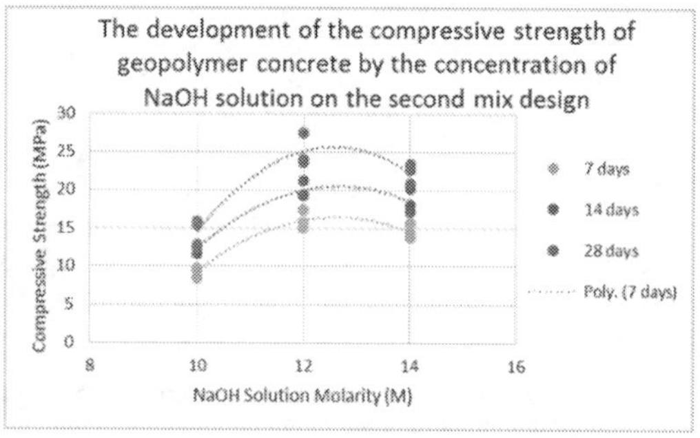

Fig. 4. The development of the compressive strength of geopolymer concrete by the concentration of NaOH solution on the second mix design

Compressive strength design of geopolymer concrete was 50 MPa with a slump of 10-15 cm. The compressive strength achieved was only 50-60% of the planned compressive strength. Hot curing on geopolymer concrete is strongly advised to get better concrete quality [5, 8]. The compressive strength in the second mix design is decreased compared to the first mix design (Figure 5). This behaviour indicates that the quality of geopolymer concrete is very sensitive to the use of extra water.

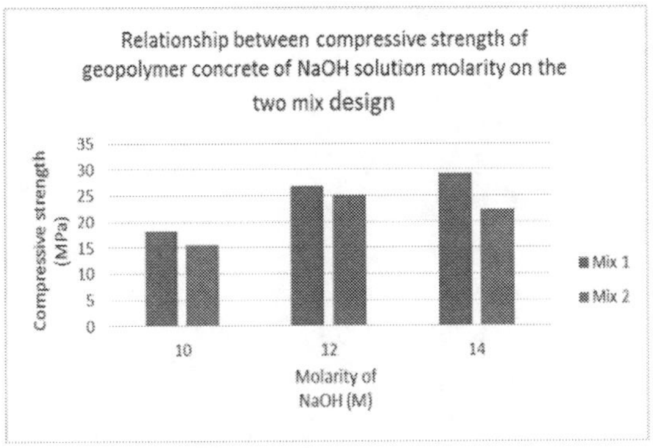

Fig. 5. The development of the compressive strength of geopolymer concrete by the concentration of NaOH solution on the two mix designs

3.3 Shrinkage of Geopolymer Concrete

Observation of shrinkage was conducted on the development of the weight of specimens. The observation was intended to see the effect of room temperature on the weight of geopolymer concrete. Figure 6 shows the development of the weight of specimen on various NaOH solution molarity. This indication was used to see the shrinkage on the weight of specimen.
The smallest shrinkage occurs in geopolymer concrete with the NaOH solution molarity of 14 M (figure 6). Increasing the amount of water used in geopolymer concrete has resulted in shrinkage. The molarity of the NaOH solution decreases, the shrink percentage increases.

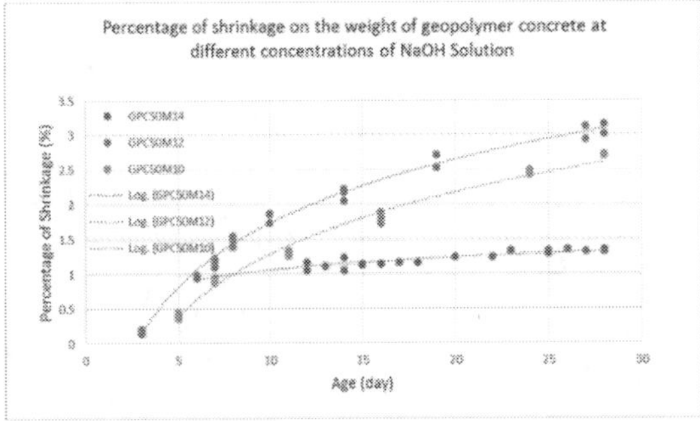

Fig. 6. Percentage of shrinkage on the weight of geopolymer concrete at different concentrations of NaOH

4. Conclusions

Fresh geopolymer concrete has a high consistency so the behaviour of mixture is rigid. This characteristic could be improved by adding super plasticizer so that the mortar becomes more elastic and workable. The setting time was faster by 150-250 minutes, observed during producing concrete mixture. The higher concentration of NaOH solution then the required setting time becomes longer Geopolymer concrete achieved an optimum compressive strength on NaOH solution molarity of 12 M. The compressive strength achieved was only 50-60% of the planned one. The smallest shrinkage took place on geopolymer concrete with NaOH solution molarity of 14 M.

Acknowledgements: This research was funded by Directorate General of Higher Education Indonesia through Excellent Research for Higher School Grant. We also thank to PLTU Suralaya for providing fly ash.

References

1 Herwani, I. Imran, I. Pane, *Studi Eksperimental Karakteristik Beton Geopolimer Berbahan Dasar Fly Ash*, Prosiding (KNPTS), Goro, G.L. (eds), pp. 87 (2015)

2 P.H. Simatupang, I. Imran, I. Pane, & B. Sunendar, *On the Development of a Nomogram for Alkali Activated Fly Ash Material (AAFAM) Mixtures*, Journal of Engineering Technology & Science, **Vol. 47**, No. 3, 231-249 (2015)

3 Siyal, A.A., Azizli, K.,A., Man, Z., & Ullah, H., *Effects of Parameters on the Setting Time of Fly Ash Based Geopolymers Using Taguchi Method*, in Proceedings 4th International Conference on Process Engineering and Advanced Materials, Procedia Engineering 148, pp. 302 – 307 (2016).

4 J. Davidovits, '*Chemistry of geopolymeric systems, terminology*', in Proceedings of Second International Conference on Geopolymer, pp. 940 (1999)

5 D. Hardjito, & B.V. Rangan, *Development and properties of low-calcium fly ash-based geopolymer concrete*, Research Report, Perth, Australia: Curtin University of Technology (2005)

6 P. Duxson, A. Fernandez-Jimenez, J.L. Provis, G.C. Lukey, A. Palomo, & J.S.J. Van Deventer, *Geopolymer Technology: The Current State of the Art*, Material Science, **42**, pp. 2917-2933 (2007)

7 F. Pacheco-Torgal, J. Castro-Gomes, & S. Jalali, *Alkali Activated Binders: A review, Part 1, Historical Background, Terminology, Reaction Mechanisms and Hydration Products*, Construction and Building Materials, **22**, pp. 1305-1314 (2008).

8 P.H. Simatupang, '*Development of Alkali Activated Material to Make Green Material of Infrastructures*', PhD Dissertation, Civil and Environment engineering, ITB, Bandung (2013)

9 N. Ganesan, R. Abraham, S. Deepa Raj, & D. Sasi, *Stress–strain behaviour of confined Geopolymer concrete*, Construction and Building Materials, Elsevier, pp.326-331 (2014)

10 A. Palomo, M.W. Grutzeck, & M.T. Blanco, *Alkali-Activated fly ash cement for future*, Cement and Concrete Research, **29** (8); pp.1323-1329 (1999)

11 H. Xu, and J.S.J. Van Deventer, *The Geopolymerisation of Alumino Silicate Minerals*, International Journal of Mineral Processing **59** (3): 247-226 (2000)

https://doi.org/10.1051/matecconf/201814701004

12 J.G.S. Van Jaarsveld, J.S.J. Van Deventer, & G.C. Lukey, *The effect of composition and temperature on the properties of fly ash and kaolinite-based geopolymers*, Chemical Engineering Journal, 4001:1-11, (2002)

13 J.J. Ekaputri, Triwulan, dan O. Damayanti, *Sifat Makanik Beton Geopolimer Berbahan Dasar Fly Ash Jawa Power Paiton Sebagai Material Alternatif* Jurnal PONDASI, volume **13** no.2 (2007)

14 R.K. Tabassum, A. Khadwal, *Effect of Sodium Hydroxide Concentration on Various Properties of Geopolymer Concrete*, IJETR pp. 2454-4698, Vol.3, Issue-10, October (2015)

15 R. M. Hamidi, Z. Man, K.A. Azizli, *Concentration of NaOH and the Effect on the Properties of Fly Ash Based Geopolymer*, 4th International Conference on Process Engineering and Advanced Materials, Procedia Engineering 148, pp. 189 – 193, (2016)

Shear Stress Distribution in the Opening Chords of Hybrid R/C T-Beams with Opening

Jonie Tanijaya[1]

[1]Civil Engineering Department, Paulus Christian University of Indonesia

Abstract. This study is carried out to evaluate the potential of three hybrid T-beams with web openings theoretical shear stresses distribution. The shear stresses at the opening edges were plotted at the working stage, yielding stage and collapse stage for these three tested beams. The available experimental results from the previous research was compared to the finite element results as well as the developed analytical. The shear stress distribution at the middle of the top and bottom chords of the opening in pure bending region are zero. At the upper and lower corners of the opening occurs the maximum shear stresses. The maximum shear stress occurs at the right lower corner chord at the high moment edge and at the left upper corner chord at the low moment edge in beams with openings at high shear and high flexural – shear region. Furthermore, an extensive parametric study is performed on these beams to find the distributing ratio of the shear force between the opening chords. The shear force at an opening in hybrid R/C T-beam is carried by the top and bottom chords of the opening according to the area – moment of inertia root ratio with the correction factor 0.70.

1 Introduction

Web opening in beams occur quite often in practice to provide convenient passage of environmental services to provide passage for utility duct and pipes, it also translates into substantial economic savings in the construction of a multi-storey building [1]. It has been practiced that pipes and ducts are usually hanged below the floor beams, and covered by a suspended ceiling for its aesthetic purpose. These openings can be of different shapes and sizes as circular, square or rectangular. As a results, story heights in buildings can be reduced resulting in major cost savings. Some openings generally give rise to excessive deflections and stresses which have to be taken into consideration. Utilization of high strength concrete in construction sector has increased due to its improved mechanical properties compared to ordinary concrete [2]. One such mechanical property, shear resistance of concrete beams is an intensive area of research. Providing an opening in beam develops cracks around the opening due to stress concentration [3]. Therefore, it is important to predict accurately the distribution of shear between the top and bottom chords of the opening. Beam openings may be of different shapes, sizes and are generally located close to the supports where shear is dominant. The behavior of such beams is affected by the size and location of the openings [4]. In practical life it is quite often use to provide convenient passage of environmental services which reduce the story heights of buildings and weight of concrete beams as it improves the demand on the supporting frame both under gravity loading and seismic excitation which results in major cost saving.

Openings should be positioned on the concrete beams to provide chords with sufficient concrete area for developing ultimate compression block in flexure and adequate depth for providing effective shear reinforcement [5]. The presence of openings will transform simple beam behaviour into a more complex behaviour, as we induce a sudden change in the dimension of the beam's cross section. The ultimate strength, crack width shear strength and stiffness may also be seriously affected [6]. Dinakaran and Sastry [7] suggested that the amount of the transverse shear distributed to each chord is depending on its respective cross-sectional areas. On the other hand, Barney et al. [8] proposed that the total shear force should be proportioned according to the moments of inertia of the chords for un-cracked section. Once a chord(s) has cracked, the distribution of shear will depend on the extent of cracking.

More recently, Kennedy and Abdalla [9] formulated a root ratio, dependent on the area and moment of inertia of each chord, for distributing the shear force between the top and bottom chords in beams with opening. The design for a beam with a large opening had been proposed that the top and bottom cross members of the opening could be assumed to behave like a *vierendeel* panel. The top and bottom chord members were expected to carry shear and axial force acted from the bending moment [10]. The openings should be provided so that chords have enough concrete area to develop the ultimate compression block in flexure and sufficient

* Corresponding author: jonie.tanijaya@gmail.com

 https://doi.org/10.1051/matecconf/201814701005

depth to provide efficient shear reinforcement. They should not be deeper than one-half the beam depth and should be located not closer than one-half the beam depth from supports or concentrated loads. In continuous beams that generally occur in practice, reduction in stiffness due to the provision of opening through webs causes a redistribution of internal forces and moment, the amount of which needs to be evaluated before a design can proceed. In some researches, generally the opening are applied on normal concrete beams with conventional aggregates. Light weight concrete application has enormous benefits in an effort to reduce the weight of structure, so theoritically it could minimize structure dimension which also means construction cost savings can be achieved. This research is using light weight concrete utilizing locally expanded clay aggregates applied on T-beams with web opening.

Finite element analysis is more superior to the conventional design approach in terms of robustness and applicability for such severe nonlinear structural behaviour [11]. Though, the development of a finite element model (FEM) may need intensive material testing to incorporate into the material model in any of the finite element (FE) packages available [12]. In the present paper, using the 'ABAQUS' computer program [13], theoretical analyses were carried out on three hybrid R/C T-beams with rectangular web opening. The opening was provided in the web of the beams, i.e. in the pure bending region, the high shear region, and the high flexural-shear region, respectively (Figure 1).

The finite element method was used to conduct an extensive parametric study on the static response of simply supported beam with rectangular opening. The shear stresses at the opening edges at the working stage, yielding stage and collapse stage were plotted for the three tested beams. A parametric study was undertaken to detemine the ratio of the shear force carried by each chord of the opening. The available experimental results from the previous research have been compared to the finite element results as well as with results obtained from the developed analytical approach. Good agreement was observed [14].

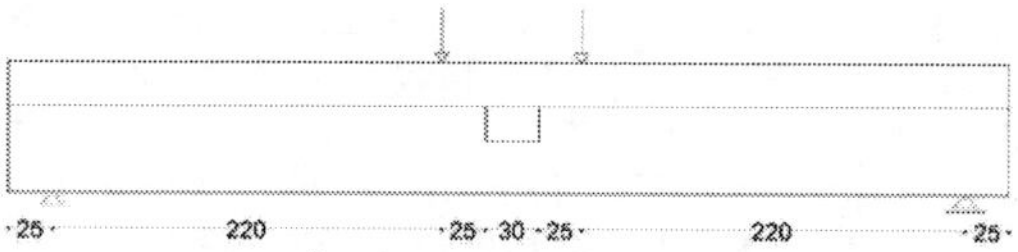

a. Opening in pure bending region BLT-1

b. Opening in high shear region BLT-2

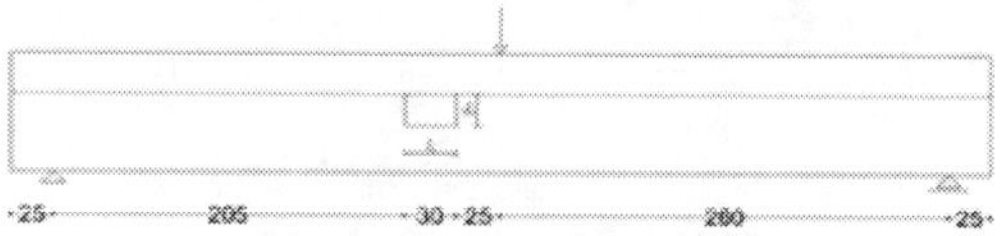

c. Opening in high flexural-shear region BLT-3

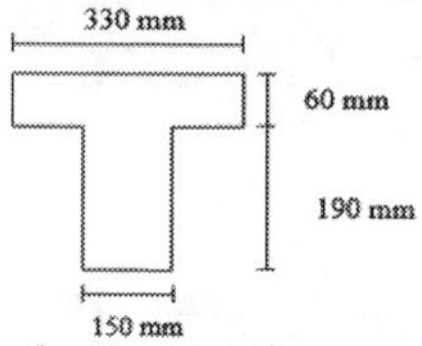

d. Cross section

Fig. 1. Beam identification

2 Distribution of Shear Stress in the Vicinity of the Opening Chords

Distribution of shear stresses in the vicinity of the opening chords from BLT-1. BLT-2, and BLT-3 with the difference loading are shown in Figure 2, 3 and 4. Figure 2(a) shows that the shear stress distribution at the middle of the top and bottom chords are zero at the working, yielding and collapse stages. This is due to the position of the beam opening (pure bending zone). From the Figure 2(b), it can be observed from these results that at collapsed stage (ultimate load) the maximum shear stress occurs at the opening corners with different stages. At the upper corner of the opening occurs positive shear stress and at the lower corner occurs negative shear stress. Figure 3(a) shows that prior to the beam collapse (close to the ultimate load) the maximum shear stress at middle of the top and bottom chords increase. It can be seen that the top chord carries only about 1/3 of the bottom chords. Figure 3(b) and 3(c) show the maximum shear stress occur at the right lower corner chord at the high moment edge and at the left upper corner chord at the low moment edge.

Figure 4(a) shows shear stress distributions at the lower middle chord occurs different sign. Shear force at the lower chord decrease with the applied load increasing, this is caused reducing cross dimension of beam. After cracking (Figure 4 (b) and (c)), with the increasing load, the shear stress distributions at the edge of opening are similar to those for BLT-2, the maximum shear stress occur at the right lower corner chord at the high momen edge and at the left upper corner chord at the low moment edge. Shear force at the opening chords cause secondary moment at the vicinity opening, which can caused crack at the corners of the chord.

https://doi.org/10.1051/matecconf/201814701005

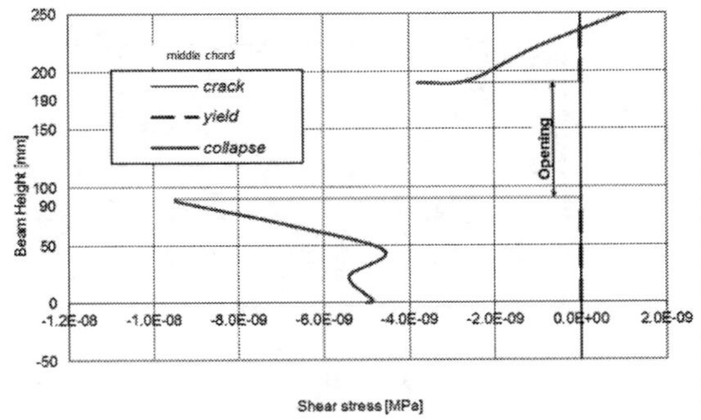

a. Middle chords

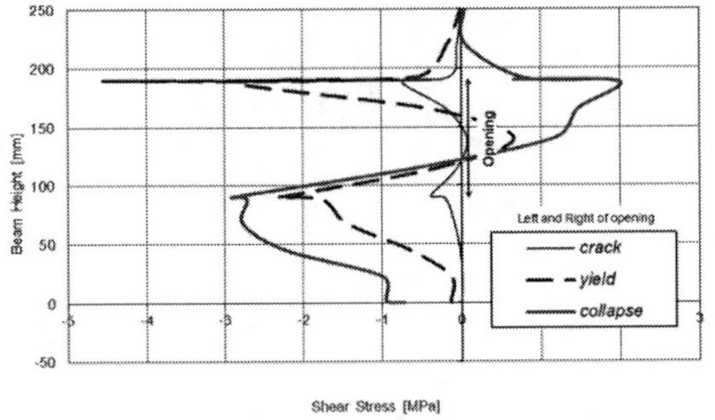

b. Opening edges

Fig. 2. Shear stress distribution fo BLT-1

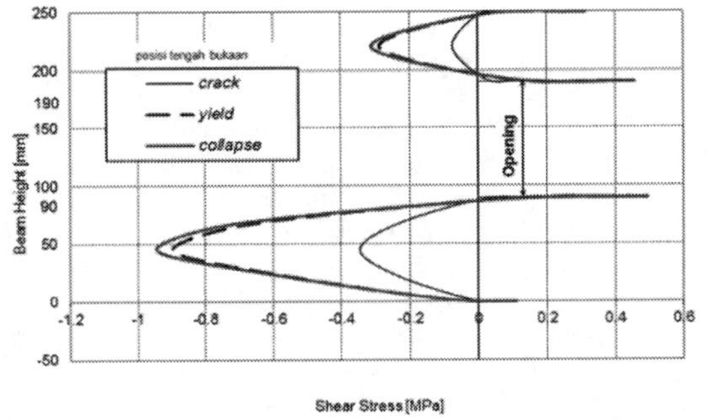

a. Middle chords

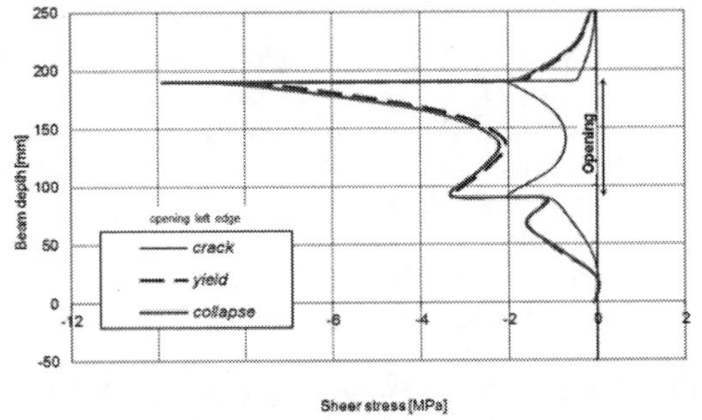

b. Left-opening edges

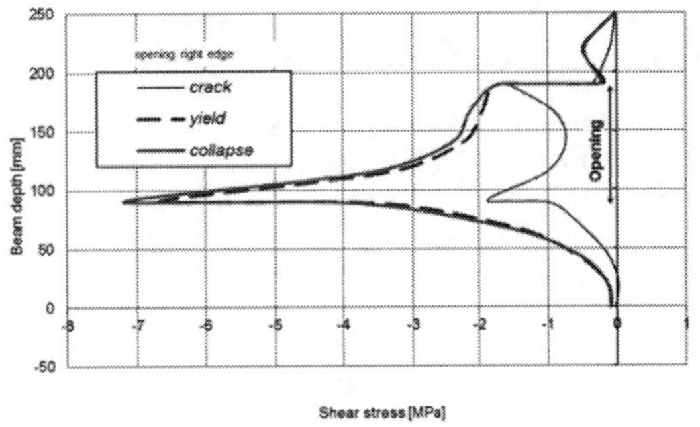

c. Right-opening edges

Fig. 3. Shear stress distribution fo BLT-2

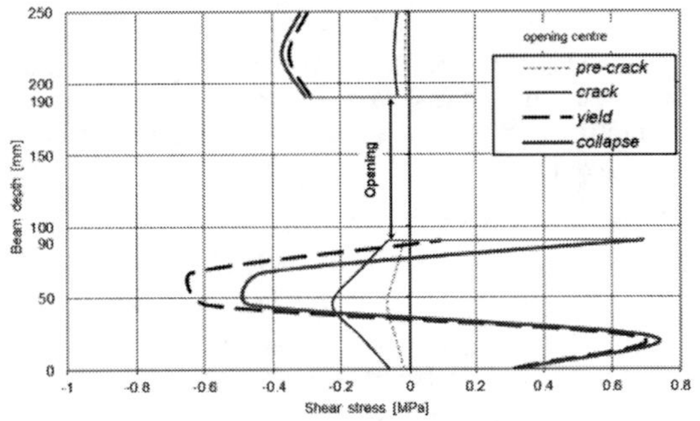

a. Middle chords

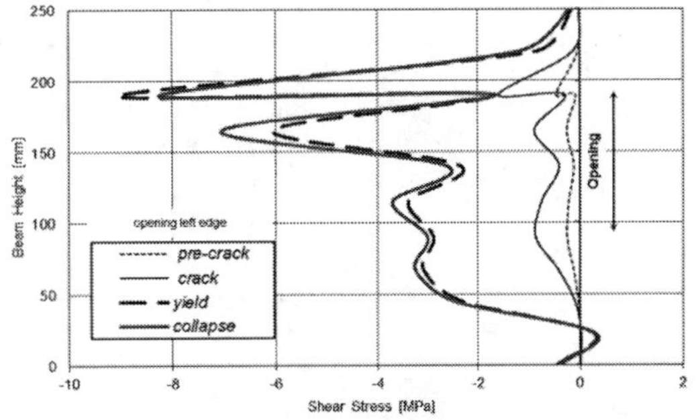

b. Left-opening edges

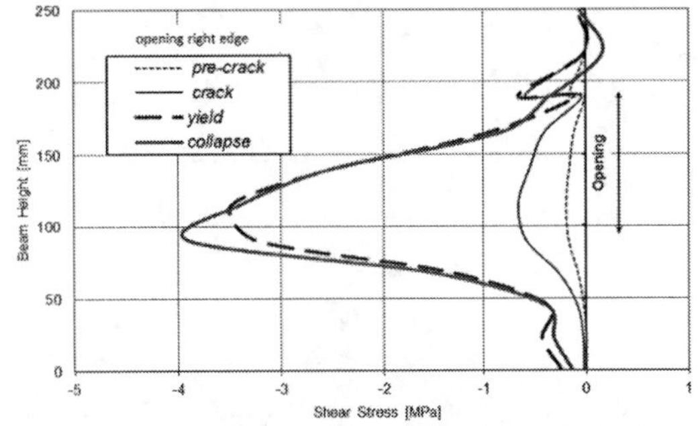

c. Right-opening edges

Fig. 4. Shear stress distribution fo BLT-3

https://doi.org/10.1051/matecconf/201814701005

3 Distribution of Shear Force between Top and Bottom Chords of Opening

The area ratio $[A_b/(A_t+A_b)]$ named Area-1, the moment of inertia ratio $[I_b/(I_t+I_b)]$ named Inertia-1, the area root ratio $[\sqrt{A_b}/(\sqrt{A_t}+\sqrt{A_b})]$ named Area-2, the moment of inertia root ratio $[\sqrt{I_b}/(\sqrt{I_t}+\sqrt{I_b})]$ named Inertia-2 and the area-moment of inertia root ratio $[\sqrt{A_bI_b}/(\sqrt{A_tI_t}+\sqrt{A_bI_b})]$ named Roof-1 were compared to the results of the finite element analysis (Figure 5).

It was found that the results lie in between moment of inertia and the area – moment of inertia root ratio, which means that the shear force carried by each chord depends on both the area and moment of inertia of this chord. Applying statistical methods to the finite element results of the investigated beams, the shear force carried by each chord of opening was determined ($[\sqrt{A_bI_b}/0.7(\sqrt{A_tI_t}+\sqrt{A_bI_b})]$ named Roof-2). Figure 6 shows the shear force carried by the bottom chord of an opening in T-section beams, using the area – moment of inertia root ratio given by:

$$\frac{V_b}{V} = \frac{(\sqrt{A_b I_b})}{0.7(\sqrt{A_t I_t} + \sqrt{A_b I_b})} \tag{1}$$

in which:

V_b = shear force carried by the bottom chord,

V = total shear force at the opening,

A_b, A_t = cross sectional areas of bottom and top chords, respectively,

I_b, I_t = moments of inertia of bottom and top chords, about their centroidal axes, respectively.

One can observe that the results based on the proposed root ratio given by Eq. (1) is the most suitable one to determine the proportion of shear carried by each chord. Therefore should be corrected by a factor of 0.7 for hybrid reinfoced concrete T-beams.

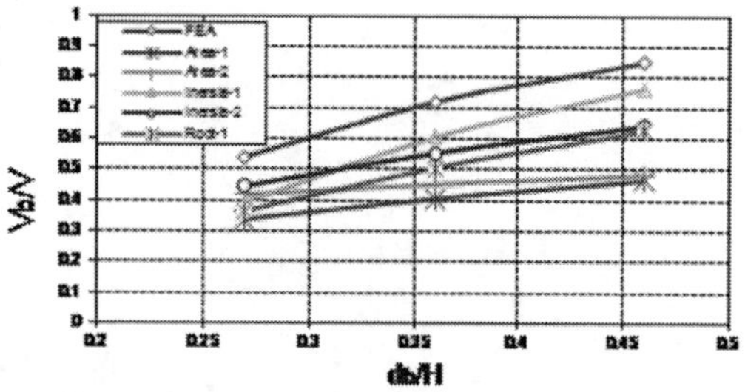

a. BLT-2

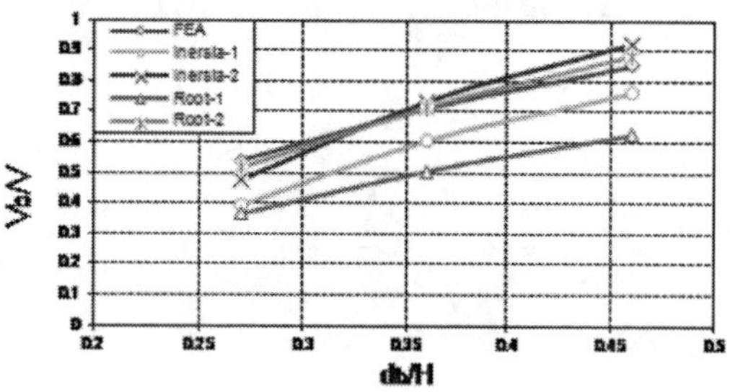

b. BLT-3

Fig. 5. V_b/V vs d_b/H

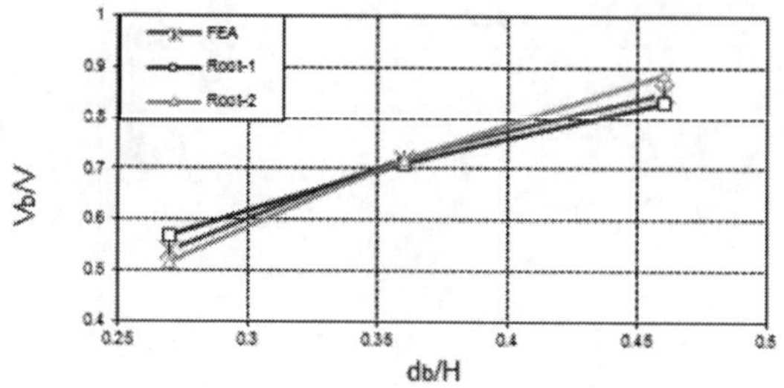

a. BLT-2

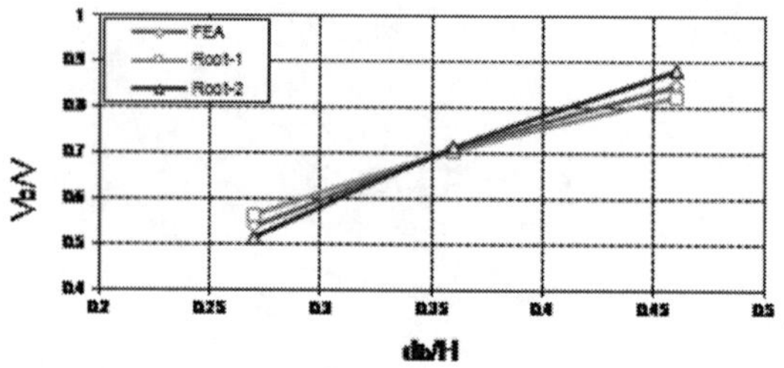

b. BLT-3

Fig. 6. Vb/V vs db/H

4 Conclusion

The shear stress distribution at the middle of the top and bottom chords of the opening in pure bending region are zero. At the upper and lower corners of the opening occurs the maximum shear stresses. The maximum shear stress occur at the right lower corner chord at the high moment edge and at the left upper corner chord at the low moment edge in beams with openings at high shear and high flexural – shear region. The shear force at an opening in hybrid R/C T-beam is carried by the top and bottom chords of the opening according to the area-moment of inertia root ratio with correction factor 0.70.

This research was partially supported by Paulus Christian University of Indonesia. This support is gratefully acknowledged by the author.

References

1. Fawzy, K., et al., *Performance of RC Beams with Web Opening Subjected to Pure Torsion Strengthened with CFRP*, International Journal of Engineering and Innovative Technology (IJEIT) Vol. 4, July 2014, pp. 197-204)
2. Ali, A.Y., and Zghair, M.G., *Experimental Investigation and Nonlinear Analysis of Hybrid*, Al-Qadisiyah Journal For Engineering Sciences, Vol. 8, No. 2, 2015, pp. 99-119
3. Al-Sheikh, S.A., *Flexural Behavior Of Rc Beams With Opening*, Concrete Research Letter Journal ISSRES, Vol. 5 (2) June 2014, pp. 812-824
4. Amiri, S., et al., *A review of design specifications of opening in the web for simply supported RC beams*,

Journal of Civil Engineering and Construction Technology Vol. 2(4) April 2011, pp. 82-89

5. Hafiz, R.B., et al., *Effects of Opening on the Behavior of Reinforced Concrete Beam*, IOSR Journal of Mechanical and Civil Engineering (IOSR-JMCE) Vol. 11, Issue 2 Ver. VII (Mar- Apr. 2014), pp. 52-61
6. Vivek, R., and Madhavi, T.C., *Behaviour Of Reinforced Concrete Beam With Web Openings*, ARPN Journal of Engineering and Applied Sciences Vol. 11 no. 9 May 2016, pp. 5993-5995
7. Barney, G.B., Corley, W.G., Hanson, J.M., and Parmelee, R.A., *Behavior and Design of Pre-stressed Concrete Beam with Large Web Openings*, PCI Journal, Vol. 22, No.6, Nov.-Dec. 1997, pp. 32-61
8. Dinakaran, V., and Sastry, M.K. (1984), *Behaviour of post-tensioned prestressed concrete T-beams with large web openings*, Indian Concrete J., Bombay, India, Vol. 58, No. 2, 1984, pp. 34-38
9. Hibbitt, H.D., Karlsson, B.I., Sorensen, E.P., ABAQUS *Version* 5.7, *Finite Element Program*, Hibbitt, Karlsson and Sorensen Inc., Providence, Rhode Islang 02906-4402, 1997
10. Chen, C.C., et al., *Experimental Study Of Steel Reinforced Concrete Beams With Web Openings*, The 14th World Conference on Earthquake Engineering October 2008
11. Long, X., and Lee, C.K., *Modelling of Two Dimensional Reinforced Concrete Beam-Column Joints Subjected to Monotonic Loading*, Journal of Advances in Structural Engineering September 2015, pp. 1461-1474
12. Sinaei, H., et al., *Evaluation of reinforced concrete beam behaviour using finite element analysis by ABAQUS*, Scientific Research and Essays Vol. 7(21) June 2012, pp. 2002-2009
13. Tanijaya, Jonie, *Parametric Study of Web Opening Dimensions on Hybrid Reinforced Concrete T-Beams*, International Journal of Structural Analysis and Design, Vol. 1 : Issue 3, September 2014, ISSN : 2372-4102, SEEK Digital Library, pp. 60 – 64
14. Kennedy, J.B., and Abdalla, H.A., *Static Response Of The Pre-Stressed Girder With Openings*, J.Struct. Engrg., ASCE, Vol. 118, No. 2, 1992, pp. 488-504

Effect of Pressure and Heat Treatments on the Compressive Strength of Reactive Powder Concrete

Masdar Helmi[1,*], *Matthew* R Hall[2,3], and *Sean P Rigby*[4]

[1]Department of Civil Engineering, Faculty of Engineering, University of Lampung, Jalan S. Brojonegoro No. 1 Bandar Lampung, Indonesia 35145
[2]GeoEnergy Research Centre, Faculty of Engineering, University of Nottingham, University Park, Nottingham NG7 2RD, UK
[3]British Geological Survey, Environmental Science Centre, Keyworth, Nottingham NG12 5GG, UK
[4]Division of Manufacturing and Process Technologies, Faculty of Engineering, University of Nottingham, University Park, Nottingham NG7 2RD, UK

Abstract. This paper presents the corresponding compressive strength of RPC with variable pressure combined with heating rate, heating duration, and starting time of heating. The treatments applied were 8 MPa static pressure on fresh RPC prims and heat curing at 240 °C in an oven. The compressive strength test was conducted at 7-d and 28-d. The images of RPC morphology were captured on the surface of a fractured specimen using Scanning Electron Microscopy in Secondary Electron detector mode to describe pore filing mechanism after treatments. The results show that a heating rate at 50 °C/hr resulted in the highest compressive strength about 40 % more than those at 10 or 100 °C/hr. A heating duration of 48 hours led to the maximum compressive strength. Heat curing applied 2 days after casting resulted in the maximum compressive. Heat curing had a signicant effect on the compresssive strength due to the acceleration of both reactions (hydration and pozzolanic) and the degree of transformation from tobermorite to xonotlite. It is concluded that the optimum condition of treatments is both pressure and heat curing at 2-day after casting with a rate of 50 °C/hr for 48 hours.

1 Introduction

Reactive powder concrete (RPC), categorized as a type of ultra-high performance concrete (UHPC) due to having compressive strength > 140 MPa, has a different composition to ordinary concrete. RPC has no coarse aggregate and contains a higher proportion of fine particles (<600 μm) comprising silica fume, quartz sand and other admixtures. RPC can be developed by controlling three main variables: composition, pressure during setting period, and post-set heat curing [1].

Firstly, composition can be improved through a microstructural engineering approach by using fine aggregate only (no coarse aggregate), reducing water-binder ratio, lowering the $CaO–SiO_2$ ratio through the introducing of silica-rich pozzolona [2, 3] and adding fibre reinforcement to increase ductility [4]. Secondly, pressure applied during the setting period also influences the strength of concrete. Pressure induces three main effects: reducing the entrapped air, removing the excess water, and eliminating the porosity caused by chemical shrinkage [1]. Meanwhile, heat curing induces the acceleration of the pozzolanic reaction between amorphous silica and calcium hydroxide as well as the silicate and aluminate hydration reaction, and also the modification of hydrate microstructure [5, 6].

An RPC mixture typically contains a high amount of cement at a dosage of 800-1000 kg/m^3. The use of cement in a high amount impacts on the heat of hydration and leads to potential risks associated with shrinkage and creep-induced strain problems, and also the production cost. To solve the cost problems, Yazici *et al.*, [7] conducted research on the composition of RPC mixtures to find the optimal amount of finer materials from industrial waste (SF, PFA, and GGBS) as cement replacement. They suggested that RPC comprising 35wt% SF and up to 40 % GGBS (as cement replacement) could result in a compressive strength of more than 200 MPa. However, the presence of mineral admixtures can slow down the hydration reaction, lower the heat of hydration, and decrease shrinkage. This is often a difficult problem to overcome in mixtures with high cement content and low water/cement ratios [8].

Most scholars have argued that the strength of RPC is mainly developed by the pozzolanic reaction and hydrate transformation after heat curing. Heat curing, which is one of the most important treatments in RPC, can be applied by various methods. Yazici *et al* [7] applied heat curing on RPC prism specimens (after demoulding) in an autoclave at 2.0 MPa and 210 °C. These conditions were achieved within a 2.5 h heating period, and the heat curing was then applied for 8 hours. Cwirzen [6] used a method of steam curing (95 %RH, 90 °C) to optimise the conditions for heat curing by varying two factors: the starting time of heating after demoulding, and the duration of heating. Heat curing was applied to the RPC cube specimens with durations of 24, 48 and 168 hours, and starting times at 1, 2, and 4 days after casting the samples.

* Corresponding author: masdar.helmi@eng.unila.ac.id

He suggests that the optimal conditions of heating can be reached after 2 days of casting with a total duration of 48 hours producing the highest degree of hydration. Another heat curing method was applied by Sadrekarimi [9] and Tam *et al.* [10] in drying oven equipment with different rates and maximum temperatures. Sadrekarimi [9] placed samples in the oven with a constant heat rate of 10 °C/hr until it reached a temperature of 240 °C. Tam *et al.* [10] set first a drying oven at 100 °C then placed samples into the oven with a temperature rate of 50 °C/hr until it reached 250 °C. It should be noted that the duration and heating rate are important factors in heat curing since they affect the crystallisation of hydrates, pozzolanic reaction, microstructure development and associated bulk mechanical properties [11].

Heat curing treatment, which is generally applied after final set of cement at a temperature between 90 °C and 250 °C, contributes to both the progression of cement hydration and the acceleration of the pozzolanic reaction between silica fume (amorphous) and quartz flour (crystalline) [12,5,6,11]. Heat curing also modifies the micro structure of hydrates by changing the C-S-H chain length from trimer to pentamer [6]. When the temperature of curing increases to more than 200 °C, this condition leads to the transformation of crystalline hydrates and dehydration of the hardened paste. The transformation hydrate will be formed spontaneously when water extracted from hydrates is trapped in the centre of samples during transient heating [12] becoming tobermorite at a temperature of 150 °C and xonotlite at temperatures of both 200 and 250 °C [11].

It is clearly understood that the treatments of pressure and heat curing affect the strength of RPC which depend on three factors; the method of application, the temperature and duration of treatments [13]. Cwirzen [6] and Tam & Tam [11] have studied the effect of treatments on RPC containing high cement content with those variables. They used the composition of RPC with high cement content and not yet in low cement content due to using industrial waste as cement replacement. Therefore it is necessary to evaluate the optimisation of treatments on RPC mixtures.

2 Experimental method

2.1. Materials

The main binder was CEM1 52.5 cement (Rugby, Cemex UK), in accordance with BS EN 197-1:2011. The silica fume was used along with microsilica grade 940-D (Elkem, Switzerland) and Ground Granulated Blastfurnace Slag (GGBS) (Hanson, UK) was used as a partial cement replacement. Quartz sand (David Ball Ltd. UK) was used where, $G = 2.65$ and water absorption = 1.5 %, with gradation conforming to BS 1881-131. This study used three grades of sand: A (2.36 mm - 1.18 mm), C (0.60 mm - 0.30 mm) and E (0.15 mm - 0.09 mm). These grades were mixed in a ratio of 2A:1C:1E (by mass) to produce a suitable particle size distribution as suggested by Yazici et al. [7].

2.2 Sample fabrication

The composition used industrial waste due to the fact that this technique can reduce the use of cement by up to 40 %, while giving compressive strength of more than 200 MPa [7]. A trial mix was done in order to get enough workability of fresh RPC mixture. Table 1 presents the result of the final composition.

The mixing of RPC containing very fine materials and low water-cement ratio is different from a conventional concrete mixing process. The process of mixing in this study was adopted from [8]. Initially, dry materials, including binder and quartz sand, were put in a stainless mixer bowl, mixed at a speed of ~120 rpm for 2 minutes and then followed by mixing with a speed of up to ~450 rpm for another 2 minutes. The half volume of water and SP premixed were then added and mixed for about 5 minutes until the mixture had a consistent appearance. After the mixture became consistent, it was cast into oiled steel moulding which contained three prisms with dimension of 40 x 40 x 160 mm.

Table 1. Composition of RPC per cubic meter

PC (kg)	SF (kg)	GGBS (kg)	Quartz (kg)			Water (kg)	SP (litre)
			A	C	E		
498	208	332	488	244	244	200	55

2.3 Sample treatments

2.3.1 Pressure treatment

Pressure treatment was used to compact RPC samples in the setting period. Static pressure was applied on the top of fresh mixtures using a compressive machine for about 5 hours after casting. The load was increased and controlled by a computer at a rate 1 kN/s to reach at 150 kN, which resulted in a stress of 8 MPa for each sample. The compressive machine was held in this load position for about 2 minutes in order to tighten four corner bolts in the moulding. Using this method will allow the load to continue pressing the samples during the setting period [13]. Releasing the pressure and demoulding the samples was done after 48 hours of casting in order to let the mixture harden properly and not stick to the moulding [10]. Figure 1 presents the illustration of the mould and the samples.

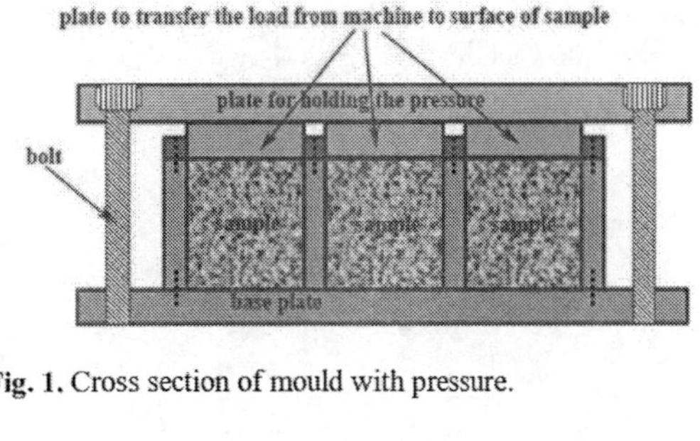

Fig. 1. Cross section of mould with pressure.

2.3.2 Heat curing treatment

Heating the sample in the setting period is a unique treatment for RPC, which is different from treatments in common concrete [13]. The heat curing process in this study uses a drying oven similar to an experiment by [10]. The samples are put on a rack and preheated at 40 °C for 2 hours, and then increased to 240 °C. The rate, durations, and starting time of heating were used as variables in the preliminary study to get the optimum condition which were indicated by compressive strength. There were three different levels in every set; rates were 10, 50, and 100 °C/hour; durations were 12, 24, and 48 hours; and starting times were 1, 2, and 3 days. After heat curing finished, the temperatures of the oven was decreased at 2 °C/min rate to 40 °C, then samples were cooled at room temperature. The samples were then submerged in water at a temperature of 10 °C until the day of testing, either 7-d or 28-d.

2.4 Method of testing

2.4.1 Compressive strength

The compressive strength test was conducted at 7-d or 28-d using one part of the two broken pieces of prism conforming to BS EN 196-1:2011. The contact pressed area of the sample was 40 x 40 mm and the height was 40 mm. The samples were tested at 7-d and 28-d under compressive machine with a maximum capacity of 500 KN and load rate of 2400 N/s. This test results in the maximum load, which was then used to calculate compressive strength by dividing it by the surface of the contact area. The average of compressive strength is taken from three samples.

2.4.2 Paste morphology

Samples for SEM tests were made from fractured bulk specimens of RPC cubes. Three representative sub-samples were mounted fracture face down and cold mounted under vacuum in 2-part epoxy resin. Mounted samples were ground using SiC paper (400, 600, 800 and 1200 grit) on 20cm diameter wheels at a rotational speed of 250 rpm, followed by polishing using 6μm then 1μm diamond pastes. Polished samples were washed using acetone and dried under a hot air blower, followed by sputter coating with ~15nm thick carbon using an Edwards 306 vacuum Coater, as shown in Figure 2.

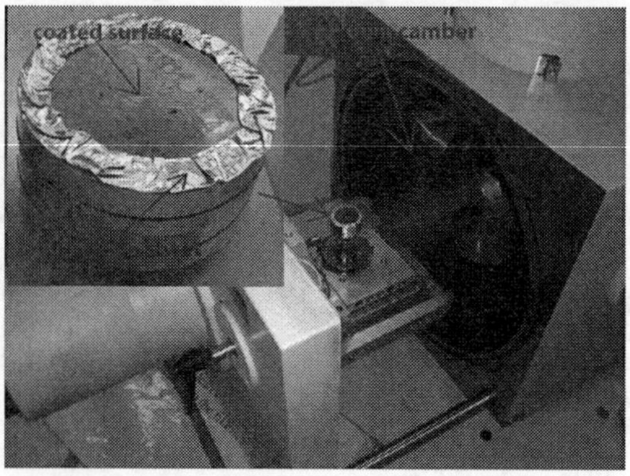

Fig. 2. Placing the mounted and coated sample in vacuum camber.

3 Result and Discussion

3.1 Compressive strength due to heating rate

The heating rate has a correlation with the speed of evaporation, this in turn influences the extent of air voids and hydration reaction in the concrete (Sadrekarimi, 2004). A low heating rate may decrease the dehydration rate by allowing sufficient time for thermal diffusion to occur, which influences the final compressive strength of the concrete. Therefore three different rates, namely 10, 50 and 100 °C/hr were evaluated in this study. Samples were tested in compression at 7-d and 28-d using two parts of a broken prism. The result is shown in Figure 3.

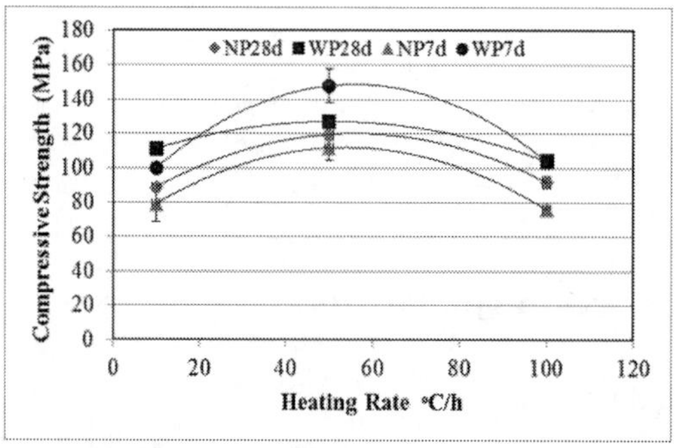

Fig. 3. Compressive strength-heating rate curves for a heating duration of 48 hours after 2 days of casting.

The compressive strength, at 7-d and a heating rate of 10 °C/hr, is 79 MPa for a non-pressured (NP) sample and 100 MPa with pressure (WP) [14]. This is stronger than the qualification of high strength concrete performance (HSCP) according to Neville [15]. At 28-d, the compressive strength continuously increases up to 89 MPa for NP and 111 MPa for WP. When the heating rate is increased to 50 °C/hr, the compressive strength at 7-d increases by 41 % (NP) and by 48 % (WP) compared to the numbers achieved for the 10 °C/hr samples.

At 28-d, however, compressive strength remains relatively stable for NP and drops about 14 % for WP compared to the same samples is at 7-d. It is supposed that the hydration and pozzolanic reactions have been accelerated with the application of heat but that micro cracks are also produced. Increasing the heat application level to 100 °C/hr has less effect on compressive strength compared to using 10 °C/hr. Compressive strength changes at this heat rate, either at 7-d or 28-d for both NP and WP, are within the calculated range of error for the measured results and can be ignored.

The lack of change in compressive strength with high temperature could mean that the high rate heating induces immediate evaporation of water from the C-S-H gel and causes a rapid increase in the pressure of pores which in turn increases the tensile stress in the cement paste. If this tensile stress is higher than the tensile capacity of RPC at 2-d (the point at which pressure is applied) microcracks are induced around the pores. Microcracks, which have been formed during the setting period due to natural shrinkage and grain movement after pressure application, could increase in number and width after heat curing. Consequently, the reason for achieving lower compressive strength in samples with heating rates of 10 and 100 °C/hr could be less effective hydration and pozzolanic reactions and the development of microcracks respectively [1].

3.2 Compressive strength due to heating duration

For this experiment, samples were heated to a maximum temperature of 240 °C with a ramp rate of 50 °C/hr in a drying oven. The maximum temperature was applied for durations of 12, 24 and 48 h. The measured compressive strengths in the samples at 7-d and 28-d are presented in Figure 4.

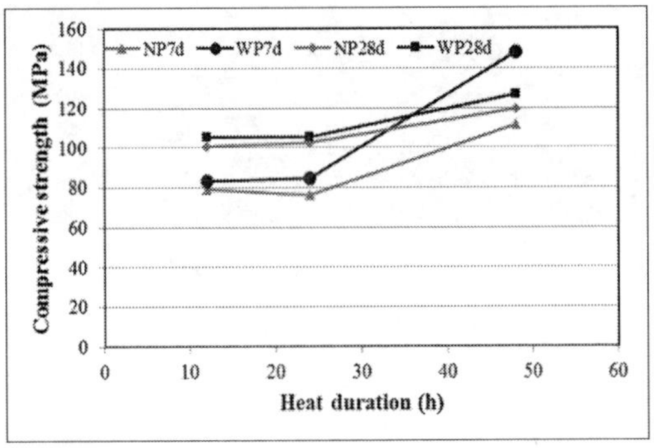

Fig. 4. Compressive strength-heating duration curves for a heat rate of 50 °C/hr after 2 days of casting.

As can be seen in Figure 4, the curve shows that there is no optimum compressive strength from the three variations of heat duration tested. However, the compressive strength tends to increase with an increase of duration for all samples. At 7-d, the compressive strength increases progressively and not linearly with heating duration. It is found that the compressive strength of samples with heating duration of 12, 24 and 48 h are 79.02, 76.34, 111.60 MPa for NP and 83.20, 84.62, 147.94 MPa for WP, respectively. The difference is less significant at 12 h compared to 24 h and becomes very significant to 48 h as there is increase of around 45 % (NP) and 75 % (WP). For the 28 day samples the trend is more gradual with the difference in compressive strength of both types of sample at around 20 %.

This result shows that the duration of heating is an important factor which affects the compressive strength of RPC and the properties of the microstructure [6]. The effect shows correlation with the acceleration of the rate of cement hydration and pozzolanic reactions for both silica fume and quartz sand. In addition, the duration of heat condition results in an increase in the strength of the RPC due to the amount of C-S-H produced and the change in crystal formation from tobermorite to xonotlite [11].

3.3 Compressive strength due to heating start time

For this experiment, three samples were heated under identical temperature conditions of 240 °C at a ramp rate of 50 °C/hr in a drying oven. The time of sample placement in the oven after casting was varied between samples with delays of 1, 2, and 3 days. The comparison of compressive strengths for the three different delay times at 7-d and 28-d is presented in Figure 5.

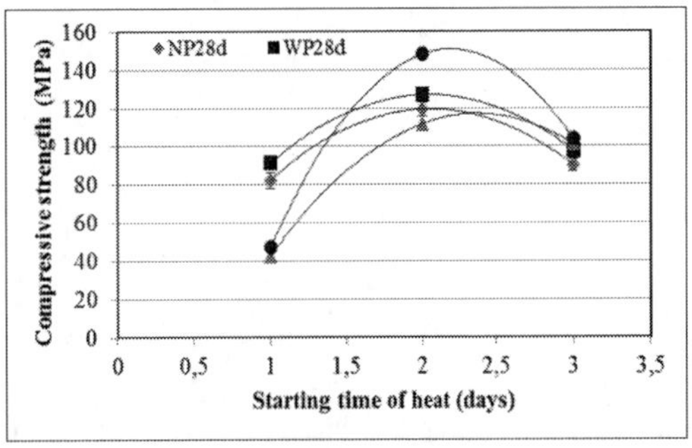

Fig. 5. Compressive strength-heated starting time curve for heating rate of 50 °C/hr and duration of 48 h.

Figure 5 shows that the optimum compressive strength is found when the heat condition is applied after a delay of 2 days from casting, both at 7-d and 28-d, for both NP and WP samples. Although the rate of strength development in concrete under normal curing conditions is influenced by the rate of hydration after one or two days [14], it will still increase gradually with time [16]. The results from applying heat curing in RPC samples at 3 days show disagreement with this theory, with a contradictory result of a 45 % reduction in compressive strength compared to the one with applied heat curing after 2 days. In any condition, the compressive strength of WP samples was always higher than NP samples for all conditions because pressure condition changes the physical microstructure of the mixture before final setting was achieved by a densification mechanism [1].

The reason for optimum compressive strength being achieved by heating at 2 days may be explained by hydration theory [16]. On the first day of hardening, tobermorite [C-S-H] gel was formed in much higher quantities than ettringite and portlandite [$Ca(OH)_2$] and starts to form C_4AF hydrate. During the second day, the volume of hydrated phases increased and the volume of ettringite decreased as it converted to monosulphate, and on the third day the density of hydrate phases began to increase significantly. Moreover the majority of ettringite converted to monosulphate prior to the onset of the hydrate phase densification at 2 days. Given that the above was a valid description. This suggests that the highest compressive strength, while heat curing was applied at 2 days, could be due to acceleration of hydrate phase densification whilst the risk of microcracks initiation/ enlargement resulting from thermal expansion of partially-hardened hydrate phases was avoided.

3.4 Morphology of cement paste

It is well known that pressure setting applied to fresh RPC mixtures can increase their density due to the reduction of entrapped air pores, both in terms of diameter and number, and the associated removal of free water content. On the other hand, the use of high cement content and low water-cement ratio results in a mixture of anhydrous and hydrated material. In addition, applying high temperature curing affects not only the prevention of the hydration process of cement grains but also induces microcracks in the hardened paste due to evaporation and the transformation of CSH gel [10].

3.4.1 Entrapped air pores

In general RPC has a denser matrix phase and is more homogeneous than normal concrete. However it still contains entrapped air pores with a wide range of diameters between 10 and 300 μm, which are formed by using high amounts of superplasticizer in the RPC mixture [8]. Entrapped air pores are generally spherical and much larger than any other type of pore.

In SEM images of cementitious materials, several kinds (and length scales) of pores can be observed as shown in Figure 6 - 9. Polished cross-section RPC specimens were prepared and imaged, and analysed using the secondary electron (SE) detector at 10 mm working distance.

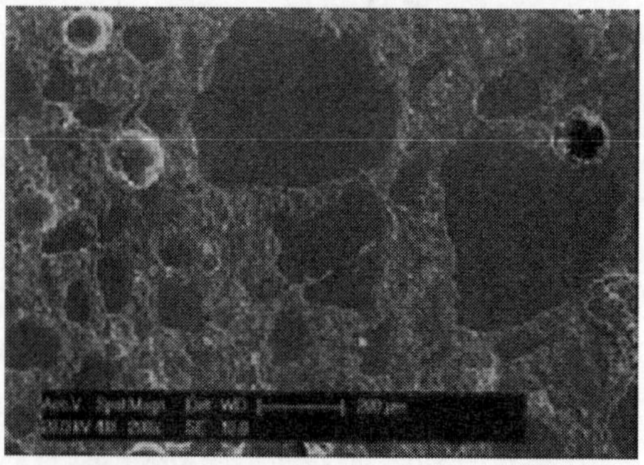

Fig. 6. Entrapped air on samples following no treatment,

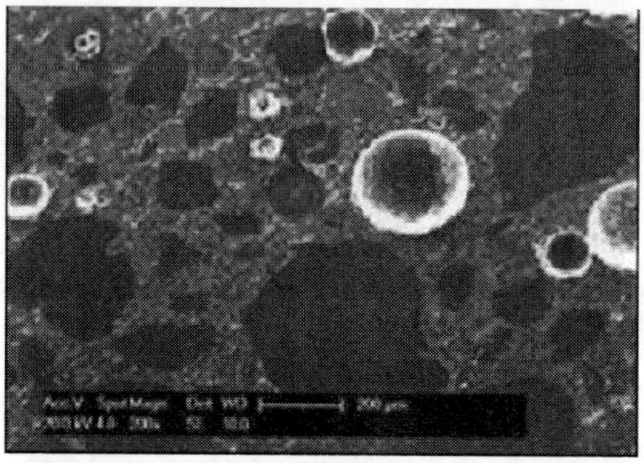

Fig. 7. Entrapped air on samples following heat curing,

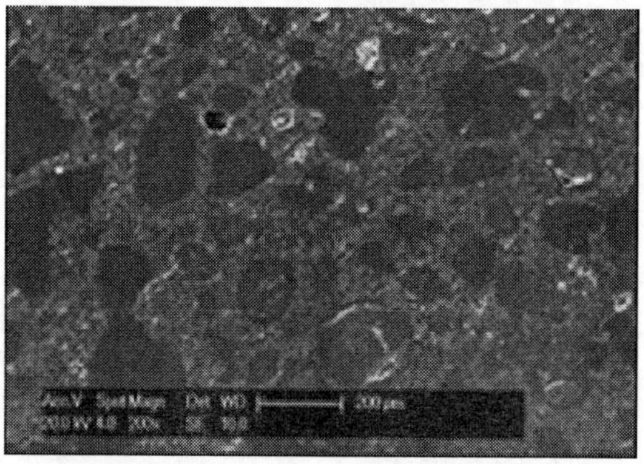

Fig. 8. Entrapped air on samples following pressure,

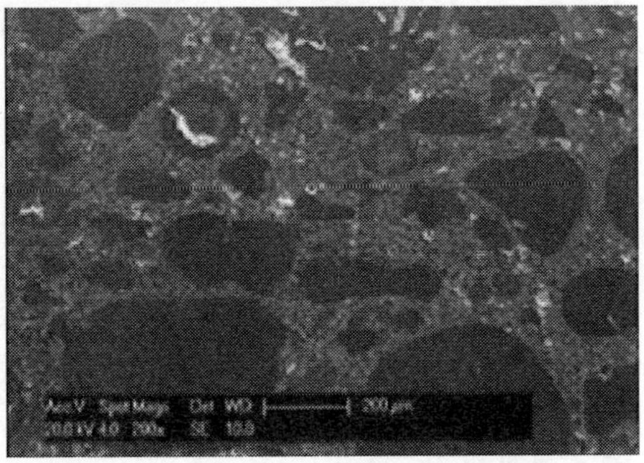

Fig. 9. Entrapped air on samples following pressure and heat curing.

Figure 6 and 7 show that RPC without pressure treatment had some entrapped air pores with a maximum diameter of about 200 μm. When pressure was applied to fresh mixtures, the number and diameter of large entrapped air pores were reduced by around one order of magnitude (Fig 8 and 9). In addition, the matrix of specimens appeared denser than those without pressure. It is suggested here that some of pores have been closed altogether with spill out of free water, whilst others were changed physically becoming smaller in diameter and greater in number. This finding supports most scholars' statements that applied pressure on fresh mixtures can reduce the pore volume and mean diameter, and consequently increase matrix density [1,8].

3.4.2 Pore filling after heat curing

Most scholars agree that heat curing induces a pore filling mechanism caused by the formation of pozzolan-derived C-S-H phases, the rate of formation for which is exothermically driven [1,6,11]. At a temperature of 90 °C it can accelerate the hydration reaction, and therefore transform both the microstructure and microchemistry of RPC [6]. Tam & Tam [11] found that tobermorite was characterised by plate-like crystals formed at a temperature of around 150 °C, whilst xonotlite was observed as needle-shaped crystals that occur at temperatures above 200 °C. At higher temperatures, between 200 and 250 °C, the chain length of C-S-H can be modified from trimer to pentamer or hexamer, with some dehydroxilation of cement gel due to both acceleration of hydration and the pozzolanic reaction [5,12]. The pore filling mechanism of RPC in this study is shown in Figure 10 - 11.

Figure 10 presents the pore inside appeared nearly clear with an irregular coarse surface, and small pores with a diameter of about 1 μm. The small platy-shaped hydrate, possibly found in RPC at ambient temperature curing, was identifiable as C-S-H [8]. The structure of this C-S-H typically was like a rigid gel which was amorphous or with poorly crystalline structure [10].

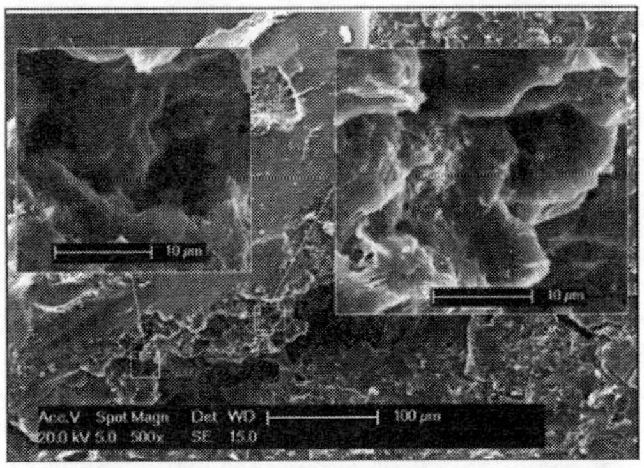

Fig. 10. Morphology on pore without heat curing. Insert: group of ettringite crystal located in pores.

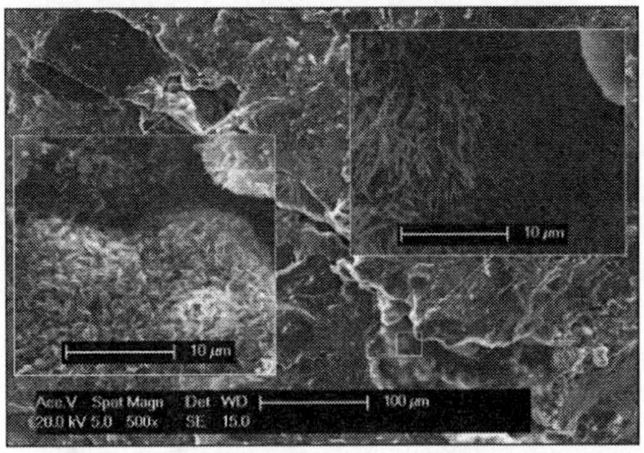

Fig. 11. Morphology on pore with heat curing. Insert: groups of tobermorite (platy-shape) and xonotlite (needle-shape) filled in pores. Heat curing induces pore filling mechanism.

Figure 11 shows some pores of specimens after heat curing, and they were filled by some platy-shaped and needle-shaped crystals. The platy-shaped crystals were identified as tobermorite ($Ca_{4+x}(H_{2-2x}Si_6O_{17}.5H_2O)$ or ($C_5S_6H_5$) which has properties such as: a hardness of 2.5 on the Mohs scale, relatively stable with a CaO/SiO_2 molar ratio of 0.83. The needle-shaped crystals were called xonotlite ($Ca_6(Si_6O_{17})(OH)_2$ or $C_6S_6H_2$), with properties similar to those of tobermorite but having a CaO/SiO_2 molar ratio of 1.0 [11].

This observation finds agreement with those of previous scholars [8,11], where high temperature curing could transform C-S-H to tobermorite or xonotlite and fill the pores inside the specimen [17]. The mechanism of the transformation was that heat treatment extracted water from the hydrates, which is then trapped inside the pores. It was firstly transformed to tobermorite, then secondly into xonotlite as a final structure [10,12]. This transformation mainly depended on both the temperature level and the duration of heating [10].

4 Conclusion

This study has verified the properties of RPC mixtures with reduced cement content (due to replacement by GGBS). The effects of pressure and high temperature curing treatment on compressive strength and paste morphology have also been investigated in three ways, i.e. heating rate, heating duration and heating start time. The conclusion of this study are as follows:

1. A heating rate at 50 °C/hr results in the highest compressive strength about 40 % more than these at 10 or 100 °C/hr due to the proportional effect on the pozzolanic reaction rate and the volume of microcracks formed in the paste.
2. A heating duration of 48 hours led to a maximum compressive strength (compared to the samples at 12 and 24 hours) due to the acceleration of both reactions; hydration and pozzolanic.
3. Heat curing applied at 2 days after casting resulted in the maximum compressive strength due to heat-induced acceleration of hydrate phase densification whilst avoiding the risk of microcrack initiation/enlargement resulting from thermal expansion of partially-hardened hydrate phases.
4. High temperature curing could transform C-S-H to tobermorite or xonotlite and fill the pores inside the specimen due to the extraction of water from the hydrates, which is then trapped inside the pores.
5. The optimum condition of heat curing is starting at 2-day after casting with a rate of 50 °C/hr for 48 hours due to the increase of pozzolonic and hydration reaction rates and the amount of crystal transformation.

References

1. P. Richard, M. Cheyrezy, Cem. and Conc. Res. 25 (1995).
2. O. Bonneau, C. Vernet, M. Moranville, P.C. Aïtcin, Cem. and Conc. Res, **30** 12 (2000).
3. V. Matte, M. Moranville, Cem. and Conc. Comp. **21** 1 (1999).
4. Y. W. Chan, S. H. Chu, Cem. and Conc. Res. **34** 7 (2004).
5. H. Zanni, M. Cheyrezy, V. Maret, S`Philippot, P. Nieto, Cem. and Conc. Res. **26** 1 (1996).
6. A. Cwirzen, Adv. in Cem. Res. **19** 1 (2007).
7. H. Yazıcı, H. Yiğiter, A.S. Karabulut, B. Baradan, B.,. Fuel, **87** 12 (2008).
8. H. Yazıcı, M. Y. Yardımcı, S. Aydın, A.S. Karabulut,. Cons. and Buil. Mat. **23** 3 (2009).
9. A. Sadrekarimi,. Jou. of Adv. Conc. Tech. 2 (2004).
10. C. M. Tam, V.W.Y. Tam, K.M. Nga, Mag. of Con. Res. **62** 10 (2010).
11. C. M. Tam, V.W.Y. Tam, Mag. of Con. Res. **64** 3 (2012).
12. M. Cheyrezy, V. Maret, L. Frouin, Cem. and Conc. Res. **25** 7 (1995).
13. M. Helmi, (Thesis, University of Nottingham, (2016)
14. M. Helmi, M.R. Hall, S.P. Rigby, *33rd Cement and Concrete Science Conference*, pp 288-333 (2013).
15. A. M. Neville, *Properties of Concrete*, (Atlantic Publications Inc, 2011)
16. P. L. Domone, *Concrete*, (E & FN Spon, 1996).
17. M. Helmi, M.R Hall, L. A. Stevens, S.P. Rigby, Cons. and Buil. Mat. J. **105** 15 (2016).

Acknowledgements

The authors wish to acknowledge the support of the Directorate General of Higher Education (DIKTI) in Indonesia for the research scholarship enabling this work, which was conducted as part of a larger project.

Physical, chemical, and mineralogical characteristics of blast furnace slag on durability of concrete

*Elakneswaran Yogarajah *, Toyoharu Nawa, Toshifumi Igarashi*

Division of Sustainable Resources Engineering, Faculty of Engineering, Hokkaido University, Kita 13, Nishi 8, Kita-ku, Sapporo, 060-8628, Japan

Abstract. A partial replacement of Portland cement (PC) by ground granulated blast furnace slag (GGBFS) is an effective method to improve the durability of concrete due to its lower diffusivity and higher chemical resistance compared to PC. Further, the microstructure of GGBFS blended cementitious materials controls the physicochemical properties and performance of the materials in concrete. Therefore, understanding of cement hydration and cementing behavior of GGBFS is essential to establish microstructure property relationship for predicting performance. In this study, hydration, microstructure development, and chloride ingress into GGBFS-blended cement have been investigated. Solid-phase assemblage and pore solution chemistry of hydrating PC and cement blended with GGBFS were predicted using thermodynamic model and compared with experimental data. A mathematical model integrating PC hydration, GGBFS reaction, thermodynamic equilibrium between hydration products and pore solution, ionic adsorption on C-S-H, multi-component diffusion, and microstructural changes was developed to predict chloride ingress into GGBFS blended cementitious materials. The simulation results on chloride profiles for hydrated slag cement paste, which was prepared with 50% of replacement of PC with GGBFS, were compared with experimental results. The model quantitively predicts the states of chloride such as free, adsorbed on C-S-H, and chemically bound as Friedel's salt.

1 Introduction

Cementitious materials have been using in various reinforced concrete structural components as well as in nuclear waste repositories [1]. Extensive use of supplementary cementitious materials, such as ground granulated blast furnace slag (GGBFS), fly ash, and silica fume, is anticipated not only to reduce CO2 footprint of Portland cement (PC) but to achieve better performance of concrete in numerous environments [2]. Concrete made with GGBFS shows better performance than conventional concrete in chloride-rich environments [3]. The slag incorporation into cement reduces the total porosity as well as significantly influences pore size distribution, increases fine pores. Furthermore, it has effect on chloride binding and diffusion. Several factors such as porosity, pore size distribution, tortuosity, mineralogical distribution, and the characteristics of calcium alumino silicate hydrate (C-A-S-H) influence on the ionic transport into slag-blended cementitious materials [4]. Therefore, further study on the properties and performance of the materials is necessary for their successful application in concrete structures. The purpose of this study is to calculate hydrate assemblage and pore solution chemistry of slag containing cementitious materials to predict chloride ingress into the materials. Geochemical code, PHREEQC [5], coupled with empirical expressions for dissolution of clinker minerals and reaction of slag is used to calculate hydration products and pore solution composition, and the determined properties are given to a reactive transport model together with microstructure properties to predict ionic transport into the materials.

2 Materials and methods

2.1 Materials, specimen preparation, and experimental methods

Hardened cement paste with 100 % of PC and GGBBFS cement paste with slag replacement ratio of 50 % and 70 % were prepared with the water to binder ratio of 0.5 by weight. Backscattered electron image (BEI), energy dispersive X-ray spectrometry (EDX), electron probe microanalyzer (EPMA), micro-elastic modulus, and thermoporometry measurements were conducted on the cured and exposed specimens. The details of the composition of raw materials and their physical properties as well as the experimental techniques are described in our previous paper [6].

2.2 Modelling approach and calculation procedure

* Corresponding author: elakneswaran@eng.hokudai.ac.jp

The hydration and reactive transport calculations were performed as presented schematically in **Fig 1**. Hydration reaction of slag-blended cementitious materials were performed based on the coupled model, which couples geochemical code PHREEQC and Excel which is used to calculate dissolution rate of clinker minerals and slag reaction. Pore solution chemistry, hydrate assemblage, and the composition of C-A-S-H were calculated for the input of PC composition, binder proportions, composition of slag, and mixing conditions. The results are transferred to reactive transport model together with boundary conditions for thermodynamic equilibrium and multispecies transport calculations. The basic theory behind thermodynamic equilibrium reactions, cement hydration, and multi-ionic transport are described in authors' previous works [7-8].

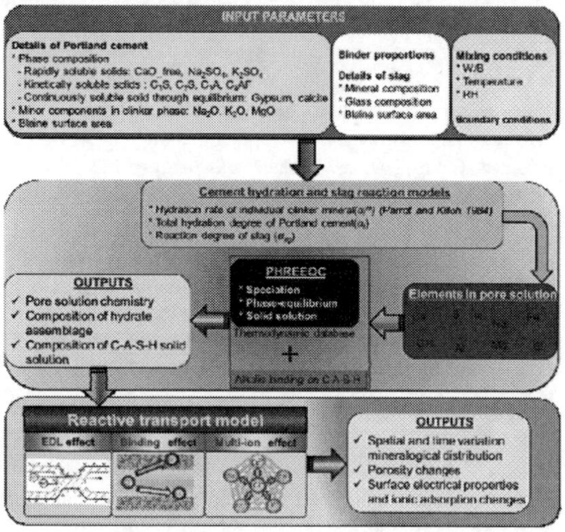

Fig. 1. Calculation procedure of an integrated model

3 Results and discussion

3.1 Hydration products and pore solution chemistry of slag cement system

The coupled model described in ref [7] with the same database is used in here. The thermodynamic calculations were performed for the hydration reaction of cement clinker minerals and the reaction of slag. It is assumed that only the glass phase of the slag is dissolved uniformly and this reaction does not influence the hydration reaction of the clinker minerals. The reaction of the glass phase and clinkers contribute the kind and amount of hydration products as well as pore solution chemistry. The predicted phase changes in terms of volume and concentration of ions in the pore solution as a function of hydration time for slag-blended cement, which consists of 60% of PC and 40% of slag, are shown in **Fig. 2**. The reaction of clinkers and slag contribute to continuous increase of hydration products and thus decrease of porosity. The main hydration products are C-A-S-H, portlandite, ettringite, hydrotalcite, monosulfoaluminate, and mono and hemicarboaluminate which are due to the presence of calcite in the cement. Further, the ettringite are converted to monosulfoaluminate as the hydration progress. The ionic concentration in the pore solution was calculated based on the release and uptake of alkalis due to the hydration reaction of both PC and slag. The Na, K, OH, Ca, S, Al, and Si are the major elements in the pore solution. The concentration of alkalis increases with hydration and contributes to high pH in the pore solution. The mature slag containing cementitious materials have very low concentration of calcium and sulfate ions than alkalis.

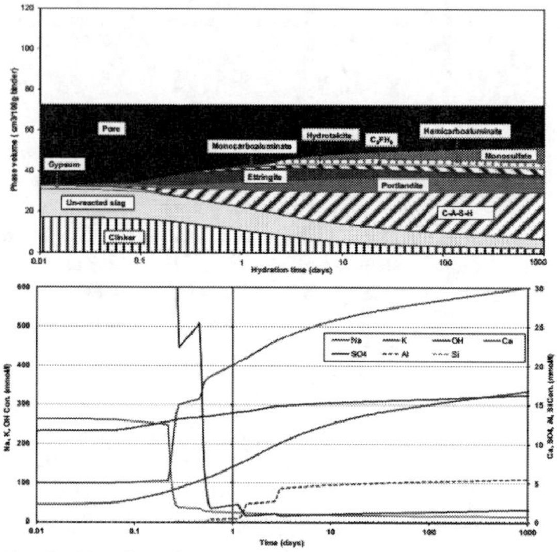

Fig. 2. Simulated phase assemblage in terms of volume and pore solution chemistry of slag-blended cement (PC: Slag = 60:40)

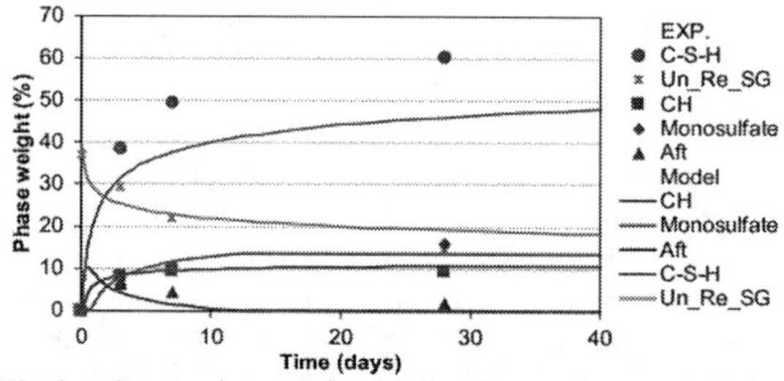

Fig.3. Comparison of simulated portlandite (CH), monosulfate, ettringite (Aft), C-S-H, and un-reacted slag (Un_Re_SG) changes with experimental data [9] for cement blended with slag (OPC: slag = 60:40) [7].

The simulated hydration products of slag-blended cement were compared with experimental data available in the literature, shown in **Fig. 3**. The same experimental conditions and the composition of slag and PC were adopted for the simulation [9]. The simulated hydration products agree well with experimental data both qualitatively and quantitatively. However, the model underestimates the amount of C-S-H formed as the hydration of slag and PC. The difference could be attributed by the consideration of the type of end-members for C-A-S-H solid solution. The hydration model for PC and slag cement system are detailly described in our previous work with more experimental verification on hydration products and pore solution chemistry as well as the importance of C-A-S-H solid solution model for cementitious materials [7].

3.2 Microstructure development of slag cement system

The pore size distribution of hydrated slag cement paste as determined by thermoporometry is shown in **Fig. 4**. The pore in cementitious materials are classified as small gel pores (< 5 nm) and large gel pores (5–10 nm) [10]. The amount of small gel pore increases with slag substitution and it is related to the pores of different C-S-Hs: low-density C-S-H (LD C-S-H) consists of large gel pores and high-density C-S-H (HD C-S-H) has small gel pores [11]. In the hydrating slag cement system, HD C-S-H and LD C-S-H are respectively similar to the inner- and outer-products of hydrating cement and/or slag [12]. The tortuosity of slag-blended cement paste was determined by fitting the porosity and diffusivity of C-S-H, predicted through a three-dimensional spatial distribution model, to Archie's law [13]. The determined tortuosity together with total porosity is tabulated in **Table 1**. The increase in tortuosity with slag replacement ratio indicates that the diffusion path of chloride ions becomes more complex as more slag is added. Based on the pore size distribution and tortuosity, the critical pore diameter which controls the diffusion is assumed and given in Table 1. The measured total porosity, tortuosity, and assumed pore radius are used in the prediction of chloride ions diffusion.

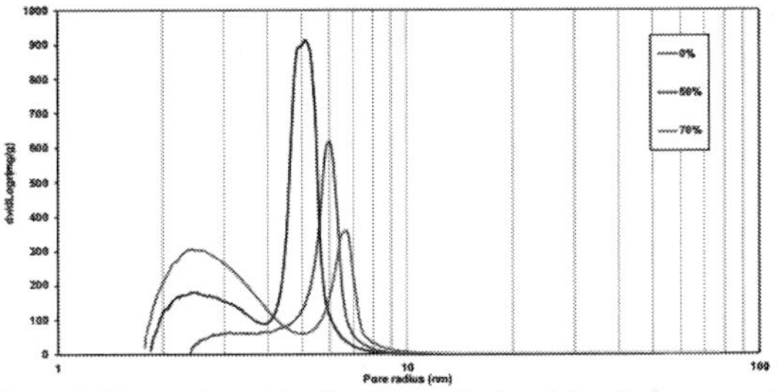

Fig. 4 Pore size distribution of slag-blended cement paste cured at 91 days [6]

Table 1. Porosity, tortuosity, and pore radius used in simulations.

Slag replacement ratio (%)	Porosity	Tortuosity	Pore radius (nm)
0	0.200	2.39	3.2
50	0.214	2.63	2.3
70	0.226	2.85	2.1

3.3 Chloride ingress into slag cement system

The integrated model developed in a previous work [8] was used to simulate multi-species transport in hydrated slag-blended cement paste sample exposed to 1 L of 0.5 mol/l NaCl solution for 91 days. The main ions in the pore solution and chloride were considered in the simulation. Initial and final boundary conditions in the simulation were set as exposure solution concentration and zero flux, respectively. As shown in **Fig. 1**, the pore solution concentration and mineral assemblage were calculated from the hydration model and other microstructure parameters for the simulation are given in Table 1. The simulated total chloride profile is compared with experimental data in **Fig. 5**. The predicted amounts of chloride in the free water of the pore solution, diffuse double layer, chemically bound as Friedel's salt, and adsorbed on C-S-H surface are also presented. The simulation result agrees well with experimental data indicates that the integrated model could reproduce the experimental data. As can be seen, the physical adsorption of chloride controls the total content, which supports the findings of previous experimental studies in blended cements [14]. Slag-blended cementitious materials have large electrical double layer thickness due to low ionic strength of pore solution as well as high tortuosity and small threshold pore size. They contribute to lower the chloride diffusion in the materials as compared to conventional Portland cement and thus show high durability performance in chloride-rich environments.

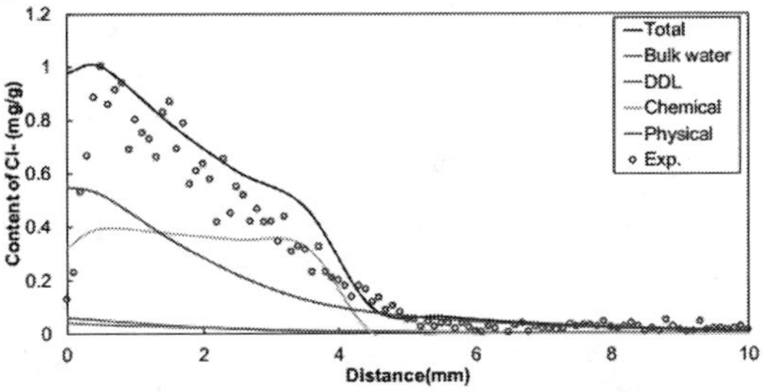

Fig. 5. Comparison between measured and simulated chloride profiles for hydrated slag-blended cement containing 50% of slag exposed to 0.5 M NaCl for 91 days. Exp.: Experimentally measured total Cl-; Total: Predicted total Cl-; Bulk water: Predicted free Cl- in bulk pore water; DDL: Predicted free Cl- in diffuse double layer; Chemical: Predicted chemically bound Cl-; Physical: Predicted physically adsorbed Cl- on C-S-H [6]

4 Conclusions

The chemical thermodynamic equilibrium calculations were performed to predict the solid-phase assemblage and pore solution chemistry of hydrating PC and slag using the coupled model. The agreement between simulation results and experimental data on hydrate assemblage indicates that the model can be used to predict mineralogical distribution and pore solution concentration in slag-blended cement. Integrated model incorporates phase-equilibrium reaction, surface complexation, and multi-ionic transport was applied to predict chloride ingress into hydrated slag cement paste.

The microstructural properties such as porosity, tortuosity, and threshold pore diameter were determined through experimental results. The chloride profiles predicted by the integrated model agree well with EPMA experimental data for slag-blended cement paste exposed to 0.5 M NaCl for 91 days. Most of the transported chloride is adsorbed by C-S-H and reduce its concentration in the pore solution as free chloride which causes the reinforcement corrosion. When cement is blended with slag, it modifies mineralogical distribution and surface electrical properties, increases tortuosity, and reduces porosity and threshold pore size to show high chloride resistance and better durability performance.

Acknowledgement

Authors wish to thank the Japan Cement Association for a part of financial support of this work and Mr. Akira Hatanaka is greatly acknowledged for his assistance in experimental works.

References

1. Malhorta, V. Durability of Concrete. Corrosion. R. W. Revie Ed.., Wiley. (2000).
2. M.S. Imbabi, C. Carrigan, S. McKenna, Trends and development in green cement and concrete technology, Int. J. Sustain. Built Environ. **1**, 194 (2012)
3. H. W Song and V. Saraswathy. Studies on the corrosion resistance of reinforced steel in concrete with ground granulated blast–furnace slag–An overview, Journal of Haz. Materials B **138**, 226 (2006)
4. Glasser, Fredrik P., Jacques Marchand, and Eric Samson. Durability of Concrete - Degradation Phenomena Involving Detrimental Chemical Reactions, Cement and Concrete Research **38**, 226 (2008)
5. D.L. Parkhust, C.A.J. Appelo, A computer program for speciation, batch- reaction, one-dimensional transport and inverse geochemical calculations, USGS Report, (1999).
6. A. Hatanaka, Y. Elakneswaran, K. Kurumisawa, T. Nawa, The Impact of Tortuosity on Chloride Ion Diffusion in Slag-Blended Cementitious Materials, Journal of Advanced Concrete Technology **15**, 426 (2017)
7. Elakneswaran Yogarajah et al. Hydration Study of Slag-Blended Cement Based on Thermodynamic Considerations, Construction and Building Materials **124**, 615 (2016)
8. Elakneswaran, Y.et al., Ion–cement hydrate interactions govern multi–ionic transport model for cementitious materials, Cement and Concrete Research **40**, 1756 (2010)
9. S. Hoshino, K. Yamada, H. Hirao, XRD/Rietveld analysis of the hydration and strength development of slag and limestone blended cement, J. Adv. Concr. Technol. **4**, 357 (2006)
10. Jennings, Hamlin M. Refinements to Colloid Model of C-S-H in Cement: CM-I, Cement and Concrete Research **38**, 275 (2008)
11. Tennis, Paul D. and Hamlin M. Jennings. Model for Two Types of Calcium Silicate Hydrate in the Microstructure of Portland Cement Pastes, Cement and Concrete Research **30,** 855 (2000)
12. Maekawa Koichi, Tetsuya Ishida, and Toshiharu Kishi. Multi-Scale Modeling of Concrete Performance" Journal of Advanced Concrete Technology **1,** 91 (2003)
13. Kurumisawa Kiyofumi, Toyoharu Nawa, and Hitoshi Owada. Prediction of the Diffusivity of Cement-Based Materials Using a Three-Dimensional Spatial Distribution Model, Cement and Concrete Composites **34,** 408 (2012)
14. Saeki, T, Sasai, K, Chloride Binding Capacity of Calcium Silicate Hydrate" Cement Science and Concrete Technology. **60**, 322 (2006)

The use of hazardous sludge solidification and green-lipped mussel shells in cementitious material: a case study of ngcc power plant of priok

Samuel Dwima Halim[1,*], *Edbert* Rainer[1], *Chris* Ryantonius[1], *Bagas* Panandito[1], *Doni* Wardoyo[1], *Mhd* Reza Fahlevy[1], and *Ivan* Sandi Darma[2]

[1]Civil Engineering Division, Faculty of Civil and Environmental Engineering, Bandung Institute of Technology, Indonesia
[2]Structural Engineering Division, Faculty of Civil and Environmental Engineering, Bandung Institute of Technology, Indonesia

Abstract. This paper aims to introduce the utilization of hazardous sludge with green-lipped mussel shells as concrete mixture components. The hazardous sludge is a side product of wastewater treatment plant operated in NGCC Priok, while green-lipped mussel shells are side product of community activity in Kalibaru, North Jakarta, Indonesia. Utilization is dedicated to preserve the environment, reducing both hazardous sludge generations from wastewater treatment plant and potential waste from social activities in the coastal area. Three different groups of concretes were prepared. The first group contains normal concretes with W/C ratio of 0.5. Subsequently, the second group contains concretes with dry sludge and mussel as a replacement of fine aggregate at different contents, 5% (C1), 10% (C2), and 15% (C3), with W/C ratio of 0.5. Finally, the last group contains concretes with dry sludge as cement replacement and the green-lipped mussel as fine aggregate replacement at different contents of 5% (C4), 10% (C5), and 15% (C6). The results show that dry sludge and green-lipped mussel provide better results when used as a substitute for fine aggregate rather than as cement replacement. In addition, the rate of strength development of concretes containing waste additions are slow compared to normal concretes.

1 Introduction

Sewage sludge production from waste water treatment plants are increasing all over the world. This kind of sludge includes the solid material left from sewage treatment processes. Generally, sludge still has high enough water content, therefore it is necessary to treat sludge which is part of the wastewater treatment. The percentage of water content can reach 95-99%. The main purpose of sludge treatment is to reduce the volume of sludge by separating water from the sludge before disposal, in order to facilitate the transport process. Therefore, the reduction of water content and sludge volume is important.

However, it should be noted that sludge may cause more harmful environmental disturbances than wastewater by considering that sludge contains more concentrated pollutants and may contain new types of contaminants that are not contained previously in the wastewater resulting from the addition of chemicals and from the decomposition of the compounds contained in the sludge. In this regard, waste treatment measures are required from the planning stage so that steps and strategies can be prepared to address negative impacts on the environment and bring about positive impacts. Sometimes, waste still has value and can be used as raw materials or other useful materials.

Furthermore, pollutants not only come from sludge but also many other waste materials that can be a source of pollutants. One of them is shell waste. For this waste, shells need a container to place, manage and develop its potential to not only be a mere waste, but to be something useful and profitable in addition to reducing the waste of the shell itself.

Many studies have been conducted to determine how far we can optimize the use of waste, especially in the field of construction materials. Valls et.al [1] confirmed that up to 10% of treatment plant sludge can be added to concrete for use in certain very specific applications. Meanwhile, by applying dry sludge as a replacement of fine aggregate, Jamshidi et.al [2] proposed as non-constructional concrete such as paving and flooring. The use of higher volume of dry and wet sludge in concrete as a replacement of cement content has been explored by Rabie [3].

However, the common thread that can be drawn from those studies is a decrease in the strength of concrete. Recent investigations show that the partial replacement of cement by sewage sludge ash led to substantial retardation of hydration rates [4]. The next problem that would arise is how much the mechanical properties of the concrete will decrease if two types of waste are used as a mixture of concrete.

* Corresponding author: sam.dwima@gmail.com

https://doi.org/10.1051/matecconf/201814701008

In this paper, we explored the effect of the concretes containing two types of waste addition i.e. dry sludge and green-lipped mussel shells. With the proposed mixture proportions coupled with various curing periods, the behavior of multiple addition of waste materials in concrete mixture would be further understood.

2 Materials and method

2.1 Sewage sludge and green-lipped mussel

Fig. 1. Waste Addition (a). Raw sewage sludge. (b) Oven dried sludge. (c) Green-lipped mussel

Figure 1 shows the waste addition used in this research. The sludge and green-lipped mussel were prepared by PT. Indonesia Power UPJP Tanjung Priok. The sludge was then oven dried for 24 hour with the temperature of 105 °C. Furthermore, the sludge was grinded and sifted out in the sieve with the size of 2.36 mm (**Figure 1(b)**). As for the green-lipped mussel, pounding process was applied in order to reduce the size of the mussel. The maximum size of mussel used in the concrete mixture is around 15 mm (**Figure 1(c)**).

2.2 Mix proportion and curing

Table 1. Mixture composition

Mixture	Cement (kg)	Water (kg)	Gravel (kg)	Sand (kg)	Sludge (kg)	Mussel (kg)
C0	450	225	915	610	0	0
C1	450	225	915	549	30.5	30.5
C2	450	225	915	488	61	61
C3	450	225	915	427	91.5	91.5
C4	427.5	225	915	579.5	22.5	30.5
C5	405	225	915	549	45	61
C6	382.5	225	915	518.5	67.5	91.5

In order to study the effects of treated dry sludge and green-lipped mussel on concrete, mix designs with water to cement ratios of 0.5 was prepared. In this research, mixture compositions can be classified into three different groups as shown in **Table 1.** First composition contains normal concretes as a control specimen (C0). The next group contains concretes with waste addition (dry sludge and mussel) as a replacement of fine aggregate at different contents of 5% (C1), 10% (C2) and 15% (C3). Finally, the last group contains concretes with dry sludge used for secondary binder (cement replacement) and the green-lipped mussel used as fine aggregate replacement at different contents of 5% (C4), 10% (C5) and 15% (C6). In addition, to increase the rate of solidification of the concretes due to the use of sludge, water-based cement admixture (ECO5000) was utilized. Furthermore, after casting, all specimens were left for a day in molds in humid condition at laboratory condition. Then they were removed and immersed in water for 6, 13 and 27 days.

3 Results and discussion

3.1 Compressive strength evolution with the increasing curing periods

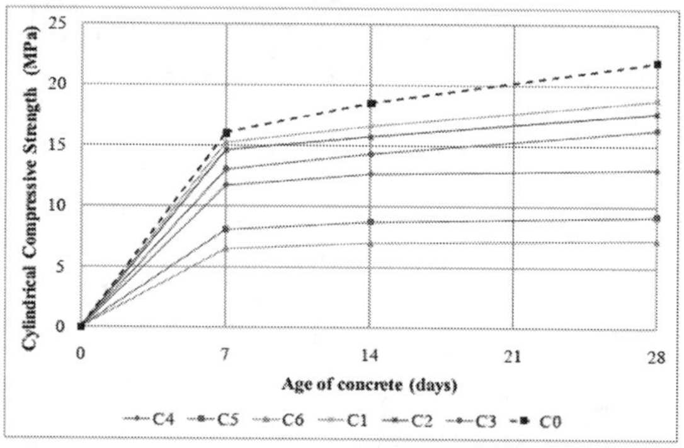

Fig. 2. Evolution of compressive strength of concrete

Figure 2 shows the evolution of compressive strength of concretes. As shown in the Figure 2, the resistance to compression of samples are always increasing with the increasing curing periods. However, it can be seen that the use of higher amount of waste additions as fine aggregate replacement resulting in the reduction of compressive strength of concrete having the same amount of cement content at the same curing period. Meanwhile, the compressive strength of the samples in which dry sludge used for secondary binder (cement replacement) and the green-lipped mussel used as fine aggregate replacement show the lowest as compared with that of normal as well as the concretes containing waste additions as fine aggregate replacement.

Sludge may form a coating on aggregate particles, resulting in weakened bond between the aggregate and the mortar. Excessive amounts of these fine materials may also unduly increase the water demand of the concrete, resulting in loss of concrete strength and an increase in its permeability. Subsequently, it can be worsened by reducing of cement content in the mixture.

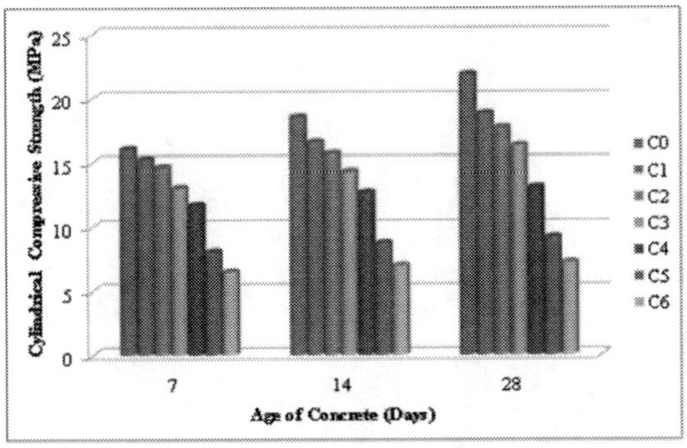

Fig. 3. Comparison of compressive strength at 7, 14, and 28 days

Figure 3 shows the comparison of strength at 7, 14 and 28 days. From **Figure 3**, it can be seen that the usage of waste additions of 5%, 10%, and 15%, as fine aggregate replacement is reducing the compressive strength of concrete at 28 days up to 14.1%, 19.1% and 25.4%, respectively. As for samples in which sewage sludge used for secondary binder (cement replacement) and the green-lipped mussel used as fine aggregate replacement, the compressive strength in 28 days reduce significantly up to 40.3%, 58.0% and 66.9% for 5%, 10%, and 15% replacement, respectively.

3.2 Rate of strength development with the increasing of waste materials content

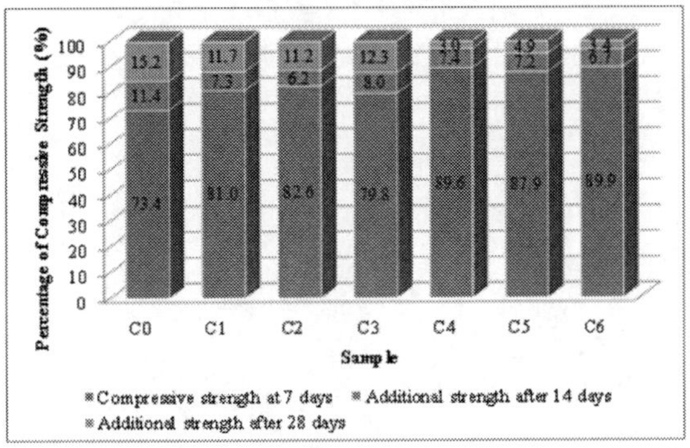

Fig. 4. Concrete strength development rate

The rate of concrete strength development of each samples are shown in the **Figure 4.** As shown in the **Figure 4**, the rate of development of waste additions concretes in 7 days curing periods are comparable to the normal concretes. It was 73.4% for normal concretes (C0). Meanwhile, for the samples of C1, C2 and C3, the rate of development varies from 79.8% to 82.6%. As for samples in which sewage sludge used for secondary binder (cement replacement) and the green-lipped mussel used as fine aggregate replacement (C4, C5 and C6) those rate varies from 87.9% to 89.9%.

Furthermore, as the hydration reaction continues, the strength of the concrete increases. The additional strength of concrete after 14 days curing periods as compared with that of 28 days of C0 is 11.4%. For the group of samples using waste additions in the concrete mixture, the additional strength that can be achieved are similarly vary from 6.2% to 8.0%.

Moreover, from 14 days to 28 days, there is still additional strength of concrete due to the amount of unhydrated cement grain undergone hydration reaction. However the amount of additional strength is vary for each group of sample. The group of samples in which sewage sludge used for secondary binder (cement replacement) and the green-lipped mussel used as fine aggregate replacement (C4, C5 and C6) were recorded to have the smallest increase in the strength during 14 to 28 days curing periods.

The rate of strength development of group samples C1, C2, and C3 makes them almost reaches the same compressive strength of the normal mixture, especially until day 14. This could be attributed to a minor pozzolanic effect as reported by other authors [5]. Meanwhile, as for the other samples of C4, C5, and C6, the reduction of cement content and low pozzolanic activity are causing the waste additions act as filler or particle packing in the microstructure of concrete. However, sludge itself has a tendency to coat the aggregate rather than fill the gap presence in the mortar. Therefore, the function of the sludge as filler might not be achieved.

4 Conclusion

In this research, the use of multiple waste materials in the concrete mixture has been presented while several new findings were also provided. There is still a potential for concrete containing multiple waste materials to be used as structural and constructional element. The used of multiple waste materials as fine aggregate provided better performance as compared with that of cement replacement. However, the rate of strength development of concretes containing waste materials should be of concern. Time required from initial setting to day 7 becomes crucial periods for concrete containing waste materials in developing its strength. After 7 days, the rate of strength development significantly decrease.

5 Acknowledgement

This research was fully supported by PT. Indonesia Power UPJP Tanjung Priok (162.SPK/061/TGP/17). The first author also express gratitude for the facilities support from Structure Engineering Laboratory, Bandung Institute of Technology.

References

1. S. Valls, A. Yague, E. Vazquez, C. Mariscal, Cem. Conc. Res., **34**, 2203- 2208, (2004)
2. A. Jamshidi, N. Mehrdadi, M. Jamshidi, J. Envi. Stud., **Vol. 37**, No. 59, (2011)
3. Rabie, J Civil Environ Eng, **6**, 209, (2016)

4. N. Rodriguez, S. Ramirez, M. Blanco Varela, M. Guillem, J. Puig, E. Larrotcha, J. Flores, Cem. Conc. Res., **40,** 778-786, (2010)
5. M. Cyr, P. Coutand, P. Clastres, Cem. Conc. Res., **37,** 1278–1289, (2007).

Performance level analyses of public transportation using importance-performance analysis method

*Nursyamsu*Hidayat[1*]

[1]Civil Engineering Diploma Program, Vocational School, GadjahMada University, Indonesia

Abstract. Yogyakarta, one of the middle cities in Indonesia, has been developing BRT system named Trans Jogja to reduce growth of motorization. However, the performance tends to decrease year by year, therefore the system and management upgrading must be carried out continuously to restore public trust. This research tries to investigate public satisfaction, their expectation, and some attributes considered important based on public's perspective using IPA method that can be used as useful information systems to evaluate service quality by prioritizing or focusing on areas where improvement is needed. The results reveal that decision maker must focus and allocate their resources to improve Trans Jogja system on the following attributes: a) keep punctuality, b) shorten waiting time in the bus stop, c) improve bus comfort, d) provide safety for child passenger, e) improve bus stop comfort, f) create new bus route to enlarge accessibility, and g) shorten bus travel time.

1Introduction

Most of the cities in developing countries have been facing big problem with the public transportation. Also, Indonesian people tend to use private vehicle rather than public transportation. Along with increasing of welfare, economic growth, and easiness of buying private car, the motorization is increasing rapidly. On the other side, quality of public transport serviceability tends to decrease year by year. Yogyakarta, one of the middle cities in Indonesia, is facing the same problems of domination of private cars on the roadways and therefore improvement of public transportation serviceability must be carried out.

Idris [1] conducted the study on city bus in Yogyakarta, using Customer Satisfaction Index (CSI) the study revealed that the performance of bus system is quite poor. Bus operator must pay much attention in some indicators related to passenger's safety and comfort. Factors related to bus availability and punctuality are considered important based on passenger point of view.

At the same time, Yogyakarta local government has established a new system Bus Rapid Transyt (BRT), named Trans Jogja, for around 12 years to provide an alternative of public transportation for the residents. At first, the Trans Jogja level of service is rated very high by the public. However, public's expectation on Trans Jogja services is increasing year by year. In addition, increasing of the number of passengers is insignificant. Performance upgrading must be undertaken continuously to attract more users. This research tries to investigate public satisfaction, their expectation, and determine attributes considered important on Trans Jogja performance based on passenger and non passenger point of view.

2 Methodology

2.1 Study Design

In this study, interviewing survey was undertaken to Trans Jogja passenger and non-passenger. As a result, 200 passengers and 200 non-passengers are taken as respondents during data collection process in Yogyakarta. Data collections are conducted in October 2014.

Non-passenger respondents are chosen from the respondent that never or rarely using BRT, and non-regular passenger. They have their own car or motorcycle as main mode of transportation.

2.2 Questionnaire

A set of survey questionnaire is designed based on literature review. The first part of the questionnaire measures passenger/non-passenger characteristics. Respondent's socioeconomic and travel characteristics are also gathered, covering gender, age, occupational status, monthly transportation cost, and education level. The second part of the survey asks respondent's perception about trip characteristics, trip frequency, origin and destination, and trip purpose. The third part is divided into two sub-parts to covers respondent's perception on Trans Jogja performance and importance. A five-point Likert scale is used for both sub-parts with

*Corresponding author: nursyamsu.h@ugm.ac.id

https://doi.org/10.1051/matecconf/201814702001

"one" representing very unsatisfied and "five" representing very satisfied for BRT performance and "one" representing very unimportant and "five" representing very important for BRT importance level.

Respondents characteristic can be declared firstly as follow: passenger respondent's characteristics of BRT Trans Jogja are elaborated based on gender and occupation. The percentage of female passenger is greater than male (51% vs. 49%, respectively). Most of the passenger respondents are students (42%) and private sector worker (23%), the remains are government employee (11%), housewife (10%), self-employed (7%), pensioner (2%), and jobless (5%).

For the non-passenger respondents, male respondent are dominant (58%) compare to female (42%). Based on the occupation, it can be elaborated as students (49%), self-employed (15%), private sector worker (16%), government employee (9%), housewife (5%), pensioner (1%), and jobless (5%).

Besides that, trip characteristics are captured in terms of monthly transport cost, trip frequency, and trip purpose.

2.3 Analytical Tools

Importance-Performance Analysis (IPA) is a statistical method to compare between performance/service experienced by user and level of satisfaction that their intended [2]. The first step of the IPA is the questionnaire establishment to investigate level of importance and performance in a five-point Likert scale.

Importance and performance analysis reveal level of suitability among level of importance and present service performed by producer. Level of suitability is comparison among scores of performance and importance of the product, therefore by this instrument could be determined priorities rank to agree with the user suitability/expectation as can be performed by following formula,

$$T_k = \frac{x}{y}100\% \quad (1)$$

$$\bar{x} = \frac{\sum x}{n} \quad (2)$$

$$\bar{y} = \frac{\sum y}{n} \quad (3)$$

where,

Tk = level of suitability
x = score of performance provided by producer
y = score of importance by users
$\bar{x}$ = average scores of performance
$\bar{y}$ = average scores of importance
n = total respondents

Wong et al. [3] stated that the two dimensional IPA model is divided into four quadrants with performance on the x-axis and importance on the y-axis. As a result of this, four quadrants are created. The quadrants can be used to generate suggestions for BRT operator by differentiating between them.

1. Quadrant I (High Importance/Low Performance). Attributes that fall into this quadrant represent key areas that need to be improved with top priority.
2. Quadrant II (High Importance/High Performance). All attributes that fall into this quadrant are the strength and pillar of the organizations, and they should be the pride of the organizations.
3. Quadrant III (Low Importance/Low Performance). Thus, any of the attributes that fall into this quadrant are not important and pose no threat to the organizations.
4. Quadrant IV (Low Importance/High Performance. It denotes attributes that are overly emphasized by the operator; therefore, operator should reflect on these attributes, instead of continuing to focus in this quadrant, they should allocate more resources to deal with attributes that reside in quadrant I.

Above explanations can be resumed in Figure 1 that shows IPA grid.

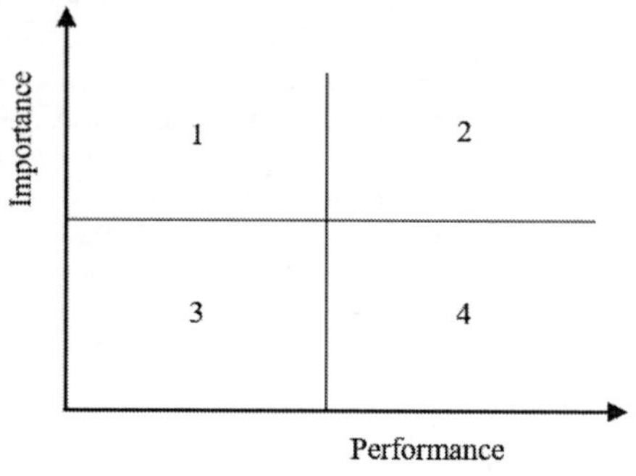

Fig. 1. The IPA Framework

The IPA evaluation tool can help to evaluate BRT attributes and to provide guidance to formulate BRT system strategy to allocate resources accurately to maximise a high advantage. The IPA model is used to evaluate the level of importance citizens attach to each of the attributes and also to examine the level of satisfaction perceived by them in the delivery of those attributes.

3 Importance Performance Analyses of Trans Jogja Based on Passenger's Perception

The amounts of 200 passengers are chosen to reveal suitability level from comparing between Trans Jogja performance score and passenger's importance score. Based on suitability level, it can be established the rank of attributes improvement to agree with passenger satisfaction. The mean importance for all the 19 attributes was rated 4.273, whereas the mean satisfaction was just an average of 3.226. Therefore, if we were to plot the importance and performance on the IPA grid Trans Jogja leaders and managers would be able to quickly identify areas in which operators should allocate their resources to improve Trans Jogja performance. This is reflected in the IPA shown in Table 1 and Figure 2.

https://doi.org/10.1051/matecconf/201814702001

Table 1.IPA Analyses of Trans Jogja Passenger

Variables	Attributes	Mean		Quadrant
		Performance	Importance	
1	Walking distance to bus stop	3.345	4.100	4
2	Easiness to buy the ticket	3.915	4.155	4
3	Ticket price	3.825	4.135	4
4	Availability of time table, bus route in the bus stop	3.395	4.440	2
5	Bus stop comfort	3.030	4.460	1
6	Waiting time in the bus stop	2.720	4.345	1
7	Bus-stop'screw service	3.535	4.080	4
8	Bus stop cleanliness	3.230	4.275	2
9	Keepingpunctuality	2.690	4.495	1
10	On buscomfort	2.915	4.385	1
11	Bus crew service	3.585	4.145	4
12	Secure from crime	3.725	4.610	2
13	Trip safety	3.470	4.645	2
14	Travel time	2.935	4.310	1
15	Route/bus transfer	3.205	4.015	3
16	Operating time	3.415	4.080	4
17	Accessibility	2.920	4.300	1
18	Governmentinstantiate by using public transportation	2.535	3.870	3
19	Safe for childpassenger	2.900	4.335	1
Total		61.290	81.180	
Mean		3.226	4.273	

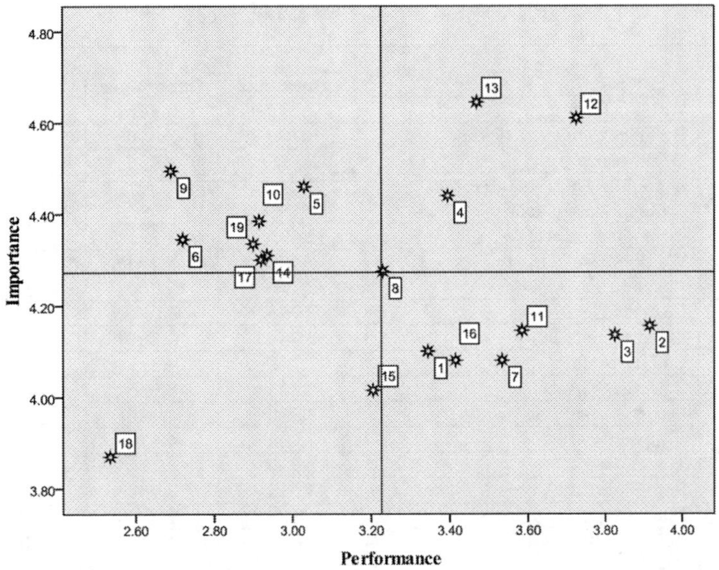

Fig. 2. Trans Jogja Passenger IPA Grid

The intersection in the IPA is made available using the mean level of importance at 4.273 and the mean level of performance 3.226. In quadrant I, passengers perceive the attributes as very important, but the perceptions of Trans Jogja performance levels are below average. Thus, further improvement efforts should be concentrated here. There are seven attributes that fall into this quadrant.

1. Keeping punctuality (9)
2. Waiting time in the bus stop (6)
3. On bus comfort (10)
4. Safe for child passenger (19)
5. Bus stop comfort (5)
6. Accessibility (17)
7. Travel time (14)

Waiting time in the bus stop and bus punctuality are the two main parameters to attract BRT user. Because previous bus system in the city is lack in reliability in terms of punctuality, waiting time, and comfort as well.

However, some attributes are important to build passengers' trust in using Trans Jogja. These attributes have already well done and must be kept in high level performance because passengers satisfy with the recent performance. Attributes such as availability of time table, bus route in the bus stop (4), bus stop cleanliness (8), secure from crime (12), and trip safety (13), situated in quadrant II, are perceived to be very important and satisfied from the passenger's perspective. All these attributes are the strengths of the Trans Jogja operation and the bus's operator should keep up the good work in maintaining or delighting their passengers, as otherwise, these benefits might risk falling into the first quadrant.

Some of the attributes that literally categorized as low priority in quadrant III are route/bus transfer (15) and government instantiate by using public transportation (18).

In quadrant IV, there are six attributes namely, walking distance to bus stop (1), easiness to buy the ticket (2), ticket price (3), bus-stop's crew service (7), bus crew service (11), and operating time (16) that are rated by passengers as of low importance with high performance. Here, the passengers are very satisfied with these Trans Jogja attributes performance, but all the continuous effort in delivering those attributes to the passenger will be futile if passengers are already satisfied. Therefore, present efforts on these attributes in this quadrant are overstated and Trans Jogja operators should consider allocating resources (i.e., money, effort, and time) elsewhere, especially on those attributes in the first quadrant, to yield a higher benefits and performance.

4 Importance Performance Analyses of Trans Jogja Based on Non Passenger's Perspective

Two hundred citizens are sampled as non-passenger respondents for gathering their perception on Trans Jogja performance. Also, their expectation are needed to identify attributes must be improved for possibility to attract new comer passengers. Based on non-passenger point of view, IPA analyses can find some attributes which need improvement to correspond with their satisfaction. The mean importance for all the 19 attributes was rated 4.113, whereas the mean satisfaction was just an average of 2.902. This is reflected in the IPA shown in Table 2 and Figure 3.

Table 2.IPA Analyses of Trans Jogja Non-Passenger

Variables	Attributes	Mean		Quadrant
		Performance	Importance	
1	Walking distance to bus stop	2.690	3.955	3
2	Easiness to buy the ticket	3.395	3.910	4
3	Ticket price	3.505	3.830	4
4	Availability of time table, bus route in the bus stop	3.130	4.215	2
5	Bus stop comfort	2.695	4.220	1
6	Waiting time in the bus stop	2.405	4.105	3
7	Bus-stop'screw service	3.160	3.960	4
8	Bus stop cleanliness	2.955	4.070	4
9	Keepingpunctuality	2.345	4.385	1
10	On buscomfort	2.745	4.325	1
11	Bus crew service	3.205	3.945	4
12	Secure from crime	3.350	4.425	2
13	Trip safety	3.190	4.415	2
14	Travel time	2.630	4.295	1
15	Route/bus transfer	3.025	3.930	4
16	Operating time	3.175	3.920	4
17	Accessibility	2.680	4.215	1
18	Governmentinstantiate by using public transportation	2.350	3.805	3
19	Safe for childpassenger	2.505	4.225	1
Total		55.135	78.150	
Mean		2.902	4.113	

The intersection in the IPA is determined using the mean level of importance at 4.113 and the mean level of performance 2.902. In quadrant I, non-passengers reveal the attributes as very important, but the performances of Trans Jogja are must be improved significantly. Thus, further improvement efforts must be focused here. There are six attributes that fall into this quadrant:

1. Keeping punctuality (9)
2. Safe for child passenger (19)
3. Travel time (14)
4. On bus comfort (10)
5. Accessibility (17)
6. Bus stop comfort (5)

 https://doi.org/10.1051/matecconf/201814702001

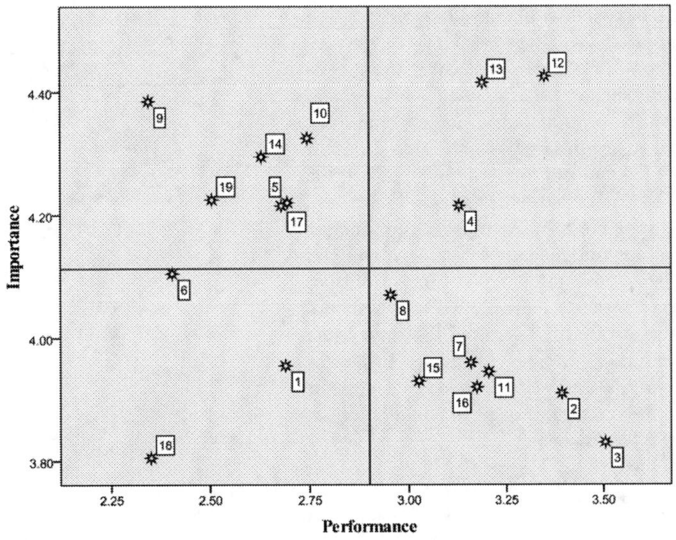

Fig. 3. Trans Jogja Non Passenger IPA Grid

However, some attributes are important to build non passengers' trust in moving to Trans Jogja. These attributes have already performed excellently and must be kept in high level performance because non passengers satisfy with the recent performance. Attributes such as availability of time table, bus route in bus stop (4), secure from crime (12), and trip safety (13), placed in quadrant II, are perceived to be very important and satisfied from the non-passenger's point of view. All these attributes must be kept up the good work in maintaining or delighting their passengers, as otherwise, these benefits might risk falling into the first quadrant.

Some of the attributes that literally categorized as low priority in quadrant III are walking distance to bus stop (1), waiting time in the bus stop (6), and government instantiate by using public transportation (18). These attributes are considered unimportant, however, the performance is deemed in average level.

Seven attributes located in quadrant IV, i.e. easiness to buy the ticket (2), ticket price (3), bus-stop's crew service (7), bus stop cleanliness (8), bus crew service (11), route/bus transfer (15), and operating time (16) that are considered as of unimportant with excellent performance. Non passengers are very pleased with these Trans Jogja attributes performance, therefore, all of the effort to improve these attributes will be useless because non passengers are already satisfied. Trans Jogja management must consider allocating their resources to develop those in Quadrant I to obtain higher benefit, rather than taking into account quadrant IV.

5 The Main Priorities Enhancing Trans Jogja Based on Passenger's and Non Passenger's Perception

This research study managed to provide useful recommendations for Trans Jogja operator, government or policy makers for developing public transportation strategies in the future. The IPA grid reveals areas of strategic focus, i.e., first quadrants with findings in this area that require the greatest attention. Also, to be highlighted in this research that pay attention on non-passenger suggestion will attract new comer and can be expected reducing private car population. Seven attributes from passenger and six attributes from non-passenger are revealed as main priority for improving Trans Jogja serviceability. Operator should pay more attention in these attributes to keep present passenger's trust and attract new passengers. The main priorities for developing Trans Jogja can be seen in Table 3.

Table 3. The Main Priorities for Enhancing Trans Jogja

PerformanceAtributtes	Passenger'sPriority	Non Passenger'sPriority
Keepingpunctuality	1	1
Waiting time in the bus stop	2	-
On buscomfort	3	4
Safe for childpassenger	4	2
Bus stop comfort	5	6
Accessibility	6	5
Travel time	7	3

Both passengers and non-passenger primary concern is punctuality. They are not satisfied with this attribute. Therefore, passengers stated that waiting time in the bus stop is another attribute which must be improved. Time scheduling should be launched to control arriving-departing Trans Jogja on the bus stop. Also, time scheduling will be expected to create trust to Trans Jogja to both passengers and non passengers, because passengers are not tolerant with unpredictable waiting time in the bus stop.

Besides that, bus stop's serviceability should be improved by size/dimension enlargement, addition and deployment in to main points, and improvement of facilities for person with physically disabilities. In order to cut the waiting time and improve on bus comfort, the number of buses must be added with the new buses, and also the project to repair old buses must be conducted. Recently, 64 buses are operated to cover 4 routes and 32.5 km2 areas of Yogyakarta city, it is considered not sufficient. The routes should be developed to reach sub-urban and some supporting areas around the city. Most of people come from country sides during week-day for working and studying. Besides that, as one of popular tourist destination, the bus routes need to connect tourist destination areas located in country-side.

Conclusion

In this paper, the IPA technique is discussed and applied to measure Trans Jogja performance from the perspective of passenger and non-passenger to find main priority for the allocation of resources. A set of survey was developed to survey both passenger's and non-passenger's perceptions to identify areas of importance and areas of satisfaction/dissatisfaction, which then enable government/providers to strategically identify focus areas to improve urban public transportation services. IPA is a strategic tool that can quickly enable Trans Jogja operator and government to understand public's needs and wants (important prioritization areas) and also to assess public's satisfaction rather than relying on performance indicators alone.

The results of this research reveal that Trans Jogja operator and decision maker/government must focus and allocate their resources to improve Trans Jogja system on the following attributes that fall in the first quadrant: a) keep punctuality, b) shorten waiting time in the bus stop, c) improve bus comfort, d) provide safety for child passenger, e) improve bus stop comfort, f) create new bus route to enlarge accessibility, and g) shorten bus travel time. In the fourth quadrant, there is an overemphasis in the operator's allocation of resources on attributes that are deemed to be unimportant and it is considered to reallocate their resources to first quadrant. This research considers investigating public's perspective on the importance and performance of attributes in terms of Trans Jogja services (demand side), but it can be performed a similar research to study attributes considering important from operator's point of view (supply side).

References

1. Z. Idris, J. Din.Tek. Spl, **9**, 189-196, (2009)
2. J. Martilla,J. James, J. Market, **41 (1)**, 77-79, (1977)
3. M. S. Wong, N. Hideki, P. George, J. Theo. and Appl. Elect. Com. Res., **6**, 17-30, (2011)

Exploring physical attributes of walkability from perspective of blind pedestrians

Russ Bona Frazila[1] and *Febri* Zukhruf[1*]

[1]Department of Civil Engineering, Institut Teknologi Bandung, Jl. Ganesha No. 10, Bandung, 40132, Indonesia

Abstract. Access has become a crucial issue for the persons with disabilities (PWDs), in which the inconvenient transportation facilities is contributed to the dependent living issue of PWD, specifically the visual impaired person. As a primary aspect of transportation, the walking facilities further needs to be carefully considered for facilitating their moving activities. Recently, the effect of built environment on walking behavior has obtained significant attention, which is generally constructed within the framework of walkability concept. The walkability has been extensively used for evaluating the physical attributes of pedestrian facilities by comparing it to the walker perception. Hence, the evaluation result can directly answer the needs of pedestrian. Despite of their recent research achievements, the walkability concept is mostly unsuccessful to take into account the blind walker characteristics. This paper then explore physical attributes of walkability environment in order to provide the friendly pedestrian facilities for the blind pedestrian, which is rarely explored. The research is established based on the blind walker perspectives within the micro-level analysis that incorporates a smaller unit of measurement (i.e., the street-level physical attributes). The physical attributes result are thus potentially to be utilized for analyzing the required pedestrian facilities for the blind pedestrians.

1 Introduction

People with disabilities (PWDs) practically meet some difficulties for accessing the transportation facilities ([1]), which directly influence their mobility. The inaccessibility of such facilities then limits their opportunities to participate in various types of social and economic activities, and hence, the access improvement becomes a necessary element for increasing the quality life of PWDs. Practically PWDs can be divided into several categories, namely, wheelchair users, blind and visually impaired people, deaf and hearing impaired people, walking impaired people, people with other handicaps, elderly people, children and people of short or tall statue. The researches on transportation accessibility generally pay attention to the first two categories of PWDs (i.e., wheelchair user, blind and visually impaired people (VIP)).

For tackling such issues, several studies have been conducted. Pecchini and Giuliani [2] surveyed the crossing behavior of PWDs and they proposed the design recommendation. The other experiments have been also implemented for capturing the PWDs behavior, such as, walking speed of VIPs [3], stair evacuation speed [4], doorways capacity by considering PWDs and elderly people ([5]). The specific research relating to the blind and VIP is also conducted, which is mainly focused on the autonomously navigation, such as, accessible smartphones for blind user [6], Acoustic way finding [7], ZebraRecognizer [8], traffic lights detection on smartphones [9], and sonification of guidance data [10].

Despite its recent development of PWDs and VIP pedestrian studies, the elaboration of built environment on walking behavior of VIPs are still left behind, compare to the non-PWDs studies (e.g., [11], [12]). Such elaborations are practically implemented within the framework of walkability concept, where the built environment evaluation have been constructed in the meso-level or the micro-level. The level is differed by its detail aggregation, for example the meso level evaluated the walkability in the area level ([13], [14]), whereas the micro-level measure the street-level physical attributes, such as the presence of trees, the width of sidewalks, and the quality of streets (e.g., [15]).

This paper then present the evaluation of physical attributes, which govern the walkability environment in order to provide the friendly pedestrian facilities for the blind and VIPs pedestrian, which is rarely explored. Furthermore, the blind walker perspectives is elaborated within the micro-level analysis for investigating a smaller unit of street-level physical attributes. The attributes result are thus potentially to be involved for analyzing the required pedestrian facilities for the blind and VIP pedestrians.

In the following section, the method is described, while in the third section, the blind walker characteristics are firstly resented, before elaborating the physical attributes of pedestrian environment in the fourth

* Corresponding author: febri.zukhruf@ftsl.itb.ac.id

section. Finally, in the fifth section, the methodologies, results, and analyses in the paper are summarized.

2 Walkability Concept

The main objective of paper is to elaborate the physical attribute that govern the walkability perception of blind pedestrian. The walkability analysis is constructed based on the micro-level analysis, in which two different surveys are designed, namely, a blind walker perception survey, and a physical survey for gathering the physical attributes that governing the walkability perception.

Since the study incorporates a micro-level analysis, the analysis is considered by measuring the micro scale of street physical attributes, for instance, width of side walk, number of commercial uses, average building width, and other physical attributes ([16]). The micro-level is possibly obtained a huge attribute combination, which is not practical to be investigated. Therefore, the physical attributes is elaborated by assuming there is a linear correlation between the attributes and the walkability perception (see Eq. (1)). Furthermore, the important attributes are selected based on the step wise based approach.

$$Y^n = c_1 y_1{}^n + c_2 y_2{}^n + \ldots + c_m y_m{}^m \quad (1)$$

where :

Y : perception value of walkability component-n

y_m : physical attributes-m that possibly govern the walkability perspective,

c_m : relative importance of each physical attributes-m.

Since the walkability is certainly influence by several components, it is importance to discover the walkability components that strongly affect the walkability perception of blind pedestrian. Therefore, the important walkability component is determined based on the blind walker perception about the importance facilities that need to be provided. In order to get more comprehensive view of walker need, the question is designed based on the pair wise comparison approach. The physical attributes are explored based on the interview and the street physical survey in the four pedestrian routes. The routes are selected based on the experience of respondents, where its several performance indicators are assessed by respondents. The assessment is then compared to the current situation of physical attribute to investigate the important attribute that constructs the walkability perception.

3 Blind Walker Characteristics

Before turn into the deep discussion relating to the physical attributes of walkability for the blind walker, this paper firstly elaborate the walking characteristic of blind persons. The characteristics is corresponded to the possible walking time, the group walking behavior, the transportation modes used and its transportation cost. Figure 1 illustrates the different perception of possible walking time between blind and non-blind pedestrians. The blind pedestrians perceive the less walking time than the non-blind pedestrians, in which the most of blind pedestrian only possible walk around 10 minutes, below the non-blind pedestrians that can averagely walk in around 20 minutes. Furthermore, the seeing pedestrians believe that they can reach up to 80 minutes walking on the farthest, where the blind pedestrians considers only a half walking time of non-blind pedestrians.

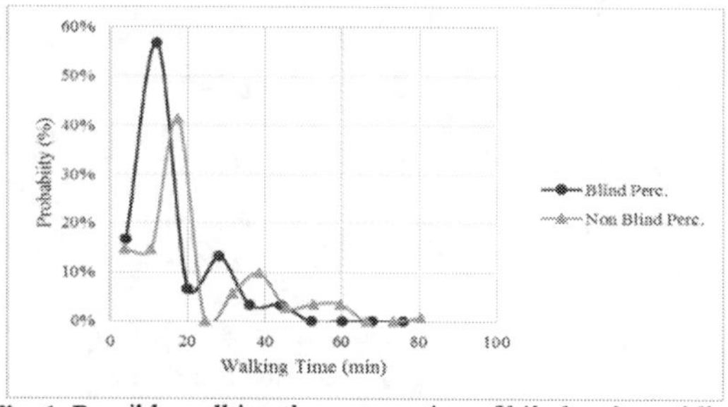

Fig. 1. Possible walking time perception of blind and non-blind pedestrians.

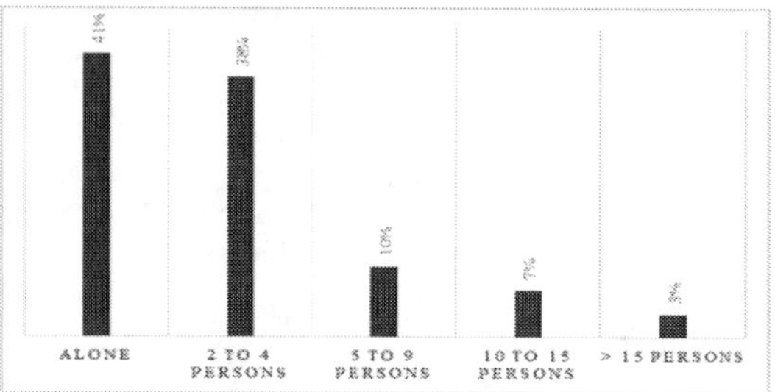

Fig. 2. Group Walking of Blind Pedestrians

The survey results also indicate that the blind pedestrian frequently walks within the group than conducts the alone walking trip (see Figure 2). To support their travel trip, the public transportation mode (i.e., Bus and Paratransit) is selected as the main transportation modes for the blind walkers. However, the end to end walking trip still gains a big portion, which can be interpreted as the inconvenient public transport service.

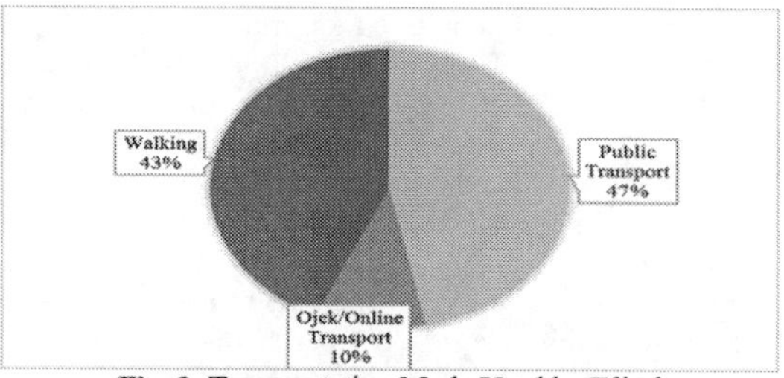

Fig. 3. Transportation Mode Used by Blind

In term of transportation cost, the blind mostly spends less than Rp. 10,000 for single trip, where for the longer trip they can be incurred up to Rp. 85.000 per trip by using online transportation modes.

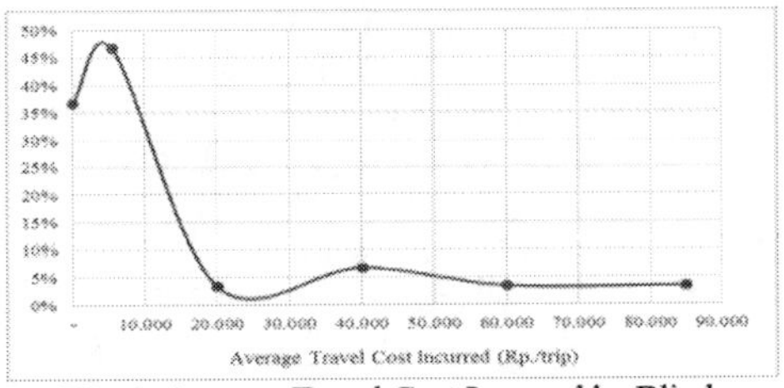

Fig. 3. Average Travel Cost Incurred by Blind.

4 Physical Attributes of Walkability for Blind Pedestrian

The further stage is conducted by elaborating the walkability components for the blind pedestrian. Due its physical limitation, the walkability components may differ to the non-blind pedestrians. This paper then explores the walkability components, which is derived from the required facilities.

The required facilities are examined by conducting the pair wise comparison interview to the blind pedestrians. The width of sidewalk, its pavement condition, crossing facilities, sidewalk level, tactile paving, braille information, safety and security facilities are selected as the main facilities. The results shows that the security and safety facilities is perceived as the most important facilities to the blind pedestrians.

In addition, the tactile paving and braille information in the walking built environment is significantly needed, which could provide the direction and guidance to the blind walker. As it can be inferred from the table above, the blind pedestrians also pay more attention to the pavement condition than the width of side walk, which can easily be understood by taking into consideration their physical shortcomings.

Table 1. Required Facilities for Blind Pedestrians.

Facilities Components	Weight of Importance
Width of Sidewalk	0,11
Sidewalk Pavement Condition	0,15
Crossing Facilities	0,16
Sidewalk Level	0,14
Tactile Paving and Braille Information (Physical Guidance)	0,22
Safety and Security	0,23

The interview and facilities surveys are also implemented for investigating the built environment of pedestrian facilities that construct the walkability components of blind pedestrians. To obtain more comprehensive view, the current performances of walkability components, which are assessed by the blind pedestrians, are compared to the physical attribute of pedestrian facilities. The attributes is listed by Park et al. [16], for instances, width of side walk, number of commercial uses, average building width, and etc. Such comparisons allow the further investigation for discovering the attributes that form the walkability components for the blind pedestrians. By assuming that there is a linear correlation between physical attributes and the blind pedestrian perception, and hence, it is possibly used for prioritizing the required pedestrian facilities to smoothly flow the blind pedestrians.

The width of sidewalk is firstly discussed by counting the total and effective width of side walk. The effective width can be easily defined as the total width reduces by the obstacle that decrease the usage of total width. This study thus studied how the perception of blind walker to the sidewalk width by considering those factors. Despite its less statistic performance, which is merely caused by number of respondent, by comparing the perception of blind walker to the physical indicator, it is interesting to find that the blind walker more considers the effective width rather than the total width. As it is illustrated in Table 2, the wider effective width is positively correlated to the satisfaction perception of blind pedestrians.

Table 2. Regressions result of sidewalk width component

Variables	Blind Walker Perspectives			
	Coeff.	Std. Err.	t	Sig
Total width (m)	-0,20	0,54	-0,37	0,72
Effective width (m)	0,37	0,91	0,41	0,69
Constant	2,34	0,36	6,58	0,00
F Statistic				0,26
Significance F				0,62

For examining the factor that build the walkability component of pavement condition, the number of potholes in the sidewalk and the condition of tactile pavement is associated to the convenient perception from the blind pedestrians. Table 3 indicates that the pavement satisfaction condition is more contributed from the tactile pavement condition than the number of sidewalk potholes. This phenomena is merely caused by the behavior of blind pedestrian that practically follow the tactile pavement as the guidance, and hence, the damage of tactile pavement toward to reduce their perception about the condition of sidewalk pavement.

Table 3. Regressions result of sidewalk pavement condition

Variables	Blind Walker Perspectives			
	Coeff.	Std. Err.	t	Sig
Number of sidewalk potholes/100 m(unit)	0,12	0,47	0,25	0,80
Tactile paving condition*)	0,82	0,37	2,19	0,04
Constant	1,00	0,66	1,51	0,14
F Statistic				2,75
Significance F				0,08

*) : 1: poor condition ; 2 : moderate condition ; 3 : good condition

The factor that build the perception about crossing facilities is also analyzed. Since most of traffic light in Indonesia is not installed by the sound equipment, this study only consider the number of zebra crossing for seeking the facilities that associate with the crossing components. Table below infers that there is a positive correlation between the numbers of zebra crossing facilities with the easiness perception of crossing for blind walkers.

Table 4. Regressions result of crossing components

Variables	Blind Walker Perspectives			
	Coeff.	Std. Err.	t	Sig
Number of zebra crossing/300 m (unit)	0,33	0,13	2,48	0,02
Constant	1,83	0,33	5,52	0,00
F Statistic				6,16
Significance F				0,02

Table 5 describes the regression result which compares the blind pedestrian perspective about the physical guidance satisfactions with the tactile existences and its condition. Since the braille information and the sound guidance equipment are not available in the study area, the physical guidance is more concentrated to the tactile guidance. The result imply that the existence of pavement tactile should be united with its good condition in order to fulfill the blind pedestrian needs.

Table 5. Regressions result of physical guidance components

Variables	Blind Walker Perspectives			
	Coeff.	Std. Err.	t	Sig
Tactile condition	0,78	0,33	2,35	0,03
Existence of pavement tactile *)	0,10	0,29	0,33	0,74
Constant	1,09	0,47	2,35	0,03
F Statistic				3,36
Significance F				0,05

*): 1: tactile pavement exist ; 0 : tactile pavement is not exist

The sense of security in the pedestrian facilities is practically difficult to be assessed, since it is mostly affected by the other aspects. However, this paper try to draw the security issue by correlating it with the existence of others. The number of nearby house and commercial are then investigated for illustrating the existence of others. The results show that the increasing number of nearby commercial probably improve the sense of security for the blind pedestrians.

The other issues, which need to be carefully considered for the blind pedestrian, is the side walk level changes. The changes within the side walk facilities will bring some difficulties to the blind pedestrian. Therefore, the study try to capture the factor that influence the perception of blind walker about the level changes. Table 7 show that the number of intermediaries, which force the walker to consider the change of sidewalk level, are negatively related to the blind walker perception. The similar situation occurred for the changing elevation. Thus, it can be simply concluded that the blind walker will find more difficulties if the number of intermediaries increase, as well as the increment of elevation changing. Furthermore, the existence of chair is interestingly related to the positive perception about the side walk level. This fact may be seen as the blind walker prefer to use chair for taking a rest.

Table 6. Regressions result of security issues

Variables	Blind Walker Perspectives			
	Coeff.	Std. Err.	t	Sig
Number of nearby house/100 m (unit)	0,09	0,08	1,24	0,23
Number of nearby commercials/100 m (unit)	0,01	0,01	1,44	0,16
Constant	1,52	0,49	3,09	0,00
F Statistic				1,73
Significance F				0,20

Table 7. Regressions result of side walk level

Variables	Blind Walker Perspectives			
	Coeff.	Std. Err.	t	Sig
Avg. number of intermediaries (per 100 m)	-0,04	0,16	-0,26	0,80
Changes in elevation (per 100 m)	-0,02	2,45	-0,01	0,99
Existence of chair	0,03	0,87	0,03	0,97
Constant	2,81	1,32	2,12	0,04
F Statistic				0,06
Significance F				0,98

Putting together the analysis of walkability component, table below summarize the important built environment factors for forming the walkability of blind walker. As can be inferred from Table 8, it is very important to provide several physical indicator for smoothly flowing the blind walker, for instance, the sufficient effective width, the existence of tactile pavement including its good condition, the existence of zebra crossing equipped by the warning tactile, the small number of intermediaries, and the sidewalk pavement without potholes.

Table 8. Physical Factor Construct the Walkability Components

Walkability Component	Important Physical Factors For Blind Walker
The width of side walk	Effective width of side walk (m)
The Sidewalk Pavement Condition	Tactile pavement condition Number of sidewalk potholes/100 m (unit)
Sense of Safety in Crossing	Number of zebra crossing/ 300 m (unit)
Physical Guidance for Blind	Existence of pavement tactile Tactile pavement condition

Walkability Component	Important Physical Factors For Blind Walker
Sense of Security	Number of nearby house/100 m (unit) Number of nearby commercials/100 m (unit)
Sidewalk Level	Avg. number of intermediaries/100 m Changes in Elevation/100 m Existence of chair

5 Conclusion

This paper discover the physical attribute that construct the walkability perception of blind pedestrian. As different with the previous research that mostly focuses on the non-blind walker perception, the paper thus takes into account the blind walker perception for investigating the physical attribute of walkability perception. The result shows that the blind walker has a specific physical attributes, which needs to be carefully handled in the pedestrian facilities planning. Therefore, the result is potentially to be used for analyzing the required pedestrian facilities to serve the blind walker.

This work is supported by the P3MI ITB research project fund. We thank our colleagues from PSBN Wyta Guna Bandung who provided insight and information that greatly assisted the research.

References

1. T. Wagener, and S. Spek, Accessibility for all – Eliminating barriers across Europe, NovaTerra, Connected Cities, Netherland, (2006).
2. D. Pecchini, F.Giuliani, Street crossing behavior of people with disabilities, J. Transp. Eng. **141**, (2015).
3. M.S. Wright, G.K. Cook, G.M.B. Webber, Emergency lighting and wayfinding provision systems for visually impaired people: phase I of a study, Lighting Res. Technol. **3** (2), pp. 35–42, (1999).
4. E.D. Kuligowski, R. Peacock, E. Wiess, B. Hoskins, Stair evacuation of older adults and people with mobility impairments, Fire Saf. J. **62**, 230–237, (2013).
5. W. Daamen, S.P. Hoogendoorn, Emergency door capacity: influence of door width, population composition and stress level, Fire Technol. **48** (1), 55–71, (2011).
6. M.C. Rodriguez-Sanchez, M.A. Moreno-Alvarez, E. Martin, S. Borromeo, J.A. Hernandez-Tamames, Accessible smartphones for blind users: A case study for a wayfinding system, Expert Systems with Applications **41**, pp. 7210–7222, (2014).
7. S. Secchi, A. Lauria, G. Cellai, Acoustic wayfinding: A method to measure the acoustic contrast of different paving materials for blind people, Applied Ergonomics **58**, pp. 435-445, (2017).
8. S. Mascetti, D. Ahmetovic, A. Gerino, C. Bernareggi, ZebraRecognizer: Pedestrian crossing recognition for people with visual impairment or blindness, Pattern Recognition **60**, pp. 405–419, (2016).
9. S. Mascetti, D. Ahmetovic, A. Gerino, C. Bernareggi, M. Busso, A. Rizzi, Robust traffic lights detection on mobile devices for pedestrians with visual impairment, Computer Vision and Image Understanding **148**, pp. 123–135, (2016).
10. S. Mascetti, D. Ahmetovic, A. Gerino, C. Bernareggi, Sonification of guidance data during road crossing for people with visual impairments or blindness, Int. J. Human-Computer Studies **85**, pp. 16–26, (2016).
11. S. Handy, Critical assessment of the literature on the relationships among transportation, land use, and physical activity, Transportation Research Board and the Institute of Medicine Committee on Physical Activity, Health, Transportation, and Land Use. Resource Paper for TRB Special Report, 282, (2005).
12. S. Kalakou, and F. Moura, Bridging the gap in planning indoor pedestrian facilities, Transport Reviews, **34**(4), 474–500, (2014).
13. R. Cervero, and K. Kockelman, Travel demand and the 3Ds: density, diversity and design. Transportation Research Part D: Transport and Environment, **2**(3), 199–219, (1997).
14. R. Cervero, O. L. Sarmiento, E. Jacoby, L.F. Gomez, and A. Neiman, Influences of built environments on walking and cycling: lessons from bogotá, International Journal of Sustainable Transportation, **3**(4), 203–226, (2009).
15. R., Ewing, & S. Handy, Measuring the unmeasurable: urban design qualities related to walkability. Journal of Urban Design, **14**(1), 65–84, (2009).
16. S. Park, E. Deakin, and J.S. Lee, Perception-based walkability index to test impact of microlevel walkability on sustainable mode choice decisions, Transportation Research Record: Journal of the Transportation Research Board, **2464**(1), 126–134, (2014).

MATEC Web of Conferences **147**, 02003 (2018) https://doi.org/10.1051/matecconf/201814702003
SIBE 2017

Is the indonesian regulation of standard level of service of a pedestrian path fulfill pedestrians' convenience

Amirotul Mahmudah[1,*], *Slamet* Legowo[2], *Agus* Sumarsono[3], *Studyamadyakomunika* Linta[4], and *Benny* Irawan[5]

[1,2,3] Lecturer of Civil Engineering Study Program, Universitas Sebelas Maret, Surakarta 57126, Indonesia
[4,5] Graduates of Civil Engineering Study Program, Universitas Sebelas Maret, Surakarta 57126, Indonesia

Abstract. Indonesia published Regulation No: 03/PRT/M/2014 about the Pedestrian Level of Service (LOS), consist of pedestrian movement characteristics and sidewalk capacity, as like the Bangkok standard, there is no parameter that represents pedestrian interest. Australian standards develop a Level of Service with the main based on categories that relate with pedestrian convenience. Within this study, identifying Indonesian pedestrian interests done by measuring pedestrian perception of convenience. This study aims to check the suitability between this standard pedestrian LOS and pedestrian perception of convenience in a commercial area, Pasar Gede market. The study results, based on Indonesia regulation LOS, the pedestrian path around Pasar Gede market is categorized as level A at all sections, and this is the same as the Bangkok standards. According to the Australian's standard, the LOS is an A for the West section, B for the North section, and C for the South section. The pedestrians have perceptions of moderately discomfort for all sections. This means that although the pedestrian path fulfills the level of service for Indonesian Government regulation, it does not fulfill user satisfaction. Therefore, the Indonesian LOS standard for pedestrian paths doesn't provide equity for the pedestrian.

1 Introduction

Walking is a significant mode of transport to promote sustainable transport, since it doesn't need fuel does not create any pollution, and improves health [1,2]. One parameter to value pedestrian path performance is Level of Service. Countries have formulated standards of level of service for pedestrian paths, and each country has its own factors and values, which depend on pedestrian characteristics.

A standard Level of Service (LOS) is really important for planning, evaluation, and improvement processes. LOS is used as a standard for the planners to design to, and a suitable LOS and user perception is key to the successful implementation of the design. Pedestrian LOS have been discussed since around 1983 by A. Polus [3], but is still a hot topic of discussion [4, 5, 6, 7, 8]. Discussion about the LOS for pedestrian path starts with the flow, speed characteristics, and sidewalk capacity, and continues with pedestrian facilities, and then pedestrian perception. Considering the right factors for the LOS should meet the satisfaction of planners and users. At this time, there are various approaches to pedestrian level of service models. Some of them emphasize pedestrian flow and volume, as well as sidewalk capacity, and others consider pedestrian path facilities [4]. Most of the guidelines consider that pedestrian flow and volume, as well as sidewalk capacity is enough to measure LOS, but this consideration is criticized because pedestrians are then treated like vehicles [4].

Other studies discuss pedestrian perception for the LOS [5, 6], and mention that having a good understanding of the pedestrian perception of LOS on a pedestrian path in an urban area can be useful in developing strategies for providing a comfortable and safe walking environment [5]. Kang et al. [5] concludes that the most influence factor on LOS value is pedestrian flow rates, however, many other factors have a significant influence on LOS perception, including sidewalk width, the presence of a barrier separating the sidewalk from motor-vehicle traffic, the presence of parking next to the sidewalk, and so on. Jena [6] suggests that pedestrian flow and volume, and te sidewalk capacity are most important and so it is reasonable for these to be the main point to value in the LOS. But there are many other facilities that pedestrians also consider important when they evaluate LOS. Other researchers use a rating system to analyze the LOS, based on road users' 5-factor perception: safety, comfort level, vendor encroachment, accessibility and sidewalk performance, and climate conditions.

Comparison and evaluation among methodologies at international level for the estimation of pedestrian levels of service conclude that current methodologies are unable to analyze the entire spectrum of the

Corresponding author: amirotulmhm@staff.uns.ac.id

walking experience, and also conclude that no method is universally applicable [7,8], so there is a need to define a specific model as a standard pedestrian LOS for each country. Based on the identification of the weakness of existing methodologies, Christopoulou and Latinopoulou [7] developed a model whose factors reflect the pedestrian LOS along Greek urban streets. The factors are developed from 3 categories: that is, traffic parameters, geometry/environmental/sidewalk parameters, and pedestrian movement parameters. Then the model is calibrated with a questionnaire survey to pedestrians, residents, and transport experts. Shingh and Jain [8] conclude that assessment methods should not only be driven by data that can be easily measured and manipulated quantitatively, but should also include the walking experience of pedestrian and planners.

In the Indonesian case, pedestrian facilities are not as good as in developed countries. The climate and weather conditions (tropical country, high temperature and humidity), public transport performance, and private vehicle use discourage the willingness to walk, and define a specific pedestrian characteristic. At this time, walking is a choice at most for captive people. Some research on pedestrian LOS discusses a walkability index [9], which can be used to identify improvements to the pedestrian way according to pedestrian perceptions. Sutikno et al. [9] measured a walkability index of Sukarno-Hatta street corridor, business corridor, in Malang city, East Java, Indonesia. They used 9 factors: physical condition of the pedestrian way; conflict with the street/accident; ease of crossing; maintenance and completeness of supporting facilities; effective width of pedestrian way; buffer with the street; accessibility of the pedestrian way; physical aesthetic of pedestrian way; and shade. Based on these factors, the respondents' value given to the walkability index of the Sukarno-Hatta street corridor was medium. The study also discussed pedestrian perception that the pedestrian facilities need to be improved; i.e. continued improvement and addition of pedestrian ways, planting plans that can reduce air pollution, availability of crossings, the addition of shade-providing trees, traffic management to reduce congestion, and parking design. Other studies discuss the pedestrian LOS model incorporating pedestrian perceptions of sidewalks with vendor activities [10]. One model was developed for a commercial area in Jakarta, based on 4 dependent variables—comfort, vendor problems, pedestrian volume, interaction with vendors—with a multiple linear regression model. Comfort has the highest influence on pedestrian LOS, followed by vendor problems, vendor interaction, and pedestrian volume. As far as the author knows, until now there has been no discussion about suitability between the existing Government regulation and pedestrian perception. This is the aim of this study.

In 2014 the Public Works Minister of Indonesia published a regulation about pedestrian level of service in Regulation No: 03/PRT/M/2014 [11]. This regulation uses pedestrian movement as its parameter. The regulation will be applied to a commercial area, since the commercial area is an important land use to support economic growth. One of important supporting infrastructures in the commercial area is a pedestrian path, which is there for the municipality to address pedestrian satisfaction on the pedestrian level of service. This regulation divides LOS into 6 levels: A, B, C, D, E, and F [11], as well as other standards [12, 13]. A is the most comfortable pedestrian environment, whereas and F is the least comfortable pedestrian environment. The Indonesian standard LOS is based on pedestrian space, average speed, flow rate, and volume capacity ratio (VCR). Within this study the pedestrian LOS and comfort based on pedestrian perception of the pedestrian path facilities at Pasar Gede market will be measured. During the survey, this facility is a new facility. Based on survey data will be analyzed and discussed the value of LOS based on area, flow rate, and volume capacity ratio is fulfil the planner design. And also measured does the facility fulfil the pedestrian convenience. Then the LOS based on Indonesian regulation also will be compared with the pedestrian path LOS for Thailand and Australia.

The study results hope to show the differences among the LOS based on Regulation No. 03/PRT/M/2014, Bangkok, and Australian standards, and also the value of pedestrian perception on pedestrian path facilities in Pasar Gede Market. The discussion results can be show the difference level of satisfaction between planner and pedestrian. This proof can be based on consideration that it is really important to use factors on pedestrian convenience in LOS.

2 Research Method

2.1 Field Location

Pasar Gede market is the oldest traditional market that serves the daily needs of downtown Surakarta. The market has multiple functions, and as well as a commercial area it has also become a tourist destination. To support this strategic function, the Government needs to provide good infrastructure facilities such as the pedestrian paths that fulfill the desired level of convenience for the pedestrian.

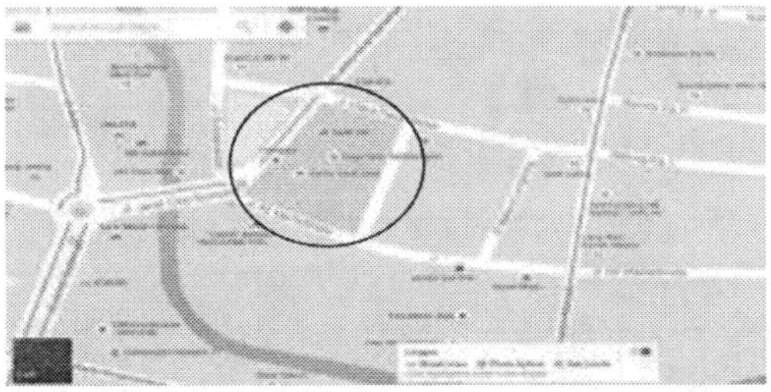

Fig 1 Pasar Gede Market

Corresponding author: amirotulmhm@staff.uns.ac.id

As be seen in Figure 1, Pasar Gede market is surrounded by Urip Sumoharjo Street, RE Martadinata Street, and Utara Pasar Besar Street. Pedestrian path is located between these streets, and the market consists of 4 sections of pedestrian path: North, West, South, and east sections. Each section has a different geometrical design and facilities so they need to be analyzed separately. The operational hours of the market are 07:00–16:00 Western Indonesia Time (WIB). During the survey was conducted the east side used as warehouse, so we do not analyze this section.

2. a. West section

2.b. North section

2.c. South section

Fig 2. Pedestrian path in Pasar Gede market

Table 1. Geometry and surface conditions of pedestrian path in Pasar Gede market

Section	Effective Width (meters)	Quality of path surface
North	4	Continuous, smooth, skid-resistant surface, without cracks and bumps
South	2,5	Continuous, smooth, skid-resistant surface, without cracks and bumps
West	2,7	Continuous, smooth, skid-resistant surface, without cracks and bumps

2.2 Pedestrian Flow Survey

First, the preliminary survey is conducted. This survey aims to determine the observation points to represent the flow and speed conditions along the section, to define the survey method, survey time, and also to design the number and tasks of the surveyors. Second, based on a preliminary survey, the main survey is conducted during operational hours to collect pedestrian flow and speeds, with an observation span of 10 m for each section. The data were collected during operational hours at 5-minute intervals. The manual survey is done by 9 surveyors for all sections. The total number of pedestrians during survey periods was 8559 persons. Based on this data, it can be calculated pedestrian density, speed, and flow.

2.3 Greenberg model

The capacity of a pedestrian path can be calculated based on the Greenshield, Greenberg, or Underwood models. The appropriate application of these models depends on the pedestrian movement characteristics. Analog with pedestrians behavior in Yogyakarta, we assume that in Surakarta the appropriate model for pedestrian flow characteristics is the Greenberg model [14].

2.4 LOS Standard

The performance of the facilities is usually measured in the Level of Service and the standard of LOS is defined by the Government. In Indonesia the regulation of standard LOS for the pedestrian path is done by the Ministry of Public Works, on "Pedoman Perencanaan, Penyediaan, dan Pemanfaatan Prasarana dan Sarana Jaringan Pejalan Kaki di Kawasan Perkotaan," according to regulation number No: 03/ PRT/M/2014 [7]. The LOS standard includes 4 parameters: Area, Speed, Flow, and the Volume Capacity Ratio, as mentioned in Table 1. The manual for the planning process, whereby the facilities fulfill the LOS-A, should fulfill the planning standard and satisfy the users; in this case, the users are pedestrians.

Table 2. Standard of Pedestrian Path LOS in Ministry of Public Works Regulation No: 03/PRT/M/2014

LOS	Area (m^2/ped)	Speed (m/minute)	Flow (ped/meter/ minute)	Volume Capacity Ratio (VCR)
A	≥ 12	≥ 78	≤ 6.7	≤ 0.08
B	≥ 3.6	≥ 75	≤ 23	≤ 0.28
C	≥ 2.2	≥ 72	≤ 33	≤ 0.40
D	≥ 1.4	≥ 68	≤ 50	≤ 0.60
E	≥ 0.5	≥ 45	≤ 83	≤ 1.00
F	< 0.5	< 45	variable	1.00

Bangkok developed its standard LOS for walkways based on space and flow. Comparison between Bangkok and Indonesian standards shows that space in Indonesia standards is higher than Bangkok, and Flow for Indonesian standards is lower than Bangkok standards. This means that Bangkok pedestrians can tolerate and occupy less personal space than Indonesian pedestrians.

Corresponding author: amirotulmhm@staff.uns.ac.id

Table 3. Developed LOS Bangkok standard for walkways in Bangkok

LOS	Space (m^2/ped)	Flow (ped/m/min)
A	≥ 2.38	≤ 28
B	1.60–2.38	28–40
C	0.98–1.60	40–61
D	0.66–0.98	61–81
E	0.37–0.66	81–101
F	≤ 0.37	101 or variable

The Government of Western Australia composed guidelines for assessing the pedestrian level of service [13]. Factors affecting the pedestrian level of service are physical characteristics, location factors, and user factors. Methods of assessment are explained in Table 3. The guidelines also provide a pedestrian level of service assessment sheet which explains the weight and definition of each point. Each factor is assessed and then multiplied by the score, then the final weight score of each factor is summed and plotted, as shown in Table 4.

Table 4. Methods of Assessment

Category	LOS Factor	Method of Assessment
Physical Characteristics	Access	Examine on site
	Path Width	Measure from plans or during site visit
	Surface Quality	Examine during site visit
	Crossing Opportunities	Assess during site visit
	Support Facilities	Examine during site visit
Location Factors	Connectivity	Assess from street directory maps and during site visit
	Path Environment	Assess from street directory maps and during site visit
	Potential for Vehicle Conflict	Count during site visit (number per km)
User Factors	Path User Volume	Local Government statistics or count during site visit
	Mix of Path Users	Count during site visit
	Personal Security	Assess during site visit, preferably at night

Table 5. LOS grade scale

LOS Grade	Description	Range of Scores
A	Ideal	132 or higher
B	Reasonable	101–131
C	Basic	69–100
D	Poor	37–68
E	Unsuitable	36 or lower
F	Problems with Access	Score of less than 15 for Access

Corresponding author: amirotulmhm@staff.uns.ac.id

2.5 Questionnaire survey

Within this research, the survey method used to collect pedestrian perceptions about pedestrian path facilities is a questionnaire survey. The error tolerance is 10%, To fulfill the data adequacy test, from the population number (for the number of visitors in 1- day during operational hours) is 8559 persons, the total respondents needed is 288: 97 respondents for the West section; 98 for the North section; and 93 respondents for the South section.

Reference [15] mentions that pedestrian comfort is influenced by the following factors: circulation, weather, noise, smell, landscape furniture, safety, cleanliness, and appearance. Within this study, safety is divided into 2 variables: safety from crime and safety of the pedestrian path (slippery, sloping, potholes), and 1 variable is added: completeness of supporting facilities. The response in the questionnaire for each factor is designed using a Likert scale. Percentage class interval is built based on score assessment.

Table 6. Comfort Criteria based on score assessment

Interval Kelas Persentase (%)	Criteria
100% ≥ Percentage > 84%	Strongly comfort
84% ≥ Percentage > 68%	Moderately comfort
68% ≥ Percentage > 52%	Slightly comfort
52% ≥ Percentage > 36%	Moderately discomfort
36% ≥ Percentage > 20%	Strongly discomfort

To guarantee that the questions in questionnaire are valid and reliable, the questionnaire is analyzed the valid and reliable based on Pearson Correlation and Spearmen Brown theory, as in [16].

3. Result And Discussion

Geometric and pavement conditions of the pedestrian path at Pasar Gede market are in good condition since it was built in the same year as the study. Each section has a different effective width, as seen in Table 1, so it is reasonable to calculate the pedestrian path LOS separately for each section.

3.1 Pedestrian Flow, Speed, and Density

The pedestrian path at Pasar Gede Market has specific characteristics since its location is really close to shops, so as pedestrians walk to their destination, they usually also glimpse the merchandise in the shops. Based on this condition it is reasonable to omit speed for categorizing the level of service (LOS).

Table 7. Movement characteristics of pedestrians in pedestrian path at Pasar Gede market

	North Section	South Section	West Section
A (m^2/ped)	59.88	135.14	79.37
V (m/min)	8.695	7.46	8,00
Q (ped/m/minute)	0.145	0.0552	0.108

A, V, Q is a pedestrian area, speed, and flow. Pedestrian area in all sections is more than or equal to 12 m^2/ped, so LOS for all sections is at the A level. For flow parameter, all sections have flow ≤ 6.7 ped/m/minute in LOS-A.

3.2 Volume Capacity Ratio (VCR)

Correlation among flow, speed, and density, based on the Greenberg model for the North section, is shown in Table 5: Vs is Velocity, D is density, and Q is flow. From the Greenberg model, the maximum flow or capacity can be calculated [17]. Pedestrian path capacity can be calculated from the equation of relationships between Q and D, or Q and Vs. Capacity is the maximum flow that can pass along the pedestrian path. Based on the Greenberg model, the capacity of the North section is 1.8056 ped/m/minute. Flow for the North section, based on Table 7, is 0.145 ped/m/minute, so the VCR is 0.08. Thus the level of service for the North section is A.

Table 8 Relationship of Flow, Speed, and Density, based on Greenberg Model

North section	South section	West section
Vs = 35.648–6.718 LnD	Vs = 36.694–5.471LnD	Vs = 34.935–6.991LnD
Q = 35.648 D–6.718 D Ln D	Q = 36.694 D–6.471 D Ln D	Q = 34.935 D–6.991 D Ln D
$Q = 0.085\ V_s . e^{-0.062\ Vs}$	$Q = 0.105\ V_s . e^{-0.025\ Vs}$	$Q = 0.091\ V_s . e^{-0.045\ Vs}$

A similar calculation shows the capacity of the South and West sections is 2.06 ped/m/minute and 1.93 ped/m/minute, respectively, so the VCR value for the South section is 0.03, and for the West section it is 0.06. Therefore the LOS of the South and West sections based on VCR parameter are at level A.

3.3 Level of Service

Recapitulation of calculation of LOS for the pedestrian path at Pasar Gede market is shown in Table 9. Based on pedestrian movement all sections have level A, but based on pedestrian path facilities (Western Australian Standard) the level of service varies within sections.

Table 9. LOS value of pedestrian path at Pasar Gede market

Section	Regulation		
	Peraturan Menteri Perkerjaan Umum Nomor : 03/PRT/M/2014		
	Flow	**Space**	**VCR**
West	A	A	A
North	A	A	A
South	A	A	A
	Bangkok, Thailand		
	Flow	**Space**	
West	A	A	
North	A	A	
South	A	A	
	Standar LOS Australia		
West	A		
North	B		
South	C		

3.4 Comfort level of Pedestrian Path Facilities

Pedestrian path facilities in Pasar Gede market are not uniform: for example, width, cleanliness, smell etc. For this reason, the analysis is divided into West, North, and South sections. The data from respondents who pass the West section are described in Table 10. The responses are: not at all good (NG); slightly good (SG); moderately good (MG); very good (VG); and extremely good (EG).

Table 10. Score weight for West Section pedestrian path at Pasar Gede Market

Corresponding author: amirotulmhm@staff.uns.ac.id

	Pedestrian Path Facilities Condition	NG	SG	MG	VG	EG
1	Circulation	1	14	189	100	5
2	Weather (temperature, moisture)	2	66	120	88	0
3	Noise	2	62	150	50	5
4	Smell	6	58	120	84	0
5	Landscape furniture	0	8	183	112	20
6	Safety from crime	1	8	150	136	40
7	Safety of pedestrian path (slippery, sloping, pot hole)	0	14	150	144	25
8	Cleanliness of pedestrian path	0	28	150	128	5
9	Appearance of landscape furniture	0	6	162	152	10
10	Completeness of supporting facilities	0	48	165	62	10
	Mean	1	31	155	106	12
		0.4 %	10. 2%	50.8 %	34. 8%	3.8 %

Based on all pedestrian path facility conditions only 0.4% of respondents stated Not at all Good, 10.2% stated Slightly Good, 50.8% stated Moderately Good, 34.8% stated Very Good, and 3.8% stated Extremely Good. The highest mean plot (Table 6) thus concludes that pedestrian perception of the pedestrian path categories is one of moderate discomfort. A similar step is done for other sections.

Table 11. Pedestrian perception of pedestrian path at Pasar Gede market

No	Pedestrian Perception on Pedestrian Path	Score (%)	Value
1	West section	50,8%	Moderately discomfort
2	North section	48.7%	Moderately discomfort
3	South section	49,4%	Moderately discomfort

3.5 Discussion

Performa and user's perception are different things, but they should be considered simultaneously within an infrastructure facility planning. Ideally, the LOS standard of a facility also meets with user perception, in order to guarantee that the planners' valuation similar with user perception of a facility performance.

As far as author know there is not an equity comparison on LOS and User perception yet. This study also does not want to make a comparison on it, but shows that there is different between LOS and user's perception valuation on a facility when in a LOS standard only consider the pedestrian flow characteristic (Indonesia and Bangkok standard).

This study also analyzes the LOS of Australian's standard, it has been considered the physical characteristics, location factors, and users factors. Pedestrian volume is the only variable of pedestrian flow characteristic that is considered. It will show how much the different value between Australian's standard and the user perception.

The pedestrian path is an important transportation facility to support the performance of the road network, transit areas, commercial areas, residential areas etc. Factors that influence pedestrian perceptions of pedestrian paths normally have different weights for each transportation facility: e.g. the pedestrian considers the travel time of the road network and transit area, whereas for the commercial area pedestrians pay much more attention to the physical characteristics of the pedestrian path. This study compares the standard LOS for pedestrian path and road user perception in a commercial area.

Based on the regulations of Indonesia and Thailand, the LOS of the pedestrian path for all section at Pasar Gede market is in level A, but pedestrian satisfaction is "moderately discomfort." However, based on Western Australia guidelines for the West, North, and South sections, the LOS is A, B, and C.

3. Conclusion

In commercial areas, Pasar Gede Market, the pedestrian path has LOS-A based on the Ministry of Public Works Regulation of No. 03/PRT/M/2014, but the user satisfaction is "moderately discomfort". This study could be the one information that there is different valuation between planners and pedestrian, so it is needed to analyze the valuation of the weight of the pedestrian flow characteristic performance and the weight of pedestrian's perception on a convenience matter. Finally these two main variable, -pedestrian flow characteristic performance and pedestrian's perception-, should be the main variables of LOS, so the planners have the similar valuation with pedestrian for a LOS on a pedestrian path.

REFERENCES

[1] World Health Organization, *Health economic assessment tools (HEAT) for walking and for*

Corresponding author: amirotulmhm@staff.uns.ac.id

cycling: Methodology and user guide (2014 update), Economic Assessment of Transport Infrastructure and Policies, ISBN: 978-92-890-0251-6, WHO Regional Office for Europe Scherfigsvej 8 DK-2100 Copenhagen Ø, Denmark, (2014)

[2] Olabarria, M,. et al., Health impact of motorized trips that could be replaced by walking, *European Journal of Public Health*, 23(2), 217–222. doi:10.1093/eurpub/cks015 (2013)

[3] A. Polus, J.L. Schofer, A. Ushpiz, *Pedestrian flow and level of service*, Journal of Transportation Engineering, Vol. **109** No. 1 January (1983)

[4] Z. Ashadi-Shekari, M. Moeinaddini, M.Z. Shah, *A Pedestrian level of service method for evaluating and promoting walking facilities on campus streets*, Land Use Policy **38**, 175 - 193 (2014)

[5] L. Kang, Y. Xiong,, F.L. Mannering, *Statistical analysis of pedestrian perceptions of sidewalk level of service in the presence of bicycles*. Journal of Transportation Research Part A **53,** 10-21 (2013)

[6] S. Jena, *Perception based pedestrian level of service*, Thesis for diploma in Department of Civil Engineering National Institute of Technology Rourkela, India.

[7] P. Christopoulou, M. Pitsiava-Latinopoulou, *Development of a model for the estimation of pedestrian level of service in Greek urban areas*, Procedia - Social and Behavioral Sciences 48 1691 - 1701 (2012)

[8] Singh K., Jain P.K., Methods of assessing pedestrian level of service, Journal of Engineering Research and Studies Vol ***II*** January-March -116-124, E-ISSN 0976-7916 (2011)

[9] F. R. Sutikno, Surjono, E. B. Kurniawan, *Walkability and pedestrian perceptions in Malang City emerging business corridor*, Procedia Environmental Sciences **17** 424 – 433 (2013)

[10] N. Hidayat, K. Choocharukul, K. Kishi, *Pedetrian level of service model incorporating pedestrian perception for sidewalk with vendor activities,* Proceedings of The Eastern Asia Society for Transportation Studies , **8** (2011)

[11] Ministry of Public Works, Regulation of Ministry of Public Works No. 03 / PRT/ M / 2014 (2014)

[12] Y. Tanaboriboon, J. A. Guyano, *Level-of-Service Standarts for Pedestrian Facilities in Bangkok*, ITE Journal, November, pp. 39 - 41 (1989)

[13] Government of Western Australia, *Guidelines For Assessing Pedestrian Level Of Service,* Western Australia (2006)

[14] Fika I. Pratiwi, *Studi Karateristik Pergerakan Pejalan Kaki Di Pedestrians Road Stasiun Tugu Yogyakarta*, Thesis for Undergraduate UGM, Yogyakarta (2011)

[15] R. Hakim, H. Utomo, *Komponen Perancangan Arsitektur Lansekap, Bumi Aksara*, Jakarta (2003)

[16] Sugiyono, *Metode Penelitian dan Pengembangan*, Alfabeta, Bandung (2015)

[17] O. Z. Tamin, *Perencanaan Pemodelan dan Rekayasa Transportasi*, Teori contoh soal dan aplikasi, ITB Publisher, Bandung (2003)

Corresponding author: amirotulmhm@staff.uns.ac.id

Relationship between train users' perceptions of walkability with access and egress mode choice

Romeiza Syafriharti [1,*], *B.* Kombaitan[2], *Iwan P.* Kusumantoro[3], and *Ibnu* Syabri[2]

[1]PhD Student of Transportation, SAPPK, Institut Teknologi Bandung, Bandung, Indonesia
[2]Urban and Regional Planning, SAPPK, Institut Teknologi Bandung, Bandung, Indonesia
[3]Transportation, SAPPK, Institut Teknologi Bandung, Bandung, Indonesia

Abstract. The purpose of this study is to understand whether there is a relationship between train users' perceptions of walkability in built environment of trip origin with access mode choice and between train users' perceptions of walkability in built environment of trip destination with egress mode choice. Train users are who ride from Cicalengka station Bandung Regency, West Java, Indonesia. To analyze the relationship is used crosstab method. The perceptual factors about walkability are those perceived by the train users consisting of walking distance, safety, comfort, and secure from crime, both in origin and destination of the built environment. The mode choice consist of walking, paratransit, motorcycle taxi, and own vehicles (or others for egress mode). To better understand the relationship is used several control variables, that are trip purposes, train usage, gender, and age. For access trip there is another control variable, that is vehicle ownership. Train users' perceptions of walkability have a relationship with both the access and the egress mode choice, except for the security aspect. The influence of control variables on the relationship between perceptions of walkability with access/egress mode choice varies for walking distance, safety, and comfort.

1 Introduction

Walking as transport mode has been extensively researched from various fields of science, not only directly related to transportation itself, but also from the urban design, urban planning, and even linked to health. The decision to walk as a transport mode can be influenced by many factors, among which many of the research findings are a good walkability of the built environment.

To decide whether to walk or not to/from transit, the measurement of walkability that is researched varies considerably. Among these measures there is a focus on walking distance [1-3]. Distance to transit together with density, diversity, (street) design, and distance to destination are an urban form attributes known as 5 D's used when verify the effect of built environment on walking and cycling [4]. In addition, there is also evidence of the effects of 3 D's urban forms, that are density, diversity, and design (street network design), for walking to transit [5]. Meanwhile, also there are the effects of urban design attributes on walking as access mode to the station [6]. Only walking distance is not enough to analyze walking to transit required more than that, ie walking effort or walking accessibility [7]. Later, psychological factors are used to measure walkability and verify the effect on walking or using vehicles to public transport facilities, such as enjoyment of walking and social pressure [8]. From previous studies, it appears that, in examining the effect of built environments walkability in the choice of walking as access/egress mode to/from transit there is a variety of walkability measures.

In recent years, research on the mode choice of transport, including walking, tends to consider soft variables or subjective variables [9]. The residential environment determines the choice of mode rather than the objective characteristics of the built environment [10]. It is aligned with other studies that include self-selection variables in analysis the relationship between the built environment and travel behaviour, especially walking [11, 12].

Research that take into account travelers' perceptions of built environment walkability and examine their relationship with mode choice is still limited in number. Most studies use objective walkability measurements, such as by using urban form or urban design attributes. Research using subjective walkability measurement are still limited to the perceptions of built environment walkability in residence neighborhood only and just few about built environment walkability in place of activity, like at work place.

For public transport users, good walkability is not only important in the precincts of the residence to/from public transport facilities, but also around the site of activities to/from public transport facilities. Based on the findings of previous studies, that the subjective measurement of walkability relative to the greater effect on the mode choice than the objective measurement, the perceptions of

* Corresponding author: romeiza.syafriharti@email.unikom.ac.id

 https://doi.org/10.1051/matecconf/201814702004

travelers is used as a measure of walkability. Walkability at the start of the journey (start miles) may be different from walkability at the end of the journey (end miles), as well as the mode used may vary between access mode and egress mode.

The object of this research is Bandung Raya railway users who ride from Cicalengka train station in Bandung Regency West Java Province Indonesia and reside around the station. The destination railway stations are spread from Rancaekek to Padalarang. Therefore, this research aims to explore the relationship between railway users' perceptions about walkability with the choice of modes, specifically to answer the following research questions:

- Is there a relation between rail users' perceptions of the built environment walkability from residence to the origin railway station (during the access trip) with the access mode choice; and
- Is there a relation between railway users' perceptions of the built environment walkability from the destination railway station to the activity place (during egress trip) with the egress mode choice.

Basically, what is meant by access mode is a mode used to get from a trip origin (in this case is home) to a main mode (rail station origin), and vice versa (returning trip). While egress mode is a mode used to get from the main mode (destination train station) to the trip destination, and vice versa (returning trip) which can be seen on Fig. 1. The mode used on the returning trip may vary with the mode on the access trip. The same condition can also occur in egress trips. Therefore, in this research, access trip are limited only to travel from home to origin station, while the egress trip is a trip from the destination station to the final destination (Fig.2), excluding the returning trip. Access trip is distinguished by egress trip. On access trip there is a chance to use private vehicle, because the journey starts from home. While on egress trip there is no possibility of using private vehicles. With good walkability then walking is expected to be chosen as access and egress mode.

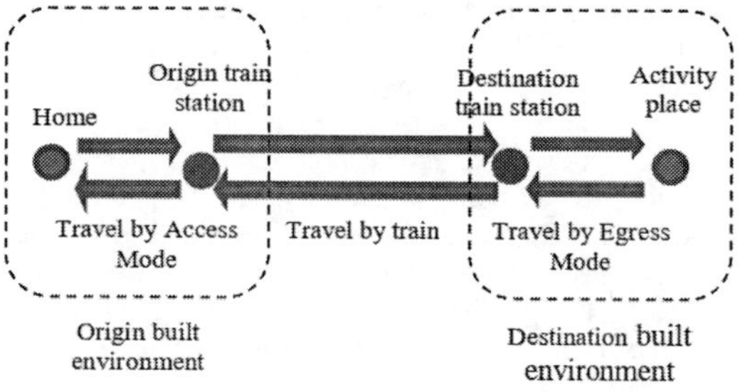

Fig. 1. General scheme of railway user travel.

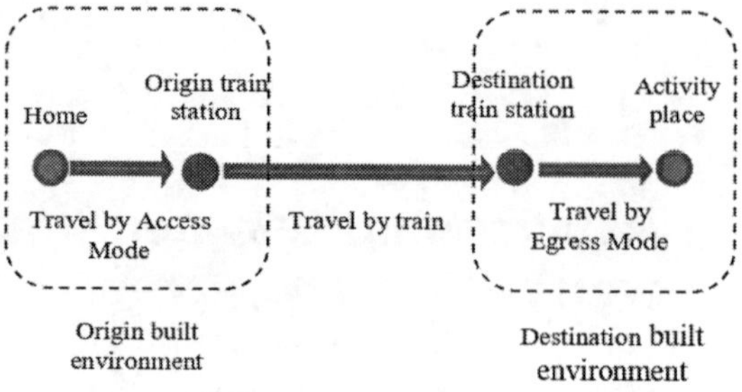

Fig. 2. Railway user travel scheme for this research.

2 Literature Review

In the process of this research, it's conducted literature review of related research on walking behavior that are influenced by the existence of psychological factors in assessing the built environment. Not much literatures are found which deals with walking as access mode and also as egress mode. Therefore, the reviewed research are walking as the whole transport mode, not only as access/egress mode but also walking as main mode. In the following table can be seen some previous research that explore the relationship between perceptions of built environment walkability with walking behavior.

From the previous research, there are some variables that are most widely analyzed. One of them is variable related to walking distance. This variable is defined as one that will be assessed by the train users who will ride from Cicalengka train station. Other variables to be considered are safety, comfort, and security.

- Walking distance
  Walking distance contains complex notions, so it is not enough to be represented only by physical distance, but more precisely defined as walking effort [7] or walking accessibility. The perceived walking distance can vary for the actual distance [2]. Therefore, for walkability, walking distance is more appropriate if using pedestrian perceptions.
- Safety from motorized-vehicle traffic
  Safety from vehicle traffic is also one of the variables whose judgments can be different for everyone. Especially if that walk are children or old people, safety standards will be different from adults or who do not have certain limitations. Safety perceptions from vehicle traffic in the built environment significantly influences walking as transport mode [11, 16]. Although traffic volume can be calculated to get the actual amount of traffic, still the perceptions of safety can vary for each person.
- Comfort
  Comfort is an abstract measurement, although it can be realized through the urban design attributes. The meaning of comfort is also contained in the 'spaciousness' [11], 'convenience' [8], 'aesthetics' [14], 'walkway width' and 'walkway quality' [15], 'quietness' [10], and even 'green streets' [16]. Such a broadness of meaning is contained in 'comfort'. 'Comfort' is something that is perceived, then for this

variable is more precisely measured subjectively from pedestrian.

- Security from crime
  Security is one of the basic requirements of transportation. However, in research on walkability, not many use it as the main variable. Most are part of attitudes [11].

Table 1. A summary of some previous research on the relationship between perceptions of built environment walkability with walking behaviour.

Researcher, Trip Purposes, Location	Methods	Walking Behavior Variables	Perceptions of Built Environment Walkability Variables	Results
Handy et al., 2006 [11] Traditional dan suburban neighborhood Northern California	Negative binomial regression Cross-sectional analysis Quasi-longitudinal analysis	Walking as transport mode	Accesibility, physical activity option, safety, socializing, outdoor spaciousness, attractiveness	Built environment has an impact on walking behavior even after accounting for attitudes and preferences.
Mehta, 2008 [13] All purposes Main street in Boston	Qualitative studies for microscale analysis	Walking as transport mode	Usefulness, safety, comfort, pleasurability, sense of belonging	Usefulness, sense of belonging and pleasurability were most important.
Walton & Sunseri, 2010 [8] All purposes Auckland & Wellington	ANOVA & ANCOVAs	Walking as access mode, besides drive	Fear of crime, distance/time, parking charges, inconvenience, and non built environment variables	Distance, fear of crime and time do not affect walking.
Panter & Jones, 2010 [14] Adults for active transport	Quantitative studies → review previous studies by examining the influence of psychological and environmental factors	Walking or cycling as transport mode	Functional (infrastructure & street connectivity), Safety, Aesthetics, Destinations (facility provision, residential density, land usemix, coastal proximity, walkable environment	Walking or cycling associated with only psychological factors or with only built environment factors or with both of them
Lin & Yu, 2011 [15] Leisure travel Students of elementary schools In the Shilin District of Taipei	Negative binomial regression model Multinomial logit model	Walking as transport mode, besides cycling and public transport	Walkway width, walkway quality, safety, and other variables (non-perceptions variables)	From perceptions variables, walkway quality encouraged a child to walk.
Adkins et al., 2012 [16] All purposes Residential neighborhood in Portland, Oregon	OLS	Walking as transport mode	Street characteristics, Adjacent land use, Pedestrian environment, Safety interventions, Green streets	Green street facilities, safety, and pedestrian network connectivity contribute to walking environment attractiveness.
Van Acker et al., 2013 [10] Leisure trips Belgium: urban, suburban, rural	Factor and cluster analysis	Walking (and cycling) as transport mode, besides car and public transport	Close to public transport, Close to shop, Quietness, Presence of green areas, Close to leisure activities, Close to family and friends, Close to work	Walkability perceptionss affect the population in the suburb.
Hernández & Witter, 2015 [2] All purposes Santiago de Chile	Comparative analysis → Perceived vs. Actual distance to transit	Walking as access mode	Walking distance to public transport station (buses and metro)	Metro supply has a higher overestimation rate than buses.

3 Methods

Respondents are train users who ride from Cicalengka train station and who live around that train station (1000 m radius). Of the nearly 300 completed questionnaires only 203 sets of data can be processed. There are other requirements set for the selection of respondents, which are restricted to those aged 17 years or older. The consideration is to be consistent with the classification of trip purposes, one of which is to work. The survey was conducted on weekdays (Monday - Wednesday) for train departures at 06.15, 07.25, 09.00, and 09.45. Questionnaires were mostly filled by respondents on trains with the help of surveyors, few succeeded in filling out questionnaires at railway stations.

Each respondent was asked the usual mode of transportation for access trip from home to Cicalengka train station and for egress trip from destination train station to activity location (final destination). Each respondent also asked about the perceptions of walkability from home to Cicalengka train station, which is related to walking distance, safety from motorized vehicle, comfort, and security from crime. Perceptions of walkability from station of destination to final destination are also asked. For all questions relating to walkability perceptions, ie 'distance', 'safety', 'comfort', and 'security', respondents are required to provide an assessment of the requested agreement on 5-point Likert Scale from "strongly agree to strongly disagree". Thus, the perceptions of walkability in this research is what train users perceived about distance, safety, comfort and security in the built environment at the beginning of the journey and at the end of the journey.

In addition, respondents also filled out data on travel characteristics and socio-economic characteristics. These variables are as control variables when analyze the relationship between walkability perceptions from home to origin station with access mode and between walkability perceptions from destination station to final destination with egress mode. Characteristics of the trip consists of trip purposes and frequency of train usage. While socio-economic characteristics consist of gender, age, and vehicle ownership.

All data from all variables are ordinal data, including control variables. Because the focus of research is walkability and walking as transport mode, then for all control variables, the largest value as ordinal data is given to the most likely to choose walking as transport mode. For example, in the ownership of vehicle variable, the value 1 is given to those who do not have the opportunity to use own vehicles, so the chances of walking to the railway station are greater than those that have the option of using own vehicles. While the value 0 is given to who owns a private vehicle. For egress trips, ownership vehicle variables are not tested as control variables because there is no opportunity to use own vehicles from destination stations to final destinations.

The same assumption is used for other variables, which is assumed to have a greater chance of choosing walking on the access or egress trip, given a larger value (see Table 2). These assumptions are for trip purposes 'work', train usage 'very often', gender 'male', age group '17 -25 years', and ownership vehicle 'no' is assumed to be most likely to choose walking as access or egress mode compared the other groups.

For train usage characteristics used 5 point scale, from very often to very rarely. The grouping of train usage is very rarely = 1 (once in more than 2 months), rarely = 2 (once in 2 months), not too often = 3 (once or twice a month), often = 4 (once or twice a week) , and very often = 5 (almost every day).

Table 2. Variables.

Main Variables
Walkability Walking distance Safety from motorized vehicle traffic Comfort Security from crime (5 point scale from strongly agree to strongly disagree)
Access Mode Walking (4), Paratransit (3), Motorcycle taxi (2), Vehicles have their own (1)
Egress Mode Walking (4), Paratransit (3), Motorcycle taxi (2), Others (1)
Control Variables
Travel Characteristics Trip purposes Work (4), School (3), Shopping (2), Others (1) Train usage (5 point scale from very often to very rarely)
Socio-economics Characteristics Gender (Male = 1, Female = 0) Age (More than 17-25 years old = 3, 26-50 = 2, more than 50 = 1) Ownership of vehicles (have vehicle(s) = 0, no = 1)

Data is divided into two sets of data, which are related to access trip and related to egress trip. To find the relationship between all aspects of walkability perceptions with access/egress mode, cross tabulation analysis was used, both for walking distance, safety, comfort, and security, with significance level (α) = 5%. SPSS is used for data processing and analysis. Thus, H0 will be rejected if the probability (value of Asymp .Sig. (2-sided) < 0.05, which means there is a relationship between the variables tested. For evaluate the strength of correlation is used Spearman Correlation value because all data are ordinal.

4 Results

4.1 Access and egress mode

From percentage of mode share, paratransit is the most preferred mode, both for access trip and egress trip. The difference is if the access trip is also dominant mode in addition to the paratransit is own vehicle, while the egress trip is paratransit and walking. Can be seen the difference of mode share on access trip with egress trip. The percentage of walking on the egress trip is greater than in the access trip. Apparently, own vehicle and motorcycle taxi users on access trip prefer walking and paratransit on egress trip.

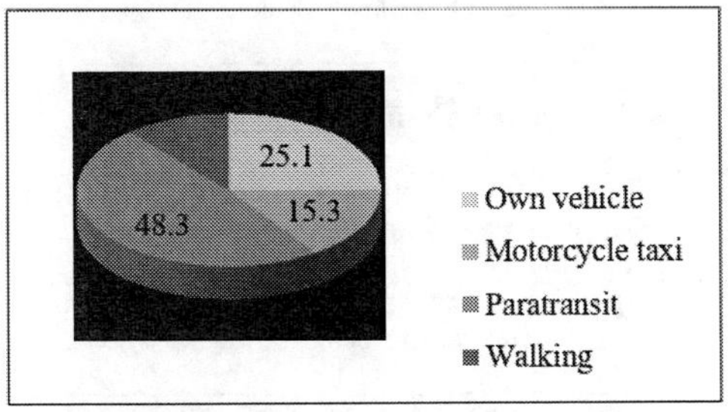

Fig. 3. Mode share percentage of access trip.

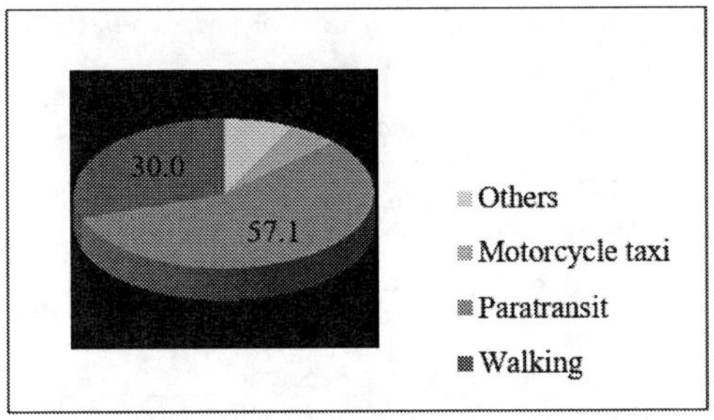

Fig. 4. Mode share percentage of egress trip.

4.2 Perceptions of walkability

In general, the perceptions of walkability is good, both for walkability from home to origin rail station and from destination rail station to final destination. The 'strongly disagree' answer is relatively small in number, even for walkability in destination built environment the 'strongly disagree' answers are not available for safety from motorized-vehicle traffic and security from crime.

Table 3. Percentage of respondents based on perceptions of built environment walkability.

	1	2	3	4	5
From home to origin station					
Walking distance	4.93	28.57	25.12	25.12	16.26
Safety from motorized-vehicle traffic	1.97	7.88	29.56	45.81	14.78
Comfort	1.97	8.87	31.03	47.29	10.84
Security from crime	0.99	9.85	29.06	41.38	18.72
From destination station to end destination					
Walking distance	7.88	18.72	25.12	36.95	11.33
Safety from motorized-vehicle traffic	0.00	7.88	33.99	46.31	11.82
Comfort	0.49	2.46	34.48	46.31	16.26
Security from crime	0.00	10.34	31.53	41.38	16.75

Note:
1 = Strongly disagree
2 = Disagree
3 = Ordinary
4 = Agree
5 = Strongly agree

Except for walking distance in the origin built environment, the perceptions of safety, comfort, and security are mostly agree. While the perception of walking distance in the origin built environment is almost balanced between disagree, ordinary, and agree. So, among the four components of the perceptions of walkability assessment, 'walking distance' earns the lowest rating compared to the other components.

4.3 Travel characteristics

Among the classification of trip purposes the most is to 'work', then 'others'. As for 'school' and 'shopping' is not too dominant for travel using the train from Cicalengka station. When viewed from the number of respondents with trip purposes other than for 'work', 'school', and 'shopping' (others), reaching 29.06%. This number is large enough that it seems that a more detailed classification of trip purposes is needed for future research.

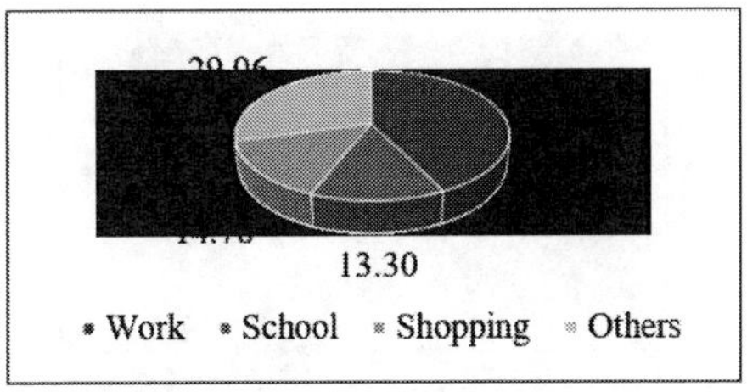

Fig. 5. Percentage of trip purposes.

When viewed from the railway usage level, about 60% of respondents answered 'often' and 'very often' which can be categorized as regular train users. However, as many as 30% of rail users are users who belong to the category 'rarely' and 'very rarely'.

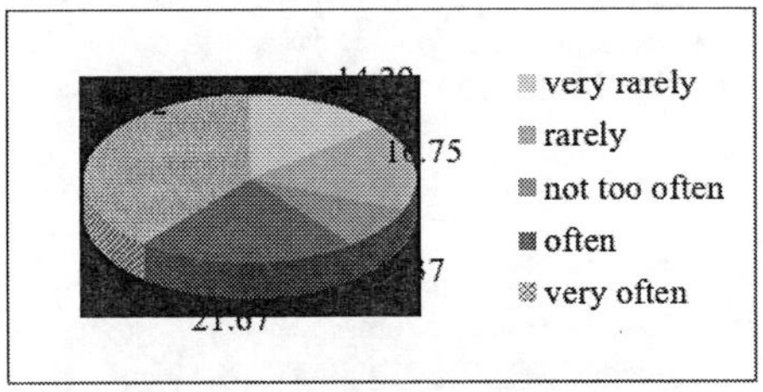

Fig. 6. Percentage of train usage.

4.4 Socio-economics characteristics

The number of male passengers is slightly higher than that of female. Whereas by age group, only a few are aged 50 years and over. Most of them have vehicles, whether they are motorcycles or cars, which are likely to be used for travel from home to the origin station.

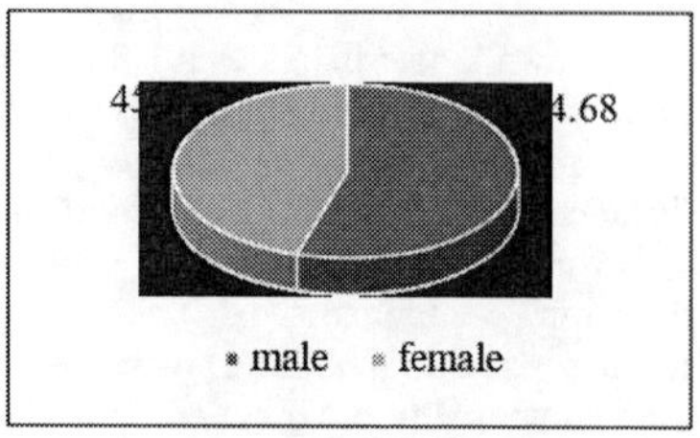

Fig. 7. Percentage of gender.

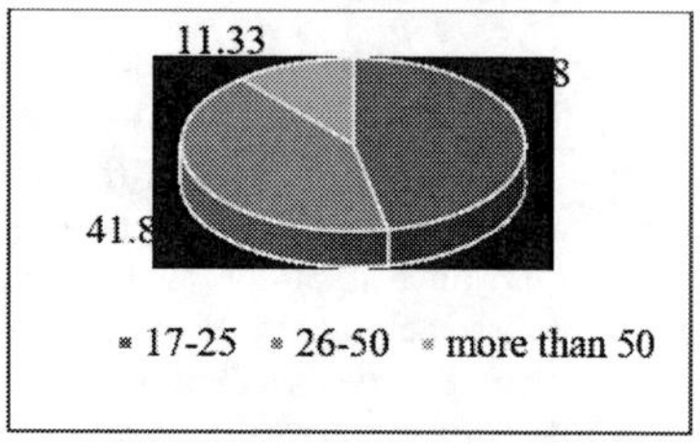

Fig. 8. Percentage of age group.

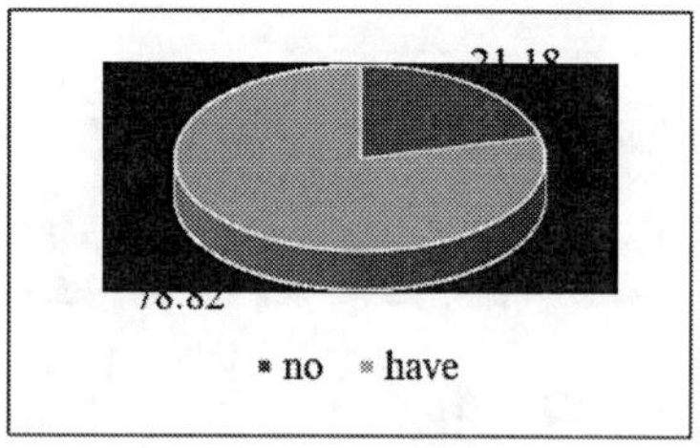

Fig. 9. Percentage of vehicle ownership.

4.5 Relationship analysis

Furthermore, it will be analyzed whether there is a relationship between walkability with mode choice, and also analyzed whether there is influence of each control variable on the relationship. Relationship analysis is performed for each walkability component, ie walking distance, safety from motorized vehicle traffic, comfort, and security from crime, with hypotheses such as Table 4.

Table 4. Hypotheses.

No.	H_0
H_0 for Access Trip	
1.	There is no relationship between perceptions of walking distance and access mode choice.
2.	Trip purposes have no effect on the relationship between perceptions of walking distance and access mode choice.
3.	Train usage have no effect on the relationship between perceptions of walking distance and access mode choice.
4.	Gender have no effect on the relationship between perceptions of walking distance and access mode choice.
5.	Age have no effect on the relationship between perceptions of walking distance and access mode choice.
6.	Vehicle ownership have no effect on the relationship between perceptions of walking distance and access mode choice.
7.	There is no relationship between perceptions of safety from motorized vehicle traffic and access mode choice.
8.	Trip purposes have no effect on the relationship between perceptions of safety from motorized vehicle traffic and access mode choice.
9.	Train usage have no effect on the relationship between perceptions of safety from motorized vehicle traffic and access mode choice.
10.	Gender have no effect on the relationship between perceptions of safety from motorized vehicle traffic and access mode choice.
11.	Age have no effect on the relationship between perceptions of safety from motorized vehicle traffic and access mode choice.
12.	Vehicle ownership have no effect on the relationship between perceptions of safety from motorized vehicle traffic and access mode choice.
13.	There is no relationship between perceptions of comfort for walking with access mode choice.
14.	Trip purposes have no effect on the relationship between perceptions of comfort for walking with access mode choice.
15.	Train usage have no effect on the relationship between perceptions of comfort for walking with access mode choice.
16.	Gender have no effect on the relationship between perceptions of comfort for walking with access mode choice.
17.	Age have no effect on the relationship between perceptions of comfort for walking with access mode choice.
18.	Vehicle ownership have no effect on the relationship between perceptions of comfort for walking with access mode choice.
19.	There is no relationship between the perceptions of security from crime and the access mode choice.
20.	Trip purposes have no effect on the relationship between perceptions of security from crime and access mode choice.
21.	Train usage have no effect on the relationship between perceptions of security from crime and access mode choice.
22.	Gender have no effect on the relationship between perceptions of security from crime and access mode choice.
23.	Age have no effect on the relationship between perceptions of security from crime and access mode choice.
24.	Vehicle ownership have no effect on the relationship between perceptions of security from crime and access mode choice.
H_0 for Eggres Trip	

No.	H_0
1.	There is no relationship between perceptions of walking distance and egress mode choice.
2.	Trip purposes have no effect on the relationship between perceptions of walking distance and egress mode choice.
3.	Train usage have no effect on the relationship between perceptions of walking distance and egress mode choice.
4.	Gender have no effect on the relationship between perceptions of walking distance and egress mode choice.
5.	Age have no effect on the relationship between perceptions of walking distance and egress mode choice.
6.	There is no relationship between perceptions of safety from motorized vehicle traffic and egress mode choice.
7.	Trip purposes have no effect on the relationship between perceptions of safety from motorized vehicle traffic and egress mode choice.
8.	Train usage have no effect on the relationship between perceptions of safety from motorized vehicle traffic and egress mode choice.
9.	Gender have no effect on the relationship between perceptions of safety from motorized vehicle traffic and egress mode choice.
10.	Age have no effect on the relationship between perceptions of safety from motorized vehicle traffic and egress mode choice.
11.	There is no relationship between perceptions of comfort for walking with egress mode choice.
12.	Trip purposes have no effect on the relationship between perceptions of comfort for walking with egress mode choice.
13.	Train usage have no effect on the relationship between perceptions of comfort for walking with egress mode choice.
14.	Gender have no effect on the relationship between perceptions of comfort for walking with egress mode choice.
15.	Age have no effect on the relationship between perceptions of comfort for walking with egress mode choice.
16.	There is no relationship between the perceptions of security from crime and the egress mode choice.
17.	Trip purposes have no effect on the relationship between perceptions of security from crime and egress mode choice.
18.	Train usage have no effect on the relationship between perceptions of security from crime and egress mode choice.
19.	Gender have no effect on the relationship between perceptions of security from crime and egress mode choice.
20.	Age have no effect on the relationship between perceptions of security from crime and egress mode choice.

As an alternative hypothesis, for every H0 is formulated H_1. H_1 is a hypothesis that states there is a relationship between variables tested. H_1 is also formulated to test the influence of control variables on the relationship of the variables tested.

From the results of relationship analysis, for access trip there are 13 pairs of related variables and some of which have influence from control variables. Meanwhile, for egress trip only 3 pairs of related variables. Overall, it can be seen that the perceptual factors of walkability in the built environment at the origin of the journey are pretty much related to the mode choice, in this case the access mode choice. In contrast, the perceptions of walkability in the built environment at the end of the trip is not much related to egress mode choice (see Table 5). Although the correlation between related variables is not very strong, but the relationship is significant.

Table 5. The result of analysis.

No.	Variables that are Related	Asymp. Sig. (2-sided)	Correlation (Spearman Correlation
Access Trip			
1.	Walking distance – access mode choice	0.002	0.188
2.	Walking distance – access mode choice, CV: work	0.000	0.438
3.	Walking distance – access mode choice, CV: very often	0.001	0.420
4.	Walking distance – access mode choice, CV: male	0.036	0.334
5.	Walking distance – access mode choice, CV: 17-25 years	0.016	0.081
6.	Walking distance – access mode choice, CV: have (vehicle ownership)	0.008	0.174
7.	Safety – access mode choice, CV: female	0.032	0.110
8.	Comfort – access mode choice	0.004	0.186
9.	Comfort – access mode choice, CV: work	0.004	0.277
10.	Comfort – access mode choice, CV: very often	0.005	0.258
11.	Comfort – access mode choice, CV: female	0.001	0.324
12.	Comfort – access mode choice, CV: 26-50 years	0.006	0.246
13.	Comfort – access mode choice, CV: have (vehicle ownership)	0.010	0.185
Egress Trip			
14.	Walking distance – egress mode choice	0.024	0.230
15.	Safety – egress mode choice, CV: very rarely	0.000	0.331

No.	Variables that are Related	Asymp. Sig. (2-sided)	Correlation (Spearman Correlation
16.	Comfort – egress mode choice, CV: very rarely	0.049	0.293

Note: CV = control variable

Among the four walkability measures tested, walking distance and comfort are those variables that not only have direct relationships with access mode choice but also have relationships when there are control variables that affect those relationships. Control variables that affect the relationship of walking distance and comfort with access mode choice are trip purpose 'work', train usage 'very often', and 'have' (vehicle ownership). There is a difference of influence of control variable on walking distance relationship with access mode choice and between comfort with access mode choice. If the walking distance, which affects the control variable 'male', while the affecting comfort is 'female'. In addition, walking distance is influenced by age group '17 -25 ', while the comfort is influenced by age group '26 -50'.

The walking distance and comfort relationship pattern from the destination station to the final destination with egress mode choice is somewhat different. There are no control variables that affect walking distance relationship with egress mode choice. Safety has a relationship with the access mode choice only when tested with the control variable 'female'. Meanwhile, for egress mode choice is related to safety when tested by control variable 'very rarely'.

Only security has no relationship with the mode choice, either on access or egress trip. It appears that the security variable is not something that the train users who departure from Cicalengka Train Station consider choosing the access or egress mode.

5 Discussion and conclusions

There are some interesting results to discuss. First of all, the security is not the measurement of walkability into consideration to choose the mode. This means that even if someone feels that the security from home to the origin station or from the destination station to the final destination is good, it does not necessarily want to walk. If returned to the data the number of respondents walking from home to the origin stations is relatively small (11.3%), a good assessment of the security built environment has no effect at all. They are more likely to choose paratransit than walking.

There is a difference in the component of walkability that affects male and female in choosing access mode. If males are more affected by walking distance, then females are influenced by comfort. A similar thing happens in the control variable age group. Walking distance is chosen by the age group '17-25 years', while comfort is the choice of age group '26-50 years'.

For people who travel for 'work', walking distance and comfort for walking become the decider in choosing access mode. Meanwhile, there is no relationship between walking distance with access mode choice for train users who are travel for 'school', 'shopping', and for other activities. Likewise for comfort, not something important for choosing the access mode except for travelling to work. The same pattern also occurs in respondents who very often use the train. Walking distance and comfort for walking are only important for the very frequent use of trains, otherwise it does not affect anything for other train users to choose the access mode.

Another interesting finding is the existence of a relationship between walking distance with access mode choice and between comfort with access mode choice for train users, even though they have private vehicles. This means that for private vehicle owners, walking distance and comfort remains a consideration for choosing to walk or not. Despite having a vehicle, if walking distance is close and comfort is nice, they are more likely to walk than using a private vehicle.

The last interesting finding is that for respondents who are very rare using train, safety and comfort of the built environment from the destination station to the final destination are important for them when choosing a mode for egress trip. Meanwhile, for this group of respondents, none of the walkability measures were decisive in the selection of access modes.

Overall, the conclusions that can be obtained from this research are:

- Using the crosstabs analysis, found a relationship between train users' perceptions of the built environment walkability factors with the mode choice, both access and egress mode.
- Security is the only measure of walkability that has no association with the access and the egress mode choice. This finding is similar to previous study [8].
- Each variable control has a different influence in the relationship between walkability factors with access and egress mode choice.

The authors would like to thank PT KAI to allow for collecting data. The authors also would like to thank the students of Urban and Regional Planning, Unikom, Bandung, who have helped carry out the survey.

References

1. R. Daniels, C. Mulley, JTLU, **6(2)**, 5-20 (2013)
2. D. Hernández, R. Witter, J. Public Transp., **18(4)**, 16-30 (2015)
3. C. P. Durand, X. Tang, K. P., Gabriel, I. N. Sener, A. O. Oluyomi, G. Knell, A. K. Porter, D. M. Oelscher, H. W. Kohl III, J. Transp. Health, **3(2)**, 154-160 (2016)
4. R. Cervero, O. L. Sarmiento, E. Jacoby, L. F. Gomez, A. Neiman, Int. J. of Sust. Transp., **3**, 203-226 (2009)
5. A. Ozbil, *Eighth International Space Syntax Symposium*, 8030:1-8030:15 (2012)
6. S. Park, K. Choi, J. S. Lee, Int. J. of Sust. Transp., **9(8)**, 529-541 (2015)

7. S. S. Wibowo, P. Olszewski, J. of the EASTS, **6**, 147-156 (2005)
8. D. Walton, S. Sunseri, Int. J. of Sust. Transp., **4**, 212-226 (2010)
9. V. Van Acker, P. L. Mokhtarian, F. Witlox, EJTIR, **11(2)**, 115-146 (2011)
10. V. van Acker, B. Derudder, F. Witlox, JTLU, **6(1)**, 53-62 (2013)
11. S. Handy, X. Cao, P. L. Mokhtarian, J. APA, **72(1)**, 55-74 (2006)
12. X. (J) Cao, P. L. Mokhtarian, S. L. Handy, Transport Reviews, **29(3)**, 359-395 (2009)
13. V. Mehta, J. of Urbanism, **1(3)**, 217-245 (2008)
14. J. R. Panter, A. Jones, J. of Physical Act. & Health, **7**, 551-561 (2010)
15. J. Lin, T. Yu, Transport Policy, **18**, 246-258 (2011)
16. A. Adkins, J. Dill, G. Luhr, M. Neal, J. of Urban Design, **17(4)**, 499-510 (2012)

Model petri net of adaptive traffic lights and its collaboration with a special event

Tomi Tristono[1,*], *Setiyo Daru* Cahyono[2],Sutomo[3], and *Pradityo* Utomo[1]

[1]Informatics Department, Universitas Merdeka Madiun, 63133Madiun–East Jawa, Indonesia
[2]Civil Engineering, Universitas Merdeka Madiun, 63133 Madiun–East Jawa, Indonesia
[3]Mechanical Engineering, Universitas Merdeka Madiun, 63133 Madiun–East Jawa, Indonesia

Abstract. Traffic lights have an important role as the system control of vehicles flow on the urban network. Commonly, most countries still using fixed time strategy. Our research proposes the adaptive traffic lights model to response the traffic demand. It uses basic Petri net as a general modeling framework. Foractuating method of minimum and maximum green signal time interval, the green traffic lights have three-time extension units. Next, we collaborate on a case of the existence of railways that crosses on the southern arm of an intersection. We introduce both of collaboration model design of traffic lights and the railway's gate which always closes while a train passing. Verification and validation of the model are based on the simulation result of vehicles queue. The collaboration model design of traffic lights has excellent performance, and it can resolve the congestion problem better than conventional schedule.

1Introduction

The primary role of the traffic lights is to improve the capacity and performance of junctions/ intersections in urban network area [1]. Traffic lights are installed to reduce congestion and travel delays, to make the main road crossable, to establish the time sharing fairly, to improve the safety of traffic vehicles and pedestrians, and so efficiency [2],[3].

Based on the time strategy, traffic lights are classified into two main categories, i.e., Fixed time strategy and traffic responsive strategy. The setting of the first strategy assumes that the traffic demand is constant. The second strategy is traffic lights that have the real-time control signal measurement [4]. It uses detectors to respond the fluctuation of traffic demand or special events [3].

The standard traffic lights have the sequence of green, yellow, and red signals that turn on in the specific time interval [5]. This paper's aim proposes modeling approach of adaptive traffic lights. This model is very beneficial to adjust to fluctuations in vehicles volume that passes on the road. Primarily in connection with the existence of a railway crossing. Of course, the number of railway's gate closures in the day will increase significantly during the beginning of double track of Railways era. It uses Petri Net (PN) to model traffic light behavior because they offer representation of conflict, sharing, synchronous and asynchronous, and priority [5]. Petri Net (PN) consisting of a four-tuple elements, i.e., places, transitions, arcs and tokens [6],[7].

The Marking places in a Petri Netcontaining a discrete number of tokens that the distribution represents a reachability of the net. An enabled transition of a Petri net may fire states while there are sufficient tokens in all of its input places. The firing of a transition consumes the required input tokens and deposits tokens in its output places [5].

A place may be refined into a subnet with many places and transitions resulting an expanded Petri Net [8]. That becomes the controller of the time interval of traffic lights.

2 Traffic Lights Model

This section reveals a green signal with a minimum interval time and maximum Petri Net model. It also explains the model of traffic lights that have two phases and P-invariants as the structural analysis to prove some properties of Petri Net.

A P-invariant indicates the tokens number in a set of places that satisfies some linear invariant.In a special case, all entries of a P-invariant are either 0 or 1 in which the number of tokens remains unchanged in all reachable markings [5].

2.1 Petri Net and The Properties

A Petri Net (PN) is a directed, weighted, and bipartite graph. The formal definition is written in the following: Petri net structure is the five-tuple elements N=(P, T, Pre, Post, M_0) [5],[6],[7].

$P = \{p_1, p_2, p_3, \dots p_n\}$ is a places finite set, $P \neq \emptyset$, $n \in N+$. Places are ilustrated by circles. Places can represent signals, controlers, or events.

*Corresponding author: tomitristono@unmer-madiun.ac.id

$T = \{t_1, t_2, t_3, \dots t_m\}$ is a transitions finite set, $T \neq \emptyset$, $m \in N+$. Transitions are illustrated by bars. The task of enabled transitions is as the trigger tools for signals, or events change. These are given the term fire.

Pre: $(P \times T) \rightarrow N+$ is an input function that defines directed arcs from places to transitions.

Post: $(T \times P) \rightarrow N+$ is an output function that defines directed arcs from transitions to places. N+ is the set of nonnegative integers. M_0 is the initial marking.

Based on the reason that the model design must correct, the Petri Net (PN) desires some properties [5], [6], i.e., Reachability, Reversibility, Boundedness, Safeness [9], Liveness or deadlock free [5].A Marking M is to be reachable from the initial if there exists a sequence of transitions firing which transform a marking from the initial. The Marking must able to return to the initial marking, It is called as reversible. Boundedness is a control strategy to prevent overflow in the model. A place p is said to be k-bounded if the number of tokens in p is always less than or equal to k($k \in N+$) for every marking M that reachable from the initial marking. It is safe if it is 1-bounded.

2.2 Green minimum and Maximum Time Interval

To reduce the top-down Petri net model complexity, the traffic lights design must be performed in a modular way. It may be presented in parts of the system, the subsystems, and a part of the subsystems. The modularity can be specified and taken into account separately.

P_Green [α, β] at Fig. 1 (left) means that the minimum time interval of green light is α and the maximum is β. In this paper, the unit of time is second. A token exists in P_Green at least α unit of time, and It is not more than β second. A token exists in P_Green means that green signal turns on. It depends and adapts to the volume of traffic demand. P_Green_min [α, α] means it has the minimum equal to the maximum time interval. P_Green_min [α, α] has a fixed time interval, i.e.,the α unit of time. Each place P_Green_ext(1), P_Green_ext(2), and P_Green_ext(3) have time interval 4 seconds [5]. The Traffic Detector Handbook [10]gives a recommendation unit extension of 3.5 seconds for roads where speeds are higher than 30 km/h. For the simplicity, it is used 4 seconds.

Thistype of adaptive traffic light system has a simple algorithm.Manysystems implement complex algorithms and require new technologywhich is often an error. This minimum and maximum systems are easy to implement and durable [5].

2.3 Two Phases Petri Net Traffic Lights Model

A signalized intersection of the urban network consists of four arms. The traffic light setting has two phases i.e. SN (south-north) and EW (east-west) [11], [12]. The phase SN provides signals for traffic from the north and south arms simultaneously. The east arm traffic moves in one direction, entering the intersection and going to the south, west, and north arms. The traffic movement in the south and north arms have two directions, i.e., approaching and away the intersection. The traffic of vehicles come from the southern arm allowed to move towards the west and the north arm. The flows entering the intersection from the north arm can move towards the south arm only. Traffic flows in the western arm moves in one direction away from the intersection. Referring to government regulations, vehicular traffics is left-hand traffic (LHT).

The traffic light model can adapt to the traffic demand. The time interval specified from a minimum to a maximum value. A special situation existing at that intersection is the railways crossing on the south arm. The presence of the train on the railways arise as an obstacle to the traffic movement causing a congestion when a heavy traffic occurs. The railway's gate closure always longer than the time interval of a traffic lights cycle [11],[12].

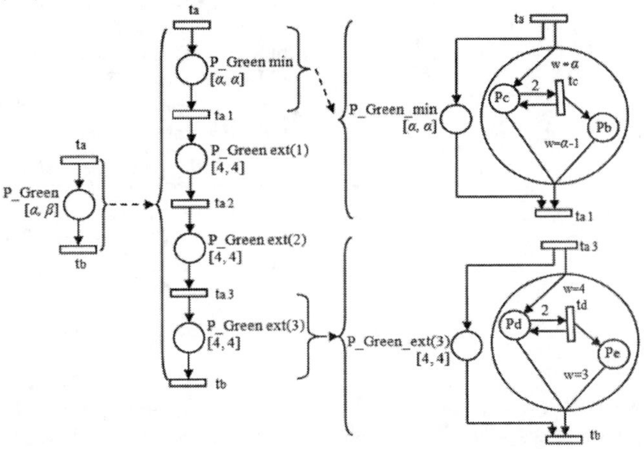

Fig. 1. At the left is the Place Green (P_Green is a signal model as Green signal). The refinement of P_Green at the center and the big circle as Place controller (P controller) of the time interval at the right side. The controller is embedded in the P_Green.

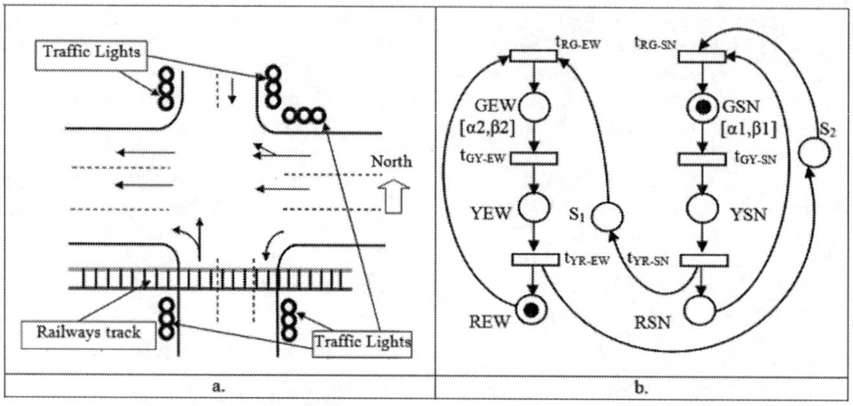

Fig. 2. a.The traffic flow of the intersection. b. The top-down Petri net model of a two-phases of traffic lights in normal schedule.

It refers to Fig. 2.b. Each GSN (Green South-North), and GEW (Green East-West) have refinement signal to adapt to the traffic demand. It defines signals YSN (Yellow South-North), RSN (Red South-North), YEW (Yellow East-West), and REW (Red East-West). Place S_1, and S_2 are the intermediation place to build the time order of Phase 1 (Phase South North) and Phase 2 (Phase East-West). A cycle in each the traffic light phase represents the whole of the green, yellow, and red time interval [5].

2.4 Signals Invariant of Two Phases Traffic Lights

Signal invariants indicate set of signal places which they guarantee the number of tokens remains unchanged in all reachable markings. All entries of signal place invariants are either 0 or 1. A signal turn on while a token exists in the signal place and the marking equal to 1. It will be off when it is empty and the marking equal to 0.

$$M(GSN)+M(YSN)+M(RSN)=1 \tag{1}$$

$$M(GEW)+M(YEW)+M(REW)=1 \tag{2}$$

$$M(GSN)+M(YSN)+M(RSN)=M(REW) \quad while\ M(REW)=1 \tag{3}$$

$$M(GEW)+M(YEW)+M(REW)=M(RSN) \quad while\ M(RSN)=1 \tag{4}$$

$$M(GSN)+M(YSN)+M(GEW)+M(YEW)+M(S_1)+M(S_2)=1 \tag{5}$$

The invariant (1) presents that can be a token available in one of three signal places GSN, YSN, and RSN only of south-north traffic lights. It is similar for signal Places invariant (2).

The signal Places invariants (3) and (4) hint that the model provides the assurance of safety traffic movement. It refers to (3) while the red signal (REW) of east arm turns on, the traffic signals on the south and north arms are either GSN, YSN, or RSN. It is similar for (4) while red signals of the south and north arms go on.

The signal place invariant (5) uses intermediation places S_1 and S_2. These are the artificial signal places. While a token exists in places S_1 or S_2, the traffic lights turn on red signal for all phases.

3 Method

The method for actuating minimum and maximum green signal time interval applies three-time extension units. Each unit time interval is 4 seconds. The total extension is 12 seconds. The detectors inform to the controller for one unit of the first extension at least 2 seconds before the ending of the minimum green signal. The vehicular detectors collect the information for the second additional while first green extra going on, and it accumulates the instruction for the third additional while the second extension takes place. Sometimes while the low demand, the traffic lights do not require the green extension. The traffic light changes directly from Green minimum time interval to the yellow and redsignals.

The traffic lights collaborated to a case of the existence of railways that cross on the southern arm of an intersection. Weintroduce a collaboration model design of traffic lights. It employs two phases only. The model is the both collaboration of traffic lights and the railway's gate which always closes while a train is passing. Special events such as the closure of the railway's gate, entirely are always longer than the full-time interval of a traffic light cycle.

The analysis and verification of traffic lights model use invariants, occurrence graphs, and Petri Net Simulator 2.0 for simulation [13]. At last, the performance test is also presented.

4 A Case Study

Acase is shown in Figure 2.a. The heavy vehicles (HV) dominate the flow entering the intersection of the eastern arm. Onthe contrary, the low vehicles (LV) and motorcycles (MC) are the largest parts of vehicles coming from the north and south arms. Bothconflicts of vehicle flow will be very dangerous.Updatingthe traffic flow schedule is very urgent and can not be delayed.

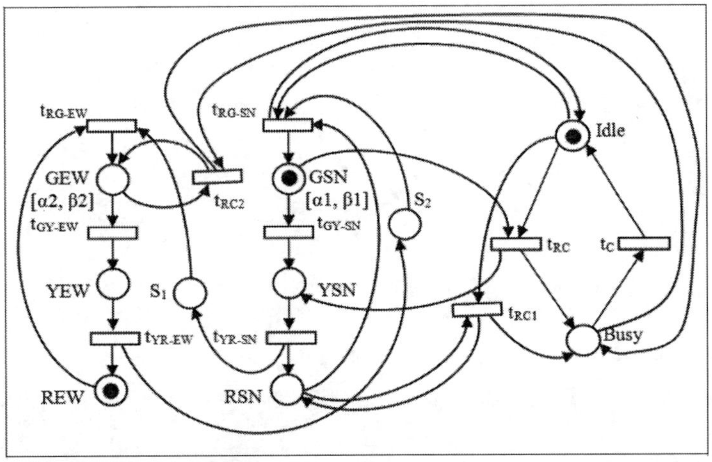

Fig. 3. The Petri net model of a collaboration of traffic lights and the railway's gate. It illustrated without control places of normal of traffic lights, the train ready to come, and completed passes.

4.1 Design Model of The Railway's Gate and Its Collaboration

Model of railway's gate in Fig. 3. has two places, namely Idle (when it open) or Busy (while closed). The place busy includes circumstances of the preparation for closing the railway's gate that must be longer than or equal to the time interval of YSN. Based on the reason of model simplification, it is not shown.

Transition t_{RC} (ready to come) and t_{RC1} represent that the train ready to come and the railway's gate ready to close. The transition can be enabled if there are tokens in all its input signal places. The enabled transition can fire a token of the previous signal places to the next signal places. Transition t_{RC} (ready to come) enabled if the train arrives while the Green South-North signal turns on. Otherwise, transition t_{RC1} enabled if the railway's gate ready to close while red South-North signal turns on. It bases on the information of the train detectors. Transition t_C (completed) informs that the train had crossed completely and the railway's gate ready to reopen.

The train is the priority. The signal model design in Fig. 3. has GSN minimum and maximum time interval. For the road users safety reason, It must have unlimited time interval connecting to the time interval of railway's gate closure. Sometimes while the train ready to come, it forces the green signal to turn yellow and then red, and it maintains the red signal until the train passes completely.

4.2 Analysis and Verification of Petri Net Model Design

While the train is passing and the railway's gate is closing, this means that the busy state is going on. Conversely, while the railway's latch is not closing, it means in the Idle state. In this state applies invariant (6), (7), and (8).

$$M(Busy) + M(Idle) = 1 \tag{6}$$

$$M(Busy) + M(GSN) = 1 \tag{7}$$

$$M(Busy) + M(YEW) + M(REW) = 1 \tag{8}$$

Places invariant (6) presents that the railway's gate state are either Busy or Idle. Places invariant (7) means that the south-north traffic lights may not turn on green signal while the railway's gate is busy. Places invariant (8) notes that while the railway's gate closes the east-west traffic lights must turn on green signal only. Occurrence Graph (OG) shows that there is no possibility of a deadlock in the model. It is not presented here.

5 Simulation Results and Discussion

The Fig. 4. is the result of the verification and validation of the east arm vehicles flow. It uses "Passenger Car Equivalence" (PCE) to measure the traffic flow volume [14].

While the traffic flows is low, the frequency of minimum green time interval on the east arm is high. Very low traffic occurs in the midnightcausing traffic lights to always be lit with minimum interval time.

When medium traffic time interval in the early morning, the green time interval can be on additional time ext 1, ext 2, or maximum. The time ext 1 (28 seconds) is 21 times, time ext 2 (32 seconds) is 25 times, and the maximum green time interval (36 seconds) is five times.

On the peak hours, it can be ext 2 or the maximum green time interval. While the heavy traffic and a congestion after the railway's gate reopened again, the traffic lights are always lit at the maximum green interval time.

Basedon the reason the system algorithm is simple, the system is easily repaired when a little damage occurs.Thefour-second extension on the green time interval proved precisely to the actual demand of traffic on each arm.

* Corresponding author: tomitristono@unmer-madiun.ac.id

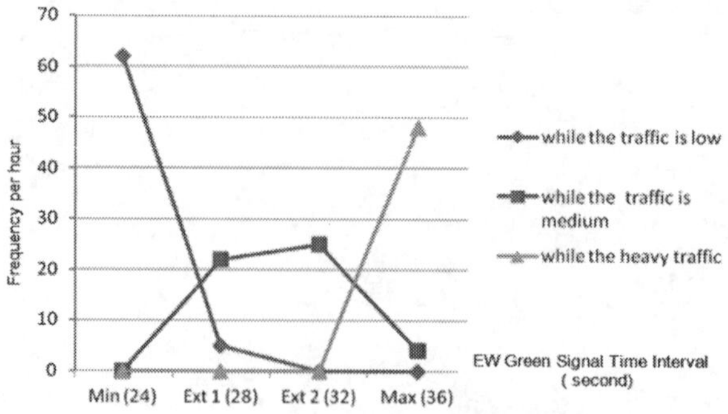

Fig. 4. The occurrencefrequency of each time interval of the green signal on three different traffic volumes of the east arm.

Fig. 5. The real case that requires the way out of the problem. The location of the intersection is in the eastern of the Jombang City railway station. The photograph of intersection shooting is from the direction of the north arm. a. when a low vehicles traffic. b. When a low vehicles traffic and the train is passing in the southern arm. c. The heavy vehicles dominate the traffic flow from the east arm.

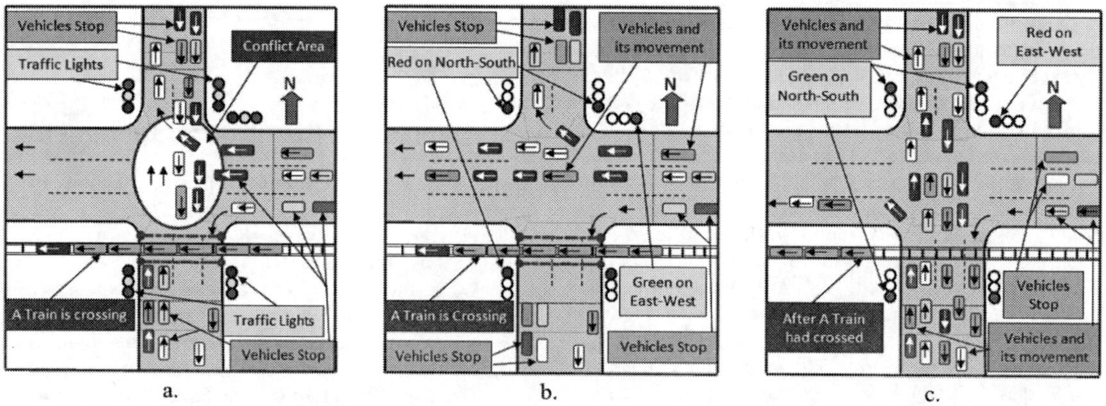

Fig. 6.a. The visualization of the congestion while the railway's gate closure. The traffic lights schedule applied without collaboration. A chaotic situation occurs causing congestion in conflict areas of traffic flow. b. and c. present the vehicles flow settings that apply the collaboration traffic lights schedule to the railway's gate closure. b. while the train is crossing, EW green signal turns on, and c. after it completed, SN must turn on green signal at the first time.

Table 1. The schedules of two phases of traffic lights on the time interval of the maximum green signal

Phases	*Green*	*Inter Green*		*Red*	*Cycle*
		Yellow	*All red*		
SN	27	3	3	45	75
EW	36	3	3	36	75

There has been a railway's gate closure for 135 seconds during the peak hours. The time interval is longer than a cycle of traffic lights. The vehicles arriving has normal distribution on a measurement every 3 seconds. The vehicles leaving the intersection while the signal turned green has a negative exponential distribution. The simulation began when the traffic light started to turn red on the phase of the south-north/ SN.

In this paper, we present a graph of a queue of vehicles based on time. Thesecond form of presentation is the comparison of the number of vehicle queues of each arm by time-invariant.Inthe second form simultaneously displayed the pattern of each arm queue.

*Corresponding author: tomitristono@unmer-madiun.ac.id

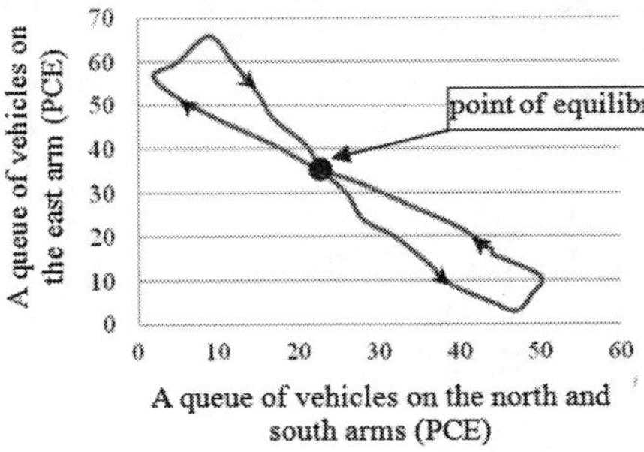

Fig.7. Vehicles queue in Normal Schedule (Presented using time-invariant)

Fig. 7 is the queue of vehicles due to the setting of the traffic lights. The vehicles queue goes up while the yellow or red lights are being lit and decreases when the green signal is on. A cycle of traffic lights is illustrated using time-invariant with an inclined lemniscate.

Fig. 8. presents the total sum of vehicles queue on the southern and the northern arm for 135 seconds of the railway's gate closure. The reopening of the railway's gate occurs at the ending of the north-south green signal. The setting traffic lights and the railway's gate are not collaborated. The congestion arising solved within six cycles. Each traffic lights cycle time interval is 75 seconds.

Fig. 9. describes a congestion in the southern and northern arm. The jamming on the east arm occurs also. A chaotic pattern illustrates the real situation at the intersection.

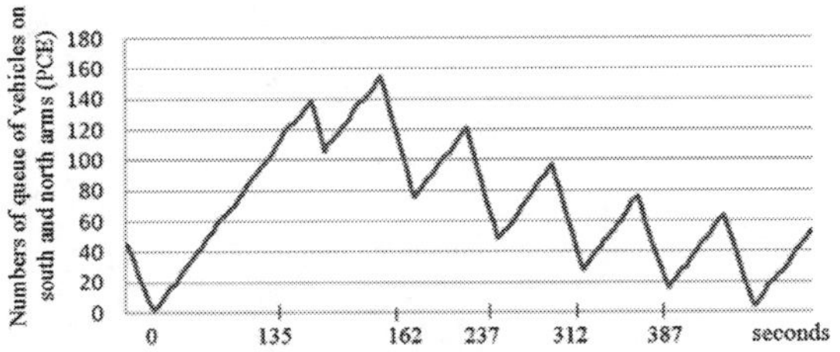

Fig. 8. The queue of vehicles of the south and north arms when the railway's gate closed and after it reopened. The settings without collaboration between light traffic and railway's gate. (Chart illustrates six cycles of traffic lights)

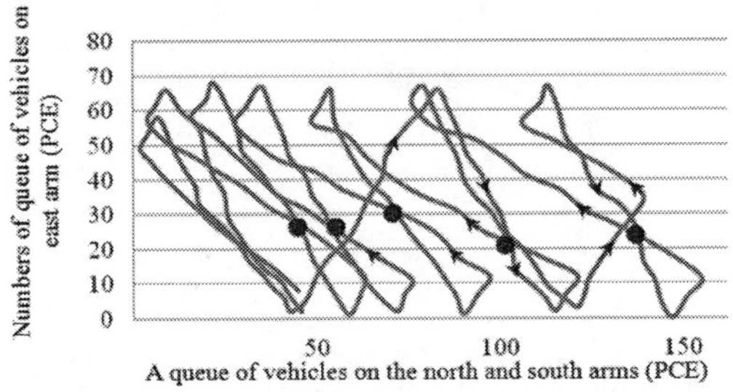

Fig. 9. The queue of vehicles of south-north arms and east arm when the railway's gate closed and after it reopened. The settings without collaboration between light traffic and railway's gate. (Chart illustrates six cycles of traffic lights). it uses time invariant.

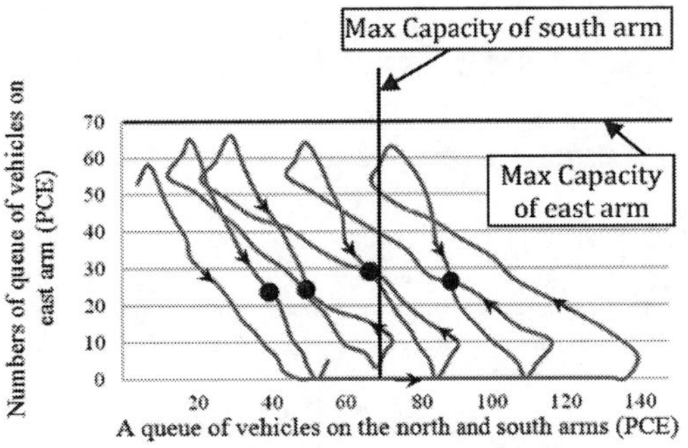

Fig. 10. Queue number of the vehicle when the train was passing and when the railway's gate has been reopened. Setting traffic lights implement collaboration with the railway's gate. (Illustrated four cycles of the traffic light after the railway's gate closure).

Fig. 10. shows the queue pattern implementing collaboration. At the beginning of the railway's gate closure, the queue of vehicles increase in accordance with the direction of the arrow. After the train has finished passing, the railway's gate reopened. The graph moves to left back. This means that the queue of vehicles decreased toward normal circumstances. It has a beautiful pattern. This illustrates the flow of vehicles in real situations that remain controlled. The bottlenecks are resolved in four cycles. It is two cycles faster when compared to the uncollaborated schedule.

The traffic lights setting applies collaboration able to reduce queues of vehicles coming from the east arm. The reopening of the railway's gate connected to the initial of the green signal for the north-south arm. The setting in this way is more efficient to ensure safety for all road users. The south-north traffic tends become impatient waiting after a long stop.

Fig. 10. is a visualization of the performance of the schedule of the collaboration of traffic lights to railway's gate. It looked better results to tackle congestion after an interruption. Whenthe railway's gate reopens, and peak hour, the traffic light applies the maximum green signal time.

The patterns while a low traffic flows has a similar shape, but smaller. The response of minimum and the maximum time interval of the traffic light are not visible, this due to traffic volume fluctuations that impossible occurred in a short period.

6 Conclusion

It has introduced an adaptive traffic lights model design with Petri net. This study also introduced a traffic light model that applies an interruption. This model is a collaboration between traffic lights and railway's gate closure. The simulation results show that this model has a better performance when compared to ordinary traffic light model to overcome the bottleneck after the closure of the railway's gate.

Acknowledgements. The research was funded by the Ministry of Research, Technology and Higher Education of Republic Indonesia on College Applied Research Scheme (PTUPT) in the year 2017. The contract number is No. 013/SP2H/HB-PT/Unmer.Mdn/LPPM/ V/ 2017.

References

1. B. Yulianto, Setiono., FLATLTDBDR., KoNTekS **7**, UNS,(2013).
2. M. Papageorgiou, C. Diakaki, V. Dinopoulou,A. Kotsialos, Y. Wang, RRTCS., *Proc.of IEEE.*, 2043-2067, (2003).
3. Y.S. Huang, and T.H. Chung, MATLCSUTCPN, *Intechopen.com*, (2010).
4. F. Kurniawan, R.A. Al Hasibi,TCATCSDSMDWV, *The Sci. J*, UMY**10**,126-135, (2007).
5. M. Soares, ADIMLDSIS,*Inf. F., Delft*, **7**, 99-133,(2010).
6. C.G. Cassandras, S. Lafortune, DES, *Kluwer Ac. Publisher*, (1999).
7. D. Adzkiya, Subiono., MTLUPNIS,*ITS-Sby*, (2008).
8. C. Choppy, M. Mayero, L. Petrucci, EFPPNR, *El. T. Comp. Sci.* **214,** 231- 254, (2008).

The analysis of overloaded trucks in indonesia based on weigh in motion data (east of sumatera national road case study)

Jongga Jihanny[1,*], *Bambang Sugeng* Subagio[1] *and Eri Susanto* Hariyadi[1]

[1]Department of Civil Engineering, Institut Teknologi Bandung, Jl. Ganesha No. 10, Bandung, 40132, Indonesia

Abstract. Overloaded trucks phenomena generally common in developing countries where the traffic control is poor. In Indonesia, the percentage of overloaded trucks can reach more than 60% in the total number of trucks and may be one of the substantial factors that reduce the service life of the road pavements. This paper presents the analysis results of the weigh in motion survey data at East of Sumatera National Road (Jalintim) in Indonesia and the impact of overloaded trucks on the pavement. For the analysis the simplified approach was used, the axle loads were converted into representative single-axle loads based on 4th power formula by AASHTO 1993 equation. The vehicle damage factor of vehicles is presented and will be compared with the Highways National Standard to estimate the remaining service life of pavement and IRI value prediction. The analysis showed that the vehicle damage factor that determined from weigh in motion data is extremely greater than vehicle damage factor of the national standard in Indonesia which may lead to accelerated deterioration, reducing the service life of the pavement structures and significantly influence the IRI value.

1 Introduction

The East of Sumatera National Road or known as Jalan Lintas Timur Sumatera/Jalintim is one of the strategic national roads in Indonesia. This road stretch from north to south of Sumatera Island and connects 6 provinces from a total of 8 provinces in Sumatera. This national road is part of the overall Asian Road Network route. Figure 1 shows the maps of Jalintim national road.

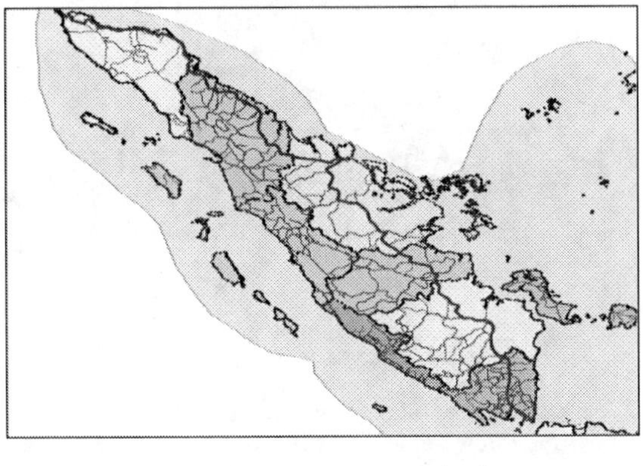

Fig. 1. Jalintim national road in Sumatera Island.

As one of strategic national road in Indonesia, Jalintim plays an important role so it needs to be maintained properly. The problem is the condition of Jalintim Road nearly always in poor condition. The pavement structure deteriorates rapidly compared their design life causing the preservation cost increased.

Main reasons that caused this problem is overloaded vehicles. A study by INDII (The Indonesia Infrastructure Initiative) stated that overloading is the main factor (nearly 50%) that contributed to pavement failure [1]. This overloaded vehicles phenomena became common problems in developing countries [2]. This phenomena contributes a significance impact on pavement performance. As stated in the previous study of overloaded vehicles and their damage on pavement, show that they have a significant role on reducing the service life of pavement structures [3,4].

The overloaded vehicle normally dominates by truck. An overloaded truck has a load or gross weight exceed their maximum legal loads. In Indonesia, main factor behind the overloaded trucks is economic issues, for example, the owner of commodities or truck owner attempt to minimize transportation cost by carrying an overload. On the other hand, law enforcement has not been optimal yet. Overload truck restrictions through weigh station failed to prevent it. Without further intervention by the government, the continuous use of overloaded trucks causes a serious problem on pavement preservation and planning policy in Indonesia.

Some of previous study in Indonesia related to overloaded trucks focus on the vehicle damage factor and remaining service life analysis [5,6). While the other studies analyze the loss cost due to overloaded trucks [7,8]. These studies based on weigh station data and the total gross weight of the vehicle. This paper further involves the axle load analysis based on weigh in motion

* Corresponding author: jongga_jihanny@itb.ac.id

survey. By using these analysis, the traffic data can represent the actual load and can present a better result comparing the previous study.

Weigh-in-Motion survey (WIM) is a survey that can be used for pavement and bridge engineering, traffic data collection, traffic controls and etc [9]. Slightly different with weigh static station or other weight survey methods, WIM data has a higher accuracy data and have no interruption to traffic flows [10].

In Indonesia, since WIM survey is very limited and costly, it is usually conducted annually by Directorate General of Highway (DGH), Ministry of Public Works and Housing at several strategic national roads. The duration of survey for each location normally between several days into weeks. Although the WIM survey quite often conducted, the data has not been studied properly in Indonesia. The database of axle load and VDF value analysis has not been used well yet related to pavement design and planning policy in Indonesia. Further analysis needed to improve and update the information for the purpose of pavement design.

One of pavement condition indicator that used as a parameter in pavement management system in Indonesia is International Roughness Index (IRI). As an International standard, IRI is practically used in pavement management system. Traffic conditions, especially ESAL, have the highest significance in contributing to the IRI value because of ESAL numbers greatly affect the changes in the surface conditions of flexible pavements [11]. Thus, the prediction of IRI value demanding an appropriate design of traffic.

The objective of this present study are to analyze the traffic condition in East of National Road of Sumatera, Indonesia and to analyze the impact of overloaded truck on pavement structure. This paper present the impact of overloaded trucks on pavement structure by considering the service life and IRI value based on WIM survey

2 Data and methods

This section discussed about data, methods and analysis that used to reach the objectives of study.

2.1. Weigh in motion data

In this study, the data from 2 segments of Jalintim Road were collected since 2007 till 2014. The location is part of south segment of Jalintim Road. These 2 segment considered as the most congested than other segment since these segment heading to main port in Sumatera Island that connect to Java Island, the main island in Indonesia

the data have already been verified and filtered to remove invalid data records. Nearly more than 80,000 total number of truck that recorded in these 2 location. The detail of survey location presented in figure 2.

Figure 3 present an information about type of truck, class and VDF standard that evaluated. The truck information was adopted from National truck configuration by Directorate General of Highway (DGH). There are 5 types of trucks that considered in the analysis of this study. This VDF standard will be compared with the VDF value that generated from WIM data survey.

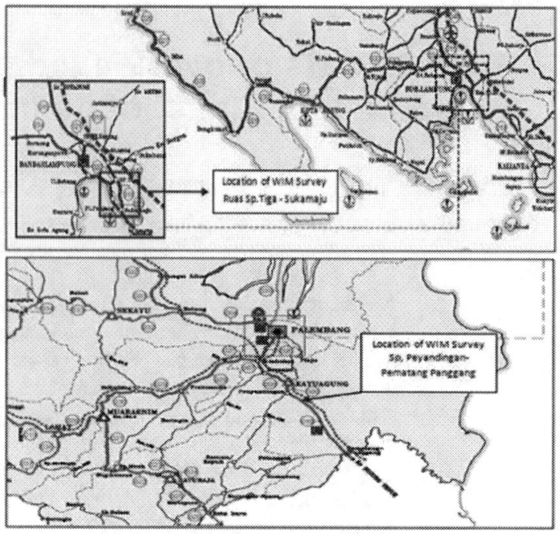

Fig. 2. Survey Location

Table 1 is a resume data of WIM survey report. Truck type of 2 axis or type 6A is the most type of truck that recorded in these WIM survey report, followed by truck type of 3 axis or type 7A.

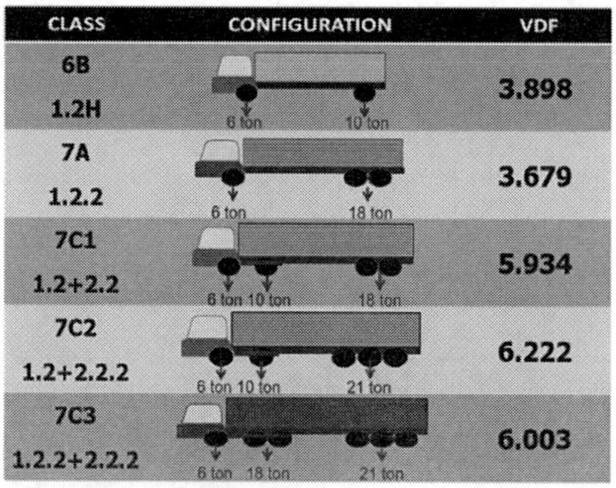

Fig. 3. Vehicle configuration

2.2 Overloaded trucks

The percentages of overloaded truck is estimated by identifying truck axle load that greater thanthe maximum legal limit (10 tons). Each of type truck has a different percentage number of overloaded. In figure 4, the percentage of the overloaded trucks varies for each type of truck, year of survey and location. The values is varies between 30% - 98%.

Table 1. Resume of data

Survey Location		Time of survey	Data Records (type vehicle and total truck)					
Road Segment	Province		6B	7A	7C1	7C2	7C3	Total trucks
Sp. Peyandingan - Pematang Panggang	South Sumatera	2007, 2012, 2013	35171	5946	323	44	203	41,687
Sp. Tiga - Sukamaju	Lampung	2007, 2010, 2013, 2014	29330	8912	211	220	525	39,198

The result show that the majority of truck is overloaded and the traffic control in that area is very poor. Compared to other countries, this number is extremely high. Other study in Poland, the percentage of overloaded vehicles vary between 6% to 16.5% which is quite low when compared to this results [12].

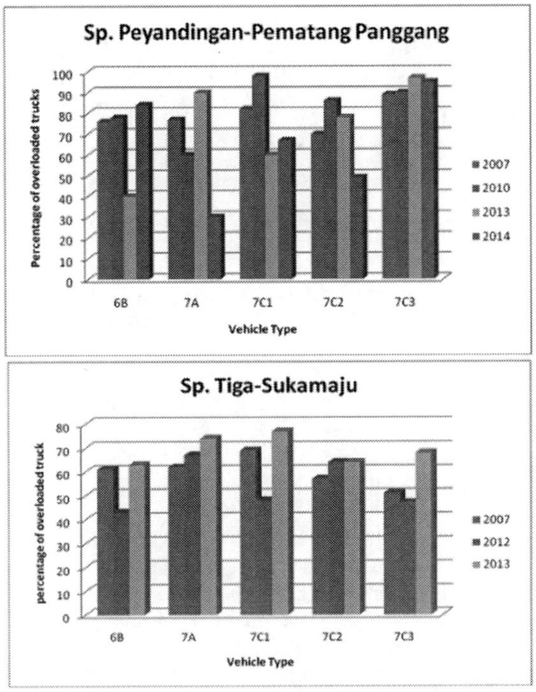

Fig. 4. The percentage of overload truck

Vehicle Damage Factor used as representation of the trucks that encountered on the pavement structure that defined as equivalent number of standard axles per truck. The VDF estimated by using the empirical approach through the 4th power formulas derived by Liddle [13]

$$DF = k\,(Q_j/Q_s)^4 \qquad (1)$$

DF : Damage Factor
Q_j : Actual Load
Q_s : standard single axle load 80 kN
k : 1 for single,
0.086 for tandem and
0.053 for tridem axle load

In this study, the vehicle damage factor was estimated from WIM survey. Since the WIM survey was conducted in 2 segment and in different years, VDF value estimated by analyzing for each year and location of survey. This was done to analyze the traffic condition and the tendency of overloaded truck.

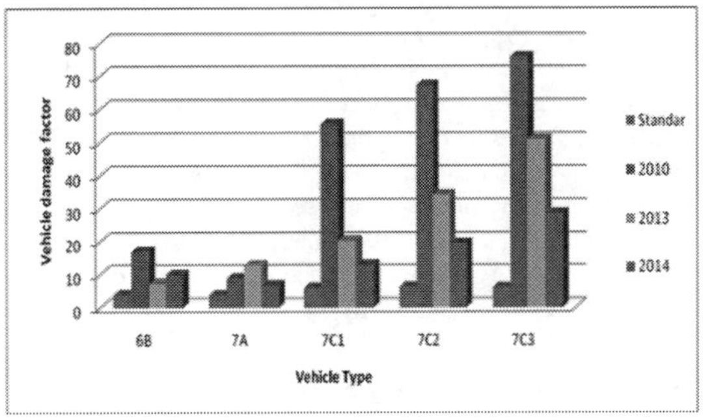

Fig. 5. VDF value at Sp. Tiga-Sukamaju

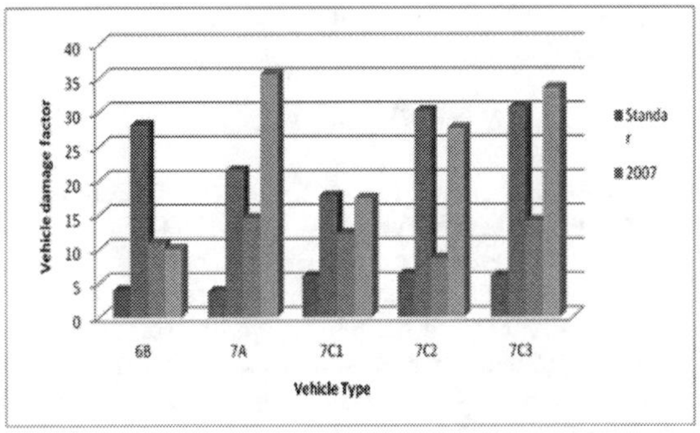

Fig. 6. VDF value at Sp. Peyandingan-Pematang Panggang

Figure 5-6 present the calculation of VDF value. The result show that the VDF value have a tendency to decreased especially in 6B truck although in other types of trucks the VDF values is ups and down.

Further analysis done by using the VDF value of truck type 6B to check the tendency of the overloaded truck. Truck type 6B chosen since this kind of truck is the majority in the daily traffic data. The result can be seen in table 2. The tabel show about a mapping from VDF value comparison between overloaded VDF and standard VDF. The percentage describe the total number of 6B truck that have a VDF value more than VDF standards. The overage number of VDF divided into several grup to descibe the magnitude of the overage compare to Standard VDF.

Table 2. VDF value of truck type 6B

Year	Percentage of Overloaded VDF (%)											Total of Overloaded Truck
	0-10	10-20	20-30	30-40	40-50	50-60	60-0	70-80	80-90	90-100	>100	
2007	15.99	18.07	18.38	18.55	11.34	8.60	5.21	2.00	1.35	0.52	0.52	2302
2010	13.37	15.84	15.35	12.11	13.32	10.31	7.85	5.91	4.21	1.74	18.93	2065
2013	13.04	10.88	3.99	4.40	12.12	23.89	21.75	8.44	1.16	0.31	0.15	9040
2014	26.13	23.12	21.11	17.09	8.04	4.52						199

From table 2, we can conclude that the overage of overloaded VDF can reach about more than 190% compare to standard VDF especially in the earlier time survey. Then in latter time of survey there is a tendency a decresing overage value of overloaded VDF. Not only in the percentage number, but also in the range of the overage. The overage of overloaded VDF simply reach below 60% compare to VDF standards.

The VDF result (overloaded VDF) then compared with the national standard VDF for each truck that stated in figure 3. From the analysis the VDF result was extremely greater than standards vehicle damage factor. The range comparison between overloaded VDF and standard VDF about 1.38 to 12.63 times. This is the reason why the pavement structures in the location tend to rapidly deteriorates during its service life.

2.3 Cumulative Single Axle Loads

AASHTHO (1993) gives formula to estimates the total number of traffic during the service life by using the following Cumulative Equivalent Single-Axle Loads (CESAL) formula:

$$W_{18} = \sum_{j=1}^{n} N_j \times VDF_j \times D_D \times D_L \times 365 \qquad (2)$$

W18 is design traffic in design life of pavement, N_j denotes number of vehicles while DD and DL are distribution factor for direction and lanes.

To analyze the impact of overloaded truck on design life, VDF value and the percentage of overloaded truck from WIM survey analysis was used to estimate the CESAL and the number of design traffic. The daily traffic data collected from Integrated Indonesia Road Management System (IIRMS). IIRMS is road maintenance management program that developed by DGH and contain important data related to road network.

The CESAL result (Overloaded CESAL) will be compared with CESAL estimation using standard load (Standard CESAL) to calculate the impact caused by overloaded trucks. Then the reduction value of service life of pavement structures can be measured.

CESAL value comparison is to compare the Overloaded and Standard CESAL. Figure 7 is a representation of CESAL value comparison. As shown in that figure, the CESAL value from WIM data analysis is greater than the Standard CESAL value. In the end of design life (10 years), total number of standard CESAL is 35.02 million ESAL. Whilst the total number of Overloaded CESAL is 96.22 million ESAL. The Overloaded CESAL value is nearly 2.75 times more than Standard CESAL.

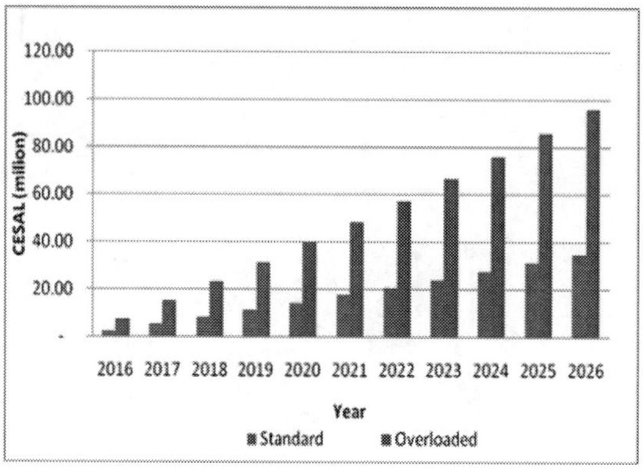

Fig. 7. CESAL comparison

3 Overloaded truck impact on pavement structure

3.1 Service life

To calculate the impact of overloaded vehicles on pavement structure, we can calculate the remaining service life of pavement. Remaining service life (RSL) has been defined as the estimation of total years that a pavement will be functionally and structurally in a normal condition by with only routine preservation [14]. The RSL was calculated using the formula:

$$RSL = \frac{CESAL\ Standard}{CESAL\ Overloaded}\ x\ DL \qquad (3)$$

Where:

RSL is remaining service of pavement (years) and DL is design life (10 years)

By using the this equation, the RSL of pavement in this study is 3.64 years. From this result, we can conclude that there is a reduction in service life approximately about 6.36 years from the design life of pavement. Which means the service life of the pavement that design in 10 years of design life, will ended in 3 years and 8 months.

3.2 International Roughness Index (IRI)

The IRI is calculated from the road profile or pavement surface. This study present IRI determination based on Indonesian Integrated Road Management System (IIRMS) that adopted from simplified methods by Paterson and Attoh-Okine [15], where the formula is:

$$IRI_n = 1.04e^{0.023n}\,[IRI_0 + 263\,(1+SNC)^{-5} \times CSA_n] \qquad (4)$$

where :

IRI_n = roughness index at n, IRI (m/km)

IRI_0 = initial roughness index, IRI (m/km)

SNC = Modified Structure Number Capacity depend on pavement type

CSA_n = CESAL at n (per 1 million ESAL)

Table 3. Road condition based on IRI value (DGH Standard)

Road Condition	IRI value	Pavement Maintenance
Good	IRI ≤ 4	Routine
Medium	4 < IRI ≤ 8	Periodic
Minor Damaged	8 < IRI ≤ 12	Rehabilitation
Major Damaged	IRI > 12	Reconstruction

Using equation (4), IRI value can be predicted in design life of pavement both for standard condition and overloaded condition. By estimate both of these condition, the impact of overloaded truck on pavement can be measured.

From table 3, there are 4 types of pavement condition based on IRI value and types of pavement maintenance. Routine maintenance and periodic maintenance needed to maintain the pavement condition to be able serve the traffic in design life. The IRI value comparison can be seen in figure 9. This figure show a cumulative value of IRI without pavement treatment.

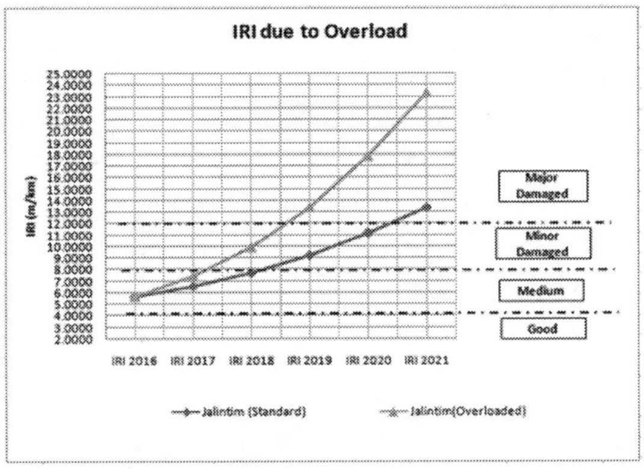

Fig. 8. IRI value comparison

By considering the pavement maintenance into IRI value comparison, the significant impact of overloaded truck can be seen in table 4. In overloaded condition, the number of periodic maintenance needed is bigger than in standard condition. In other words, the government should prepare extra budget for maintenance and proved

Table 4. IRI comparison value and pavement maintenance

Years	Standard		Overloaded	
	CESAL	IRI Prediction	CESAL	IRI Prediction
2016	2.73	5.68	7.51	5.68
2017	5.55	2.00	15.25	2.00
2018	8.45	2.89	23.22	4.21
2019	11.44	4.10	31.43	2.00
2020	14.52	2.00	39.89	5.70
2021	17.69	3.71	48.60	2.00

Where:

Routine maintenance	Periodic Maintenance

4 Conclusion

This study outlines weigh in motion survey data consideration in traffic data analysis at East of Sumatera National Road (Jalintim) in Indonesia to estimate the impact of overloaded trucks on the pavement structures. Nearly more than 80,000 total number of truck that recorded and from the analysis found that the total of overloaded truck is varies between 30% - 98% for different type of truck, time and survey location.

The specific result of this study show that the VDF values form WIM survey report is extremely greater than the standard VDF. The overloaded trucks contributed greatly in reducting the service life of pavement. On the other hand, by using IRI as pavement condition indicator, the overloaded truck proved to be costly in pavement management system's budgets.

From this study we can conclude that the overloaded truck has a severe damage to the pavement especially on service life. The percentage of overloaded truck that found in this study signifies the traffic control is poor. The government need to enforce a restriction regarding this problem.

The study is supported by ITB research program. The data were provided by Directorate General of Highway (DGH), Ministry of Public Works and Housing

References

1. Directorate General of Highway, Handling the vehicle overloading. Indonesia Transport, Supply Chain and Logistics. (2016).
2. L.G. Fuentes, L.F. Macea, A. Vergara, G.W. Flintsch, A.E. Alvarez, O.J. Reyese, Evaluation of trucks factors for pavement design in developing

countries. Procedia of Social and Behavioral Sciences **53**,1141-1150. (2012).

3. J. Mohammadi, and N. Shah, Statistical evaluation of truck overloads. J. of Trans. Eng. **118** (5), 651–665. (1992)
4. J.C. Pais, S.I.R. Amorim, and M.J.C. Minhoto, Impact of traffic overload on road pavement performance. J. of Trans. Eng. **139** (9), 873–879. (2013).
5. L. Sentosa, and A.A. Roza, Analysis of the Impact of Overloading Vehicle on Design Life of Rigid Pavement Structures. Civil Eng. Journal. **19**, 161-168. (2012). (written in bahasa)
6. Syafriana, S.M. Saleh and R. Anggraeni, Evaluation of road service life with overloading at east of sumatera road in aceh province. Proceeding of The 18th FSTPT International Symposium. (2015). (written in bahasa)
7. A.T. Mulyono, and M. Antameng, Analysis of loss cost of road pavement distress due to overloading freight transportation. J of the EASTS. **8**, 706–721. (2010).
8. S.M. Saleh, O.Z. Tamin, A. Sjafruddin, and R.B. Frazila, Reducing road maintenance cost caused of overloading trucks with multimodal freight transportation policy. Proceedings of the EASTS. **7**, 159-159. (2009).
9. B. Jacob, V.F.L. Beaumelle, Improving truck safety: potential of weigh in-motion technology. IATSS Res. **34**(1):9–15. (2010).
10. R.B. Machemehl, C.E. Lee, and C.M. Walton, Truck Weight Surveys by In-Motion Weighing, Research Report No. 181-1F, Center for Highway Research, Univ. of Texas. (1975)
11. A.I. Rifai, S.P. Hadiwardoyo, A.G. Correia, P.A. Pereira and P. Cortez, The Data Mining applied for the prediction of highway roughness due to overloaded trucks. Inter. J. of Tech. **6**(5), 751-761. (2015).
12. D. Rys, J. Judycki And P. Jaskula, Analysis of effect of overloaded vehicles on fatigue life of flexible pavements based on weigh in motion (WIM) data. Inter. J. of Pavement Eng. **17** (8), 716-726. (2016).
13. AASHTO, *The AASHTO guide for design of pavement structures*. AASHTO. (1993).
14. D. Gedafa, M. Hossain, R. Miller, and T. Van, Estimation of Remaining Service Life of Flexible Pavements from Surface Deflections. J. of Transp. Eng. 136 (4), 342-352. (2010).
15. W. D. Paterson and A. Busby, Simplified Models of Paved Road Deterioration Based on HDM-III. TRB, 71st Annual Meeting, Washington, D.C. (1992).

Application of gis on determination of flood prone areas and critical arterial road network by using chaid method in bandung area

Darwin[1], *Benecditus* Kombaitan[2], *Gatot* Yudoko[3], *Heru*Purboyo[4]

[1]PhD Student of Transportation, SAPPK, InstitutTeknologi Bandung, Bandung, Indonesia
[2]Urban and Regional Planning, SAPPK, InstitutTeknologi Bandung, Bandung, Indonesia
[3]School of Business and Management, Institut Teknologi Bandung, Bandung, Indonesia
[4]Transportation, SAPPK, InstitutTeknologi Bandung, Bandung, Indonesia

Abstract. Floods in Bandung area often occur when the rainfall is high then the water volume exceed the capacity of Citarum watershed. Floods cause economic and social losses. The purpose of this research is to get the GIS application model in the estimation of puddle area and road network in Bandung Metropolitan Area has disturbed.Geospatial map preparation methodology used statistical data from 11041 flood points, which divided into two groups, 7729 flood points to estimate the decision tree model and 3312 flood points to validate the model. The process of making flood vulnerability maps is approached by Chi-square Automatic Interaction Detection (CHAID) method, and validation using Receiver Operating Characteristic (ROC) method. Validation results in the area under the curve with a value of 93.1% for success rate and 92.7% for the prediction level.Chaid result is class 0 - 0,047 covering 76,68% area; Grades of 0.047-0.307 include 5.37%; Grades 0.307 - 0.599 (Low) covering 5.36%; Grades 0.599 to 0.4444 include 5.31% and grade 0.844-1 (high) covering 7.27% of the research area. Flood-prone road network is Link from Rancaekek (Area of PT Kahatex), link from Solokan Jeruk (Cicalengka-Majalaya), Link Baleendah, and linkDayeuhkolot (M.Toha - Andir)

1. Introduction

Indonesia has 17,508 islands located between two continents (Asia and Australia) and between two oceans (Indian Ocean and Pacific). As the world's largest archipelago country, Indonesia, is at the meeting of three major plates of the world that is the Indo Australia, Eurasian and Pacific Plate, and 9 small plates that have the potential to cause earthquakes if the plates collide so that the Indonesian territory becomes prone to various disasters, both earthquakes, Forest fires, tsunamis, floods, landslides, storms, and disasters caused by technological failures. In addition, Indonesia also has 129 active volcanoes, 80 of which are dangerous as shown in figure 1.

The National Disaster Management Agency has recorded disaster statistics as shown in Figure 2 on the following page. Based on the statistics of disaster occurrence in Figure 2 it shows that floods are frequent, whereas earthquakes and tsunamis cause the most deaths. Flood incidents have the highest intensity, causing traffic disruption, both congestion and broken roads, and bridges damage, housing, and some shops and industrial activities are disrupted hence causing huge economic losses. This condition as a result of global climate change hence the incidence of this flood will not decrease in the near future, [11].

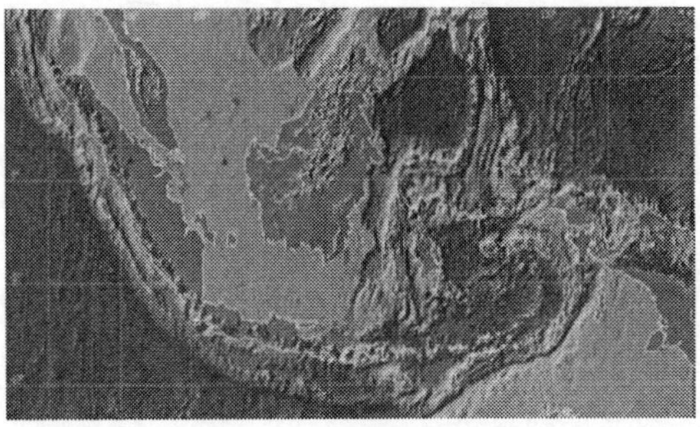

Figure -1. Active volcanoes Distribution in Indonesia

Bandung metropolitan area covers 71 districts, especially in the area of South Bandung which experienced flood disaster every rainy season, as shown in figure- 3 in the following page.

The major floods of 2010 in southern Bandung covered an area of 6,420 ha and caused economic losses of up toRp 615 billion [5] and the floods in 2016 more bigger than 2010.

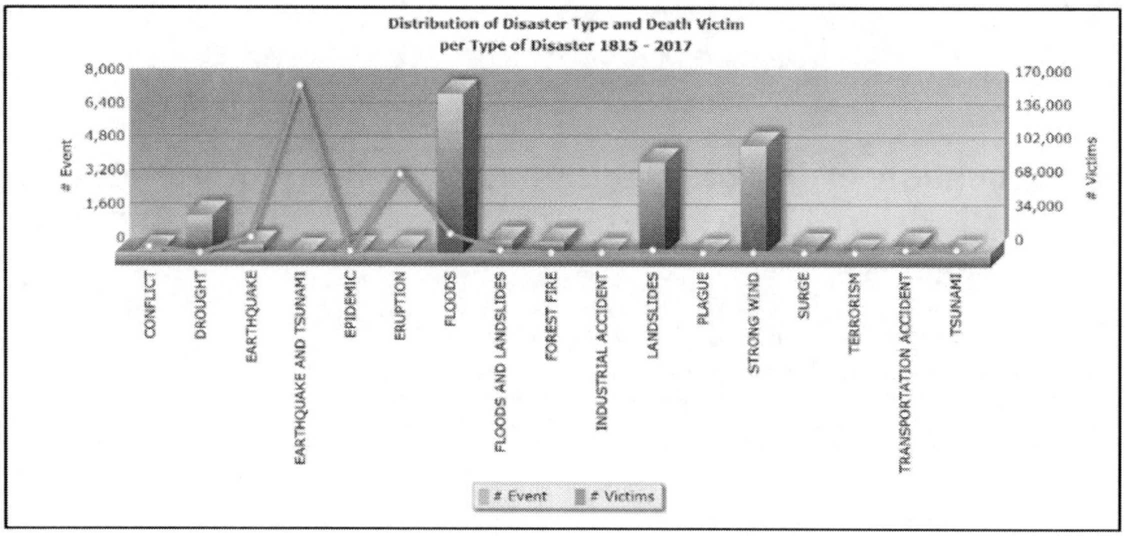

Figure - 2 The statistics of Indonesian disaster occurrence

Figure – 3 Floods on arterial roads in Bandung Area

Flood modeling (including hydrological process modeling) is carried out with various techniques, including rainfall modeling techniques - rain runoff, data-based techniques, and a combination of both [12] Physical modeling techniques are more difficult because of the dynamics of change in water catchment areas. There is a model that requires an understanding of the physical processes occurring in the watershed so that the data-driven approach to the flood model is increasingly explored [16].

Theory of machine learning methods to be the primary choice for disaster modeling with data-driven approach. Decision tree method derived from the theory of machine learning is a highly efficient tool for classification and estimation. Unlike other statistical methods, decision tree methods do not take statistical assumptions, capable of handling data that is represented by the various measurement scale, and it only takes a short computation [15]

The utilization of GIS for flood mapping is conducted in Jambi[7], for green open space in Depok area [7], but they do not discuss the flood-affected road network. In the Bandung area there has been no research on floods and road networks that are affected so as to disrupt traffic flow. The analysis of traffic disturbances caused by flooding is very important in transportation planning, especially to conduct traffic engineering so that greater losses can beavoided

The purpose of this research is to get the GIS application model to estimate flood area and road network disturbed by flood. Flood analysis model using advanced decision-based rule tree by utilizing Geographic Information System (GIS) software to process thematic data and then using module that provided by SPSS software togenerate decision tree.Then decision trees are used to generate flood prone areas and re-use GIS software to illustrateflood-prone maps. The benefit of this research is to obtain a map of flood-prone areas in Bandung so that the government can conduct good disaster management, such as planning of evacuation system, flood logistic needs, and traffic engineering management.

2. Literature Review

2.1. Research Area

The area of Bandung metropolitan includes 71 districts, which entered Into 5 cities, namely Sumedang District, Bandung City, Cimahi City, Bandung Regency, and West Bandung. The shape of Bandung's highlands consists of a plain of about 40 km from west to east and 30 km from north to south, with a population of 7 million in 2005 and is projected to increase to 12.8 million by 2025, while the city of Bandung population is 2,470,802 People in 2014 [1].

The geological conditions of the Citarum watershed are divided into several morphologies such as those in the Atlas of Water Resources of the Department of Water Resources Management (PSDA).

Upstream watersheds in the Citarum Basin are formed from volcanic morphology with relief of soil and gravel, with a height of 750 - 2300 m above sea level, intersected 5-15% feet, in the middle of 15-30. %, and at the peak of 30-90%. The pattern of parallel river flow and river flow is the main sink area of shallow air and in the air where the slope of the slope is released. Stone constituents are the sediments of young and old volcanoes, which consist of tuff, breccia, and lava.

The river flows in Bandung quite a lot, such as the river Cikapundung, Cidurian, Cinambo, Cisangkuy, and others that empties into Citarum. In extreme weather conditions floods often occur because the capacity of drainage and Citarum rivers is lower than the volume of rainwater.

2.2. Geographic Information Systems

Geographic Information System is an application designed to perform various operations on geographic information. Geographic information is defined as information about the location on or near the earth's surface, and can be organized in various ways [4]. GIS presents the information in a graphical form by using the map as an interface. GIS is composed of the concept of multiple layers (layer) and relationships. Each layer in GIS represent data and specific information according to the geographical location and the relationships that are defined [6]

Geographical phenomena is generally conceptualized using continuous fields and discrete objects. But these two concepts problematic when digitized as a result of the information contained therein was not clearly quantified. So if you want to describe a geographic phenomena that can be processed by a computer, the format should be changed using the two methods, the vector and raster. Basically vector and raster can be used to codify the continuous field and discrete objects, but there is a strong correlation between raster and continuous fieldand vector with discrete objects [13]

GIS database is usually conceptualized as a collection of layers, each layer bound to the coordinates on the earth's surface. Alayer can contain representations of fields, such as land cover maps, or may contain multiple discrete objects, such as buildings. Layer-based raster usually describe only the geographical variation of one property only, such as the name of the district or land cover types, so the raster layeris usually conceptualized with a continuous field. Layer-based vectors on the other hand can represent a collection of discrete objects with a number of diverse attributes, or a variation of a property which is conceptualized by a continuous field [4]. Database in GIS application known as the "geodatabase" because the data in the database of this type are usually equipped with geographic co-ordinates.

2.3 Decision Tree and CHAID Algorithm

The decision tree is a data mining technique that breaks down the heterogeneous data set into the group ofmore homogeneous data by using directed knowledge discovery. Directed knowledge discovery explains the relationship between the target variable with other variables to find certain patterns, which are then used to predict future events by using the chain of decision rules. Chain of decision rules generated using the patterns found in the analyzed data. In this way, a decision tree model provides an accurate and clear where a decision tree model can explain the reason of certain decisions using decision rules. The decision tree can be used in classification problems and also in the estimation problem in which the output is a continuous value.

CHAID stands for "Chi-Squared Automatic Interaction Detection". This method follows the three stages in the decision tree generation process, namely: merging, splitting, and stopping. As the name implies, CHAID using a p-value of the chi-square as merging and splitting criteria. Splitting the CHAID method was not consider limited the binary split, but also consider multi-way split. CHAID algorithm is able to receive input nominal and ordinal and able to accommodate different scale measurements of each variable

3.Research Methods

3.1. Model GIS

relates to the standardization of spatial maps. The coordinate system used is W GS 1984 UTM Zone 48S with Transverse Mercator projection, and with the geographic coordinate system GCS WGS 1984 datum D WGS 1984. The Linear units used in this coordinate system is meter.

Geographical Information Systems (GIS). GIS is a generic term denoting the use of computers to create and depict digital representations of the Earth's surface. From humble beginnings in the 1960s, GIS has developed very rapidly into a major area of applicationand research, and into an important global business. [14]. GIS models used in this study

3.2. Decision Tree

https://doi.org/10.1051/matecconf/201814702007

There are various models of decision tree that can be used to process the data in this study. Models of the decision tree include: Classification and Regression Tree (CART), Quick, Unbiased, Efficient Statistical Tree (QUEST), Chi-square Automatic Interaction Detector (CHAID) and C4.5 algorithms. The model used in this study are CHAIDbecauseCHAID analysis able to accommodate categorical data and continuous data at a time. This factor is suitable for use in the data in this study, because there is a categorical data (soil type, land cover, and rainfall) as well as continuous data (altitude, slope, curvature, and the distance from the river). Besides, CHAID analysis is not limited to do a binary split only, but be able to do the splits with the result more than two nodes. CHAID analysis thus produces a decision tree nodes that are more specific.

3.3. Research Steps

GIS data processing carried out in accordance with the model analysis described in next section. Processing begins with a projection of spatial maps into the selected coordinate system. The first spatial map projected is DEM because these maps will be used as a base map to facilitate the processing of other spatial maps. Projections into the coordinate system and then proceed to map soil types, land cover maps, maps Citarum river and tributaries Citarum, rainfall map, a map of past floods, and a map of the study area.Then do cutting the area of spatial maps with a map of the study area Metropolitan Bandung Area. This is done to focus the map region be analyzed only in the area of research only. The next stage in the form of DEM data processing into slope map and curvature maps.

Then the whole spatial maps that are in vector format (maps of soil type, land cover maps, rainfall maps, map distance from the river, and maps of past flood) is converted into a raster format. The process of conversion from vector to raster or rasterization is done by the method of maximum combined area. The entire map in vector format is converted into a raster format with a resolution of 92.72 m, the same as the map resolution DEM.

Map of flooded areas that have been in rasterization is divided into two, with the division of 70% of cells and 30% of the cells. Cell division is done at random and will be used in conjunction with other data at the stage of processing the decision tree. Other data to be used for processing the decision tree is a data cell that is not subject to flooding of a number of cells flooded area. The cells was not affected by the floods also were randomly divided into sections of 70% and 30%.

The generation of a decision tree using a data from extraction result at the processing stage GIS. 70% of cell data is used to perform decision tree training and 30% cell data is used to perform testing a decision tree as a result from generation.Before performing the generation of a decision tree, there are several criteria that must be set. These criteria is the dependent variable, independent variables, limiting criteria, level of significance, and the size of the independent variable interval. This phase will generate decision trees along with the rules of each terminal node in the decision tree. These Rules along with the data overlaying the entire spatial map was used in the processing stage flood-prone maps.

Rules node terminal as a result from generation decision tree along with each overlaying data cell from GIS process become an input to generate flood-prone maps for the Metropolitan area of Bandung Area. Rules node terminal are used to generate the probability of flood-prone in every overlaying maps cell. The probability of flooding already known is then used to visualize them in the form of a map using GIS software. Flood probability map is then classified into several classes to determine the level of danger of flood-prone throughout the Metropolitan area of Bandung Area.

4. Data Collection

The data used in this research is secondary data collected from various sources. Here is described a method of collecting individual data points.

Flood maps is obtained by processing remote sensing imagery or aerial photographs that were recorded during the flood. Using algorithms, to identify the surfaces that are flooded from the recorded image. The results of identification using algorithms then check back to verify the entire surface of the flood. The map of past Flood shown in figure 4.

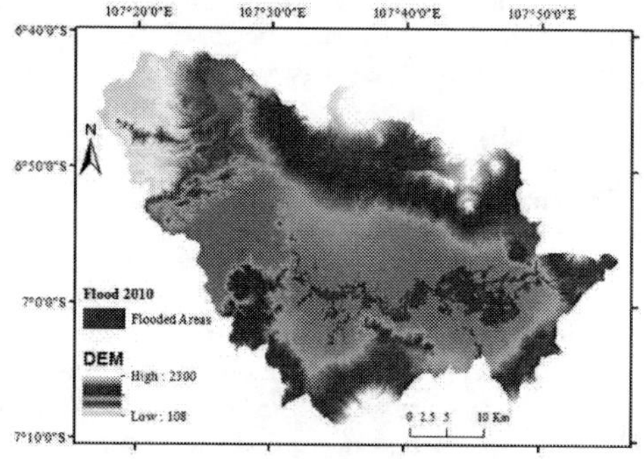

Figure 4. Flood in the past

Map land cover was obtained in a manner similar to the flood maps, which isby processing the remote sensing imagery or aerial photography. The algorithm used to identify the type of land cover more advanced because the object to be identified more as compared to only one object for identification flooding. Just as the flood maps, to map land cover, need to be verified by the man against the identification is done by algorithms and the result shown in figure 5.

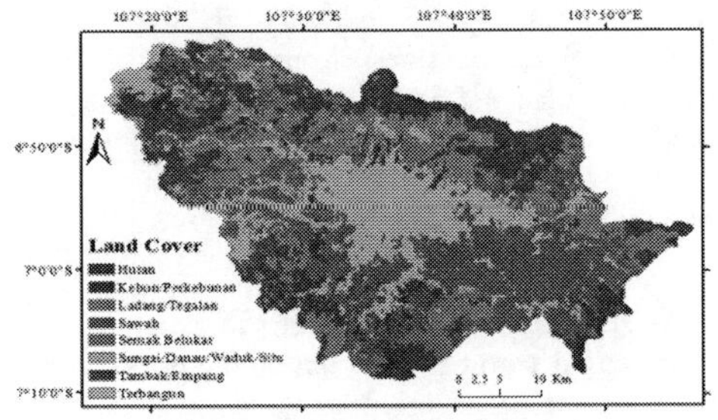

Figure 5. Covering Area

Rainfall map was obtained by collecting rainfall data from various measuring stations rainfall in the region studied. Location of rainfall stations accurately mapped using the coordinates and is associated with rainfall data. Then the rainfall data and the location of measuring stations are interpolated using Kriging method. Interpolated into the rainfall map is used as shown in figure 6.

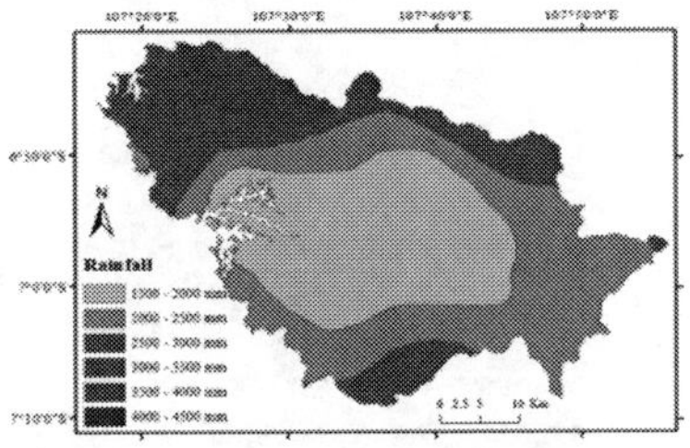

Figure 6. Rainfall map

Map of soil types is obtained by performing soil sampling at locations spread out evenly. Soil sample data along with the coordinates of the sampling is stored for processing. The data is then interpolated using Kriging method, resulting in a soil type map for the area sampling as shown in figure 7.

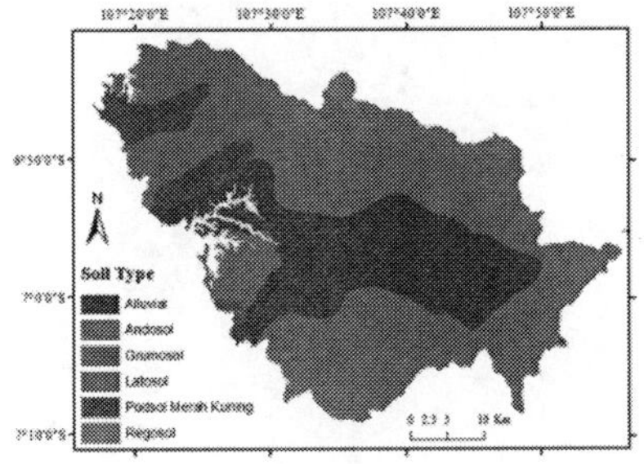

Figure 7 Soil Type

DEM data source is from the National Aeronautics and Space Administration (NASA) generated by an international study known as the Shuttle Radar Topography Mission (SRTM). SRTM data in the form of topographic world map with a resolution of 90 m. Data results are distributed by NASA SRTM have holes where the pixels have the value "no-data". So the data must be cleaned with one of them re-interpolation. Data cleansing is done by Jarvis [10] and the results of such processing is open to the public. Data is distributed by the Consortium for Spatial Information (CGIAR-CSI). DEM map for the area of West Java and the surrounding shown in DEM map with latitude 10S - 5S and longitude 105E - 110E.as shown in figure 8

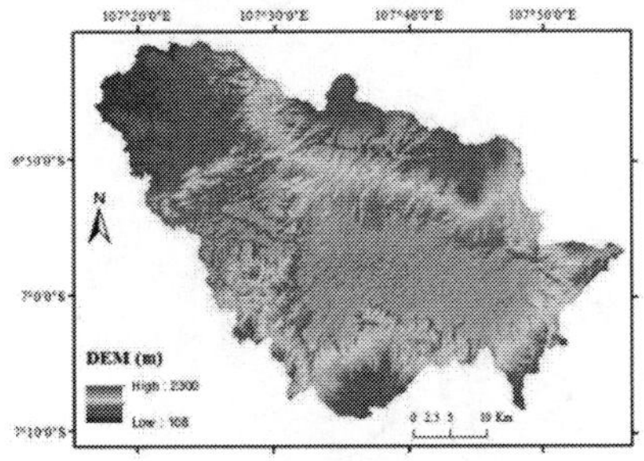

Figure 8. DEM

Slope and curvature maps of the soil surface is obtained by performing data processing DEM. Map of the slope is the first derivative of DEM map, and map the curvature is a derivative of the slope or the second derivative map from DEM map.

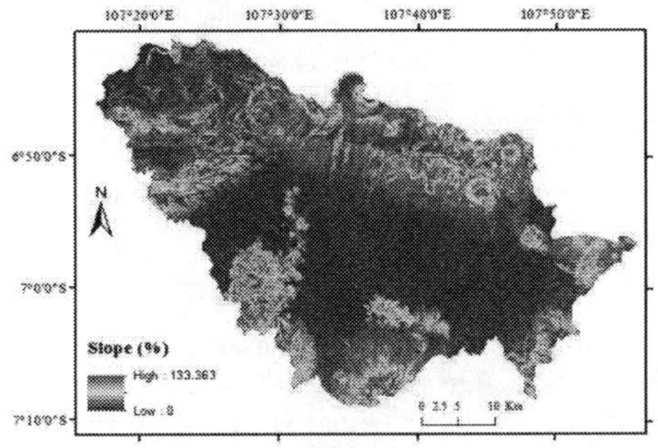

Figure 9. Slope of soil surface

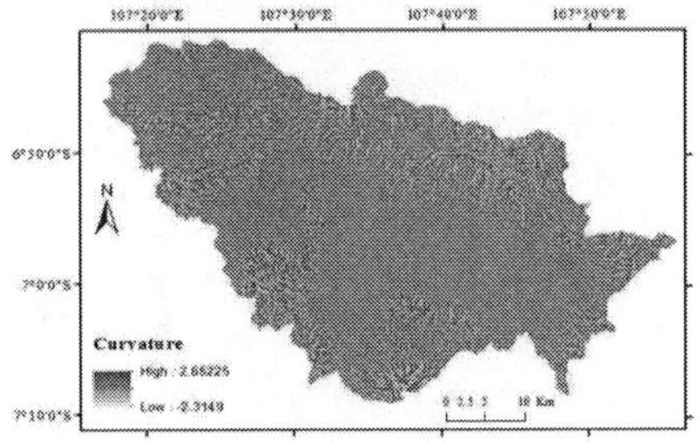

Figure 10. Curvature of soil surface

Map of the Citarum river and tributaries Citarum obtained by making use of remote sensing imagery or aerial photography. Citarum river imagery is used to determine the boundaries of the river. The boundaries of the river described using appropriate lines and then saved as a polyline features in the GIS software for further processing.

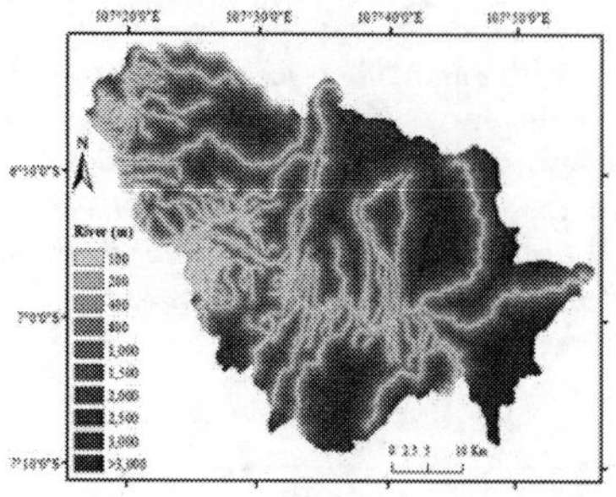

Figure 11. Distance from River

Map of the road network in the Metropolitan Bandung Area obtained from non-profit institutions Open Street Map (OSM). The institute aims to provide a complete map of the whole world with open access. Figure polyline path obtained by processing remote sensing imagery or aerial photography.

All data obtained are fed into statistical input data to be processed with Chaid to obtain the probability that an area will experience floods based on past experience. The data in the input format of the statistics processing is shown in the figure below.

All XYPoints.sav [DataSet1] - IBM SPSS Statistics Data Editor

File Edit View Data Transform Analyze Direct Marketing Graphs Utilities Add-ons Window Help

14 :

	OBJECTID	DEM	Slope	Curvature	LandCover	SoilType	Rainfall	Flood	River	POINT_X	POINT_Y
1	1	110	1.348190	.011633	8	4	1	0	100	762194.799970000	9259932.80955000
2	2	327	6.092250	-.174490	8	4	2	0	100	766181.639083000	9259932.80955000
3	3	337	5.382620	.046531	5	4	2	0	100	766274.356272000	9259932.80955000
4	4	338	4.181550	-.058163	5	4	2	0	100	766367.073460000	9259932.80955000
5	5	335	6.346510	-.162857	5	4	2	0	100	766459.790649000	9259932.80955000
228895	228895	1881	20.27310	.348980	1	1	1	0	10000	787877.461230000	9208938.35578000
228896	228896	1891	22.73040	.255919	1	1	1	0	10000	787970.178419000	9208938.35578000
228897	228897	1914	27.14110	.442042	1	1	1	0	10000	788062.895607000	9208938.35578000
228898	228898	1937	27.22730	.186123	1	1	1	0	10000	788155.612796000	9208938.35578000
228899	228899	1965	28.71410	-.034898	1	1	1	0	10000	788248.329985000	9208938.35578000
228900	228900	2002	26.92320	.174490	1	1	1	0	10000	788341.047173000	9208938.35578000
228901	228901	2029	14.50160	.314082	1	1	1	0	10000	788433.764362000	9208938.35578000

Figure 12 Input Data Statistik for Chaid analysis

5. The results of the study

There are criteria for the generation of a decision tree was defined in this study. The dependent variable was the incidence of floods (**Flood**) or no flood (**No Flood**). The independent variable is DEM, rainfall, soil type, distance from the river, slope and curvature.Limiting criteria used are minimum of 100 cases for parent nodes and a minimum of 50 cases for child nodes. The significance level (alpha) is 0.05 or 95% confidence level.From the results of the process output module Decision Trees on SPSS software with the method of calculation CHAID produced decision tree (Figure 13) with overall 98 nodes and 57 terminal nodes. All independent variables are acceptable for use in generating the decision tree and decision trees have maximum depth of as much as 6 levels.

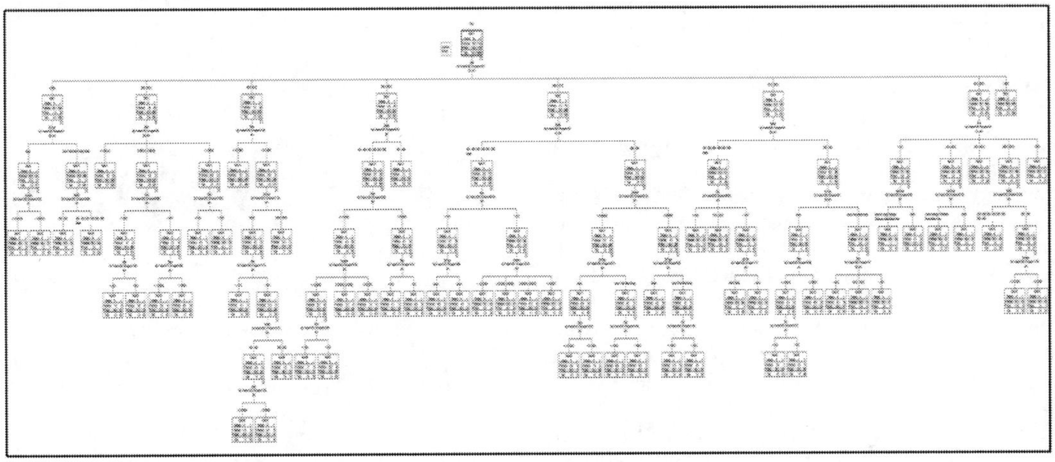

Figure 13The Tree of floodingprobability

The decision tree is generated to produce rules that are then used to analyze the new data set. There are 57 pieces of Rules decision tree, this rules are used to create SPSS syntax and then will be used in determining the probability of flooding. The model validation is done by using the method of Receiver Operating Characteristic curve (ROC). Validation is done by using the success rate curve and the curve prediction rate. Success rate curves are generated using 70% the training data set (7729 points floods). Success rate curve produces the value Area Under the Curve (AUC) which is equal to 0.931, or representingthe success rate accuracy of 93.1%.While the prediction rate curves are generated using 30% data sets of the flood point (3312 points) that have been saved to do the validation. Prediction rate produces curve value (AUC) which is equal to 0.927 or representing the prediction rate accuracy of 92.7%.The final result of the generation of the flood vulnerability maps based on the rules derived from a decision tree in the form of a map of the probability of flooding with a range of values of 0 to 1. Where the closer a number to 1, the greater the probability of flooding and otherwise getting close a number to 0, the smaller the probability of flooding.

In the making of flood vulnerability maps in this study, there are four types of classification methods considered for use. Fourth methods such as quantiles, natural breaks, equal intervals, and standard deviation. Then each method tested and selected the most appropriate method to the scope of this study and with the kind of information wish to display.

Natural breaks method suitable for use if there is a significant leap between the values of which are classified [2]. In this study, there were significant leap between the probability value was not flood with a probability value flooding, so use natural breaks classification method shown in Table 1

Tabel 1. Classification of the probability of flooding area

Probability	Class	%
0 – 0.047	Extremely low	76.68
0.047 – 0.307	Very low	5.37
0.307 – 0.599	Low	5.36
0.599 – 0.844	Moderate	5.31
0.844 - 1	High	7.28

Map of flood vulnerability of this research is beneficial for a variety of things. One was to identify the main street in the Metropolitan Bandung Area who have a high risk of flooding affected. This analysis is done by using the main road map in overlaying position to the maps of vulnerability flooding.

The duration of traffic disturbance depends on the time of flood receding. From the results of interviews with flood-affected communities, the following regression equation is obtained:

$$Y = 8.830 + 0.575X1 + 28.524X2 \quad (1)$$

Where:

Y :Predicted flood time recedes (hours)

X1: Duration of rainfall with intensity> 60 mm in Bandung area (hour)

X2: Depth of flood (m)

The coefficient of determination r2 = 0.734, which shows that the equation can only explain as much as 73.4% that the length of time receded is influenced by the depth of flood and the period of heavy rainfall in Bandung area..Traffic disturbances occur during flooded roads. Losses resulting from traffic congestion and lost productive time and decreased vehicle utility. In addition to the loss of delay in delivery of goods, especially for logistics vehicles. In this study not yet calculated all social losses, but the duration of traffic disturbance is indicated by the length of time receded in the arterial road until the road can function again.

6. Discussion

The decision tree results from the data processing has 6 levels of depth and is very wide relative to the side, due to the many separations in some independent variables. The land altitude variable becomes the most significant variable in flood event since it has the smallest p value adjusted to the root node among other variables. Thus the land altitude variable becomes the first branch of the root node. As the first branch of the root node, variable land altitude automatically become predictors for all terminal nodes, thus always affecting flood prediction or not.

There are several things that should be discussed in the model produced in this study. Model decision trees and model prediction capabilities are strongly influenced by training data used at the start of the study. Moreover, training data is an example of historical data that is static. So if using different training data, decision trees and predicted results are likely to change. So the decision tree model is less powerful. This is one of the major disadvantages of Chaid's decision tree method. Due to this deficiency, this model is only suitable for making predictions according to current circumstances. Predictions of future floods with certain rainfall could be done, but it is feared the results are not accurate.

The next thing is related to the decision making process. The decision tree making algorithm that applies to SPSS software states that flood or non-flood categorization depends on the percentage of most terminal nodes. Cutting categorization is greater than 50%. So if one category value is greater than 50% on the terminal node, then the category will be the flood status in each terminal node.

It should be remembered that the model developed in this research is a simplified model of physical flood model that tends to be very complex. So the flood vulnerability map from this study can be considered as a simplification of the actual flood vulnerability. But still an accurate prediction based on the data source used to make predictions. The advantage gained is that it simplifies the modeling time much faster thanks to the data-driven approach and the ease of data collection, which only uses the existing secondary data.

When pay attention to Figure 14 in the form of flood prediction and 4 in the form of the past flood area, looks rough correlation between the incidence of past flooding with the flood prediction. So in the visible, itlooks a relatively strong correlation between predictions of flooding to the real conditions are illustrated on the maps spatially. Seen a few predictions of floods that shaped the line snaking out of the main collection. The lines illustrate the vulnerability of flood higher on the banks of rivers region.

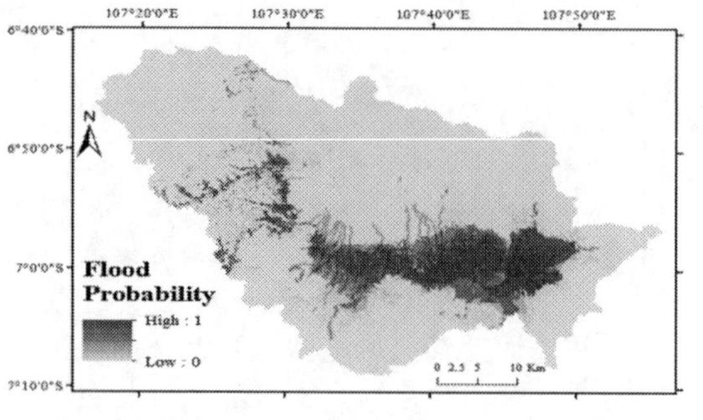

Figure 14. Map of the probability of flooding

Figure 15 shows that the flood vulnerability map dominated by classification extremely low (very low). When pay attention to Table 1 shows that the classification of the extremely low probability ranges from 0 to 0.047 covered 76.68% of the whole area which is reflected in Figure 15. It is far different from the four other classification with details very low 5.37%, low 5.36% , moderate 5.31% and 7.28% high. This indicates that most of the Metropolitan area of Bandung Area has a very low vulnerability of flood. Flood events are concentrated in certain areas only.

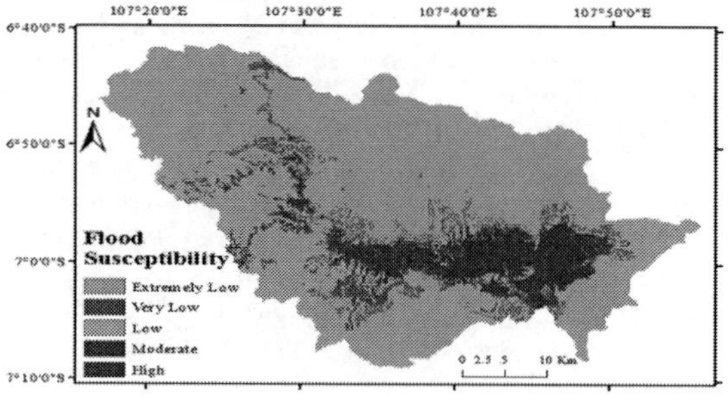

Figure 15. Map of flood vulnerability

Figure 16 shows an arterial road network that can experience flooding during the rainy season, so the road is susceptible to flooding. The analysis does not include the possibility of the flood water level, so it can not be determined which path is completely disconnected or just stagnant. Figure 17 shows a map of the Bandung area, with some arterial road names, and combined with Figure 16 it can identify potential critical road names when floods arrive.

The model developed in this research still has some shortcomings, mainly the set of data used to have differences in the year of collection. The development of future research should start by updating the sets of data with new data sets. Besides the accuracy of pixels should be increased in order to improve the accuracy of predictions resulting flood vulnerability maps are much more accurate. Suggested accuracy used was 15 m x 15 m or 10 m x 10 m. Accuracy like this can allow the use of data Stream Power Index (SPI) and Topographic wetness index (TWI) which requires precision pixels high.

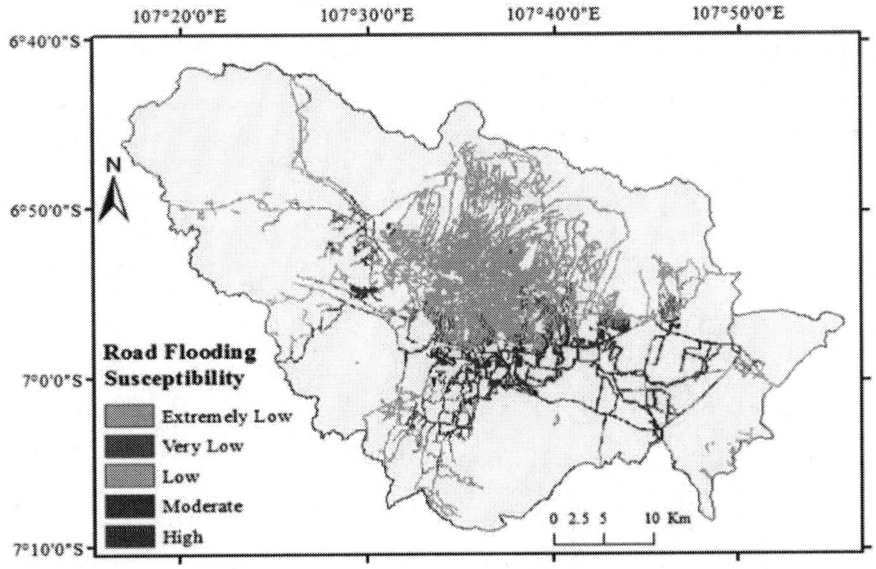

Figure 16. Map of road network flood vulnerability

Figure 17. Map of the road network with identity

7. Conclusion

Preparation of the Bandung area highway map is done by using the application of geographic information system. Flood event data in that area was processed by using CHAID classification analysis obtained by flood risk map, and then overlayed with Road Network to get flood road network map.

Digital Elevation Model (DEM) data, land cover, rainfall, soil type, distance from Citarum river, slope and curvature, Road identification was analyzed by Tree Diagram model and CHAID resulted in decision tree, which has 97 nodes, and 57 nodes Terminal node. A possible map of flood is generated using rules based on the terminal node. In predicting the flood, the DEM variable shows the most significant. The validation process of decision tree prediction model shows identification of 93.1% flood point with prediction rate of 92.7%. Classification of flood-prone areas obtained by Chaid result are class 0 - 0,047 covering 76,68% area; Grades of 0.047-0.307 include 5.37%; Grades 0.307 - 0.599 (Low) covering 5.36%; Grades 0.599 to 0.4444 include 5.31% and grade 0.844-1 (high) covering 7.27% of the research area. Based on the result of overlay map processing, the critical arterial road network is the road segment of Rancaekek - ParakanMuncang (Area of PT Kahatex), Solokan Jeruk(link ofCicalengka - Majalaya). Baleendah (Link of Bojongsoang - Baleendah), and link ofMochamadToha - DayeuhKolot, which will experience flooding if it rains more than 3 hours in all areas of Bandung. The flood vulnerability maps in this study make it easier for stakeholders to identify areas that need more attention in reducing flooding.

8. Suggestion

Further research is recommended to use the higher pixel precision such as 15 x 15 m, or 10 x 10 m in order to increase the accuracy of the data processing. In addition, higher data accuracy allows the use of SPI and TWI variables in research.

Additionally further research may use a decision tree method such as CART method, QUEST, C4.5 or other machine learning methods like ANN, Logistic Regression, or Naïve Bayesian. Then the results were compared with the results of this study to analyze the accuracy of each method in predicting flood vulnerability in the Metropolitan Bandung Area.The research results can be used to analyze other data that associated with vulnerability to flooding. Such as calculating the potential loss due to floods, the number of vulnerable populations affected by the floods, the area of agricultural potential damaged by flooding, and a variety of other uses.

Floods and landslides occur every year, especially in the rainy season, therefore the Regional Disaster Management Agency needs to redesign the disaster database system, by incorporating 3-dimensional coordinates on the geographic information system map so as to predict the potential of flood areas, network of critical road, time recede of flood and time of traffic disruption. Therefore it is necessary to coordinate disaster mitigation with Dinas Bina Marga and DinasPerhubungan and Police to develop contingency plan in case of flood disaster so that social losses can be minimized by educating the public to work together to prevent flooding earlier and to conduct traffic engineering management.

We would like to thank Mr. Oki (Pusair) and Mr. Dedi (BWS Citarum) for assisting in providing data on the characteristics of Citarum and ZhilalYusa that have assisted the preparation of GIS maps and the flood affected communities who have been willing to be interviewed to provide information on flood behavior in the area of Bandung

9.Reference

1. Abidin, H. Z., Gumilar, I., Andreas, H., Murdohardono, D., & Fukuda, Y. (2013). On causes and impacts of land subsidence in Bandung Basin, Indonesia. *Environmental earth sciences*, *68*(6), 1545-1553.
2. Ayalew, L., & Yamagishi, H. (2005). The application of GIS-based logistic regression for landslide susceptibility mapping in the Kakuda-Yahiko Mountains, Central Japan. *Geomorphology*, *65*(1), 15-31.
3. Dam, M. A. C., Suparan, P., Nossin, J. J., &Voskuil, R. P. G. A. (1996). A chronology for geomorphological developments in the greater Bandung area, West-Java, Indonesia. *Journal of Southeast Asian Earth Sciences*, *14*(1), 101-115.
4. Goodchild, M. F. (2009). Geographic information system. In *Encyclopedia of Database Systems* (pp. 1231-1236). Springer US.
5. Gumilar, I., Abidin, H. Z., Hutasoit, L. M., Hakim, D. M., Andreas, H., Sidiq, T. P., & Gamal, M. (2012). Pemetaankarakteristikpenurunanmukatanahberdasarkanmetode geodetic sertadampaknyaterhadapperluasanbanjir di cekungan. *MajalahIlmiah Globe*, *14*(1).
6. Hanifah, R., Isnanto, R. R., &Christyono, Y. (2010). Simulasisisteminformasigeografis (SIG) pemantauanposisikendaraan via SMS gateway.*TRANSMISI*, *12*(2), 45-49.
7. HendiHamdani, SulwanPermana, AdiSusetyaningsih (2014),Analisadaerahrawanbanjir
8. Menggunakanaplikasisisteminformasigeografis (studikasuspulaubangka),*JurnalKonstruksi, ISSN: 2302-7312 Vol. 12 No. 1 2014.*
9. Irina Mildawani, Diana Susilowati, Lia Rosmala Schiffer (2016), Aplikasisisteminformasigeografis (sig) DalamanalisispemanfaatandanpengelolaanRuangterbukahijaukota (rthk) Studikasus: kotadepok, LembagaPengembanganFakultasTeknikSipildanPerencanaan (LePTeSP), UniversitasGunadarma.
10. Jarvis A., H.I. Reuter, A. Nelson, E. Guevara, (2008). Hole-filled seamless SRTM data V4, International Centre for Tropical Agriculture (CIAT)
11. Khan, S. I., Hong, Y., Wang, J., Yilmaz, K. K., Gourley, J. J., Adler, R. F., ... & Irwin, D. (2011). Satellite remote sensing and hydrologic modeling for flood inundation mapping in Lake Victoria basin: Implications for hydrologic prediction in ungauged basins. *IEEE Transactions on Geoscience and Remote Sensing*, *49*(1), 85-95.
12. Kia, M. B., Pirasteh, S., Pradhan, B., Mahmud, A. R., Sulaiman, W. N. A., &Moradi, A. (2012). An artificial neural network model for flood simulation using GIS: Johor River Basin, Malaysia. *Environmental Earth Sciences*, *67*(1), 251-264.
13. Longley, P. A., Goodchild, M. F., Maguire, D. J., &Rhind, D. W. (2005).Geographic information systems and science. John Wiley & Sons Ltd.
14. Paul A Longley, Michael F goodchild, David J Maguire and David w rhind,1999, Geographical information system, Principles and Techniaal issues, 2nd edition,Jhon Wiley.
15. Saito, H., Nakayama, D., & Matsuyama, H. (2009). Comparison of landslide susceptibility based on a decision-tree model and actual landslide

occurrence: the Akaishi Mountains, Japan. *Geomorphology*, *109*(3), 108-121.

16. Varoonchotikul, P. (2003). *Flood forecasting using artificial neural networks*. CRC Press

Porous concrete basic property criteria as rigid pavement base layer in indonesia

*Frisky*Ridwan Aldila Melania Care[1], *Bambang*Sugeng Subagio[2],and *Harmein*Rahman[3]

[1]Transport Engineering Research Group, Faculty of Civil Engineering and Environment, Bandung Institute of Technology, Jalan Ganeca 10 Bandung, Indonesia

[2]Transport Engineering Research Group, Faculty of Civil Engineering and Environment, Bandung Institute of Technology, Jalan Ganeca 10 Bandung, Indonesia

[3]Transport Engineering Research Group, Faculty of Civil Engineering and Environment, Bandung Institute of Technology, Jalan Ganeca 10 Bandung, Indonesia

Abstract. Base layer with grained material tend to occur in saturated condition at the moment of rainfall or after rainfall. In that condition, the present of traffic might also create movement below the foundation layer and reduce its bearing capacity. Nevertheless, the foundation material that contain large fine creates base layer with low permeability and slow water movement. This paper presented the procedure and result of laboratory experimental study to obtain basic property criteria for rigid pavement base layer.The porous concrete mixture was successfully designed and surpassed the minimum requirements required by Directorate General of Highways [11] [12]. It was found that there was very strong correlation between void content and permeability which created fine likelihood to use void content to represent porous concrete permeability behavior. High void content with good connection resulting weaker aggregate interlocking inside the mixture so that permeability rose whereas compressive strength fell. Solid fresh density and more filled by cement paste created a smaller void that would reduce permeability and improve compressive strength. Strong correlation between fresh density towards void content and permeability opened potency to use the fresh density as a reference to set the desired porous concrete void content on the field.

Keywords: Porous Concrete, Base Layer, Rigid Pavement, Fresh Density, Compressive Strength, Void Content, Permeability.

1 Introduction

In this couple of year, concrete technology has been developed significantly. One of the latest product is a modern concrete mix-design called porous concrete. Porous concrete, sometimes referred to as no-fines, gap-graded, permeable, or enhanced porosity concrete, is an innovative approach for controlling, managing, and treating stormwater runoff. When used in pavement applications, pervious concrete can effectively capture and store stormwater runoff, thereby allowing the runoff to percolate into the ground and recharge groundwater supplies. (FHWA [13]).

In Indonesia, most of rigid pavement structures generally use grained material such as class A or Class B aggregate as base layer. The foundation of rigid pavement might also serve as drainage layer to prevent pumping and intrusion at joint locations, cracks dan the edge of plate (Center of Transportation Infrastructure Research and Development [9]). However, base layer with grained material tend to occur in saturated condition at the moment of rainfall or after rainfall. In that condition, the present of traffic also create movement below the foundation layer and reduce its bearing capacity. In several cases, base layer is employed to grant a uniform support for rigid pavement structure. The increasing of traffic load will aim to the use of denser, more robust and anti-erosion base material as well. The foundation material that contain large fine creates base layer with low permeability and slow water movement. The combination of trapped infiltration water and repeated traffic loading will generate void between base layer surface and surface layer base (erosion). The majority of infiltrated water that accumulated inside the void comes from pavement surface by infiltration through cracks, joints and gaps alongside the edge of pavement. Erosion phenomenon could lead to the loss of foundation support, pavement premature distress and short-term service life issue (Kozeliski [15]).

Survey results related to rigid pavement distress had been collected from several toll roads and highways in Indonesia. The amount of pumping, faulting and longitudinal cracking distress were very plentiful, especially in highway. These types of distress usually happened due to loss of foundation support, poor drainage and erosion problem in base layer (Huang [14]). Hence, a base layer with drainage type like porous concrete was required to tackle those problems. Researches have shown that a permeable base layer must meet three requirements. First, it must have enough permeability to quickly free itself from any water that has entered and then is trapped in the pavement structure. Secondly, it must have sufficient strength to

* Corresponding author: *friskyramc@gmail.com*

serve as a pavement construction platform. Finally, the design should meet the strength requirements to carry the traffic loads of the pavement structure (Zheng [24]). In addition, the porous concrete mixture should fulfill the base layer criteria which was required in Indonesian Guidance, Manual or Specification as well (**Table 1**).

Table 1. Rigid pavement base layer criteria (Center of Transportation Infrastructure Research and Development [9]; Directorate General of Highways[11] [12]).

Base Layer Type	Minimum f'c value 7 days (MPa)	Minimum f'c value 28 days (MPa)	Minimum k value (cm/sec)
Cement Treated Base	4.5 – 5.5	-	-
Lean Concrete	5.6 – 7.7	8.0 – 11.0	-
Aggregate Class A	-	-	0.1

2 Research Methodology

The laboratory study of porous concrete basic property criteria as rigid pavement base layer consists of following three main steps: (a) porous concrete mix-design proportion; (b) measurement of basic design parameters based on fresh density, compressive strength, void content and permeability; (c) Statistical correlation test between the parameters. These three steps are described in detail in the following sub-sections. All of the material properties are tested based on ASTM. The minimum criteria for mix-design will be attained by varying the aggregate cement ratio, water cement ratio and aggregate size value along with sand addition, fly-ash replacement and chemical additive addition such as HRWR, VMA and retarder.

2.1 Mix Design Proportion

Porous concrete mixtures were designed using ACI Committee 522 standard and some references from Meininger's [18] experimental test result. These mixtures will consists of uniform graded coarse aggregate, portland cement type I, water, sand, fly-ash class F, HRWR, VMA and retarder. The chemical additive dosage was used based on reference from Anderson [3]. External vibration combined with iron stick stabbing 25 times was used as compaction method. Polyethylene bag curing method was conducted to gain optimum compressive strength (Yusak [23]). Compressive strength specimens test were cast into cylinders with diameter 6" and length 12" (ASTM C-39 [5]), whereas void content and permeability specimens test were cast into cylinders with diameter 5" and length 5.24" (ASTM D-5084 [8]). **Table 2** and **Table 3** show the overall of porous concrete mix-design proportion in this research.

Table 2. Porous concrete mix-design proportion for preliminary test (1 m^3 concrete).

w/c	A/C	Cement (kg)	Water (kg)	Aggregate (kg)	Retarder (kg)
*0.27	4.0	338.2	91.3	1367.0	2.0
0.27	4.0	338.2	91.3	1367.0	2.0
*0.30	4.0	345.1	103.5	1367.0	2.1
0.30	4.0	345.1	103.5	1367.0	2.1
*0.33	4.0	343.9	113.5	1367.0	2.1
0.33	4.0	343.9	113.5	1367.0	2.1
*0.27	6.0	228.4	61.7	1367.0	1.4
0.27	6.0	228.4	61.7	1367.0	1.4
*0.30	6.0	228.1	68.4	1367.0	1.4
0.30	6.0	228.1	68.4	1367.0	1.4
*0.33	6.0	227.9	75.2	1367.0	1.4
0.33	6.0	227.9	75.2	1367.0	1.4

*curing at 7-days ; curing at 28-days ; aggregate size 12.7 mm

Table 3. Porous concrete mix-design proportion (1 m^3 concrete)

Cement (kg)	Water (kg)	Aggregate (kg)	Sand (kg)	Fly-ash (kg)	HRWR (mL)	Retarder & VMA (mL)
*342.8	99.4	1367.0	-	-	0.7	2.1
342.8	99.4	1367.0	-	-	0.7	2.1
*342.8	99.4	1367.0	-	-	0.7	2.1
342.8	99.4	1367.0	-	-	0.7	2.1
*342.8	99.4	1367.0	-	-	0.7	2.1
342.8	99.4	1367.0	-	-	0.7	2.1
*328.7	93.9	1307.6	66.5	-	0.6	2.1
328.7	93.9	1307.6	66.5	-	0.6	2.1
*312.1	89.1	1248.1	126.9	-	0.6	2.0
312.1	89.1	1248.1	126.9	-	0.6	2.0
*341.2	96.9	1367.0	-	37.9	0.7	2.4
341.2	96.9	1367.0	-	37.9	0.7	2.4

*curing at 7-days ; curing at 28-days ; aggregate size 9.5 mm ; aggregate size 12.7 mm ; aggregate size 19.0 mm

2.2 Measurement of Basic Design Parameter

Four basic parameters were applied to represent the required criteria for pavement base layer: (1) fresh density; (2) compressive strength; (3) void content; (4) permeability. All of the explanation about the variable in the equations are attached in Nomenclature along with their units.

Fresh density describes the density of freshly mixed pervious concrete. The test was conducted in accordance with ASTM C-1688 [6] and calculated using formula:

$$D = \frac{M_c - M_m}{V_m} \quad (1)$$

$$T = \frac{M_s}{V_s} \quad (2)$$

where:

D = density of concrete (kg/m^3)
M_c = mass of the measure filled with concrete (kg)
M_m = mass of the measure (kg)
V_m = volume of the measure (m^3)
T = theoretical density of concrete computed on an airfree basis (kg/m^3)
M_s = total mass of all materials batched (kg)
V_s = sum of the absolute volumes of the component ingredients in the batch (m^3)

Compressive Strength (f'_c) describes the typical concrete ability to receive compressive force per unit area. The compressive strength test was conducted in accordance with ASTM C-39 [5] and calculated using formula:

$$f'c = \frac{F}{A} \quad (3)$$

where:

f'_c = compressive strength (MPa)
F = maximum compressive force (N)
A = cross-sectional area of the specimen (m^2)

Void content (V_r) describes the total open space within the porous concrete. It is a comparison of the volume of voids, to the total volume of cement paste and aggregate. The void content test was conducted in accordance with ASTM C-1688 [6] and ASTM C-1754 [7] and calculated using formula:

$$U = \frac{T-D}{T} x\, 100 \quad (4)$$

$$V_r = \left[1 - \left(\frac{M_w - M_d}{\rho_w x Vol}\right)\right] \quad (5)$$

where:

U = percentage of voids in the fresh porous concrete, includingentrained and entrapped air voids in the cement paste
T = theoretical density of concrete computed on an airfree basis (kg/m^3)
D = density of concrete (kg/m^3)
V_r = void content (%)
M_w = specimen mass in water (kg)
M_d = dry specimen mass (kg)
P_w = density of water (kg/m^3)
V_{ol} = volume of specimen (m^3)

Permeability describes the easiness of concrete to be penetrated or passed by liquid or gas. Normally, the material permeability is indicated by its hydraulic conductivity or permeability value. Darcy's Law defines hydraulic conductivity as the ratio of fluid average velocity that passing through a transverse area against applied hydraulic gradient. The test was conducted using modified falling head permeameter apparatus which developed by McCain [17]. Montes and Haselbach [19] showed that under these head values laminar flow conditions are expected to exist in pervious concrete pavement and Darcy's law can be used to determine the hydraulic conductivity. The permeability test was conducted in accordance with ASTM D-5084 [8] and calculated using formula:

$$k = \frac{a.L}{A.t} \ln\left(\frac{h_1}{h_2}\right) \quad (6)$$

where :

k = permeability coefficient (m/s)
a = cross-sectional area of the standpipe (m^2)
L = length of the specimen (m)
A = cross-sectional area of the specimen (m^2)
t = time for water to drop from h_1 to h_2 (s)
h_1 = initial water level (m)
h_2 = final water level (m)

2.3 Statistical Correlation Test

The correlation between the basic design parameters was described using simple correlation test. This test is a statistical technique which is used to measure the relationship level between 2 variables and the shape of their relationship. The relationship level is stated in number wherein bigger number shows stronger relationship (**Table 4**). The simple correlation test was computed utilizing formula:

$$R^2 = \frac{\sum XY}{\sqrt{\sum X^2 \sum Y^2}} \quad (7)$$

where:

R^2 = simple correlation coefficient
X = independent variable
Y = dependent variable

Table 4. Guidance of correlation coefficient interpretation.

Value of Correlation Coefficient	Relationship Interpretation
0,000 – 0,199	Very Weak
0,200 – 0,399	Weak
0,400 – 0,599	Moderate
0,600 – 0,799	Strong
0,800 – 1,000	Very Strong

3 Test Results and Discussion

This section covers the result of basic design parameter and correlation of basic parameter. Several discussions were attached as well to illustrate the laboratory test result.

3.1 Statistical Correlation Test

Preliminary test was conducted to obtain the optimum compressive strength and void content by varying water cement ratio and aggregate cement ratio. The preliminary test results are showed on **Figure 1** to **Figure 3**.

As can be seen from **Figure 1** mix-proportion with A/C = 4.0 had a higher compressive strength value compared to A/C = 6.0 at the age of 7-days and 28-days. In addition, the porous concrete mixture also had an optimum compressive strength value at specific water cement ratio wherein 0.30 was showed as an optimum value in this case. Both of these conditions nearly close to the results which were gained by Meininger [18] and Amde [2].

From**Figure 2** and **Figure 3**, therewere a dramaticdecreasefrom the void content and permeability value due to the raise of water cement ratio and the subtraction of aggregatecement ratio McCain [17]. This condition alsoshowed a potency of compressive strengthenhancement by reducing the aggregatecement ratio value. Therefore, w/c = 0.30 and A/C = 4.0 wouldbechosen as reference to create the nextspecimen composition usingaggregate variation and additive inclusion. The following compressive strength test results are presented on **Figure 4** and **Figure 5**.

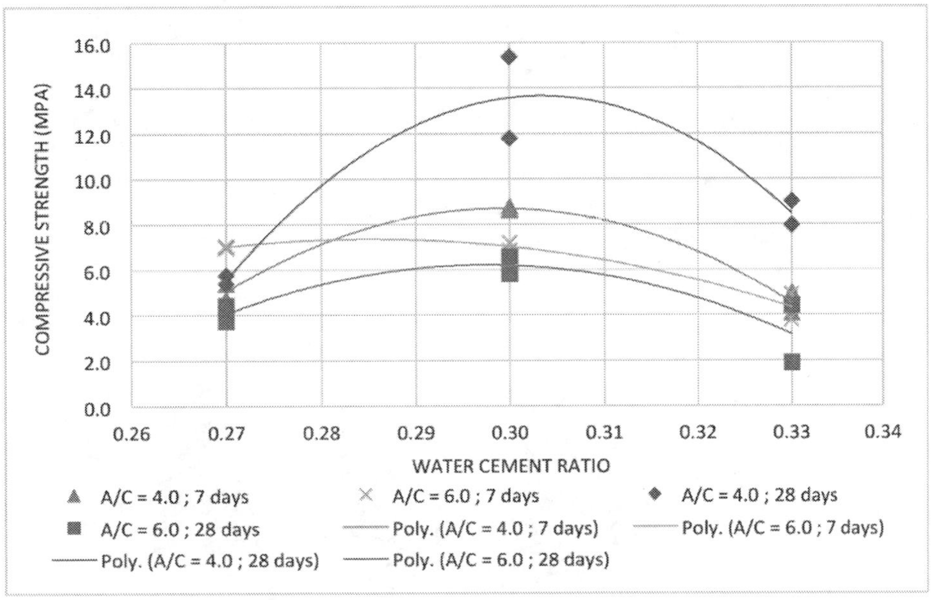

void content range: 28% - 40 %
permeability coefficient range: 2.5 cm/sec – 4.7 cm/sec

Figure1. Relationship between water cement ratio and compressive strength.

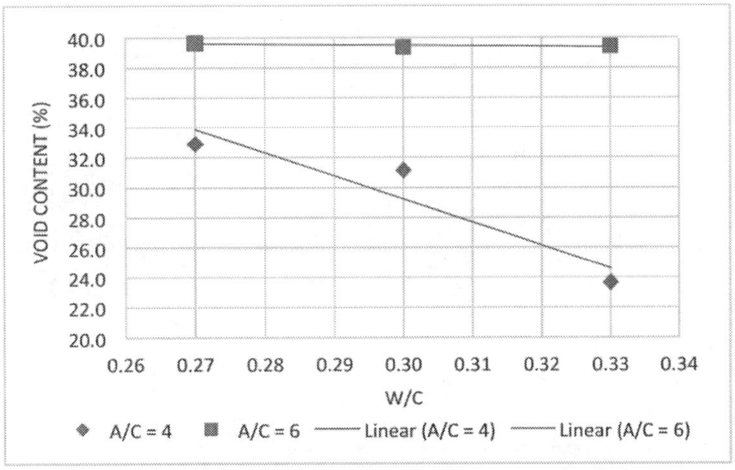

compressivestrength range: 1.9 MPa – 15.4 MPa
permeability coefficient range: 2.5 cm/sec – 4.7 cm/sec

Figure2. Relationship between w/c and void content at the age of 28-days.

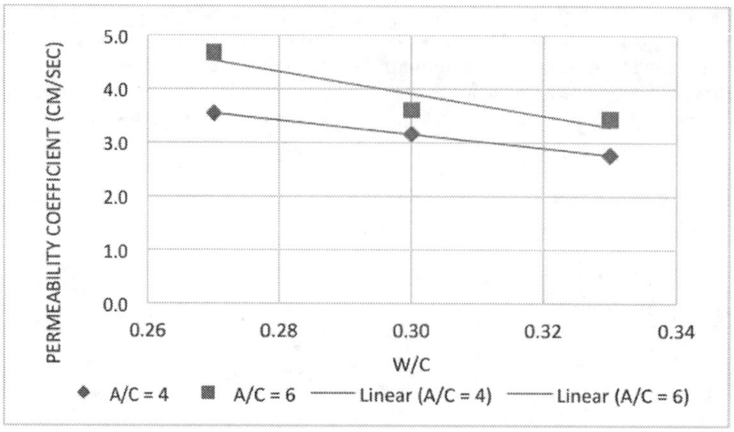

compressivestrength range: 1.9 MPa – 15.4 MPa
void content range: 28% - 40 %

Figure3. Relationship between w/c and permeabilityat the age of 28-days.

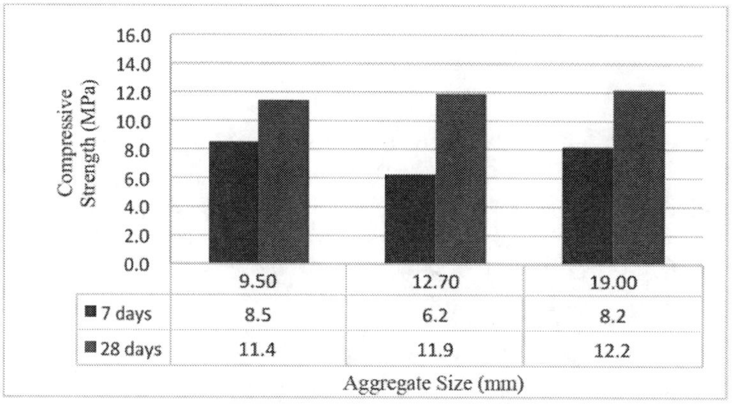

void content range: 27% - 32 %
permeability coefficient range: 1.4 cm/sec – 2.6 cm/sec

Figure 4. Relationship betweenaggregate size and compressive strength.

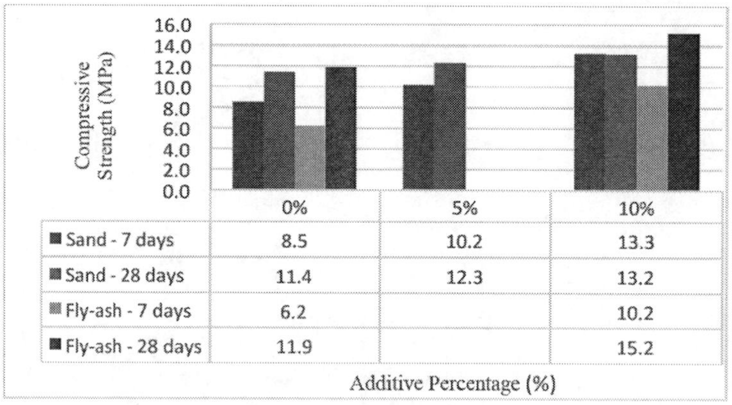

void content range: 25% - 32 %
permeability coefficient range: 1.0 cm/sec – 2.6 cm/sec

Figure 5. Relationship between additive use and compressive strength.

Based on **Figure 4** the increase of aggregate size did not give a significant influence against compressive strength value. This situation wasdifferentwhencompared to Crouch [10] whichsaidthatfineraggregate size produced more superior compressive strength value. Turning to **Figure 5**, the sandincreaseuntil 10% rose the compressive strengthgradually. However, when the sandwasadded more than 10%, the compressive strength value mightdroppedalongwith the growth of time. Because of sand addition, contact area that must becovered by cementpastewouldincrease and the cementpastethatcovering the materialwouldalsodiminish. Moreover, thinnercover by cementpastewouldreduce the bondingstrengthbetweenaggregate and cementpastewhichresultinglower compressive strength. Still on the same figure, the rise of fly-ash as cement replacement until 20% increased the compressive strength as well. Nevertheless, this condition wasdifferentfromAoki [4] and Muthaiyan [20] statementwhichconcludedthat the compressive strength value wasdeclinedwhenfly-ash content wasincreased. Futhermore, the porousconcrete mixtures withsand and fly-ash inclusion surpassed the minimum compressive strengthrequirements, i.e. 8.0 MPa, for base layer onlyin 7-days although the accepted minimum value isin 28-days required by Directorate General of Highways[12].

The followingvoid content and permeability test results are presented on **Table 5** and **Figure 6** to **Figure 9**.**Table 5**exhibitsvarious values of void content and permeabilityat the age of 28-days. Eventhough, twospecimenshadquite far void content value from the targetvoid content, their value stillremainedabove 20% whichwastargeted in thisstudy. **Figure 6** and **Figure 7**showthat the increase of aggregate size decreasedvoid content and permeability. Biggeraggregate size causedcementpastethatcovered the contact area becamelessersothatthis excessive cementpastewouldfill the voidinside the mixture. When the void content decreased, the permeabilitywoulddecreased as well. Evenso, this situation wasdifferentwhilecompared to Crouch [10] whichsaidthat the rise of aggregate size producedsimilarvoid content.

Table 5 Void content and permeability value of test specimensat the age of 28-days.

Target Void Content (%)	PorousConcreteSpecimens		
	SpecimenNumber	Void Content (%)	Permeability (cm/sec)
29	14D	29.99	2.56
29	16D	29.45	2.39
29	18D	29.21	2.37
29	20D	28.62	1.48
30	22D	26.07	1.38
27	24D	21.91	1.02

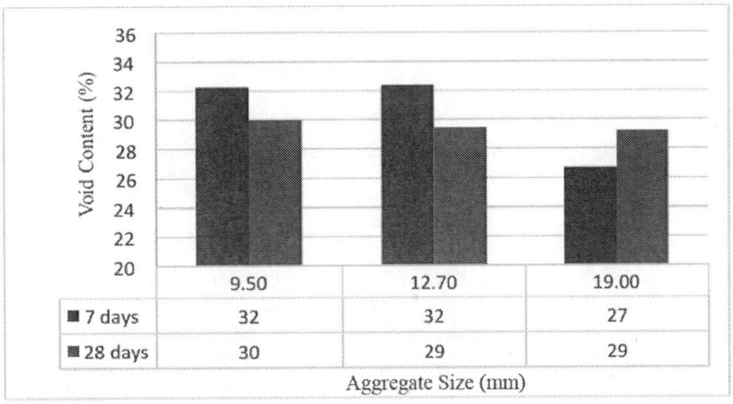

compressivestrength range: 6.2 MPa – 12.2 MPa
permeability coefficient range: 1.4 cm/sec – 2.6 cm/sec

Figure 6. Relationship between aggregate size and void content.

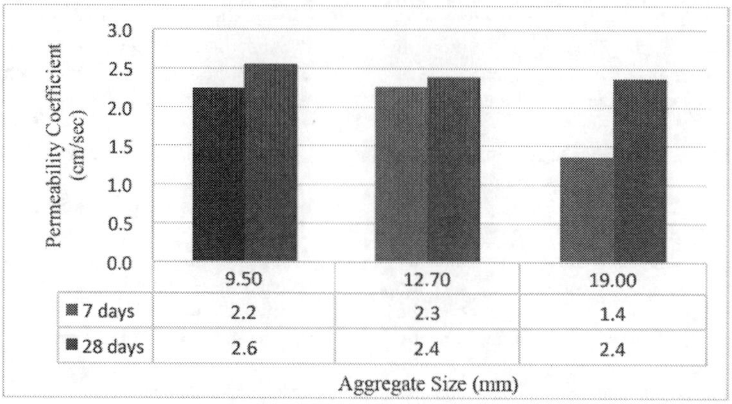

compressivestrength range: 6.2 MPa – 12.2 MPa
void content range: 27% - 32 %

Figure 7. Relationship between aggregate size and permeability.

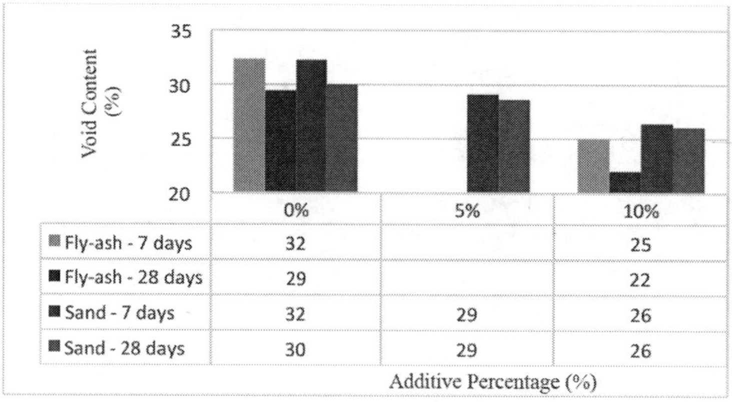

compressivestrength range: 6.2 MPa – 15.2 MPa
permeability coefficient range: 1.0 cm/sec – 2.6 cm/sec

Figure 8. Relationship between additive use and void content.

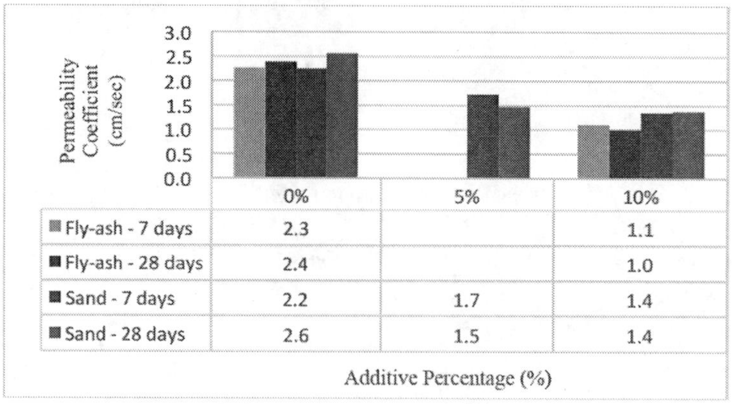

compressivestrength range: 6.2 MPa – 15.2 MPa
void content range: 25% - 32 %

Figure 9. Relationship between additive use and permeability.

Turning to **Figure 8** and **Figure 9**, sand addition up to 10% from cement content dropped the void ratio and permeability values (Schaefer [21]). Utilization of fly-ash up to 20% also lowered the void ratio and permeability values. This condition walked the same way with Aoki [4] and Muthaiyan [20] which concluded that permeability values tend to drop while fly-ash content was rose. Fly-ash addition lowered void content significantly compared to sand because fly-ash possess typical grain size 100 times smaller and contact area 10 times smaller rather than sand. Additionally, the use of greater aggregate size also produced greater void content. As a result, fly-ash with a smaller grain size had a higher chance to fill the void space rather than sand. All mixtures also produced void content above 20%, wherein this condition could maintain a relatively high level of permeability throughout the service life of pavement (Lim [16]). Futhermore, the porous concrete mixtures with sand and fly-ash inclusion stayed above the minimum permeability requirements, i.e. 0,1 cm/sec, for base layer in 7-days and 28-days required by Directorate General of Highways [11].

3.2 Correlation of Basic Parameter

Correlation test was conducted to observe which variables among the basic parameters were connected and how strong the relationship through the correlation coefficient that was gained from correlation test. Strong relationship was indicated by a high coefficient of correlation value, and vice versa.Several basic properties that were used in this study tried to represent each of the requirements which given by Directorate General of Highways [11] [12], i.e. permeability coefficient to represent run-off performance, compressive strength to represent strength performance. Fresh density and void content were utilized as control variable or independent variable.The results of correlation test are presented on **Figure 10** to **Figure 14**.

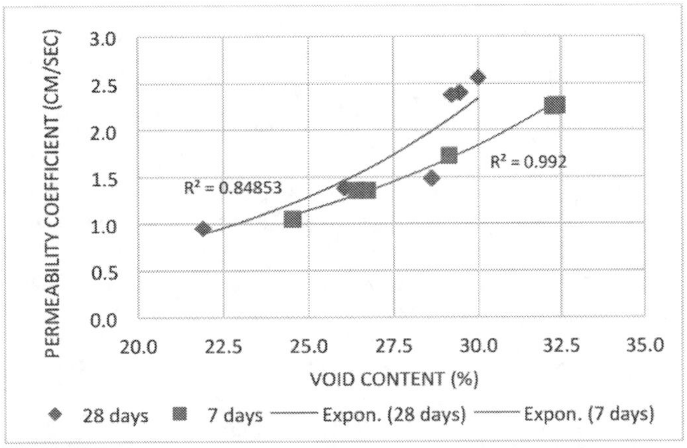

Figure 10. Correlation between porous concrete void content and permeability.

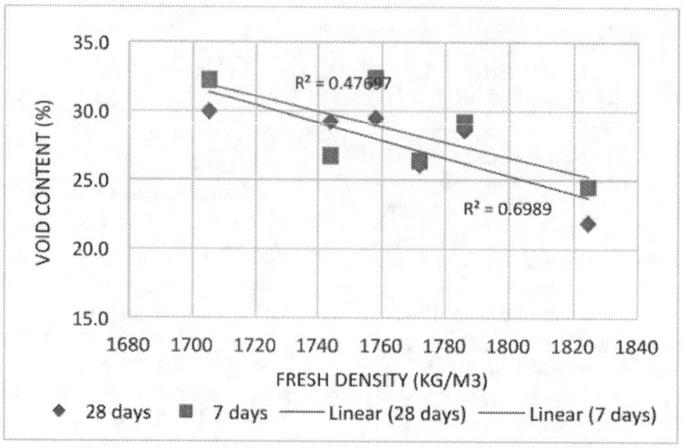

Figure 11. Correlation between porous concrete fresh density and void content.

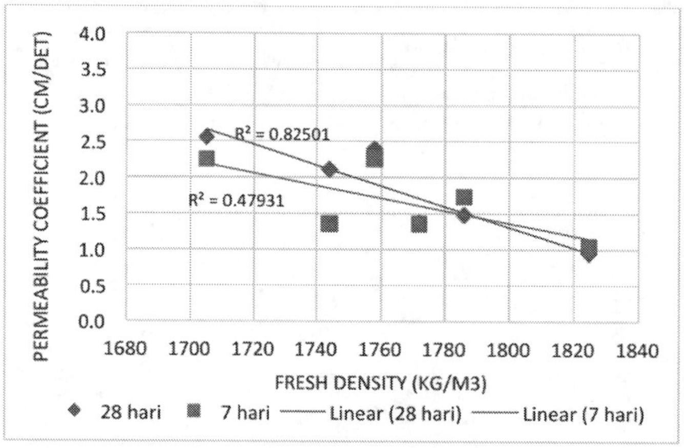

Figure 12. Correlation betweenporousconcretefreshdensity and permeability.

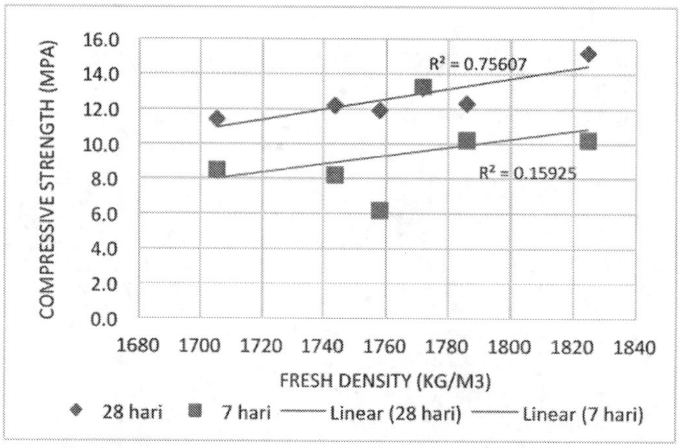

Figure 13. Correlation between porousconcretefreshdensity and compressive strength.

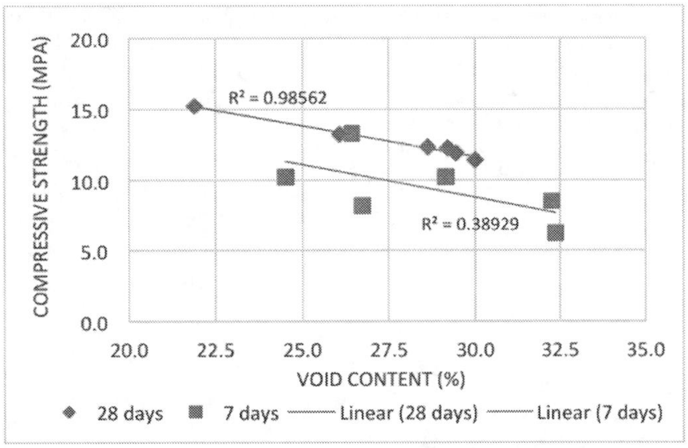

Figure 14. Correlation between porous concrete void content and compressive strength.

Based on correlation test results, the following discussion can be made:

a. **Figure 10** illustrated that permeability had a direct and very strong correlation with void content at the age of 7-days and 28-days which was evidenced by a dramatic rise exponentially from permeability value while void content was increased (ACI Committee 522 [1] and Shaefer [21]). Based on this condition, there was fine likelihood to use void content to represent porous concrete permeability behavior.

b. As can be seen on **Figure 11** and **Figure 12**, there were strong negative correlation between fresh density towards void content and permeability at the age of 28-days. Greater fresh density created smaller void and small void would reduce permeability. It was happened because the void inside fresh concrete mixture with a bigger density was filled by cement paste and aggregate. This result strengthened previous test results which were explained by Sonebi [22] and created a fine potency to use the fresh density as a reference to set the desired porous concrete void content on the field.

c. Based on **Figure 13** and **Figure 14,** there were strong correlation between fresh density and void content against compressive strength at the age of 28-days. The increase of fresh density rose the compressive strength significantly. On the contrary, the increase of void content declined the compressive strength gradually. High void content with a good connection resulting weaker aggregate interlocking inside the mixture so that the compressive strength value decreased. On the other hand, compact fresh density with a good aggregate interlocking inside the mixture increased the compressive strength value. These situations showed a close result to the prior results which been obtained by Made [2], Meininger [18] and ACI Committee 522 [1].

d. Looking at **Figure 11** - **Figure 14,** most of age 7-daysshowedweak relationship among the parameters.Weak coefficient of correlation values indicated that one parameter had no connection with other parameter and the variability level between two parameters still high. There was lower possibility where if the independent variable increased or decreased, the dependent variable would be increased or decreased as well. This situation might happen because most of pasta semen inside the porous concrete had not fully hydrated yet.

4 Conclusion

This paper has presented the procedure and result of laboratory experimental study to obtain basic property criteria for rigid pavement base layer. Those basic parameters were fresh density, compressive strength, void content and permeability. Porous concrete mixture was successfully designed using ACI Committee 522 [1]standard and some references from Meininger's [18] experimental test result. It also surpassed the minimum compressive strength and permeability requirements for base layer which are required by Directorate General of Highways [11] [12]. The compaction method and curing method also gave significant contribution for the porous concrete mix-design. Minimum criteria for mix-design was attained by using water cement ratio 0.30, aggregate cement ratio 4.0 and varying aggregate size value along with sand addition, fly-ash replacement and chemical additive addition.

It was found that there was very strong correlation between void content and permeability at the age of 7-days and 28-days which created fine likelihood to use void content to represent porous concrete permeability behavior. Greater fresh density created smaller void and small void would reduce permeability. High void content with a good connection resulting weaker aggregate interlocking inside the mixture so that the permeability increased and the compressive strength decreased. On the other hand, compact fresh density with a good aggregate interlocking increased the compressive strength value. It was happened because the void inside fresh concrete mixture with a bigger density was filled by cement paste and aggregate. Strong correlation between fresh density towards void content and permeability at the age of 28-days opened a fine potency to use the fresh density as a reference to set the desired porous concrete void content on the field. Weak relationship among the parameters at the age of 7-days might happen because most of pasta semen inside the porous concrete had not fully hydrated yet.

5 References

1. ACI Committee 522, Report on Pervious Concrete, *American Concrete Institute*, 7-15 (2010).
2. Amde, A., M., Rogge, S., Development of High Quality Pervious Concrete Specifications for Maryland Conditions, State Highway Adminstration Maryland Department of Transportation, 13-54 (2013).
3. Anderson, Walsh, Oka, Dewoolkar, Limberg, Sevi, Schmeckpeper, Laboratory Performance of Pervious Concrete Subjected to Deicing Salts and Freeze-Thaw, University of Vermont Transportation Research Center, 20-20 (2015).
4. Aoki, Y., Development of Pervious Concrete, *M.Eng Thesis*, Faculty of Engineering and Technology, University of Technology, Sydney, (2009).
5. ASTM Standards C39/C39M-05, Standard Test Method for Compressive Strength of Cylindrical Concrete Specimens,*ASTM International*, (2005).
6. ASTM Standards C1688/C1688M-13, Standard Test Method for Density and Void Content of Freshly Mixed Pervious Concrete, *ASTM International*, (2013).
7. ASTM Standards C1754/C1754M-12, Standard Test Method for Density and Void Content of Hardened Pervious Concrete, *ASTM International*, (2012).
8. ASTM Standards D5084-03, Standard Test Methods for Measurement of Hydraulic Conductivity of Saturated Porous Materials Using a Flexible Wall Permeameter, *ASTM International*, (2003).
9. Center of Transportation Infrastructure Research and Development, *Pedoman Perencanaan Perkerasan Beton Semen*, Department of Regional Settlement and Infrastructure, 7-9, 20-20(2003).
10. Crouch, L., K., Pitt, J., Hewitt, R., Aggregate Effects on Pervious Portland Cement Concrete Static Modulus of Elasticity, *Journal of Materials in Civil Engineering*, 566-566(2007).
11. Directorate General of Highways, *Manual of Hydraulic for Road and Bridge Works – Hydraulic Design*, Department of Public Works,96-96, 99-99 (2005).
12. Directorate General of Highways, *Spesifikasi Umum Revisi 3*, Ministry of Public Works, 5.66-5.66 (2010).
13. Federal Highway Administration, Pervious Concrete, *The Advanced Concrete Pavement Menttechnology (ACPT) Techbrief*, 1-1 (2012).
14. Huang, Y., H., *Pavement Analysis and Design*, 2nd Edition, Prentice-Hall, Inc., 376-388(2004).
15. Kozeliski, F., A., Permeable Bases Help Solve Pavement Drainage Problems, the Aberdeen Group, 1-2(1992).
16. Lim, E., Fwa, T., F., Tan, K., H., Laboratory Evaluation of Clogging Behaviour of Pervious Concrete Pavements, *Journal of the Eastern Asia Society for Transportation Studies***Vol. 11**, 1611-1611(2015).
17. McCain, G. N., Dewoolkar, M., M., Porous Concrete Pavements: Mechanical and Hydraulic Properties, *TRB 2010 Annual Meeting*, 7-18.
18. Meininger, R., C., No-Fines Pervious Concrete for Paving, *Concrete International***Vol. 10 Issue 8**, 22-25(1988).
19. Montes, F., Haselbach, L., Measuring Hydraulic Conductivity in Pervious Concrete, *Journal of Environmental Engineering and Science*, 968-968 (2006).
20. Muthaiyan, U., M., Thirumalai, S., Studies on the Properties of Pervious Flyash-Cement Concrete as a Pavement Material, Cogent Engineering,15-16 (2017).
21. Schaefer, V., R., Wang, K., Suleiman, M., T., Kevern, J., T., Mix Design Development for Pervious Concrete in Cold Weather Climates, National Concrete Technology Pavement, 41-41 (2006).
22. Sonebi, M., Bassuoni, M., T., Pervious Concrete: Mix Design, Properties and Applications, *Research Gate*, 110-113 (2016).
23. Yusak, M., I., M., Jaya, R., P., Hainin, M., R., Ismail, C., R., Ibrahim, M., H., W., Strength of Porous Concrete Pavement at Different Curing Methods, *Jurnal Teknologi*, 103-103(2015).
24. Zheng, M., Chen, S., Wang, B., Mix Design Method for Permeable Base of Porous Concrete, *International Journal Pavement Research Technology***Vol. 5 no.2**, China, 102-102(2012).

Pedestrian Perception about Facility of Pedestrian Crossings

Agah Muhammad Mulyadi

Institute of Road Engineering, Ministry of Public Works and Housing. Jl. A.H. Nasution No. 264 Bandung, 40294, Indonesia

Abstract. Pedestrians are vulnerable road user which have potential conflicts with vehicles, especially when pedestrians crossing the road. The number of accidents which was involving pedestrians crossing on the roads were relatively high. Data showed that accidents which were involving pedestrians in Indonesia were 3878 events1. The purpose of this study is to determine the needs of pedestrians on the crossing facility and how pedestians desired types of crossing facilities. Data were collected by questionnaire distribution and interviews to the pedestrians. Collecting data and information were conducted in Bandung. Perception of comfort on each type of crossing facility, the majority of respondents choose pelican crossing and crossings with creative road marking with the highest comfort 50% and 51% respectively. Respondents' perception on the secure of each types of crossing facility is to assume that overpass/underpass has the highest secure pedestrian crossing by 48%. In addition, the recommendation of types of pedestrian crossing on Jalan Asia Africa Rd and Aceh Rd are pelican crossing on weekday and overpass/underpas at the weekend based on PV^2 formula.

1 Introduction

Walking is the simplest transportation mode in daily activities. Every road which used for public shall be equipped with facilities for bicycles, pedestrians, and persons with disabilities1. Pedestrians is a vulnuarble road user, therefore the safety factor must be improved. Accidents that occur between pedestrians and motor vehicles are common nowadays. Pedestrians accidents usually occurs whereas there are no crossing facility for pedestrians on the road, thus pedestrian cross the roads without crossing facility and the they facing the risk hit by vehicles. Pedestrians who hited by vehicles with travel speed of more than 64.4 km/h have a death risk by 85% [2].

Pedestrians need facilities that serve to protect pedestrians from conflicts with motor vehicles, thereby increasing the sense of security for both pedestrians and motorists.. The crossing pedestrian facility is part of road which shows by difference surface path, difference textures, or different colors in order to make it contrast so it is easily to recognize by pedestrians or vehicles rider.[3]. Basically, pedestrian crossing facilities are divided into two types, there first is pedestrian crossing at grade and the second is overpass/underpass pedestrian crossing. The choice of pedestrian crossing facilities should also consider pedestrian needs factors, which consist of internal factors such as age, gender, or willingness to using pedestrian facilities.[4].

2 Figures and tables

Roads are road transport infrastructures covering all sections of roads, including auxiliary buildings and equipment intended for traffic, located on soil surface, above ground level, below ground level and/or water, and above water level, except railroads, lorries, and cable roads.[5]

Pedestrians are part of the physical element of urban design.[6] Each pedestrian needs to walk on the road safely, and is fun, hence its needed pedestrian infrastructure to walking along the road i.e pedestrian sidewalk and pedestrian crossing . The pedestrian facilities were builded to provide for pedestrians in order to improve the smoothness, safety, and comfort of pedestrians. Pedestrian facilities provided for pedestrians are classified into two types:[7]

1.1 At Grade Pedestrian Crossing.

1.1.1 Zebra Cross

its distinguishing feature is alternating dark and light stripes on the road surface, resembling the coat of a zebra as shown in Figure 1.

* Corresponding author: author@e-mail.org

Fig. 1. Zebra Cross

1.1.2 Pelican Crossing

Pelican Crossing features a pair of poles each with a standard set of traffic lights facing oncoming traffic, a push button and two illuminated, colored features a pair of poles each with a standard set of traffic lights facing oncoming traffic, a push button and two illuminated, colored pictograms facing the pedestrian from across the road as shown in Figure 2.

Fig. 2. Pelican Crossing

1.1.3 Crossings with creative markers

Crossings with creative markers declared with creative road marking and show by contrast colors that can attract attention from road users as shown in Figure 3.

Fig. 3. Crossings with creative markers

1.2 At Grade Pedestrian Crossing.

Pedestrian overpasses and underpasses allow for the uninterrupted flow of pedestrian movement separate from vehicle traffic as shown in Figure 4.

Fig. 4. Pedestrian overpasses and underpasses

The determination of a non-level crossings facility shall be in accordance with the conditions of road traffic being reviewed. The following method of determining the crossing facility is not-level[11]. On the road with a plan travel speed above 75 km / hour. In strategic areas where waders are not allowed to cross the road, except only on the bridge. PV^2> 5 x 108 with P> 100 person / hour and V 5000 pcu/ hour. The V value is taken from the average current value during the 4 hrs. Selection of Pedestrian Crossing Facility based on PV^2 Formula is shown in Table 1.

Table 1. Setting Word's margins.

PV^2	P	V	Crossing Facility Recommendation
$> 5 \times 10^8$	100-1250	2,000-5,000	Zebra Cross
$> 10^{10}$	3,500-7,000	400-750	Pelican Cross
$> 5 \times 10^9$	100-1,250	> 5,000	Pelican Cross
$> 5 \times 10^9$	> 1,250	> 2,000	Pelican Cross
$> 10^{10}$	100-1,250	> 7,000	Overpass
$> 10^{10}$	> 1,250	> 3,500	Overpass

3 Research methods

In this research method of data analysis to be used is quantitative method. Meanwhile data collection techniques that using in this research community perception of pedestrian path in Bandung by interviewing, observation, and distribute the questionnaires. This research to be conducted by using quantitative analysis will be further divided into several analysis based on the needs and field conditions. Analysis to be conducted in this research includes descriptive statistical analysis. Descriptive statistics are the process of collecting and summarizing the data, as well as attempts to describe the various important characteristics of the organized data.[8]

The data collection activities in this study mainly were collected by distributing questionnaires. Questionnaires will be given to respondents related to this research are pedestrian path users who have experiences using pedestrian path. Sampling technique to be taken in this research is accidental systematic sampling. The accidental sampling technique is performed when the selection of sample members is made against any person or object that happens to exist or be encountered. The advantage of this sampling technique is fast and easy.[9] While the weakness is less representative. Then the number of samples to be used is

120 respondents. This amount is determined based on the consideration of the accuracy of research results to be representative and can represent the existing population in the study area. Furthermore, the number of respondents is divided based on certain times between weekend and weekdays to obtain data distribution. Primary survey data is shown in Table 2.

Table 2. Primary Survey Data.

Day	Time Session	Time	Locations	Number of respondents	City
Sunday and Monday	Morning	06.00-08.00	Asia Africa Rd and Aceh Rd.	20	Bandung
	Day	12.00-14.00		20	
	Noon	15.00-17.00		20	

Analysis to be used is the frequency distribution analysis for each parameter. The frequency distribution in principle is to organize the data into several classes of the same data, therefore that each class can describe the characteristics of the existing data.[10] At this stage the public opinion of the pedestrian path is summed and percentage is made for each community choice.

4 Pedestrian perception analysis

The result of perceptual analysis of the questionnaire results, the most widely found pedestrian path users are in productive age. This is because most of users in the productive age are still able to do many walking activities when compared to road users who in the old age or children. The classification of productive age is aged 15-30 years old and 31-45 years old, the result of the distribution of data obtained by domination at productive age is 86.67%. The distribution data by age of respondents is shown in Table 3.

Table 3. Distribution of data by age of respondents

Age Clasification	Frequency	Percentage
<15	1	0.83%
15-30	86	71.67%
31-45	18	15.00%
46-60	12	10.00%
>60	3	2.50%
Total	**120**	**100%**

In addition, the analysis based on the education background, found that most respondent have education background are high school and college/universities is 86.67%. The distribution data base on education background is shown in Table 4.

Table 4. Distribution of data based on the education background

Education Background	Frequency	Percentage
Never been Studied	0	0.00%
Primary school	4	3.33%
Junior high school	16	13.33%
Senior High School	85	70.83%
College	15	12.50%
Total	**120**	**100%**

Respondents characteristics are users who are in the ages that tend to perform recreational activities or shopping, consequently that in line with the condition of the pedestrian path which close with shopping centre and public park. Analysis level of income show most of respondents have income below 1 million rupiah is 81.67%. This is correlation because most of respondents still study in the school or college, accordingly, they do not have their own income, or if they work with that level of education, they will have an average income less than 1 million rupiah. Distribution of respondents income is shown in Table 5.

Table 5. Distribution of data based on respondents income

Respondents' Income (Rp.)	Frequency	Percentage
<1,000,000	98	81.67%
1,000,000-2,500,000	10	8.33%
2,500,001-4,000,000	8	6.67%
4,000,001-7,000,000	3	2.50%
>7,000,000	1	0.83%
Total	**120**	**100%**

Activities near the pedestrian facility at the time of distributing the questionnaire should be known to measure the typical journey of a pedestrian. Most of respondents came to Asia Afrika Road to recreation and leisure. At the location of data collection in Bandung as much as 48% of respondents who have a reason accidentally visited the location of pedestrian facilities. Distribution of respondents activity in the area around the pedestrian crossing is shown in Table 6.

Table 6. Distribution of trip purpose

Trip Purpose	Frequency	Percentage
Sightseeing	25	20.84%
Shopping	10	8.33%
Waiting for Public Transport	5	4.17%
Accidentally Visited this place	57	47.50%
Meal time	8	6.67%
Working	15	12.50%
Total	**120**	**100%**

https://doi.org/10.1051/matecconf/201814702009

The results of pedestrian knowledge about types of pedestrian crossing facilities i.e zebra cross, pelican cross, creative road marking cross, and overpass and underpass. Crossing facilities of zebra cross and Overpass/ Underpass pedestrian crossing is the most familiar crossing facility that 100% of respondents know this facility. Meanwhile, the type of pelican cross and crossings with creative markers have already known by 53.33% and 100% respectively. Distribution of respondents knowledge is shown in Table 7.

Table 7. Distribution of data by knowledge about types of pedestrian crossing

Pedestrian Crossing Facility	Respondent has already known	Percentage (%)	Respondent did not know	Percentage (%)
Zebra Cross	120	100.00 %	0	0.00%
Pelican Crossing	64	53.33%	56	46.67%
Crossings with creative markers	89	74.17%	31	25.83%
Overpass/Underpass Pedestrian Crossing	120	100.00 %	0	0.00%

The question that states whether the respondent always cross the road by using the pedestrian crossing facilities, the majority respondent's answer is sometimes as 34%, often as 30%, rare as 19%, very often as 9% dan very rare as 8%. These results are needed to know the respondent's awareness about road safety. In addition, by using pedestrian crossing facility, By crossing through pedestrian fasilities will encourage others people to do the something. Figure 5 shown the piechart about frequency of cross the road by using the pedestrian crossing facilities.

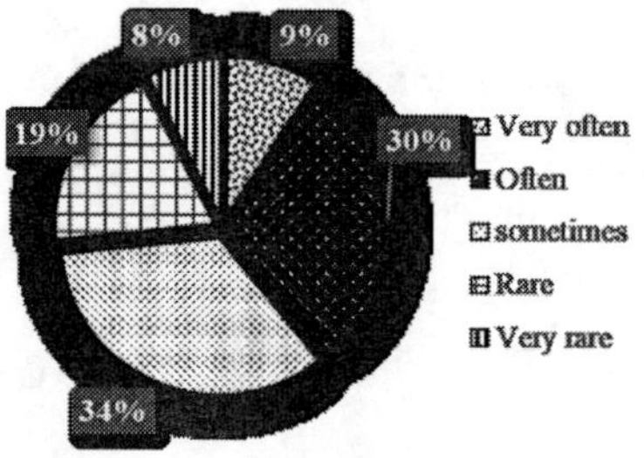

Fig. 5. Respondents cross the road by using the pedestrian crossing facilities

Data analysis with statistical analysis method is data response to performance of each crossing facility on the Asia Africa Rd and Aceh Rd based based on secure and comfort feeling with score as shown in Table 8.

Table 8. Classification fot Secure and Comfort Pedestrian Crossing

Score	Classification
1	Very High
2	High
3	Netral
4	Low
5	Very Low

Distribution of secure feeling using of pedestrian crossing is shown in Figure 6.

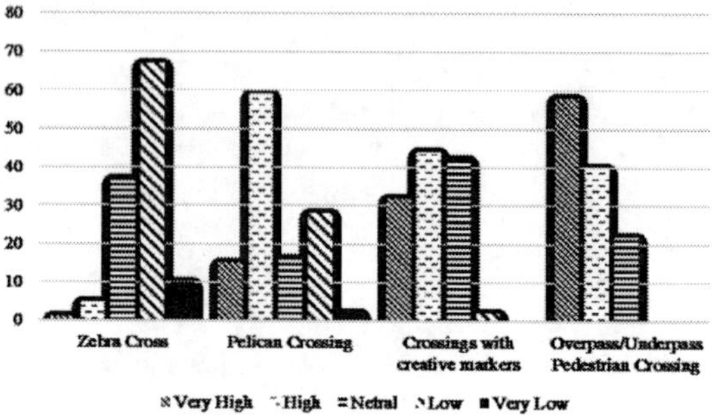

Fig. 6. Distribution of Secure Feeling Using of Pedestrian Crossing

Based on the results of data processing at each pedestrian crossing facility it is obtained for the perception of secure feeling itself. At the zebra cross facility the majority of respondents felt a low secure by 55.83%. Furthermore, pelican crossing facility has a perception of high secure by respondent is 49.17%. The result of high secure perception also obtained by pedestrian crossings with creative markers facility with 36.67%. Meanwhile, pedestrian crossing overpass/ underpass facility has very high secure perception with majority of respondents choose 48.33%. So that, they feel very safe and secure when crossing by using pelican cross and underpass/overpass crossing. Distribution of comfort feeling using of pedestrian Crossing is shown in Figure 7.

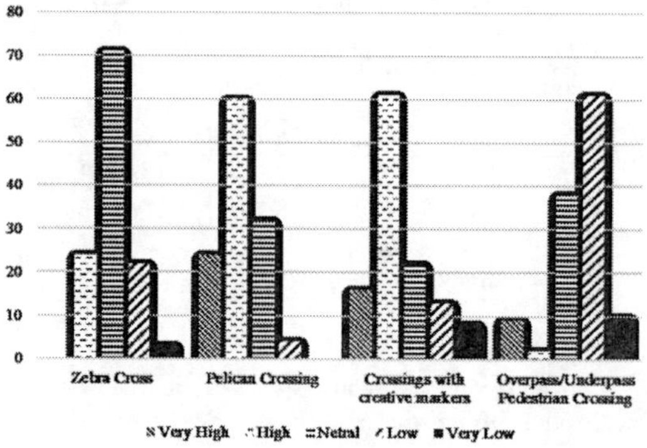

Fig. 7. Distribution of Comfort Feeling using of Pedestrian Crossing

Respondents tend to using crossing facilities if it can be achieved easily, free of obstacles and there is separator in the middle of the roas as waiting area.

The result of respondent's perception about comfort feeling for each types of pedestian crossing facility is the zebra cross facility the majority of respondents feel the normal by 59.17% while Pelican Crossing facility has a perception of comfort are high comfort by 50.00%. On the other hand, high comfort perception is obtained by crossings with creative markers facility with 50.83% while Crossing Overpass/Underpass facility has low comfort perception with majority of respondents by 50.83%.

The analysis for the questionnaire data is done by calculating the average value of the respondent's answer in every variable in the questionnaire. In addition, average values and mode values can be obtained by using the statistical analysis method on each survey zone. Table 9 show the average value and mode value for each types of pedestrian crossing.

Table 9. Average Value and Mode Value

Crossing Facility	Secure		Convenience	
	Average Value	**Value mode**	**Average Value**	**Value mode**
Zebra Cross	2.13	2	3.87	3
Pelican Crossing	4.10	4	4.03	4
Crossings with creative markers	3.60	2	4.03	4
Overpass/Underpass Pedestrian Crossing	4.30	4	2.57	2

To determine the types of crossing facility in each survey zone, the determination of pedestrian crossing facilities is used, thus that the types of pedestrian crossing facilities can be determined as shown in Table 10.

Table 10. Recommendation Crossing Facility

Time Survey	**P (Pedestrian/ hour)**	**V (vehicle/hour)**	**PV^2**	**Crossing Facility Recomendation**
Week day	203	7010	9.97E +09	Pelican Crossing
Weekend	562	5994	2.01E +10	Overpass

The analysis result, it is found that the recommendation of crossing facility on weekday is pelican crossing with PV^2 result of 9.97E + 09. Further recommendation of crossing facility at the weekend is pelican crossing with PV^2 result of 2.01E + 10.

5 Conclusion

Based on the analysis, followings are the conclusions:

- The type of crossing facility that is most familiar with the respondent is the type of Zebra Cross and Overpass/Underpass. Whereas the majority of respondents sometimes using pedestrian facilities.
- Respondents' perception on the secure of each types of crossing facility is to assume that overpass/underpass has the highest secure pedestrian crossing.
- Perception of comfort on each type of crossing facility, the majority of respondents choose pelican crossing and crossings with creative road marking with the highest comfort level compared with overpass/underpass is less comfortable.
- Based on the analysis PV^2 result, the recommendation of types of pedestrian crossing on Jalan Asia Africa Rd and Aceh Rd are pelican crossing on weekday and overpass/underpas at the weekend.

6 References

1. Pemerintah Republik Indonesia, 2009, *Undang-undang Nomor 22 Tahun 2009 tentang Lalu Lintas dan Angkutan Jalan*, Jakarta.
2. Korlantas-IRSMS Police, 2017. Jumlah Kecelakaan yang melibatkan pejalan kaki. Jakarta.
3. AASHTO, 2004. Guide for the Planning, Design, and Operation of Pedestrian Facilities. American Association of State and Highway Transportation Officials.
4. Hidayat. Edwin, 2012. Selection of Crossing Facilities Based on Critical Gap (Case Study Dharmawangsa Street, Surabaya). Cibinong.
5. Pemerintah Republik Indonesia, 2004, Undang-Undang Nomor 38 Tahun 2004 tentang Jalan, Jakarta.
6. Iswanto, Danoe. 2006. Jurnal Ilmiah Perancangan Kota dan Pemukiman Vol.5 No 1. Maret,2006. *Pengaruh Elemen – Elemen Pelengkap Jalur Pedestrian Terhadap Kenyamanan Pejalan Kaki (Studi Kasus : Penggal Jalan Pandanaran, Dimulai Dari Jalan Randusari Hingga Kawasan Tugu Muda)*, Semarang.
7. Direktorat Jenderal Bina Marga, 1995. Tata Cara Perencanaan Fasi- litas Pejalan Kaki di Kawasan Perkotaan. Jakarta.
8. Supangat, Andi. 2007, Statistika Dalam Kajian Deskriptif, Inferensial dan Nonparametik, Kencana perdana Group, Jakart
9. Santoso, Singgih. 2003. Statistk Deskriptif. Yogyakarta. Penerbit ANDI.
10. Putri, Pinkan Amelinda K dan Denny Zulkaidi. 2013. "Faktor yang Mempengaruhi Permint aan Terhadap Penyediaan Jalur Pejalan Kaki Pada Suatu Kawasan" dalam jurnal Perencanaan Wilayah dan Kota B

 https://doi.org/10.1051/matecconf/201814702009

SAPPK V2N3. Hlm: 595-603. Bandung: Sekolah Arsitektur, Perencanaan dan Pengembangan Kebijakan ITB.

11. Direktorat Jenderal Bina Marga. 1999. Pedoman Perencanaan Jalur Pejalan Kaki Pada Jalan Umum. Jakarta.

Model of Ciliwung River Flood Diversion Tunnel Using HEC-RAS Software

Joko Nugroho[1,*], *Indratmo* Soekarno[1], and *Dhemi* Harlan[1]

[1]Institut Teknologi Bandung, Faculty of Civil and Environmental Engineering, Ganesha 10, Bandung, Indonesia

Abstract. As a coastal city which lies in lowland area, Jakarta is prone to flooding. One major river which flow through Jakarta is Ciliwung River. There are alternatives to reduce flood risk, such as: river capacity improvement, existing natural reservoir and polder system improvement, upstream reservoir construction, city drainage improvement, flood channel construction and flood diversion. This paper presents capacity analysis of a proposed flood diversion of Ciliwung River to Cipinang River. Cipinang River has its downstream end at Eastern Flood Canal (Kanal Banjir Timur, KBT). This diversion is based on the available capacity of KBT. A 1-D numerical hydraulic model using HEC-RAS based on a proposed design is used to assess the performance of the diversion system in any combination of upstream and downstream boundary condition. Simulations were done for steady condition. The results show that capacity of the system can be achieved for certain condition at upstream and downstream boundary. The effects at the downstream reach of Ciliwung and Cipinang River due to the diversion are also obtained.

1 Introduction

Jakarta is a coastal city with 13 rivers pass through its area. Flood is one of main problem at Jakarta. Ciliwung River is one of the rivers which has a significant contribution on flooding of Jakarta region. Ciliwung River has a length of 117 km. The catchment area of Ciliwung River is 382.6 km^2.

Cipinang River has 48 km^2 catchment areas, approximately. This river has its downstream end at Kanal Banjir Timur (KBT, Eastern Flood Canal). KBT receives flow from Cipinang, Sunter, Buaran, Cakung and Jati Kramat rivers. Total catchment of KBT is 207 km^2, approximately.

Based on previous feasibility study, Kanal Banjir Timur has an extra capacity which could be provided for diverted partial discharge from Ciliwung River at any possible condition. It is expected to utilize this capacity to reduce flood risk at the downstream reach of Ciliwung River.

The purpose of research is to study the capacity of flood diversion tunnel by using HEC-RAS software and explore its application in this case.

2 Literature Review

Flood is frequently occurred in Jakarta, due to drainage problem and river flood. In the last twenty years there are at least five major floods event occurred, i.e. in 1996, 2002, 2007, 2013 and 2014. This flood caused severe inundation, public utilities and service disruptions and fatalities. For example, flood event in 2013 resulted in about 40 death tolls, 45,000 people evacuated, and severe economic lost [1]. The estimated economic loss due to flood in early 2007 was 566 million dollar, approximately [2].

HEC-RAS (Hydrologic Engineering Software-River Analysis System) was developed by Corps' Civil Works Hydrologic Engineering Research and Development Program. HEC-RAS is a one dimensional flow model which capable to perform calculation of steady flow, unsteady flow, movable boundary sediment and water quality analysis for a network of natural or constructed channel. HEC-RAS was developed based one dimensional energy, momentum and continuity equations. The equations are solved numerically by finite difference approach [3]. HEC-RAS is equipped with graphical user interface to facilitate data input and result visualization in tables and graphic [3]. HEC RAS has been applied in design of flood control [4], flood waves analysis [5] and river system modelling [6].

3 Ciliwung River – Kanal Banjir Timur (Eastern Flood Channel) Diversion System

The diversion system consists of inlet at Ciliwung River, transition, tunnel and outlet at Cipinang River. The connection between Ciliwung River, diversion tunnel and Cipinang River is presented in Figure 1. Inlet invert level elevation is at +9.150. Outlet invert elevation is at +7.00. The length of inlet transition is 53 meters. The horizontal length of diversion tunnel is 1,200 m, with

* Corresponding author: joko@si.itb.ac.id

longitudinal slope of 0.586% for the first 670 meter of tunnel and 0.711% for the rest. The tunnel consists of two circular barrels with inside diameter of 3.5 meter.

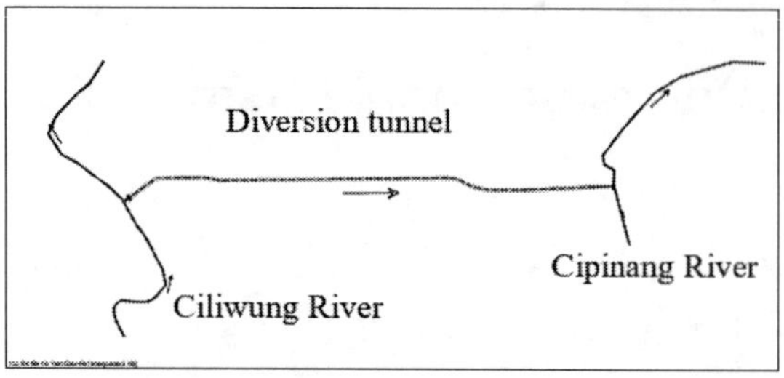

Figure 1. Layout of diversion system in HEC-RAS.

4 Methodology

In this study, hydrological data is obtained from previous study. River cross sections are available for Ciliwung and Cipinang river reach. In this study, normalized cross section is used in the modelling. Steady discharges applied for the model are presented in Table 1.

Table 1. Flood discharges and water level of Ciliwung and Cipinang River after normalization.

Return Period	Ciliwung		Cipinang	
(years)	Discharge (m^3/s)	Water Level (m)	Discharge (m^3/s)	Water Level (m)
Normal flow	n/a	n/a	7	6.481
2	301	11.086	84	7.951
5	400	11.929	106	8.233
10	452	12.342	114	8.330
25	509	12.773	122	8.425
50	547	13.048	127	8.483
100	570	13.227	130	8.517

The model consists of three main reach, i.e. Ciliwung river reach, diversion tunnel reach, and Cipinang river reach. The downstream of Cipinang river reach is connected with Kanal Banjir Timur (KBT, Eastern Flood Channel). Steady flow condition is assumed to obtained diversion capacity of the tunnel.

Typical normalized Ciliwung and Cipinang River are presented in Figure 2. Ciliwung and Cipinang River were modelled as open channel, and the diversion tunnel was modelled as open channel with lid. Example of a cross section with lid is presented in Figure 3.

The flow in the diversion tunnel may be in the state of flow with free surface, as flow in a closed conduit or flow with free surface at upstream part and closed conduit at the downstream. This will depend on the upstream and downstream flow boundary conditions. An example of flow condition in the diversion tunnel is presented in Figure 4.

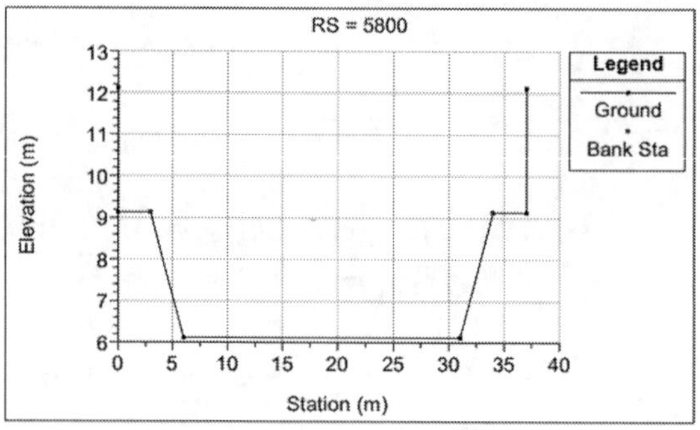

Figure 2. Typical cross section of Ciliwung River at upstream of the diversion inlet.

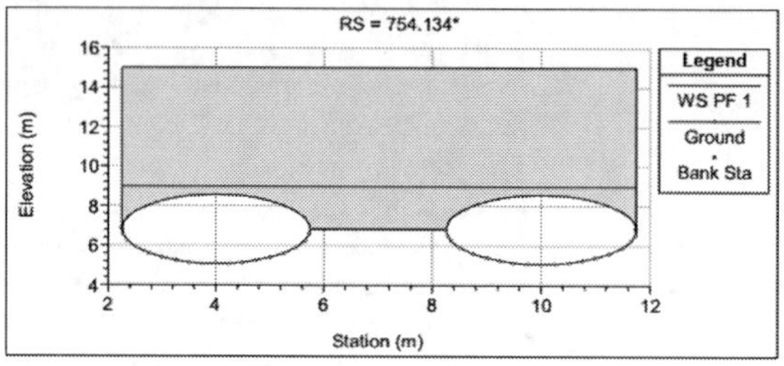

Figure 3. Cross sections of Ciliwung river diversion tunnel.

Boundary conditions of steady discharges were applied for upstream boundary of Ciliwung and Cipinang River. Boundary condition of normal depths was applied for downstream end of Ciliwung and Cipinang River.

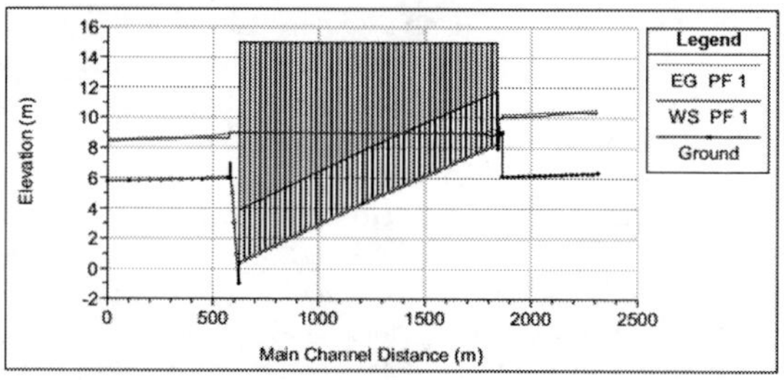

Figure 4. Long section of Ciliwung River (upstream)-diversion tunnel-Cipinang River (downstream).

For every scenario of discharge condition at Ciliwung and Cipinang River, the discharge of diverted flow was assumed first. After a calculation of a scenario in HEC RAS is completed, the energy gradient before and after every junction were checked. If energy level upstream and downstream of a junction is equal, the assumed diversion discharge is acceptable. Otherwise, another assumption of diverted discharge must be taken until equality of energy level at the junction is satisfied. The results of all scenarios are presented in Table 2.

5 Results and Discussion

Simulations were conducted for different combinations of Ciliwung and Cipinang river discharge. The discharge which flow through the diversion tunnel for these combinations are presented in Table 2.

Simulation showed that the discharges through diversion tunnel ranges from 10 to 75 m^3/s, or 5% to 13% of the respective discharges at Ciliwung River. Based on rating curve of Ciliwung River, the maximum reduction of flow depth is 0.52 m. This reduction will reduce risk of river bank overtopping, and also add more time for evacuation of people and their valuable properties.

Table 2. Flow discharge (m^3/s) through diversion tunnel.

				Cipinang								
	Return Period (year)			**Normal**	**2**	**5**	**10**	**20**	**25**	**50**	**100**	**100***
		Discharge (m^3/s)		7	84	106	114	122	127	130	135	152
			Elevation (+m)	6.481	7.951	8.233	8.330	8.425	8.483	8.517	8.574	8.740
Ciliwung	**1.01**	200	10.006	10	10	10	10	10	10	10	10	10
	2	301	11.086	38	39	38	38	38	38	38	38	35
	5	400	11.929	59	59	58	58	57	56	56	55	56
	10	452	12.342	66	66	64	63	62	62	61	61	58
	25	509	12.773	70	70	69	68	67	66	66	66	63
	50	547	13.048	73	73	71	71	70	69	69	69	66
	100	572	13.227	75	74	73	72	72	71	71	70	66

*) : After Cipinang river improvement.

6 Conclusion

System of diversion from Ciliwung River to eastern flood canal was successfully modeled and simulated in HEC-RAS. According to the simulation result, the diversion capacity ranges from 10 to 75 m^3/s. Reduction of flow at the downstream of diversion point will result in reduction of flood water level and reduce inundation area. Hence the diversion tunnel will reduce impact of flood along the downstream reach of Ciliwung River and can be proposed as alternative in flood disaster mitigation.

However, the presented results are limited to 1 D assumption. Three dimensional flow eventually will affect flow condition at inlet and outlet where change of flow direction occurred. Further study is required to accommodate these considerations.

The authors thank to Water Resources Engineering Research Group, Faculty of Civil and Environmental Engineering, Institut Teknologi Bandung for the support in this research.

References

1. Moe, I.R., Kure, S., Farid, M., Udo, K., Kazama, S., Koshimura, S., *Numerical Simulation of Flooding In Jakarta and Evaluation Of A Countermeasure To Mitigate Flood Damage*, Journal of Japan Society of Civil Engineers Ser. G, **71**, No. 5, I_29-I_35 (2015)
2. Kementerian Negara Perencanaan Pembangunan Nasional (BAPPENAS), *Laporan Perkiraan Kerusakan dan Kerugian Pasca Bencana Banjir Awal Februari 2007 di Wilayah JABODETABEK (Jakarta, Bogor, Depok, Tangerang, dan Bekasi)*, (2007)
3. United States Army Corps of Engineers (USACE): *Hydraulic Reference Manual v.4.1. Hydraulic Engineering Center*, USA, (2010)
4. Modak, S., Nagarnaik, P., *Flood Control And Prediction of Flood Using HEC-RAS – A Review*, International Journal of Science and Research, ISSN (online): 2319-7064, **6** - Issue 3, (2017)
5. Kovacs, S., Kiss, A., Szekers, J., *Experiences In Application Of HEC-RAS Model Under Circumstances Of Flood Waves*, Hazards, Vulnerability and Mitigation Measures, 47–58. J. Schanze et al. (eds.), Flood Risk Management, Springer (2006)
6. Kowalczuk, Z., Świergal, M., Wróblewski, M., *River Flow Simulation Based on the HEC-RAS System*, Advances in Intelligent Systems and Computing, **635**, (to be published)

Erosion index formulation with respect to reservoir life in the upper Citarum watershed

Bakhtiar[1,*]

[1]Civil Engineering Department, Sangga Buana University, Bandung, Indonesia

Abstract. This study aimed to formulate erosion index in the upper Citarum watershed with respect to the Saguling reservoir life. Soil and Water Assessment Tool model was incorporated to simulate hydrological processes in the catchment. From the calibration and validation results, the model is considerably of good performance. The simulated sediment inflow at Nanjung outlet was then extrapolated to determine the sediment inflow into the reservoir. The study revealed that the average value of sediment inflow into the reservoir is 29.24 tonnes/ha/year just below the tolerable erosion limit of 30 tonnes/ha/year assumed by Hammer (1981). It was also found that the relationship between sediment yield and sediment inflow is non linear. Erosion index is formulated as the ratio between the mean annual sediment yield generated in the watershed and the mean annual sediment yield that leads dead storage to be full in the designated life of the reservoir. Erosion index equals to 1.0 indicates that the dead storage will be full in the designated life of the reservoir. A classification of erosion index can be subsequently be made based on erosion index and reservoir life relationship.

1 Introduction

The upper Citarum watershed covers the fountain of the Citarum river down to Saguling with an area of 1744.98 km2 as a part of the Citarum watershed, one of the biggest watershed in West Java. The observed monthly rainfall in 2008 is averaged to be 172 mm with the total annual rainfall of 2,070 mm. The topography of the region is dominated by mountains along the borders and a broad plain in the middle. The sedimentation rate in the Upper Citarum watershed in the last decade is reported to be increasing. It is indicated by the mean sedimentation rate of 4,048,132 tonnes per year (PT Indonesia Power [1]). This fact is due to the ecosystem degradation and the decreasing forest area over the watershed and the catchment of the Saguling reservoir. The aim of this study is to analyse the effect of rainfall and land use scenarios on sediment yied in the Upper Citarum watershed.

2 Method

In this study the erosion and sedimentation behaviour is modeled using ArcSWAT tool (Arc GIS Interface for SWAT Model). SWAT model (Soil and Water Assessment Tool) is a semidistributed watershed model with ArcGIS interface delineating subbasins and river networks from digital elevation model and calculating the daily water balance from meteorological, soil and land use data.

The erosion modeling in SWAT is based on the following Modified Universal Soil Loss Equation (MUSLE) equation (Neitsch [2]):

$$sed = 11.8 \cdot \left(Q_{surf} \cdot Q_{peak} \cdot area_{hru}\right)^{0.56} \cdot K_{USLE} \cdot C_{USLE} \cdot P_{USLE} \cdot LS_{USLE} \cdot CFRG \quad (1)$$

where:

sed	=	sediment yield (tonnes)
Q_{surf}	=	surface runoff (mm/ha)
Q_{peak}	=	peak runoff rate (m^3/s)
$area_{hru}$	=	area of the HRU (hydrologic response unit (ha)
K_{USLE}	=	USLE soil erodibility factor (0.013 tonnes m^2 hr/[m^3 tonne cm])
C_{USLE}	=	USLE cover and management factor
P_{USLE}	=	USLE support practice factor
LS_{USLE}	=	USLE topographic factor
$CFRG$	=	coarse fragment factor

3 Soil and water assessment tool (SWAT)

SWAT models chosen for this study given the ability to simulate the processes of land management on a watershed-large watershed. SWAT Model is physically based simulation model (physically based) that was developed to simulate the processes of landscapes and watersheds for continuous time with a high degree of spatial details that allow the watershed was divided into sub-watershed. Each sub-watershed is divided again into some combination of land use and soil type called the hydrologic response unit (HRU) based on the percentage

*bakhtiarab8@gmail.com

https://doi.org/10.1051/matecconf/201814703002

threshold used in selecting land use and soil types (Arnold et al. [3]).

4 Data preparation and analysis

The data prepared in the simulation process is data in the form of data hidroklimatologi rainfall, evaporation, climate and discharge, as well as physical data of Citarum upstream in the form of topographic maps, land use maps, river networks, the type of soil and sediment there. More detailed analysis and clear of each data will be described in the following. Data input for the model preparation derived from various sources are shown in Table 1 below.

Table 1. Model input data and corresponding data sources.

Type of data	Scale	Source	Description
Digital elevation model	90 m resolution	Shuttle Radar Topographic Mission (SRTM)	Elevation, length and slope
River networks (hydrography)	Projection: UTM/WG S84	West Java Water Resources Agency	
Geology	Zone 48 S	Soil and Agroclimate Research Centre	Soil physical properties (bulk density, texture, organic matter, hydraulic conductivity, etc.)
Landuse/ Landcover	1:250.000	West Java Spatial Planning Agency	Landuse classification
Rainfall	Projection: UTM/WG S84	Saguling reservoir authority	
Streamflow and sediment	Zone 48 S	Saguling reservoir authority	

5 Model setup

After all the input data is prepared and added to the SWAT database, SWAT models hereafter devised a new project with ArcSWAT, i.e. SWAT interface and ArcGIS. The main steps of the SWAT procedure include delineating sub-watershed and HRU and writing all input files. Figure 1 below shows the steps SWAT modeling using ArcSWAT (Winchell et al, 2009 [4]).

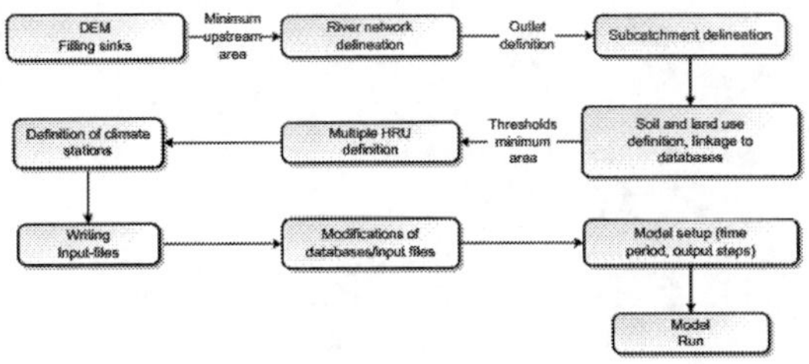

Fig. 1. Steps for setting up a new SWAT project with ArcGIS User Interface.

The first step required to build a model of SWAT is to define parameters related to elevation as: elevation above sea level, slope aspect, network flow, distance to the nearest river, and divide the watershed into sub-sub-watershed. DEM data with a resolution of 90 m were obtained from the Shuttle Radar Topographic Mission (SRTM) is used in this stage.

ArcSWAT automatically delineates the boundaries of the watershed and river network based on DEM data. The number of sub-watershed is determined based on the level of accuracy we have set previously. The higher level of accuracy, the greater the number of sub-watershed generated. In this study the number of sub-watershed produced are 92 sub-watersheds (see Figure 2).

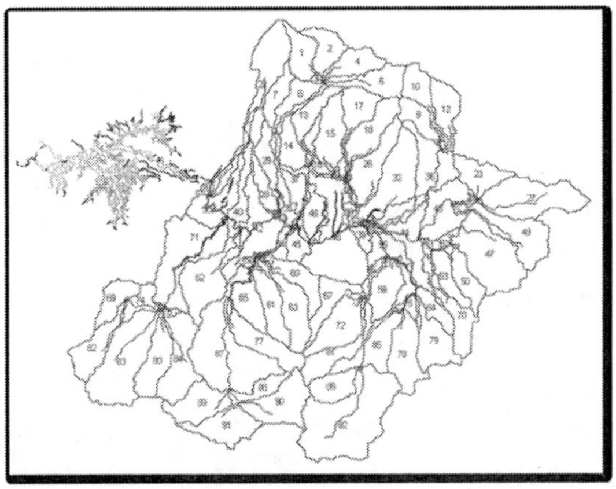

Fig. 2. Watershed delineation using ArcSWAT.

6 Model calibration

Calibration was perfomed on water yield, average annual water inflow, and annual sediment inflow at the watershed outlet i.e. Nanjung gauge station (see Figure 3) for the period of 11 years i.e from 1987 to 1997. The model performance is then evaluated on their statistical parameters i.e. coefficient of correlation (r), coefficient of determination (R2), model efficiency (ME), and index of agreement (IA) (Spiegel, 1961). Table 2 shows the observed and simulated mean annual values as well as their respective model performance.

https://doi.org/10.1051/matecconf/201814703002

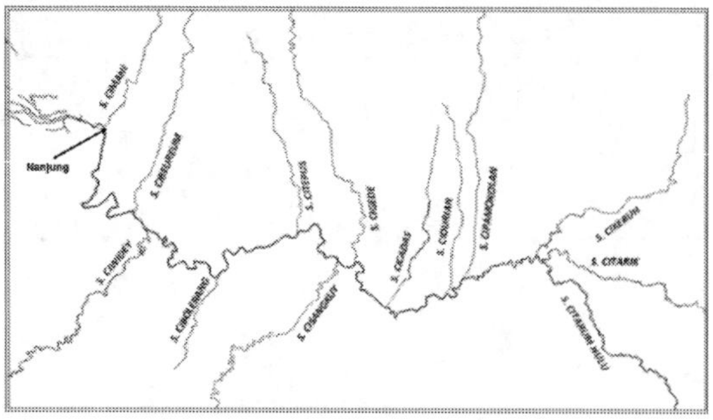

Fig. 3. Location of Nanjung gauging station.

Table 2. Summary of the calibrated annual values.

Parameter	Water yield	Water inflow	Sediment inflow
Unit	mm	m^3/s	tonnes
Observed	1,503	82.77	3,862,941
Simulated	1,513	83.57	3,678,091
r	0.938	0.941	0.668
R^2	0.876	0.886	-0.497
ME	0.874	0.878	-0.497
IA	0.968	0.969	0.755

7 Model validation

Validation was perfomed on the calibrated model using the consecutive 11 year data i.e. from 1998 to 2008. Based on the validation on the annual observed values, it can be seen that the model is considerably of good performance as shown by the coefficient of correlation, coefficient of determination, model efficiency and index of agreement which are close to 1 except for the sediment inflow for the same reason as the case of sediment inflow calibration.

Table 3. Summary of the validated annual values.

Parameter	Water yield	Water inflow	Sediment inflow
Unit	mm	m^3/s	tonnes
Observed	1,628	89.61	5,668,783
Simulated	1,626	89.74	5,328,818
r	0.987	0.988	0.742
R^2	0.974	0.975	-70.500
ME	0.974	0.976	-70.565
IA	0.993	0.994	0.272

8 Determination of sediment inflow into the reservoir

Since the inflow of sediment being modeled by SWAT only to outlets Citarum upstream, ie heading suspect water Nanjung, then the data volume of sediment observations from Saguling to be used for calibration and validation of models must be transformed into data inflow of sediment at Nanjung assuming a volume of sediment proportional of the watershed area as follows:

$$sed^*_{out(obs)} = \frac{A_{DAS}}{A_{DTW}} \cdot V_{sed} \cdot \gamma_{sed} \qquad (2)$$

where:

$sed_{out(obs)}^*$ = observed sediment inflow at Nanjung (tonnes/year)

A_{DAS} = area of the watershed upstream of Nanjung = 1,745 km^2

A_{DTW} = area of Saguling catchment = 2,313 km^2

V_{sed} = observed annual sediment entering the reservoir (m^3)

γ_{sed} = initial specific gravity of sediment = 1,229 tonnes/m^3

The sediment inflow into the reservoir is obtained from the inflow of sediments in the watershed outlet resulted from the simulation model SWAT (sedout*) using the following equation:

$$sed_{out(sim)} = \frac{A_{DTW}}{A_{DAS}} \cdot sed^*_{out(sim)} \qquad (3)$$

where:

$sed_{out(sim)}$ = simulated sediment inflow into the reservoir (tonnes/year)

A_{DAS} = area of the watershed upstream of Nanjung = 1,745 km^2

A_{DTW} = area of Saguling catchment = 2,313 km^2

$sed_{out(sim)}^*$ = simulated sediment inflow at the catchment outlet (tonnes/year)

The simulated sediment inflow at Nanjung obtained from the SWAT model simulation is then used to determine the sediment inflow into the Saguling reservoir using Equation 3. It is found that the average value of sediment inflow into the reservoir is 29.24 tonnes/ha/year which is close to the tolerable erosion limit of 30 tonnes/ha/year assumed by Hammer (1981) in Arsyad [5].

9 Relationship between sediment yield and sediment inflow

The simulated sediment yields and sediment inflows are tabulated in the following table.

Table 4. Relationship between sediment yield and sediment inflow.

Rainfall (R)	**Sediment yield (Sed)**	**Water inflow (I_w)**	**Sediment inflow (Sed_{out})**
1,021	165.90	1,750,878,720	3,256,757
1,062	170.66	1,841,387,040	3,112,277
1,174	197.45	2,020,196,160	3,319,056
1,236	210.23	2,140,663,680	3,808,166
1,359	202.14	2,440,255,680	4,514,658
1,403	213.64	2,521,618,560	4,444,406
1,423	247.35	2,364,569,280	4,641,906
1,452	177.75	2,586,582,720	4,481,520
1,494	201.44	2,476,206,720	4,575,631
1,507	257.19	2,626,633,440	5,328,516
1,510	214.64	2,630,733,120	4,702,879

Based on Table 4 above, the relationship between sediment yield and sediment inflow can be formulated as:

$$Sed_{out} = 20.394 Sed^{1.009} \quad (4)$$

where:

Sed_{out} = mean annual sediment inflow (tonnes/year)

Sed = mean annual sediment yield (tonnes/ ha/year)

10 Determination of reservoir life

Reservoir life can be predicted by the trap efficiency method. Gunner Brune (1953 in the Bureau of Reclamation [6]) suggests that the trap efficiency depends on the ratio between the reservoir storage capacity (C) and the annual water inflow (I_w). The trap efficiency of a reservoir reduces with time since the capacity of the reservoir is reduced by the accumulated sediment. To determine the trap efficiency curve equation median Brune (Bureau of Reclamation [6]) in Figure 4 is used.

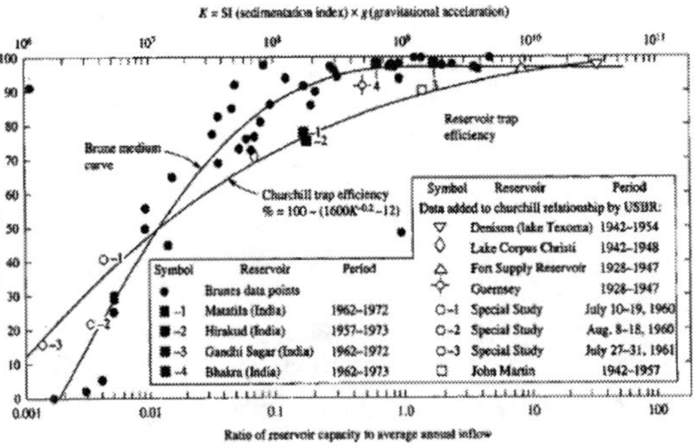

Fig. 4. Trap efficiency curves (Churchill, 1948; Brune, 1953).

The reservoir life is achieved when the volume of the dead storage has been full of sediment that blocks the intake.

For example, according to Table 4 in the previous section a mean annual rainfall of 1,062 mm results in the sediment yield of 170.66 tonnes/ha, the sediment inflow of 3,112,277 tons/year and the average water inflow of 58.39 m³/s which is equal to:

$I_w = 58.39 \cdot 60 \cdot 60 \cdot 24 \cdot 365 = 1{,}841{,}387{,}040\ \text{m}^3\text{/year}$

According to data from PT Indonesia Power Service Business Unit Saguling [1] the total capacity of the reservoir is 875 million m³ with a 167.689 million m³ dead storage. Assuming that the sediment in the reservoir is located in a continuously submerged zone and consists of 15% clay, 35% silt and 50% sand, then using Equation (5) the initial specific gravity of sediment can be determined as follows:

$$\gamma_{sed} = \gamma_c p_c + \gamma_m p_m + \gamma_s p_s \quad (5)$$

where:

γ_{sed} = specific gravity of sediment (tonnes/m³)

p_c = percentage of clay = 15%

p_m = percentage of silt = 35%

p_s = percentage of sand = 50%

γ_c = specific gravity of clay = 0.416 tonnes/m³

γ_m = specific gravity of silt = 1.120 tonnes/m³

γ_s = specific gravity of sand = 1.550 tonnes/m³

Hence:

$\gamma_{sed} = 0.416 \cdot 0.15 + 1.120 \cdot 0.35 + 1.550 \cdot 0.50 = 1.229\ \text{tonnes/m}^3$

Furthermore, the average specific gravity of sediment after t years of reservoir operation can be determined using the following equation:

$$\gamma_{sed,t} = \gamma_{sed,o} + 0.4343 \cdot K \left[\frac{t}{t-1} (\ln t) - 1 \right] \quad (6)$$

where:

$\gamma_{sed,t}$ = average specific gravity of sediment after t years of reservoir operation

$\gamma_{sed,o}$ = initial specific gravity of sediment = 1.229 tonnes/m³ (from the previous calculation)

K = constant which is based on the type of reservoir operation and sediment size analysis obtained by the following equation:

$$K = K_c \cdot p_c + K_m \cdot p_m + K_s \cdot p_s \quad (7)$$

where:

K_c = constant for clay = 0.256 tonnes/m³

K_m = constant for silt = 0.091 tonnes/m³

K_s = constant for sand = 0

Hence:

$K = 0.256 \cdot 0.15 + 0.091 \cdot 0.35 + 0 \cdot 0.50 = 0.070\ \text{tons/m}^3$

Entering $\gamma_{sed,o}$ = 1,229 tonnes/m³ and K = 0.070 tonnes/m³ into Equation (6), the average specific gravity of sediment after t years of reservoir operation can be determined as follows:

$$\gamma_{sed,t} = 1.229 + 0.4343 \cdot 0.070 \left[\frac{t}{t-1} (\ln t) - 1 \right] \quad (8)$$

Reservoir life can subsequently be determined implicitly using the following equation:

$$DEAD_{t=n} = \overline{sed}_{out} \sum_{t=1}^{t=n} \frac{0{,}96^{-0{,}25 \log\left(\frac{C_o - DEAD_{t-1}}{I_v}\right)}}{\gamma_{sed}} \quad (9)$$

where:

$DEAD_{t=n}$ = specific gravity of sediment (tonnes/m³)

The reservoir life can eventually be determined implicitly using Equation (9). By interpolation the reservoir life is predicted to be 70.49 years.

11 Erosion index

Erosion index is a function of sediment yield and reservoir life. Sediment yield generated by the SWAT model simulation is the total sediment that consists of bed load and suspended load. Reservoir sedimentation is only influenced by suspended load and not affected by bed load. It is for this reason, the correlation between reservoir life and sediment yield becomes less significant as compared to the effect of sediment inflow in the dead storage. This is proved by the evidence that the value of the coefficient of correlation (r) is -0.80 which is far from 1.0. The relationship between sediment yield and reservoir life is shown in Figure 5 below.

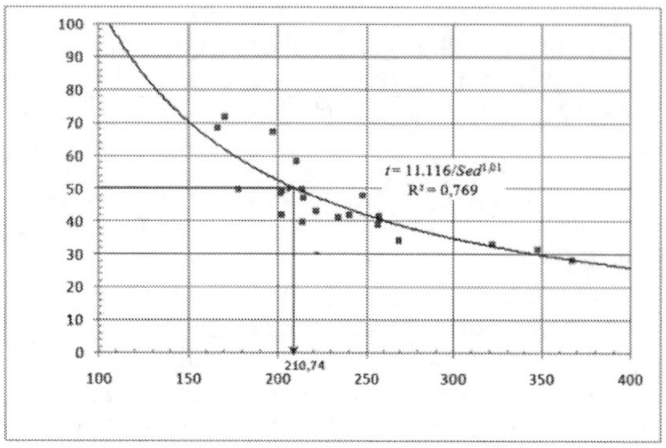

Fig. 5. Relationship between sediment yield and reservoir life.

Erosion index (*IE*) is formulated as the ratio between the mean annual sediment yield generated in the watershed (*Sed*) and the mean annual sediment yield that leads dead storage to be full in the designated life of the reservoir (Sed_t). Erosion index equal to 1.0 indicates that the dead storage will be full in the designated life of the reservoir. In this study the designated life of the reservoir (*u*) is set to be 50 years. Thus, in this case, erosion index equal to 1.0 refers to the sediment yield that causes the dead storage to be full in the period of 50 years.

Hence, for this case, erosion index (*IE*) is formulated as the ratio between the mean annual sediment yield that takes place in the watershed (*Sed*) and the mean annual sedimen yield that causes the dead storage to be full in the period of 50 years (Sed_{50}).

According to the curve in Figure 5 above, the reservoir life (*u*) of 50 years is correlated with the sediment yield of 210.74 tonnes/ha/year. Therefore, erosion index can be formulated as:

$$IE = \frac{Sed}{Sed_u} = \frac{Sed}{Sed_{50}} = \frac{Sed}{210.74} \qquad (10)$$

where:

IE = erosion index

Sed = actual mean annual sediment yield (tonnes/ha/year)

Sed_u = mean annual sediment yield that leads dead storage to be full in the designated life of the reservoir (tonnes/ha/year)

Sed_{50} = mean annual sediment yield that leads dead storage to be full in 50 years = 210.74 tonnes/ha/year

Erosion index values for various sediment yield are shown in Figure 6.

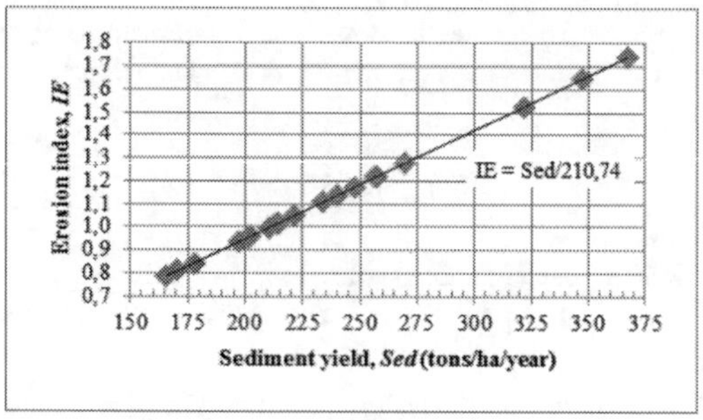

Fig. 6. Erosion index determination based on sediment yield and reservoir life.

The relationship between erosion index and reservoir life is shown in Figure 7.

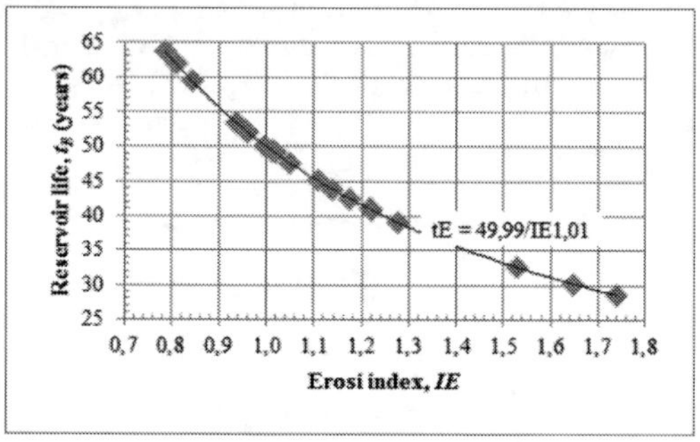

Fig. 7. Relationship between erosion index and reservoir life.

Based on the relationship between erosion index and reservoir life above, a classification of erosion index can be made as shown in Table 5 below.

Table 5. Erosion index classification.

Sediment yield, *Sed* (tonnes/ha/year)	Erosion index, *IE*	Category	Reservoir life, t_E (*years*)
$Sed < 210{,}74$	$IE < 1$	Non critical	$t_E > 50$
$Sed = 210{,}74$	$IE = 1$	Critical	$t_E = 50$
$Sed > 210{,}74$	$IE > 1$	Very critical	$t_E < 50$

12 Comparison between erosion index and erosion hazard index

If the values of erosion index found in this study are compared to the values of erosion hazard index (Hammer, 1981 in Arsyad [5]), the same sediment yield will fall into a different category (see Table 6). The difference resulted because each index has different ranges for their respective categories.

Table 6. Erosion hazard index (*IBE*) assessment criteria.

Erosion hazard index (IBE)	Category
IBE ≤ 1.0	Low
1.0 < IBE ≤ 4.0	Moderate
4.0 < IBE ≤ 10.0	High
IBE > 10.0	Very high

The comparison between erosion index and erosion hazard index is shown in Figure 8.

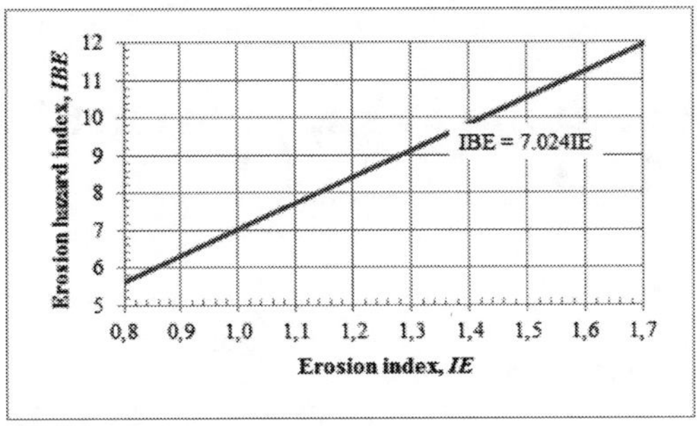

Fig. 8. Relationship between erosion index and reservoir life.

13 Conclusion

From the previous discussion it can be concluded that:

1. The mean sediment inflow into Saguling reservoir is 29.24 tonnes/ha/ year which is close to the tolerable erosion rate of 30 tonnes/ha/year (Hammer, 1981).
2. The relationship between sediment yield and sediment inflow is nonlinear with an exponential equation.
3. Erosion index (IE) is formulated as the ratio between the mean annual sediment yield generated in the watershed (Sed) and the mean annual sediment yield that leads dead storage to be full in the designated life of the reservoir (Sedt).
4. Based on the relationship between erosion index and reservoir life, erosion index can be classified into three categories i.e. non critical, critical and very critical.
5. If the values of erosion index found in this study are compared to the values of erosion hazard index, the same sediment yield will fall into a different category. The difference resulted because each index has different ranges for their respective categories.

14 Recommendation

Based on this study it is recommended that:

1. Regarding a great number of parameters involved in the modelling process, it is recommended to conduct further research on those parameters to obtain the appropriate values in order to obtain acceptable results.
2. In Indonesia modelling using SWAT model is a new thing and scarce. It is for this reason, the implementation of SWAT model into the future research will be very much welcomed.
3. The above resulted formula are only valid for the location under consideration. However, the methods can be applied to produce a similar formula for other locations.

References

1. The Saguling Service Business Unit, Monitoring Report of The Saguling Reservoir Sedimentation (2008)
2. S. Neitsch, J. Arnold, J. Kiniry, J. Williams, Soil and Water Assessment Tool: Theoretical Documentation Version 2005, 225 (2005)
3. J. G. Arnold, R. Srinivasan, R. S. Muttiah, J. R. Williams, Large Area Hydrologic Modeling and Assessment Part I: Model Development, Journal of the American Water Resources Association, **34(1)**, 73-89 (1998)
4. M. Winchell, R. Srinivasan, J.M. Di Luzio, ArcSWAT 2.3.4 Interface for SWAT2005: User's Guide (2009)
5. S. Arsyad, Konservasi Tanah dan Air (1989)
6. Bureau of Reclamation, Erosion and Sedimentation Manual (2006)

Development Land Erosion Model Using Model Builder GIS (Case Study: Citepus Watershed)

Dina PA Hidayat[1,*], *Sih* Andajani[2]

[1]Laboratorium Keairan, Faculty of Civil Engineering and Planning, Trisakti University, Jakarta, Indonesia
[2]Laboratorium Keairan, Faculty of Civil Engineering and Planning, Trisakti University, Jakarta, .Indonesia

Abstract. Land erosion is the impact of increasing runoff discharge and land use conversion to impervious areas. Land erosion usually calculated by formula called USLE (Universal Soil Loss Equation) then modified as MUSLE (Modified Universal Soil Loss Equation). These formula calculate average annual soil loss in tons/areas depends on rainfall erosivity (R), soil erodibility factor (K), topographic factor (LS), cropping and conservation factor (CP). GIS (Geographic Information System) is a system designed to capture, manipulate, and analyze spatial/geographic data. There are some tools related water resources analysis in ArcGIS such as: watershed analysis and also have a tools for user to create their own model called model builder. This research was aim to create a model to calculate land erosion using MUSLE formula by model builder in ArcGIS. The output for this research is the model which can be used to calculate annual soil loss in watershed area based on GIS systems. For the model trial and case study, we use Citepus watershed located on Bandung West Java, that has 5 river branches: Cibogo, Cikakak, Cilimus, Cipedes and Ciroyom. As the result of the model, the value of average annual soil loss in Citepus watershed can be calculated automatically by developed model.

1 Introduction

1.1. Background

The erosion process of soil particle on the surface watershed usually called land erosion. Sediment discharge which is resulted by land erosion is the one of sediment input in the river beside bank erosion. As a result of runoff from rainfall, soil particles on the surface of a watershed can be eroded and transported through the processes of sheet, rill, and gully erosion. Once eroded, sediment particles are transported through a river system and are eventually deposited in reservoirs, in lakes, or at sea.

Soil erosion is a serious problem from agricultural intensification, land degradation and other anthropogenic activities. Assessment of soil erosion is useful in planning and conservation works in a watershed or basin [1]. Beside that, flood and inundation are getting more complex especially in urban areas, and it can be worst by water pollution. Flood and inundation has occurred because of incapability of drainage channel/river to collect runoff flow. This lack of capacity can caused by sedimentation in river bank/drainage channel, that sourced by land erosion.

Since the last century, soil erosion accelerated by human activities has become a serious environmental problem. Soil erosion risk varies from case to case depending on the configuration of the watershed (topography, shape), the soil characteristics, the local climatic conditions and the land use and management practices implemented [2].

Because of the risk of soil erosion, the estimation of soil erosion rate is very important to predict the value of annual soil erosion in a watershed. After soil erosion is predicted, this information is needed to provide pre-requisite information for effective watershed management, including soil conservation strategies. Appropriate watershed management plans for sustainable development need reliable long-term soil loss information at various parts of the watershed [3].

Soil erosion prediction technology began over 70 years ago when Austin Zingg published a relationship between soil erosion (by water) and land slope and length,followed shortly by a relationship by Dwight Smith that expanded this equation to include conservation practices [4]. The general equation of land erosion is Universal Soil Loss Equation (USLE) (Wischmeier and Smith, 1965). This equation is the multiplication of rainfall factor, soil erodibility factor, topography factor, cropping management factor and erosion control factor. There are continued research of land erosion equation such as: Revised Universal Soil Loss Equation (RUSLE) (Wischmeier and Smith, 1978) and Modified Universal Soil Loss Equation (MUSLE) (Williams, 1975).

In the last decade, good amount of work has been done on soil erosion assessment in large catchments and at a regional scale by using lumped approaches and very few

* Corresponding author: dinaparamitha06@gmail.com or dina.hidayat@trisakti.ac.id

on micro-units and gridcell basis [3]. Today, GIS (Geographical Information Systems) has been used for engineering field, and many research used GIS for predict/estimate soil erosion on watershed.

This research aim to calculate soil annual loss rate by MUSLE equation and GIS application. GIS application on this research are generating Citepus watershed, MUSLE factor calculation and developing annual soil loss model by model builder GIS.

This research use Citepus watershed as the case study, located on Bandung, West Java, Indonesia (7° 52' 15" – 7° 57' 50" S and 107° 34' 15" – 107° 36' 45"E). The catchment area of Citepus watershed is 1820 ha, consist of 8 sub district (Sukasari, Sukajadi, Cicendo, Andir, Bojongloa Kidul, Astana Anyar and Bojongloa Kaler). The Citepus watershed has elevation between 670 meters until 950 meters above sea level and different slope between north and south area. In the north area of Citepus watershed, it is steep area, has 7% slope. On the other hand, the middle and south area of Citepus watershed is relatively flat area, has 0-6% slope. Rainy season dominates this area with average daily precipitation for 10 years (1997-2007) of 95mm and large variation for each year. Citepus River is one of the main river, flow to Citarum River, has 10 km of length and 5 river branches (Ciroyom,Cikakak,Cilimus,Cibogo and Cipedes).

One of factor that cause flood in Bandung is sedimentation at the section of the river. Sedimentation cause the less of river capacity for retaining stream discharge. The area is covered by housing area 76%, funeral area 1%, green area 2%, rice field 1% and others 21%. Furthermore, less of river capacity can cause overtopping in some location of Citepus river section, such as: overtopping in Pagarsih street section that has 5 hours duration and 0.8 ha flooding area. As a result of identification at the section, there is sediment stack that causes river width changing from 12 meters to 6 meters.

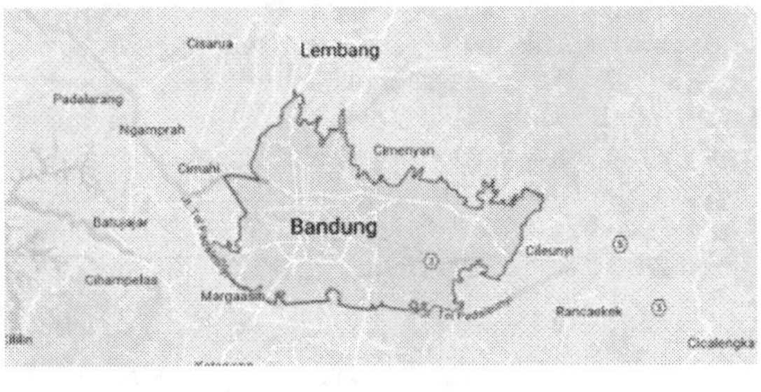

Fig. 1. Map of Location Study *(maps.google.com)*

1.2. Objectives

This research is aim to: develop MUSLE equation model with model builder GIS and calculate soil annual loss for Citepus watershed by developed model.

1.3 Scope

This research located in Citepus Watershed with Citepus River as the main river that has 5 branch (Cilimus, Cikakak, Cibogo, Ciroyom and Cipedes). GIS for this research is used for watershed boundary analysis, MUSLE factor analysis and develop MUSLE equation with model builder GIS

2 Materials and Method

2.1. Methodology

This research was done by following step:

a. Data collection

Raw data was collected consist of: hydrology data, land use, topography, conservation data,etc. All data sourced from related government agency (such as: bakosurtanal, Ministry of Public Works), books, journal, paper and related project.

Hydrology Data

Hydrology data was used for estimate rainfall erosivity factor. This research used Cemara station as the sources of rainfall data.

Topography Map

Digital Elevation Model (DEM) is one of the important parameters for soil erosion assessment. Notable uncertainties are observed in this study while using three high resolution open source DEMs [5]. For topography data, this research used SRTM with open source download facility.

b. MUSLE factor mapping on GIS

For this research, GIS use for generate watershed boundary and analyze MUSLE factor (soil erodibility, topography factor and cropping factor). Watershed boundary was generating by watershed menu on GIS toolbox, with SRTM (Shuttle Radar Topography Mission) as the input.

c. Develop MUSLE equation using Model Builder GIS

After all factor was generated, develop land erosion model for calculating soil loss annual rate using Model Builder GIS. The model can be used for another case, not only for Citepus watershed.

d. Calculate annual soil erosion rate by developed model and classify erosion rate

After model was generated on model builder GIS, try to running the model for Citepus Watershed data, then classify Citepus erosion rate by following table (Wiharta et al,1997).

Table 1. Erosion Rate Classification

Annual Soil Loss Rate (Tons/ha/yr)	Erosion Rate Classification
0-15	Very Light
15-60	Light
60-180	Medium
180-480	Heavy
>480	Very Heavy

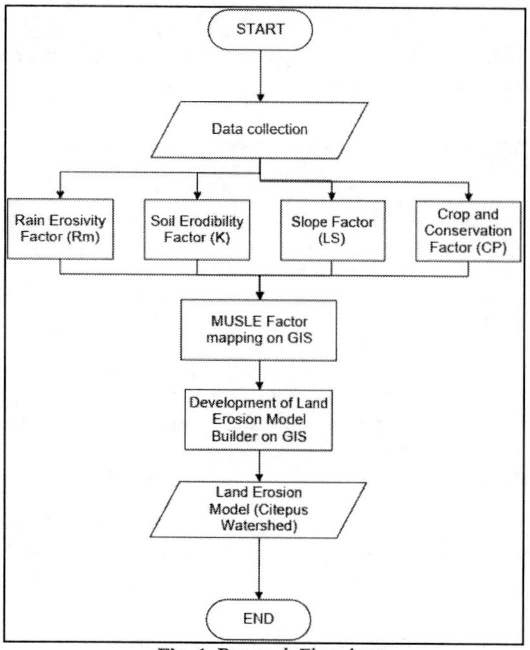

Fig. 1. Research Flowchart

2.2. Land Erosion Modelling

Soil erosion is an important item for consideration in the planning of watershed development works. It reduces not only the storage capacity of the downstream reservoirs but also deteriorates the productivity of the watershed. Erosion involves the detachment, transport and deposition of soil particles and aggregates. Among available soil erosion and sediment yield models, the universal soil loss equation (USLE), the revised version, Revised universal soil loss equation (RUSLE), and its modified version (MUSLE) are widely used in hydrology and environmental engineering for computing the potential soil erosion and sediment yield [6].

Modified Universal Soil Loss Equation (MUSLE) is an empirical model, which gives average soil loss produced by rainfall events (Williams, 1975). This equation used runoff energy, soil erodibility, slope length and steepness vegetative cover and management practices as factors. MUSLE (Modified USLE) equation was developed using rainfall erosivity factor as the different factor than USLE.

$$A = R \, x \, K \, x \, LS \, x \, C \, x \, P \quad (1)$$

A = long term average annual soil loss, ton/acre/year, ton/ha/year
Rm = rainfall erosivity
K = soil erodibility
LS = topography factor
C,P = cropping and conservation factor

The rainfall erosivity factor was developed by Williams dan Berndt (1977), consist of runoff volume and peak discharge. Peak discharge is calculated using rasional equation which assume that rainfall has uniform intensity on entire watershed.

$$R = 11{,}8 \, (Qv \, x \, Qp)^{0{,}56} \quad (2)$$

Qv = runoff volume (m^3)
Qp = peak discharge (m^3/sec)

For the model accuracy, MUSLE increases sediment yield prediction accuracy and as well as it eliminates the need for delivery ratios [6]. The MUSLE equation has been used previously by many researchers (Tripathi et al. 2001) and, in some cases, the equation was subjected to different modifications. The sediment yield model like MUSLE is easier to apply because the output data for this model can be determined at the watershed outlet (Pandey et al. 2009)

2.3. GIS Application and Model Builder

Geographic Information System is a tools to analyze, visualize and interpret data with spatial and geographic data. Nowadays, many research on engineering field especially on water resources engineering using GIS as a tools. There are several menu/tools that helped us to analyze water resources data, such as: thiessen polygon, watershed generating menu, etc. For soil erosion estimation, LWIS is one of tools that can be used. LWIS (the Integrated Land and Water Information System) is a a GIS software package from the International Institute for Geo-Information Science and Earth Observation in the Netherlands, has moved from paid (100 euros) proprietary status to free open source status as of today [11].

Based on Chandramohan (2015), ILWIS (Integrated Land and Water Information System) GIS was used to estimate MUSLE parameters and average soil loss. The topo-sheets were geo-referenced and base maps of watershed boundary, contours and drainage were generated. Different thematic maps, such as land use map, soil map, etc were prepared using the base maps and the data collected from field work. The digitized contour and spot height information were used to obtain the DEM (Digital Elevation Model) with the help of various interpolation routines available in ILWIS. This DEM was used for the creation of slope map.

Beside ILWIS, ArcGIS support user to make their own model with a tools called model builder GIS. Model Builder is a visual programming language for building geoprocessing workflows. Geoprocessing models automate and document your spatial analysis and data management processes.

Model Builder in ArcGIS support us to: Build a model by adding geoprocessing tools, map layers, datasets, and other data types, and connecting them to a process; Iteratively process every feature class, raster, file,or table in a workspace; Visualize your workflow sequence as an easy-to-understand diagram; Run a model step-by-step, up to a selected step, or run the entire model; Make model into a geoprocessing tool that can be shared with others or can be used in Python scripting and other models [10]. By model builder GIS, we can explore our creativity to make own model and developed model can be used for another case.

In fact, models have been more heavily influenced by their use in participatory contexts where visualization is of course important but where the predominant mode is to simply pick and choose from available software and engage in a loose coupling wherever such a coupling is required [9]. Several studies showed the potential utility of GIS technique for quantitatively assessing soil erosion hazard based on various models [2].

There are several research that used GIS for estimating soil erosion, such as:

a. GIS has been used to estimate soil loss in the Nethravathi Basin located in the southwestern part of India. Soil erosion in India has a major effect on the agricultural sector, siltation of reservoirs, degradation of soils, etc. in the nation (Ganashri,2016)
b. GIS environment used to create soil erosion maps of the specific years using data from meteorological stations, soil surveys, topographic maps, landsat satellite images in Jordan. Jordan is dominated by arid ecosystems that are vulnerable to human interventions and activities. It receives little rainfall with high intensity, which renders land degradation, soil erosion and desertification imminent threats of the ecosystem balance. Several changes in land cover have also occurred during the last decades (Minwer,2013)
c. GIS has been used to visualized the process of the dynamical change of water level in Nam Co Lake and dynamically demonstrate the evolution situation of Nam Co lake water level in three-dimensional space (Liu Bing,2011)

3 Result and Discussion

3.1. Watershed Boundary Analysis

Based on Mondal (2017), there is a significant variation in the rate of soil erosion only by using different DEMs of the same resolution and then by aggregating the resolution of all DEMs. the SRTM DEM performs better than the ASTER and the CARTOSAT DEMs in assessing the accuracy. With the increased resolution also, SRTM shows better performance.

Watershed boundary analysis was done by 2 step: cropping SRTM topography map using Mapper and watershed boundary analysis using ArcGIS 10.0. Cropping topography map using watershed coordinates as the boundary. The result can be seen on **Fig. 3**.

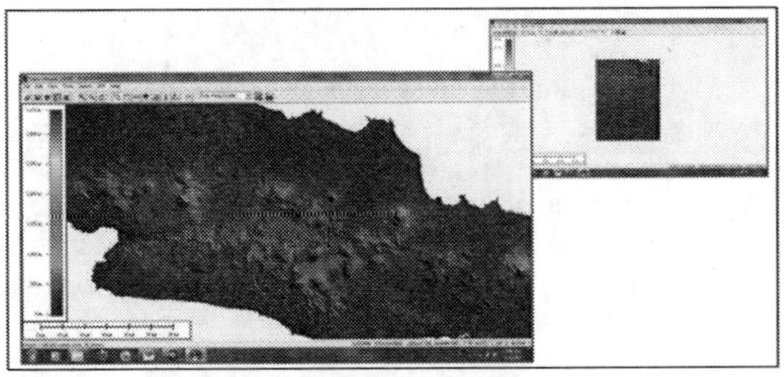

Fig. 2. Cropping Location Study

The cropped map was used for analyzing watershed boundary in ArcGIS 10.0. Firstly, analyze stream pattern and find Citepus river by using flow accumulation and flow direction menu. Citepus river location can be seen on **Fig. 4**. The next step is analyze the watershed boundary with determining the outlet point of watershed and subwatershed. The outlet point was determined on the downstream of Citepus River and for sub watershed was determined on the downstream of each river branch. The Citepus watershed and 5 sub watershed can be seen on **Fig. 5**.

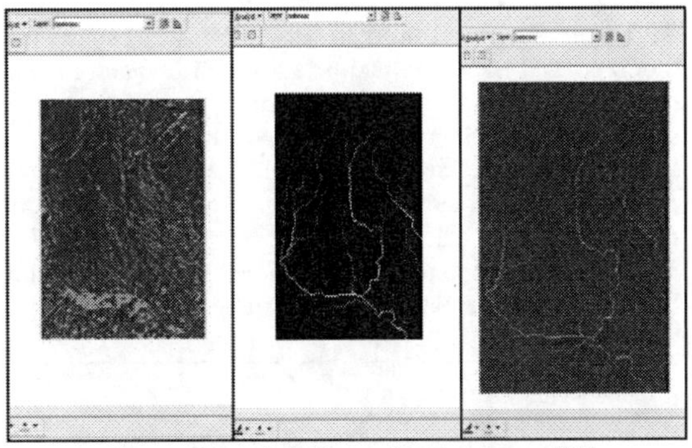

Fig. 4. Watershed Generating Process on GIS

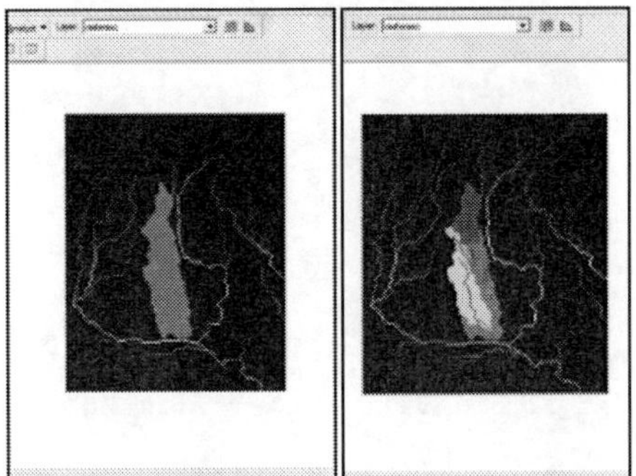

Fig. 3. Citepus Watershed and Sub Watershed on GIS

3.2. MUSLE factor calculation result

Before generate the model using model builder, the average value of MUSLE factor (such as: cropping factor, topography factor and soil erodibility factor) for each sub watershed was determined by percentage of areas. Percentage of areas can be estimated by calculate areas on GIS attribute. Soil erodibility, cropping factor and topography factor of Citepus watershed seen on **Fig. 6**.

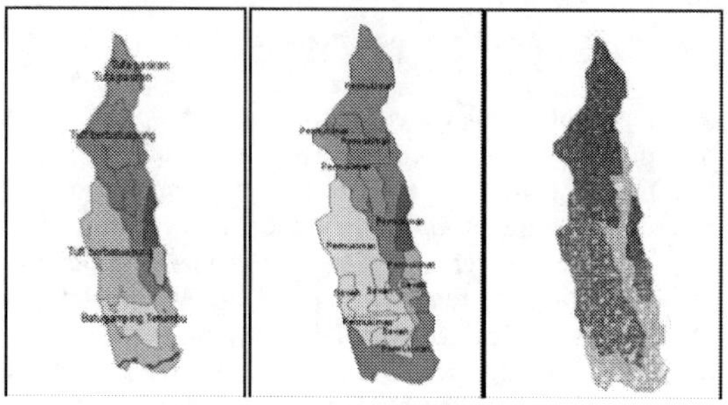

Fig. 4. MUSLE factor on GIS

The annual soil loss rate was calculated for each sub watershed using developed model. The model was developed by model builder GIS and raster calculation menu consist of multiplication of MUSLE factors. This model can be used for another study case/location by saving model file in arctoolbox format **(Fig. 7)**. Developed model by model builder is quite easy to rerunning for another case/location, by copying model toolbox file and attached related layers on the new file. Model builder is quite simple to develop a model and it has varying command/menu that can be used.

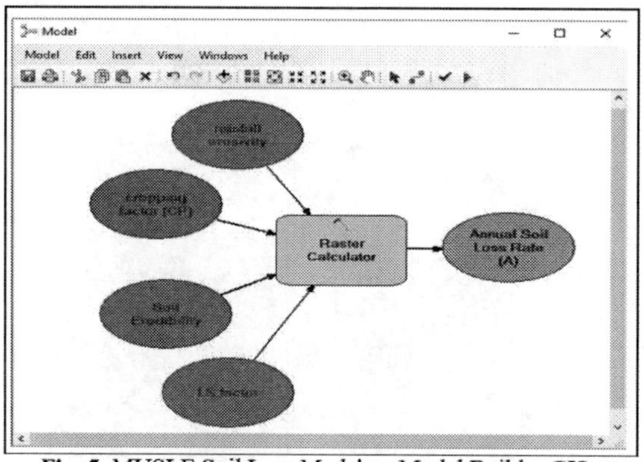

Fig. 5. MUSLE Soil Loss Model on Model Builder GIS

By using developed model, the annual soil loss can be calculated for each sub watershed. The calculation result can be seen on **Table 2**. From 5 sub watershed, Cilimus has the biggest value of annual soil loss rate (230 tons/ha/yr), while the smallest value is Ciroyom (129 tons/ha/yr). After MUSLE calculation, the next process was classify erosion rate by **Table 1** as reference, and the result can be seen on **Table 3.** From 5 sub watershed, erosion rate result varying from medium to heavy classification. MUSLE calculation usually compared with sediment measured data for calibration. For Citepus watershed, there are no available sediment data that can be used for calibration because in Indonesia, small watershed such as Citepus usually doesn't have sediment measured station.

Table 2. Annual Soil Loss Result Calculation

Sub Watershed	MUSLE Factor Calculation					Annual Soil Loss Rate
	R	C	P	LS	K	A
						tons/ha/yr
Cibogo	827	0.637	1	4.6	0.09	218
Cikakak	947	0.53	1	3.5	0.09	158
Cilimus	1048	0.637	1	3.9	0.09	234
Cipedes	601	0.637	1	6.1	0.09	210
Ciroyom	653	0.48	1	3.5	0.13	143

Table 3. Erosion Rate Classification

Sub Watershed	Annual Soil Loss Rate	Erosion Rate Classification
	tons/ha/yr	
Cibogo	218	Heavy
Cikakak	158	Medium
Cilimus	234	Heavy
Cipedes	210	Heavy
Ciroyom	143	Medium

4 Conclusion and Recommendation

From this research, it can be conlude that Cilimus sub watershed has the biggest value of annual soil loss rate (234 ton/ha/yr), while the smallest value is Ciroyom (143 ton/ha/yr). All Citepus sub watershed (Cibogo, Cikakak, Cilimus,Cipedes and Ciroyom) have serious problem in land erosion (with erosion rate classification medium-heavy). This problem must be handled as soon as possible, by technical management (implementation of ecodrain concept, such as: land conservation) and non technical managemet (regulation implementation, community participation,etc). Developed model by model builder is quite easy to rerunning for another case/location. Model builder is quite simple and has varying command/menu that can be used.

References

1. B.P. Ganasri, H. Ramesh, Geoscience Frontiers **7**, 953 – 961 (2016)
2. M.M. Alkharabsheh, T.K. Alexandridis, G. Bilas, N. Misopolinos, N. Silleos, Procedia Environmental Sciences **19**, 912 – 921 (2013)

3. G. Singh, R.K. Panda, International Soil and Water Conservation Research **5**, 202 – 211 (2017)
4. J.M. Laflen, D.C. Flanagan, International Soil and Water Conservation Research **1**, 1 – 11 (2013)
5. A. Mondal, D. Khare, S. Kundu, S. Mukherjee, A. Mukhopadhyay, S. Mondal, Geoscience Frontiers **8**, 425 – 436 (2017)
6. P.S. Kumar, T.V. Praveen, M.A. Prasad, Aquatic Procedia **4**, 1291 – 1298 (2015)
7. T. Chandramohan, B. Venkatesh, A.N. Balchand, Aquatic Procedia **4**, 1227 – 1234 (2015)
8. L. Bing, J. Yao, Q. Liqing, Procedia Environmental Sciences **10**, 1513 – 1518 (2011)
9. M. Batty, Procedia Social and Behavioral Sciences **21**, 10 – 17 (2011)
10. http://pro.arcgis.com/en/proapp/help/analysis/geoprocessing/modelbuilder/what-is-modelbuilder-.htm
11. http://freegeographytools.com/
12. http:// maps.google.com

Study of the Relation between Sediment Characteristics and Multiphase Flow to the Presence of Sediment Oxygen Demand (SOD) in Open Channels

Dhimas Dwinandha[1*], and *Haryo Satriyo* Tomo[2]

[1,2]Department of Environmental Engineering, Institut Teknologi Bandung, Jalan Ganesha No.10 Bandung 40132, Indonesia

Abstract. A laboratory study was conducted with the aim to determine the correlation between sediment characteristics and multiphase flow analysis of carrier fluid to the presence of SOD. Six sediment samples were tested on their physical and chemical characteristics along with particle size distribution of sediments to find the indication of oxygen consumption and to classify the soil sediment class. For multiphase flow analysis, there were two transition velocities calculated: the transition between a pseudo-homogenous flow and a heterogeneous flow and the limit deposit velocities at the onset of solid particle bed. The SOD test was done in laboratory-scale by using a 600-mL reactor. According to tests, the amount of organic carbon content (TOC) in the samples were ranging from 34.58 to 81.27%, with the sediments' textures categorised as silt loam, silty clay loam, and sand. In the channels, heterogeneous flow occurred in two channel segments, while the other segments' regime was classified as homogeneous flow. The obtained SOD values were varied from 0.2427 to 0.8487 g/m2/day with K3 values obtained ranged from 8.6537 to 12.4028 m-1. Based on all analysis, the organic characteristic of sediment holds a key role in the presence of SOD value.

1 Introduction

SOD is the rate at which DO is removed from the overlying water column by biochemical processes in the streambed sediments [5]. SOD includes oxygen consumption of biological activity (e.g. by microorganisms in the sediment) and chemical oxidation reactions to certain chemical species, including Fe^{2+}, Mn^{2+}, and S^{2-} [6].

Research on the SOD is not a new in determining water quality. Rounds and Doyle [8] found that the rate of SOD by 1-4 g/m2/day (typical SOD value of sandy sediments with little organic material) was still high enough to be an important component in a DO depletion. Seiki, et al [9] found the similar result that a minute amount of biodegradable organic compound in the sandy environment could also consume oxygen significantly. The decrease in concentration DO can cause a condition called hypoxia (DO < 2 mg/L) and anoxia (no DO at all) that can affect aquatic ecology and human population due to the loss of resources and water pollution [7].

Based on the explanation above, conducting a research about SOD is advantageous for several purposes. First, it can be used to develop mathematical models for predicting the allocation of water pollution load. In addition, this study can be a basis for planning the revitalization of the open channel, either natural or artificial channel. This study, however, was focusing on the relation between sediment characteristics and multiphase flow to the existence of SOD in open channels in Indonesia, particularly in the Province of West Java. This study is expected to be a trigger for conducting the other research related to the SOD in Indonesia, so there will be more references about sediment and its effect to the water ecosystem.

2 Methodology

2.1. Sampling and Field Observation

Six sediment samples from six channel segments were collected and tested in this study. Sampling sites located in the Citarum River (Kota Baru Parahyangan, Buah Batu, and downstream to Saguling Reservoir segment), Citarik River (downstream to the Citarum River segment), Cikapundung River (upstream segment), and West Tarum Channel (downstream near the Cikeas River segment). The sample number that corresponds to the sampling location is given in Table 1. Sediment samples were collected using sediment grab method and sealed in plastic bags. All the samples were stored and kept in a cool box. DO concentrations of each segments were also measured to be defined as DO target in SOD measurement.

* Corresponding author: author@e-mail.org

Table 1. Sampling location.

No.	Sampling Location
1	Citarum River (Kota Baru Parahyangan)
2	Citarum River (Buah Batu)
3	Citarum River (Saguling Reservoir)
4	Citarik River
5	Cikapundung River
6	West Tarum Channel (WTC)

2.2 Sediment Characteristic Tests

There were physical and chemical characteristic test conducted as sediment characterisation. Tests of physical characteristics were carried out in Soil Mechanics Laboratory of Civil Engineering Programme ITB. The test included densities (bulk and dry density), water content, and particle size distribution. TOC test method, as a chemical property of sediment samples, was based on the principles of FAS titration toward the excess potassium dichromate (K2Cr2O7) while being destructed under acidic conditions and heat. The testing was done twice (duplo) to obtain more accurate and precise results.

2.3 Multiphase Flow Calculation

Multiphase flow analysis was conducted to observe the behaviour of flow occurred above the sediment. The analysis included settling velocity and flow regime determination. Settling velocity (vs) was calculated using Equation [1], initiated by assuming CD value, calculating Reynolds number, and then calculating vs value. Iteration was stopped when the calculation result was convergent (less than 10^{-4} error).

$$v_s = \sqrt{\frac{4}{3} \cdot \frac{g \cdot d}{C_d}\left(\frac{\rho_p - \rho_w}{\rho_w}\right)} \quad (1)$$

Determination of flow regime was done by using Equation [2] developed by Newitt, et al. (written in Giguere, et al. [3]) to differentiate between the pseudo-homogenous and heterogeneous flow, while to separate between heterogeneous and stationary bed flow, Equation [3] and [4] were used [3].

$$v_{LH} = (1800 \cdot g \cdot d \cdot v_s)^{1/3} \quad (2)$$

$$v_{LD} = F_{LD}\left[2 \cdot g \cdot d \cdot \left(\frac{\rho_p - \rho_w}{\rho_w}\right)\right]^{1/2} \quad (3)$$

$$F_{LD} = 2 + 0{,}3 \cdot \log_{10}\left(\frac{d}{D \cdot C_D}\right) \quad (4)$$

The value of v_{LH} and v_{LD} was compared to the flow velocity in the observed channel. Pseudo-homogenous flow occurred if $v_{flow} > v_{LH}$, heterogenous flow occurred if $v_{LH} > v_{flow} > v_{LD}$, and stationary bed condition occurred if $v_{LD} > v_{flow}$.

2.4 SOD Measurement

SOD was assessed in laboratory scale by measuring oxygen depletion over time. DO concentrations were measured every 10 minutes, 30 minutes, 60 minutes, and 120 minutes until DO concentration remained constant (steady). The reactor used in the test was 600 mL reactor as shown in Figure 1.

Fig. 1. SOD test reactor design.

SOD test was started by filling aquadest to the reactor that had been filled with approximately 300 mL sediment until full. The DO meter probe then mounted on the lid of the reactor, and then that set of the lid was used to seal the reactor until there were no more air exchanges, both from inside and outside the reactor. The equation used to determine the SOD value was derived from Gin, et al. [4], which is showed in Equation [5]. The obtained SOD value then corrected by temperature using Equation [6].

$$SOD = \frac{\left(1000 \frac{V}{A} \cdot (C_f - C_i)\right)}{\Delta t} \quad (5)$$

$$(SOD)_{20} = \frac{(SOD)_T}{\theta^{(T-20)}} \quad (6)$$

2.5 Determination of Deoxygenation Coefficient (K_3)

Determination of deoxygenation coefficient due to SOD (K3) was done by: (a) calculating the SOD value; (b) determining the regression equation of the obtained DO decline profile; (c) conducting simple linear correlation test to determine the significance of the regression equation; and (d) calculating K_3. Determination of the regression equation was performed by using Microsoft Excel. DO concentration was defined as variable X and time as variable Y. In addition, determinant coefficient value (R^2) was taken, as well, for simple linear correlation analysis tests. After the statistical test, the K_3 value was calculated using Equation (7) adapted from the general equation of the first order reaction.

$$K_3 = \frac{\left(\frac{C_{initial} - C_{final}}{\Delta t}\right)}{SOD} \quad (6)$$

3 Result and Discussion

3.1 Particle Size Distribution

The sediment size composition or distribution is showed in Figure 2. The data indicates that most of the particle size in the sediment sample was silt, except sample 5 (Cikapundung River), which dominated by the sand. Distribution of particles in sample 5 caused by the sample was taken in the upper reach of the river with high flow velocity, so the smaller particles had not been formed yet or had been washed out. On the contrary, other sampling locations had a low flow velocity. That condition allowed the sedimentation of suspended particles in water, therefore the range of particle size was larger.

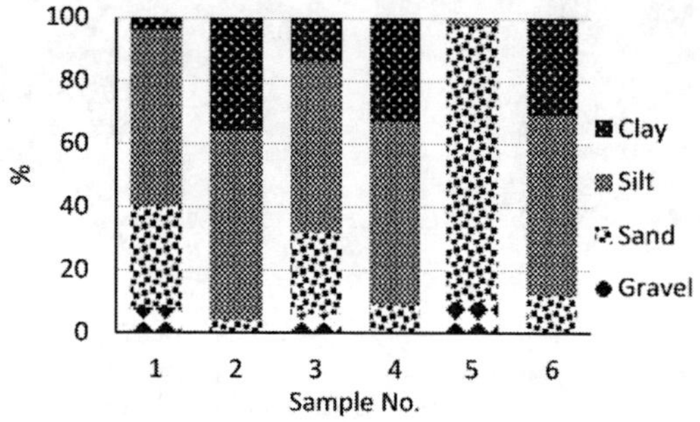

Fig. 2. Distribution of sediment particle size.

The plot of particle size composition in soil texture triangle according to USDA (US Department of Agriculture) is illustrated in Figure 3. The figure reveals that the sample 1 and 3 was categorised as silt loam soil, the class of sample 2, 4, and 6 are silty clay loam, and sample 5 classified as sand, as expected.

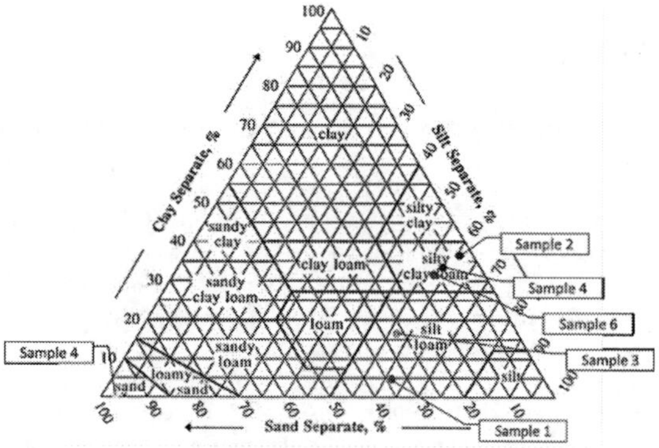

Fig. 3. Soil texture triangle for sediment samples.

3.2 Index Properties of Sediment

Index properties of the observed sediment samples (water content along with bulk and dry density) are given in Table 2.

Table 2. Index properties of sediment samples.

No.	γ_m (kN/m^3)	γ_d (kN/m^3)	Wn (%)
1	15,40	9,30	66
2	13,70	6,33	116
3	15,20	9,07	68
4	13,40	5,81	131
5	17,20	11,89	45
6	13,60	6,76	101

The result reveals that the sediment in Cikapundung River (sample 5) had the highest density, while the Citarik River sediment (sample 4) had the lowest compared to other samples. The magnitude of particle density of the sediment could be influenced by the chemical composition of the sediments. The higher amount of mineral content in the sediment (commonly found in sand particles) is, the denser the sediment will be. Meanwhile, the density of sediment will be low if the composition of the sediments mostly contains light organic compounds, as found in clay and silt particles.

The data showed that the sediments were highly saturated by water. High water content could be resulted by sediment particle size composition dominated by clay and/or silt. Clay and silt have a high adhesion force that can accommodate and retain water, resulting in high water holding capacity of the sediment. On the other hand, sand particle has a low adhesion force, thus the water tends to be escaped when passed through a sand layer.

3.3 Total Organic Carbon (TOC) Content

The presence of organic carbon might indicate the consumption of dissolved oxygen in the water body because organic carbon is a source of energy for the microorganisms. Table 3 gives the test result of total organic carbon content (TOC) in sediment samples taken.

Table 3. Percentage of total organic carbon (TOC) in sediment.

No	TOC (%)
1	46,44
2	49,84
3	81,27
4	54,64
5	37,63
6	34,58

Sediment in Citarum River–Reservoir Saguling had the highest percentage of total organic carbon (TOC) content with 81.27%. The high amount of TOC could be caused by organic compounds accumulation from previous river segments. In addition, the particle size composition that dominated by silt and clay particles made high level of organic carbon compounds were adsorbed from the water.

The sample with the lowest percentage of TOC was a sample 6 (West Tarum Channel). This might be due to the low concentration of suspended organic compounds found in the water, hence only few solid particles containing organic compounds settled to the bottom of the channel. As the result, less oxygen consumption would be occurred in the channel.

There was a large variation of concentrations in each sample that can be caused by the significant difference in location sampling. Besides, the level of pollution of the water body at sampling locations can also affect the amount of TOC percentage in the sediment.

3.4 Settling Velocity of Sediment Particles

Sediment settling velocity calculated using the cut size diameter (d_{50}) of the sediment. The result of the settling velocity calculation is shown in Table 4. Five of six samples have d_{50} included in the range of 0.002–0.05 mm, which indicated that the particle was classified as silt. Only sample 5 (Cikapundung River) that have d_{50} on the sand category (range of 0.05 – 2.00 mm).

Table 4. Settling velocity of sediment particle.

No.	d_{50} (mm)	V_s (m/s)
1	0,0650	0,01458
2	0,0079	0,00677
3	0,0400	0,01198
4	0,0100	0,00684
5	0,3700	0,05070
6	0,0140	0,00753

The settling velocity was the parameter that is directly associated to the size and density of the particles. This showed that the settling velocity of particles would increase in line with the increasing of particle diameter and mass. This phenomenon could be seen in the data provided in Table 3. Sample 5 (Cikapundung River) has the highest settling velocity due to the largest diameter and highest density.

3.5 Slurry Flow Regime

The calculation result in determining slurry flow regime is shown in Table 5. There were two types of flow regime in observed channels, which were homogeneous (samples 1, 2, 4, and 5) and heterogeneous (3 and 6). The difference of this regime can be explained by the variation in flow velocity of the carrier fluid and the turbulence of the flow as the effects of the flow velocity.

Table 5. Slurry flow regimes.

No.	V_{LH} (m/s)	V_{LD} (m/s)	V_{flow} (m/s)	Flow Regime
1	0,25576	0,01717	0,3942	Homogenous
2	0,09811	0,00231	0,6722	Homogenous
3	0,20375	0,01094	0,0178	Heterogenous
4	0,10652	0,00292	0,1429	Homogenous
5	0,69190	0,07698	1,2800	Homogenous
6	0,12301	0,00380	0,1197	Heterogenous

Multiphase flow regime can show the difference of solid concentration distribution in a carrier fluid. Heterogeneous flow regime indicates that there are more particles at the base of the channel than in the middle or in the surface. It showed a strong cohesion force between particles of sediment at the bottom of the channel so that the carrier fluid flow has a little effect on sediment particles movement.

On the contrary, homogeneous flow shows the concentration of solids evenly distributed, thus resulting in similar concentration of solids throughout the flow. Unlike heterogeneous regime, the cohesion force between sediment particles is weak, but the adhesion force between sediment particles with water is strong, therefore the solid particles move followed by the water as carrier fluid. Additionally, lightweight sediment particles (small diameter or low density) can also lead to the occurrence of a homogeneous flow regime because lightweight sediment particles have low settling velocity. However, the flow velocity of carrier fluid has a greater influence than the setting velocity of the particle. The statement can be proved by looking at the data for sample 5 (Cikapundung River), which had the highest settling velocity among other samples but its flow regime was homogeneous.

The process of sedimentation in a homogeneous flow regime should not have occurred because the solid particles are carried away until the flow regime changes by the flow velocity difference. The presence of sediment in sampling location indicated that there was an unnatural process of sedimentation. The unnatural sedimentation process that possibly occurred was solid particles stuck on the edge of a channel that has vegetation and was accumulated in the base of the channel.

3.6 Sediment Oxygen Demand (SOD) Analysis

Figure 4 depicts the dissolved oxygen (DO) concentration depletion, while Table 6 represents the calculated SOD values.

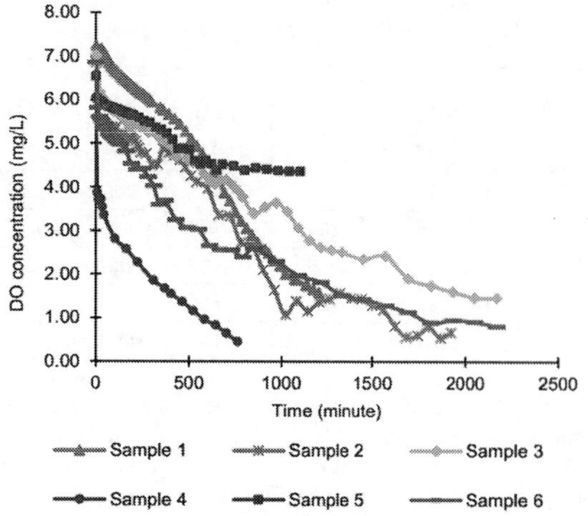

Fig. 4. DO concentration depletion profile.

Table 6. Obtained SOD values in sediment samples.

No.	SOD @ 20°C g/m²/day
1	0,5805
2	0,3835
3	0,3143
4	0,8487
5	0,2427
6	0,3317

The overall decrease in DO concentration profiles showed a similar trend, which was a substantial decrease of DO concentration in early test. According to Barcelona and Wang [1], it is most likely to occur because of the presence of chemical SOD on sediment. Chemical SOD is composed of sulfidic demand and chemical–sulfidic demand. Chemical–sulfidic demand indicates the oxidation of organic compounds and iron-manganese minerals, while sulfidic demand describes the oxidation of sulphide compound, such as $[HS^-]$ and $[S^{2-}]$. Therefore, the greater decline of DO concentration at the beginning of the test, the greater chemical SOD in these sediments.

Based on the data in Table 5, it shows that the SOD value in sample 4 (Citarik River) was the highest among other samples. The high amount of SOD value can be caused by the huge amount of organic matter that can be degraded. In addition, sediment particles that categorised as silty clay loam accommodate high adsorption of organic compounds from water (especially suspended organic compounds) because of the large surface area. Another cause was the major fraction of chemical SOD presented in the sample (showed in rapid drop of DO concentration in early test).

The channel segment that had the lowest SOD value was the Cikapundung River (sample 5). This was because the sampling location laid on the upper reaches of the river with sand-class sediment particles; therefore, it is less capable of adsorbing organic content as much as in the Citarik River sediment. The flow velocity of the river might also be the reason; while the velocity was high enough, most of the organic content in the sediment particles could be washed out and dissolved again.

Although the TOC percentage in sample 3 was the highest of the overall sample, the SOD value of the sample did not follow. This shows that the amount of degradable organic carbon compounds in the sample holds a significant role in affecting SOD value. Sample 3 might have a low amount of degradable organic compound, and so did its SOD value. The amount of degradable organic compounds can be affected by the duration of sediments present in the water. Leutheuser (written in Barcelona and Wang [1]) opined that organic compounds that have been accumulated in the sediment for a long time would react slower than the fresh organic compounds.

3.7 Deoxygenation Coefficient due to SOD (K_3)

First step in determining K_3 value was conducting a statistical test called simple linear correlation analysis. This was done to observe if there was any alteration in DO concentration gradient. The alteration is important to analyse DO consumption kinetics. The complete result of the test presents in Table 7. Based on the results, it could be concluded that all regression equations were significant in 5% and 1% level. Thus, there was no change in the gradient of decreasing concentration of dissolved oxygen during testing.

After statistical tests, the calculations of K_3 was performed. The results of the calculation of the value of the coefficient of deoxygenation due to SOD in samples are listed in Table 8.

Table 7. Results of simple linear correlation analysis.

No	n	R^2	r-calculated	r-table		Conclusion
				5%	1%	
1	61	0,9914	-0,9957	0,250	0,325	Significant for α = 1%
2	58	0,9436	-0,9714	0,260	0,340	Significant for α = 1%
3	60	0,9624	-0,9810	0,250	0,325	Significant for α = 1%
4	18	0,8267	-0,9092	0,468	0,590	Significant for α = 1%
5	44	0,9083	-0,9530	0,304	0,393	Significant for α = 1%
6	41	0,8835	-0,9399	0,312	0,403	Significant for α = 1%

Table 8. The value of K_3 in sediment samples.

No.	K_3 @ 20°C m^{-1}
1	12,4028
2	11,2640
3	10,9970
4	8,6537
5	11,2713
6	10,8544

Empirical calculation of K_3 values conducted in this research shows the value of K_3 had an inversely proportional relationship to the SOD value in sediment and was directly proportional to the rate of decrement in DO (indicated by the value of the gradient of DO concentration). This relationship is an auto-correlation obtained from the equation to calculate the value of K_3. Further research using a variation of the SOD value will reaffirm the statement. However, factors that affect the magnitude of the value K_3 cannot yet to be analysed with certainty. Therefore, further research needs to be conducted about the factors that affect K_3 value in the sediment.

Recapitulation of the SOD and K_3 value based on sedimentary soil texture shows in Table 9.

Table 9. Recapitulation of the SOD value and K_3 based on sediment texture.

No.	SOD @ 20°C g/m²/day	K_3 @ 20°C m^{-1}	Soil Texture Class
1	0,5805	12,4028	*silt loam*
2	0,3835	11,2640	*silty clay loam*
3	0,3143	10,9970	*silt loam*
4	0,8487	8,6537	*silty clay loam*
5	0,2427	11,2713	*sand*
6	0,3317	10,8544	*silty clay loam*

3.8 The Implication of SOD Value

The presence value of SOD in the sediment showed that the consumption of DO in the water is not only derived from BOD (biochemical oxygen demand) and COD (chemical oxygen demand) that have suspended or dissolved phase, either for natural or artificial channel channels. The consumption of DO will still occur in the channel that has huge amount of sediment although the polluters with the high BOD and/or COD rate have been stopped. Therefore, the management of disposed wastewater is not enough to revitalise the watercourse and improve its water quality. The good management of sediments must also be established to give an effective result of channel revitalization and the water quality of the watercourse can remain good.

There are several methods in managing sediment in the channel. One of them is to build an aeration structure that allows sediment particles to accumulate so they can be removed, either manually or automatically. Another option is an aeration structure that facilitates sediment particles to dissolve into water as dissolve or suspended organic compounds, so the removal process can be done by using a same method as removing BOD or COD in the water.

4 Conclusion

The existence value of SOD in the open channel was indicated by the carrier flow regime. Heterogeneous flow regime showed a concentration of solids in the bottom of the channels that can form a sediment bed and therefore the oxygen consumption. In addition, sediment that categorised as clay and/or silt (indicated by small particle diameter, low density, and high percentage of water content) and a high degradable organic content can increase value of SOD in the sediment.

5 Nomenclature

C_d = discharge coefficient
d = particle diameter, selected diameter *cut size* (d_{50})
g = gravitational acceleration, m/s²
ρ_p = sediment particle density, kg/m³
ρ_w = water density, kg/m³
C_D = drag coefficient
D = channel diameter, m
F_{LD} = limit factor of deposition velocity
v_{LD} = limit velocity of particle deposition, m/s
v_{LH} = limit velocity of homogenous flow, m/s
$(SOD)_{20}$ = the number of sediment oxygen demand (SOD) flux at 20°C, mg/m²/day
$(SOD)_T$ = the number of sediment oxygen demand (SOD) flux at T°C, mg/m²/day
A = cross-sectional area of overlying water, m²
C_f = DO concentration at the end of measurement, mg/L
C_i = DO concentration at the beginning of the measurement, mg/L
T = average testing temperature, °C
V = overlying water volume, m³
θ = temperature constant = 1,065
C_{final} = DO concentration at t (calculated based on the regression equation), mg/m³
$C_{initial}$ = DO concentration at t_o (calculated based on the regression equation), mg/m³
K_3 = deoxygenation coefficient due to SOD, m^{-1}
SOD = the number of sediment oxygen demand (SOD) flux, mg/m²/day

Authors would like to express their gratitude to Institut Teknologi Bandung (ITB) who had provided financial support in this research through P3MI programme (Program Penelitian Pengabdian Masyarakat).

References

1 M. Barcelona and W. Wang, *Usefulness of Sediment Oxygen Demand as a Tool for Impoundment Management* (Water Resources Center, Univ. of Illinois, 1982)
2 C.R. Ellis and H.G. Stefan, Wat. Res. Bull. **25**, 1169-1176 (1989)
3 R. Giguère, L. Fradette, D. Mignon and P. Tanguy, J. Chem. Eng. Rsch. Dsgn., 943–950 (2009)
4 K.Y.H. Gin and A.P. Gopalakrishnan, J. Env. Eng. **136**, 78-85 (2010)
5 K. Hatcher, *Sediment Component of Oxygen Demand in Streams* (Institute of Natural Resources, University of Georgia, Athens, 1980)

6 C.B. Price, C. Cerco and D. Gunnison, *Sediment Oxygen Demand and Its Effects on Dissolved Oxygen Concentration and Nutrient Release: Initial Laboratory Studies* (U.S. Army Corps of Engineers, Washington DC, 1994)

7 T.A. MacPherson, *Sediment Oxygen Demand and Biochemical Oxygen Demand: Patterns of Oxygen Depletion in Tidal Creek Sites* (University of North Carolina, 2003)

8 S. Rounds and M. Doyle, *Sediment Oxygen Demand in the Tualatin River Basin, Oregon* (U.S. Geological Survey Water Resources Investigations, 1997)

9 T. Seiki, H.Izawa, E. Date and H. Sunahara, Wat. Rsch. **28,** 385-393 (1994)

The Performance of Equalization Model of Water Allocation Inter Irrigation Areas in River System

Anang M. Farriansyah[1], *Pitojo* Tri Juwono[2], *Ery* Suhartanto[2], *Very* Dermawan[2], and *Rahimatus* S. Z. Alyaminy[3]

[1]Civil Engineering- Doctoral Program, University of Brawijaya, Malang, Indonesia
[2]Water Resources Engineering, University of Brawijaya, Malang, Indonesia
[3]Youth Professional Engineer, Mataram, Indonesia

Abstract. In Indonesia, water is public goods so it is necessary to control water allocation. Inequity of water allocation between water users is expanding largely, including irrigation as the largest user, while the density of headworks in the river are getting higher. Considering that water is limited, the practice of irrigation water allocation needs to be refined, from the traditional equity to volumetric equity. MEQAA (*Model Equalisasi Alokasi Air*/Equalization Model of Water Allocation) plays a role in determining water sharing between headworks in order to meet the maximum-equal K-factor in river. MEQAA-Generic is a calculation machine with: analog-deterministic dynamic model; network equation according to mass balance and linear optimization; independent-based system; sustainability-efficiency-equity constraints; Ms. Excel-VBA. The inputs are: scheme system, local inflow, and irrigation demand. The outputs are: K-factor, release and ecosystem quote. The model performance is identified by comparing the output to the class of K-factor based on treatment of water distribution. The model test is performed in an uncontrolled and complicated system in Kukusan Tanggek watershed with 24 headworks in Lombok river basin. As long as it is adequate for water sharing, MEQAA-G can always produce maximum-equal K-factor. The output model is used to operation control.

1 Introduction

System is a collective work between components for a particular purpose [3]. Since every main structure node of water withdrawal (headwork/HW) in the river is hydraulic-gravity connected as a system unit, therefore one integrated management is needed [32]. Water withdrawal must be based on a proportional, efficient, and equitable water allocation plans [13], considering in Indonesia water is public goods, and limited.

Practically, water allocation is not based on system, and disoriented, resulting in inequity [22]. Water allocation gap occurs mostly in water deficit areas, resulting in conflicts [12], [25], and [27]. In [26], competition case between water users in South Bali were caused by irrigation water deficit which makes it a complicated task. Regarding to domination of irrigation (usage rate over 90%) [7], this sector needs to be controlled. This effort is aligned with the emphasis of equity in water allocation by IWRM (Integrated Water Resources management) [29].

To create equity, a proportional water sharing is needed [8], [9], [12], and [25]. In [11], proportional equalization is an equal K-factor (release demand ratio) between same users. Also, K-factor method in Indonesia has been known in irrigation operations since the Dutch era. K-factor indicates volumetric reliability in system [16], which is Release Demand Ratio (RDR) [5], [4], [11], and [30].

Inequity due to the failure of water sharing will trigger conflict, even threaten food security [18]. Inequity also hit the ecosystem. For ecosystems, rules are required and must be taken as other water demands in the calculation [17]. The rule of ecosystem allotment is regulated in Government Regulation RI No. 38 Year 2011 regarding River, 95% probability of water availability and 5 % of water availability (Directorate General of Water Resources RI, Irrigation Planning Criteria or KP-02-revised edition, 2013: 28).

Water allocation control between nodes (headwork) in the complex river systems should be supported by models, such as Decision Support System (DSS) [11]. However, DSS produces many alternatives, other than optimal solutions, that professionals are required to make decisions [3]. From a series of DSS models in Table 1, upon local models it can be concluded as follows: i) limited ability in rigid systems, ii) does not consider ecosystems, and iii) does not produce water release according to volumetric equity. Besides, upon import models, it can be summarized that: i) priority oriented (with penalty), thus not suitable in Indonesia, and ii) integrated hydrology/water quality model.

Alternative model is expected to work multisystem. The output of the model has to fulfill the triangle constraints of sustainability-efficiency-equity [22] or SEE.

anangmf62@gmail.com

Table 1. Recapitulation of water allocation model

No	Model/Origin/ Year	Programming Language	Orientation/ Function	Integrated Analysis	Type/Scale/ Type of Nodes	Result	Negotiaton
1	HLD/ LOMBOK RB/ England Expert/ 1991	Fortran	Equality / crop production, real time	Hydrology	Specific, scheme (interdependent), Limited nodes, Diversion & reservoir	Release, Unequal K-factor	Yes
2	MS EXCEL MODEL/ Indonesia/ 1998 & 2006	Lotus123 and Excel-VBA	Equality, real time	No	Specific, simple scheme, Very limited nodes, Diversion	Release, Unequal K-factor	Yes
3	PAA/ Indonesia/ 2010	Excel	Equality, real time	Irrigation water demand	Specific, simple scheme, Very limited nodes, Diversion	Water balance, Operation rule "adjusted"	Yes
4	WRMM/ Canada/ 1981	Visual Basic	Priority, planning	Hydrology	Generic, watershed, Diversion & reservoir	Water balance	Yes
5	RIBASIM/ Delft-Belanda/ 1985	Fortran	Priority, planning	Hydrology & water quality	Generic, watershed, Limited nodes, Diversion & reservoir	Water balance	Yes
6	REALM/ Australia/ 1980th	Fortran	Priority, planning	Hydrology & water quality	Generic, system scheme, Nodes >>>, Diversion & reservoir	Water balance	Yes
7	WRAP/ Texas-USACE/ 1980th	Fortran	Priority, planning	Hydrology & water quality	Generic, watershed, Nodes >>>, Diversion & reservoir	Water balance	Yes
8	WEAP/ SEI-Swedia/ 2000th	Javascript	Priority, planning	Hydrology & water quality	Generic, watershed, Diversion & reservoir	Water balance	Yes
9	MIKE BASIN/ DHI-Denmark/ 2008	Link with Excel- Visual Basic	Priority, equally deficit, and crop stress. Planning	Hydrology, groundwater, & water quality	Generic, watershed, Diversion & reservoir	Water balance	Yes

If water is public goods, then "equal for equity" will be accomplished by MEQAA-G. The target of the model is maximum-equal K-factor which is to allocate water proportionally, efficiently, and equally between HW in the system. In the beginning, this model was called MEQAA-E (Embryo), a conceptual model with a linear simulation-optimization based on mass balance equation.

The optimization concepts of MEQAA-E [5] is: i) to control the utilization of local inflow [23], ii) based on node in independent system [14] and [2], and iii) refers to the principle of mass balance [30] and [9]. Based on the concepts, linear network equation (NE) can be formed according to the system configuration. [5] continued, MEQAA-E's experiment used system scheme and hypothetical data, with NE being arranged manually, resulting in: i) max-equal K-factor between same users (irrigation), ii) SEE constraints are fulfilled, and iii) policy rules about K-factor gap between wet-dry system can be applied.

MEQAA-E's ability is limited and prone to human errors if it applied in various independent system states. The current research is the development of MEQAA-E to MEQAA-G (Generic). Generic model applies multisystem [20] and [30]. The principle of K-factor equalization of MEQAA-G follows MEQAA-E. The difference is MEQAA-G automatically builds up a specific model according to system scheme on Excel sheet.

Since MEQAA-E/G is conceptual, the validity of the model is determined based on the theory and assumptions used to construct the model structure [3]. With input variability in the form of system scheme, and also inflow and demand value, it is expected that every specific model formed by MEQAA-G in specific system can produce the optimal output.

As long as water can be allocated based on gravity flow water system, MEQAA-G will produce max-equal K-factor or equal K-factor class. Output model is used to control water allocation proportionally, efficiently, and equally between HW of same users (irrigation) in the river system.

2 Study Area

Lombok River Basin (RB) in Fig. 2 (4.738 km^2, 3.5 million populations, rainy season Nov-Mar, dry season Apr-Oct, normal rain 1189-1505 mm) with irrigation as the dominant user (94%) with system capacity in Table 2. High Level Diversion (HLD) suppletion canal was built in the 1980s to distribute water from the wet watershed in the west Lombok to dry watershed in the middle-east-south with 92 nodes in 6 watersheds interconnection for 60.000 ha of irrigation areas (Kartabrata & Marjanto, 1994: 40). And today, in HLD interconnection with 398 nodes in 12 watersheds for more than 97.000 ha [33].

Meanwhile, Kukusan Tanggek watershed is a dependent system (185 km^2, density of 1 node/5 km^2, 30 units of weir, 1 unit of small reservoir (local name is embung), 6446 ha of irrigation areas, 2 units of 30 lt/s of water domestic).

anangmf62@gmail.com

Table 2. Capacity of water resources system in Lombok RB.

No	Item	Unit	Independent	Interdependent
1	Total watersheds	unit	40	12
2	Total area of effective watersheds	km^2	1.152	1.454
3	Total nodes: - Weir - Small reservoir (embung) - Reservoir - Diversion in HLD canal	unit	145 118 27 -	371 308 60 3 60
4	Node density	unit/ km^2	0.13	0.26
5	Total irrigation area	ha	26.221	97.037

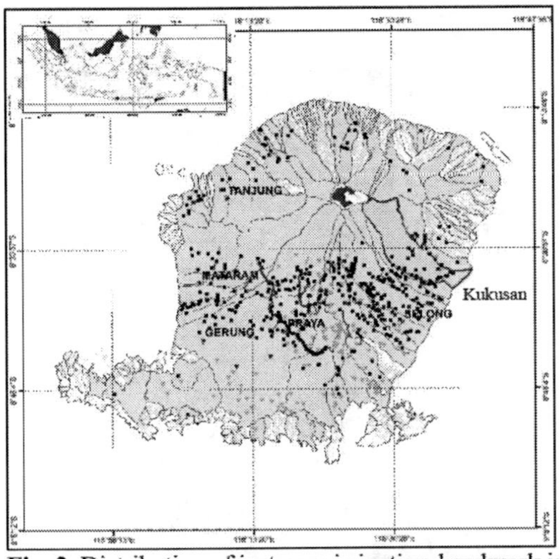

Fig. 2. Distribution of instream irrigation-headwork in Lombok RB and research site

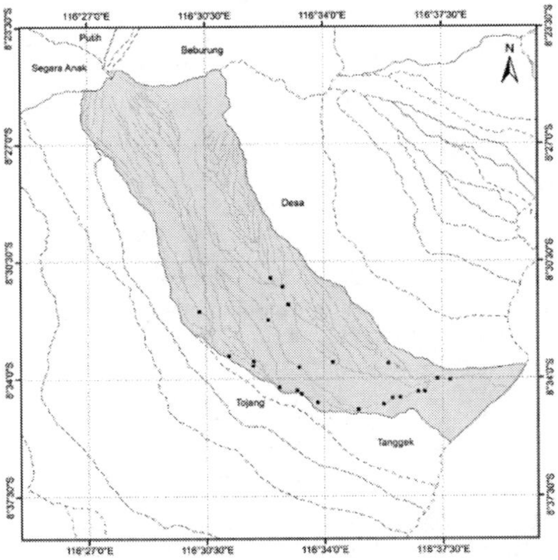

Fig. 3. Kukusan Tanggek watershed

The deficit of irrigation water in Lombok RB is caused by continuity and uneven distribution of water potential [21]. National rice granary areas, such as Lampung, Sulawesi Selatan, Bali and Lombok are experiencing high water vulnerability [10]. Based on the survey of water distribution in irrigation areas in Lombok RB (March IV and May I 2016, crop pattern: paddy (P) and paddy-soybean (P-S), in 114 nodes in 14 main rivers), there are: i) inequity in dry watershed, and ii) K-factor in the upstream is K1/K2, while in the downstream is deficit. The gap is in sync with [27], that in Lombok RB there are 386 conflict cases/year between water users, and drag the government as one of its actors. The gap occurs in Kukusan Tanggek as shows in Table 3.

Table 3. Cass of K-factor in the main river of Kukusan Tanggek

Headwork	Irrigation Area (ha)	Crop Start	Crop Pattern	Crop Area (ha)	Class of K-Factor	
					Mar IV	May I
B Meloang	168	Sep II	P-S	162	K1	K2
B Madang 1	138	Sep II	P-S	130	K1	K2
B Madang 2	46	Sep II	P	46	K1	K2
B Bagek Nyake	189	Okt II	P-S	172	K2	K2
B Mamben	477	Okt II	P-S	453	K2	K2
B Kukusan	1117	Okt II	P-S	1055	K2	K2
B Jowet	100	Okt II	P-S	96	K2	K2
B Tegaron	202	Okt II	P-S	192	K2	K2
B Reban Aji	129	Okt II	P-S	124	K3	K2
B Kerumut	58	Okt II	P	58	K3	K2
B Sukamulia	513	Okt II	P-S	480	K3	K3
B Batu Yang	532	Okt I	P-S	517	K2	K2

3 Method

3.1 Equalization Method

Water distribution between HWs need to be controlled, considering that water is limited and unequal. Gravitationally, water in the upstream is allocated to the downstream due to their own needs so K-factor gap does not happen, surely with emphasis on SEE constraints. K-factor could not be measured, therefore K-factor class which is analogized based on water distribution pattern in irrigation area is used. Practically, K-factor class is influnced by the adequacy of water supply and affects the inter-block rotation within the irrigation areas.

Table 4. Optimization constraint within MEQAA-G

	Sustainability	Efficiency	Equity
Indicator	- According to the demands and priority. - Ecosystem. - Both are fulfilled from time to time.	- Water release does not exceed the water demand. - Minimum water sharing is as much as the ecosystem allotment.	Equal K-factor between the same users after the fulfillment of domestic and ecosystem needs.
Goal	To maximize water release proportionally, equally between the same users (max-equal K-factor)		

K-factor class in Table 5 was inspired by [1], [4], [28] and Minesterial Regulation PU-PERA No. 12/PRT/M/2015). This K-factor's class classification

 https://doi.org/10.1051/matecconf/201814703005

considers inflow fluctuations, weir/diversion dominance, operational intensity, and error measuring tool.

Table 5. K-factor class of irrigation

Class	Range of K-Factor (%)	Water Deficit	Operational Categories	Interlude of Water Distribution
K1*	100	-	Continuously	-
K1	80 - 99	Very low	Continuously (limited)	-
K2	60 - 79	Low	Rotation (low)	Short
K3	40 - 59	Medium	Rotation (medium)	Medium
K4	20 - 39	High	Rotation (high)	Long
E	< 20	Very high	Emergency	Withered crop is prioritized

3.2. Model Structure

The purpose of water allocation is to distribute water optimally between nodes in the system [9]. For that purpose, MEQAA-G was built with characteristics of: i) as a "calculating machine" of water allocation with a max-equal K-factor, after drinking water and ecosystem are fulfilled [23], dan ii) using the technique of "tracking" system scheme to create NE. MEQAA-G is supported by Excel-VBA, because it is popular/practical for non-professional programmers and its code format is relative simple to run such a great program [28].

Solutions on water resources system are often using optimization [9], [23], [24] and [30], such as linear programming (LP). In [9], [15], [20] and [30], optimation/simulation model is used to attain system operation decisions. In [24] both combinations produce an optimal solution, although it contains many variables. Concluded by [30], optimal/nearly optimal value is obtained from execution of ad hoc model with iteration. Iteration requires a set up initial value [9], which is K-factor draft 100%-0 with step 0,1 – 1% [5].

The principle of mass balance [9], [30], [31] in every HW is: with the number on the right-hand side.

$$I - O = \Delta V \qquad (1)$$

With inflow (I) includes local inflow (QL) and water sharing (QS) from upstream, outflow (O) includes release (QR) and water sharing (QS) to downstream, and ΔV= a change in volume. It is assumed that the loss of water (LL) between HWs mutually negates with return flow (RF) from irrigation area. The loss value is hard to quantify therefore it is often ignored [30].

MEQAA-G with its purpose of maximizing water distribution is:

$$\max Z = \sum_{i=1}^{n} K_i.QD_i \qquad (2)$$

If every HW is weir/diversion, then ΔV=0 [30]. Based on formula 1 and 2, NE is arranged based on independent system. NE follows domino effect flow [30] and [2]. The simple structure of NE as shown in Figure 3 combines 5 functions of constraint, such as:

- mass balance: $QL_i + QS_{i-1} - QR_i - QS_i = 0$ (3)
- water availability: $0 \leq QR_i \leq (QL_i + QS_{i+1} - QE_i)$ (4)
- ecosystem sustainability: $QS_i \geq QE_i$ with

$$QE = 0.05\ QA \qquad (5)$$

- efficiency: $0 \leq QR_i \leq QD_i$ which can substituted into

$0 \leq K_i \leq 1$ or $QR_i = K_i \,.\, QD_i$ [5] (6)

- equity from upstream to downstream of HW:

$K_i - K_{i+1} = 0$ [5] (7)

QD = irrigation demand, QE= ecosystem allocation

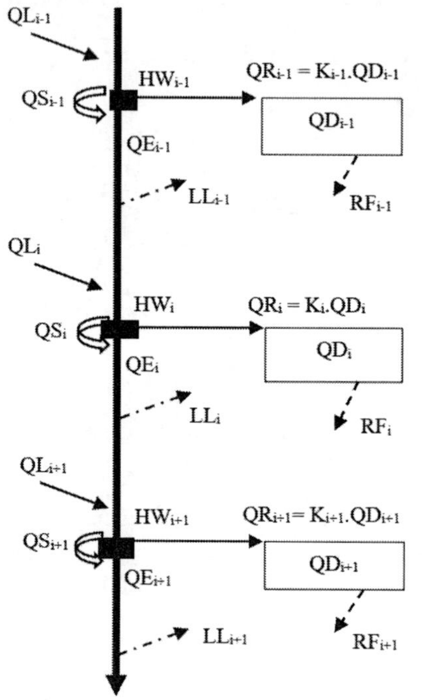

Fig. 3. Network flow sketch in system segment with weir

As for reservoir ($\Delta V \neq 0$), volume calculation is applied [30], [6] and [31]:

$$V_{end} = V_{beg} + V_{inlfow} - V_{release} - WL \qquad (8)$$

With $0 \leq V_{end} \leq V_{eff}$, V_{eff} = effective volume and WL = water loss because of evaporation and seepage. Since in Kukusan Tanggek system there is Embung Senang, therefore reservoir rule curve is required as constraint adjusment factor [31]. Operational volume limit is:

$$VRC = V_{beg} * RCC \text{ on condition that } V_{end} \geq VRC \qquad (9)$$

Regarding to formula 8, rule curve coeficient ($0 \leq RCC \leq 1$) is spesific in every reservoir. This coefficient is based on inflow-demand, reservoir capacity and operating policy. In this study, the equation of RCC line is assumed to follow functional equation of Batujai reservoir.

For more practical [33], WL ($=0.01.V_{beg}$) is assumed to be 1% of the beginning volume of Batujai Reservoir

 https://doi.org/10.1051/matecconf/201814703005

(V_{eff} = 18.000.000 m^3, with evaporation of 4-6 mm/day and seepage of 0.08 – 0.70 lt/s).

NE's development applies the numbering of streams and juctions respectively such as main river 100, and tributary 200, 300 and so on, while for junction number is {(tributary number: 100) – 1} (Figure 5). The output of model performance test will prove that the field conditions is uncontrolled, which means existing K-factor class is unequal between upstream and downstream.

In MEQAA-G, there is about 1000 lines of syntax code, including: i) tracking (including NE) and ii) equalization. Trial of the code uses hypothetical data, including river scheme, inflow, dan demand.

4 Results and Discussion

4.1. System scheme tracking

In Figure 5, there are: i) 24 HWs with B code for 23 weirs/diversions and EM code for 1 small reservoir, ii) number 100 in the main river, then number 200 – 500 in every branch river, and iii) junction number 1 at tributary 200, 2 at 300, etc.

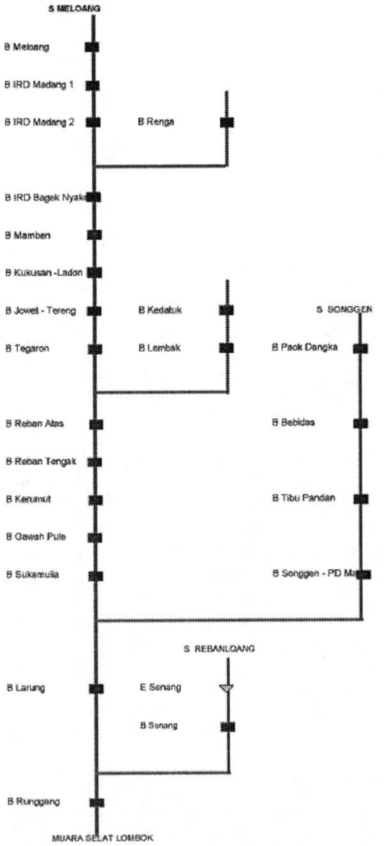

Fig. 5. Node codes of HW and junction in Kukusan Tanggek system scheme

4.2 River group and network equation structure

Based on system scheme track (Figure 5), the model automatically produces river grouping (Table 6). In this case, there are 5 river groups with numbers 100-500 (Table 6). NE structure is on the Excel sheet (Table 7), generally described that: i) NE translates the mass balance principles and optimation that detailed into 3 to 10 equations to form computational equations in every HW, ii) If a HW has demand =0, then link of gap will go towards HW in the downstream, and iii) there is no link of gap in the most downstream HW, due to the estuary.

Table 6. River grouping

43	48	53	58	63
RIV_100	RIV_200	RIV_300	RIV_400	RIV_500
B MELOANG	B RENGA	B KEDATUK	B PAOK DANGKA	EM SENANG
B MADANG 1	B BAGEK NYAKE	B LEMBAK	B BEBIDAS	B SENANG
B MADANG 2	B MAMBEN	B REBAN ATAS	B TIBU PANDAN	B RUNGGANG
B BAGEK NYAKE	B KUKUSAN-LADON	B REBAN TENGAK	B SONGGEN-PDMARE	
B MAMBEN	B JOWET-TERENG	B KERMUT	B LARUNG	
B KUKUSAN-LADON	B TEGARON	B GAWAH PULE	B RUNGGANG	
B JOWET-TERENG	B REBAN ATAS	B SUKAMULIA		
B TEGARON	B REBAN TENGAK	B LARUNG		
B REBAN ATAS	B KERMUT	B RUNGGANG		
B REBAN TENGAK	B GAWAH PULE			
B KERMUT	B SUKAMULIA			
B GAWAH PULE	B LARUNG			
B SUKAMULIA	B RUNGGANG			
B LARUNG				
B RUNGGANG				

Table 7. Some specific MEQAA-G worksheeet format in Kukusan Tanggek system

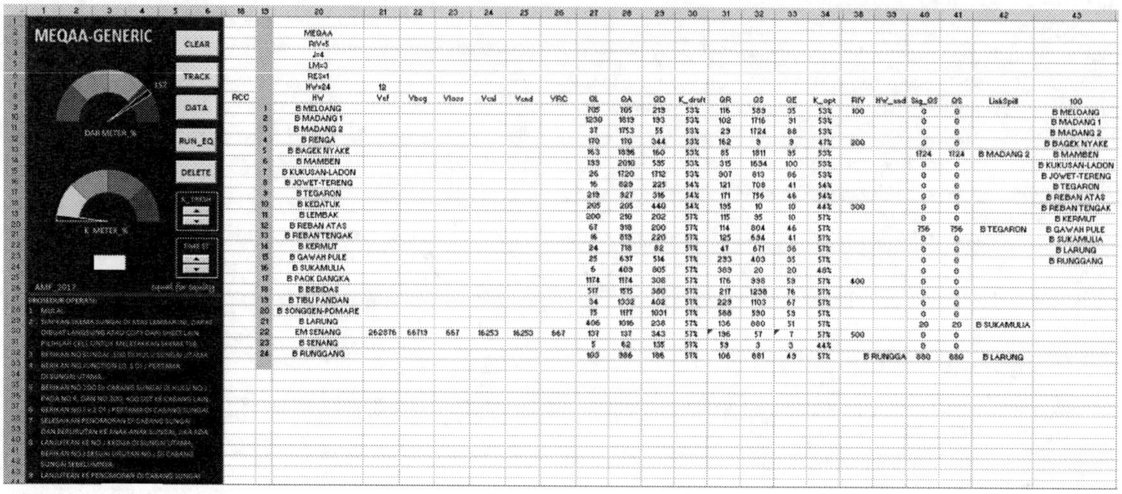

RCC		HW	Vef	Vbeg	Vloss	Vcal	Vend	VRC	QL	QA	QD	K_draft	QR	QS	QE	K_opt	RIY	HW_end	Sig_QS	QS	LinkSpill	100
		MEQAA																				
		RIV=5																				
		J=4																				
		LM=3																				
		RES=1																				
		HW=24	18																			
	1	B MELOANG							105	105	213	53%	116	589	35	53%	100		0	0		B MELOANG
	2	B MADANG 1							1230	1819	193	53%	102	1716	31	53%			0	0		B MADANG 1
	3	B MADANG 2							37	1753	55	53%	29	1724	88	53%			0	0		B MADANG 2
	4	B RENGA							170	170	344	53%	162	9	9	47%	200		0	0		B BAGEK NYAKE
	5	B BAGEK NYAKE							163	1836	160	53%	85	1811	35	53%			1724	1724	B MADANG 2	B MAMBEN
	6	B MAMBEN							133	2010	535	53%	315	1634	100	53%			0	0		B KUKUSAN-LADON
	7	B KUKUSAN-LADON							26	1720	1712	53%	907	813	66	53%			0	0		B JOWET-TERENG
	8	B JOWET-TERENG							16	829	225	54%	121	708	41	54%			0	0		B TEGARON
	9	B TEGARON							219	927	316	54%	171	756	48	54%			0	0		B REBAN ATAS
	10	B KEDATUK							205	205	440	54%	195	10	10	44%	300		0	0		B REBAN TENGAK
	11	B LEMBAK							200	210	202	57%	115	95	10	57%			0	0		B KERMUT
	12	B REBAN ATAS							67	918	200	57%	114	804	46	57%			756	756	B TEGARON	B GAWAH PULE
	13	B REBAN TENGAK							16	813	220	57%	125	684	41	57%			0	0		B SUKAMULIA
	14	B KERMUT							24	718	82	57%	47	671	36	57%			0	0		B LARUNG
	15	B GAWAH PULE							25	637	514	57%	233	403	35	57%			0	0		B RUNGGANG
	16	B SUKAMULIA							6	409	805	57%	369	20	20	46%			0	0		
	17	B PAOK DANGKA							1174	1174	308	57%	176	998	58	57%	400		0	0		
	18	B BEBIDAS							517	1515	360	57%	217	1238	76	57%			0	0		
	19	B TIBU PANDAN							34	1332	402	57%	229	1103	67	57%			0	0		
	20	B SONGGEN-PDMARE							75	1177	1001	57%	588	590	53	57%			0	0		
	21	B LARUNG							406	1016	238	57%	136	880	51	57%			20	20	B SUKAMULIA	
	22	EM SENANG	262876	66719	667	16253	16253	667	137	137	343	57%	136	57	7	57%	500		0	0		
	23	B SENANG							5	62	135	57%	59	3	3	44%			0	0		
	24	B RUNGGANG							103	986	186	57%	106	881	49	57%		B RUNGGA	880	880	B LARUNG	

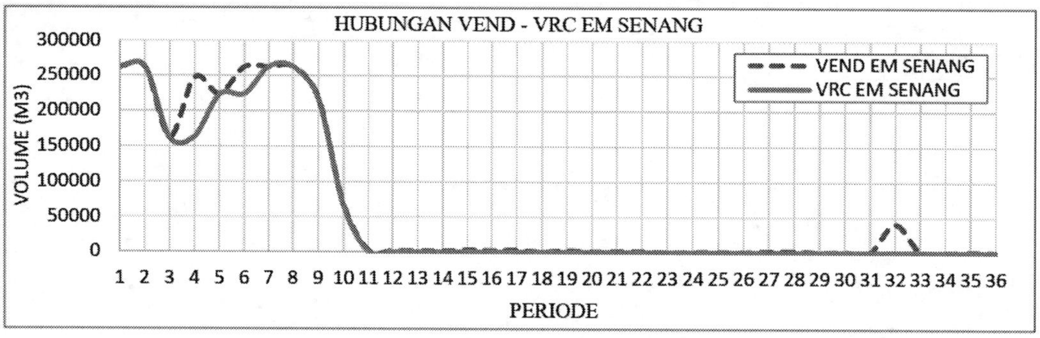

Fig. 5. Vend and VRC of Senang Small Reservoir

4.3 Result and discussion

Based on the input data of QL and QD (Table 8), MEQAA-G output is obtained as shown in Table 9 (K-factor) dan Figure 6a and 6b. It is generally described: i) on trial with hypothetical data and ends with secondary data of 36 periods of inflow and demand, there is ouput model with release that is compatible with max-equal K-factor, ii) K-factor value is strongly influenced by inflow, iii) spillout/water sharing is always the same as or more than the ecosystem rations, iv) water sharing from Senang small reservoir (embung) to Senang Weir can be controlled ($V_{end} \geq$ VRC), and v) if the inflow is high, then the iteration will be fast, and vice versa. In this case, there are about 1.298 – 844.240 iteration (K-factor's step = 0.1%), with total time around 7 minutes.

Table 8. Some local inflow (QL, lt/s) and irrigation demand (QD, lt/s) data

NO	HEADWORK	QL	QD	QL	QD
		JAN I	JAN I	JUL I	JUL I
1	B MELOANG	1057	0	337	99
2	B MADANG 1	1849	0	588	22
3	B MADANG 2	62	0	18	6
4	B RENGA	270	0	82	160
5	B BAGEK NYAKE	267	0	78	103
6	B MAMBEN	332	0	95	402
7	B KUKUSAN-LADON	49	11	13	1146
8	B JOWET-TERENG	31	27	8	172
9	B TEGARON	418	40	105	252
10	B KEDATUK	304	34	98	267
11	B LEMBAK	325	20	95	140
12	B REBAN ATAS	132	25	32	161
13	B REBAN TENGAK	31	30	7	183
14	B KERMUT	48	11	11	68
15	B GAWAH PULE	52	71	12	444
16	B SUKAMULIA	12	84	3	581
17	B PAOK DANGKA	1713	0	561	153
18	B BEBIDAS	746	0	247	193
19	B TIBU PANDAN	50	0	16	236
20	B SONGGEN-PD MARE	125	58	52	736
21	B LARUNG	698	0	283	165
22	EM SENANG	236	0	96	252
23	B SENANG	10	0	4	94
24	B RUNGGANG	194	0	72	134

Table 9 shows some model outputs (8 of 36 periods), referring to equal K-factor in every river group. The flow follows hydraulic link from HW in the upstream to HW in the downstream in each river. Downstream in river 100 will experience more than once equalization, due to its part of many rivers (200 – 500) as shown in Table 6. Under conditions of demand = 0 for the period of Jan I and Oct III (Table 9), it is written "FALSE" which means the HW is not included in the calculation of the K-factor gap (grey colour in Figure 6a). The calculation of gap is forwarded to the downstream HW.

In Figure 5, $V_{end} \geq VRC$ is fulfilled. RCC=100% occurred in Jan 1 (rainy season) that the reservoir was full. Menwhile, in Apr I - Des III (dry season), RCC=1%, the volume reservoir was minimum (≈0).

Table 9. Some MEQAA-G output (K factor)

NO	HEADWORK	1	3	10	12	19	21	28	30
		JAN I	JAN III	APR I	APR III	JUL I	JUL III	OKT I	OKT III
1	B MELOANG	FALSE	100%	88%	53%	37%	55%	14%	14%
2	B MADANG 1	FALSE	100%	88%	53%	37%	55%	14%	FALSE
3	B MADANG 2	FALSE	100%	88%	53%	37%	55%	14%	FALSE
4	B RENGA	FALSE	100%	88%	47%	37%	41%	11%	14%
5	B BAGEK NYAKE	FALSE	100%	88%	53%	37%	55%	14%	14%
6	B MAMBEN	FALSE	100%	88%	53%	37%	55%	14%	14%
7	B KUKUSAN-LADON	100%	100%	88%	53%	37%	55%	14%	14%
8	B JOWET-TERENG	100%	100%	88%	54%	38%	55%	15%	14%
9	B TEGARON	100%	100%	88%	54%	38%	55%	15%	14%
10	B KEDATUK	100%	100%	88%	44%	35%	49%	15%	15%
11	B LEMBAK	100%	100%	89%	54%	38%	55%	15%	15%
12	B REBAN ATAS	100%	100%	89%	54%	38%	55%	15%	15%
13	B REBAN TENGAK	100%	100%	89%	54%	38%	56%	15%	15%
14	B KERMUT	100%	100%	89%	54%	38%	56%	15%	15%
15	B GAWAH PULE	100%	100%	89%	54%	38%	56%	15%	15%
16	B SUKAMULIA	100%	100%	88%	53%	38%	55%	14%	14%
17	B PAOK DANGKA	FALSE	100%	100%	83%	65%	95%	22%	24%
18	B BEBIDAS	FALSE	100%	100%	83%	65%	95%	22%	24%
19	B TIBU PANDAN	FALSE	100%	100%	83%	65%	96%	22%	24%
20	B SONGGEN-PD MARE	100%	100%	100%	82%	64%	95%	21%	23%
21	B LARUNG	FALSE	100%	100%	100%	100%	100%	69%	89%
22	EM SENANG	FALSE	63%	62%	30%	28%	60%	21%	28%
23	B SENANG	FALSE	62%	61%	29%	27%	59%	20%	27%
24	B RUNGGANG	FALSE	100%	100%	100%	100%	100%	68%	89%

Still in Table 9, in the period of Oct I and Oct II, there is an equal K-factor with a very small value (11-15%) or emergency (E). In this condition, it is difficult for the water to flow along the river and also the operations can be hampered. To overcome this, it is necessary to allocate water between irrigation area in the river segment with specific HW. In this on-off system, water allocation has to be prioritized according to crop age or urgent demand. This effort is made to increase the K-factor into class K4 or K3.

Figure 6a and 6b are the examples of output machine for the period of Jan I and Jul I. The colours will be plotted on the scheme, such as grey (HW with demand = 0), red (K4), yellow (K3), green (K2), light blue (K1) and dark blue (K1*). These colours are suitable with the equalization results as shown in Table 5.

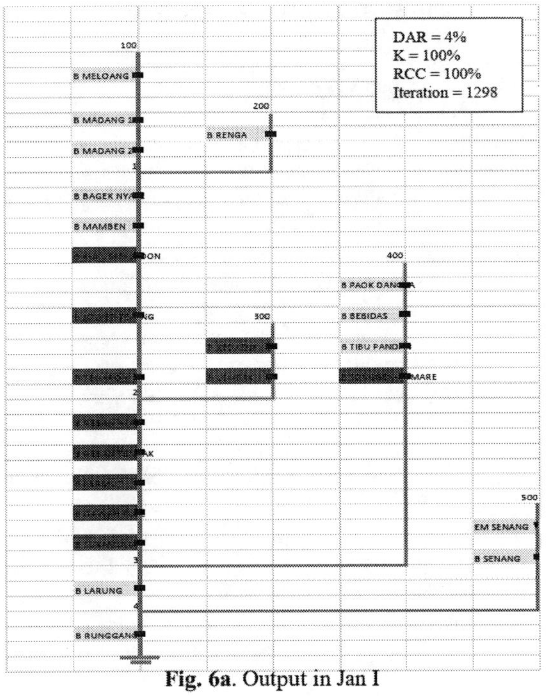

Fig. 6a. Output in Jan I

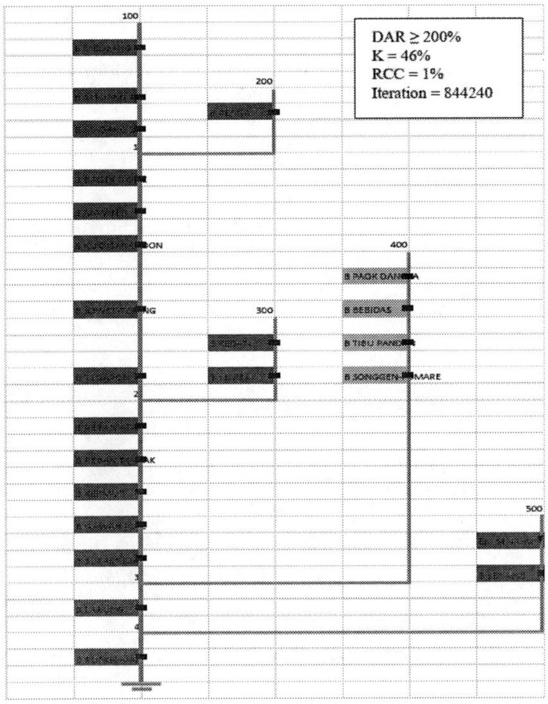

Fig. 6b. Output in Jul I

3 Conclusions and recommendation

Based on the trials in the hypothetical system and Kukusan Tanggek system, it can be concluded and recommended as follows:

1. MEQAA-G simplifies the computation of water allocation in the independent river systems with the final result always meet the max-equal K-factor, and also the ecosystem quota,
2. For the reservoir that produces a result as mentioned above in point a, the water sharing will follow the V_{end} $\geq$VRC criteria, with RCC value always varying between operating periods. Rule curve is specific that it always be different in every reservoir.
3. MEGAA-G is not influenced by types of data (hypothetical and or secondary data), since it is a conceptual model and also because NE shapes and links are always dynamic following HW configuration in the system,
4. The number of iterations depends on the density/configuration of HW (including reservoir), water balance and K-factor step,
5. MEQAA-G output is very contrastive with on the field reality (Table 3), i.e. from inequal classification of K-factor (reality) become equal.
6. It is necessary to make priority on water allocation (on-off system) if K-factor < 20% (emergency) occurs in some HWs in the system, and
7. MEQAA-G needs to be applied in a more complex independent system with cascade/multi reservoirs or dependent system (2 watersheds) to interdependent systems (more than 2 watersheds) in Lombok RB and another.

Thanks are due to individual consultant as youth professional engineer in Water Allocation Unit – Balai Wilayah Sungai Nusa Tenggara I, Lombok River Basin.

References

1. R. Azuan, AH. Wahyudi, Sobriyah, PKOWDR, *Media Teknik Sipil,* **IX**, 71-75 (2009)
2. D. Dutta, S. Kima, J. Vazea, J. Hughesa, A. Yanga, WASWRM, *21st International Congress on Modelling and Simulation. Australia,* 2040-2046 (2015)
3. Eriyatno, *Ilmu Sistem-Meningkatkan Mutu dan Efektivitas Manajemen*, Jilid I E **IV** (2012)
4. A.M. Fagi, PSDA, *Buletin Iptek Tanaman Pangan*, **2** No 1, (2007)
5. A.M. Farriansyah, G.R. Novelia, B. Husnan, DEMWA, *Prosiding Seminar Internasional HATHI-V*, 722-732 (2016)
6. S.S. Fayaed, A. El-Shafie, O. Jaafar, RSSOT, *Stoch Environ Res Risk Assess*, 27:1751–1772 (2013)
7. M.A. Fulazzaky, CIWRM, *Water*, **6**, 2000-2020 (2014)
8. S.D. Gorantiwar, I.K. Smout, *Performance Assessment Of Irrigation Water Management Of Heterogeneous Irrigation Schemes: 1. A Framework For Evaluation. Irrigation and Drainage Systems*, Loughborough University's Institutional Repository (2005)
9. D. Haro, J. Paredes, A. Solera, J. Andreu, MSOWAPWNFP, *Water Resour Manage,* 26:4059–4071 (2012)
10. W. Hatmoko, IKAI, *Paper Conference, Seminar INAICD*, **1** (2014)
11. W. Hatmoko, W. Triweko, D. Yudianto, SPKAA, *Jurnal Teknik Hidraulik Puslitbang SDA*, **3**, 71-86 (2012)
12. G. Irianto, PWS, *Kompas* (2004)
13. M. Kartabrata, W.D. Marjanto, PCMDAA, *Jurnal Informasi Teknik, Ditjen Pengairan-Dep PU,* **14**, (1994)
14. T.J. Kim, R.A. Wurbs, MRRSM, *KSCE Journal of Civil Engineering*, 15(8):1457-1467 (2011)
15. Koch, Grunewald, CMSDRWRMP, *Water Resour Manage*, 23:1403-1422 (2008)
16. Z.W. Kundzewicz, J. Kindler, MCERA, *Modelling and Management of Sustainable Basin-scale Water Resource Systems (Proceedings of a Boulder Symposium, July 1995), IAHS Publ. 231*, 217 – 224 (1995)
17. K.S. Meijer, W.N.M. Krogt van der, E. Beek van, NAIEFR, *Water Resour Manage*, 26:1271–1286 (2012)
18. J.K. Mutiga, S.T. Mavengano, S. Zhongbo, T. Woldai, R. Becht, WAPTMWUC, *Water Resour Manage*, 24:3939–3959 (2010)
19. H. Pawitan, B.I. Setiawan, B. Kartiwa, K. Subagyono, P. Rejekiningrum, MPAPBKL, *Laporan Akhir Penelitian Kerja Sama Kemitraan Penelitian Pertanian dengan Perguruan Tinggi (KKP3T)* (2009)
20. B.J.C. Perera, B. James, M.D.U. Kularathna, REALM, *Journal of Environmental Management*, **77**, 291–300 (2005)
21. D. Rohmat, UKKKSDA, *talk show Hari Air Dunia* (2010)
22. R. Roozbahani, B. Abbasi, S. Schreider, A. Ardakani, MATRWA, *Water Resour Manage*, 28:5447–5463 (2014)
23. R. Roozbahani, B. Abbasi, S. Schreider, OAWCS, *Ann Oper Res*, 229:657–676 (2015)
24. A. Singh, CMMWR, *Hydrogeology Journal*, 23: 1217–1227 (2015)
25. H. Sosiawan, K. Subagyono, Strategi Pembagian Air Secara Proporsional Untuk Keberlanjutan Pemanfaatan Air. Balai Besar Penelitian dan Pengembangan Sumberdaya Lahan Pertanian, *Pengembangan Inovasi Pertanian* 2 (4) (2009)
26. S. Strauß, WC, *Hum Ecol*, 39:69–79 (2011)
27. S. Suryadi, Kekeringan dan Hak Guna Air, *Koran Kompas 09 Agustus* (2006)
28. B. Tongongar, C.E. Kan, J.H. Chen, DFT, *American-Eurasian J. Agric. & Environ. Sci*., **3** (3) (2008)
29. UN Water, World Water Development Report 2015, "Water for a Sustainable World", *UNESCO* (2015)
30. R.A. Wurbs, Comparative Evaluation of Generalized Reservoir/River System Models, Technical Report No. 282 Texas Water Resources Institute (2005)

31. Y.B. Yazdeli, O.B. Haddad, E.F. Mehdipour, M.A. Mariño, ERSO, *Water Resour Manage*, 28:715–729 (2014)
32. B. Yulistiyanto, B.A. Kironoto, PPSDA, *Media Teknik*, **2** (2008)
33. BWS-NT I, Rencana Alokasi Air Tahunan (RAAT) WS Lombok, *Balai Wilayah Sungai Nusa Tenggara I* (2016)

Diversion Canal to Decrease Flooding (Case Study : Kebon Jati-Kalibata Segment, Ciliwung River Basin)

Dian Indrawati[1,*], *Iwan* K. Hadihardaja[2], M. *Bagus* Adityawan[2], *Syambali* F. Pahrizal[1] and *Fajar* Taufik[1]

[1]Faculty of Engineering, Jenderal Achmad Yani University, Jalan Terusan Jenderal Sudirman, PO BOX 148, Cimahi, West Java, Indonesia
[2]Faculty of Civil and Environmental Engineering, Institut Teknologi Bandung, Jl. Ganesha No.10 Bandung 40132, West Java, Indonesia

Abstract. The flood in Jakarta has become a national concern in Indonesia. It is a haunting disaster, with a high probability to happen when heavy rainfalls in Jakarta and/or its upstream area. Based on data that was provided by Public Work Agency of DKI Jakarta, there are 78 vulnerable points of inundation in which, most of them are located in Ciliwung river basin, commonly in the meandering segments. One of the worst flooding occurs in Pancoran, at Kebonjati to Kalibata segment in particular. The river discharge in this segment is much higher as compared to the carrying capacity. In addition, this area has a high density of population and thus, difficult to increase the *river capacity* by enlarging the river dimension. In this research, a closed diversion canal is proposed as a solution. The effectiveness of the solution is evaluated using a numerical model, HEC-RAS 4.1. The diversion canal is designed as two culverts, with 2.0 m in diameter. Nevertheless, hydraulic jump may occur at the outlet of the canal due to the relatively steep slope. Therefore, the canal outlet should be designed accordingly. A Hydraulic structure such as a stilling basin can be employed to reduce the energy. The results show that the diversion canal has a good performance in decreasing water level and flood discharge in the study area. The canal has the capacity of 17,72 m3/sec and succesfully decreases the water level by 4.71 – 5.66 m from flood level for 2 – 100 years returned period.

1 Introduction

Based on data from Indonesian National Board for Disaster Management (BNPB), since 1815, flood become the most frequently disaster with 31.5% of events compare to others such as puting beliung, terorism, fire and earthquake. Specially for DKI Jakarta, during period 2011-2015, there are 98 flood events occur and most of them were happened on Ciliwung river basin and make these a national concern in Indonesia. It is a haunting disaster with a high probability hazard in every heavy rainfall occurs in DKI Jakarta area and/or its upstream of river basin.
Flood in the Ciliwung river basin had presented annually since the 1600's and in order to manage its disaster, several projects have been developing such as Amanus straits digging in North Bandengan in 1647, Bageracht and Mookervart from 1678 to 1686, and The Jakarta Flood Canal i.e. West Flood Canal (KBB) and East Flood Canal (KBT) recently.
Furthermore, a numerous sources of this disaster had been identifying, and one of them is the insufficient of the carrying capacity of the river specially in meandering segments. One of the worst flooding occurs in Pancoran area whileas in a year several spots in this area reach 2 meter-height.
Several studies related to this flood had conducted i.e Formanek [1] studied about Regional Flood Map Index for Ciliwung river flood based on FESWMS modelling, Emam [2] using HEC-HMS modelling in order to analyse land-use changes and assesment of climate in upper Ciliwung river basin, and Murniningsing [3] shows that the developing of Ciawi, Sukamahi and Katulampa weir DAM takes less significant values compared to its huge cost.
This study will analyze the effectiveness of diversion canal in order to decrease the flooding specially in Kebonjati- Kalibata segment which indicated as the source of problem in Pancoran area.
Although in some of researches show lackness of this method because of its impacts regard to sedimentation [4] and [5], the river diversion recomended for river on flattened area [6] with dense population.

2 Literature Study

The flow in diversion chute canal should be occoring in supercritical scheme as the result of the deployed flow on the upstream structure which is requiring quick acceleration. So in the design flow, its depth in this section should be in uniform depth. Thus, it is necessary to define which condition the flow occur along the canal. Also whileas a possibility of occuring the hydraulics jumps inside, if it is happen, the location of it should be determined. The last thing, back water curves probably occur inside and need to be analyzed along the canal [7].

* Corresponding author: author@e-mail.org

In HEC-RAS [8], the diversion chute canal is modeling as the culvert condition. Which is the assumption of inlet control is that the flow passes through critical depth near the inlet and transiting to supercritical flow. If the hydraulics jump occured in this barrel, it is assumed its cause the pipe to pressuirize along the length and flow will act like an orrifice type. If this occurs, then the outlet control will become the answer.
For computing inlet control headwater with submerged inlet, the equation are :

$$\frac{HW_i}{D} = c\left[\frac{Q}{AD^{0.5}}\right]^2 + Y - 0.5S \quad (1)$$

Whereas Hwi is the headwater energy depth above the invert of the culvert inlet (m), D is the interior height of the culvert barrel (m), Q is the discharge through the culvert (m3/s), A is the full cross sectional area of the culvert barrel (m2), S is the slope of culvert barrel (m/m) and K,M,c,Y is equation constants, which vary depending on culvert shape and entrance conditions.
On contrary, for the outlet control flow using Bernoulli's equation in order to compute the change of energy through the barrel under outlet condition is following:

$$Z_3 + Y_3 + \frac{a_3V_3^2}{2g} = Z_2 + Y_2 + \frac{a_2V_2^2}{2g} + H_L \quad (2)$$

Where Z3 is the upstream invert elevation of the culvert (m), Y3 is the depth of water above the upstream culvert inlet (m), V3 is the average velocity upstream of the culvert (m2/s), a3 is the velocity weighting coefficient upstream of the culvert (dimensionless), g is the acceleration of gravity (m/s2), Z2 is the downstream invert elevation of the culvert (m), Y2 is the depth of water above the downstream culvert inlet (m), V2 is the average velocity downstream of the culvert (m2/s), a2 is the velocity weighting coefficient downstream of the culvert (dimensionless) and HL is the total energy loss through the culvert (m).
The head loss is computed using the formula :

$$H_L = h_{en} + h_f + h_{ex} \quad (3)$$

Which hen is the entrance loss (m), hf is the friction loss (m) and hex is the exit loss (m). The friction loss in the culvert is computing using Manning's formula derived as follow :

$$h_f = L\left(\frac{Qn}{1.486AR^{2/3}}\right)^2 \quad (4)$$

Where hf is the friction loss (m), L is the culvert length (m), Q is the flow rate in the culvert (m3/s), n is the Manning's roughness coefficient (dimensionless), A is the flow area (m2) and R is the hydraulic radius (m).

3 Study Site and Methodology

3.1 Study Site

In order to manage the flood in Pancoran area, specially at Kebonjati to Kalibata segment, Public Work Regional Office of DKI Jakarta Province will develop the diversion canal which located on 6°15'16.99"S - 6°15'14.51"S and 106°51'37.96"E - 106°51'40.28"E describes on Figure 1 below:

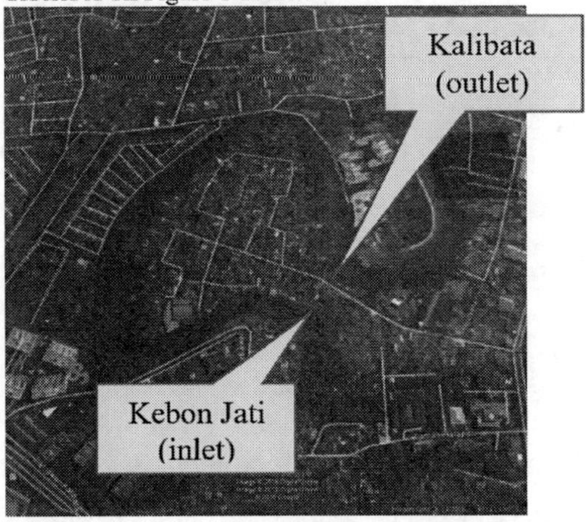

Fig. 1. Diversion Canal Location at Kebon Jati and Kalibata segment

The diversion, which located before the Manggarai water gate, is a subsystem of flood management in DKI Jakarta province and being a part of KBB.

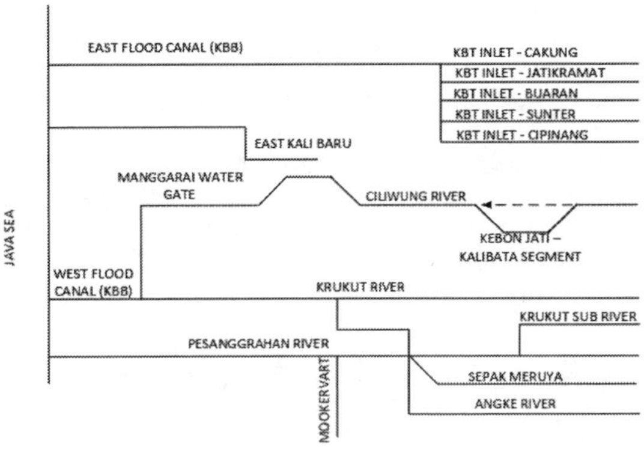

Fig. 2. Schematization of The Diversion Canal of Kebonjati – Kalibata Segment on Flood Management in DKI Jakarta

3.2 Methodology

This study consists of hydrological and hydraulics model which are hydrological model was used to obtain flood hydrograph and hydraulics model was used to simulate the flow on the river. In order to estimate several rainfall returned period, the data was extrapolated using analysis frequency method.
For the hydraulics model, its been calibrated by comparing 2 year flood return period with capacity of river bankfull in the inlet of diversion canal. And HSS Nakayashu has closed enough values compares to Rational, Haspers and HSS Snyder-Alexjev. Moreover, the schematication of the modeling shows in Figure 3 below.

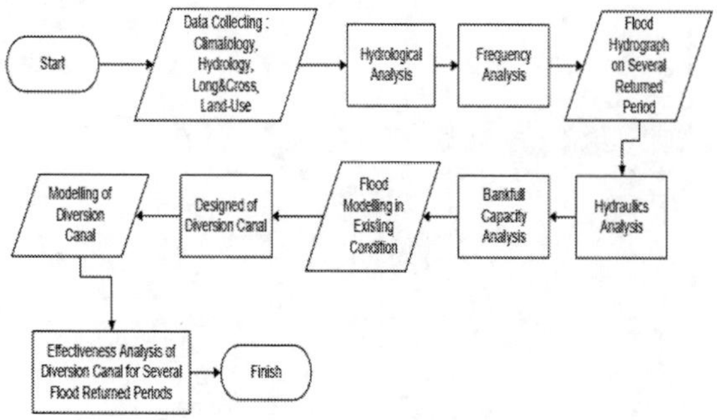

Fig. 2. Flowchart Modelling

4 Results and Discussion

4.1 Boundary and Initial Condition

The model was extended about 11.824 km which is divided into 210 cross sections namely by P1 to P210. From the upstream to the canal, i.e P1 to P169, has 10,271.58 meter distance for mufflling the model dissipation and analyze the backcurve current. And there are 739.35 meter along after the tunel in order to analyze the effect of canal on the downstream river which defined as P188 to P210.

For the boundary condition, the model using flood hydrograph on the upstream and water elevation in the downstream, which are the water surface will compute through existing (without canal) and design (with canal) condition, each running 2, 5, 10, 25, 50, and 100 of the return periods of flood hydrograph.

In order to validation, the 2 years of hydraulics river model is comparing by bankfull capacity of the river to adjust the appropriate flood hydrograph for the HEC-RAS model as mention in the previous section.

4.2 Diversion Canal Data and Coefficient

The diversion canal will be held on P169 for the inlet and P188 for the outlet with 813.75 meter length. Based on the multiple choice analysis, circle culvert has been selected compare with box culvert or other shape of culvert.

The shape of the culvert itself defined by 2 circle culverts with diameter 2 meter each. The value of Manning's is 0.013, the entrance loss coefficient is 0.5, and the exit loss coefficient is 0.5. The culvert also has bottom elevation on inlet +12.81 meter and outlet +9.36 meter.

4.3 Hydrological Model

These hydraulics model held by several hydrograph with 2, 5, 10, 25, 50, and 100 return periods, whereas these discharges are being derivated from 11 (eleven) year of maximum daily rainfall data since 2005 until 2015 from selected stations and was transformed to the rainfall area using Polygon Thiessen which has influences area each describe on Table 1 below.

Table 1. Influence surfaces of Rainfall Station in Ciliwung River Basin

Stations	Infuence Area (km^2)
Bendung Gadog	89.61
Cawang	64.10
FT UI	114.04
Gunung Mas	101.61

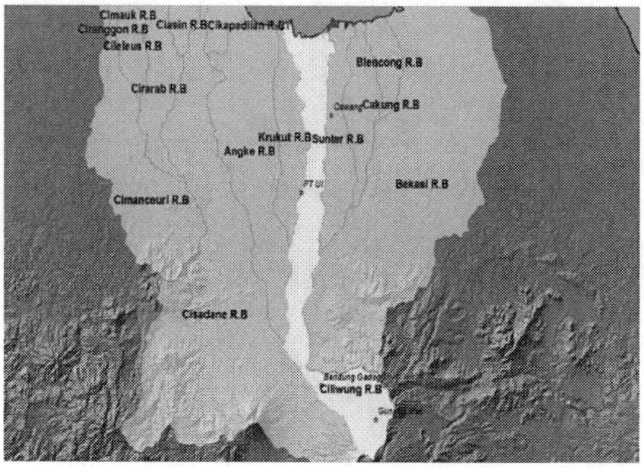

Fig. 3. Ciliwung River Basin

Table 1. Rainfall Area in Ciliwung River Basin using Polygon Thiessen

No.	Year	Bendung Gadog	Cawang	FT UI	Gunung Mas	Average
1	2005	17.32	66.96	49.13	39.96	41.98
2	2006	12.27	12.69	17.65	14.44	14.60
3	2007	11.98	14.23	34.84	19.83	21.59
4	2008	14.53	15.12	43.63	19.57	25.00
5	2009	17.62	20.81	27.34	16.08	20.75
6	2010	17.87	16.19	20.58	25.58	20.53
7	2011	15.09	8.84	18.45	12.11	14.22
8	2012	19.70	13.45	21.92	11.92	17.16
9	2013	21.50	14.68	20.27	17.98	18.97
10	2014	20.95	19.38	26.46	22.71	22.86
11	2015	15.17	15.31	18.50	10.44	14.92

Source : Puslitbang SDA, 2016

Based on the frequency analysis, Log Pearson III distribution was accepted with the deviation standar value is 0.13, the skewness coefficient (Cs) is 1.25 and the Kurtosis coefficient (Ck) is 2.44. Further, fit tests using Chi square and Smirnov Kolmogorov that show the acceptance of Log Pearson III distribution.

The values of calculated Chi-Square is 3.09, The Critical Chi-Square value is 5.99, the freedom of degree is 2.00 and the significant degree is 0.05. On contrary, for the smirnov kolmogorov test results the D max value is -0.428, the significant degree is 0.050 and the Critical D is 0.396.

Because of the lack of hourly rainfall data in its location, the rainfall pattern assumed by using average pattern rainfall in Indonesia specially in Jawa island with distribution 55%, 14%, 10%, 8%, 7% and 6% for six hours of rainfall. Wheras calculation of the returned period of rainfall is :

Table 2. Rainfall Distribution in Ciliwung River Basin

Parameter		Returned Period							
		2	**5**	**10**	**25**	**50**	**100**	**200**	**1000**
Rainfall (R)		21.45	25.21	30.47	38.49	45.59	53.75	63.17	91.18
Netto Rainfall [mm/day]		10.61	12.47	15.07	19.04	22.55	26.58	31.24	45.10
Hour	Distribution	Hourly Rainfall [mm/hour]							
1	55 %	5.84	6.86	8.29	10.47	12.40	14.62	17.18	24.80
2	14 %	1.49	1.75	2.11	2.67	3.16	3.72	4.37	6.31
3	10 %	1.06	1.25	1.51	1.90	2.25	2.66	3.12	4.51
4	8 %	0.85	1.00	1.21	1.52	1.80	2.13	2.50	3.61
5	7 %	0.74	0.87	1.05	1.33	1.58	1.86	2.19	3.16
6	6 %	0.64	0.75	0.90	1.14	1.35	1.60	1.87	2.71

Furthermore, the flood discharge calculated by using UH Nakayashu based on fittest analysis. The results of Nakayashu unit hydrograph are shown in Table 3.

Table 3. Figure 5 Flood Discharge using UH Nakayashu

Returned Periods	Discharge [m^3/sec]
2	55.44
5	63.50
10	78.40
25	102.59
50	125.17
100	152.27
200	184.89
1000	290.36

4.4 Modelling of Existing Condition

Based on its contour, Ciliwung river has steep and tight topography in the upstream and become more mild and widen follows the downstream. Because of the lack of observation discharge data, the result from hydrology modelling is being compared with analysis of bankfull capacity at P146. The modelling results for two year return period of rainfall will lead 102.59 m3/s of discharge with 5.172 m of water elevation and the rating curve resulted 103.64 m3/s of discharge with 5.172 m of the water elevation, respectively.

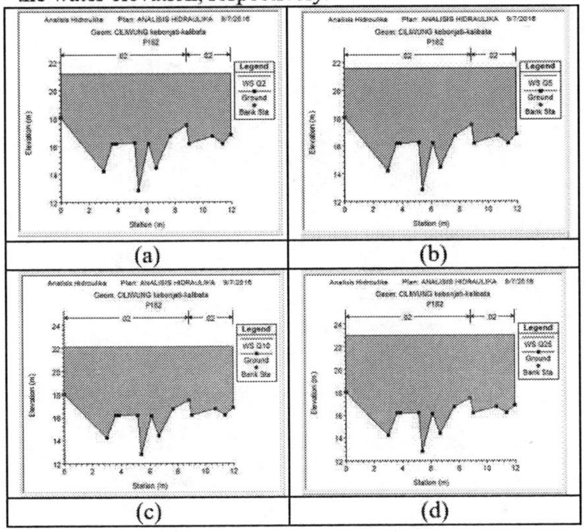

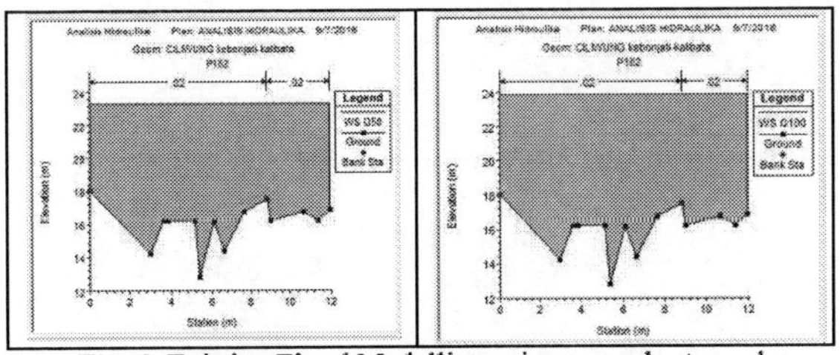

Fig. 4. Existing Flood Modelling using several returned periods. (a) 2 year; (b) 5 year; (c) 10 year; (d) 25 year; (e) 50 year; and (f) 100 year

Figure 5 shows the overtopping of Ciliwung river specially in the inlet location for all years of return periods. The discharge of each is 55.44 m3/s, 63.50 m3/s, 78.40 m3/s, 102.59 m3/s, 125.17 m3/s and 152.27 m3/s for 2, 5, 10, 25, 50 and 100 year, respectively.

4.5 Modelling of Diversion Canal

Based on model, the diversion canal has super critical flow with Froude number 1.418 inside and has pressure difference 43654.50 kg/m.s2. This modeling using two condition in order to calculate head loss i.e normal and flooding. Total head loss in flood condition is 0.402 meter and on normal condition and 0.492 meter on flooding condition which the water temperature is 30°C. The diversion canal obviously making a significant impact for decreasing water elevation in Kebonjati-Kalibata segment and clearly reducing the floodi in Ciliwung river basin. With fully submerged flow, the diversion canal could carried 35.44 m3/s discharge from inlet to outlet and shortcut the way along meandering segment in Ciliwung river as shown on Table 4.

Table 4. Recapitulation of Influence of the Diversion Canal to The Water Level Elevation on Ciliwung River

Return Periode	Flood Discharge [m^3/s]	Diversion Canal Discharge [m^3/s]	Total Discharge [m^3/s]	Water Elevation [m]		Note
				Before Diversion Canal	After Diversion Canal	
2	55.44	35.44	20.00	21.17	16.46	Down 4.71 m
5	63.50	35.44	28.06	21.54	16.76	Down 4.78 m
10	78.40	35.44	42.96	22.16	17.08	Down 5.08 m
25	102.59	35.44	67.15	23.02	17.52	Down 5.50 m
50	125.17	35.44	89.73	23.33	17.82	Down 5.51 m
100	152.27	35.44	116.83	23.87	18.21	Down 5.66 m

5 Conclusions

Results of the simulation of its diversion canal shows on Figure 7 below, whereas on every returned period, water level becomes decreased significantly from 4.71 – 5.66 meter. Its means the diversion canal achieved 22.25% - 23.71% decrease of flood in this segment.

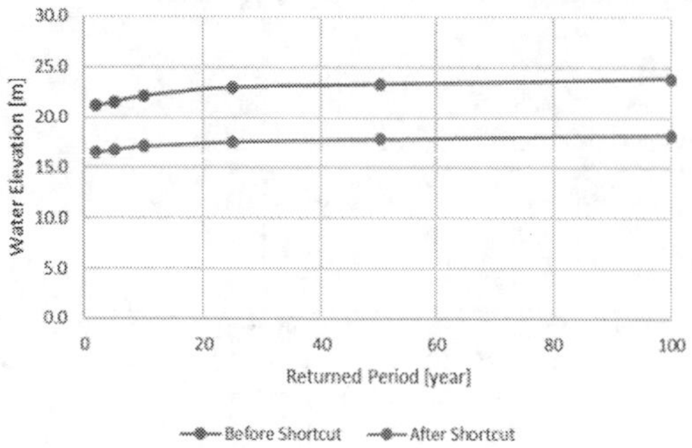

Fig. 5. Water Elevation Before and After Diversion Canal Development

Eventhough the diversion canal obviously making a significant impact for reducing the flood in meandering segment of Kebonjati-Kalibata area. But there are some circumtances related to supercritical flow along the canal and hydraulic jump on the canal outlet. They will make a scouring and sedimentation problem in the canal downstream and these complex interacting system will influences the river morphology.

Further, research about sedimentation need to be held seriously and developing of stilling basin on the canal outlet is necessary in order to muffle the hydraulic jump.

This study was supported by Public Work Regional Office of DKI Jakarta Province for project title as "Planning and Detail Designed of Management Ciliwung River from Manggarai Watergate to its Upstream" specially for the topography data and particularly indebted to the UNJANI Competitive Research 2017 for some part of financial sharing and support.

7 References

1. Formanek. Adam, Rasmiaditia Silasari, M. Syahril B.K, Kardhana.. H, J. Eng. Technol. Sci. E **45**(3), 307-325 (2013)
2. Emam A.R, Mishra B.K, Kumar P, Masago Y, Fukushi K, Water. J.E **8**(12), 559 (2016)
3. Murniningsih S, Anggraheni E, Int. Journal of Environment, Chemical, Ecological, Geological and Geophysical Eng. J.E **10**, 2 (2016)
4. Sharp A.J, Heath. R.E, Clifton N.D, Journal of Water Res and Protection, E **5**, 16-29 (2013)
5. Bowles D.S, Grant J.L, Humphries W.E, O'hayre A.P, WR Bulletin EWRA, E. **21**(6), 995-1003 (1985)
6. BUREC, *Design of Small Dams* (WR Tech Publication, Denver, 1987)
7. Silva S.A.O, *River Diversion Structures: Conception, Design and Adaption to Definitive Structures* (Instituto Superior Tecnico)
8. Brunner G.W, *HEC-RAS River Analysis System Hydraulic Reference Manual Version 4.1* (U.S. Army Corps of Engineers: Davis, CA, 2010)

Hydraulic and Groundwater Chemical Parameters of the Aquifer in Malakasari, Bandung

Jemi S. Ahnaf[1,*], *Zufialdi Zakaria*[1], and *Aton Patonah*[1]

[1]Geology Engineering Department, Padjadjaran University, 45363 Sumedang, Indonesia

Abstract. In order to reveal the physical condition of the aquifer, the pumping test using Cooper-Jacob (1946) principle has conducted at well SM5. The observation data of the test then processed to generate various value of hydraulic properties i.e. 3.241x10-4 cm^2/sec for transmissivity (T), 8.103x10-6 cm/sec for conductivity (K), 0.05055 for storativity (S), and 3.852x10-3 ft^{-1} for specific storage (Ss). These data show that the aquifer composed of unconsolidated sedimentary rocks ranged from coarse sand to silt. In addition, also performed the feasibility test of groundwater by using Multimeter which produces chemical parameter data. The chemical parameter of eight well samples have average values of 6.62, 766.25 µs/cm and 376.25 mg/L for pH, electric conductivity (EC), and total dissolved solid (TDS) respectively, while physical observation shows no turbidity and odor.

1 Introduction

1.1 Research background

Human needs for fresh water will continue to increase as the population increases, and indirectly causes studies on hydrogeology continue intensively in the future. This not surprising in behalf of water is one of the eminent human needs both for rural and urban communities, specifically it stated in [1] that surface water and groundwater are the main water sources that people used to meet various needs. Thus, it would be useful if groundwater studies were conducted, at least to perceive potential areas for water exploration. Furthermore, [2] reinforced that the need for water by humans is endless, especially clean water that feasible for household uses such as bathing, cooking, even the most crucial is to drink.

This study aimed to determine the physical properties of aquifer and water chemical parameters in Malakasari, it is expected that the results of this research will be a preliminary study for exploration or further research. Meanwhile in the framework of exploration, studying hydrogeological aspects such as aquifer or water chemical parameters are particularly important as it supports the sustainability of the exploration. The aquifer plays an important role in storing and passing water so that it can be assessed its physical properties to know the values of transmissivity (T), conductivity (K), storativity (S), and specific storage (Ss). While water chemical parameters can be used to determine groundwater ground quality before further advance water chemistry analysis, by observing it, parameters such as pH, Electric Conductivity (EC), and Total Dissolved Solid (TDS) are known.

1.2 Research location and geology condition

This research located in Malakasari Village, Baleendah district, Bandung Regency, West Java Province. Malakasari has an average air temperature of 24°c-30°c and fairly fluctuating rainfall average of 1,856 mm per year which administratively bordered by Bojong Malaka in the north, Rancamanyar in the west, Baleendah in the south, while the eastern part is bordered by the Citarum river and Andir. Based on the Geological Map of the Garut and Pamuengpuek quadrangle [3], the research area included in the quaternary deposits resulting from volcanic activities, the area included in Qd unit which is a sediment lake with rock composition such as clay, silt, fine-coarse sand, and gravel, generally these rocks are tuffaceous and has young age (Quaternary). Furthermore, in a view of geomorphology, this area classified as highland category that located at altitude of 600-715 msl and belongs to the Bandung basin.

* Corresponding author: jemi14001@mail.unpad.ac.id

Fig. 1. Location of research area in Malakasari Village, District Baleendah, Bandung Regency, West Java Province. Uncoordinated and unscaled, image sourced from Google Map Pro.

1.3 Aquifer test

Aquifer is a layer of rock that has a cavity between the granules or cavities due to fractures, where these cavities contain water and are able to drain the water in sufficient quantities [4], according to Halik & S. (2008) in [5], before groundwater drilling is carried out, a preliminary ground investigation is done to determine the presence of carrier layer (aquifer). In addition, this aquifer investigation can also determine its physical condition by performing an aquifer test.

An aquifer test is a test to forecast aquifer physical condition by determining its hydraulic properties. This test do through pumping test, while pumping test itself is a practical and reliable method of estimating well performance, well yield, the zone of influence of the well and aquifer characteristics (i.e., the aquifer's ability to store and transmit water, aquifer extent, presence of boundary conditions and possible hydraulic connection to surface water). The analysis of the hydraulics of wells for the evaluation of groundwater potentialities by pumping tests falls in the category of groundwater hydrology [6]. Actually, aquifer test could also perform through slug test, but it provide limited information on the hydraulic properties of the aquifer and often produce estimates which are only accurate [7], thus pumping test more common used than the slug test.

2 Method

Pumping test by principle of Cooper-Jacob (1946) conducted to acquire hydraulic properties condition of the aquifer at SM5, this pumping conducted with a constant debit (long-term constant rate) by aid of Jet Pump. From this test, the data produced are changes in the drawdown to its elapsed time, these data are then processed using AQTESOLV (an aquifer test program) to generate hydraulic values as well as transmissivity (T), conductivity (K), storativity (S) and specific storage (Ss). In addition, all samples from eight wells analyzed by Multimeter to give picture related to water chemical parameters such as pH, EC and TDS, besides the physical parameters of water as temperature, color, and odor were also observed.

3 Result and discussion

3.1 Pumping test

Pumping test conducted in well SM5, which is a representative of the 8 wells mapped. The well is located on coordinates S 07°00'22.77" and E 107°36'30.55" at an elevation of 679 m above sea level with watertable saturated at 677.60 m, this well also has 3.80 m in depth and 0.75 m and 1.1 m for inner and outer diameter respectyively. Pumping test carried out by draining water out aided by Jet Pump, thus debit of water taken is constant amounted to 0.214 liter/s or 0.01284 m^3/m. The test is single-well test, as consequence, drowdown measured in pumped well only and induce observation glued to well SM 5. The data that produced from observation during pumping test then processed by AQTESOLV using the Cooper-Jacob (1946) principle.

Table 1. Observation data of pumping test from SM 5, displacement is the change in head relative to the static condition (drowdown), displacement = h(0) - h(t), where h(0) is the initial (static) head and h(t) is the head at time t.

Time (min)	Displacement (m)	Time (min)	Displacement (m)	Time (min)	Displacement (m)
1	0.01	14	0.12	60	0.18
2	0.02	16	0.12	70	0.20
3	0.04	18	0.13	80	0.22
4	0.07	20	0.12	90	0.23
5	0.07	25	0.13	100	0.24
6	0.08	30	0.14	110	0.25
7	0.09	35	0.16	120	0.25
8	0.09	40	0.16	135	0.26
9	0.10	45	0.17	150	0.26
10	0.10	50	0.17	165	0.26
12	0.11	55	0.18	180	0.26

Pumping test was conducted non-stop until water level shows steady state (no change displacement within a certain time), this situation occurs when the amount of water pumped equal to the entering water from the aquifer. In this observation, steady state began to occur in minute 135, therefore, in minute 180 the test was stopped and continue to observe recovery well by measuring the rise of water level to its elapsed time. Observation of recovery well done until the water level back to its original state before the well pumped. This state accomplished within a period of 240 minutes, thus recovery time is 60 minutes longer than pumping time, this situation indicates that the amount of debit taken during pumping, i.e. 0. 0128 m^3/m is greater than entering water from aquifer to the well.

The displacement-time plot shows a linear trend that corresponds to the Cooper-Jacob (1946) graph should be, it means the retrieval and processing data of the test are compatible. A derivative ds/d(log(t))-time plot provide a useful diagnostic tool for detecting deviation in the rate of displacement change, for example this plot can help in identification of wellbore storage and aquifer boundaries, unfortunately, other needed data to determine the aquifer boundaries is still lacking, so it could not be determined. But certainly from this derivative plot is known that the largest change displacement occurred in the minute 20 (lowest) to minute 100 (highest), but even then, the average change displacement occur quite negligible at around 0.05 m.

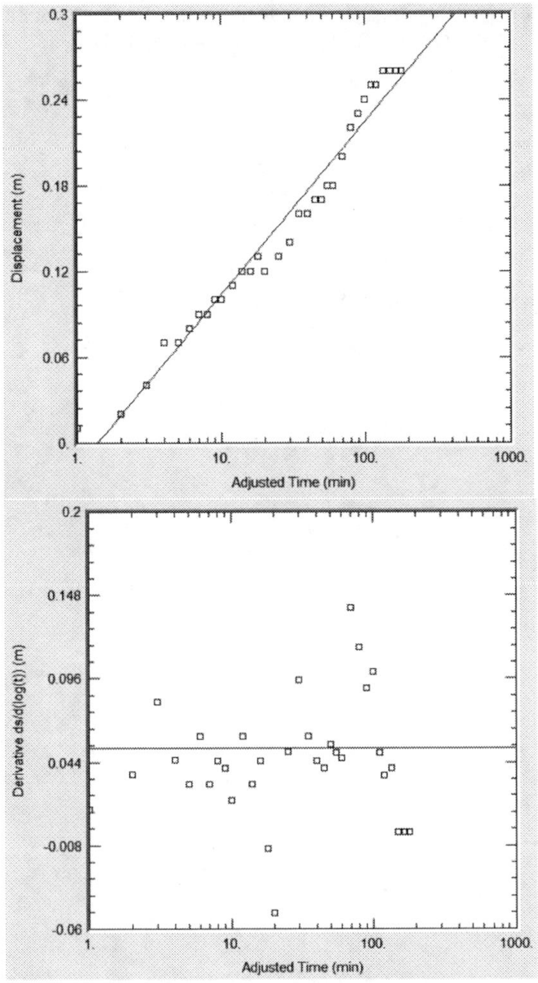

Fig. 2. These graphs are interpretation of the observational data were processed using AQTESOLV, (above) displacement-time plot, and (below) derivative ds/d(log(t))-time plot.

3.2 Hydraulic properties

The hydraulic properties determined from this research are transmissivity (T), conductivity (K), storativity (S), and specific storage (Ss), which has a value 3.241×10^{-4} cm^2/sec, 8.103×10^{-6} cm/sec, 0.05055, 3.852x10-3 ft^{-1}, respectively. These values then examined to describe the type of material making up the aquifer. In short, transimissivity used to define conductivity which then interpreted to seek the material making up the aquifer based on Domenico & Schwartz (1990) table [8]. Since the aquifer in the research area possibly sediments have not been litified (Quaternary lake deposit), used unconsolidated sedimentary rocks table to narrow down the possibilities. With the value of hydraulic conductivity 8.103x10-6 cm/sec, then material making up the aquifer should be unconsolidated coarse-fine sand to silt.

As transmissivity used to define the conductivity value, so does the storativity (S) to specific storage (Ss). The specific storage then interpreted to describe geological material of the aquifer as conductivity did. By value 3.852x10-3 ft^{-1} of specific storage and using Domenico & Mifflin (1965) table [9], the possibility constituent materials of the aquifer are dominantly plastic clay, this does not correspond to the conductivity that indicates unconsolidated coarse-fine sand to silt for the aquifer.

Table 2. Representative values of hydraulic conductivity for various unconsolidated sedimentary materials (Domenico & Schwartz, 1990) [8].

Unconsolidated sedimentary material	
Material	**Hydraulic conductivity (cm/sec)**
Gravel	3×10^{-4} to 3×10^{-2}
Coarse sand	9×10^{-7} to 6×10^{-3}
Medium sand	9×10^{-7} to 5×10^{-4}
Fine sand	2×10^{-7} to 2×10^{-4}
Silt, loess	1×10^{-9} to 2×10^{-5}
Till	1×10^{-12} to 2×10^{-6}
Clay	1×10^{-11} to 4.7×10^{-9}
Unweathered marine clay	8×10^{-13} to 2×10^{-9}

3.3 Water chemical parameters

Eight well water samples were taken and tested by using a multimeter to determine its chemical parameters. This test can be used for rough water quality analysis before further analysis related to ion-cation of water, in fact this parameter is only useful as the preliminary data and can not be used as the main data for determining the feasibility of groundwater to be drinking water, consequently it will be discussed in outline.

From the test, the average value for pH is 6.62, EC is 766.25 μs/cm, and TDS is 376.25 mg/L. All samples showed normal pH and near-acid values for the SM6 and SM8 samples. These values are quite good, because according to World Health Organization (WHO) drinking water standards in Oyem (2014) [10] recommends a pH value above 6.5 to avoid corrosion. The EC value is usually twice from its TDS, so the interpretation can be represented by TDS. The TDS shows an average value of 376.25 mg/L, WHO in [10] recommends a TDS value of 1000 mg/L, whereas Queensland Health Forensic Scientific Services (QHFSS) recommends a value below 500 mg/L for taste reason. Thus the WHO recommendation value for TDS does not meet for these samples, but however these samples are complied to TDS standard of QFHSS. In addition, it needs to be analyzed further since these samples are not directly taken from the aquifer.

Table 3. Chemical and physical parameters of water eight well samples. The measuring of physical parameters such as T, turbidity, and odor conducted at the time of sampling.

Sample	pH	EC (μs/cm)	TDS (mg/L)	T (°c)	Turbidity	Odor
SM1	6.6	1.070	520	25	no	no
SM2	6.6	900	470	25	no	no
SM3	6.8	700	340	26	no	no
SM4	6.9	830	410	24	no	no
SM5	6.6	800	390	25	no	no
SM6	6.5	580	280	25	no	no
SM7	6.6	780	380	25	no	no
SM8	6.4	470	220	25	no	no

4 Conclusion

Data interpretation of pumping test using AQTESOLV generated various values of hydraulic properties i.e. 3.241x10-4 cm2/sec for transmissivity (T), 8.103x10-6 cm/sec for conductivity (K), 0.05055 for storativity (S), and 3.852x10-3 ft-1 for specific storage (Ss), hence predicted that the constituent material of the aquifer is in the form of unconsolidated sediments which ranged from coarse-fine sand to silt. Chemical parameter analysis of well samples shows average values of 6.62, 766.25 μs/cm and 376.25 mg/L for pH EC and TDS respectively. Thus, according to WHO and QHFSS, groundwater in the Malaksari complied to standard of drinking water, making it possible to further groundwater exploration, but it needs more detailed analysis to other water chemistry such as total hardness, temporary hardness, alkalinity, anion-cation content, the content of trace element and so forth.

References

1. D.A. Enggarwati, A. Susilo, D.D. Wardhana, Physics Student Journal Univ. Brawijaya (2015)
2. Z. Lubis, N.A. Affandy, Jurnal Teknika, **6**, 577-588 (2014)
3. M. Alzwar, N. Akbar, S. Bachri, Geological Map of the Garut and Pameungpeuk Quadrangle, Java, Geological Research and Development Centre, Indonesia (1992).
4. R.S. Waspodo, jTEP Jurnal Keteknikan Pertanian, **25**, 29-35 (2011)
5. G. Halik, J. Widodo, Media Teknik Sipil, 109-114 (2008)
6. J. Bruin, H.E. Hudson, Report of Investigation, Department of Registration and Education, Illinois, (1961)
7. P.S. Osborne, Ground Water Issue-United State Environmental Protection Agency, (1993)
8. P.A. Domenico, F.W. Schwartz, Physical and Chemical Hydrogeology, John Wiley & Sons, (1990)
9. P.A. Domenico, M.D. Mifflin, Water Resources Research, **1**, 563-576 (1965)
10. H.H. Oyem, I.I. Oyem, D. Ezaweali, Research Journal of Environmental Science, **8**, 444-450 (2014).

Study on the Effectiveness of Infiltration Wells to Reduce Excess Surface Run Off In ITB

Afifah Muhsinatu Mardiah[1,*], *Cherish* Nurul Ainy [2], *Mohammad* Bagus [3], and *Dhemi* Harlan [3]

[1] Study Program of Civil Engineering, Faculty of Civil and Environmental Engineering Institut Teknologi Bandung, Jln. Ganesha No. 10 Bandung, Indonesia
[2] Master Program of Landscape Architecture, School of Design, Planning and Development Policy Institut Teknologi Bandung, Jln. Ganesha No. 10 Bandung, Indonesia
[3] Water Resources Engineering Research Group, Faculty of Civil and Environmental Engineering Institut Teknologi Bandung, Jln. Ganesha No. 10 Bandung, Indonesia

Abstract. Institut Teknologi Bandung (ITB), Ganesha Campus, Indonesia, has an area of 28.86 hectares. The campus is located in Bandung. Starting from 2012, new buildings were constructed within the area, reducing the area of permeable surface significantly. In the past few years, there were several excess run off incidents in the campus. The insufficient area of permeable surface as well as the inadequate capacity of the drainage system contributes to the excess surface run off. The drainage system has only two outlets. Moreover, in some areas, the drainage systems are disconnected. Thus, most the surface run off are stored within the drainage system. The purpose of this study is to evaluate the effectiveness of infiltration wells for reducing the local excess run off in ITB. Precipitation data and drained service area are used to estimate the design discharge from each building in ITB. In order to avoid the excess surface run off of certain locations in ITB, then the infiltration wells are proposed to balance the area of impermeable surface. The effectiveness of the infiltration wells are evaluated by assessing their number to their contribution in reducing the excess surface runs off.

1 Introduction

Precipitation in Bandung area has increased in the past few years. The rainfall intensity in the last 25 years ascends by 10-20 mm every year [1].Based on Regional Regulation No. 26 of 2007 on Green Area (Ruang Terbuka Hijau/RTH), – from 661.4 thousands hectare area of Citarum River Basin, 198 thousands is supposed to be reserved area. In accordance with Regional Regulation No. 2 of 2016 on controlling the utilization of North Bandung Area (Kawasan Bandung Utara/KBU), the local government – tries to protect KBU by limiting physical development to only 20% of land area. Area of KBU Bandung is 38 thousands, but 80% have been intervened by constructions. The increase in industrial activities eventually increases the degree of groundwater withdrawal from the aquifers in Bandung [2]. Thus, most the surface run off are stored within the drainage system. Furthermore, land use and land cover change affect hydrological components such as interception, infiltration and evaporation. They influence the soil moisture content, runoff generation (both process and volume) and stream flow regimes [3]. Previous surfaces decrease the amount of runoff as precipitation infiltrates [4]. These also affect the groundwater table [5].

As well as the city of Bandung, new buildings have been constructed within the area of ITB. These building significantly reduce the permeable area. Soil type in the area is mostly clay up to 6 meters depth, which has a slow infiltration rate. The construction was not followed by increasing the capacity of the drainage system. Thus, excess runoff occurs quite frequently during the last few years in a few internodes such as Labtek V,VI,VII,VIII, Aula Timur and Lapangan Seni Rupa.

One of the solutions to reduce excess surface run off is infiltration wells. Infiltration wells are structures that highly allow surface runoff to drain through underground [6]. Infiltration well was initiated in Indonesia by Sunjoto [7] and developed by some other researchers [6,8-10]. The benefits of infiltration wells are a quantity of unconfined groundwater can be conserved, the surface level of unconfined ground water stays stable, the area of ponding water is minimized, and the dimension of drainage networks is minimized. In addition, the groundwater quality is improved, land subsidence is prevented, and salt water intrusion in coastal area is avoided [11]. The disadvantage of infiltration wells is the infiltration takes time depending on the permeability. The purpose of this study is to evaluate the affectivity of infiltration wells in order to reduce the excess surface run off in ITB.

2 Methodology

The results in this paper are based on two well-established and complementary hydrology analysis and infiltration wells standard design. They are SNI T-02-02-2006 for Local Drainage Design and SNI 03-2453-2002 for Infiltration Wells Design. The design is based on the available topography, hydrology, and soil data. The hydrology data are as follows:

1. Precipitation of Metropolitan Bandung Raya Data (2017-2015)

* Corresponding author: mardiahafifah@gmail.com

2. Precipitation of ITB station (AWS November 2011-December 2016)

3. Drainage system map

2.1. Hydrology

Hydrology analysis was conducted based on the daily maximum precipitation. The maximum precipitation in a return period can be estimated through 4 distribution methods such as normal, log normal, log pearson III and gumbel [12]. The methods concepts are fitted to probability concepts. SNI T-02-02-2006 states that the local drainage system design is based on a 5 years return period.

2.2 Standard Design for Infiltration Wells

SNI 03-2453-2002 states the general requirements of infiltration wells are as follows:

1. Infiltration wells are placed relatively in flat contour;
2. The water entering the catchment is unpolluted rainwater;
3. Decision of infiltration wells shall take into safety building;
4. It must pay attention to local regulations;
5. Things that do not meet these requirements shall be approved by authorized agencies.

Technical requirements that must be included are in the following:

1. To ground water minimum 1.5 meter in the rainy season;
2. The usable soil structure should have soil permeability ≥ 2.0 cm / hour.

The shape and type of the infiltration well can be either an infiltration wells made in quadrilateral or cylindrical with a certain depth and the bottom of the well is located above the groundwater level. Based on the Regional Disaster Management Agency of Jakarta, types of infiltration wells construction are:

1. Wells without couple walls, bottom without filled stone or fibers (empty).
2. Wells without couple walls, bottom filled with stones and fibers
3. Wells with brick, stone or stone compartments on the wells wall, bottom filled with stones and fibers or empty
4. Wells use blawong (rock formed specifically for wells wall).

3 Results and Discussions

3.1 Topography and Impermeable Surface

Topography map data with a 50 cm contour interval is provided to continue the study. ITB has a relatively mild slope, almost flat, with the north is higher than the south.

The map of constructed area in ITB is also given to by Infrastructure Facilities Division of ITB. Most of areas in ITB are impermeable surface, such as building, pavement area (parking, plaza, and corridor) and roadway thus ITB has more impermeable surface. The total building area in ITB is 99,733 m^2. Open space are garden, median, and island. ITB has less of infiltration area and the excess surface run off would drastically increase at the time of heavy precipitation.

3.2 Drainage Systems

The existing drainage systems were analyzed based on the drainage network data as shown in figure 1. In addition, field survey was conducted to assess the flow direction. In general, the condition of existing drainage in ITB: (1) The drainage systems are disconnected in some places, (2) The drainage system has 2 outlet points but the remaining functional outlet is Dago. Furthermore, the drainages in several points were clogged as shown in figure 1. In some places, the channels dimensions were relatively small and might be in insufficient during a heavy rain. Additionally, the flow directions in some sections were not heading to any outlet causing excess surface run off within the area.

Fig. 1. Existing Drainage of ITB

Depth	
	0.00-1.5 m **Silty Clay**, dark brown, medium stiff, high plasticity
1.50-2.00	
2.00-2.45	
	2.45-3.00 m **Silty Clay with some gravel**, brown, medium stiff, high plasticity
3.00-3.50	
	3.50-4.00 m **D.O**
4.00-4.45	
	4.45-5.00 m **D.O**
	5.00-6.00 m **Clayey Silt with some gravel**, brown, spoted yellow, low plasticity
6.00-6.45	
	6.45-8.00 m **Sandy Gravel with some clay**, brown, well graded, fine grained
8.00-8.45	
	8.45-10.00 m **Gravelly Sand**, grey brown, well graded, coarse grained
10.00-10.45	

3.3 Soil Characteristics

Fig. 2. Soil Characteristics in ITB

The characteristic of soil in ITB depend on the depth of the soil (Soil investigation FTTM Building, 2012). Between 0-1.5 meter depth is silty clay with dark brownish, medium stiff and high plasticity. This kind of soil takes time to infiltrate water [12]. The ground water table reached >20 m. The coefficients permeability of silty clay is 0.48 (m^3/m^2/day) or 1.3 x 10-4 m/sec.

3.4 Hydrology Analysis

The data are ITB station and Dago Pakar Station in 06° 52' 51'' South Latitude and 107° 36' 52'' East Longitude. Maximum daily precipitation design for the area is calculated using arithmetic mean method. The average maximum daily precipitation data of both stations are shown in the table 1.

Table 1. Maximum Daily Precipitation Design

Year	Maximum Precipitation (mm)
2007	113
2008	80
2009	73
2010	104
2011	45
2012	61.6
2013	70.75
2014	95.65
2015	65.3
2016	71.4

The design precipitation was statistically analyzed using the Normal, Log Normal, Log Pearson III, and Gumbel methods. The area design is the area of ITB which is 28.865 ha. According to SNI T-02-2006, the design precipitation return periods are 5, 10, 15, 20 year. In this paper, the return period used is 5 years because the scope of the area is university. Based on the results, the smallest error is Log Normal Method with a 5 years return period daily precipitation of 98.15 mm.

3.5 Infiltration Wells Design

3.5.1 Infiltration Wells Capacity

The design of infiltration wells has 1 meter diameter with the depth of 10 meter and infiltration efficiency is 100%. Soil clay has coefficient permeability 0.48 m3/m2/day and course sand has 0.864 m^3/m^2/day.

1. Surface run off volume (V_{AB})

For every 500 m^2 of service area, the surface run off volume is given as.

$$V_{AB} = 0.855\times0.75\times500\times98.15\times10^{-3}=31.47\ m^3 \quad (1)$$

2. Storage volume after precipitation (V_{STO})

The infiltration wells have 1 meter diameter and 10 meter depth along the vertical permeability (Kv) is calculated as below:

$$K_V = (0.48\times1.5+0.864\times4.95+8.64\times3.55)/10 \quad (2)$$

$$Kv= 3.57\ m^3/m^2hr \quad (3)$$

The infiltration wells is not impermeable, thus the water could flow as an infiltration or percolation (flowing through 2 axes). The horizontal permeability (Kh) is twice the vertical permeability, so the value is 7.13 m^3/m^2/day.

The precipitation duration (Tc) is the time for the precipitation to prolong.

$$Tc = 0.9 \times R^{0.92} \quad (4)$$

$$Tc= 0.9 \times 98.15^{0.92}=1.02\ hours \quad (5)$$

The total permeability (Kaverage) of the infiltration wells for permeable walls is

$$K_{average} = (Kv\times Ah + Kh \times Av)/(Ah+Av) \quad (6)$$

Where Ah is area of base wells, Av is area of wall wells. For the 10 meter depth of wells,

$$Ah = \Pi D^2/4= \Pi 1^2/4=0.785\ m^2 \quad (7)$$

$$Av = \Pi DH= \Pi\times1\times10 = 31.4\ m^2 \quad (8)$$

D is diameter of wells and H is height of wells

$$K_{average} = (3.57\times 0.785 + 7.13 \times 31.4)/(0.785+31.4) \quad (9)$$

$$K_{average} = 7.0468\ m^3/m^2/day \quad (10)$$

Volume of infiltration water during the precipitation with Tc=1.02 hours is

$$V_{rsp} = Tc/24\ A_{total} \times K_{average} \times efficiency \quad (11)$$

$$V_{rsp} = 1.02/24\times(0.785+31.4) \times 7.0468\times100\% \quad (12)$$

$$V_{rsp} = 9.64\ m^3 \quad (13)$$

$$V_{STO}= V_{AB} - V_{rsp}=31.47-9.64=21.82\ m^3 \quad (14)$$

3. Infiltration Wells Total

The total of infiltration wells is depends on storage volume.

$$H=V_{STO}/Ah=21.82/0.7854 = 27.79\ m \quad (15)$$

For 10 meter depth of infiltration wells, with a service area of 500 m2, the number of infiltration wells needed is:

$$Total\ wells = H/Hdesign=27.79/10 \approx 3\ wells \quad (16)$$

3.5.2 Infiltration Wells Application for ITB

If the totals of infiltration wells are designed according to roof area per building, with the total roof area 99,733 m^2, ITB is supposed to have 770 wells with the total area of wells is 604.757 m^2. The number of infiltration wells per building is given in table 2.

Table 2. Number of infiltration wells per building

No	Building	Area (m^2)	n_1	n_2
1	Kriya	153	0.4	1
2	STJR	190	0.4	1
3	FSRD	294	1.2	2
4	FTSL	341	1.6	2
5	PPFK	409	2.4	3
6	LAB RADAR	452	2.8	3
7	KANTIN BENGKOK	460	2.8	3
8	PSDA II	583	3.6	4
9	PSDA I	611	4.0	4
10	KIMIA BARU	741	4.8	5
11	DOPPING	794	5.2	6
12	BANKER	807	5.6	6
13	BSC B	847	5.6	6
14	OKTAGON	1036	7.2	8
15	TVST	1036	7.2	8
16	BSCA A	1066	7.6	8
17	LAB PLN	1098	7.6	8
18	GKU TIMUR	1116	8.0	8
19	CC TIMUR	1177	8.4	9
20	COMLAB TPB	1157	8.4	9
21	KONVERSI	1291	9.2	10
22	PPTI	1394	10.0	10
23	CADL	1634	12.0	12
24	KIMIA LAMA	1647	12.0	12
25	GKU BARAT	1675	12.4	13
26	CC BARAT	1776	13.2	14
27	AULA BARAT	1879	14.0	14
28	RME	1888	14.0	14
29	LAB HIDRAULIKA	1993	14.8	15
30	CRCS	2028	15.2	16
31	LABTEK CIII	2134	16.0	17
32	CAS	2172	16.4	17
33	SBM	2312	17.6	18
34	PERPUS PUSAT	2428	18.4	19
35	T.LING	2496	18.8	19
36	LABTEK VI	2732	20.8	21
37	PAU	2736	20.8	21
38	LABTEK II	2930	22.4	23
39	LABTEK VIII	2982	22.8	23
40	LABTEK I	2995	22.8	23
41	LABTEK VII	3088	23.6	24
42	LABTEK V	3158	24.4	25
43	LABTEK III	3259	25.2	26
44	LABTEK VI	3281	25.2	26
45	TEKNIK SIPIL	3354	26.0	27
46	LAB MESIN	3423	26.4	27
47	AULA TIMUR	3572	27.6	28
48	CADL LAMA	4076	31.6	32
49	LABTEK IX	4147	32.0	33
50	FISIKA	5265	41.2	42
51	LABTEK X&XI	5620	44.0	45
TOTAL		99733	745.5	770
TOTAL AREA OF WELLS (m^2)			585.5	604.8

Nb. n_1 is number of wells and n_2 is number of wells round up.

Infiltration wells are expected to reduce the amount of water that flow from the outlets to the roadside public drainage. This means that ITB run off contribution to the city drainage system will be reduced significantly.

Infiltration wells are prioritized in the area with known history of excess surface run off such as Aula Timur, Lapangan Seni Rupa and Aula Barat. The wells can either be an infiltration well, connected to the drainage system (drainage well), or not (RTH well). The locations are shown in figure 3.

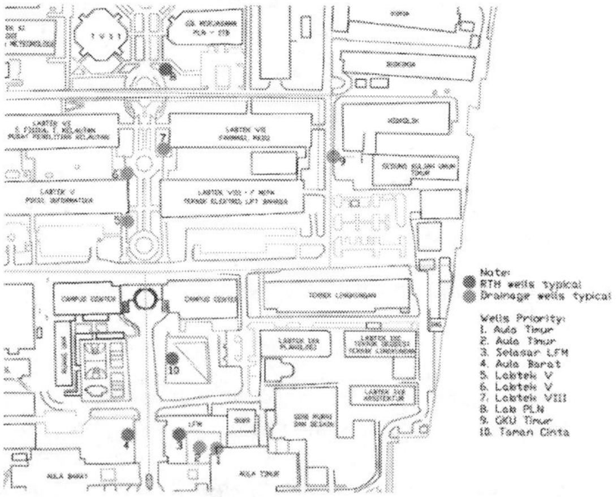

Fig. 3. Priority Wells of ITB

4 Conclusions

In this study, the effectiveness of infiltration wells to reduce the excess local run off in ITB has been assessed. The ground water table is located 20 meter depth and in general, the soil type is clay thus the infiltration wells may not be the best answer to the excess run off problem for ITB.

Dimension of the infiltration well is 1 meter diameter and 10 meter depth. It was found that for a service area of 500 m^2, 3 wells are needed. Therefore, to retain 100%

precipitation from the buildings into wells, ITB needs approximately 770 wells.

Due to the large number of wells needed, it is recommended to construct these wells only in several locations, at internodes that have high excess run off. In addition, it is recommended to develop a new master plaster plan for the ITB drainage system.

The authors would like to express their gratitude to the Infrastructure Facilities Division of ITB for providing the data.

References

1. A. Arif, *Jangan Salahkan Cuaca Berubah*, *Kompas*, https://www.pressreader.com/indonesia/kompas/20170320/281857233354137, (17 June 2017)
2. I. Gumilar, et al, Land subsidence in Bandung Basin and its possible caused factors, *Procedia Earth and Planetary Science*, **12**, 47–62 (2015)
3. W. Gumindoga, *Hydrology Impact of Landuse Change in the Upper Gilgel Abay River Basin, Ethiopia, TOP MODEL Application,* Integrated Watershed Management and Modelling (2010)
4. Haselbach, M. Liv, S. Valavala, F. Montes, Permeability predictions for sand-clogged Portland cement pervious concrete pavement systems, *Journal of Environmental Management*, **81**, 42–49 (2006)
5. F. DeSmedt, O. Batelan, The impact of land use changes on the groundwater in the Grote Nete River Basin, Belgium, *Conference Future of Groundwater Resources*, 151–158 (2001)
6. Y. Arafat, Reduksi Beban Aliran Drainase Permukaan Menggunakan Sumur Resapan, *SMARTek*, **6**, 144–153 (2008)
7. S. Sunjoto, Optimization of infiltration well to avoid salted water intrusion in coast areas, *IUC-UGM*, Yogyakarta (1988)
8. B. Soenarto, Rekayasa Peresapan Buatan untuk Mengatasi Susutnya Peresapan Air Alami Akibat Konservasi Lahan Alami Menjadi Lahan Permukiman, *Seminar Sehati "Air Tanah Cekungan Bandung"*, Satgas Penelitian Sumberdaya Air ITB Bandung, II-11–II-58 (1995)
9. K. Iriani, A. Gunawan, B. Besperi, Perencanaan Sumur Resapan Air Hujan untuk Konservasi Air Tanah di Daerah Permukan (Studi Kasus di Perumahan RT II, III, dan IV Perumahan Lingkar Timur Bengkulu), *Inersia*, **5**, 9–22 (2013)
10. A. Azis, Y. Hamzah, F. Zulviyah, Konservasi Airtanah Melalui Pembuatan Sumur Resapan Air Hujan di Keluarahan Maradekaya Kota Makassar, *Intek*, **3** (2), 87-90 (2016)
11. S. Sunjoto, Infiltration Well and Urban Drainage Concept, *Conference on Future Groundwater Resources at Risk*, IAHS Publ. no. 222, Helsinki, 527–532 (1994)
12. M. Sobolewski, Various Method of The Measurement of The Permeability Coefficient in Soils – Possibilities and Application, *Electronic Journal of Polish Agricultural Universities*, **8**, 2 (2005)

2D Modeling of Flood Propagation due to the Failure of Way Ela Natural Dam

Bagus Pramono Yakti[1,*], *Mohammad Bagus* Adityawan[2], *Mohammad* Farid[2], *Yadi* Suryadi[2], *Joko* Nugroho[2] and *Iwan Kridasantausa* Hadihardaja[1]

[1]Graduate School of Civil Engineering, Institut Teknologi Bandung, Jalan Ganesha 10, Bandung 40132, Indonesia
[2]Water Resources Engineering Research Group, Institut Teknologi Bandung, Jalan Ganesha 10, Bandung 40132, Indonesia
[3]Water Resources Development Center, Institut Teknologi Bandung, Jalan Ganesha 10, Bandung 40132, Indonesia

Abstract. A dam break induced-flood propagation modeling is needed to reduce the losses of any potential dam failure. On the 25 July 2013, there was a dam break generated flood due to the failure of Way Ela Natural Dam that severely damaged houses and various public facilities. This study simulated the flooding induced by the failure of Way Ela Natural Dam. A two-dimensional (2D) numerical model, HEC-RAS v.5, is used to simulate the overland flow. The dam failure itself is simulated using HECHMSv.4. The results of this study, the flood inundation, flood depth, and flood arrival time are verified by using available secondary data. These informations are very important to propose mitigation plans with respect to possible dam break in the future.

1 Introduction

Way Ela Dam was a dam located in Negeri Lima Village, Leihitu District, Ambon Island. This dam was formed naturally by landslide due to high rainfall on 13 July 2012 [1]. Avalanche material was carried in the form of soil and rocks with a very large volume, thus closing the flow of the Way Ela River. The avalanche caused the formation of a dam on Way Ela River.

A year after it was formed, on 25 July 2013 flooding occurred due to an extreme rainfall that caused the failure of the natural dam. According to Badan Nasional Penanggulangan Bencana (BNPB), the generated flood depth was about 6-7 m and took the lives of 1 person while the number of refugees reached to 3,000 people. The flood caused by the failure of the natural dam at least destroyed more than 350 housing units and other public facilities.

The climate change that causes extreme rainfall, has increased the frequency of floods, landslides, and increasing disaster risks caused by a dam failure [2]. Modeling the dam failure and the impact of the flood is necessary to reduce losses from the potential for dam failure [3]. A numerical model is often used to predict the flood propagation due to a dam failure, but the model used is generally still 1-D, so the distribution of inundation can not be predicted accurately. However the 1-D model approach can be used for certain cases, mostly used in channel planning, but it has limitations on runoff analysis [4]. At the time the water starts to overflow to surface, 2-D phenomenon occurs. Therefore, it is necessary to have the calculation of the area of inundation, water level inundation, and rapid flood flooding (flood travel time). By conducting such analysis, the disaster risk of the affected area can be known [5].

Way Ela dam failure was studied and modeled using Zhong Xing HY21 [6]. The study was based on the assumption that the dam failure was caused by middle piping due to a PMF (Probable Maximum Flood) discharge, although it was mentioned that the inflow during the incident was not due to a PMF discharge. Additionally, based on investigation report [7], the cause of the dam failure was overtopping. Therefore it is necessary to conduct research based on the latest data from investigation report [7]

HEC-RAS v.5 is a numerical software model with a free license that has been widely used to simulate flood. Starting from v.5, the model is capable of simulating 2D flow, which gives it an advantage for estimating flood inundation. The model was developed based on the full 2D St. Venant equations, solved using a finite difference scheme. The model is robust and accurate. The model was applied for simulating Monticello Dam Breach with good results [8].

In this study, 2-D modeling of flood propagation caused by the failure of the Way Ela natural dam is conducted in order to predict the flood inundation, flood depth and flood arrival time so that it can anticipate and reduce the human-economic losses . The modeling is based on the failure information that has occurred and the DEM (Digital Elevation Model) data. A two-dimensional (2D) numerical model, HEC-RAS v.5, is used to simulate the overland flow. The dam failure itself is simulated by using HECHMSv.4.

2 Methodology of study

The flow chart of this study is given in Figure 1.

* Corresponding author: bagusyakti@gmail.com

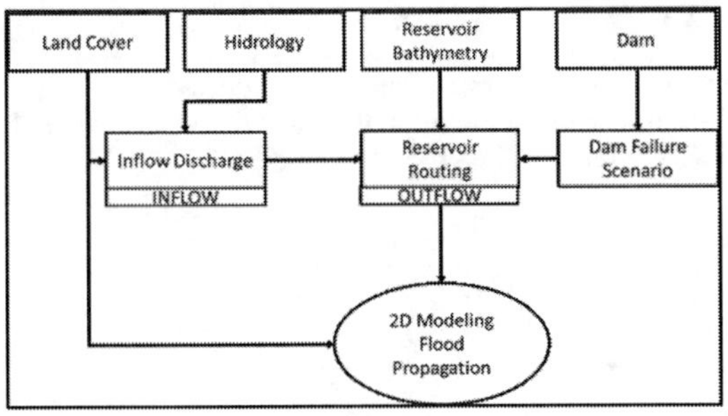

Fig. 1. Flow chart.

Land Cover was analyzed to obtain the runoff coefficient and manning's n value for the inflow discharge hydrograph. Hydrological parameter was analyzed to obtain the inflow discharge hydrograph as the input for reservoir routing the dam failure outflow. Reservoir bathymetry was analyzed for reservoir routing as the parameter of reservoir storage characteristic. Dam parameter was analyzed with the reservoir routing as the parameter for dam characteristic and dam failure scenario.

2.1. Study Area

Way Ela Natural Dam is located in Negeri Lima Village, Leihitu District, Ambon Island. The village had a mean elevation between 0-700 m above sea level. The main River are the Way Ela River which in 2012 dammed to form a natural dam due to avalanches.

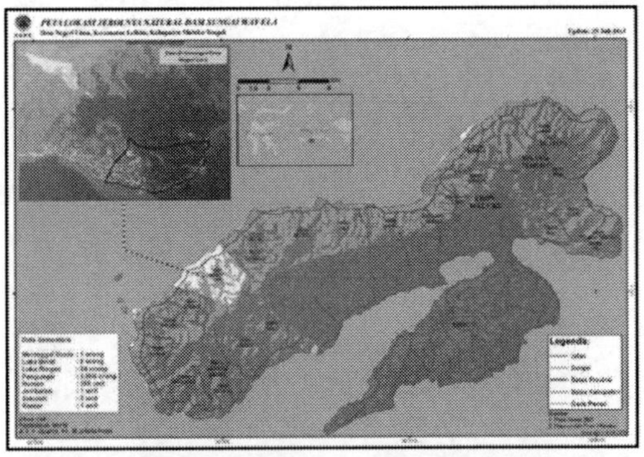

Fig. 2. Study area. BNPB [1]

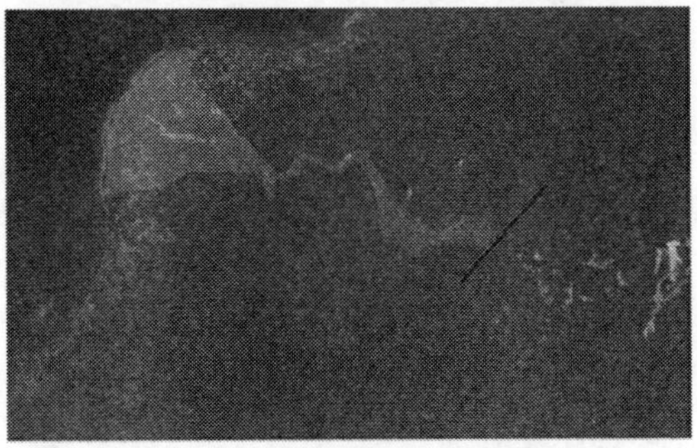

(a) Before dam failure.

(b) After dam failure.

Fig. 3. Orthophoto inunduation image.

The images before and after dam failure are shown in Figure 3. The urban areas affected by flood due to the dam failure is approximately 147,749.8 m^2.

2.2 Land Cover

In this study the land cover is used to define the Runoff coefficient (C). The Way Ela Watershed Land is covered by forest with Runoff coefficient is 0.5 [9].

Fig. 4. Land cover map of the study area

2.3 Reservoir Routing

The reservoir routing was calculated using HEC-HMS to estimate the amount of discharge when the dam failed. There are three main parameters used as the input for reservoir routing in HEC-HMS, inflow, reservoir bathymetry and the dam failure scenarios. The output from this reservoir routing is the outflow discharge due the failure of Way Ela Natural Dam.

Table 1. HEC-HMS data used.

Data	Values
Watershed Area (m^2)	12.56
River Length (m)	6.85
Land Cover (Runoff Coefficient)	0.5
Inflow Hydrograph	Figure 4
Dam Top Elevation (m)	+ 215.66

Data	Values
Dam Top Length (m)	300
Normal Water Level (m)	+ 197

2.3.1 Inflow

The inflow of the reservoir was calculated using rainfall at the time of the dam failure. The condition of this inflow at the time of dam failure was estimated with Synder-Alexeyev Synthetic Unit Hydrograph based on the characteristic of the catchment area as given in table 1.

The rainfall detected at the dam failure by Ambon Meteorological Agency (BMKG) was 432 mm. The results are shown in the Figure 5 along with the capacity of the spillway of the natural dam. It shows that inflow exceeds the spillway capacity.

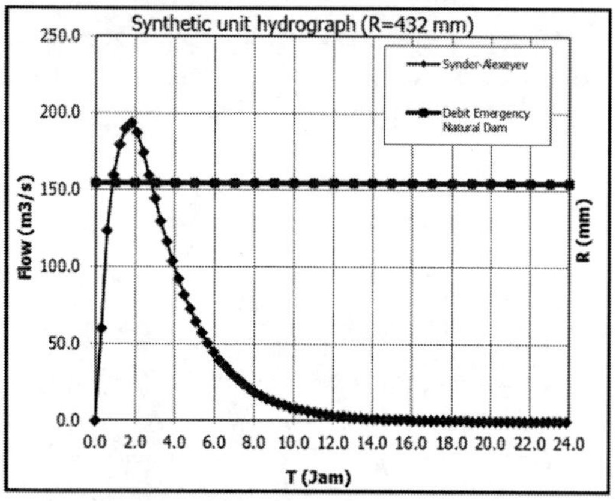

Fig. 5. Inflow hydrograph

2.3.2 Reservoir Bathymetry

Reservoir Bathymetry of the Way Ela natural dam was analyzed based on previous studies [7], the result is given in figure 5 along with the dam elevation (table 1).

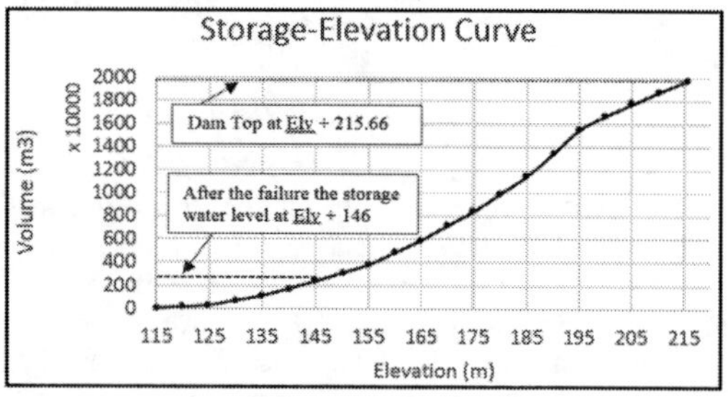

Fig. 6. Storage-Elevation curve

Figure 6 shows the maximum storage capacity of the reservoir is 19.807,000 m^3 and after the dam failure the volume of water becomes 2,500,000 m^3.

2.3.3 Dam Failure Scenario

In this study a simulation was conducted for the dam failure with the assumption that the failure was caused by overtopping based on previous studies [7].

Table 2. HEC-HMS data used.

Overtopping	
Top Elevation (m)	215.66
Bottom Elevation (m)	115
Bottom Width	201.32
Left Slope (xH : 1V)	1
Right Slope (xH : 1V)	1
Development Time (HR)	10
Trigger Elevation (m)	200

The scenarios parameter is given in table 2. The trigger elevation is set to + 200 m, is 3 m above spillway = 197 m, because based on chronology when the dam failure sliding occurs on the dam body because the spillway only finishes 70% and then collapses. The dam failure development time is 5 hours but lengthened to 10 hours due to emergency spillway [7]. The Bottom Width parameter is 2 × dam height [10].

2.4 Flood Propagation Model

The 2D flood propagation was modeled using HEC-RAS v5. The HEC-RAS v5 solves the full 2D Saint Venant equations [11] :

Continuity Equation

$$\frac{\partial \zeta}{\partial \tau} + \frac{\partial p}{\partial x} + \frac{\partial q}{\partial y} = 0 \qquad (1)$$

Momentum Equation :

$$\frac{\partial p}{\partial \tau} + \frac{\partial}{\partial x}\left(\frac{p^2}{h}\right) + \frac{\partial}{\partial x}\left(\frac{pq}{h}\right) = -\frac{n^2 pg\sqrt{p^2+q^2}}{h^2} - gh\frac{\partial \zeta}{\partial \tau} + pf + \frac{\partial}{p\partial x}\left(h\tau_{xx}\right) + \frac{\partial}{p\partial y}\left(h\tau_{xy}\right) \qquad (2)$$

where h is the water depth (m), p and q are the specific flow in the x and y directions (m^2/s), is the surface elevation (m), g is the acceleration due to gravity (m/s), n is the Manning resistance, is the water density (kg/m^3), xx, yy and xy are the components of the effective shear stress and f is the Coriolis (s^{-1}). The manning roughness is estimated for each type of land cover as given in the table and figure bellow.

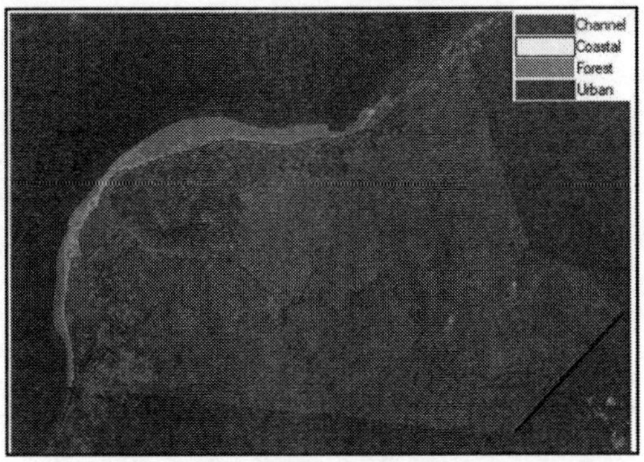

Fig. 7. Manning distribution map of study area

Table 3. Manning's Roughness

Land Cover	Manning's n
Forest	0.03
Channel	0.025
Coastal	0.025
Urban	0.172

Source : Kotani, et. al. [12], chow [13], Gibson [14]

In order to ensure the stability of the model, the time step was estimated according to the Courant-Friedrichs-Lewy condition :

$$Cr = \frac{c\Delta t}{\Delta x} = \frac{\sqrt{gh}\Delta t}{\Delta x} \leq 1 \quad (3)$$

Based on the dam failure the overtopping failure will be modeled. The model results will be compared with the past inundation from the image (orthophoto). The inundation orthophoto is given in Figure 3.

Table 4. HEC-RAS Parameters

Data	Values
Outflow Hydrograph (Upstream Boundary)	From Reservoir Routing
Stage Hydrograph (Downstream Boundary)	0
Digital Elevation Model (Terrain)	30 m × 30 m
Land Cover (manning's n)	Table 1
Δt [s]	0.5
Δx [m]	10
Simulation Period (day)	1

3 Result and Discussion

3.1 Reservoir Routing

Outflow from dam failure is simulated uses HEC-HMS. The overtopping failure scenario is selected refers to previous study [8]. Inflow Hydrograph was analyzed using the Synder-Alexeyev Synthetic Unit Hydrograph. The result of this routing process will be shown by Figure 7.

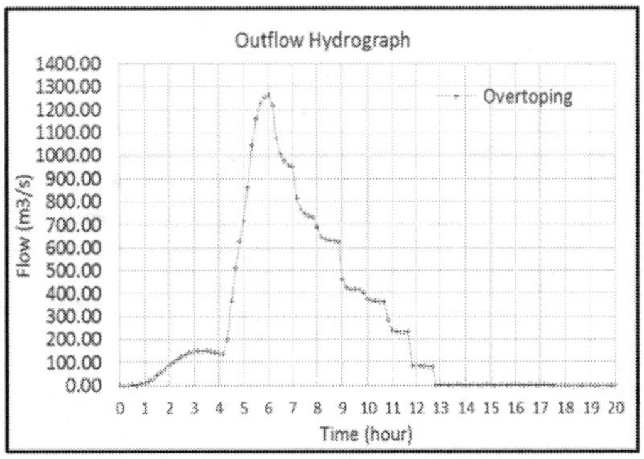

Fig. 8. Outflow hydrograph due the dam failure

Figure 8 shows the outflow peak discharge due the dam failure by the overtopping failure is 1,268.4 m^3/s.

3.2 Flood Depth

The relative error of overtopping failure shown in Table 5. Table 5 shows the result that this overtopping failure has a small relative error compared with the inunduation of the orthphoto image. It can be concluded that the previous dam failure was caused by overtopping. Time requirement to execute this model is 30 minutes.

Table 5. Flood Area

	Way Ela Village Urban Area	Inunduation	
		Orthophoto	HEC-RAS v5
			Overtopping
Area (m2)	404,945.33	147,749.38	147,132.91
Area (%)	-	36.49	36.33
Relative Error (%)	-	-	0.42

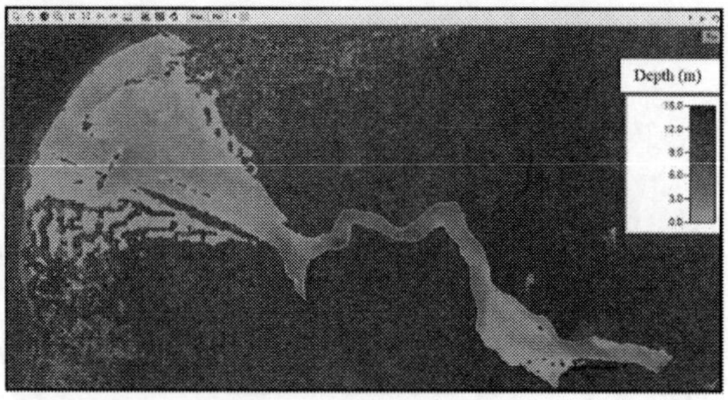

Fig. 9. Flood depth map

Figure 9 shows the maximum depth of flood inundation. In general, the depth agrees well with report provided by BNPB [1]. The maximum depth in the village area are between 3-6 m, the depth of flood in the upstream near the dam is 3-8 m, and in the middle of the channel, the depth flood inundation is quite high 3-7 m due to its topographic condition which is a cliff.

3.2 Flood Arrival

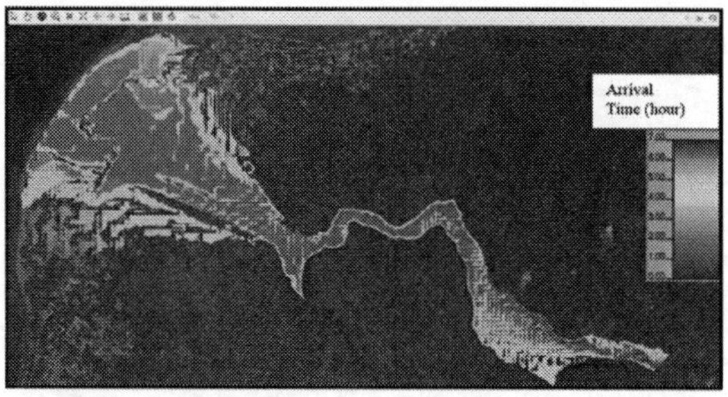

Fig. 10. Flood arrival time map

Figure 10 shows that the arrival time of the flood caused by the dam failure when it reaches the village is between 2 hours and in the downstream is between 5-5.5 hours.

4 Conclusions

A two-dimensional (2D) numerical model, HEC-RAS v.5, is used to simulate the over land flow generated by a dam failure. The dam failure itself is simulated using HECHMSv.4. The model was applied to the dam break case of the Way Ela natural dam.

The results from the study are the inundation area, depth, and arrival time. The inundation area was verified using the data from orthophoto. It was found that both areas corresponds well with a very small relative error (0.42%). The inundation depth in the village there are up to 6 m, which agrees well with the reported depth. The arrival time of the flood is 2 hours. These informations are very important to propose a mitigation plans with respect to possible dam break in the future.

References

1. S.P. Nugroho, *Sebagian Bendungan Way Ela di Maluku Jebol*, http://www.bnpb.go.id/ (25 July 2013)
2. C. S. Chen, T. W. Lin, C. Y. Chen, Eng. Geol. **188**, pp. 148-158, (2015).
3. Y. Xiong, J. Water Resource Prot. **3**, pp. 370-379, (2011).
4. K.Srinivas, M. Werner, N. Wright, Comparing forecast skill of inundation models of differing complexity-the case of Upton upon Severn, Flood risk management : research and practice, Issue ISBN 978-0-415-48507-4, (2009)
5. D. Wirustyastuko, J. Nugroho, Jurnal Teknik Sipil ITB **20**, pp 129-140, (2013)
6. Rachmadan, C.R., Juwono, P.T., Asmaranto, R., Analisa Keruntuhan Bendungan Alam Way Ela dengan Menggunakan Program Zhong Xing HY21, http://pengairan.ub.ac.id/, (1 March 2014)
7. Balai Wilayah Sungai Maluku Kementrian Pekerjaan Umum Republik Indonesia, *Final Report Investigasi Pekerjaan Bencana Alam Natural Dam Way Ela Negeri Lima*, Kab. Maluku Tengah, PT. LAPI ITB, (2014)
8. G.W. Brunner, 2D Modeling and Mapping with HEC-RAS v.5 Monticello Dam Breach. US Army Corps of Engineers, (2015)
9. R.K. Sivanappan, *Soil and water conservation and water harvesting*, **2nd ed**, Madras : Tamil Nadu Social Forestry Project, Indo-Swedish Forestry Coordination Programme, (1992)
10. F.A. Johnson, P. Illes, *A Classification of Dam Failures*. Water Power and Dam Construction, pages 43 -45, (1976)
11. V. M. Quiroga, S. Kure, A. Mano, IAHR **36th**, pp. 25-33, (2015)
12. M. Kotani, F. Imamura, and N. Shuto, JSCE **45**, pp. 356–360, (1998)
13. V.T. Chow, *Open-Channel Hydraulics*. (McGraw-Hill Book Company, New York. p.113. 1959)
14. S. Gibson, *Unsteady HEC-RAS Model of the Downtown Reach of the Truckee River*, HEC, PR-58, 39 p., (2005)

Study on Water Resources Allocation for Kertajati, Jatitujuh, and Ligung Sub-Districts to Support the Development of West Java International Airport (BIJB) and Kertajati Aerocity Area

Yadi Suryadi[2,3,*], *Asrini* Chrysanti[1], *Febya* Nurnadiati[1], *Mohammad* Bagus Adityawan[2,3], and *Arno* Adi Kuntoro[2]

[1]Graduate School of Civil Engineering, Institut Teknologi Bandung, Jalan Ganesha 10, Bandung 40132, Indonesia

[2]Water Resources Engineering Research Group, Institut teknologi Bandung, Jalan Ganesha 10, Bandung 40132, Indonesia

[3]Water Resources Development Center, Institut teknologi Bandung, Jalan Ganesha 10, Bandung 40132, Indonesia

Abstract. The Bandarudara Internasional Jawa Barat (BIJB) and Kertajati Aerocity are under construction and expected to be a center of economic activity supported by potential of natural resources and agriculture. They later will act as a driver of economic growth in West Java, especially for surrounding area such as Kertajati, Jatitujuh and Ligung Sub-districts. As an affect of the the development of BIJB and Kertajati Aerocity, the water demand of surrounding area will increase. Therefore an analysis of water demand and availability is needed. This research supports by analyzing the water balance, water demand, and also water allocation using WEAP (Water Evaluation and Planing) software tool. Water balance of Cimanuk-Tomo, Cimanuk-Monjot, and Cilutung-Dam Kamun river are analyzed, water demand of the three sub-districts are also projected and become the inputs of the water allocation model. The result of WEAP model simulation shows that the Cimanuk River can meet the water demand of the three sub-districts until 2040.

1 Background

PT Bandarudara Internasional Jawa Barat (PT BJIB) responsible for operating West Jawa Internasional Airport and developing Aerocity. In order to encourage the development of the Ciayumajakuning (Cirebon-Indramayu-Majalengka-Kuningan) region as well as the increase of economic growth and public service especially in the field of transportation, set in 2010 the development and growth of the international airport of West Java and Kertajati Aerocity. BJIB's accessibility is guaranted by having both highway and railway connected Bandung, Kertajati, dan Cirebon. BJIB plans to ensure sufficient capacity to meet future demand in air travel wich can handle 5 million passanger per year.

Development and growth of BIJB and Kertajati Aerocity is implemented by developing Kertajati Urban Area which is integrated with the development and growth of airport. The development of Kertajati Urban Area certainly affect Kertajati Area itself and also around. Among them is the Center for Local Activities Kertajati with service functions as a commercial area and services, an integrated industrial area BIJB, the development of maritime Aerocity, Central Services Area Jatitujuh with service functions as housing development, services, industry, and agriculture Serving sub-district or village-level activities, and Central Services Area Ligung with service function as security area, industrial development, social service, and agriculture serving sub district or some village. Therefore it is necessary to plan development and growth of three sub districts to support development and growth of BIJB and Kertajati Aerocity Regions.

Water demand increases along with population growth in Majalengka Regency and will continue to increase in line with the development and growth of BIJB and Kertajati Aerocity. Especially in the regions service areas in Kertajati Subdistrict (outside Region Aerocity), Jatitujuh, and Ligung. Therefore it is necessary to analyze the water demand and identify the water resources to meet the total water demand, as well as the planning of water resources management in 3 (three) sub districts to support the development and growth of BIJB and Kertajati Aerocity. With well managed water resources, water demand (domestic needs) can be fulfilled and sustainable water resources management can support infrastructure development and growth. This research using Water Evaluation and Planning (WEAP) system for water resources allocation, which is suitable for incorporates a demand priority and supply preference approach to drive the allocation of water from water supply. WEAP model using linear program algorithm, iterated to optimize demand and supply priorities. WEAP model has been used in several water related researches across the world, for example to determine the impact of a number of possible Development scenarios in Lake Tana Catchment, Ethiopia (Alemayehu et al. 2010) [7]. It is also used in Western Africa to study the impacts of hydropower production due to land use and land cover changer under climate changes (Obahoundje et al. 2017) [9].

2 Location of the study area

The BJIB and Kertajati Aerocity is located in Majalengka district, while the focus of this research are Kertajati, Jatitujuh and Ligung districts as shown in **Figure 1**. Majalengka is among the big districts such as Indramayu, Cirebon, Kuningan, Ciamis and Tasikmalaya. In urban

* Corresponding author: suryadi350@yahoo.com

areas the water demand has been fulfilled from the Cimanuk River.

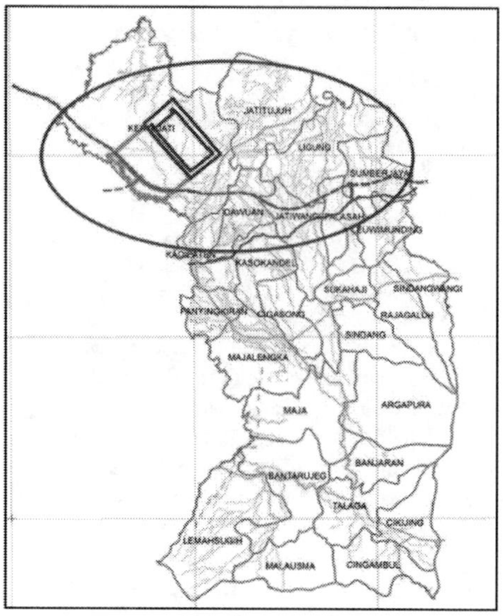

Fig. 1. Location of study area

2.1 Hydroclimatology

Majalengka and Kuningan districts are classified as wetlands with average annual rainfall ranges between 2000 - 4000 mm. The wettest area of the region with annual average rainfall of 3500 - 4000 mm is in the northwest. Average annual rainfall in Majalengka regency ranges from 2,400 mm-3,800 mm per year with an average rainy day of 11 days per month. In 2010, Majalengka regency rained every month with the highest rainfall occurred in March 2010 which reached 586 mm. During the period of 2008-2012, the annual rainfall in Majalengka was the lowest at 1,953 mm per year in 2009 and the highest at 3,459 mm per year in 2010, with an average of 5 years in year of 2,723 mm per year. There are 2 (two) big river which become heart of big enough water requirement to be utilized especially for irrigation that is Cimanuk River and Cilutung River. In 2011 the air temperature in Majalengka Regency ranged from 26.8 to 29.3 degrees celcius. Maximum air temperature occurred in October of 35.3°C, while the minimum air temperature occurred in August with a temperature of 22.7°C. Average wind speeds ranged from 3 to 5 knots per hour with a top speed in March of 28 knots per hour, making the Majalengka Regency nicknamed the City of Wind.

2.2 Supply sources

The water supply system in Majalengka Regency consists of piping systems and non-piped systems. Piping systems are managed by municipal waterworks while non-piped systems are managed by the community. The areas that have been served by water piped systems are urban areas mostly located in Majalengka Central District, while non-piped areas (community-managed) are located in Majalengka Regency in the North and South. For the South Majalengka Regency generally do not face problems in the fulfillment of their water supply because the water source is available by utilizing the water springs. However, the fulfillment of water demanddemands in the northern region of Majalengka Regency is still derived from ground water. The quantity of ground water is not sufficient and the quality of water is also not good because it contains lime.

There are many water sources utilized by municipal waterworks Majalengka district. Water source utilized by municipal waterworks Majalengka district are Cilongkrang Springs, Cisurian Springs, Cipadung Springs, Cihaneut Springs, Cibulakan Spring, Citamba Springs, Cigowong Springs, Colom Springs, Cilutung River, Cimanuk River, and Jatigede Reservoir. Based on Busines Plan municipal waterworks Majalengka District, from 26 Subdistricts in Majalengka new district 11 Sub-districts that have been served with the number of 67 villages. In addition, Majalengka district has some other potential surface water in the form of lake located in Cipadung, Umbrella, Sangiang, and Talagaherang villages. Groundwater, based on existing potential conditions, in general the Northern and Central Territories of Majalengka district have the potential of good quality ground water.

2.3 Previous Study

Several studies have been done to analyze the water balance in the study area. The total water demand has been calculated in the previous study, the total water requirement is the amount of domestic water demand covering the needs of airport and residential activities and industrial water requirements projected as an integrated industrial location. The estimated total annual requirement is presented in the **Table 1**.

No.	Types of Needs	Projected water demand (m3/s)				
		2007	2010	2025	2030	2040
1	Domestic	0.1038	0.0855	0.3306	0.3950	0.7831
2	Industry	0.0000	0.0213	0.1150	0.1400	0.3211
3	Maintenance	0.0052	0.0150	0.0223	0.0268	0.0552
4	Irigation	1.0747	1.1024	1.2516	1.3660	1.4210
	Total	1.1837	1.2242	1.7195	1.9278	2.5804

Table 1. Total water demand from previous study
Source: Final Report of Feasibility Study of Cipanasaat Reservoir Construction (PT. Sarana Bhuana Raya, 2006)

The result shows the maximum monthly average discharge occurs in January with 4.021 m^3/s and the minimum in September with 0.144 m^3/s, with the average per year is 1.771 m^3/s. The water availability can be determined as the dependable flow (80%). Maximum dependable flow occurs in February with value of 3.505 m^3/s and the minimum is in October with value of 0.106 m^3/s. Water Balance in the Cipanasaat watershed is reviewed for local hydrological conditions with respect to the inflow of each tributary and water requirements along the river channel, water balance diagram for water availibility and water demand from 2007 to 2040 shown in **Figure 2**. Based on the water availability analysis, the monthly dependable flow is range between 0.106-3.382 m^3/s, with the maximum occurring in January and the minimum occurring in October. While the water requirement of the Sub-watershed currently in 2007 is 0.109 m^3/s, short term in 2010 is 0.122 m^3/s, medium term in 2020 is 0.468 m^3/s, and long term in 2040 is 1.156 m^3/s. The results of the water development analysis show that for the water supply from Cipanasaat River in the medium and long term will experience deficiency especially in the dry season.

3 Methodology

Concept of aerocity development is an area in which there are various activities support each other with the activities

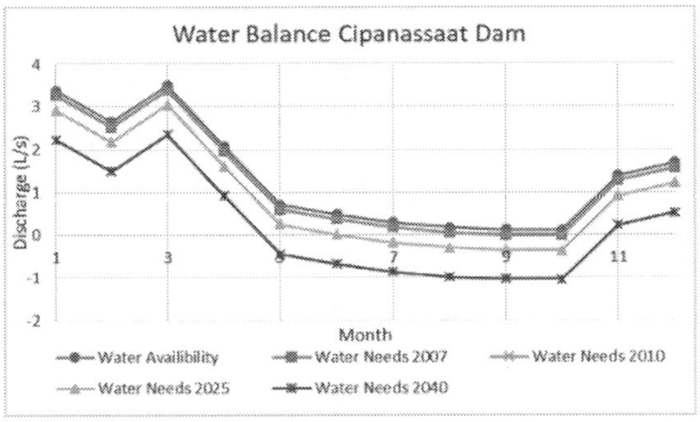

Fig. 2. Water balance
Source: Final Report of Feasibility Study of Cipanasaat Reservoir Construction (PT. Sarana Bhuana Raya, 2006)

of the airport. There are three main activities on aerocity design namely core aeronautical activities, airport related activities and airport oriented activities. Based on this concept, Kertajati aerocity has two schemes of airport activity type of airport there are activities within the aerocity and outside the aerocity. Therefore the analysis in this study was conducted on both areas within the aerocity and outside.

To allow water allocation, we need to calculate the water demand and supply on the location study. This study only take into account the quantitative factors, therefore it is assumed that all the water resources from three rivers have the same quality dan able to use as a water resource.

There are three stages of analysis used in this research; projection of water demand, water requirement, and water distribution. Projected water demand are intended to serve domestic facilities, educational facilities, health, offices, tourism, handling and industry. To meet the water demand of each activity calculated based on predictions every year until 2040. Total water demand is the amount of domestic water demand which covering the needs of airport and residential activities and industrial water demand projected as an integrated industrial location.

Water availability will be analyzed using F.J Mock method. F.J. Mock (1973) [3] proposed a monthly water balance simulation model for the drainage region in Indonesia, known as the Mock flow simulation. This model is specifically used for rivers in Indonesia (Bappenas, 2006). The Mock method refers to the water balance in which the total water volume in the earth is fixed, only its circulation and distribution varies. Mock takes into account the volume of incoming (rain), out and stored water in the soil (infiltration, percolation and evapotranspiration).

Water resources network of Kertajati Aerocity will use piping system, so analysis of capacity and water allocation is required to support the plan.

The software that supports the analysis is Water Evaluation and Planning System (WEAP). The WEAP algorithm is implemented as a series of linear programming (LP) problems, iterated over demand and supply priorities. The algorithm is documented in Yates, Sieber, et al. (2005) [6]. WEAP has been used in several water related researches across the world, for example to determine the impact of a number of possible Development scenarios in Lake Tana Catchment, Ethiopia (Alemayehu et al. 2010) [7]. It was also used to evaluate Ethiopia's policy of large dam construction in the Blue Nile River basin by simulating the impact of climate change scenario on the performance of existing and planned hydropower and irrigation schemes (McCartney and Girma 2012) [8].

WEAP operates with the basic principle of water balance and can be applied to municipal and agricultural systems, single river basins or complex watershed systems. WEAP simulates various natural and engineering components of this system, including rainfall runoff, relative steady streams, and groundwater infiltration from rainfall; Sectoral demand analysis; Water conservation; Water policies and allocation priorities, reservoir operations; Generation of hydropower; Tracking pollution and water quality; Kerentana assessment; And ecosystem requirements.

4 Results and discussion

4.1 Projected water demand

The three sub districts of the study area have function as Kertajati Local Activity Center with service function as commercial and service area, integrated industrial area, BIJB area, urban development "aerocity", and agriculture. In this research, the total water demand of those three sub-districts is analyzed, includes water demand of domestic facilities, education facilities, health facilities, office facilities, tourism facilities, trade and service facilities, and industrial estates.

Projection of water is done based on Local Regulation and Spatial Planning of Majalengka district. Water demand is calculated by considering the population growth rate of 4% and water requirement 120 L/day/person. The ratio of the population of Kertajadi district and the number of worker in each facilities and water requirement for each facilities shown in **Table 2.**

Table 2. Water demand for each facilities

Facilities	District					
	Kertajati		Jatitujuh		Ligung	
	Ratio/Area	Water Needs (L/d/p)	Ratio/Area	Water Needs (L/d/p)	Ratio/Area	Water Needs (L/d/p)
Education	10.09%	10	25.23%	10	22.93%	10
Health	7.94%	200	7.94%	200	10.64%	200
Office	6.17%	10	13.20%	10	17.80%	10
Industry	20%	60	20%	60	20%	60
Tourism	74.46 Ha	0.3	770.61 Ha	0.3	428.3 Ha	0.3
Trade and Service	1504.1 Ha	12,000	381.82 Ha	12,000	177.41 Ha	12,000

Source: Analysis Result

The total demand of three district can be seen from recent years (2015) untill projected years in 2040 in **Figure 3**. At 2040 total water demand will reach 1,450.60 L/s.

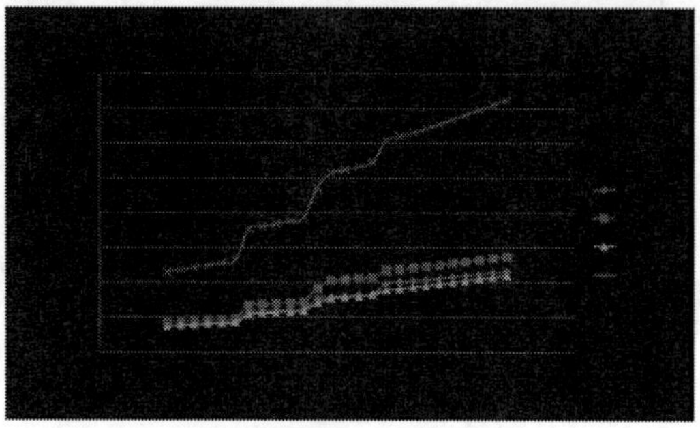

Fig. 3. Total water demand
Source: Analysis Result

4.2 Water avaibility

From the survey results, Cibeureum Lake, Nyesel Lake, Anggrarahan Lake, Tarisi Lake, Belik Lake, Jawura Lake, Cimaneuh Lake, Citamiang Lake, Cicabe Lake, Telik Lake can only be used during the rainy season because the water source are from rain water. So the analysis of water availability is only done in 3 points, namely Cimanuk Tomo River, Cimanuk-Monjot River, and Cilitung-Dam Kamun River. Water availibility for each river shown in **Figure 4**.

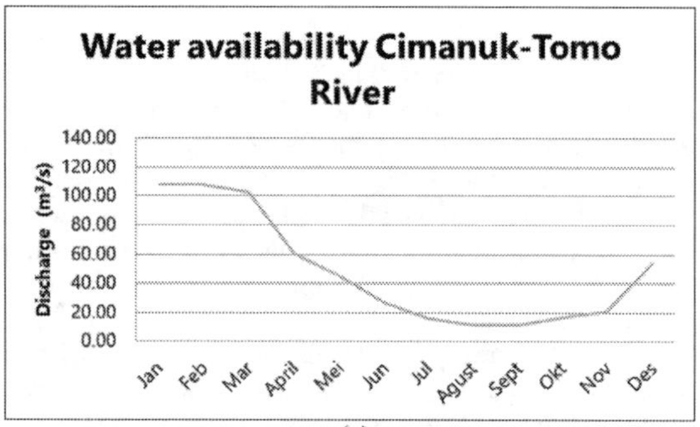

(a)

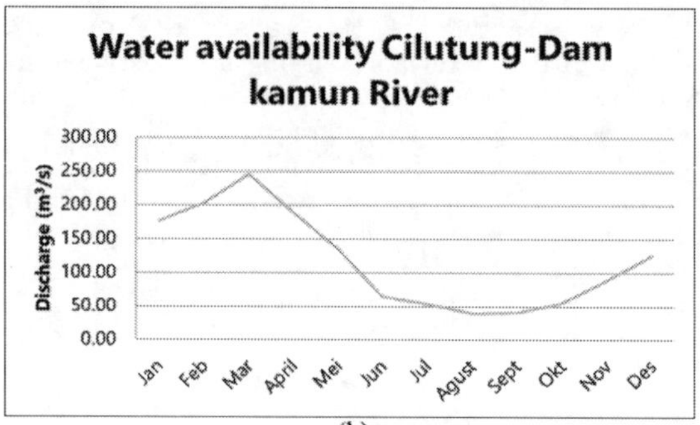

(b)

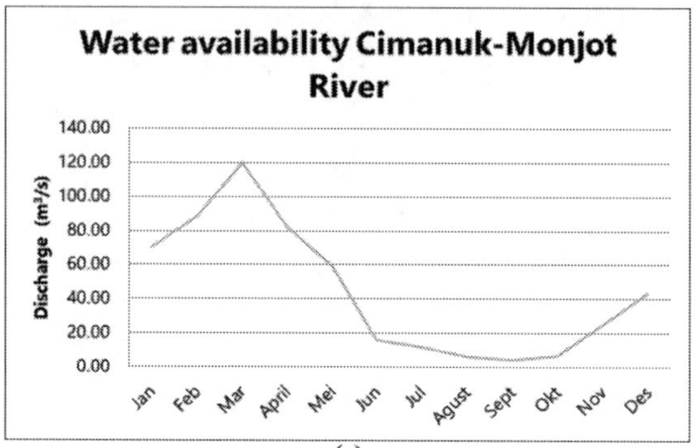

(c)

Fig. 4. Water avaibility
Source: Analysis Result

4.3 Water allocation

In this study, water balance analysis of Kertajati, Jatitujuh, and Ligung districts is conducted. GIS-based data in the form of river networks, lake points, and water use allocation can be seen in the **Figure 5**. The water allocation for each district is classified into seven in accordance with the previous analysis. Water collection for each class has been set previously. Each allocation is given the value of percentage amount of water consumed. Unconsumed water will be returned to the river. The classification of water allocation and percentage of water consumed, and water retrieval can be seen in the **Table 3**.

The water source used in this water allocation model is surface water taken from the Cimanuk-Monjot River. This is done with the following considerations:

- The location of Cilutung River is on the different side with the study location and the river empties into Cimanuk River, so it will be more efficient if water retrieval done in main river (Cimanuk River) compared to making channel from Cilutung River to study location.
- The result of water availability analysis in Cimanuk-Tomo River is greater than the availability of water in the Cimanuk-Monjot River, whereas the Cimanuk Monjot River is located downstream of the Cimanuk River and Cilutung River catchments. So it is assumed there are water retrieval in the Cimanuk-Tomo River range and Cimanuk-Monjot River.

Table 3 Water calssification and

Demand Site	2015	Scale	Unit
JATITUJUH Industri	70	Percent	
JATITUJUH Kesehatan	40	Percent	
JATITUJUH Pariwisata	30	Percent	
JATITUJUH Pendidikan	20	Percent	
JATITUJUH Perdagangan	30	Percent	
JATITUJUH Perkantoran	20	Percent	
JATITUJUH Perumahan	50	Percent	
KERTAJATI Industri	70	Percent	
KERTAJATI Kesehatan	40	Percent	
KERTAJATI Pariwisata	30	Percent	
KERTAJATI Pendidikan	20	Percent	
KERTAJATI Perdagangan	30	Percent	
KERTAJATI Perkantoran	20	Percent	
KERTAJATI Perumahan	50	Percent	
LIGUNG Industri	70	Percent	
LIGUNG Kesehatan	40	Percent	
LIGUNG Pariwisata	30	Percent	
LIGUNG Pendidikan	20	Percent	
LIGUNG Perdagangan	30	Percent	
LIGUNG Perkantoran	20	Percent	
LIGUNG Perumahan	50	Percent	

Demand Site	2015	Scale	Unit	
JATITUJUH Industri	21900		L	/person
JATITUJUH Kesehatan	73000		L	/person
JATITUJUH Pariwisata	9460800		L	/hectare
JATITUJUH Pendidikan	3650		L	/person
JATITUJUH Perdagangan	4380000		L	/hectare
JATITUJUH Perkantoran	3650		L	/person
JATITUJUH Perumahan	43800		L	/person
KERTAJATI Industri	21900		L	/person
KERTAJATI Kesehatan	73000		L	/person
KERTAJATI Pariwisata	9460800		L	/hectare
KERTAJATI Pendidikan	3650		L	/person
KERTAJATI Perdagangan	4380000		L	/hectare
KERTAJATI Perkantoran	3650		L	/person
KERTAJATI Perumahan	43800		L	/person
LIGUNG Industri	21900		L	/person
LIGUNG Kesehatan	73000		L	/person
LIGUNG Pariwisata	9460800		L	/hectare
LIGUNG Pendidikan	3650		L	/person
LIGUNG Perdagangan	4380000		L	/hectare
LIGUNG Perkantoran	3650		L	/person
LIGUNG Perumahan	43800		L	/person

- However, the availability of Cimanuk River water must be rectified after the Jatigede Reservoir operates because the availability of Cimanuk River water will depend on the water that the reservoir estimates.
- In addition, although the location of the Cimanuk-Monjot River more downstream, but closer to the location of the study.

Water demand projection from 2015 to 2040 can be seen in **Figure 6**. Based on the results of analysis using WEAP, Cimanuk River can meet the water demand in these three sub-districts until 2040.

5 Conclusions

From the results of studies it can be taken some conclusions as follows:

- Water retrieval from Cilutung River to meet the water demand in Kertajati, Jatijutuh, and Ligung district is less effective because the location of the river is on the different side of the three districts and empties into Cimanuk River. However, the river can be an option but the water availability record in the river must be validated because the existing data is not in accordance with the conditions in the field.
- Water sources that can be used to meet water demand in Kertajati, Jatijutuh, and Ligung district are Cimanuk River (Cimanuk-Monjot). However, the availability of water in Cimanuk River must be corrected after the Jatigede Reservoir operates because the availability of water in the river will be greatly influenced by the flow of water that flowed reservoir
- There is no water source in Kertajati, Jatitujuh and Ligung sub-districts that can meet the water demand of the three sub-districts.
- In this study has identified potential water resources in other areas that can be used to meet water demand in Kertajati, Jatitujuh and Ligung districts.

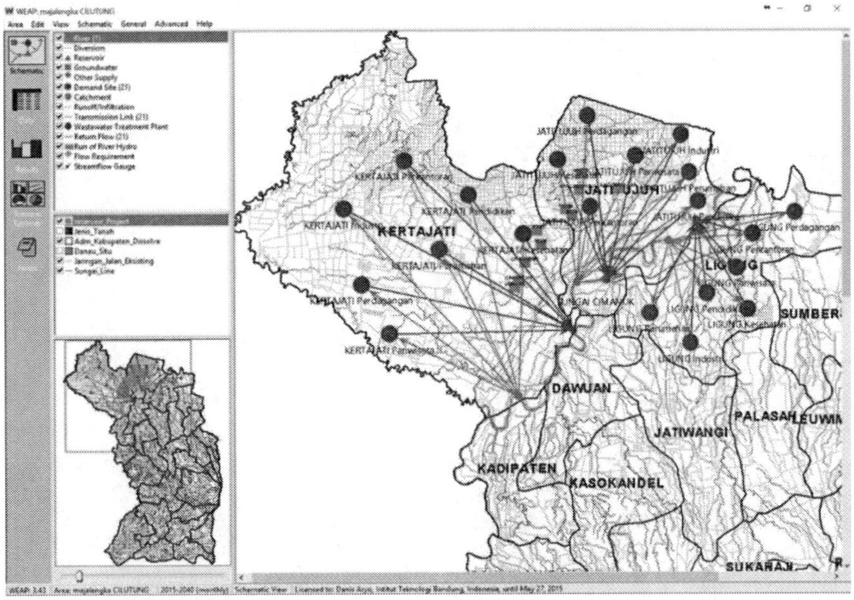

Fig. 5. WEAP Scheme

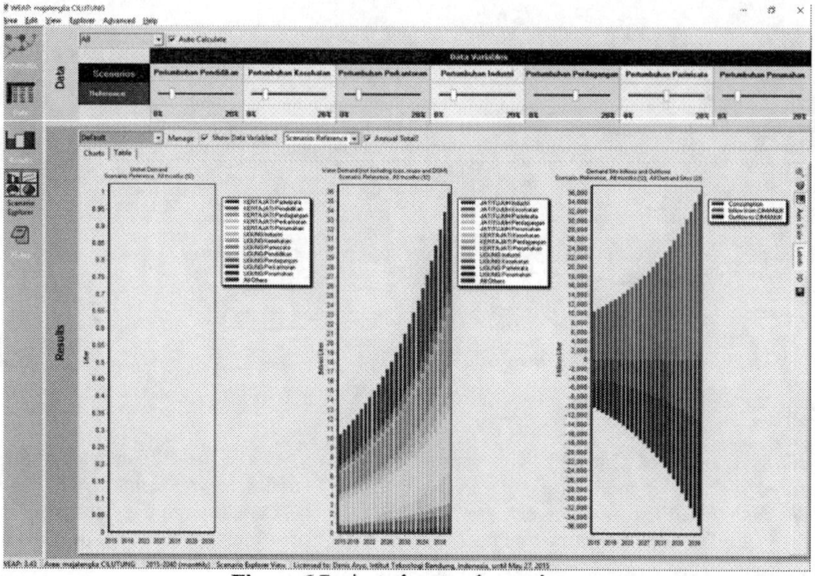

Figure 6 Projected water demand

- Based on the data obtained, the water demand in these three sub-districts can still be met up to 2040 by relying on the existing river flow.
- Required data clarification, especially river flow measurement data and other water source potentials to support the results of studies that have been done.

6 References

1. LPPM ITB, *Final Report Study of Raw Water Potential for BIJB Area and Kertajati* (2015)
2. PT. Sarana Bhuana Raya, *Final Report of Feasibility Study of Cipanasaat Reservoir Construction* (2006)
3. F.J. Mock, *Water Availibity Appraisal* (1973)
4. T.B. Adgolign et al, WEAP modeling of surface water resources allocation in Didessa Sub-Basin, West Ethiopia. *Springer Int. Publishing Switzerland* (2015)
5. A. Alfarra et al, Modeling Water Supply and Demand for Effective Water Management Allocation in the Jordan Valley. *Journal of Agricultural and Aplications (JASA)* (2012)
6. Yates D, Sieber J, Purkey D, Lee, WEAP21—a demand-, priority-, and preference-driven water planning model-part 1: model characterisitics. *Int Water Res Assoc* 30(4): 487–500 (2005)
7. T. Alemayehu, M. McCartney, S. Kebede, The water resource implications of planned development in the Lake Tana catchment. *Ethiopia Ecohydrol Hydrol* 10(2–4):211–222 (2010)
8. M.P. McCartney, M.M Girma, Evaluating the downstream implications of planned water resource development in the Ethiopian portion of the Blue Nile River. *Water Int* 37(4):362–379 (2012)
9. Obahoundje et al, Land Use and Land Covers Changes under Climater Undercertainty: Modelling the Impacts on Hydropower Production in Western Africa, *MDPI* (2017)
10. PT. Bandarudara International Jawa Barat, About PT BIJB, http://bijb.co.id/en/sejarah/ (30 July 2017).

Application of Finite Difference Schemes to 1D St. Venant for Simulating Weir Overflow

Nur Lely Hardianti Zendrato [1*], *Asrini* Chrysanti [1], *Bagus Pramono* Yakti [1], *Mohammad Bagus* Adityawan [2,3], *Widyaningtias* [2,3] and *Yadi* Suryadi [2,3]

[1]Graduate School of Civil Engineering, Institut Teknologi Bandung, Jalan Ganesha 10, Bandung 40132, Indonesia
[2]Water Resources Engineering Research Group, Institut Teknologi Bandung, Jalan Ganesha 10, Bandung 40132, Indonesia
[3]Water Resources Development Center, Institut Teknologi Bandung, Jalan Ganesha 10, Bandung 40132, Indonesia

Abstract. Depth averaged equations are commonly used for modelling hydraulics problems. Nevertheless, the model may not be able to accurately assess the flow in the case of different flow regimes, such as hydraulic jump. The model requires appropriate numerical method or other numerical treatments in order to simulate the case accurately. A finite volume scheme with shock capturing may provide a good result, but it is time consuming as compared to the commonly used finite difference schemes. In this study, 1D St. Venant equation is solved using Artificial Viscosity Lax-Wendroff and Mac-Cormack with TVD filter schemes to simulate an experiment case of weir overflow. The case is chosen to test each scheme ability in simulating flow under different flow regimes. The simulation results are benchmarked to the observed experimental data from previous study. Additionally, to observe the scheme efficiency, the simulation time between the models are compared. Therefore, the most accurate and efficient scheme can be determined.

1 Introduction

Open channel flow is a channel that drains water with free surface. There are many problems that occur in the flow in the open channel, e.g. is the failure of the structure on the weir. Therefore, early prevention is needed at the planning stage. On the planning stage, structure performance can be simulated by physical model or a numerical model, therefore structure failure can be prevented.

The physical model is created when the physical phenomenon of the problems in the prototype can be made on a smaller scale with adequate congruence. Numerical model is created when the problems to be studied can be formulated with the formulation / mathematical equations in detail.

However, physical modelling is usually more expensive than numerical models. Therefore, making a physical model is not economical if the existing problem can be solved using a numerical model and as stated in Toombes and Chanson [7], real-life situations are frequently too complex to solve without the aid of numerical models. Numerical models also offer ease of workmanship that can save time in very wide study that can be produced.

Geometry parameters, calibration process, and validation of simulation results in this study are benchmarked to the observed experimental data from previous study by Binilang [1]. On that previous study, flow is observed in multiple discharge and downstream water level.

This research develops numerical model using one dimensional shallow water equation with TVD filter applied MacCormack and Artificial Viscosity Lax Wendroff discretization scheme. Numerical model results then compared to previous experimental study for validation. Additionally, those schemes are compared to each other to determine the most accurate and efficient scheme.

2 Governing Equation

The system of partial differential equations which describe the 1-D incompressible flow in an open channel of arbitrary cross section used in this model is continuity and momentum of Saint-Venant equations. It derived from mass and momentum conservation laws shown in equations [1] and [2].

2.1 Continuity Equation

$$\frac{\partial A}{\partial t}+\frac{\partial Q}{\partial x}=0 \quad (1)$$

where A is cross-sectional area, Q is discharge, t and x are time and space respectively.

2.2 Momentum Equation

$$\frac{\partial Q}{\partial t}+\frac{\partial\left(\frac{Q^2}{A}\right)}{\partial x}+g\,A\left(\frac{\partial h}{\partial x}+S_f-S_0\right)=0 \quad (2)$$

* Corresponding author: nurlelyhardiantizendrato@gmail.com

where g is gravitational acceleration, h is water depth, Sf and S0 is energy slope and bed slope respectively.

3 Numerical Modelling

3.1 MacCormack Discretization Scheme

The application of MacCormack method to the hyperbolic equation proceeds in two steps; a predictor step which is followed by a corrector step. In predictor step, spatial and temporal derivative were made by forward difference scheme. In corrector step, forward scheme is used for temporal derivation and backward scheme for spatial derivation. Final step for (t+1) time discretization is computed as average of predictor and corrector results. In mathematical form, MacCormack discretization scheme described in equations [3] to [7].

For predictor step:

$$\frac{\partial U}{\partial x} = \frac{U_{i+1}^n - U_i^n}{\Delta x} \tag{3}$$

$$\frac{\partial U}{\partial t} = \frac{U_i^P - U_i^n}{\Delta t} \tag{4}$$

For corrector step:

$$\frac{\partial U}{\partial x} = \frac{U_{i+1}^n - U_i^n}{\Delta x} \tag{5}$$

$$\frac{\partial U}{\partial t} = \frac{U_i^C - U_i^n}{\Delta t} \tag{6}$$

And final step:

$$U_i^{n+1} = \frac{U_i^P - U_i^C}{2} \tag{7}$$

Where U is velocity, t and x are time and space respectively.

Farid et al. [8] has studied discretization of continuity and momentum equation in [1] and [2] using MacCormack scheme as shown in equations [8] to [15].

3.1.1 Predictor Step Discretization

For continuity equation:

$$\frac{A_i^P - A_i^n}{\Delta t} + \frac{Q_{i+1}^n - Q_i^n}{\Delta x} = 0 \tag{8}$$

$$A_i^P = A_i^n - \left(\frac{Q_{i+1}^n - Q_i^n}{\Delta x}\right)\Delta t \tag{9}$$

where A is cross-sectional area, Q is discharge, t and x are time and space respectively.

For momentum equation:

$$\frac{Q_i^P - Q_i^n}{\Delta t} + \frac{\left(\beta\frac{Q^2}{A}\right)_{i+1}^n - \left(\beta\frac{Q^2}{A}\right)_i^n}{\Delta x} + gA\frac{h_{i+1}^n - h_i^n}{\Delta x} + g\frac{AR}{2\rho}\frac{\rho s_{i+1}^n - \rho s_i^n}{\Delta x} = gA\left(S_0 - \frac{n^2Q^2}{A^2R^{4/3}}\right) \tag{10}$$

$$Q_i^P = Q_i^n + \left[gA\left(S_0 - \frac{n^2Q^2}{A^2R^{\frac{4}{3}}}\right) - \frac{\left(\beta\frac{Q^2}{A}\right)_{i+1}^n - \left(\beta\frac{Q^2}{A}\right)_i^n}{\Delta x} - gA\frac{h_{i+1}^n - h_i^n}{\Delta x} - g\frac{AR}{2\rho}\frac{\rho s_{i+1}^n - \rho s_i^n}{\Delta x}\right]\Delta t \tag{11}$$

where A is cross-sectional area, Q is discharge, t is time, x is space respectively, g is gravitational acceleration, h is water depth, n is manning roughness coefficient, R is hydraulic radius, Sf and S0 is energy slope and bed slope respectively.

3.1.2 Corrector Step Discretization

For continuity equation:

$$\frac{A_i^C - A_i^n}{\Delta t} + \frac{Q_i^n - Q_{i-1}^n}{\Delta x} = 0 \tag{12}$$

$$A_i^C = A_i^n - \left(\frac{Q_i^n - Q_{i-1}^n}{\Delta x}\right)\Delta t \tag{13}$$

For momentum equation:

$$\frac{Q_i^C - Q_i^n}{\Delta t} + \frac{\left(\beta\frac{Q^2}{A}\right)_i^n - \left(\beta\frac{Q^2}{A}\right)_{i-1}^n}{\Delta x} + gA\frac{h_i^n - h_{i-1}^n}{\Delta x} + g\frac{AR}{2\rho}\frac{\rho s_i^n - \rho s_{i-1}^n}{\Delta x} = gA\left(S_0 - \frac{n^2Q^2}{A^2R^{4/3}}\right) \tag{14}$$

$$Q_i^C = Q_i^n + \left[gA\left(S_0 - \frac{n^2Q^2}{A^2R^{\frac{4}{3}}}\right) - \frac{\left(\beta\frac{Q^2}{A}\right)_i^n - \left(\beta\frac{Q^2}{A}\right)_{i-1}^n}{\Delta x} - gA\frac{h_i^n - h_{i-1}^n}{\Delta x} - g\frac{AR}{2\rho}\frac{\rho s_i^n - \rho s_{i-1}^n}{\Delta x}\right]\Delta t \tag{15}$$

3.1.3 Total Variation Diminishing (TVD) Filter and Averaged Results

As stated in Tadmor [4], every discretization method is associated with an appropriate finite-dimensional projection. It is known that linear projections which are monotone, are the most first order accurate. The lack of monotonicity for higher order projections is reflected by spurious oscillations near jump discontinuities. These are evident with the second – order (and higher) centered difference, whose dispersive nature is responsible to the formation of binary oscillations. TVD schemes avoid spurious oscillations to this end they use the necessarily nonlinear projection. TVD filters, instead, suppress spurious oscillations.

Lal and ASCE [2] stated that the MacCormack method shows oscillatory behaviour near steep gradients and shocks which can be suppressed using a TVD method. Final step of numerical computation with MacCormack scheme with TVD filter applied can be described mathematically as shown in equations [16] to [24] as stated in Liang et al. [3]

Uin+1 = (UiP + UiC) / 2 + TVDi (16)

TVDi = [G(ri+) + G(r-i+1)] (ΔUni+1 - ΔUni) - [G(r+i-1) + G(ri-)] (ΔUni - ΔUni-1) (17)

$$ri+ = \frac{(h_i^n - h_{i-1}^n).(h_{i+1}^n - h_i^n) + (u_i^n - h_{i-1}^n).(u_{i+1}^n - h_i^n)}{(h_{i+1}^n - h_i^n).(h_{i+1}^n - h_i^n) + (u_{i+1}^n - h_i^n).(u_{i+1}^n - h_i^n)} \tag{18}$$

$$ri- = \frac{(h_i^n - h_{i-1}^n).(h_{i+1}^n - h_i^n) + (u_i^n - h_{i-1}^n).(u_{i+1}^n - h_i^n)}{(h_i^n - h_{i-1}^n).(h_i^n - h_{i-1}^n) + (u_i^n - h_{i-1}^n).(u_i^n - h_{i-1}^n)} \tag{19}$$

G(x) = 0,5 * Cin * [1 – φ(x)] (20)

$$\varphi(x) = \max(0, \min(2x, 1)) \tag{21}$$

$$\mathrm{Cin} = \begin{cases} \mathrm{Cr}_i^n . (1 - \mathrm{Cr}_i^n) \text{ for } \mathrm{Cr}_i^n \leq 0{,}5 \\ 0{,}25 \text{ for } \mathrm{Cr}_i^n > 0{,}5 \end{cases} \tag{22}$$

$$\mathrm{Cr}_i^n = \frac{\left(|u_i^n| + \sqrt{g.h_i^n}\right).\Delta t}{\Delta x} \tag{23}$$

$$u_i^n = \frac{Q_i}{A_i} \tag{24}$$

3.2 Lax-Wendroff Discretization Scheme

Lax-Wendroff Scheme is modification of FTCS with artificial diffusion. This scheme consists of Lax and Wendroff steps as shown in Figure 1. Both of them are using central difference to get space derivative and forward difference for time derivative. It is needed to apply, first, one step "Lax-step", and later half a time step.

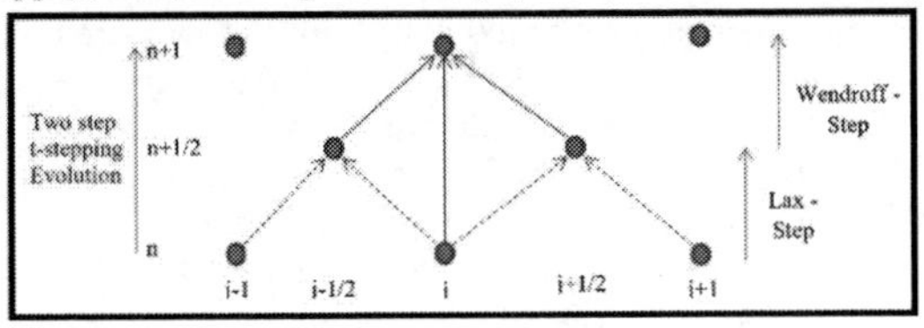

Fig. 1. Lax Wendroff Discretization Scheme

In mathematical form, Lax Wendroff discretization scheme described in equations [25] to [32] with Artificial Viscosity filter as shown in equations [33] to [36].

3.2.1 Lax Step

For continuity equation:

$$\frac{A_{i+\frac{1}{2}}^{n+\frac{1}{2}} - A_{i+\frac{1}{2}}^{n}}{\Delta t/2} + \frac{Q_{i+1}^n - Q_i^n}{\Delta x} = 0 \tag{25}$$

$$A_{i+\frac{1}{2}}^{n+\frac{1}{2}} = \frac{A_i^n + A_{i+1}^n}{2} - \frac{\Delta t}{2}\left(\frac{Q_{i+1}^n - Q_i^n}{\Delta x}\right) \tag{26}$$

where A is cross-sectional area, Q is discharge, t and x are time and space respectively.

For momentum equation:

$$\frac{Q_{i+\frac{1}{2}}^{n+\frac{1}{2}} - Q_{i+\frac{1}{2}}^{n}}{\frac{\Delta t}{2}} + \frac{\left(\frac{Q^2}{A}\right)_{i+1}^n - \left(\frac{Q^2}{A}\right)_i^n}{\Delta x} + gA\left(\frac{h_{i+1}^n - h_i^n}{\Delta x} + \frac{n^2Q^2}{A^2R^{\frac{4}{3}}} - S_0\right) = 0 \tag{27}$$

$$Q_{i+\frac{1}{2}}^{n+\frac{1}{2}} = \frac{Q_i^n + Q_{i+1}^n}{2} - \frac{\Delta t}{2}\left(\frac{\left(\frac{Q^2}{A}\right)_{i+1}^n - \left(\frac{Q^2}{A}\right)_i^n}{\Delta x} + gA\left(\frac{h_{i+1}^n - h_i^n}{\Delta x} + \frac{n^2Q^2}{A^2R^{\frac{4}{3}}} - S_0\right)\right) \tag{28}$$

where A is cross-sectional area, Q is discharge, t is time, x is space respectively, g is gravitational acceleration, h is water depth, n is manning roughness coefficient, R is hydraulic radius, Sf and S0 is energy slope and bed slope respectively.

3.2.2 Wendroff Step

For continuity equation:

$$\frac{A_i^{n+1} - A_i^n}{\Delta t} + \frac{Q_{i+\frac{1}{2}}^{n+\frac{1}{2}} - Q_{i-\frac{1}{2}}^{n+\frac{1}{2}}}{\Delta x} = 0 \tag{29}$$

$$A_i^{n+1} = A_i^n - \Delta t\left(\frac{Q_{i+1/2}^{n+1/2} - Q_{i-1/2}^{n+1/2}}{\Delta x}\right) \tag{30}$$

For momentum equation:

$$\frac{Q_i^{n+1} + Q_i^n}{\Delta t} + \frac{\left(\frac{Q^2}{A}\right)_{i+\frac{1}{2}}^{n+\frac{1}{2}} - \left(\frac{Q^2}{A}\right)_{i-\frac{1}{2}}^{n+\frac{1}{2}}}{\Delta x} + gA\left(\frac{h_{i+\frac{1}{2}}^{n+\frac{1}{2}} - h_{i-\frac{1}{2}}^{n+\frac{1}{2}}}{\Delta x} + \frac{n^2Q^2}{A^2R^{\frac{4}{3}}} - S_0\right) = 0 \tag{31}$$

$$Q_i^{n+1} = Q_i^n - \Delta t\left(\frac{\left(\frac{Q^2}{A}\right)_{i+\frac{1}{2}}^{n+\frac{1}{2}} - \left(\frac{Q^2}{A}\right)_{i-\frac{1}{2}}^{n+\frac{1}{2}}}{\Delta x} + gA\left(\frac{h_{i+\frac{1}{2}}^{n+\frac{1}{2}} - h_{i-\frac{1}{2}}^{n+\frac{1}{2}}}{\Delta x} + \frac{n^2Q^2}{A^2R^{\frac{4}{3}}} - S_0\right)\right) = 0 \tag{32}$$

3.2.3 Lax Wendroff Artificial Viscosity

The equations of hydrodynamics can be modified to include additional terms as stated in Neumann and Richtmeyer [6] which greatly simplify the procedures needed for stepwise numerical solution of the equations in problems involving shocks. A set of difference equations is given below.

Consider:

$$\frac{\partial A}{\partial t} + \frac{\partial Q}{\partial x} = 0 \tag{33}$$

Replace flux by:

$$Q' = Q - \alpha\frac{\partial A}{\partial x};\ \alpha = Dh^2\left|\frac{\partial A}{\partial x}\right| \tag{34}$$

Giving:

$$\frac{\partial A}{\partial t} + \frac{\partial Q}{\partial x} = -\frac{\partial}{\partial x}\left(-\alpha\frac{\partial A}{\partial x}\right) = \frac{\partial}{\partial x}\left(Dh^2\left|\frac{\partial A}{\partial x}\right|\frac{\partial A}{\partial x}\right) \tag{35}$$

The right term in equation (35) simulates the effect of the physical viscosity on the grid scale, concentrated around discontinuity and negligible elsewhere.

Approximation of artificial viscosity with Lax Wendroff discretization gives:

$$\frac{\partial}{\partial x}\left(Dh^2\left|\frac{\partial A}{\partial x}\right|\frac{\partial A}{\partial x}\right) = \frac{D}{\Delta x}\left[|A_{i+1}^n - A_i^n|(A_{i+1}^n - A_i^n) - |A_i^n - A_{i-1}^n|(A_i^n - A_{i-1}^n)\right] \quad (36)$$

3.3 Numerical Parameters

This model developed with cell size of 1 centimetres of 3.61 meters domain, thus resulting 362 number of cell nodes. Courant-type stability criterion is used in this model to calculate the maximum allowed time-step size. As stated in study conducted by Haun et al. [11], the Courant Number implies how fast the fluid passes through a cell. If the Courant Number is greater than 1, the velocity of a particle is so high that it passes through a cell in less than one-time step. This leads to numerical instabilities. Stability criteria and model trial lead to time step 0.001 seconds avoids instabilities in the computation and makes satisfied model result.

3.4 Initial and Boundary Conditions

Boundary condition of this model is static discharge on the upstream and static water depth on the downstream, using value from previous experimental research.

3.5 Convergence Criteria

The solution in this model is satisfy if the discharge has converged into upstream boundary value resulting water depth no longer change in time. Discharge is considered to be converged if the sum of discharge deviation of all cell is less than 0.01 m3/s. After the convergence criteria is met, the model should stop running and value of water depth and velocity are recorded.

3.6 Model Calibration

Calibrated parameter in this model is manning roughness coefficient. Three main criteria of this calibration process are hydraulic jump position, water depth in supercritical region, and upstream water depth. Calibration is taken using long section profile with discharge 0.00255 m3/s and downstream water level 0.0715 m from previous experimental research.

After calibration process is taken, it is obtained different value of manning roughness coefficient on weir body and floor. On the weir body, manning roughness coefficient varies gradually between 0.0175 on the upstream and 0.00835 on the downstream, and constant 0.00835 on the weir floor.

3.7 Previous Experimental Research

Previous experimental research which is benchmarked to this research consists of two set of discharge. Each discharge tested with 8 downstream different water depths. L1, L2, and L3 are conditions of non-submerged hydraulic jump; P1 and P2 are conditions of transitional hydraulic jump; T1, T2, and T3 are conditions of submerged hydraulic jump. Value of discharge and downstream water depth are shown in Table 1.

Table 1. Set of experimental research from previous study

Case	Discharge (m^3/s)	Downstream Water Depth (m)							
		L1	L2	L3	P1	P2	T1	T2	T3
Q2	0.0033	.045	.078	.090	.100	.115	.125	.140	.155
Q3	0.0026	.034	.072	.083	.100	.108	.128	.140	.155

4 Result and Discussion

The weir experiment consists of a 3.61 m long, 0.075 m wide flat channel with a streamline weir located on the upstream. The model is observed for two discharge values and for each discharge, there are eight downstream water depth variables. Computed water profile compared and examined to the result from previous experimental research as shown in Figure 2 and Figure 3 for MacCormack scheme, and Figure 4 and Figure 5 for Lax Wendroff scheme. Continuous lines in figure represent experimental result, and dotted line represent computed numerical results. The objective of this comparison is to examine the capability of the 1D SWE with MacCormack and Lax Wendroff numerical scheme to model weir overflow and hydraulic jump. The model results are presented in Figure 2 to Figure 5 below:

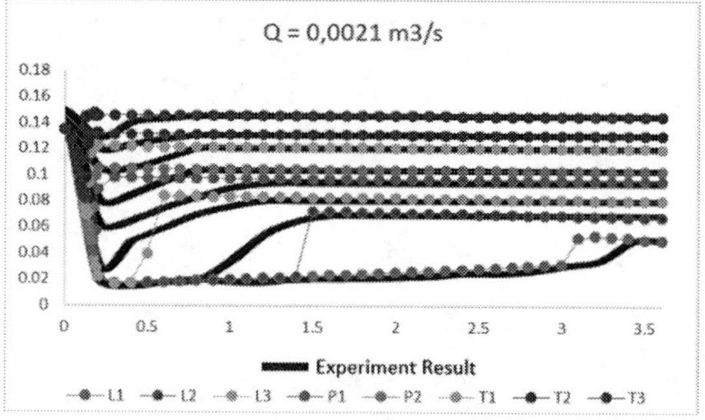

Fig. 2. Long section profile of MacCormack scheme result (Q2 case)

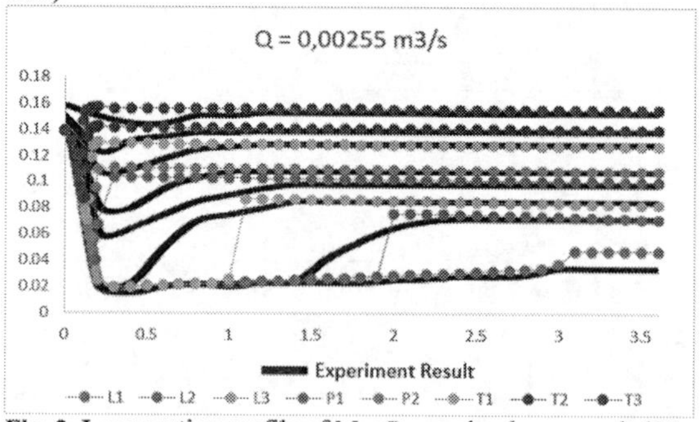

Fig. 3. Long section profile of MacCormack scheme result (Q3 case)

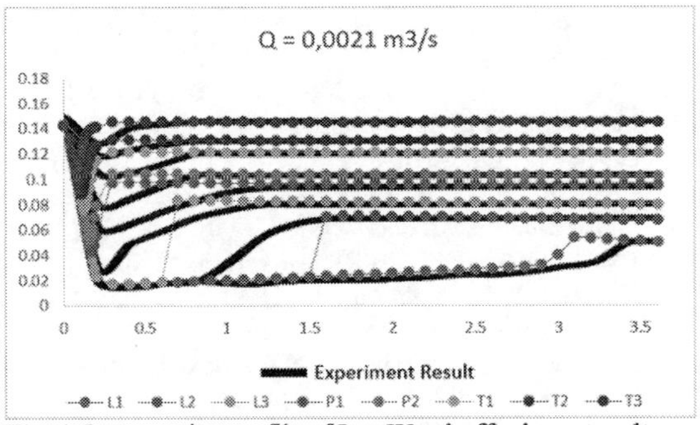

Fig. 4. Long section profile of Lax Wendroff scheme result (Q2 case)

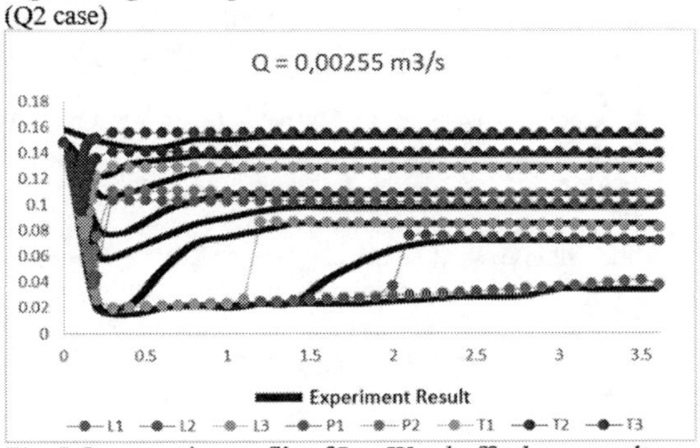

Fig. 5. Long section profile of Lax Wendroff scheme result (Q3 case)

The close agreement between the computed results and the experiment for hydraulic jump location, supercritical region and upstream water depth shows that the proposed method is comparatively accurate, which value of error shown in Table 2. However, both models fail to compute length of hydraulic jump since 1D SWE gives a linear height profile before the jump as stated in Singha et al. [10]. In addition, models are not valid for submerged hydraulic jump condition and there is significantly error on the upstream as roller effect influence flow significantly, which is 2D phenomenon.

Table 2. Maximum Error

Mac-Cormack Scheme			
Case	Maximum Water Depth Error (m)		
	Upstream	Supercritical	Subcritical
L1	0.001	0.031	-
L2	0.003	0.006	0.010
L3	0.000	0.000	0.009
P1	0.002	-	0.046
P2	0.002	-	0.035
T1	0.001	-	0.024
T2	0.009	-	0.019
T3	0.017	-	0.012
Lax Wendroff Scheme			
Case	Maximum Water Depth Error (m)		
	Upstream	Supercritical	Subcritical
L1	0.010	0.010	-
L2	0.012	0.006	0.011
L3	0.009	0.000	0.009
P1	0.007	-	0.046
P2	0.007	-	0.035
T1	0.008	-	0.024
T2	0.000	-	0.019
T3	0.008	-	0.012

Comparative result between MacCormack and Lax Wendroff schemes for L2, P2, and T2 in Q3 case as shown in Figure 6 shows that Lax Wendroff scheme provides more oscillation result relative to MacCormack. Overall for both schemes, hydraulic jump location, supercritical and upstream water depth match each other with only slight difference due to oscillation on Lax Wendroff's result.

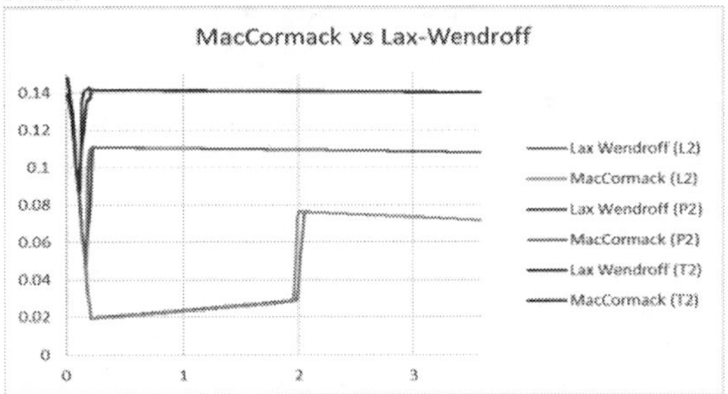

Fig. 6. MacCormack and Lax Wendroff schemes comparison (Q3 Case)

Numerical result is verified using Belanger formula as stated in Liu et al. [12] for the case of L2 (dark red line in Figure 2 to 5). Verification result satisfies with maximum error 0.003 meter, as shown in Table 3. Simulation times for TVD MacCormack scheme to obtain all results in 5 discharges and 8 downstream water depth cases are 56 minutes and 54 minutes for Lax Wendroff scheme.

5 Conclusion

Numerical model has been developed for 1D shallow water equations (Saint-Venant equations) to compute weir overflow including hydraulic jump phenomenon at downstream. The 1D shallow water equations are solved with the MacCormack and Lax Wendroff scheme discretization with addition of a symmetric total variation diminishing (TVD) term to the corrector step of MacCormack, and artificial viscosity term to Wendroff step of Lax Wendroff.

Table 3. Belanger formula verification

MacCormack Scheme				
Case: Q2 L2			Case: Q3 L2	
y_1	0.040		y_1	0.031
y_2	0.080		y_2	0.073
Q	0.003		Q	0.003

v_1	1.104	v_1	1.110
Fr_1	1.765	Fr_1	2.025
y_2/y_1	2.045	y_2/y_1	2.408
y_1	0.039	y_1	0.030
Δy_1	0.001	Δy_1	0.000
Lax Wendroff Scheme			
Case: Q2 L2		Case: Q3 L2	
y_1	0.050	y_1	0.029
y_2	0.071	y_2	0.074
Q	0.003	Q	0.003
v_1	0.881	v_1	1.189
Fr_1	1.258	Fr_1	2.245
y_2/y_1	1.348	y_2/y_1	2.713
y_1	0.053	y_1	0.027
Δy_1	0.003	Δy_1	0.001

Both schemes show good result to estimate hydraulic jump location, upstream and supercritical water depth. In Lax Wendroff's result, there is small oscillation around bed slope transition and hydraulic jump. Simulation time between both schemes shows almost same compared each other. Therefore, it can be concluded that TVD MacCormack scheme more accurate and efficient to model weir overflow and hydraulic jump compared to Lax Wendroff with Artificial Viscosity, it is because TVD is more effective to filter shocks.

Numerical simulation for the weir overflow and hydraulic jump indicates that this model is fairly effective and accurate for non-submerged condition and can be used to estimate hydraulic jump location for a planning stage of weir construction. Additionally, this model is able to simulate weir overflow and hydraulic jump with dynamic discharge and water depth. High-resolution mathematical model can be further modified and extended to multi-dimensional hyperbolic conservation laws to effectively simulate 2D and 3D weir overflow, in condition of submerged hydraulic jump occur at downstream.

6 References

1. A. Binilang, Kajian Pengaruh Hubungan Antar Parameter Hidrolis Terhadap Sifat Aliran Melewati Pelimpah Bulat da Setengah Lingkaran pada Saluran Terbuka, *Jurnal Ilmiah Media Engineering* **4(1)**, pp.55-61 (2014)
2. A.M.W. Lal, A. M. ASCE, A TVD MacCormack Method for Open Water Hydraulics and Transport, *Research Engineer, Department of Civil and Environmental Engineering, Clarkson University*, (Potsdam, New York)
3. D. Liang, B. Lin, R.A. Falconer, Simulation of Rapidly Varying Flow Using an Efficient TVD-MacCormack Scheme, *International Journal for Numerical Methods in Fluids*, **53(5)**, 811-826, (2006)
4. E. Tadmor, Approximate Solutions of Nonlinear Conservation Laws and Related Equations, *Proceedings of Symposia in Applied Mathematics*, 54, 344 (1996)
5. J.M. Murawska, S. Michal, Lax-Wendroff and MacCormack Schemes for Numerical Simulation of Unsteady Gradually and Rapidly Varied Open Channel Flow, *Archives of Hydro-Engineering and Environmental Mechanics*, **60(1-4)**, 51-62 (2013)
6. J.V. Neumann, R. D. Richtmyer, A Method for the Numerical Calculation of Hydrodynamic Shocks (Princetown, New Jersey, 1949)
7. L. Toombes, H. Chanson, Numerical Limitations of Hydraulic Models (Brisbane, Australia, 2011)
8. M. Farid, B.P. Yakti, A. Rizaldi, M.B. Adityawan, Finite Difference Numerical Scheme for Simulating Dam Break Flo, *5th International Seminar of HATHI* (Bali, 2016)
9. R. Luciano, Numerical Methods for the Solution of Partial Differential Equations, *Albert Einstein Institute, Max – Plank Institute for Gravitational Physics*, (Potsdam, Germany, 2011)
10. S.B. Singha, J. K. Bhattacharjee, A. K. Ray, Hydraulic Jump in One-Dimensional Flow, (India, 2005)
11. S. Haun, N.R.B. Olsen, R. Feurich, Numerical Modelling of Flow Over Trapezoidal Board-Crested Weir, *Engineering Applications of Computational Fluid Mechanics*, **5(3)**, 397-405, (2011)
12. Y.L. Liu, Y.X. Cai, W.L. Wei, Zh. Ma, Numerical Simulation of Hydraulic Jump Using ENO Scheme, *Journal of Chemical and Pharmaceutical Research*, **6(6)**, 603-607, (2014)

3D Numerical Modeling of Flow in Sedimentation Basin

Dhemi Harlan[1,*], *Dantje K.* Natakusumah[1], *Mohammad Bagus* Adityawan[1], *Hernawan* Mahfudz[1], and *Fitra* Adinata[2]

[1]Water Resources Engineering Research, Institut Teknologi Bandung, Jl. Ganesa No. 10 Bandung, Indonesia
[2]PT Sapta Adhi Pratama Consultant Bandung, Indonesia

Abstract. Normal operation sedimentation basin flushing systems require large volumes of water, typically up ten times of the deposited sediment volume for efficient flushing. A complete sediment removal, can only be realized by combination of mechanical removal with drawdown flushing. This operation reaches much longer operation time resulting in water loss and reducing power and energy production of Mini Hydro Power Plant (MHPP). The objective of this study is to improve the flushing system of sedimentation basin based on a numerical approach. Fluid motion is described with non-linear, transient, second-order differential equations. A numerical solution of these equations involves approximating the various terms with algebraic expressions. The resulting equations are then solved to yield an approximate solution to the original problem. The simulation result shows that the 3D numerical modeling of flow in sedimentation basin gives the reasonable result to predict the suspended load movement in the flow.

1 Introduction

Mini Hydro Power Plant has installed an 6,7 MW mini hydro power plant located in Pasir Melayu Village, Sialu Kahean Sub District, Simalungun District, North Sumatera. Unfortunately, it has experienced severe problems with silt in the head-works and subsequent damage to equipment. The problem caused very serious impact because that sediment coarser than 0.20 mm in size was harmful for turbine blades and excess of the carrying capacity of the canal. As result, the excess sediment load has to be eliminated from sedimentation basin as part of regularly scheduled maintenance activities and it has increased maintenance load and cost.

Normal operation sedimentation basin flushing systems require large volumes of water, typically up ten times of the deposited sediment volume for efficient flushing. In sedimentation basin, however, a complete sediment removal, can only be realized by combination of mechanical removal with drawdown flushing. This operation reaches much longer operation time resulting in water loss and reducing power and energy production of mini hydro power plant. Therefore, it is important to have an effective and efficient sedimentation basin.

There are various design codes for a sedimentation basin. These codes are mostly based on empirical approaches. A past study has shown that they may not provide accurate results, requiring a site specific modification [1]. However, a site specific design can be assessed using a laboratory experiment or a numerical model. The effectiveness of a retention basin was evaluated using a 1D model with good accuracy [2]. However, a 1D model may not be suitable for assessing the hydraulic condition in a sedimentation basin with a submerged hydraulic structure, e.g. a flushing gate, as in the case of this study. In this case, a 3D model is more accurate. A study was conducted using a 3D numerical model to evaluate the efficiency of a sedimentation basin with various configurations [3]. The results provide a more detail on the effectiveness of a sedimentation basin.

The objective of the study is to evaluate the flow in a sedimentation basin by using a 3D model in order to assess the effectiveness of the basin, and to improve the flushing system of a sedimentation basin system based on a numerical approach.

2 Technical Data of Karai-7 Mini Hydro Power Plant

The technical data of Mini Hydro Power Plant are obtained based on the design report [4]. The analysis in this study was focused on the sedimentation basin of Mini Hydro Power Plant in Karai-7 location. The inflow and outflow discharges of the sedimentation basin for short period were measured by the consultant [5], and some sediment samples both suspended and bed loads. We use these data in our computation and analysis. This study focuses on flow modeling for obtaining the flow characteristic in sedimentation basin. However, in this section, we explain them briefly such as sedimentation basin, inflow and outflow discharge, and sediment grain size.

* Corresponding author: dhemi@si.itb.ac.id

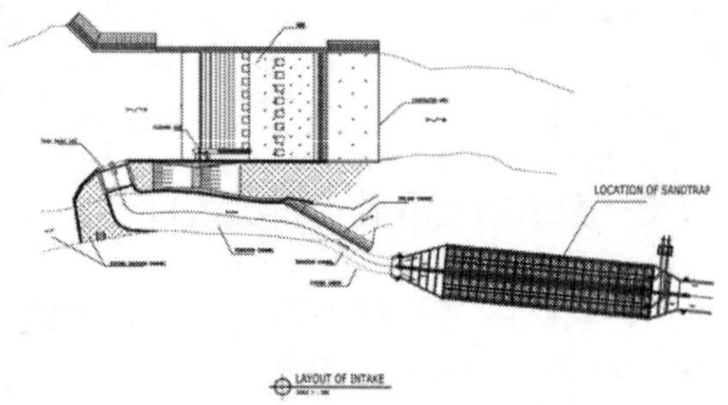

Fig. 1. Layout of Sedimentation Basin
Source: Design Report, PT BHE 2014

2.1. Sedimentation Basin

Sedimentation basin as one of mini hydro infrastructure is needed to overcome sedimentation problem in water way channel. It is a pond constructed for the purpose of removing entrained solids by sedimentation [6]. Sedimentation along water way channel can reduce channel capacity that affecting the productivity of electricity. The design of sedimentation basin based on the Design Report [4] as Figure 2 following below.

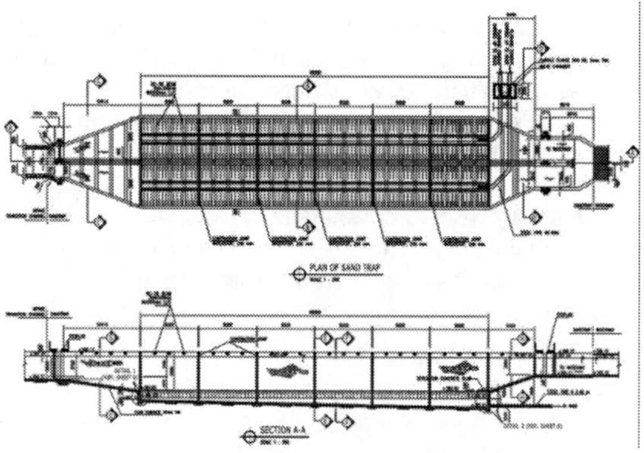

Fig. 2. Sedimentation basin of MHPP
Source: Design Report, PT BHE 2014

2.2. Inflow and Outflow Discharges

The inflow discharge and outflow discharge are used based on consultant report [5]. The average inflow discharge to sedimentation basin has the value of 13,5 m3/s and the value of 8 m3/s for the average outflow. The reducing of discharge between inflow and outflow is about 40 %. This could be caused by the sedimentation problem in the basin. This problem will affect the productivity of electricity. Based on installed capacity, the Karai-7 mini hydro just produces around 4 MW of electricity.

2.3. Sediment Grain Size

Removal of suspended particles by sedimentation depends upon the size and specific gravity of those particles. Suspended solids retained on a filter may remain in suspension if their specific gravity is similar to water while very dense particles passing through the filter may settle. Solids that can settle are measured as the visible volume accumulated at the bottom of basin [5]. Periodically this sediment is flushed every 2 days because the basin is full. All of the sediments are sand or sandy. The distribution of sand is more than 90 %. Mostly has diameter between 0.1 and 1 mm (70%-90%), Seven (7) % or less has diameter more than 2 mm. D50 is around 0.5 mm, less than 0.2 mm allowed passing the turbine.

3 Mathematical Formulation

The differential equations to be solved are written in terms of Cartesian coordinates (x, y, z) . For cylindrical coordinates (r, θ, z), the x-coordinate is interpreted as the radial direction, the y-coordinate is transformed to the azimuthal coordinate, θ, and z is the axial coordinate. For cylindrical geometry, additional terms must be added to the Cartesian equations of motion. These terms are included with a coefficient ξ, such that ξ = 0 corresponds to Cartesian geometry, while ζ = 1 corresponds to cylindrical geometry. All equations are formulated with area and volume porosity functions. This formulation is called Fractional Area-Volume Obstacle Representation (FAVOR) Method, used to model complex geometric regions [7].

3.1. Governing Equation

The flow can be modeled by using continuity equation, momentum equation, and the volume of fluid equation in Cartesian coordinate. These equations are presented in equation 1 to 3 below.

$$\frac{V_F}{\rho c^2}\frac{\partial p}{\partial t}+\frac{\partial}{\partial x}(uA_x)+R\frac{\partial}{\partial y}(uA_y)+\frac{\partial}{\partial z}(uA_z)+\xi\frac{v_\rho A_x}{x}=\frac{R_{SOR}}{\rho} \quad \text{.............. (1)}$$

$$\frac{\partial u}{\partial t}+\frac{1}{V_F}\left\{uA_x\frac{\partial u}{\partial x}+vA_yR\frac{\partial u}{\partial y}+wA_z\frac{\partial u}{\partial z}\right\}+\xi\frac{A_y v^2}{xV_F}$$
$$=-\frac{1}{\rho}\frac{\partial p}{\partial x}+G_x+f_x-b_x-\frac{R_{SOR}}{\rho V_F}(u-u_w-\delta u_s)$$

$$\frac{\partial v}{\partial t}+\frac{1}{V_F}\left\{uA_x\frac{\partial v}{\partial x}+vA_yR\frac{\partial v}{\partial y}+wA_z\frac{\partial v}{\partial z}\right\}+\xi\frac{A_y uv}{xV_F}$$
$$=-\frac{1}{\rho}\left(R\frac{\partial p}{\partial y}\right)+G_y+f_y-b_y-\frac{R_{SOR}}{\rho V_F}(v-v_w-\delta v_s)$$

$$\frac{\partial w}{\partial t}+\frac{1}{V_F}\left\{uA_x\frac{\partial w}{\partial x}+vA_yR\frac{\partial w}{\partial y}+wA_z\frac{\partial w}{\partial z}\right\}$$
$$=-\frac{1}{\rho}\frac{\partial p}{\partial z}+G_z+f_z-b_z-\frac{R_{SOR}}{\rho V_F}(w-w_w-\delta w_s) \quad (2)$$

$$\frac{\partial F}{\partial t}+\frac{1}{V_F}\left[\frac{\partial}{\partial x}(FA_xu)+R\frac{\partial}{\partial y}(FA_yv)+\frac{\partial}{\partial z}(FA_zw)+\frac{\xi FA_xu}{x}\right]=F_{SOR} \quad (3)$$

Where V_F is the fractional volume open to flow, ρ is the fluid density, c^2 is the square of the sound speed, and R_{SOR} is a mass source. The velocity components (u, v, w) are in the coordinate directions (x, y, z) or (r, R_{SOR}, z) . A_x is the fractional area open to flow in the x-direction, and A_x , A_y , A_z are similar area fractions for flow in the y and z directions, respectively. The coefficient R depends on the choice of coordinate system in the following way. When cylindrical coordinates are used, derivatives must be converted to azimuth derivatives.

In these equations, (G_x , G_y , G_z) are body accelerations, (f_x , f_y , f_z) are viscous accelerations, and (b_x , b_y , b_z) are flow losses in porous media or across porous baffle plates, and the final terms account for the injection of mass at a source represented by a geometry component.

The term $U_w = (u_w, v_w, w_w)$ in Equation (2) is the velocity of the source component, which will generally be non-zero for a mass source at a General Moving Objects Model. Meanwhile the term $U_s = (u_s, v_s, w_s)$ is the velocity of the fluid at the surface of the source relative to the source itself.

3.2. Numerical Model

3.2.1 Model Geometry

Geometry is constructed by assembling solid geometric objects to define the flow region for a simulation, as shown in Figure 3.

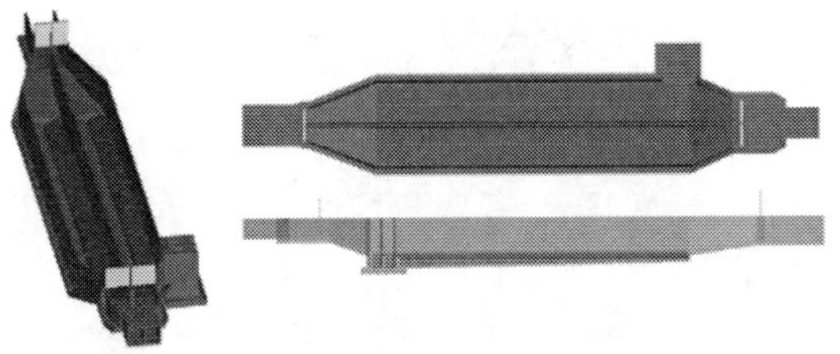

Fig. 3. 3D model of Karai-7 sedimentation basin

3.2.2 Model Mesh Block

The most basic mesh in FLOW-3D is a single-block mesh, i.e. only one mesh block is used to define the domain. The default settings for a mesh block will create a uniform mesh, which is a mesh with a uniform cell size in all directions for this model the cell size uniformly by 0.15 m each. These may be created by supplying only a cell size, a total number of cells, or a specific number of cells in each coordinate direction as shown in Figure 4. It is also possible to define multiple mesh blocks (multi-block gridding) to create more efficient meshes when modeling complex flow domains. The additional mesh blocks may be defined to be fully within another mesh block (including aligned with the containing block boundaries; this is called a nested block), they may adjoin at a boundary of another mesh block (called a linked block), or they may partially overlap.

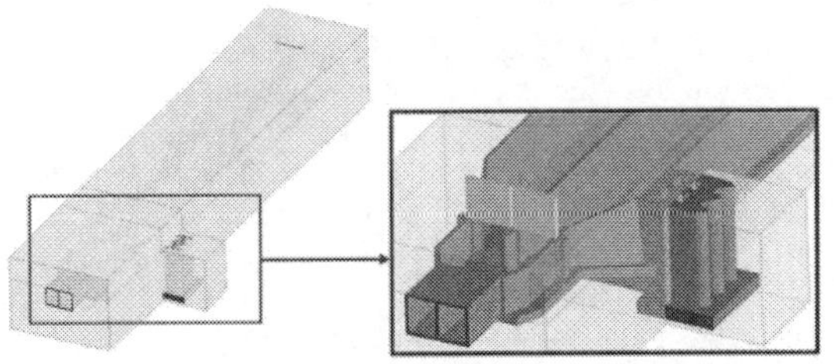

Fig. 4. Model block-grid

3.2.3 Boundary Condition

The boundary conditions associated with Geometry are handled by the choice of component type or via other component-specific properties that are associated with a particular physical model/equation (described in Model Reference). However, boundary conditions on the faces of each mesh block also need to be defined as shown in Figure 6 and Table 1 below.

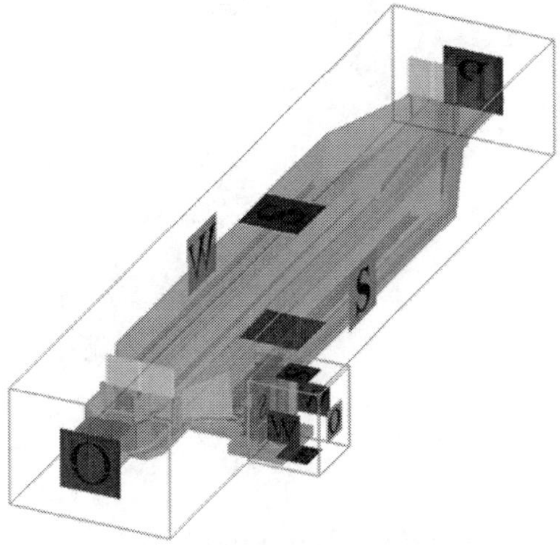

Fig. 5. Boundary conditions in model Karai-7 Sedimentation Basin

Table 1. Boundary conditions in the model

No Block	Block Name	Cell Size	Boundary Conditions					
			Xmin	Xmax	Ymin	Ymax	Zmin	Zmax
Mesh Block 1	Block 1	0.5 m	P	O	W	S	W	S
Mesh Block 2	Block 2	0.5 m	W	W	S	O	W	S

** P : Presure (Stagnations Presure) with water elevations given
S : Symetry
O :Outflow
W:Wall

3.2.4 Initial Condition

Initial conditions define the starting conditions in the simulation. As mentioned in Mesh Boundary Conditions, the initial state of the solution for transient fluid flow problems must be known in order to find a solution and, in a manner similar to what is done with boundary conditions, the initial conditions are assumed, approximating the true state at time t = 0. The accuracy of the initial conditions is somewhat less important than

https://doi.org/10.1051/matecconf/201814703012

that of the boundary conditions because their effect is reduced as time progresses. Because of this effect, it is common to define the initial geometry of the fluid with reasonable accuracy but to assume that the pressure and velocity fields are uniform. This typically produces good results with a relatively short time where the solution is affected by physical initial conditions.

Fluid regions are used to define areas of fluid in a manner similar to how sub components are defined. As such, they may be defined using limiters, coefficients for a quadratic function. In the case where the quadratic function is used the use function coefficients for box provides control over how the input coefficients (set in the Coefficients branch) are interpreted. The other options are used to define distributions of different quantities (density, temperature, velocity, etc.).

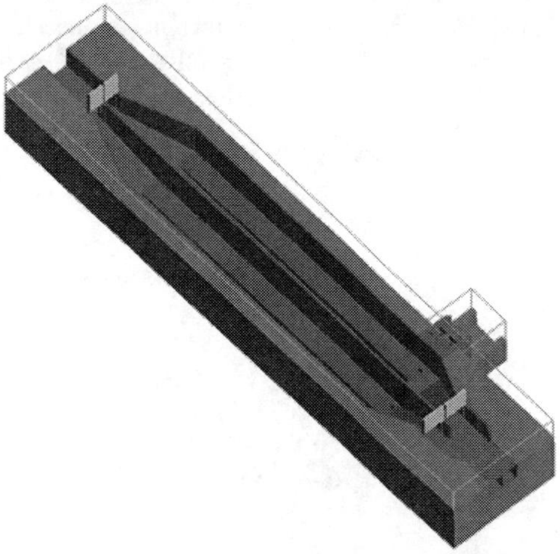

Fig. 6. Initial conditions t=0 in model Karai-7 Sedimentation Basin shows the model has no fluid

4 Result and Discussion

Figures 7 to 9 show the vertical velocity distribution along the left and right chambers at 60, 120 and 180 seconds, respectively. These results are important to assess whether the sediment basin can effectively trap the sediment. In general, the vertical velocity distribution in both chambers shows similar behavior.

The velocity distribution on the surface of the basin is given in Figures 7(a), 8(a)and 9(a), for 60, 120, and 180 seconds respectively. The sedimentation basin is divided into two chambers, left and right. The flushing gate is located in the left chamber (looking from downstream). In general, the flow distribution in the right and left chambers are similar. However, the flow magnitude tends to be higher in the left chamber. This is caused by the release square hole location of right chamber from center line of basin not similar position compared to the left one.

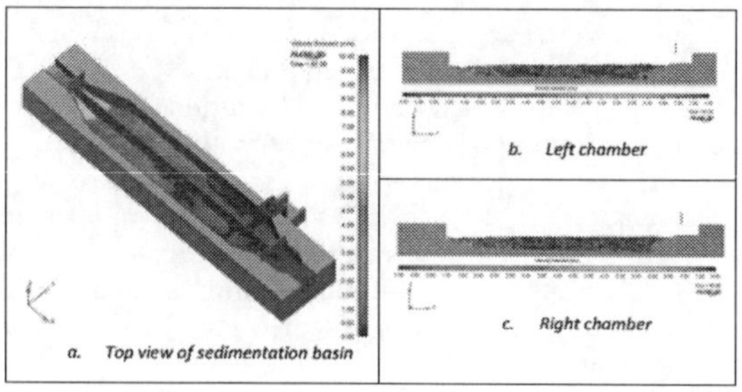

Fig. 7. Modeling result of velocity magnitude for t=60'sec

The flow velocity was reduced significantly after entering the basin. Figure 7 shows the horizontal velocity distribution on surface at 60 seconds after the gate was opened. The flow velocity at the first half of the basin is approximately 2 m/s, which is reduced to 1 m/s at the last half.

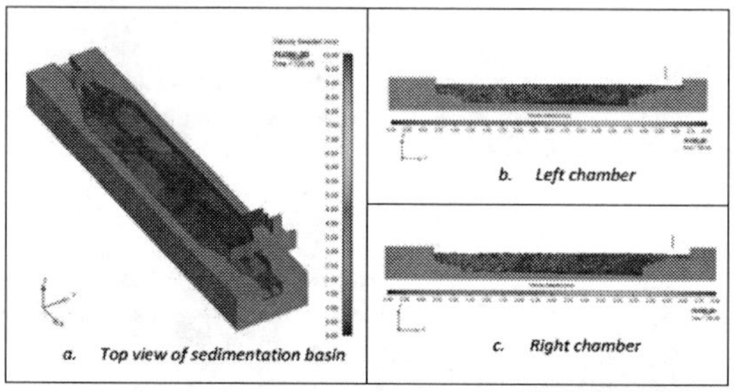

Fig. 8. Modeling result of velocity magnitude for t=120'sec

Figure 8 shows the model result at 120 seconds, the flow velocity is much calmer along the basin. It is shown here that the surface velocity is less than 1 m/s, and it is still higher than 0.3 m/s. It is expected that the velocity magnitude should be less than 0.3 m/s to allow sediment deposition in the basin or to trap the minimum size of 0.2 mm sediment. The simulation reaches a stable condition after 180 seconds as shown in Figure 9.

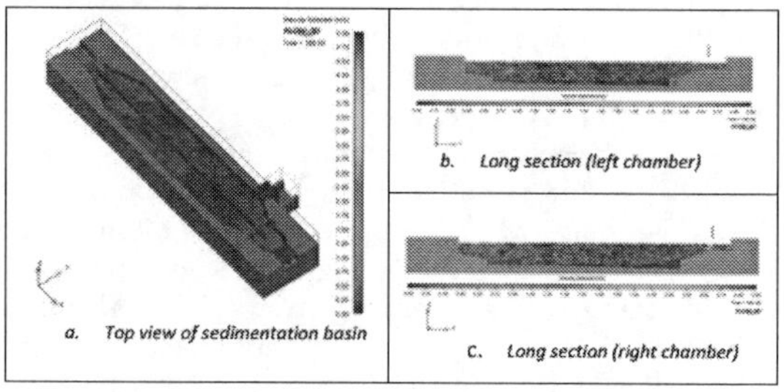

Fig. 9. Modeling result of velocity magnitude for t=180'sec

Figure 10 shows six observation locations for the left and right chambers in Cartesian coordinate that every location consists of three depths: top, middle, and bottom of water surface. Hence, observation points consist of nine point probes for every chamber.

https://doi.org/10.1051/matecconf/201814703012

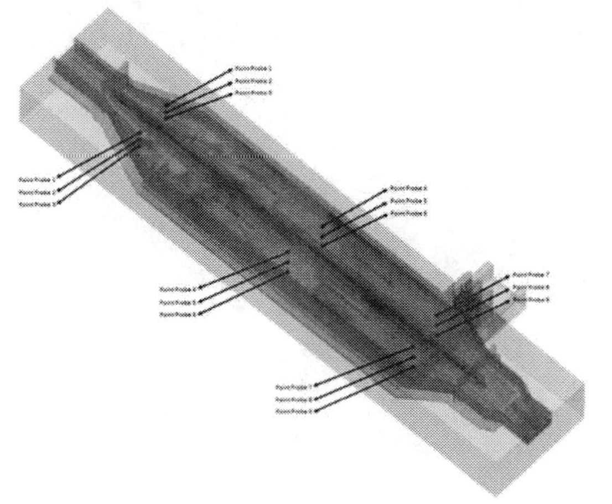

Fig. 10. Left and right sediment chambers Point probes of model observation

Left Chamber				Right Chamber			
Probe Name	X	Y	Z	Probe Name	X	Y	Z
PROBE 1	15	10.95	6.675	PROBE 1	15	6.5	6.675
PROBE 2	15	10.95	5.675	PROBE 2	15	6.5	5.675
PROBE 3	15	10.95	4.875	PROBE 3	15	6.5	4.875
PROBE 4	43	11.95	6.675	PROBE 4	43	5.45	6.675
PROBE 5	43	11.95	4.875	PROBE 5	43	5.45	4.875
PROBE 6	43	11.95	3.275	PROBE 6	43	5.45	3.275
PROBE 7	64.5	11.95	6.675	PROBE 7	66.5	5.45	6.675
PROBE 8	64.5	11.95	4.675	PROBE 8	66.5	5.45	4.675
PROBE 9	64.5	11.95	2.175	PROBE 9	66.5	5.45	2.175

Figures 11 to13 show the flow velocity at 60 seconds in the left and right chambers, accordingly. The flow velocity near the bed is high along the basin in both chambers, approximately 2 m/s, although the velocity near the surface is less than 1 m/s. The near bed velocity is not low enough to trap the sediment.

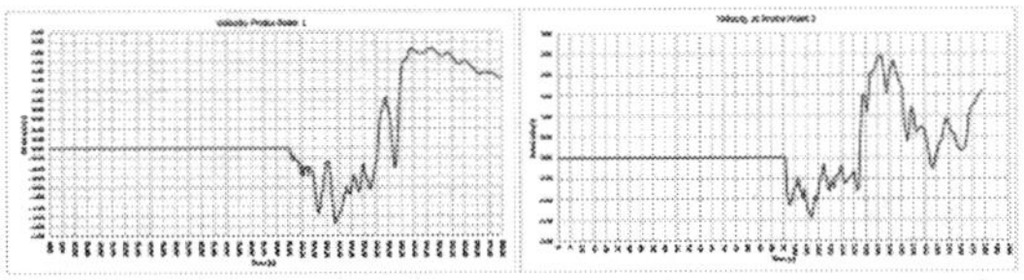

Fig. 11. Velocity at point probe 1 (left and right sediment-trap)

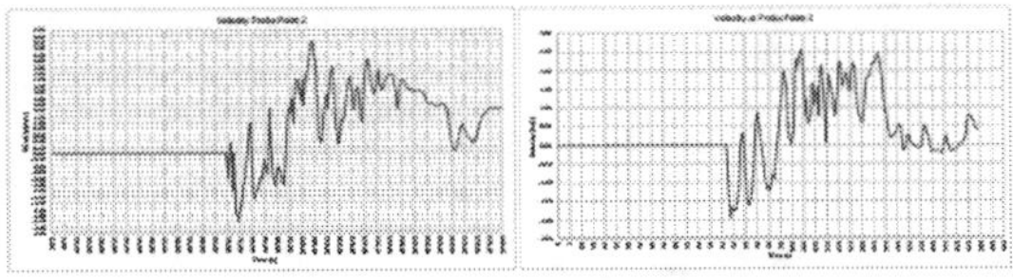

Fig. 12. Velocity at point probe 2 (left and right sediment-trap)

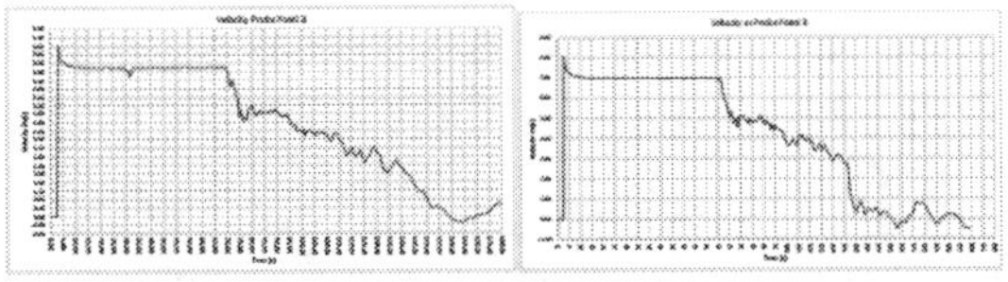

Fig. 13. Velocity at point probe 3 (left and right sediment-trap)

Figure 14 to 16 show the flow velocity at 120 seconds in the left and right chambers. The near bed velocity is significantly low, less than 0.25 m/s, which is low enough to deposit 0.2 mm sediment. However, the vertical velocity distribution at the first half of the basin is higher than 1 m/s.

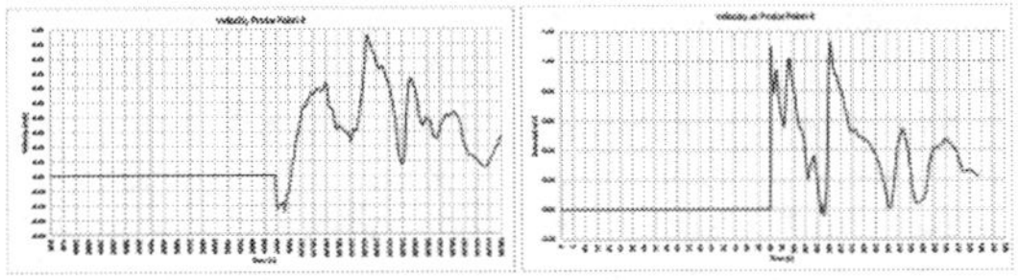

Fig. 14. Velocity at point probe 4 (left and right sediment-trap)

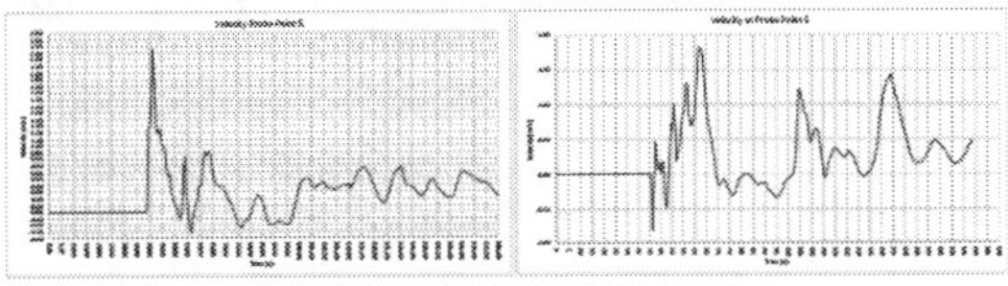

Fig. 15. Velocity at point probe 5 (left and right sediment-trap)

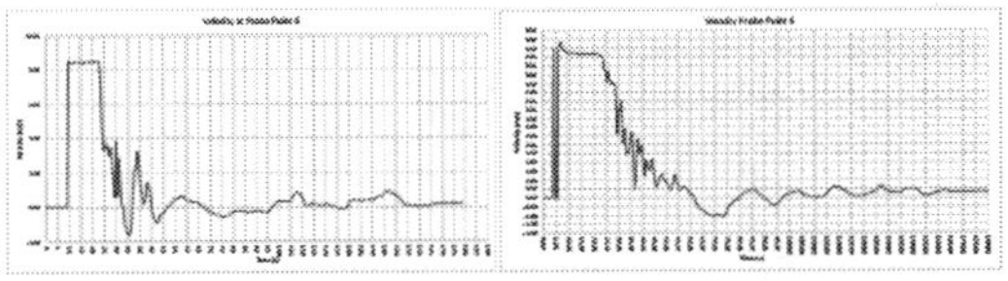

Fig. 16. Velocity at point probe 6 (left and right sediment-trap)

The stable condition is given in Figures 17 to 19, at 180 seconds for the left and right chambers. Both chambers are equally effective for trapping the sediment. In general, the velocity magnitude in both chambers is lower than 0.3 m/s. This magnitude is sufficiently low to deposit 0.2 mm sediment.

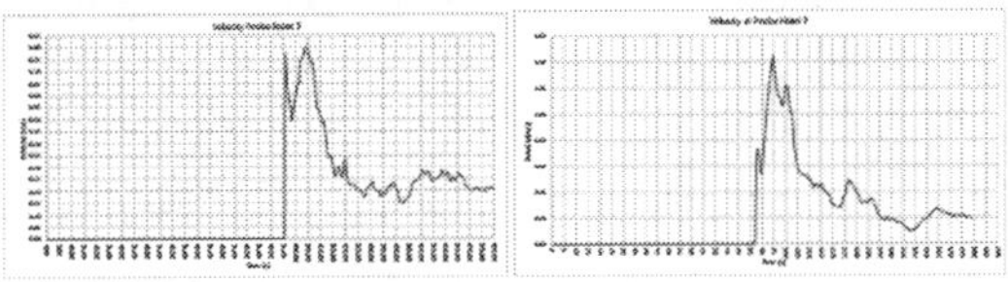

Fig. 17. Velocity at point probe 7 (left and right sediment-trap)

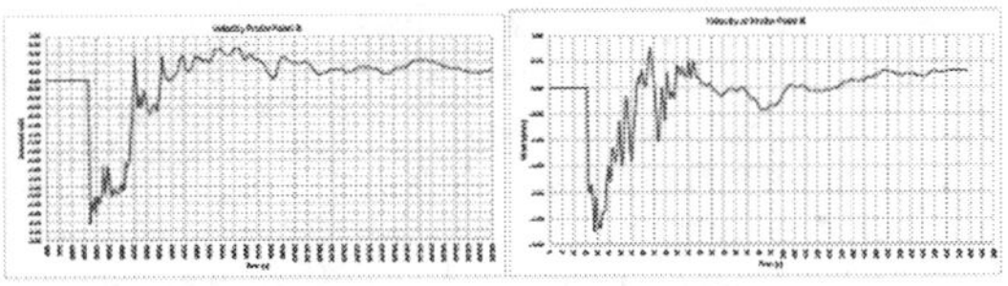

Fig. 18. Velocity at point probe 8 (left and right sediment-trap)

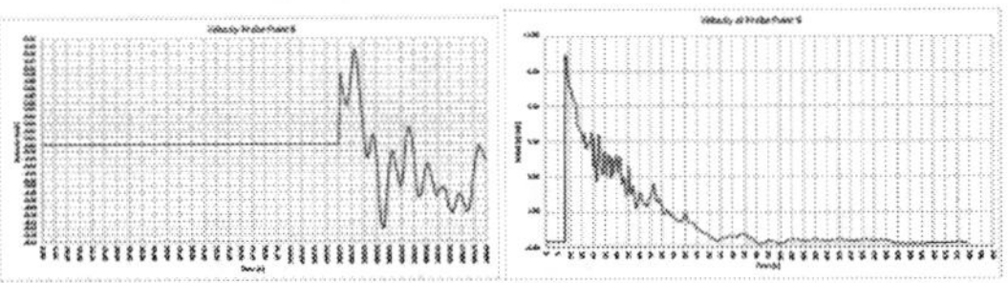

Fig. 19. Velocity at point probe 9 (left and right sediment-trap)

5 Conclusion

The sedimentation basin is expected to trap sediment with the diameter of 0.2 mm. A velocity magnitude of approximately 0.3 m/s is expected to allow this sediment to deposit. It was found from the vertical velocity distribution that the velocity magnitude in the basin is less than 0.3 m/s. Therefore, the sedimentation basin is effective for less than 0.2 mm sediment. This shows that 3D numerical modeling of flow in sedimentation basin gives the reasonable result to predict the suspended load movement in the flow. It is recommended to conduct a coupled hydraulic and sediment transport simulation to assess in more detail the effectiveness of the sedimentation basin. The simulation may also provide an in depth understanding of the sedimentation rate in the basin.

This work is supported by the Faculty of Civil and Environmental Engineering, ITB through The Program of Research, Community Service, and Innovation for Research Division ITB (P3MI ITB), year 2017.

References

1. S.E. Lewis, Z.T. Bainbridge, P.M. Kuhnert, B.S. Sherman, B. Henderson, C. Dougall, M. Cooper & J.E. Brodie. Calculating Sediment Trapping Efficiencies For Reservoirs In Tropical Settings : A Case Study From The Burdekin Falls Dam, NE Australia. Water Resources Research (49), pp. 1017–1029, 2013. doi:10.1002/wrcr.20117,
2. H. Toniolo & G. Parker. 1D Numerical Modeling Of Reservoir Sedimentation, Proceedings of IAHR Symposium on River, Coastal and Estuarine Morphodynamics, pp.457-468, 2003.
3. A. Taylor, G. Macky & A. Gasc. Sedimentation pond Performance from Three-Dimensional Modelling. Asia Pacific Stormwater Conference, 2015.
4. PT BHE. Design Drawing of Karai-7 Mini Hydro Poer Plant. PT Global Karai, 2014.
5. Lapi ITB. Watershed Level Risk Assessment for Karai-7 Mini Hydro Power Plant Development. MCA Indonesia, 2017. (Book)
6. Goldman, J. Steven, Jackson, Katharine & Bursztynsky, A. Taras. Erosion & Sediment Control Handbook. McGraw-Hill. ISBN0-07-023655-0. pp. 8.2, 8.12., 1986. (Book)
7. Flow Science Team. Flow-3D User manual. Flow Science, Inc., 2008.

MATEC Web of Conferences **147**, 03013 (2018) https://doi.org/10.1051/matecconf/201814703013
SIBE 2017

The Diurnal and Semidiurnal Patterns of Rainfall and its Correlation to the Stream Flow Characteristic in the Ciliwung Watershed, West Java, Indonesia

Edi Riawan[1,*], *Tri Wahyu* Hadi[1], *Hadi* Kardhana[2,3], *Dadang K.* Mihardja[4], *and Benyamin* Sapiie[5]
[1]Atmospheric Sciences Research Group, Institut Teknologi Bandung
[2]Water Resources Engineering Research Group, Institut Teknologi Bandung
[3]Infrastructure and Regional Development Research Centre, Institut Teknologi Bandung
[4]Oceanography Research Group, Institut Teknologi Bandung
[5]Geology Research Group, Institut Teknologi Bandung

Abstract. Based on the data analysis of 16 years of TMPA dataset, the common patterns of rainfall over the Ciliwung River Basin are diurnal and semidiurnal. Those patterns can be associated by a stationary or moving rainstorm with different magnitude and direction. Based on hydrological model simulations, both the pattern and movement have a significant role to the discharge. At the downstream area, the discharge that triggered by semidiurnal pattern of rainfall can produces higher peak discharge and longer flood duration than diurnal pattern. This result open possibility to improve our prediction on design discharge.

1 Introduction

The spatial-temporal characteristics of rainfall have significant role on characteristic of stream flow [1]–[4]. These characteristics strongly effected by moving rainstorm [2]. The greatest effect of moving rainstorms occurred when the velocity of storms is half of stream flow velocity in downslope direction [1]. de Lima and Singh simulated the moving rainstorm using numerical and physical models, and concluded the runoff characteristics are sensitive to the speed and direction of storm movement [3], [4].

An analysis of the spatial-temporal characteristic can be simplified by analyze the spatial distributions and the temporal patterns of rainfall. Based on the four severe floods that occurred in Jakarta, the spatial-temporal variation is inherent characteristics on the extreme rainfalls. Table 1 showed the distribution of rainfall and storm motions on flood events in Jakarta. The distribution of rainfall can be concentrated on upstream areas (example: 1996 and 2007 flood events), or downstream areas (example: 2002 and 2013 flood events) and the storms can moving to downslope (example: 2002) or upslope (example: 2013) or downslope-upslope (example: 2007).

Table 1. The rainstorm characteristics in the Ciliwung Watershed on extreme

Flood events		Heavy Rainfall Characteristic	
		Spatial	temporal
1	January 1996	Upstream	-
2	January 2002	downstream	downslope
		[5]	[6][7]
3	February 2007	upstream	Downslope-upslope
		[7], [8]	
4	January 2013	downstream	upslope
		[9]	

The objective of this study is to study the common characteristics of rainfall and related to peak flow and floods durations. The characteristics shown by: spatial distributions (upstream, midstream, downstream), temporal patterns of rainfall and its combinations (diurnal and semidiurnal).

2 Data and Method

2.1 Study Area

This study focuses on small river basin in the tropical area, ie: the Ciliwung River Basin (CRB). CRB is the watershed with highest flood risk in Indonesia. Ciliwung Rivers is the largest river which is through DKI Jakarta, the capital city of Indonesia. In the last 20 years, the Ciliwung River contributed to the 3 major devastating floods in DKI Jakarta, ie early February 2002 and 2007 and the end of January 2013. Floods that occurred in 2007 resulted in losses reached 5.6 trillion rupiahs [10].

* Corresponding author: ediriawan@meteo.itb.ac.id

Ciliwung River was diverted to West Flood Canal at Manggarai hydrological station (HS). This condition increasing the complexity of hydrological analysis in CRB, especially for the downstream area after the diversion. For this reason, this study is limited to the area before the diversion, precisely at M.T Haryono HS. Total area of CRB for an outlet at M.T Haryono HS reached 352 km^2 (**Fig.1**). CRB has a complex topography. The upstream area consists of steep mountains with maximum altitude above 2982 m. The downstream area is a flat floodplain.

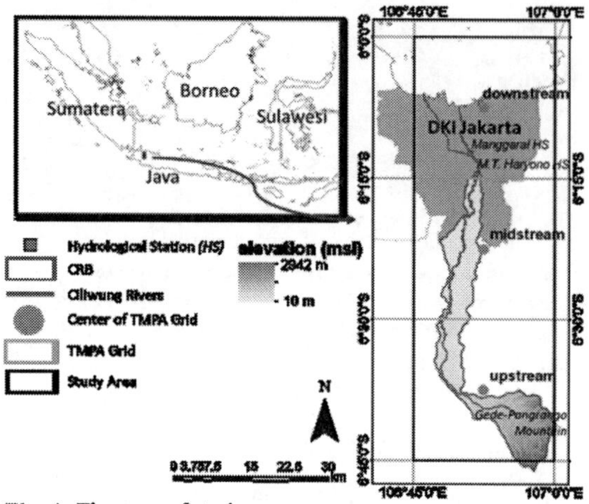

Fig. 1. The map of study area.

2.2 Rainfall Data

The Tropical Rainfall Measuring Mission (TRMM) Multisatelite Precipitation Analysis (TMPA) was used in this study. The TMPA was produced by a calibration-based sequential scheme for combines multiple satellites to produce precipitation data at fine scales, 0.25°×0.25° and 3 hourly, and available since 1998 [2]. This data was compare to other TRMM-derived products to analysis the extreme rains over the Maritime Continent [3]. Nesbitt and Zipser (2003) used the previous product of TMPA to analysis the diurnal cycle of rainfall in the tropical area [4].

This study used 16 years period of TMPA data, 1998-2013 to produce the general pattern of daily rainfall over CRB. There is 3 grid over CRB named by the upstream, midstream and downstream grid (Fig 1).

2.3 Diurnal variation analysis method

The diurnal variation of rainfall identified composite analyzes according to the method that proposed by Ruhf and Cutrim (2003). The 3-hour rainfall data from TMPA was accumulated for the same hour (hh), for each month (mm), of 16 years (yy) study periods, then divided by the number of years. This method produced the average annual accumulation rainfall for each 3-hour (mm/year). However, since monthly analysis is required, the month is considered as a calculated variable (equations (1)). The calculation result was plotted on the hour-month chart for represented the variability of diurnal variation for each month, **Fig 3** and **Fig 4** are the results of this method.

- $$P(hh, mm) = \frac{\sum_{yy=1}^{yymax} \sum_{dd=1}^{ddmax} P_{year,day}(hh,mm)}{16} \quad (1)$$

2.4 Hydrological model and simulation scnearios

The distributed hydrological model was used to investigate the influence of diurnal characteristics of rainfall to overland flow at CRB. The numerical model used in this study is Gridded Surface Subsurface Hydrological Analysis (GSSHA) [2]. GSSHA is the distributed and physics-based hydrological modeling. The model also has two dimensional (2D) overland flow and 1D stream flow that has been fully coupled.

The domain of the model is Ciliwung Watershed from the upstream area at Gede-Pangrango Mountain area to the outlet at M.T. Haryono hydrological station, DKI Jakarta Province. The domain has 250 m horizontal resolution and divided by 234 x 99 grids and 5704 grid is active cell (inside watershed). **Fig.1** and **Fig.2** showed the domain of the hydrological model.

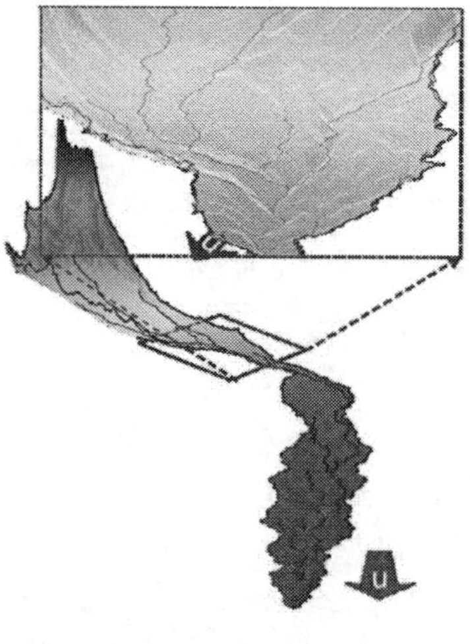

Fig. 2. Hydrological Domain in 2D visualization

The topography, rivers, and land cover data are GIS-vector dataset with scale 1:25,000 which can be downloaded from the website of Badan Informasi Geospasial (BIG) at http://portal.ina-sdi.or.id/. The contour topography was converted to DEM and used as the surface elevation of the model. The river data was used to correct the DEM data, especially in the gently slope area. The land cover dataset is used as a proxy for roughness parameterization. There are four classifications for roughness parameterization that is: forest, cultivated area, water body, and urban area. The roughness value for each land cover classification was taken into account in the calibration processes.

This model was run to simulate two following scenarios, namely (1) the combination of rainfall pattern on upstream, midstream, and downstream area that produced by TMPA data analysis (section 3.1), and (2) the stationary and moving rainfall (~5 m/s) based on previous research.

3 Result and Discussion

3.1 Variation of Diurnal Rainfall

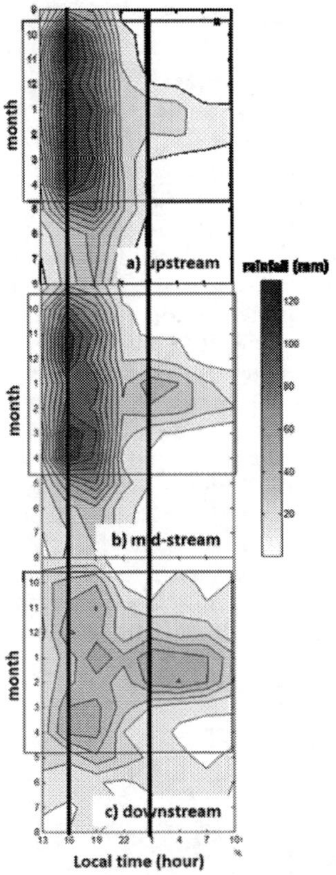

Fig. 3. Total annual rainfall for each month and each hour in (a) upstream, (b) midstream and (c) downstream area of CBR.

The diurnal variation of rainfall in CRB variated for each location and month. **Fig. 3** shows the average of the 16 years of total annual rainfall for each month and each hour in the upstream, midstream, and downstream area of CRB based on TMPA data. The figure can represent the frequency of rain for each location, hour and month. In upstream and midstream area, the rain commonly occurs in afternoon. The morning rain, the rain that occurs in the morning, is more frequent in the midstream area than upstream. In the downstream, the morning rain becomes dominant than the afternoon rain. This indicates the diurnal patterns getting more dominant in the upstream area and vice versa for the semidiurnal pattern, getting more dominant in the downstream area. Where the diurnal pattern is a pattern with single rain event in a day and the semidiurnal pattern has two rain events in a day.

The domination of semidiurnal pattern in downstream area was studied by Siswanto et al. (2016) based on long term period of observation data at the Jakarta Observatory meteorological station [3]. Semidiurnal pattern getting dominant in December-January-February (DJF) than the other month.

May to September is a dry month and August is the driest month, therefore those months will not be further analyzed in this study.

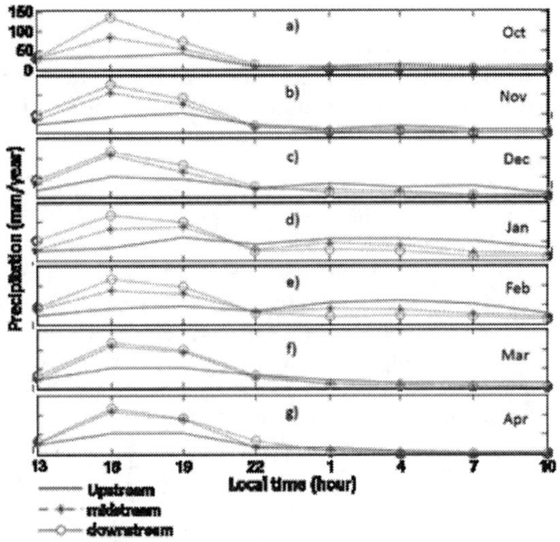

Fig. 4. The general patterns of diurnal rainfall for each month from October to April in three grid of TMPA data in CRB

Fig. 4 is a simple form of the **Fig.3** to classify the general pattern of rainfall in CRB. The diurnal patterns of rainfall dominated the whole area of CRB in March-April. In October-November, only the downstream area dominated by the semidiurnal pattern. And on the December only upstream area which dominated by the diurnal pattern. The semidiurnal pattern becomes dominant in the whole area of CRB in January and February which is the months with most frequent major floods in CRB. This result indicated correlations between semidiurnal patterns and the major floods.

The correlation of the diurnal variation of rainfall and extreme discharge conducted by hydrological model. To perform this method, simple patterns of rainfall is needed to produce the simulation scenarios but still represent the patterns that exist in CRB. **Fig. 5** shows the overlay of the topographic profile, grid of TMPA and domain of hydrological model along the Ciliwung River. The figure shows only the midstream and upstream areas within the hydrological model domain. Consequently, the hydrological simulation only involves patterns combinations in the upstream and midstream areas. There are three combinations of patterns, ie DD and SS respectively for diurnal and semidiurnal in the whole domains and DS for diurnal in upstream and semidiurnal in downstream.

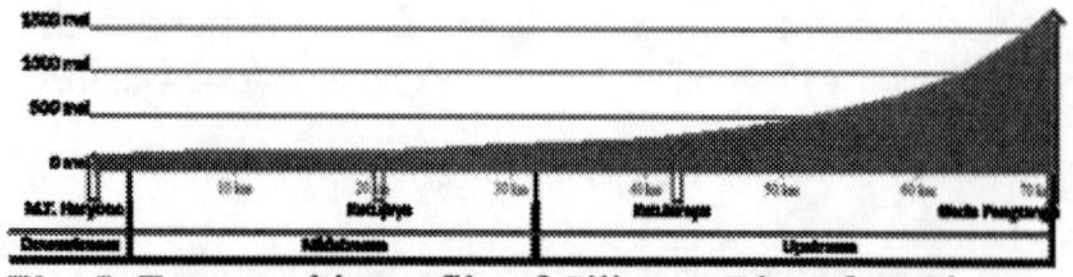

Fig. 5. Topographic profile of Ciliwung River from the upstream and the position of hydrology station (red arrows) and TMPA grid boundary (vertical black lines)

Fig. 6 shows the cumulative probability function of each rain event characteristics based on TMPA data. **Fig.6**.b shows the most rain occurs in 3-hour duration. The maximum rainfall intensity in one rain event is about 15 mm/hour (**Fig.6**.c), although the maximum intensity can reach 40 mm / hour but this condition did not occur continuously during one rain event. Based on this result, one rain event in the rainfall scenarios can be simplified by a rain with 3-hour of duration and 15 mm/hour of rainfall intensity. In afternoon, the rain occurs at 14.30 to 17.30 LT and in the morning the rain occurs at 23.30 to 02.30 LT. To clarify the difference between the effects caused by diurnal and semidiurnal patterns, rainfall volume needs to be maintained by making the rainfall intensity become 7.5 mm / hour for semidiurnal patterns.

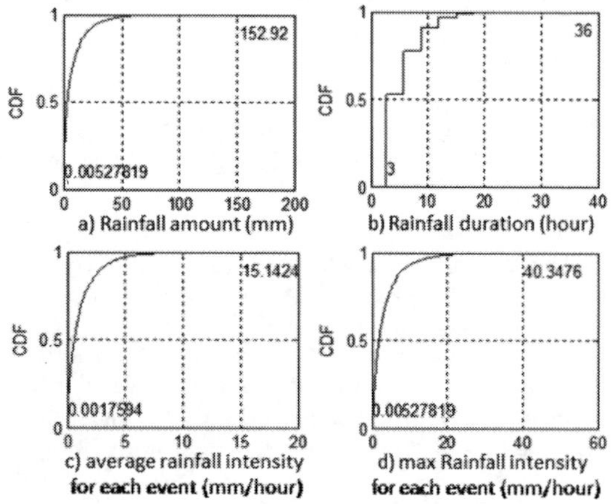

Fig. 6. The cumulative probability function of: a) amount, b) duration, c) average intensity and d) maximum intensity for each rainfall event

The diurnal rainfall analysis produced four rainfall scenarios for hydrological simulation. The scenarios named DD, DS, SS, and 2SS. DD, DS, SS is symbol for diurnal-diurnal, diurnal semidiurnal, and semidiurnal-semidiurnal respectively, where the first alphabet for upstream area and the second alphabet for the midstream area. The rainfall intensity is 15 mm/hour for the diurnal and 7.5 mm/hour for the semidiurnal with 3-hour duration for each event. Fig. 7 is the ilustration of the scenarios.

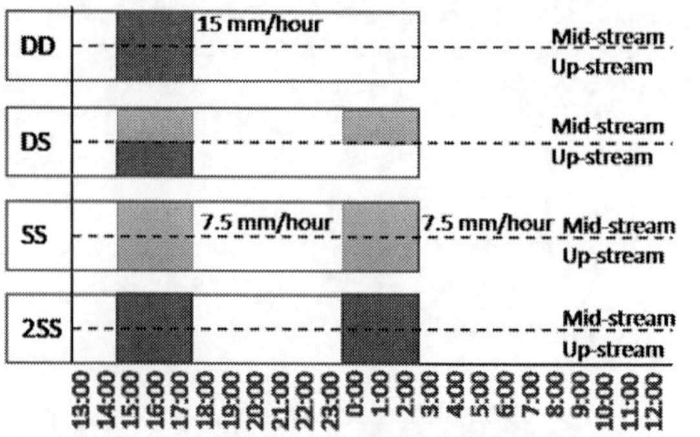

Fig. 7. Rainfall scenarios for hidrological simulation based on general patterns of diurnal rainfall

Besides of diurnal patterns, there are also other rainfall characteristics triggering the extremes discharge, which is rain propagation. Due to data limitations in Indonesia, there has few of research discuss this topic. However, several studies have mentioned their existence in CRB [6], [9], [8], [11]. Wu et al. (2013) stated the convective system that triggered major floods event in 2013 propagate to upstream by 8 m/s of speed. Sulityowati showed that the speed of rainfall propagation can vary between 2.57 - 15.42 m/s based on radar data on 9-15 February 2010[11]. Numerical model and laboratory experiment in artificial domains shows that propagated rain to downstream direction will lead to greater peak discharge. Based on this, a new parameter added to the scenario, which is the movement of rain. At this moment only considered the movement to downstream at 5 m / s of speed. The addition of this parameter produces four additional scenarios illustrated in Fig.8.

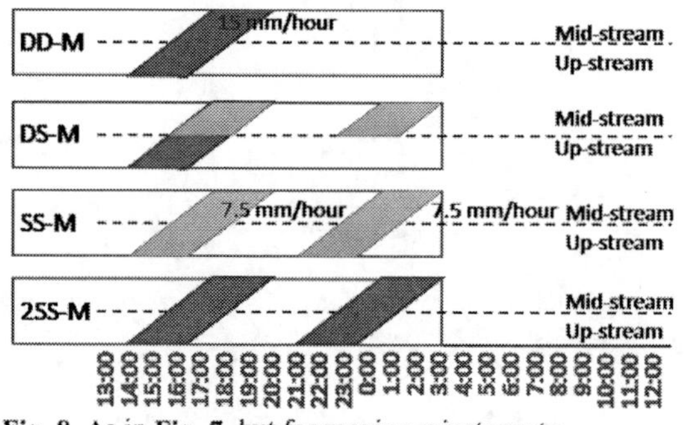

Fig. 8. As in **Fig. 7**, but for moving rainstorm to downstream directions

3.2 The role of the rainfall pattern combination

The GSSHA model was applied to produce hydrograph of discharge effected by diurnal-semidiurnal pattern of rainfall. This simulation consists of combinations of a simple spatiotemporal variability of daily rainfall over CRB that described in sections 3.1.

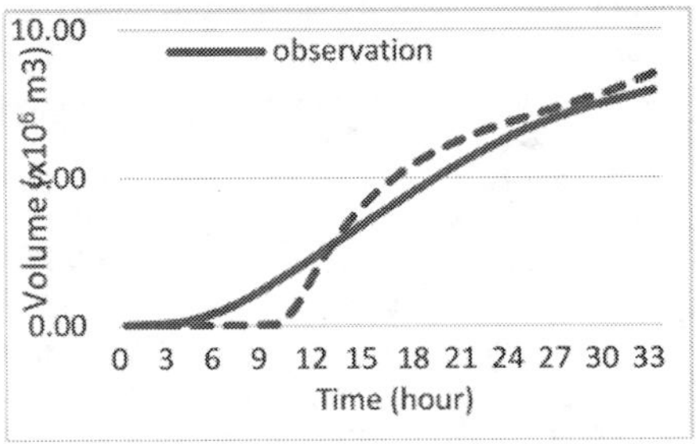

Fig. 9. Mass curve of the calibration processes

The discharge data provided by Ciliwung-Cisadane Large River Basin Agency (BBWS: Abbreviation in Indonesian) was used to calibrate the model. The calibration output showed a good agreement between observation data and simulation result for the volume of rivers discharges at outlet without any significant bias (**Fig. 9**). This result indicates the model parameterization was good enough to explain the role of rainfall volume on overland flow. The calibration process is produced by the roughness parameterization and the ratio of effective rainfall. That roughness value is 0.184, 0.01, 0.01, and 0.001 for forests, cultivated areas, urban areas, and water bodies, respectively. Effective rainfall was used in this study is 0.7.

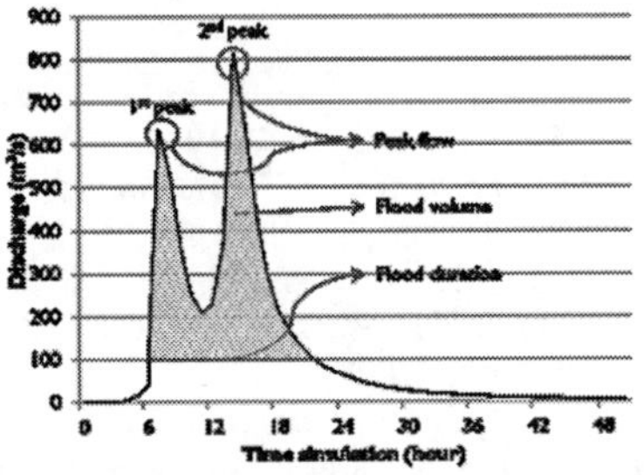

Fig. 10. The illustration of hydrological simulations results and the definition of peak flow, flood volume, and flood duration that used in this study

Fig. 10 showed an illustration of the peak flows and flood durations have been used to show the differentiation of the simulation results. There is one and two peak flow respectively for the diurnal and semidiurnal pattern of discharge. The flood durations are defined by duration where the discharge values are more than 100 m^3/s. This assumption based on the minimum value between two peaks on the semidiurnal pattern of the hydrograph. The other reason is the bank full discharge or flood discharge in M.T Haryono hydrological stations have not well defined.

Comparison of the DD, DS, SS and 2SS scenarios showed: 1) diurnal pattern of rainfall produced higher peak flow than semidiurnal, but 2) semidiurnal pattern of rainfall triggered semidiurnal pattern on discharge and prolong flood duration (Fig. 11). The rainfall with 15 mm/hour of intensity and 3 hours on the duration that occurred on the whole watershed area have produced the hydrograph with 470 m^3/s of peak flow and 8 hours of flood duration. If the semidiurnal rainfall with 7.5 mm/hour of rainfall intensity and 2 x 3 hour of durations occurs, respectively in the afternoon and early morning, the hydrograph shape will be turned into semidiurnal with the values of the first and second peak flow respectively are 0.38 and 0.5 times of the peak flow of scenario DD. Besides of that, the flood duration increased from 8 hours to 11 and 13 hours on the DS and SS scenarios, respectively.

The simulation results in **Fig. 11** showed the existence of the role of the diurnal-semidiurnal patterns of rainfall against to in the diurnal-semidiurnal patterns of discharge. Existence of diurnal-semidiurnal patterns of the rivers water level, generally has a logarithmic function to the discharge, has shown by result of spectral analysis conducted by Sulityowati et al. (2014) at Manggarai hydrological station (about 5 km to downstream area from outlet of the hydrological model in this study) [11]. The study showed the correlation between the diurnal pattern of rainfall and water level, but have no explanation about correlation for the semidiurnal pattern.

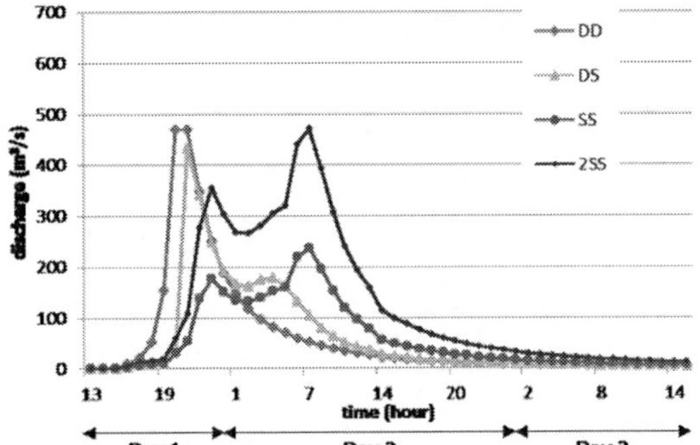

Fig. 11. Hydrographs comparison of all scenarios for stationer rainfall.

The semidiurnal pattern of discharge (**Fig. 11**, blue curve), the 2nd peak flow showed a larger value than 1st peak flow. This condition is caused by a superposition between hydrographs that produced by the afternoon and morning rain on semidiurnal rainfall. The value of the second peak will be higher when the time difference of the afternoon and morning rain is shorter. These results indicate that the semidiurnal pattern has the potential to trigger greater peak discharge for maximum rainfall intensity occur in both afternoon and morning (15 mm / hour in this study, the 2SS simulation scenarios).

The 2SS scenario has twice amount of daily rainfall than the SS scenario. This scenario accordance with the research results from Trilaksono et al. (2012) which mentions the rain that occurred in 2007 has a semidiurnal pattern that triggered by two different convective systems. This condition has a possibility to produce two rainfall events that have maximum intensity. The result opens possibility to analyze the major flood events that occurred in 2002, 2007 and 2013 known have a long flood duration [9], [8], [12], [7].

In addition to the diurnal-semidiurnal pattern, Sulityowati et al. (2014) also explains about rain propagation that can occur to upstream or downstream direction with speeds varying between 2.57 - 15.42 m/s based on radar data on 9-15 February 2010 [11]. Wu et al. (2013) stated the convective system that triggered major floods event in 2013 propagate to upstream about 8 m/s [9]. Rain propagation was also described in Hadi et al. (2006) and Trilaksono et al. (2012) which each addressed the flood events of 2002 and 2007 [6], [8]. That research result indicates that, besides of the semidiurnal diurnal pattern, rainfall propagation also plays an important role in triggering the peak discharge that causes flooding.

The DD-M, DS-M, SS-M and 2SS-M scenarios are rainfall scenarios that combine diurnal-semidiurnal patterns of rainfall and rain propagation. The simulation results show that rain propagation to downstream area will amplify peak discharge about 1.3, 1.1, 1.2 and 1.2 times for DD, DS, SS, and 2SS patterns, respectively (**Fig.12**). However, this rain propagation will result in shorter flood duration.

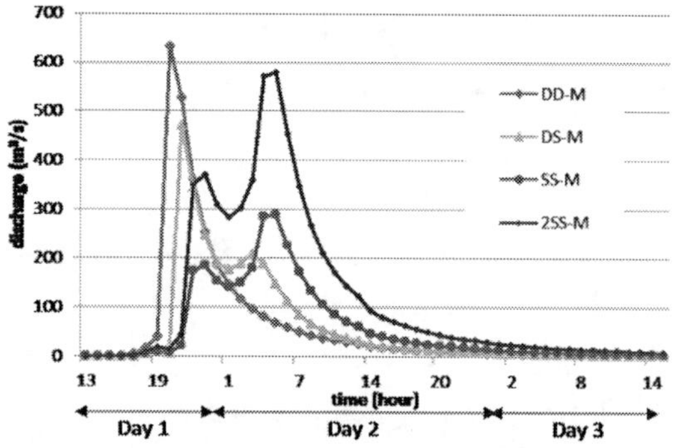

Fig. 12. As **Fig.11**, but for moving rainfall.

These simulation results are consistent with Ogden et al. (1995), de Lima and Singh (2002) and de Lima and Singh (2003) [1], [3], [4]. Those study also mentioned downstream movement direction will cause the opposite condition, ie lower peak discharge but longer flood duration. In the real case, the convective system causing major floods in Jakarta has varying speed and direction values, such as the 2013 flood event, the conventive system was propagated to upstream direction [9] and the 2007 flood event the convective system propagated to the downstream area at night but to the upstream area at morning [8].

4 Conclusions

The results of this study highlight the spatiotemporal variance of rainfall patterns and its role in peak flow and flood durations. In addition, the role of rainfall propagation also compared to the role of rainfall patterns. The result shows the rainfall pattern variated for each location and month. The semidiurnal pattern becomes dominant in the whole area of CRB in January and February which is the months with most frequent major floods in CRB. Based on the hydrological model simulation, for the equal amount of daily rainfall (a single event for semidiurnal pattern): (1) the diurnal pattern will produce higher peak-flow than the semidiurnal pattern of rainfall and (2) vice versa for flood duration. The double amount of rainfall for the semidiurnal pattern will produce higher peak-flow and flood durations. This conditions will be enlarger by the moving rainstorm to downstream directions.

The results of this research indicate there are two factors that trigger major floods are: (1) semidiurnal pattern of rainfall that occurs in the whole of the watershed, (2) the rainfall propagation to downstream area. Both of these factors need to be studied further in real cases.

This research was funded by Riset ITB 2017 and was the part of "Peranan Hujan yang Bergerak terhadap Debit Ekstrem di Sungai Ciliwung" research.

References

[1] F. L. Ogden, J. R. Richardson, and P. Y. Julien, "Similarity in Catchment Response: 2. Moving Rainstorms," *Water Resources Research*, vol. 31, no. 6, pp. 1543–1547, 1995.

[2] V. P. Singh, "Effect of spatial and temporal variability in rainfall and watershed characteristics on stream flow hydrograph," *Hydrological Processes*, vol. 11, no. 12, pp. 1649–1669, 1997.

[3] J. L. M. P. De Lima and V. P. Singh, "The influence of the pattern of moving rainstorms on overland flow," *Advances in Water Resources*, vol. 25, no. 7, pp. 817–828, 2002.

[4] J. L. M. P. de Lima and V. P. Singh, "Laboratory experiments on the influence of storm movement on overland flow," *Physics and Chemistry of the Earth*, vol. 28, no. 6–7, pp. 277–282, 2003.

[5] A. Formánek, R. Silasari, M. S. B. Kusuma, and H. Kardhana, "Two-Dimensional Model of Ciliwung River Flood in DKI Jakarta for Development of the Regional Flood Index Map," *Journal of Engineering and Technological Sciences*, vol. 45, no. 3, pp. 307–325, 2013.

[6] T. W. Hadi, N. J. Trilaksono, and I. D. G. Junnaedhi, "A Numerical Study of The Jakarta Flood Event of January / February 2002 : Simulation of Convective Rainfall Using Regional Weather Model," Bandung, 2006.

[7] P. Wu *et al.*, "The Impact of Trans-equatorial Monsoon Flow on the Formation of Repeated Torrential Rains over Java Island," *Sola*, vol. 3,

no. February, pp. 93–96, 2007.
[8] N. J. Trilaksono, S. Otsuka, and S. Yoden, "A Time-Lagged Ensemble Simulation on the Modulation of Precipitation over West Java in January–February 2007," *Monthly Weather Review*, vol. 140, no. 2, pp. 601–616, 2012.
[9] P. Wu *et al.*, "The effects of an active phase of the Madden-Julian oscillation on the extreme precipitation event over western Java Island in January 2013," *Sola*, vol. 9, no. January, pp. 79–83, 2013.
[10] Bappenas, "Laporan Perkiraan Kerusakan dan Kerugian Pasca Bencana Banjir Awal Februari 2007 di Wilayah JABODETABEK Jakarta, Bogor, Depok, Tangerang, dan Bekasi)," Jakarta, 2007.
[11] R. Sulistyowati, R. I. Hapsari, F. Syamsudin, S. Mori, S. T. Oishi, and M. D. Yamanaka, "Rainfall-Driven Diurnal Variations of Water Level in the Ciliwung River, West Jawa, Indonesia," *Sola*, vol. 10, no. 0, pp. 141–144, 2014.
[12] M. Wheeler, "The Madden Julian Oscillation and the Java floods of January/February 2002," *BMRC Modeling Workshop*, no. February, 2002.

Effective real-time forecasting of inundation maps for early warning systems during typhoons

Jhih-Huang Wang[1], *Gwo-Fong* Lin[1,*], *Bing-Chen* Jhong[1]

[1]Department of Civil Engineering, National Taiwan University, Taipei 10617, Taiwan

Abstract. Accurate forecasts of hourly inundation depths are essential for inundation warning and mitigation during typhoons. In this paper, an effective forecasting model is proposed to yield 1- to 6-h lead-time inundation maps for early warning systems during typhoons. The proposed model based on Support Vector Machine (SVM) is composed of two modules, point forecasting and spatial expansion. In the first module, the rainfall intensity, inundation depth, cumulative rainfall and forecasted inundation depths are considered as model input for point forecasting. In the second module, the geographic information of inundation grids and the inundation forecasts of reference points are used to yield inundation maps for spatial expansion. The results show that the proposed model is able to provide accurate point forecasts at each inundation point. Moreover, the spatial expansion module is capable of producing accurate spatial inundation forecasts. Obviously, the proposed model provides reasonable spatial inundation forecasts, and is able to deal with the nonlinear relationships between inputs and desired output. In conclusion, the proposed model is suitable and useful for inundation forecasting.

1 Introduction

The heavy typhoon rainfall regularly result in serious disasters. Not only is typhoon rainfall an important and precious water resource, but also the flood inundation resulting from excessive rainfall brought by typhoons frequently leads to loss of human life and property. Therefore, more accurate inundation forecasts are always required. However, the inundation process is a highly nonlinear and complex physical process. Artificial neural networks (ANNs) are a kind of information processing system with great flexibility in modeling nonlinear systems. As to inundation forecasting, applications of NNs have been presented (Chang et al. [3]; Pan et al. [7]). The ASCE Task Committee [1, 2] and Maier and Dandy [6] have presented comprehensive reviews of the applications of ANNs in hydrology. To obtain effective forecasts of inundation, a model with good generalization ability is needed.

In recent years, support vector machines (SVMs) has been paid more and more attention and used for hydrological time series forecasting (Hong [4]; Lin et al. [5]). The algorithms and the architectures of the SVMs are guaranteed to be unique and globally optimal. Moreover, SVMs can be quickly trained with real-time data. However, because of only a few researches for applying SVMs on the forecasting of regional inundation maps. Therefore, the aforementioned advantages have prompted us to improve inundation forecasting by using SVM-based models.

To provide effective real-time forecasting of inundation maps for early warning system during typhoons. Therefore, the purpose of this paper is to propose a new forecasting technique, to yield 1- to 6-h lead time inundation maps for early warning system during typhoons. The proposed model based on Support Vector Machine (SVM) is composed of two modules, point forecasting and spatial expansion. In the first module, the rainfall intensity, inundation depth, cumulative rainfall and forecasted inundation depths are considered as model input for point forecasting. In the second module, the geographic information of inundation grids and the inundation forecasts of reference points are used to yield inundation maps for spatial expansion.

2 Study area and materials

An application to Chiayi City in southern Taiwan is conducted to demonstrate the superiority of the proposed model. Chiayi City is a major urban city in southwestern Taiwan with a population of about 270 thousand people, and has an area of approximately 60 km^2. Inundation disasters have frequently been caused due to the topography characteristics and heavy typhoon rainfall in Chiayi City.

The hourly rainfall data were obtained from the Taiwan Water Resources Agency (TWRA). The inundation data are simulated and validated by a physically based two-dimensional model (FLO-2D), are used as a database for inundation forecasting in this paper. A total of 16 historical typhoon events were used to establish forecasting modules herein.

* Corresponding author: gflin@ntu.edu.tw

3 The proposed forecasting technology

In this study, the proposed forecasting technology is composed of two steps: point forecasting step and spatial expansion step. The flowchart of the proposed forecasting methodology is shown in Fig. 1.

The first step is to establish the point forecasting module for each inundation point. The 1- to 6-h ahead forecasts of inundation depth are desired. The rainfall, cumulative rainfall and inundation depth are employed as inputs to develop the point forecasting module for each inundation point. The general form of the proposed point forecasting module (hereinafter SVM-A) can be expressed as:

$$\hat{I}_{\mathrm{P},t+\Delta t}=\begin{cases} f_{\mathrm{SVM}}\big(R_{\mathrm{P},t},\ldots,R_{\mathrm{P},t-(l_R-1)},C_{\mathrm{P},t},\ldots,C_{\mathrm{P},t-(l_C-1)},\ldots, \\ \qquad I_{\mathrm{P},t},\ldots,I_{\mathrm{P},t-(l_I-1)}\big) & \text{for } \Delta t=1 \\ f_{\mathrm{SVM}}\big(R_{\mathrm{P},t},\ldots,R_{\mathrm{P},t-(l_R-1)},C_{\mathrm{P},t},\ldots,C_{\mathrm{P},t-(l_C-1)},\ldots, \\ \qquad I_{\mathrm{P},t},\ldots,I_{\mathrm{P},t-(l_I-1)},\hat{I}_{\mathrm{P},t+1},\ldots,\hat{I}_{\mathrm{P},t+\Delta t-1}\big) & \text{for } \Delta t>1 \end{cases} \quad (1)$$

where t is the current time, Δt is the lead time period (from 1- to 6-h), $R_{\mathrm{P},t}$ is the hourly rainfall of the inundation point at time t, $C_{\mathrm{P},t}$ is the cumulative rainfall of the inundation point at time t, $I_{\mathrm{P},t}$ is the inundation depth of the inundation point at time t, $\hat{I}_{\mathrm{P},t+\Delta t}$ is the forecasted inundation depth of the inundation point at time $t+\Delta t$, L_R denote the lag length of rainfall, L_C denote the lag length of cumulative rainfall and inundation, L_I denote the lag length of inundation. In addition, the SVM-A is compared with a conventional module (hereinafter SVM-B) to show the improvement in point forecasting step. The input of the conventional module are rainfall and inundation. The general form of the conventional module can be expressed as:

$$\hat{I}_{\mathrm{P},t+\Delta t}=f_{\mathrm{SVM}}\big(R_{\mathrm{P},t},\ldots,R_{\mathrm{P},t-(l_R-1)},I_{\mathrm{P},t},\ldots,I_{\mathrm{P},t-(l_I-1)}\big) \quad (2)$$

The second step, the forecasted results from the point forecasting step are expanded to spatial estimation. First, the inundation data are classified by clustering technology. For each classification, the forecasted inundation depths in the point forecasting step, rainfall, cumulative rainfall and inundation depth are employed as inputs to develop the spatial expansion module. The general form of the spatial expansion module is:

$$\hat{I}_{g,t+\Delta t}=f_{\mathrm{SVM}}\big(X_g,Y_g,H_g,R_{g,t},C_{g,t},\hat{I}_{\mathrm{P},t+\Delta t}\big) \quad (3)$$

Where $\hat{I}_{g,t+\Delta t}$ is the forecasted inundation depth of grid g at time $t+\Delta t$, X_g and Y_g are the coordinates of grid g, H_g is the elevation of grid g, $R_{g,t}$ is the rainfall of grid g at time t and $C_{g,t}$ is the cumulative rainfall of grid g at time t. Once the inundation depths of all the inundated grids are forecasted, the inundation map is then obtained.

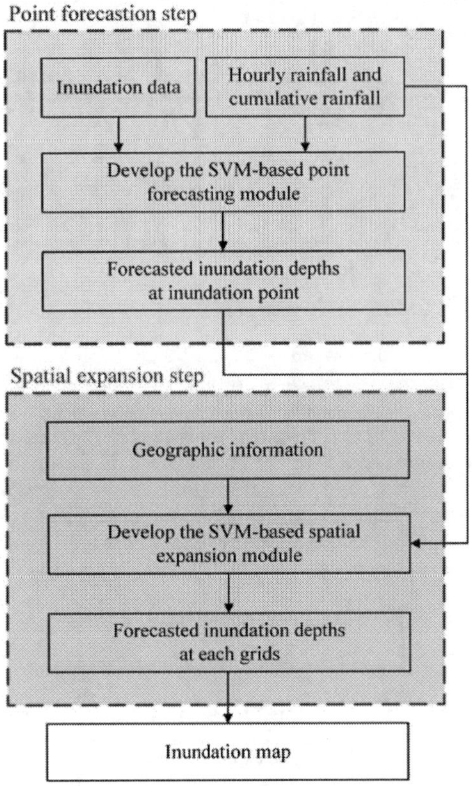

Fig. 1. Flowchart of the proposed forecasting technique.

4 Results and discussion

To demonstrate the improvement in forecasting performance using the proposed point forecasting module, we first focus on the comparison between the proposed module (SVM-A) and the conventional module (SVM-B). The performance of modules is evaluated by several criteria, such as root mean square error (RMSE) and correlation coefficient (CC). As shown in Fig. 2, the bar charts clearly show that the RMSE values of the SVM-A are lower than those of the SVM-B. Moreover, the RMSE values of the SVM-A increase more slowly than those of the SVM-B with increasing lead time. The negative impact of increasing forecast lead time has been effectively decreased by using the SVM-A. In addition, the CC values of the SVM-A are higher than those of the SVM-B. It is obvious that the performance of the SVM-B gets worse. The results confirm that the proposed module effectively improves the forecasting performance. The use of the proposed module instead of the conventional module effectively decreases the negative impact of increasing forecast lead time.

https://doi.org/10.1051/matecconf/201814703014

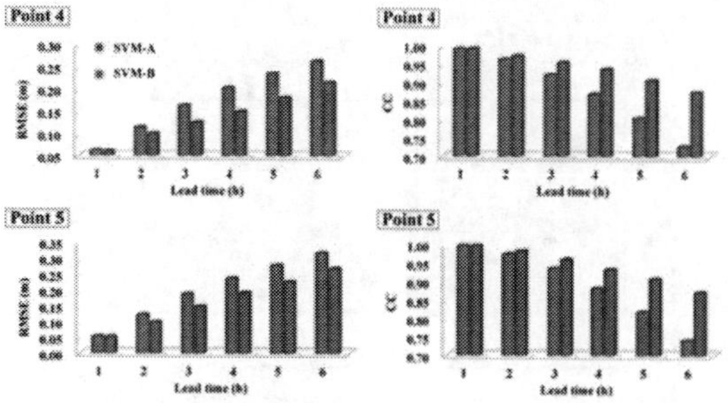

Fig. 2. RMSE and CC values of SVM-A and the SVM-B.

The comparison of the inundation data and corresponding forecasts resulting from the proposed technique for Typhoon Krosa is presented in Fig. 3. The forecasted results from the proposed technique are in good agreement with the inundation data, especially for the western and central region of the study area. Moreover, the pattern of the map obtained from the proposed technique is similar to that from the inundation data for 1- to 6-h lead time.

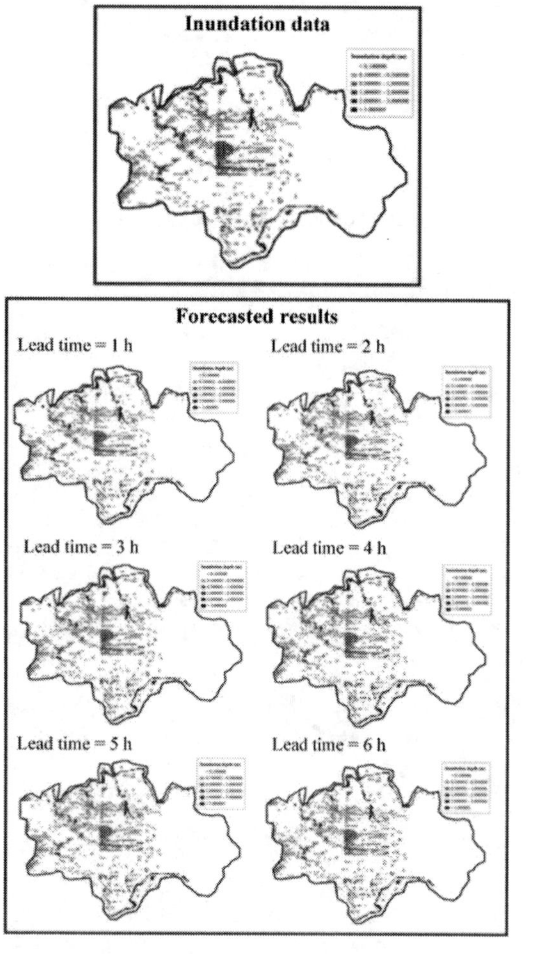

Fig. 3. Comparison of the inundation data and the forecasted results of the proposed model.

5 Conclusions

The purpose of this paper is to provide effective real-time forecasting of inundation maps for early warning system during typhoons. The results show that the proposed model is able to provide accurate point forecasts at each inundation point. Moreover, the spatial expansion module is capable of producing accurate spatial inundation forecasts. The maps of the forecasted results are similar to that of the inundation data. It is concluded that the proposed technique is suitable to be integrated with the decision support system. In the future, observation of inundation depth could be applied to the proposed forecasting approach to forecast inundation maps if the historical observed data of inundation was obtained from the 7-Eleven stores. The proposed forecasting approach is also expected to be useful to support inundation warning systems.

References

1. ASCE Task Committee on Application of Artificial Neural Networks in Hydrology, Artificial Neural Networks in hydrology I: Preliminary Concepts, Journal of Hydrologic Engineering **5**, 115–123 (2000).
2. ASCE Task Committee on Application of Artificial Neural Networks in Hydrology, Artificial Neural Networks in hydrology II: Hydrological Applications, Journal of Hydrologic Engineering **5**, 124–137 (2000).
3. L.C. Chang, H.Y. Shen, Y.F. Wang, J.Y. Huang, Y.T. Lin, Clustering-based hybrid inundation model for forecasting flood inundation depths, Journal of Hydrology **385**, 257–268 (2010).
4. W.C. Hong, Rainfall Forecasting by Technological Machine Learning Models, Applied Mathematics and Computation **200**, 41–57 (2008).
5. G.F Lin, Y.C. Chou, M.C. Wu, Typhoon flood forecasting using integrated two-stage support vector machine approach, Journal of Hydrology **486**, 334–342 (2013).
6. H.R. Maier, G.C. Dandy, Neural networks for the prediction and forecasting of water resources variables: A review of modeling issues and applications, Environmental Modelling and Software **15**, 101–124 (2000).
7. T.Y. Pan, J.S. Lai, T.J. Chang, H.K. Chang, K.C. Chang, Y.C. Tan, Hybrid neural networks in rainfall-inundation forecasting based on a synthetic potential inundation database, Natural Hazards and Earth System Sciences **11**, 771–787 (2011).

Alternative Intake Station in Saguling Reservoir for The Needs of Raw Water in Bandung Metropolitan Area

Mariana Marselina[1,*], *Arwin* Sabar[1], *Indah* R S Salami[1] and *Dyah* Marganingrum[2]

[1]Faculty of Civil and Environmental Engineering, Institut Teknologi Bandung, Indonesia
[2] Indonesian Institute of Science, Indonesia

Abstract. Bandung Metropolitan Area (BMA) region is the upper watershed of Citarum with an area of ± 2338 km^2. The status carried by BMA as a National Strategic Area from the perspective of economic encourage the increasing migration flows to BMA. These circumstances lead to an imbalance between supply and demand, in which on the one hand, demand for clean water is increasing. The potency of Saguling Reservoir as an alternative of raw water of BMA region in terms of quantity in this research was determined based on the determination of mainstay discharge. In this study, the intake site selection 11 monitoring posts will be carried out by reviewing the concentration of all parameters in Government Regulation No. 82 Year 2001 on any division of discharge grade using 5-grade Makov Discrete method (very dry, dry, normal, wet and very wet). In addition, the calculation of the value of Water Quality Index (WQI) was done at each monitoring station for each division of discharge grade that has been done. The series of data flow and concentration parameters used in this study start from the year 1999 to 2014. The allocation of raw water discharge calculation for Saguling Reservoir in order to fulfill the needs of raw water in Bandung Metropolitan Area is 46,92m^3/second (R5 dry for irrigation raw water supply and 29,53 92 m^3/second (R10 dry for drinking water supply). Based on the assessment of the concentration of measured parameters and determination of Water Quality Index, it can be found that around Muara Ciminyak location is the most qualified location to be used as drinking raw water intake for Bandung Metropolitan Area. Based on this study, it also notes that the determination of the concentration of pollutant parameters needs to be done on the each division of discharge grade occurred.

1 Introduction

Bandung Metropolitan Area (BMA) is an upstream of Citarum watershed with an area of 2338 km^2 [6]. The status held by the BMA as a National Strategic Area with the point of economic interest plus various function carried out by Bandung City as the core of the region, turned out to be investors attraction to make investments that encourage the increasing migration flow. The increasing population and economic development and production activities in this region led to successive land conversion and the increasing water demand. This condition triggers an imbalance between supply and demand, were on one hand the demand for clean water increases and on the other hand the water source has suffered from quantity and quality degradation.

The utilization of the potency of Saguling Reservoir as raw water source has an impact on the changing of reservoir operating management from single function reservoir to multi-sector reservoir. Of course this condition needs careful target and preparation. In-depth calculation of Saguling Reservoir capability in terms of quantity and quality as raw water alternatives should be performed. Based on research from Hart et al (2007), Saguling was classifed as a eutropic reservoir.

The potency of Saguling Reservoir as raw water alternative in BMA region in terms of quantity in this research was determined based on the determination of mainstay discharge. Mainstay discharge is the discharge available in a watershed at a certain time which its existence is located to be able to meet the rate of water demand. The amount of this mainstay discharge can be determined by the ranking method (statistic historic) or rational.

The analysis of reservoir water quality was conducted by Water Quality Index method. The analysis of water quality status in this research was conducted by comparing the observation results with the standard values [19], [11], [18]. Evaluation of water quality can be conducted with spatial and temporal approach [1-5], [12], [20], [21]. Therefore, the availability of periodic and regular (time series) data monitoring is the success key in performing the water quality evaluation [14-17], [8], [9]. In addition to historical data series, the selection of appropriate water quality parameters is also a very important variable in evaluating the quality of water source [18], [20].

*Corresponding author: mariana.marselina@yahoo.com

WQI shows the relative contamination levels allowed in water quality parameters or water quality parameter standards. Water quality parameter standards used in this WQI analysis refer to Government Regulation No. 82 of 2001 on the designation of water class. Saguling is categorized as class I because it is used for drinking water in downstream cascade Citarum.

Based on this research, it was understood that the mainstay discharge of drinking water of Saguling Reservoir by using the R10 concept as dry (the discharge value is less in one time within 10 years period). Based on the Water Quality Index research conducted, the monitoring location which become the best alternative as region or location of raw water intake can be found.

2 Methodology

2.1 Mainstay Dicharge of Saguling Reservoir

The reservoir discharge analysis is often associated with the reliability level of water availability in meeting the needs and is often referred to as mainstay discharge. For example, the activities that require the reliability of water availability is the provision of raw water and hydropower [13]. Mainstay discharge is determined by a certain minimum quantity that has a certain probability in meeting the needs, for example, for the needs of drinking raw water usually set as 90% or 95%, while for irrigation is set as 80%. The return period equation for probability is defined as Tr = (1/P) in which Tr is the return period and P is the expected probability. The design criteria of domestic raw water and irrigation can be seen in Table 1.

Table 1. Design Criteria of Domestic Raw Water and Irrigation

Reservoir Water Source	Raw Water Planning Design Criteria			
Successive Dry Water Discharge	Domestic		Irrigation	
	30 days	T_R=10-20 Years	30 days	T_R=5 Years

Source: Modified by Sabar (2009)

Observation of minimum discharge value and maximum discharge value is expected to provide an explanation for the water potential tendency in Saguling Reservoir. Discharge observation has been done in research by Marganingrum (2013) and Rhamdani (2009) in Saguling Reservoir in two periods, namely 1950-1985 (period of development and management planning of Saguling Reservoir) and 1986-2008 (period after the construction of Saguling Reservoir). An observation was also conducted on the minimum discharge trends from year to year as to provide an explanation on the potential base flow of watershed, while observation on the maximum discharge trends from year to year can provide explanation on floods or runoffs due to land-use changes. Furthermore, the observation of dischager in the research of Citarum Cascade Reservoir was conducted within the observation span of 1986-2013.

2.2 Determinantion of Wet, Norma, and Dry Years with Markov Model

Markov model has a formula as follows (Descombes dan Berthod, 2006):

$$q_i = d_i + e_i \quad (1)$$

where:
di = deterministic component
ei = random component
Otoregresip deterministic component has a linear formula as follows:

$$d_i = \beta_0 + \beta_1 q_{t-1} + \beta_2 q_{t-2} + \dots\dots\dots + \beta_m q_{t-m} \quad (2)$$

where d_i is a linear combination of m previous flow, for a limited m. The simplest model has the following formula:

$$d_i = \beta 0 + \beta_1 q_{t-1} + e_1 \quad (3)$$

The above model assumes that all previous influence to the flow is now reflected in the value of the previous flow. Furthermore, in lag model, the amount of β_0 dan β_1 constants are specified with the exact form of e_i. In the first place, it is considered that the flow follows the normal distribution.

If μ is the middle-value flow, ρ is the correlation of coefficient serial lag and σ is flow diversity, the following equation is established:

$$q_1 = \mu + \rho\,(q_{i-1} - \mu) + e_i \quad (4)$$

Here, it is assumed that the flow of qi is the sum of the median, some of the differences in previous flow q_{i-1} with the mean and random components e_i. If q_i has a normal distribution, then e_i should also have a normal distribution as well. The flow of q_i has median μ, so the suggested form will give the desired middle value. The flow q_i has diversity as follows:

$$E\,[\mu + (q_{i-1} - \mu) + e_i]^2 - \mu^2 = \sigma^2\rho^2 + \sigma_e^2 \quad (5)$$

where σ_e^2 is a diversity of the random components ei. The diversity q_i is related with σ_e^2 which expressed in the following formula:

$$s_e^2 = \sigma^2\,(1-\rho^2) \quad (6)$$

If t_i is a random variable that has a normal distribution, but does not serially rely on 0 median and has 1 value of standard deviation, then:

$$t_i\,\sigma\,((1 - \rho 2\,))^{0.5} \quad (7)$$

also, a normally distributed variable does not depend serially, on median 0 and variance $\sigma 2$ $(1-\rho 2)$. Thus the model is as follows:

$$q_i = \mu + \rho\,(q_{i-1} - \mu) + t_i\,\sigma\,((1 - \rho^2\,))^{0.5} \quad (8)$$

Simplification of water discharge that entered the reservoir in Markov model was conducted by dividing them into three classes. Based on the water division,

 https://doi.org/10.1051/matecconf/201814703015

matrix stochastic can be created each month that divides the historical data into three classes. The stages of Markov model processing in dividing discharge category are as follows:

a. Frequency distribution analysis
An analysis of the frequency distributions of water discharge per month in the period 1999-2013 was conducted. Then the frequency distribution of historical data was matched with the two frequency distribution models: Normal and LogNormal. Each month, the most suitable distribution of the two alternative types was chosen

b. Class division
The second phase after the selection of frequency distribution was the distribution of water discharge class. The process investigated in this study was the first order for three classes. Therefore, the amount of discharge was divided into three classes as follows:
1) Dry discharge (represented by 0)
2) Normal discharge (represented by 1)
3) Wet discharge (represented by 2)
Class intervals for each class divisions was obtained by dividing the probability curve of the distribution of the selected population into 3 equal parts, namely 0.333, 0.667, and 1 as shown in Figure 1. Range value of each class was the middle value of each class which is in probability curve 0333, 0.667, and 1. The probability determination of each data was conducted by using Weibull method as follows:

$$P\,(Xm) = \frac{m}{N+1} \qquad [9]$$

with:

P (Xm) = probability of a set of values that are expected during the observation period
N = number of observation of X variate
m = serial number of events or rate of events

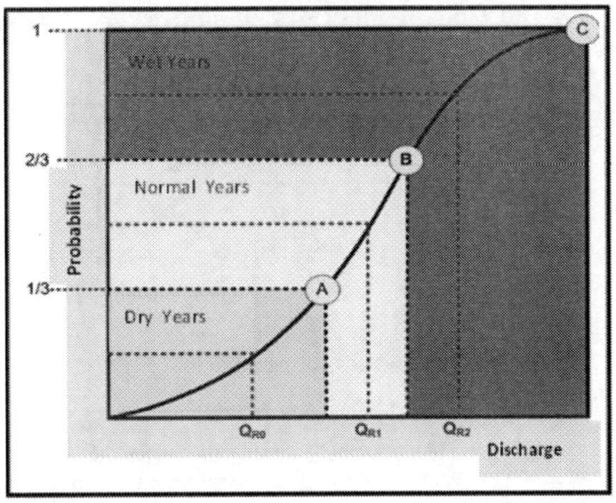

Fig. 1. Class division based on probability in Markov model

2.3 Monitoring Quality Station in Saguling Reservoir

Monitoring of water quality in Saguling was conducted regularly every three months within one year span, with each taken in March, June, September, and December respectively. There were 11 locations of Saguling quality monitoring stations: Nanjung Station (input), Batujajar, Cipatik Muara, Muara Ciminyak, Cimerang, Tjihaur, Muara Cijere, Cijambu Muara, Muara Tjihaur, Turbine Intake and Tailrace. The monitoring was conducted on forty-four (44) water quality parameters which consist of physical and chemical parameters. Main study location in this research was Nanjung Station, Muara Ciminyak Station, and Intake Turbine station as shown in Figure 2.

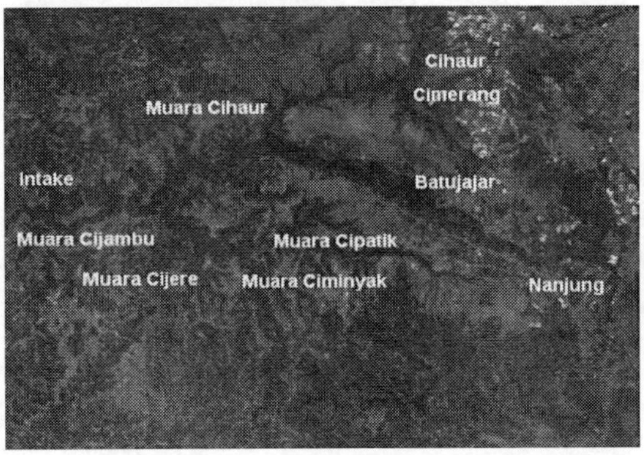

Fig. 2. Location of Monitoring Quality Station in Saguling Reservoir

2.4 The Measurement of Saguling Water Quality Index

WQI method does not distinguish between physical, chemical, and biological contaminants. All types of contaminants are given equal value [18]. WQI method is quite easy and simple with the following equation:

$$WQI = (\sum[Ci/Pi])/n \qquad (10)$$

WQI = quality index (water quality index)
Ci = concentration of the variable i
Pi= quality standards for variable i
n = number of variable

Source or basis for determining the water quality was based on WQI which was based on research by Altansukh and Davaa that conducted in Tuul River, Mongolia. Determination of water quality based on WQI value can be seen in Table 2.

Table 2. Determination of Water Quality Based on WQI Value

No	WQI Value	Water Quality		Recommendation
		Class	Status	
1	WQI ≤ 0.30	1	Very clean	No treatment necessary. Suitable for all kind of water usage.
2	0.31 ≤ WQI ≤ 0.89	2	Clean	After treatment, it can be used for drinking and agriculture. Without treatment, can be used for fishery.
3	0.90 ≤ WQI ≤ 2.49	3	Slightly polluted	Unsuitable for drinking and agriculture. If there's no choice, use it after treatment. Without treatment, can be used for livestock, recreation, and sports purposes.
4	2.50 ≤ WQI ≤ 3.99	4	Moderately polluted	Can be used for irrigation and industrial purposes after a proper treatment.
5	4.00 ≤ WQI ≤ 5.99	5	Heavily polluted	After an appropriate treatment can be used for heavy industrial activities without human body contact.
6	WQI ≥ 6.00	6	Dirty	Unsuitable for any purpose. Requires an extensive treatment.

3 Discussion

3.1 Raw Water Dicharge Allocation at Saguling Reservoir

Discharge analysis in Citarum Cascade Reservoir started from the discharge observation data at Saguling Reservoir, namely Nanjung discharge station as primary station and Safuling discharge station as secondary station. Figure 3 shows that the average monthly discharge of Nanjung Station and Saguling Station have the same monthly pattern in both data series of 1920-1980 and 1986-2013.

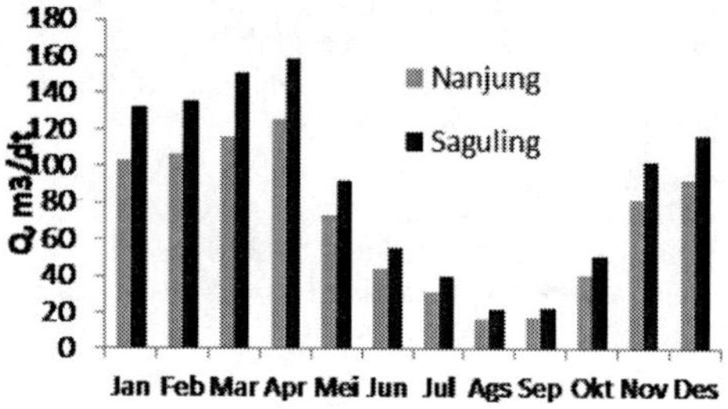

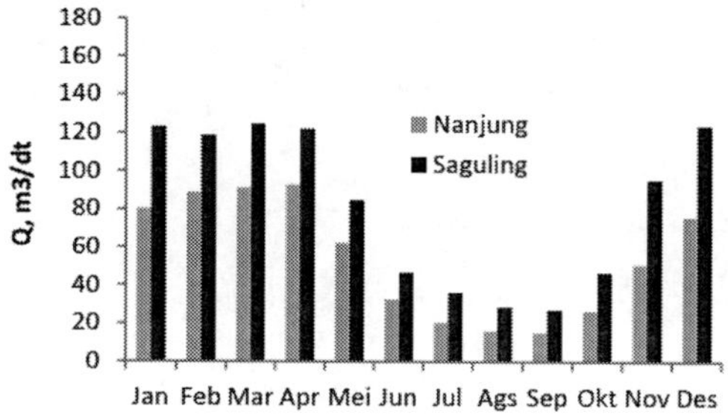

Fig.3. Monthly Average Discharge Fulucation in Nanjung Statin and Saguling Station Within Two Different Time Periods

Drinking water allocation that the Saguling Reservoir can provide is based on the discharge plan of dry R5 or dry R10 which can be seen in Table 3.

3.2 Division of Discharge Class of Saguling Water by Year

Distribution of research observation year (1999-2013) was conducted by Markov Discreet Method. It was divided into dry, normal, and wet years. The class divisions were performed in Nanjung, Muara Ciminyak, and Turbine Intake Stations. The class divisions in dry, normal, and wet years in each station can be seen in Table 4.

https://doi.org/10.1051/mateccconf/201814703015

Table 4. Determination of Dry, Normal, Wet Years according to Markov Discreet Method based on Discharge Data of Saguling Reservoir

Dry Years (0)	Normal Years (1)	Wet Years (2)
2002	1999	2002
2003	2000	2003
2004	2008	2004
2006	2009	2006
2011	2012	2011

3.3 The Average Value of WQI of Overall Monitoring Post in Saguling Reservoir

Based on Figure 4, it was found that based on the average value of WQI for each monitoring location station in Saguling Reservoir. Monitoring location that has the smallest WQI value was Muara Ciminyak Station with WQI value of 3,05. The monitoring location that has the highest WQI in this case showed the highest pollution level was Nanjung monitoring location.

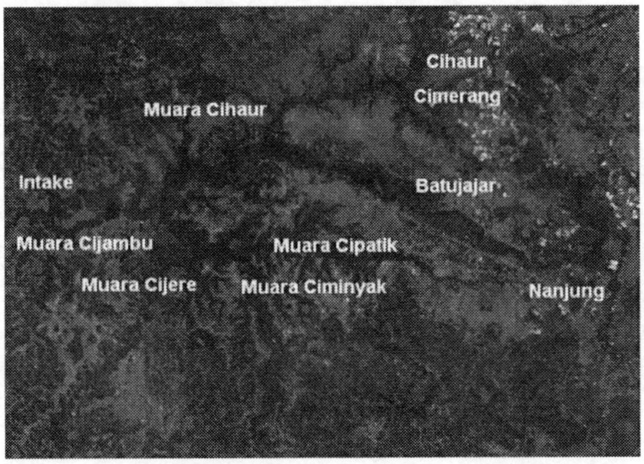

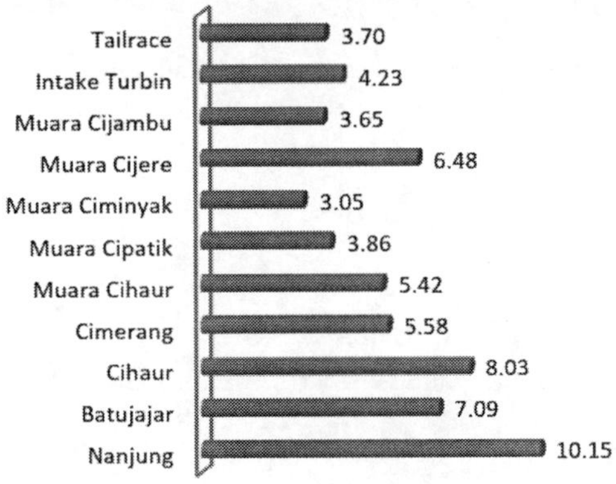

Fig.4. The average WQI value of each monitoring station in Saguling Reservoir

Determination of water quality based on WQI value (Altansukh & Davaa) can be seen in Table 5.

Table 5 Water Quality Status of Each Monitoring Station in Saguling Reservoir

Monitoring Location	WQI Value (Average)	Status
Nanjung	10,150	Dirty
Batujajar	7,092	Dirty
Cihaur	8,035	Dirty
Cimerang	5,575	Heavily polluted
Muara Cihaur	5,416	Heavily polluted
Muara Cipatik	3,858	Moderately polluted
Muara Ciminyak	3,046	Moderately polluted
Muara Cijere	6,482	Dirty
Muara Cijambu	3,650	Moderately polluted
Intake Turbin	4,227	Heavily polluted
Tailrace	3,704	Moderately polluted

The high value of WQI or the high pollution in Nanjung Station is caused by the land-use in Nanjung watershed which is dominated by industry and household. The previous surveys informed that the main types of industry in Citarum watershed are textile industry, leather tanning industry, food industry, and electroplating industry. Based on a study conducted by Environmental Agency of West Java Province in 2007, there were 359 companies that are divided into 11 different industry sectors, located in four administrative areas along the upstream of Citarum River. Among those industrial sectors, textile industry is one of the sectors that need to be considered because of its most dominant number. Other industrial sector such as electroplating, pharmaceuticals, metals, food/beverages also need attention. Detailed data on the number of industry of inventory result of Environmental Agency of West Java Province (2007) can be seen in Table 6.

Table 6 Data of the Number of Industry

No	Industry Sector	Municipality/City				
		Sumedang Regency	Bandung Regency	Bandung City	Cimahi City	Total
1	Textile	10	152	54	46	262
2	Electrop lating	0	5	9	5	19
3	Leather	1	0	1	0	2
4	Chemic al	0	4	0	0	4
5	Pharmac eutical	0	3	3	3	9
6	Paper	0	2	1	0	3
7	Paint	0	1	0	2	3
8	Integrat ed IPAL	0	2	0	0	2
9	Food/Be verage	1	3	10	10	24

https://doi.org/10.1051/matecconf/201814703015

No	Industry Sector	Municipality/City				
		Sumedang Regency	Bandung Regency	Bandung City	Cimahi City	Total
10	Garment	0	0	2	0	2
11	Metal	0	0	1	6	7
	Total	15	176	93	75	359

3.4 The Average WQI Value Based on Grouping of Monitoring Station Location

Distribution of Monitoring Station based on Location of Saguling Reservoir	
Inlet	Nanjung
	Batujajar
North of Reservoir (Middle)	Cihaur
	Cimerang
	Muara Cihaur
	Muara Cipatik
	Muara Ciminyak
	Muara Cijere
	Muara Cijambu
Outlet	Intake Turbin
	Tailrace

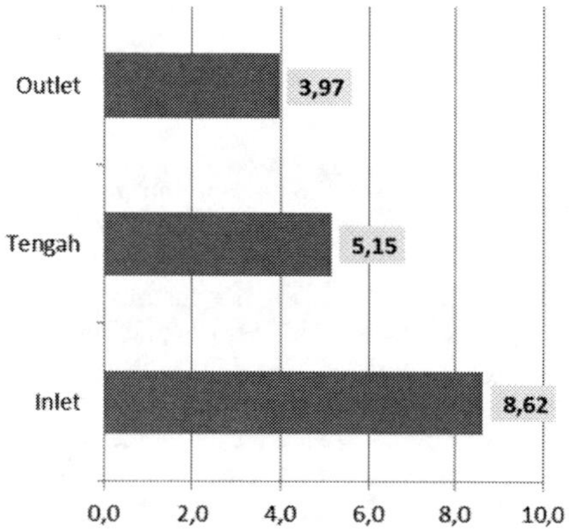

Fig.5. The average value of EQU based in the Location Grouping of Monitoring Station

Grouping of the entire monitoring station into three main location of Reservoir (inlet, middle, and outlet) and WQI value based on the location division can be seen in Figure 5. Based on Figure 5, it can be seen that the highest pollution is in the inlet area of Saguling Reservoir discharge, in this case from the upstream Citarum watershed and the lowest pollution is in outlet location. This is in accordance with the dilution phenomenon occurs along the reservoir or water body which is the increasingly away from the inlet location with high pollution, the pollution can decrease.

3.5 WQI value based on the Discharge Class Division (Dry, Normal, and Wet)

Water quality of each monitoring location cannot be separated from the volume condition occurred in Saguling Reservoir. In obtaining the actual status of reservoir, it is necessary to determine the quality of reservoir with the scenario of year condition or the volume occurred, in this case the dry, normal, and wet year. The WQI value of each monitoring location the condition series of dry, normal, and wet year can be seen in Table 7. The specific WQI value in the year condition of wet, normal, and dry for all monitoring station of Saguling Reservoir can be sen in Figure 6, Figure 7, and Figure 8.

Table 7. WQI Value of Each Monitoring Station in the Discharge Class Division of Wet, Normal, and Dry Year

Year Division Based on Discharge Class	Nanjung	Batujajar	Cihaur	Cimerang	Muara Cihaur
Drw	14,277	10,893	11,681	6,494	9,041
Normal	8,383	5,689	7,781	5,206	4,053
Wet	7,792	4,693	4,642	5,025	3,154

Table 7. WQI Value of Each Monitoring Station in the Discharge Class Division of Wet, Normal, and Dry Year (Cont.)

Year Division Based on Discharge Class	Muara Cipatik	Muara Ciminyak	Muara Cijere	Muara Cijambu	Intake Turbin	Tailrace
Dry	5,818	4,051	7,244	5,397	6,270	4,074
Normal	3,058	2,734	6,675	2,785	3,587	3,811
Wet	2,697	2,353	5,526	2,768	2,823	3,227

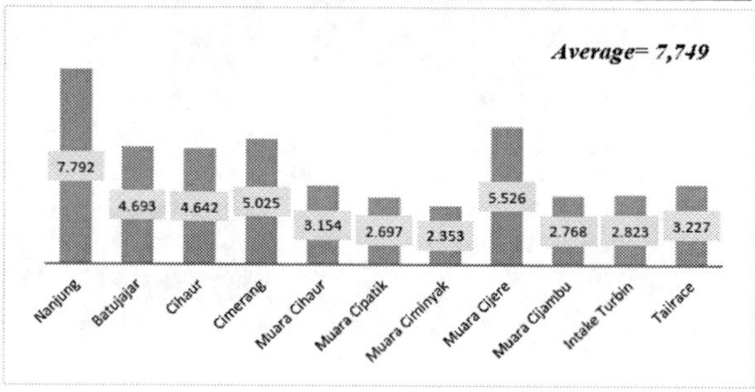

Fig. 6. WQI Value of Each Monitoring Station in the Discharge Class Division of Wet Year

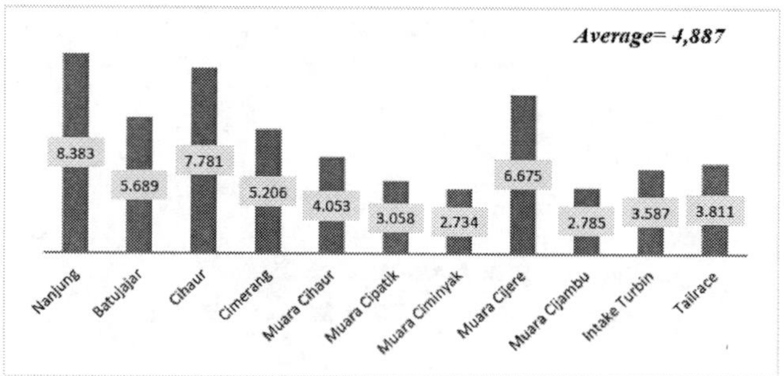

Fig.7. WQI Value of Each Monitoring Station in the Discharge Class Division of Normal Year

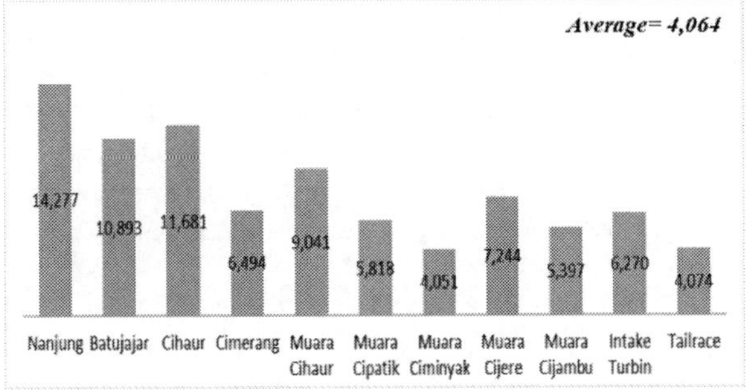

Fig.8. WQI Value of Each Monitoring Station in the Discharge Class Division of Dry Year

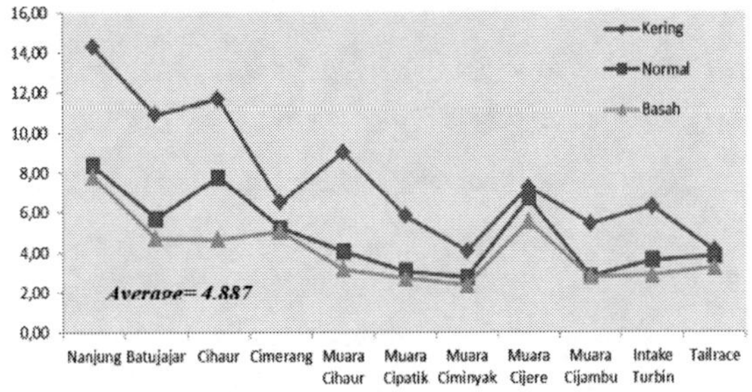

Fig.9. WQI Value Graph in Each Monitoring Station in Dry, Normal, and Wet Condition

Based on Figure 9, the highest WQI value is gained from the dry year, followed by the normal year, then, the wet years. It happened in each monitoring station in Saguling Reservoir. This was in accordance with the consistency of flow rate and concentration where it tend to be low in the dry season and the pollution concentration will increase. On the contrary, the reservoir flow rate increased in the rainy season, therefore the pollution concentration will decrease due to the dilution in the reservoir.

Based on Figure 9, the highest WQI value is gained from the dry year, followed by the normal year, then, the wet years. It happened in each monitoring station in Saguling Reservoir. This was in accordance with the consistency of flow rate and concentration where it tend to be low in the dry season and the pollution concentration will increase. On the contrary, the reservoir flow rate increased in the rainy season, therefore the pollution concentration will decrease due to the dilution in the reservoir.

4 Conclusion

The calculation allocation of raw water discharge for Saguling Reservoir in order to fulfill the needs of raw water in Bandung Metropolitan Area is 46,92 m3/second (R5 dry for irrigation raw water supply) and 29,53 92 m3/second (R10 dry for drinking raw water). The pollution load in Saguling Reservoir is a function of pollutant concentration and quantity or discharge occurred in the reservoir. This is demonstrated by the research of water pollution in Saguling Reservoir with Markov discrete approach and the determination of Water Quality Index (WQI) in the condition of dry, normal, and wet year. Based on the water quality analysis by Water Quality Index (WQI) method, it was found that the WQI Value in Muara Ciminyak Station is 4,051 in the dry year condition; 2,374 in normal year condition; and 2,353 in wet year condition. The monitoring location of this Muara Ciminyak Station is the most qualified as an option in the intake location of raw water supply from Saguling Reservoir. Utilization of Saguling Reservoir as an alternative potency of raw water supply requires careful target and preparation. In-depth calculation of Saguling Reservoir capability in terms of quantity and quality as an alternative of raw water needs to be done.

References

1. Kazi et al : Assessment of Water Quality of Polluted Lake using Multivariate Statistical Technique : A case Study. Ecotoxicol Environ Saf, 72, 301-309, (2009)
2. Singh, K.P., Malik, A., Mohan, D., Sinha, : Multivariate Statistical Techniques for The Evaluation of Spatial and Temporal Variations in Water quality of Gomti River (India): A Case Study. Water Research, 38, 3980-3992, (2004)
3. Aubert, A.H., Gascuel-Odoux, C., Merot, P : Annual Hysteresis of Water Quality: A Method to Analyse The Effect of Intra and Inter-annual Climatic Condition. Journal of Hydrology, 478, 29-39, (2013)
4. Lai et al : Development of A Water Quality Modeling System for River Pollution Index and Suspended Solid Loading Evaluation. Journal of Hydrology, 478, 89-101, (2013)
5. Merten, G. H.; Minella, J. P : The Identification of Sediment Sources in a Small Urban Watershed in Southern Brazil: 30, No. 11, October 2009, 1145–1153, (2002)
6. Djuangsih, N : Eutrofikasi di Waduk Saguling. Laporan Penelitian Hibah Bersaing. PPSDAL – Lembaga Penelitian Universitas Padjadjaran, (1996)

7. Hart, et al : Nutrient Budget for Saguling Reservoir, West Java, Indonesia. Water Research, 36, 2152–2160, (2002)
8. Andrade, E. M; Palacio, H. A. Q; Souza, I. H; Leao, R. A.O; Guerreiro, M. J. Land use effects in groundwater composition of an alluvial aquifer (Trussu River, Brazil) by multivariate techniques. Environmental Research, 106, pp. 170-178, (2008)
9. Guedes, H. A. S., Silva, D. D., Elesbon, A. A. A., Ribeiro, B. M., Matos, A. T., Soares, J. H. P : Application of Statistics Multivariate Analysis in Dove River. Jurnal Pertanian dan Teknik Lingkungan, 16, 558-563, (2002)
10. Laureano and Navar: An Assessment of Stream Water Quality of the Rio San Juan, Nuevo Leon, Mexico, 1995–1996. Journal of Environmental Quality, 31, 1256-1265, (2002)
11. Peterson et al: Process Identification by Principal Component Analysis of River Water-Quality Data. Ecological Modelling, 138, 193–213, (2001)
12. Selle et al : Understanding Processes Governing Water Quality in Catchments Using Principal Component Scores. Journal of Hydrology, 486, 31–38, (2013)
13. Vialle, C., Sablayrolles, C., Lovera, M., Jacob, S., Huau, M. C., and Montrejaud-Vignoles, M: Monitoring of Water Quality from Roof Runoff : Interpretation Using Multivariate Analysis. Water Research, 45, 3765–3775, (2011)
14. Vonberg, D., Vanderborght, J., Cremer, N., Pütz, T., Herbst, M., and Vereecken, H : 20 Years of Long-Term Atrazine Monitoring in A shallow Aquifer in Western Germany. Water Research, 50, 294-306, (2014)
15. Fan, et al : Water Quality Management Based on Division of Dry and Wet Seasons in Pearl River Delta, China. Clean - Soil, Air, Water, 40, 381–393, (2012)
16. Sebastia, et al : Analysis of The Effects of Wet and Dry Seasons on A Mediterranean River Basin: Consequences for Coastal Waters and Its Quality Management. Ocean & Coastal Management, 78, 45-55, (2013)
17. Altansukh and Davaa : Application of Index Analysis to Evaluate the Water Quality of The Tuul River in Mongolia. Journal of Water Resource and Protection, 3, 398-414, (2011)
18. Rosemond, Simone de, Duro, Dennis C. Dan Dubé, Monique : Comparative Analysis of Regional Water Quality in Canada using The Water Quality Index. Environmental Monitoring Assessment, 156, 223–240, (2009)
19. Mahmood Q, Baig SA, Nawab B, Shafqat MN, Pervez A : Development of Low-Cost Household Drinking Water Treatment System of The Earthquake Affected Communities in Northern Pakistan. Desalination, 273, 316-320, (2011)
20. Mustapha, A., and Aris, A.Z : Spatial Aspect of Surface Water Quality Using Chemometric Analysis. *Journal of Applied Sciences in Environmental Sanitation*, **6**, 411- 426, (2011)

Development of audit technology approach for performance improvement of faecal sludge treatment plant (IPLT) (Case study: IPLT Bawang in Tangerang City and IPLT Pecuk in Indramayu District)

Rydha Riyana Agustien[1,*], *Prayatni* Soewondo[2], and *Arief* Sudradjat[2]

[1]Research Institute for Housing and Human Settlements, Ministry of Public Works and Housing, Jl. Panyawungan Cileunyi Wetan Kabupaten Bandung, Indonesia
[2]Faculty of Civil and Environmental Engineering, Institute Technology Bandung, Jl. Ganesha 10 Bandung 40132, Indonesia

Abstract. One of the infrastructure to support the achievement of the 100% sanitation target is IPLT. The development of audit technology approach needed for performance improvement of IPLT. The objectives of this research are to determine the weight of components technology priority that have affect to IPLT, assessing the achievement of performance in IPLT Bawang and IPLT Pecuk, and to determine the priority of performance improvement strategy in study area. This research consist of descriptive analysis about public sanitation condition in the study area by distributing questionnaires, scoring method analysis of IPLT technology components using pairwise comparison method to determine main priority of technology component, performance analysis of IPLT Bawang and IPLT Pecuk using audit technology approach, and determining strategy of IPLT performance improvement. The weight of technology components that affect the performance of IPLT are humanware with weight 0,35, orgaware with weight 0,29, technoware with weights 0,24 and infoware with weight 0,13. The value of performance achievement IPLT Pecuk is 60,2% and IPLT Bawang is 58,1%. The priority of performance improvement strategy in IPLT Bawang are improvement strategies of humanware, technoware, infoware and orgaware, meanwhile in IPLT Pecuk are improvement strategies of technoware component, humanware, orgaware and infoware.

1 Introduction

Nowadays, the government has been conducting various program to achieve the RPJMN III (2015-2019) target, such as the achievement of 100% sanitation access. One of the sanitation aspects is the management of domestic waste water. Indicators and target of waste water development is open defecation free. To achieve the RPJMN targets, the government built sanitation facility on site and off site. In Indonesia, on site system with used septic tank is the most widely used. But, the effluent from septic tank is not yet eligible to be discharged into river because the high content of Biological Oxygen Demand (BOD) dan coliform. Without faecal sludge management, there will be contamination of pathogens into the environment and causing health damaged [1]. So, faecal sludge treatment plant needed to treat faecal sludge.

Faecal sludge treatment plant is one of the planed efforts to improve the treatment and to safe waste disposal. Based on Direktorat PPLP's data, nowadays Indonesia have 146 faecal sludge treatment plant which scattered in 23 provinces (out of 33 provinces) [2]. The condition of faecal sludge management in Indonesia is not functioning and not operating optimally due to constrains of technical and non technical aspect [2]. The main problem in faecal sludge treatment plant is a problem in the operation of faecal sludge treatment palnt. Based on Direktorat PPLP Dirjen Cipta Karya's data, the operating rate of faecal sludge treatment plant is 65% [3]. The problem in the management of faecal sludge treatment plant is lack of faecal sludge supply, lack of public knowledge about faecal sludge treatment and lack of willingness to pay of waste water retribution, lack of cooperation with private sector in term of suction faecal sludge [2].

One of the faecal sludge treatment plant that has been built is IPLT Bawang in Tangerang city dan IPLT Pecuk in Indramayu district. IPLT Bawang in Tangerang city is manage by Dinas Cipta Karya dan Penataan Ruang Tanggerang City. The physical condition of IPLT Bawang in Tangerang city is good and operated but not optimaly because the faecal sludge supply is more than the capacity [2]. IPLT

* Corresponding author: rydhariyana@gmail.com

Pecuk Indramayu district manage by Dinas Kebersihan dan Pertamanan Indramayu District. The physical condition of IPLT Pecuk Indramayu district is good and operated but not optimaly because the faecal sludge supply is less than the capacity. Both of the faecal sludge used the pond system, but the same system have the diferrent problem.

Therefore it is necessary to do research about faecal sludge management with using development of audit technology approach to determine factor which have the effect in optimizing IPLT. When main factor is already known, advice can be formulated for performance improvement of IPLT.

2 Methodology

This study has been initiated with the identification of problems. Identification of problems has been carried out by observation, direct interviews with stakeholders, and the study of literature. The study of literature has been carried out by reading the article, book, journal or related research. Then, collected primary and secondary datas. Primary datas has been obtanined from questionnaires data, observation data, laboratory test result data and interviews data with stakeholders. Secondary datas has been obtained by the study of literature from the stakeholders.

This research has been carried out by audit technology approach. Audit technology approach is a systematic and objective evaluation conducted by auditors technology against technology assets to achieve the purpose of technology so as to provide added value and increase the performance of the auditee or stakeholder. Audit technology method has been set in Peraturan Kepala Badan Pengkajian dan Penerapan Teknologi No. 004 Tahun 2011 tentang Pedoman Audit Teknologi [4]. Component technology consists of technoware, humanware, infoware, orgaware which later became known as the audit technology approach method [5]. The audit technology approach method can used for analysis of management technology [6]. In this study, audit technology approach has used as a method to analyze and evaluate the suitability of the technology IPLT with planning and needs. So that the role and function of audit technology in this research is more focused on the role and function of technology management as the management of technology assets IPLT.

The next stage is a descriptive analysis of sanitary conditions of society by distributing questionnaires, weighting analysis of technology components IPLT using pairwise comparison, performance analysis IPLT with technology audit approach, determining IPLT performance improvement strategy.

The objectives of descriptive analysis of public sanitation conditions are to determine the condition and perceptions about waste water treatment and to determine ability to pay and willingness to pay about retribution of faecal sludge suction. This analysis has been carried out by distributing questionaires. The objective of weighting analysis of technology components IPLT is to determain main priority of technology component (*technoware, humanware, infoware,* dan *orgaware* (THIO)) which influential in determining performance of IPLT. This analysis using pairwise comparison method. Pairwise comparison method is a method for determining the ratio between an object with another object [7]. Mathematical formulation used in this method [8]. The objectives of permormance IPLT analysis are to determine the performance achievement IPLT in study areas and to determine performance improvement strategy in IPLT.

Performance IPLT analysis has been carried out by technoware analysis, humanware analysis, infoware analysis, and orgaware analysis. Technoware interpreted as the physical facilities to increase energy and human capabilities in to produce products and services. Technoware analysis has been carried by quality product assessment, planning criteria and installation functions. Technoware analysis processed quality product assessment has been carried out by the approach of determining the pollution index and BOD removal efficiency. Assessment planning criteria has been carried out by calculating the volumetric load and load surface processing unit. Installation function analysis has been carried out by looking at the performance of the installation through the method of Overall Equipment Effectiveness (OEE). OEE methods is usually used to measure performance in manufacturing such as research conducted by [9], [10], [11] and [12]. OEE calculation is based on three factors: availability, performance, and *quality* [13].

Humanware defined as people who have the skills or competence without which technoware useless. Humanware analysis in terms of the adequacy of human resources indicators, human resource competencies indicators, employment status indicators and performance assessment indicators. Infoware interpreted as written or electronic documents, which allows humanware to get information and learn faster. Infoware analysis in terms of required documents indicators, and ease of access to the document indicators. Orgaware interpreted as procedures and organizational structure terimplementassi that coordinate resources and activities to achieve goals. Orgaware analysis in terms of organization indicators, cooperation indicators and finance indicators.

3 Overview IPLT in study areas

3.1 Overview IPLT Bawang in Tangerang City

Based on Dinas Kesehatan Tangerang City's data in 2013, from 504.354 household in Tangerang city, 497.577 household (97,5%) already have access to a private toilets to defecate. Tangerang city have waste water treatment plant which has built in 1982 and faecal sludge treatment plant (IPLT) which has been operate in 1999 (Anggraini, et al., 2016). IPLT

Bawang located in Bawang street Perumnas Karawaci. This capacity is ± 70 m^3/day. The average of faecal sludge supply is 85 m^3/day. IPLT Bawang managed by Dinas Cipta Karya dan Penataan Ruang. The treatment unit in IPLT Bawang are imhoff tank, anaerobic ponds, facultatif ponds, oxidation pond, maturation pond and sludge drying bed [2].

3.2 Overview IPLT Pecuk in Indramayu District

Waste water treatment especially waste water and faecal sludge is managed by Dinas Kebersihan dan Pertamanan in Indramayu district (Perda Kabupaten Indramayu No. 8 Tahun 2008). Septic tank suction servis in 2011 is 270 household/year. Facilities and infrastructure are 1 unit faecal sludge truck and 1 unit IPLT Pecuk. The area of IPLT is 176,95 m^2 with capacity 25 m^3/day. The average of faecal sludge supply is 2 m^3/day. The treatment unit in IPLT Pecuk are imhoff tank, anaerobic ponds, facultatif ponds, maturation pond and sludge drying bed [14].

4 Result

4.1 Descriptive analysis of sanitary conditions of society in Tangerang City and Indramayu District

According to the result of questionnaires, community in Tangerang city as much as 89% defecation using private toilets, 10% in public toilets and 1% in the river. The last estuary of feces is 93% in septic tank, 5% in river/draine and 2% in communal waste water treatment plant. Septic tank material which suitable with SNI 03-2398-2002 is 41% septic tank. The percentage of people in Tangerang city who are not willing to improve his septic tank according to the standard is 99%. The percentage of people who are willing to do the faecal sludge service scedule program (LLTT) is 47%. The average of ability to pay in faecal sludge suction tariff is Rp. 333.960,-/suction. The percentage of willingness to pay which suitable with retribution tariff is 42,0%. The percentage of willingness to pay with tariff Rp. 20,000,- until Rp. 30,000,- per m^3 is 13,39% and the percentage of people who are not willing to pay is 44,61%.

According to the result of questionnaires, community in Indramayu district as much as 91% defecation using private toilets, 4% in public toilets and 5% in the river. The last estuary of feces is 89% in septic tank and 11% in river/draine. Septic tank material which suitable with SNI 03-2398-2002 is 26% septic tank. The percentage of people in Indramayu district who are not willing to improve his septic tank according to the standard is 94%. The percentage of people who are willing to do the faecal sludge service scedule program is 54%. The average of ability to pay in faecal sludge suction tariff is Rp. 346.950,-/suction. The percentage of willingness to pay which suitable with retribution tariff is 53,0%. The percentage of willingness to pay with tariff Rp. 40.000,- until Rp. 50.000,- per m^3 is 4,23% and the percentage of people who are not willing to pay is 42,77%.

4.2 Weighting analysis of technology components IPLT using pairwise comparison

The data used in the weighting analysis by using pairwise comparison method is consisten data from 25 respondents. The percentage of 25 consistent respondent are 28% of respondents from the central government, 20% from the Tangerang city, 20% from professionals, 16% from the Indramayu district government and 16% from the provincial government. Based on data Roscoe in Hill (1998) in [15] explained that in the simple experimental research with tight control, the number of respondents can be done with a sample of 10 to 20 respondents. So the data from respondents consistently declared valid and can represent for research.

Based on the results of the assessment, the variable priority or have the greatest weight is humanware variable as much as 0,35. The next priority are orgaware variable with a weight of 0,29, technoware variable with a weight of 0,24 and infoware variable with a weight of 0,13. Comparison between the variable weights can be seen in Figure 1. The 10 indicators / parameters of the most influential are the human resources training parameter is amounted to 11,14%, installation function indicators 9,02%, employee performance assessment indicators is amounted to 8,14%, planning criteria indicator is amounted to 8,02%, the adequacy of human resources indicator of is amounted to 7,29%, refined quality indicators is amounted to 6,95%, operational and mainterance (OM) availability parameter is amounted to 4,70%, human resources education parameter is amounted to 4,67%, an indicator of the legal basis for financing the management IPLT is amounted to 4,42%, and employment status indicator is amounted to 3,45%. The result of this research is different from the results of research conducted by Anggraini et. al. in 2016. Based on Anggraini et. al. in 2016 research, the priority of IPLT component are technoware variable with a weight of 0,384, humanware variable with a weight of 0,250, orgaware variable with a weight of 0,212 and infoware variable with a weight of 0,154. The differences of both research is the research conducted by Anggraini et. al. in 2016, the overall respondents who filled out weighting questionnaires works as a researcher so the attention in technoware component. In this study, the respondent's professional backgrounds are more varied as described previously.

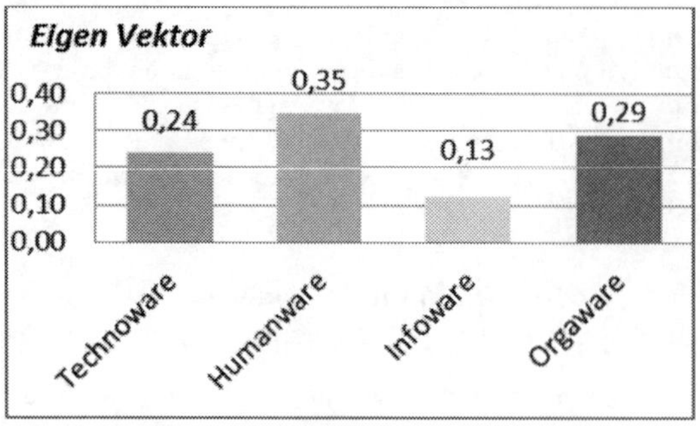

Fig. 1. The comparison between variable weights

4.3 Performance analysis IPLT with audit technology approach

4.3.1 Technoware Component Analysis of Performance IPLT Bawang and IPLT Pecuk

Technoware value is the incorporation from the value of quality product assessment, planning criteria and installation functions that have been previously analyzed. Technoware value in IPLT Pecuk is smaller than technoware value in IPLT Bawang. Technoware value in IPLT Pecuk is amount to 0,3987 while technoware value in IPLT Bawang is amount to 0,4963. Technoware perfect score is equal to 0,8797. When compared with a perfect technoware score, then the technoware value IPLT Pecuk is 45% while the technoware value IPLT Bawang is 56%.

4.3.2 Humanware Component Analysis of Performance IPLT Bawang and IPLT Pecuk

Humanware value is the incorporation from the adequacy of human resources indicators value, human resource competencies indicators value, employment status indicators value and performance assessment indicators value. Humanware value in IPLT Pecuk is smaller than humanware value in IPLT Bawang. Humanware value in IPLT Pecuk is amount to 0,5823 while humanware value in IPLT Bawang is amount to 0,5860. Humanware perfect score is equal to 0,9248. When compared with a perfect humanware score, then the humanware value IPLT Pecuk is 63,0% while the humanware value IPLT Bawang is 63,4%.

4.3.3 Infoware Component Analysis of Performance IPLT Bawang and IPLT Pecuk

Infoware value is the incorporation from the required documents indicators value and ease of access to the document indicators value. Infoware value in IPLT Pecuk is bigger than Infoware value in IPLT Bawang. Infoware value in IPLT Pecuk is amount to 0,3499 while infoware value in IPLT Bawang is amount to 0,2002. Infoware perfect score is equal to 0,4440. When compared with a perfect Infoware score, then the Infoware value IPLT Pecuk is 79% while the Infoware value IPLT Bawang is 45%.

4.3.4 Orgaware Component Analysis of Performance IPLT Bawang and IPLT Pecuk

Orgaware value is the incorporation from the organization indicators value, cooperation indicators value and finance indicators value. Orgaware value in IPLT Pecuk is bigger than orgaware value in IPLT Bawang. Orgaware value in IPLT Pecuk is amount to 0,5406 while orgaware value in IPLT Bawang is amount to 0,5248. Orgaware perfect score is equal to 0,8622. When compared with a perfect orgaware score, then the orgaware value IPLT Pecuk is 63% while the orgaware value IPLT Bawang is 61%.

4.4 Performance Achievement Analysis of IPLT

IPLT technology component values are calculated by merge technoware, humanware, infoware and orgaware value. Performance achievement value in IPLT Pecuk is bigger than performance achievement value in IPLT Bawang. Performance achievement value in IPLT Pecuk is amount to 60,2% while performance achievement value in IPLT is amount 58,1%. IPLT performance evaluation results are shown in Table 1.

Table 1. IPLT performance evaluation results

Technology Component	Perfect Score	Indramayu Score	Tangerang Score
Technoware	0,8797	0,3987	0,4963
Humanware	0,9248	0,5823	0,5860
Infoware	0,4440	0,3499	0,2002
Orgaware	0,8622	0,5406	0,5248
Totaly	3,1106	1,8716	1,8073
Performance Achievement of IPLT		**60,2%**	**58,1%**

4.4 Determining IPLT Performance Improvement Strategy

Recommendations for improved performance IPLT Bawang in Tangerang City are the implementation of human resource training for managing waste treatment plant; dredging of faecal sludge in treatment unit of IPLT; upgrading of treatment units design; need execution LLTT; supplying program of septic tank which suitable with standard; establish cooperation with the community in supporting the LLTT program; collate documents SOP operation, maintenance, repair and K3; preparing technical documents and IPLT monitoring, assessing employee performance against IPLT manager; increased ease of access to documents; and strengthening the management and financing IPLT regulation. The formulation of these recommendations based on analysis of improvement priority IPLT Bawang in accordance with Figure 2.

Recommendations for improved performance IPLT Pecuk in Indramayu district are increasing equalization tank with bar screen before imhoff tank; dredging of faecal sludge in treatment unit of IPLT; necessary implementation LLTT; supplying program of septic tank which suitable with standard; improve the competence of human resources by improving education and implementation of human resource training for managing waste treatment plant; the increase in funding sources OM; increasing the number of human resources; cooperation with private and public companies in support LLTT program; monitoring of faecal sludge truck; strengthening regulation of the management and financing of waste treatment plant; as well as complete the licensing documents and SOP K3. The formulation of these recommendations based on analysis of improvement priority IPLT Pecuk accordance with Figure 3.

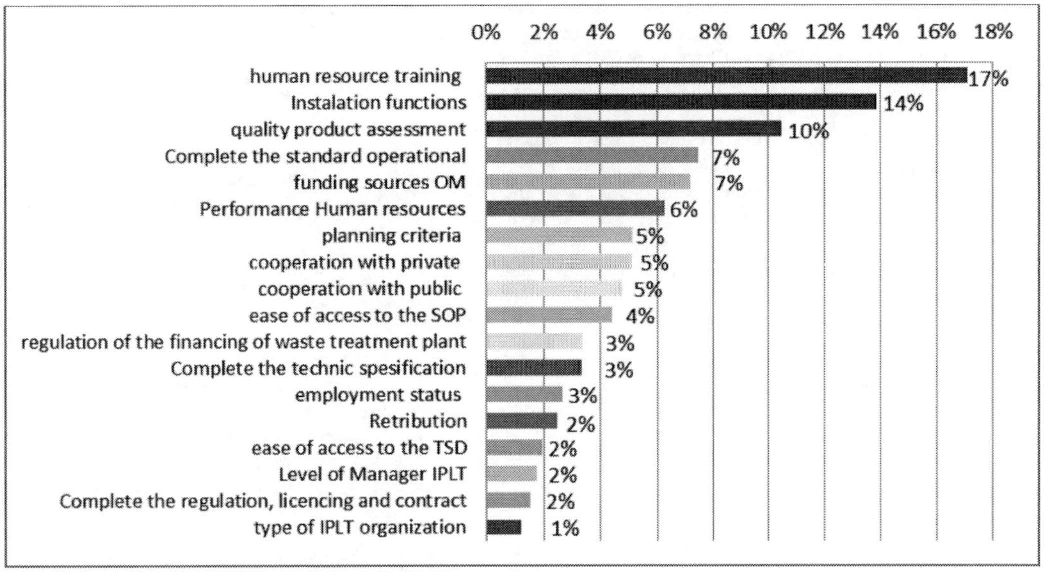

Fig. 2. Improvement priority IPLT Bawang

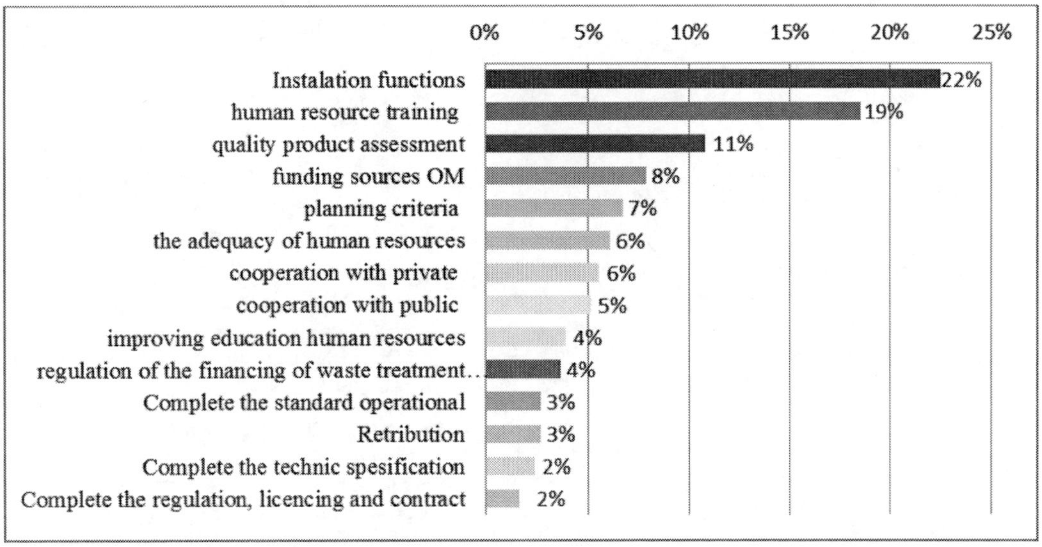

Fig. 3. Improvement priority IPLT Pecuk

5 Conclusion

1. The weight of technology components that affect the performance of IPLT are humanware with weight 0,35, orgaware with weight 0,29, technoware with weights 0,24 and infoware with weight 0,13.
2. Performance achievement value in IPLT Pecuk is amount to 60,2% while performance achievement value in IPLT is amount 58,1%.
3. The priority of performance improvement strategy in IPLT Bawang are improvement strategies of humanware component, improvement strategies of technoware component, improvement strategies of infoware component and improvement strategies of orgaware component.
4. The priority of performance improvement strategy in IPLT Pecuk are improvement strategies of technoware component, improvement strategies of humanware component, improvement strategies of orgaware component and improvement strategies of infoware component.

References

1. Strande, L, *The Global Situation*, Faecal Sludge Management, L. Strande, M. Ronteltap, & D. Brdjanovic, , London: IWA Publishing, 1-14, 2014.

2. Anggraini, F., Nuraeni, R., Yudiarti, I., Rosa, Y., Prayudi, T. R., Effendi, R. R., Pamekas, R., *Audit Teknologi Sistem Instalasi Pengolahan Lumpur Tinja (IPLT),* Bandung: PT. Kiblat Buku Utama, 2016.

3. Anggraini, F., & Nuraeni, R., Diagnosis Kinerja Instalasi Pengolahan Lumpur Tinja (IPLT), Widyariset, Vol. 1 No. 1 Hal. 87-99, 2015.

4. Pusat Audit Teknologi, Framework, Kode Etik dan Standar, Pedoman Umum Audit Teknologi, Jakarta: Pusat Audit Teknologi – BPPT, 2011.

5. Toloui, A., & Matin, Y. A., *Measuring Technological Level and Capability of the Industries in East Azerbaijan and Providing Prover Strategies for Improvement and Promotion of Technology*, Journal of Basic and Applied Scientific Research, 2 (4) 3664-3669, 2012.

6. Zaker, A. R., Tajik, H., & Bahramifaraz, F., *Analyzing the Effect of Management of Technology Transfer in Science and Technology Parks for Wealth Creation, Case Study: Science and Technology Park of Tehran University*, International Research Journal of Applied and Basic Science, Vol. 4 (10): 3427-3433, 2013.

7. Padmowati, R. d., *Analisis Teknik Indeks Konsistensi harmonik pada metode Analytical Hierarchy Process*, Bandung: LPPM-Universitas Katolik Parahyangan, 2012.

8. Febrinata, *Perancangnan Key Performance Indicators (KPI) Menggunakan Metode Balance Scorecard dan Pairwise Comparison di PT. Aston System Indonesia,* Surakarta: Universitas Sebelas Maret, 2013.

9. Samad, M. A., Hussain, M. R., & Asrafuzzaman, M., *Analysis of Performace by Overall Equipment Effectiveness of the CNC Cutting Section of a Shipyard*, ARPN Journal of Science and Technology, Vol. 2 No. 11 pp. 1091-1096, 2012.

10. Zandieh, S., Tabatabael, S. A., & Ghandehary, M., *Evaluation of Overall Equipment Effectiveness in Continuous Process Production System of Condensate Stabilization Plant in Assalooyeh*, Interdisciplinary Journal of Contemporary Research in Business, Vol. 3 No. 10 pp. 590-598, 2012.

11. Boban, B., & Joseph E., J., *Enchancing Overall Equipment Effectiveness for a Manufacturing Firm through Total Productive Maintenance*, Enchancing Overall Equipment Effectiveness for a Manufacturing Firm through Total Productive Maintenance, Vol. 3 Issue. 8 pp 425-429, 2013.

12. Nayak, D., Kumar M N, V., Naidu, G., & Shankar, V., *Evaluation of OEE in a Continuous Process Industry on an Insulation Line in a Cable Manufacturing Unit*, International Journal of Innovative Research in Science, Engineering and Technology, Vol. 2. Issue. 5 pp. 1629-1634, 2013.

13. Rahmad, Pratikto, & Wahyudi, S., *Penerapan Overall Equipment Effectiveness (OEE) dalam Implementasi Total Productive Maintenance (TPM) (Studi Kasus di Pabrik Gula PT. "Y")*, Jurnal Rekayasa Mesin, Vol. 3 No. 3 Hal. 431-437, 2012.

14. Dinas Kebersihan dan Pertamanan Kab. Indramayu, *Standar Operasional dan Prosedur IPLT Pecuk Kabupaten Indramayu*, Indramayu: Dinas Kebersihan dan Pertamanan Kabupaten Indramayu, 2015.

15. Putra, R. R., *Analisis Faktor Penentu Pengembangan Sistem Pengolahan Air Limbah Terpusat Kota Tangerang (Studi Kasus: Kecamatan Tangerang),* Bandung: Institut Teknologi Bandung, 2016.

* Corresponding author: rydhariyana@gmail.com

Analysis of Drinking Water Supply System Improvement Using Fuzzy AHP(Case Study: Subang Local Water Company)

Satrio Budi Prakoso[1] *and Suprihanto* Notodarmojo[2]

[1]Water and Sanitation Infrastructure Management, Civil and Environmental Engineerning Faculty, Institut Teknologi Bandung, 40132, Bandung, Indonesia
[2] Water and Wastewater Engineering Department, Institut Teknologi Bandung, 40132, Bandung, Indonesia

Abstract. Study of infrastructure is required to improve the drinking water supply system, in order to achieve The Long-Term Plan Targets on Water Supply (RPJMN) 2015-2019 concerning 100% universal and equitable access to drinking water services, in accordance with the purpose of Sustainable Development Goals (SDGs) in 2019. Subang district was selected as a case study of water supply improvement because of the improving development in that area and the population growth rate, influencing the need for drinking water. The purpose study is to determine the problems of drinking water supply system, to analyze alternatives for improving water supply system by performing weighted ranking factors in drinking water supply improvement with AHP method and analyzed by fuzzy logic. The method for water supply system improvement could be performed by looking at influential factors using fuzzy AHP approach to obtain some appropriate alternatives. Technical and operational factors was considered the most influential factor in the improvement of water supply system, followed by management capacity, environment, and financial aspect. Fuzzy method with rule base input approach used in this analysis was influencing the output that became the policy alternative to improve drinking water supply.

1. Introduction

There were 386 Local Water Company (PDAM) from 542 regional autonomous in Indonesia, where is only 53% that include in "Healthy" category [2] (BPPSPAM, 2015). drinking water supply in 2015 was not reached the target yet, with 68.87% proper drinking water access. The improvement of water infrastructure and tool is needed to reach 100% of drinking water universal access, as what has been expected by Sustainable Development Goals (SDGs) in 2019. The case study of this research was done in Subang District served by PDAM Tirta Rangga as local water company to provide drinking water.

Subang District continues to grow, because of the total population was more than 1.4 million people continues to increase and the economic development in the industrial, transportation, residential, and tourism aspect according to Local Goverment Spatial Plan (RTRW) 2011-2031. The potential availability of water in this area was more than enough to serve the whole population, which used streams, springs and deep wells as a source of raw water.

In this study, Fuzzy AHP approach was done in order to improve drinking water supply. This method is often used in research to improve drinking water supply as the research that was done by [16] Bojan Sradjevic and Medeiros in the year (2008). In Thailand year [18] (2015), Thungngern has done AHP approach and multi-criteria decision analysis (MCDA) in water resources management. [9] Okeola (2012) did the same approach to evaluate the distribution of urban drinking water in Nigeria.

Availability of sanitation infrastructure, in this case drinking water supply system is a basic human need [15] (Setiawati et al, 2013). A water supply management should basically provide services to the public on an ongoing basis whether quality, quantity, continuity, reliability and tariff [4] (de Souza, 2014). Improvement of water supply system requires policy alternative which is based on the factors and sub-factors that influence significantly to the improvement of water supply.

Population growth rate prediction required to know water needs in the future. The fluctuation of water used make an influance to water needs [19] (Trifunivic, 2012). Water used fulfillment should be obtain by water availability and distribution, analyzed with EPANET [13] (Rossman, 2000). Financial aspect evaluation is important in water supply plan to make this system more effective and efficient.

Fuzzy method has been used by [1] Ayunda (2013) to determine policy priorities in improving the performance of local water company. The approach is based on the weighting value by expertise respondents, where linguistics language can be converted mathematically by fuzzy logic method.

* Corresponding author: satriobprakoso@gmail.com

The main problems of water supply management in Indonesia, specially in Subang District dominated by operation technique, strategy for financial management, environment conservation and preventive aspect from development, and management capacity evaluation for resources potential optimalization.

Fuzzy version of AHP presents an approach which aims to improve real water management situation and can be considered to be flexible and robust. Sradjevic study in 2008 said, it has been recomended as a reliable support tool by decision makers in real situation, characterized by the uncertainly of both the problem and the decision maker's expertise and cognitive abilities.

Seeing the potential advantages of Subang Region, that has a lot of water resources potential, many income funds, environmental aspect not only good topography for water supply, but also the magnitude of the stream flow continously for supporting the water production and distribution, and company assets and human resorces potential. By doing this analysis and adjust it to the spatial planning document that have been made by local goverment, the water company can improve the water supply system and handle water distribution problems.

2. Methodology

This research was conducted in Subang District, West java Indonesia. Figure 1 shows the water service area in Subang District, provide by 16 water production units supplying for 23 subdistrict. This water supply system improvement was analyzed using Fuzzy AHP approach. AHP has an excess to account all important factors as well as to facilitate opinion and conflict [14] (Saaty, 1980), Figure 2 shows the research methodology flowchart.

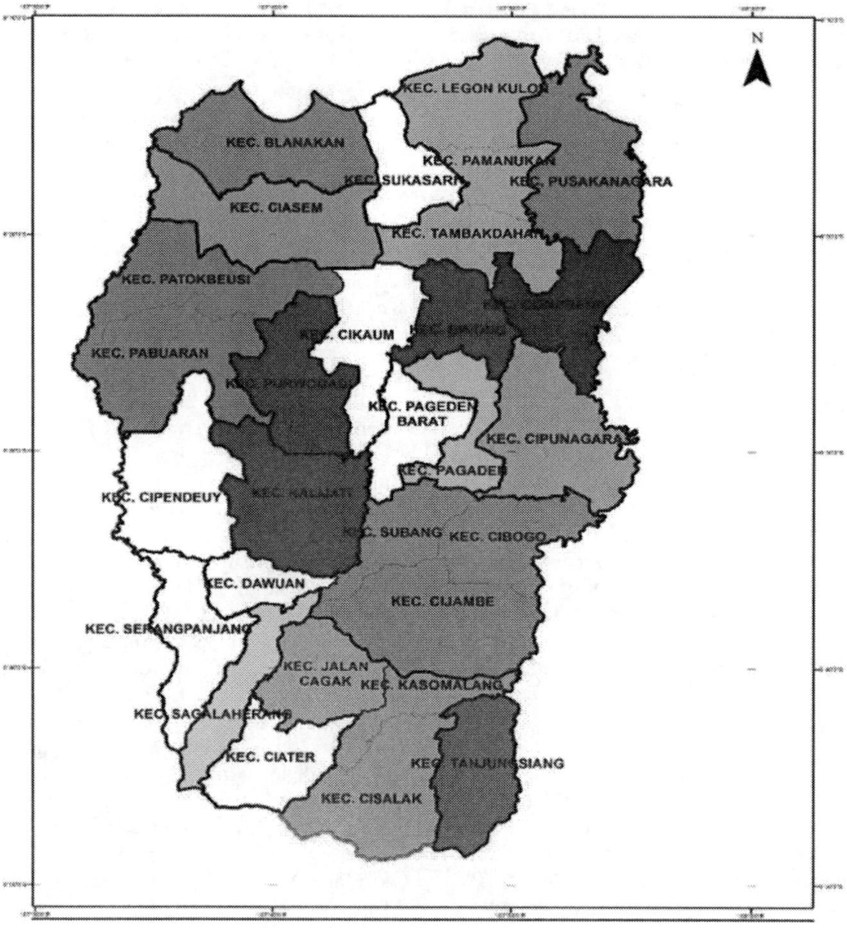

Fig 1. Drinking water supply area in Subang District

* Corresponding author: satriobprakoso@gmail.com

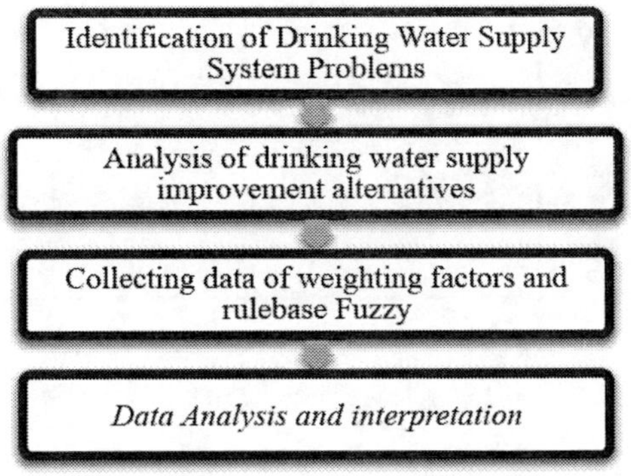

Fig 2. Research methodology

The initial stage of this research was related to the water supply system by Subang Local Water Company, including literature review, formulating the problem, and field surveys. Basic concepts of water supply system, the distribution by local water company, the company's financial viability, improvement analysis and the basic concept of AHP and Fuzzy logic method will be obtained from literature review.

EPANET 2.0 software was used to measure the piping dimensions of the piping of raw water source to service center distribution, futher find out the output of pipe dimensions and the pumps used [12] (Ramdoni, 2013). In terms of the projected water needs, population projections should be calculated. In this study, population growth rate used 4 (four) projection method. According [5] Gawatre et al (2016), arithmetic method be applied to small town, low economy, with high growth.

While the geometric method used for areas with growth rates that rise rapidly [7] (Jain et al, 2013), moreover exponential and logarithmic were used, where the projection used was the method that produces the smallest standard deviation. Projected water demand using the average daily water needs (Qrh), which is the amount of water needed for one day.

Futhermore, financial analysis of the investment costs incurred for the improvement of planed water supply system in Subang, using investment feasibility methods, which are BCR, IRR, NPV and BEP for financial planning aspect of Subang Local Water Company. Cost analysis was performed to determine company ability in generating benefit from customers.

Based on Regulation no. 23 2006 the cost could be identify by calculating the rate of Low Cost (TBR), Level of Basic Cost (TBD) and Full Cost Rate (TBP). Water tariffs is not allowed to exceed more than 4% of people's income customers [6] (Gunawan, 2016).

Drinking water supply system improvement was done by doing Analytic Hierarchy Process (AHP). AHP methodology according to [14] Saaty (2008) have 10 stages, those are: formulating the problem and determining the main objective, determining factors, forming a hierarchical structure, pair wise comparisons, measuring the level of consistency, incorporating the respondent value, weighting factors and analyzing sensitivity. The initial stage was to determine the factors and sub-factors that affect the improvement of water supply system.

Table 1. Factors affecting the improvement of water supply system [3, 4, 8-11, 15-18, 20]

Factor	References
Operational Technique	Thungngern (2015), Taylan (2014) Setiawati (2012) Vilanova (2015), Wibowo (2010), Srdjevic (2008), Okeola (2012)
Financial	Meixner (2003), Taylan (2014), Setiawati (2012) Vilanova (2015), Wibowo (2010), Srdjevic (2008), Okeola (2012)
Environment	Thungngern (2015), Meixner (2003), Setiawati (2012), Qu (2016), Srdjevic (2008), Okeola (2012)
Management Capacity	Taylan (2014), Calabrese (2012), Pangeran (2012), Setiawati (2012), de Souza (2014), Srdjevic (2008), Okeola (2012)

Factors that exist in Table 1 was developed into sub-factor of each factor. Each of these factors was included in the questionnaires for the weighting factors assessment. Meanwhile, the scale of pair wise comparisons used were as follows Table 2.

Table 2. Pair wise comparisons scale

AHP Values	Judgement
1	Equal importance
3	Week dominance
5	Strong dominance
7	Demonstrate dominance
9	Absolute dominance
2,4,6,8	Intermediate values

Respondents, who were considered to know about the problems of water supply system improvement particularly in the Subang district were 32 of the 50 expertise respondents. Respondents consisted of some stakeholders that were respondents from the local water company as facilitator, Regional body for planning and development (BAPPEDA) as the bureaucracy, and academician as consultant of public works, who were considered capable of representing stakeholders who have a different point of view.

It was according to the research conducted by [10] Pangeran et al (2012). In addition, respondents also came from a group of community leaders as stakeholders to complement the linguistic scale presentations [9] (Okeola, 2012).

* Corresponding author: satriobprakoso@gmail.com

The tabling of weighting value matrix factor 1 to factor 2, and will be divided by 1 reverse valuation of weighting factor 2 to 1, and so on. The entire weighting value of the series summed by table column, such as equation (1) below.

$$A = \begin{bmatrix} x_{11}y_1 & x_{12}y_2 & \cdots & x_{12}y_2 \\ \cdots & \cdots & \cdots & \cdots \\ x_{N1}y_1 & x_{N2}y_2 & \cdots & x_{NK}y_K \end{bmatrix} \quad (1)$$

X = matrix weighting factor assessment results
A = comparison between factor ratings

Calculation of the value of the priority vectors was done to produce the policy priorities of each factors affecting the improvement in water supply sytem. Multiplication of each value factor in the series was done and then rooted by 4 (four). Vector results between factor divided by the sum of vector, the formula in equation (2).

$$\sqrt[4]{\begin{bmatrix} x_{11} & x_{12} & \cdots & x_{1K} \\ \cdots & \cdots & \cdots & \cdots \\ x_{N1} & x_{N2} & \cdots & x_{NK} \end{bmatrix}} \quad (2)$$

V = vector between factors of results $\sqrt[4]{}$ (root four) weighting the value of each factor

Network modeling was described by the hierarchy of the factors and sub-factors to determine the weight and the order of factors priority. Weighting results were processed using Expert Choice 11 application, by considering the level of inconsistency. The next step was developing a model analysis to improve drinking water supply with a fuzzy logical approach. The modeling was done with Matlab Fuzzy Toolbox 7.11.

From rule base weight of each factors and sub-factors, as well as the results obtained from AHP, it will obtain the weight of drinking water supply improvement on each fuzzy rule base, according to the following equation (3).

$$Z = (W1Z1 + W2Z2 + \ldots. WnZn) / (W1 + W2 + \ Wn) \quad (3)$$

Z = improvement of drinking water supply system weight
Wn = the weight of each factor or sub-factor of the AHP results
Zn = the weight of the risk level of a factor or sub-factor in the rulebase

Drinking water supply system factors were converted into a fuzzy set. Membership function used was trapezoidal function Figure 3.

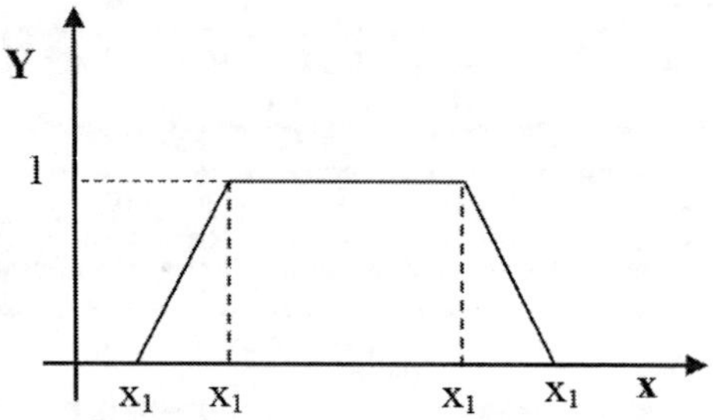

Fig 3. Fuzzy set on rulebase trapezoid

Membership function:

$$\mu(x) = \begin{cases} 0; & x \le a \text{ or } x \ge d \\ (x-a)/(b-a); & a \le x \le b \\ 1; & b \le x \le c \\ (d-x)/(d-c); & x \ge d \end{cases}$$

Modeling validation was done by looking at the alternative between sub-factors that presented the results of fuzzy modeling. By getting weight validation result between the sub-factors vectors, modeling simulation analysis can be done.. As in the equation (4) and (5) [16] (Sradjevic, 2008), data validation in Figure 4.

$$A = \begin{bmatrix} x_{11} & x_{12} & \cdots & x_{1N} \\ \cdots & \cdots & \cdots & \cdots \\ x_{N1} & x_{N2} & \cdots & x_{NN} \end{bmatrix}, k = 1, \ldots, K$$

$$xij = \sum_{k=1}^{K} \tilde{a}ik \otimes [\sum_{l=1}^{N} \sum_{m-1}^{K} \tilde{a}lm]^{-1}, i = 1, \ldots, N\,; j = 1, \ldots,$$

N =the pairwise comparison of alternative values
K =rate fuzzy matrix of output decision
X =represents the results of the assessment of fuzzy
A =matrix of alternatives membership sub-factor

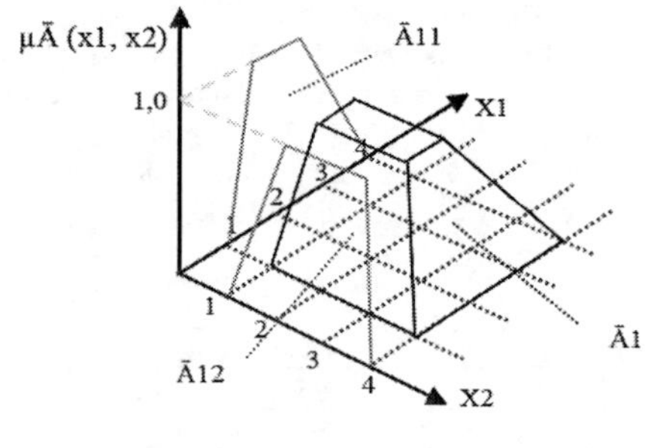

Fig 4. 3D Graphic trapezoidal validation of alternative evaluation

* Corresponding author: satriobprakoso@gmail.com

The next step was to conduct simulation and model application for drinking water supply system improvement based on the available data on the PDAM Subang existing condition, from the fuzzy inference statements (if-then). The results of the application of sub-factors alternative is represented in Equation 6.

$$Z = \begin{bmatrix} x_{11} \otimes y_1 & x_{12} \otimes y_2 & \cdots & x_{1K} \otimes y_K \\ \cdots & \cdots & \cdots & \cdots \\ x_{N1} \otimes y_1 & x_{N2} \otimes y_2 & \cdots & x_{NK} \otimes y_K \end{bmatrix}$$

y = the weight value of the sub-factors of the results of vectorization sub-factor

Fuzzy methods has been used for a variety of specific applications in decision making problems, including. The method used in this study is a group fuzzy with detailed explications of two proposed improvements, namely, a consistency analysis of experts opinions and using a difference technique regarding to handle uncertainties of applying a group decision making to substantiate the advantages and differences, by simulating result intrepetation. AHP fuzzy process, each stage seen in Figure 5.

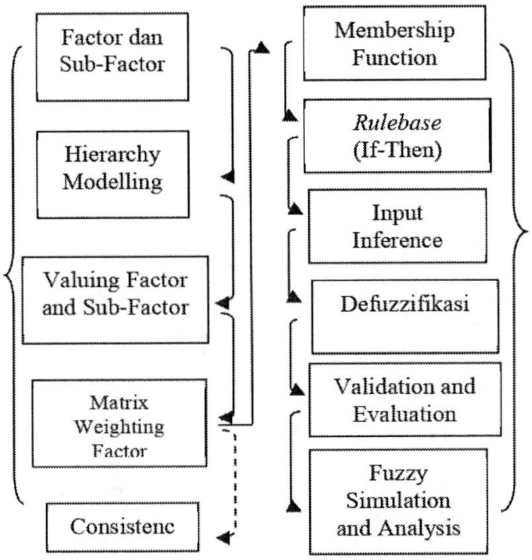

Fig 5. Stages fuzzy AHP increase in SPAM

3. Results and discussion

3.1 Identification of PDAM Subang Problems

Subang District is an area in the northern districts of West Java Province. Based on the data from Subang Department of Population and Civil Registration, Subang District total population in 2015 was 1,576,042, with 1% per year population growth rate from 2010 to 2015. Coverage of drinking water supply system by Subang local water company in 2015 was 16.97%, or for regional coverage was 52.48%. Engineering was needed for water loss that continues to increase.

3.2 Technical Analysis Distribution System Improvement Study Areas

Analysis of water distribution scheme was done based on the source of raw water, the flow of rivers, springs and deep wells. 11 districts out of 23 districts served by the unit of production sourced from streams. Most of the production units using raw water from deep wells (deep well). Springs have great potential in terms of quantity, continuity, and quality for multiple production units. Three unit from 16 units was selected for analysis of the production services and the hydraulic pressure using EPANET 2.0 software.

* Corresponding author: satriobprakoso@gmail.com

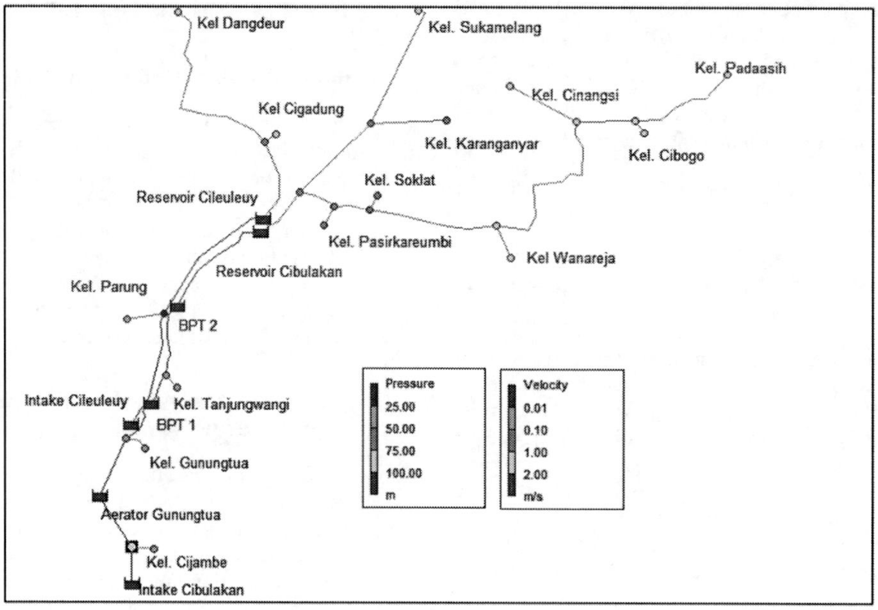

Fig 6. The Epanet simulation results of Subang Unit

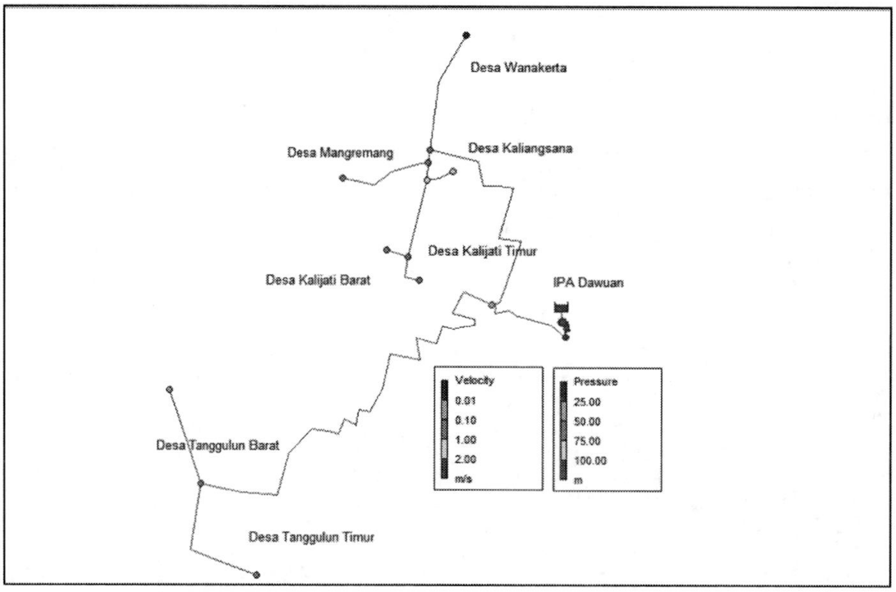

Fig 7. The Epanet simulation results of Kalijati Unit

* Corresponding author: satriobprakoso@gmail.com

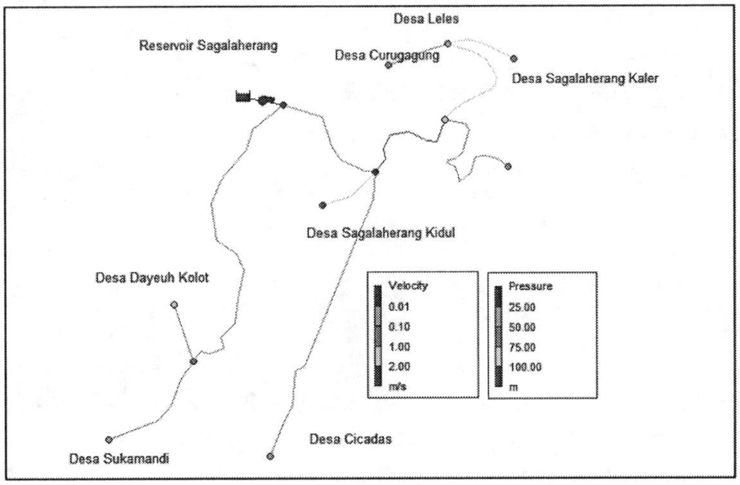

Fig 8. The Epanet simulation results of Sagalaherang Unit

In the simulation model of Epanet on the number of customers adjusted to the target of 100% service by piping systems. In Figure 6, a need to increase the production capacity of existing condition, based on the analysis of water availability was not sufficient for drinking water supply. Based on Epanet simulation results, the pressure throughout its service area met the standard in Ministry of Public Works Regulation no. 18, 2007 (more than 10 m). In Figure 7, the flow velocity in the pipe did not meet the standard of 0.3 m/s, the piping was required to be engineered. The addition of pump power was done in Figure 8, in order to achieve the specified pressure.

3.3 Calculation of Drinking Water Consumption Improvement Planning

Based on the analysis of drinking water supply planning projections for 20-year, it was discovered that at the end of the planning, drinking water needs to increase to 1,287 l/s. Domestic demand is predicted to increase from 40.420 house connection at the end of 2015 to 195.621 house connection, and non-domestic from 1214 connection to 4872 connection in the late years of planning, it is necessary to increase production capacity gradually from the existing condition in the range of 450 l/s to reach the service 100%.

3.4 Financial Aspects of Water Supply System

Anchored Ministry of Internal Affairs no. 23 in 2006, the ability to pay the basic needs of drinking water does not exceed 4% of public revenue.

Table 3. Determination of water tariff

	Tariff Alternative I	Tariff Alternative II	Tariff Alternative III
NPV	Rp 18,681,880,130	Rp 152,272,709,745	Rp 168,298,479,549
BCR	1.01	1.08	1.09
BEP	17.39	14.13	12.13
HPP	Rp 4.,117	Rp 4.117	Rp 4,117

Judging from the financial feasibility in Table 4, the alternative tariff III (three) fare better than others, with NPV and BCR's largest and fastest-year return of capital.

3.5 Selection Analysis and Weighting Method with AHP

Questionnaire was a process of verification after identifying each factor, the number of respondents was 27 expertise respondents. The entire sub-factors were influencing water supply system improvement in Subang district, with the value for each factor was above 65%.

The final results of the paired comparison questionnaire were inputted into the AHP model with Expert Choice 11 Program, which begins by describing paired comparisons matrix, as seen in Figure 9.

* Corresponding author: satriobprakoso@gmail.com

 https://doi.org/10.1051/mateconf/201814704002

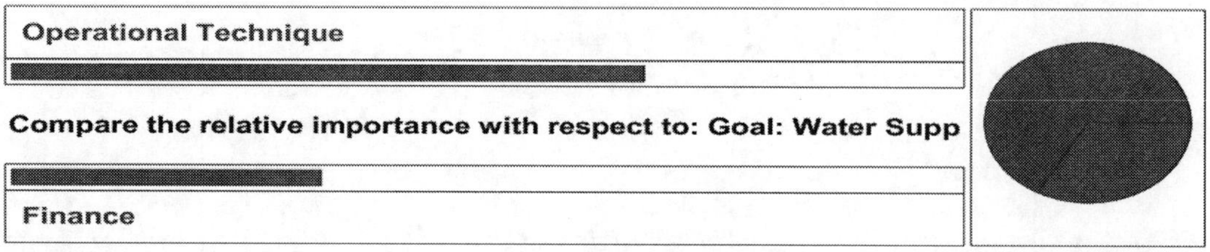

	Operational	Finance	Environmei	Managemei
Operational Technique		2,03	2,79	1,34
Finance			(1,07)	(1,33)
Environment				1,48
Management Capacity	Incon: 0,04			

Fig 9. Paired comparisons matrix between factors

Next was analyzing the weighting value between the factors and sub-factors to make a ranking by 33 respondents experts, this was the output of the matrix calculations.

Fig 10. The value of weighting among factors

The most influential factor to improve the water supply system in Figure 10 was operational technique with the weight value was 0.39, then capacity of management environment, and financial. This factor inconsistencies rate was 0.04, where the rate was below 0.1 so that results and expertise can be justified [12] (Romdoni, 2013). The management of drinking water supply systems was not considered a good operational technique, and require priority improvement.

3.6 Analysis of Water Supply Improvement Model with Fuzzy Method

The identification results of the questionnaire, resulting in 22 sub-factors that need to be examined in this modeling. Sub factors were included into four main factors, those were operational technique, financial, environment, and management capacity, which from these four factors form several sub-models of water supply system improvements. Rule base design of water supply improvement analysis model was the most important stages before performing fuzzy rule base. The concept of the model structure of drinking water supply system improvement can be seen in Figure 11.

* Corresponding author: satriobprakoso@gmail.com

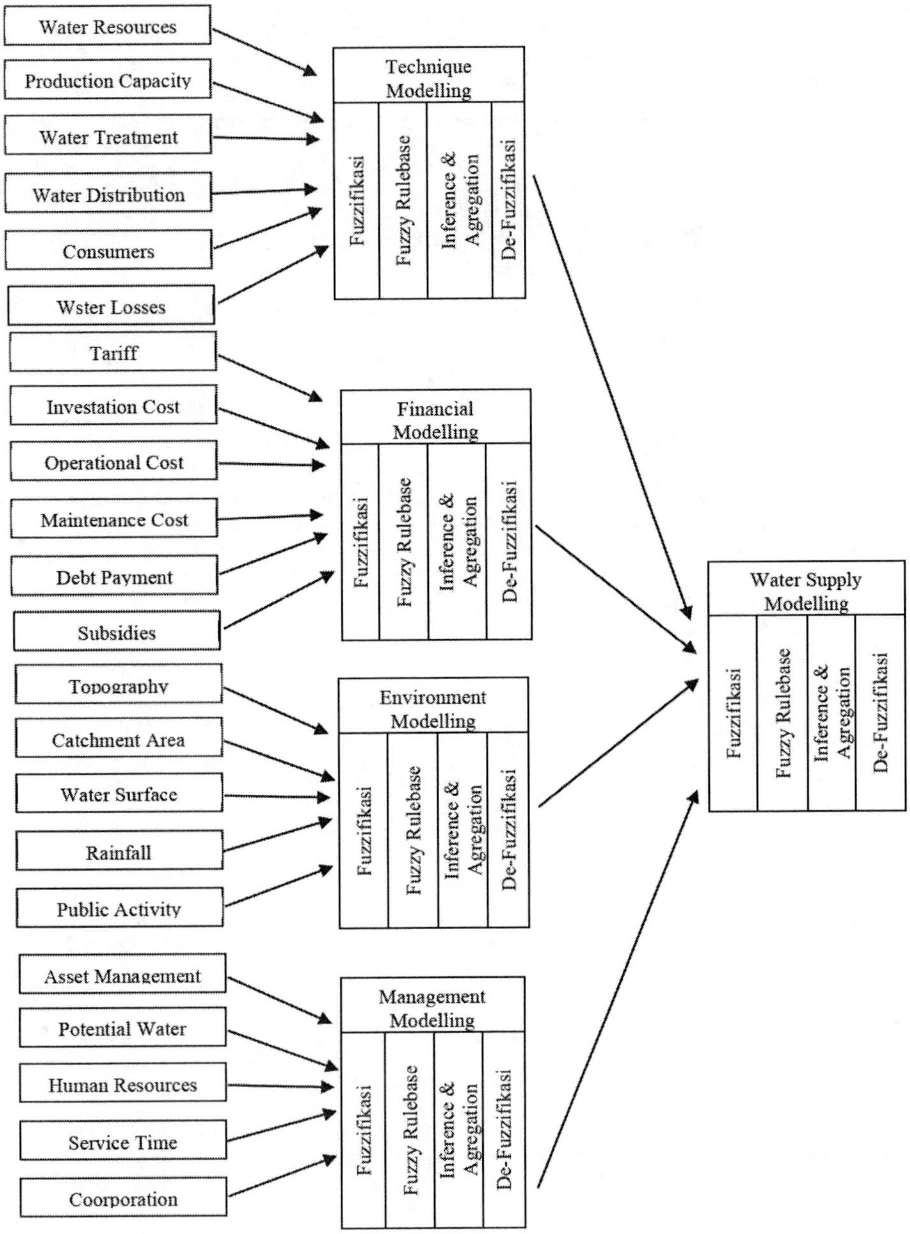

Fig 11. Structure of drinking water supply improvement fuzzy rulebase model

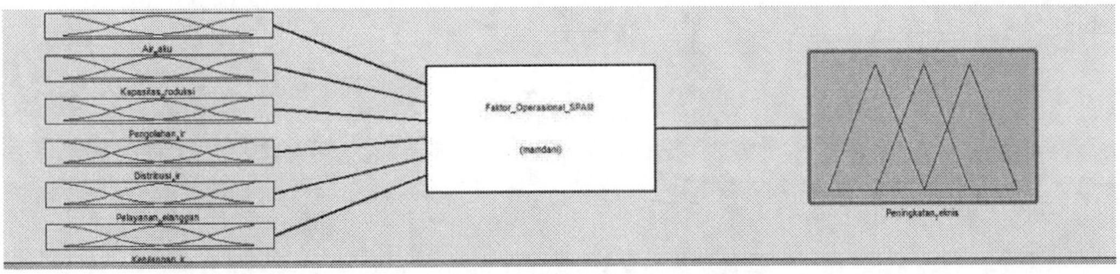

Fig 12. Operational technique fuzzy model

* Corresponding author: satriobprakoso@gmail.com

First set of fuzzy models as Figure 12. The next step was validating the model by incorporating fuzzy rulebase into sub models , to review the three-dimensional graphic, as shown in Figure 13, for example validation of modeling is done.

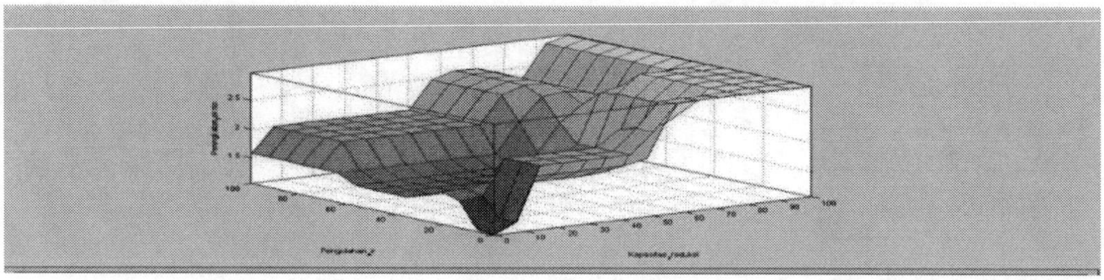

Fig 13. 3D graphic between water production and processing

It can be seen from Figure 13, that the water production capacity will continue to improve the operational technique of drinking water supply. Improvement of operational technique was initially influenced by water treatment process, but that pattern changed to be declivous after entering 30% fulfillment of water quality standards. The improvement in production capacity trend was also always followed by the improvement of water treatment, although it not sharp.

Drinking water supply system improvement model has been designed, simulated with existing data and improvements planning in Subang District Local Water Company. Simulation of the model was done by using the current scenario/existing and future scenarios that had an improved water supply system.

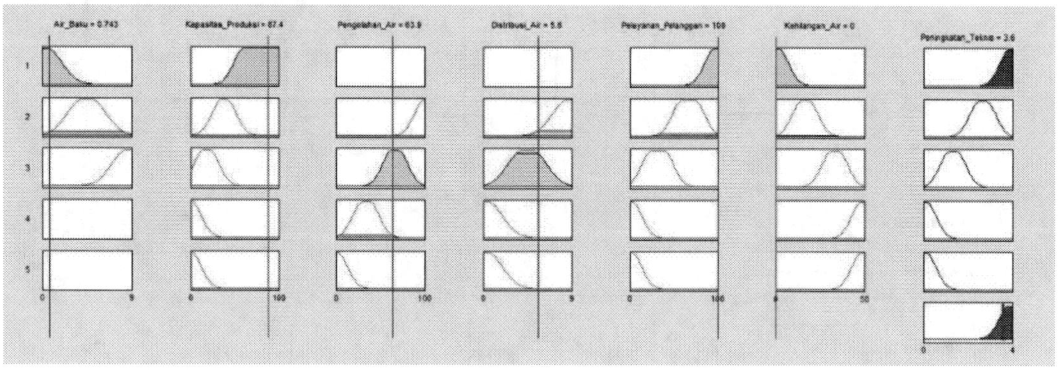

Fig 14. Operational technique improvement model simulation

In **Figure 14,** improvement of operational techniques will be able to approach the optimal number at 3.6 of 0 to 4, if drinking water supply system utilize the capacity of the water from streams, capable of producing water with a large capacity, capable of achieving 80% quality standards of raw water treatment efficiency, use gravity in the distribution system, and unavailable water losses.

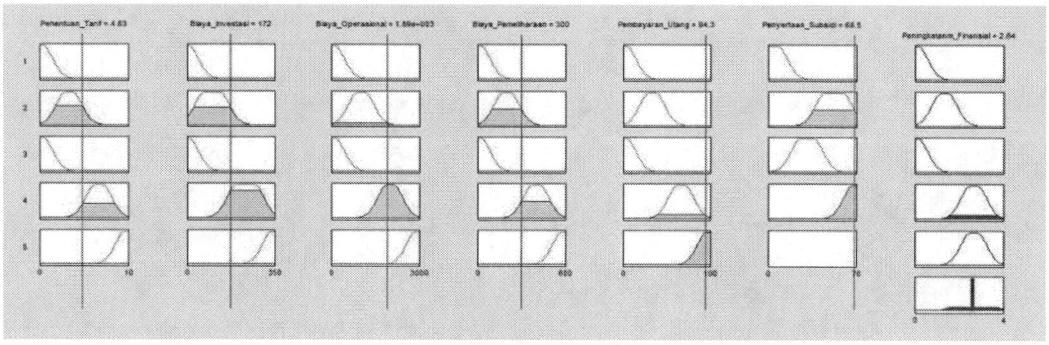

Fig 15. Financial improvement model simulation

* Corresponding author: satriobprakoso@gmail.com

The results of the simulations showed that using low tariff will be able to influence the improvement of drinking water supply system, which can increase the number of customers of local water company, as shown in Figure 15. This tariff was not maximized, futher it only able to reach 2.64. At the maximum point of the investment, operation cost, maintenance cost, if it supported the ability to pay off debt and aided by 70% government subsidies.

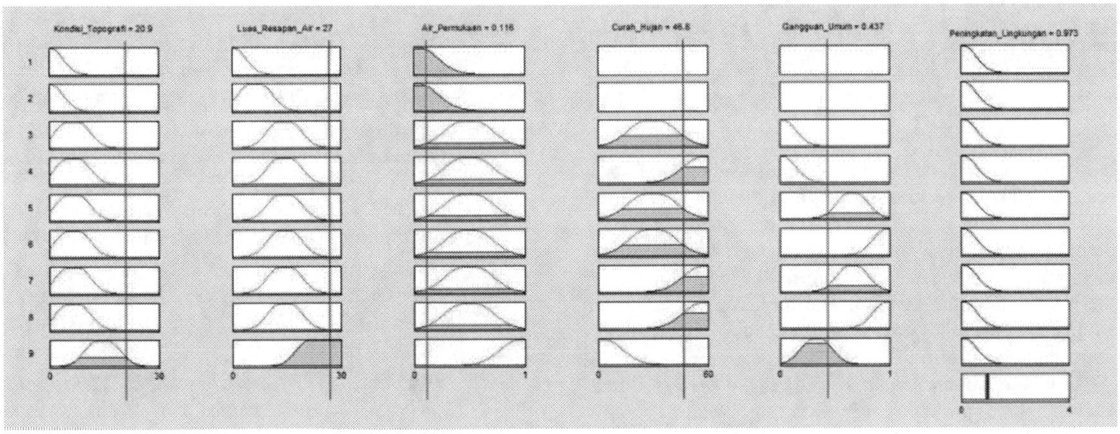

Fig 16. Environmental improvement model simulation

In the analysis of fuzzy, the improvement of environmental factors cannot be the main policies alternative to improve the water supply system. Although it has been serving the high topography area, which facilitate gravity flow system, in an area that has a high water catchment, maximizing the potential of the river flow, in the high rainfall area, and avoid disruption to the traffic and the industrial, environmental improvements didn't have the high-value it was just 0.973 as in Figure 16.

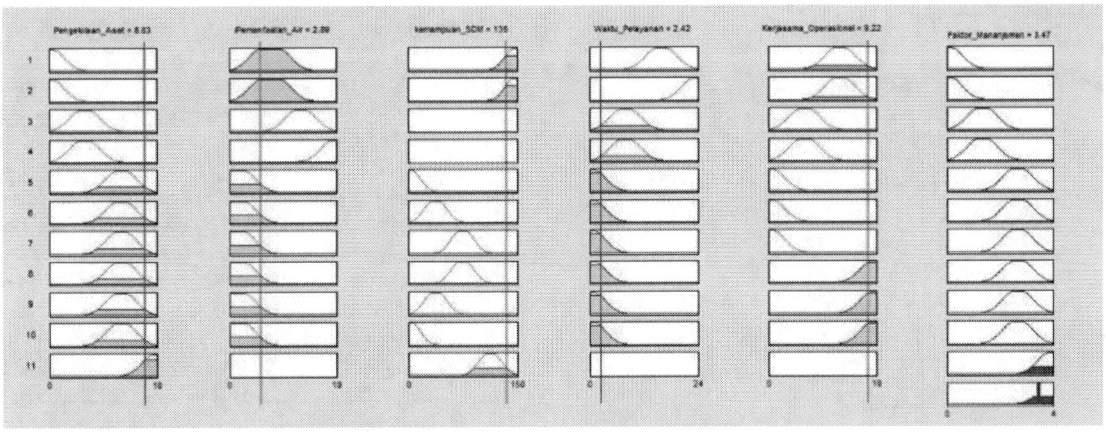

Fig 17. Management capacity improvement model simulation

Fuzzy modeling analysis in Figure 17 shows that asset management that able to provide maximum income, river water utilization, upgrading, the service time over 20 hours, and cooperation with funding supporting partners, will be able to improve water supply systems management. The Fuzzy AHP method output in this study was for water supply improvemnt, which is able to be the base of decision making in infrastructure development.

AHP method approach showed operational technique was become first priority in policy, as in the master plan of water supply (RISPAM) Subang District, that weight focusing about operational problems. Water supply improvement plan with water resource potential optimalization, water production capacity improvement, water losses reduction, and provide a wider water access, it can help achieving SDGs target in operational technique, determined by fuzzy model.

Level of financial improvement would be very high if able to investment increasing, by tarif allocation about a half and 10% for maintenance cost, and overcame the debt problem. Water distribution with gravitation used in environmental support factor, whereas management capacity would be extreamly lift by doing exercise, study, and evaluation to human resorces, for supporting not optimalization assets. Environment aspect naturally

* Corresponding author: satriobprakoso@gmail.com

as one of company management responsibility, except water supply technique operational system of local water company.

Fuzzy AHP method be able to accomplish water supply system in addition to conventional method, which are distribution pipeline network financial analysis. Improvement of water supply made by policy based on this decision making tools.

4. Conclusion

Until 2015, PDAM Tirta Rangga Subang District reached 52.48% service coverage, although it has high potential sources of raw water. The improvement can be done by the technical and financial analysis. Operational technique factors with the value of 0.3 was considered the most influential factor in the improvement of water supply system, followed by environmental, financial and management capacity.

Modeling simulation analysis of fuzzy AHP is able to answer the problem of drinking water supply development, by utilizing the capacity of raw water from streams, maximizing the production of drinking water, and water losses minimalization. Financial aspect can be very high impact, if the cost of investment, operation, and maintenance would be optimized. Fuzzy method with rule base inputs will affect the output of the drinking water supply improvement alternative, by reviewing the sub-model of factors, based exsisting condition in the case study.

References

1. Ayunda,N., P.P.K.U.P.K.P.D.A.M.M.Fu.A.H.P. *Institut Tekhnologi Surabaya. Jurnal Mahasiswa Matematika* Vol. **1**, 1. (2013)
2. BPPSPAM. Kinerja PDAM 2015. Kementrian Pekerjaan Umum dan Perumahan Rakyat. Jakarta. (2015)
3. Calabrese, A., Costa, R., Menichini, T., U.Fu.AHP.M.I.C.A. *Elseiver Academic Press. Expert Systems with Applications Vol* **40**, 3747–3755. (2013)
4. De Souza, E.V., dan Da Silva, M.A. Costa., M.S.E.U.W.S.W.Sup. *Procedia Engineering* Vol **70**, 458 – 466. (2014.)
5. Gawatre, D.W., Kandgule, M.H., Kharat, S.D., C.S.P.F.M. *IOSR Journal of Mechanical and Civil Engineering (IOSR-JMCE):*, **13**, 16-19. (2016)
6. Gunawan, I.A., S.P.P.K.P.P.U.D.A.B.T.O. *Institut Teknologi Bandung.* (2016)
7. Jain, S., Mishra N., Bokade, S., Chaturvedi, S., A.R.P.F.D.F.S.D.M.M. *(IJSR) ISSN*: **6**.14 (2013)
8. Meixner,O., Fu.AHP.G.D.A.ITS.A.E.E.S. *University of Natural Resources and Life Sciences. Vienna, Austria,* 1-14. (2003)
9. Okeola, O.G., Sule, B.F., E.Man.A.U.W. Sup.S. M.C.D.M. *University of Ilorin. Nigeria Journal of King Saud University – Engineering Sciences* Vol **24**, 19–24. (2012)
10. Pangeran, M.H., Pribadi, K.S., Wirahadikusumah, R.D., Notodarmojo, S., A.R.M.Cap.Pu.S.O.PPP.S.D.W.Sup.Ind. *Civil Engineering Dimension,* Vol. **14**, 26-35. (2012)
11. Qu,J.,Meng,X.,You,H.,M.S.Ra.E.Tec.A.W.S.P. A.Fu.Gr.D.M.T. *Elseiver Academic Press Journal of Hazardous Materials* Vol **310**, 68–81. (2016)
12. Romdoni, Y., *Analisis Risiko Kegagalan Jaringan Pipa Distribusi Air Minum dengan Pendekatan AHP-FUZZY.* Intitut Teknologi Bandung. (2013)
13. Rossman, L.A., *Epanet 2 Users Manual. Environmental Protection Agency. United States.*(2000)
14. Saaty, T.L., *Pengambilan Keputusan bagi Para Pemimpin.* Jakarta: P.T. Pustaka Binaman Pressindo. (1980)
15. Setiawati, E., Notodarmojo, S., Soewondo, P., Effendi,A.,J.,Otok,B.,W.,I.De.S.S.W.S.U.SEM. M. *Jakarta. Elseiver Academic Press. Procedia Environmental Sciences* Vol **17**, 685 – 692. (2013)
16. Sradjevic, B. Medeiros Y., D., P., Fu.AHP.Ass. Wa.M.P.*University of Novi Sad. Serbia. Water Resources Management* Vol **22**, 877-894 (2008)
17. Taylan, O., Bafail, A.O., Abdulaal, R.M.S, Kabli,M.R., C.P.R.A.Fu.AHP.F.TOPSIS Met. *Elseiver Academic Press. Applied Soft Computing* Vol **17**, 105-116. (2014)
18. Thungngern, J., Wijitkosum, S., Sriburi, T., SukhsriC.,A.R.T.An.H.P.:A.App.Wa.R.M.Tha. *Chulalongkorn University. Bangkok. App. Envi. Res.* Vol **37**, 13-32. (2010)
19. Trifunovic,N.,P.R.R.A.W.D.N.,*PhDmanuscript, Taylor & Francis Group. London.UK* (2012)
20. Vilanova, M.R.N, Filho, P.M., Balestieri, P.M., P.M.I.Wa.Su.M. *Elseiver Academic Press. Renewable and Sustainable Energy Reviews* Vol **43**, 1–12. (2015)

* Corresponding author: satriobprakoso@gmail.com

Screening The Effect of Cu, Mn, and Mg on Ethanol Formation in Degradation Process of Palm Oil Mill Effluent (POME) under Anaerobic Condition Using Two-Level Factorial Design Method

Tia Widya Puteri[1*] and *Mindriany* Syafila[2]

[1]Department of Environmental Engineering, Institut Teknologi Bandung, 40132 Bandung, Indonesia
[2]Department of Environmental Engineering, Institut Teknologi Bandung, 40132 Bandung, Indonesia

Abstract. Anaerobic digestion can be used for the treatment of POME to reduce organic content and generate some substances, such as volatile acids, ethanol, and various gasses. During anaerobic fermentation process, microorganism requires the presence of trace elements for growth to improve their performance. This research will be carried out as a study of the presence of trace elements, such as metals (Cu^{2+}, Mn^{2+}, and Mg^{2+}) that has significant effects on ethanol formation. Circulating Bed Reactor was used and operated in a batch system for 48 hours. Metal ions were screened and analyzed by using two-level factorial design method whether there is any correlation effect between the addition of Cu^{2+}, Mn^{2+}, and Mg^{2+} and ethanol formation. Several parameters which consists of Total Volatile Acids (TVA), dissolved Chemical Oxygen Demand (CODs), Volatile Suspended Solid (VSS), pH, Dissolved Oxygen (DO), and ethanol were measured every sampling. Mn metal is proven statistically affect both TVA and ethanol formation while Mg metal only affect TVA formation. Cu^{2+} and Mg^{2+} metals combination affect ethanol formation with largest detected ethanol concentration is 7,483.07 mg/L. The result from this study had identified the metal ions which has significant effect as a foundation for optimization ethanol formation.

1 Introduction

The energy used today almost entirely fossil-derived oil was known as non-renewable energy. As the production of energy from these fuels is associated to resource depletion, the identification of new and renewable sources of energy has become a matter of priority in many countries. One of those alternative energy is a biofuel derived from biomass conversion. Alternative energy can be obtained by using various types of substrates, such as raw materials containing saccharides (sugar cane, molasses), starch (corn, wheat, potatoes, tubers), dan biomass cellulose (wood, agricultural waste, agri-industrial waste) [17] that will process by anaerobic fermentation. The use of substrate selection based on availability of resources, economic efficiency considerations, as food materials availability, and other factors.

Palm oil is one of the biggest agricultural products produced in Indonesia. Oil palm is the only fruit that can give two types of oil, extracted from the fruit of the oil palm (crude palm oil, CPO) which are chemically different. Palm oil comes from the mesocarp (flesh of the fruit) and the lauric palm kernel oil from the kernel at the fruit's core (crude palm kernel oil, CPKO). The two oils are separated in the oil palm bunches by the thick shell of palm kernels. The world demand for palm oil has increased in the past few decades [25]. It causes the oil palm industry to increase production capacity to meet the demand. Along with the increased in the palm oil production, it will also generate in contaminant the production of waste. There were some types of wastes produced in this process, which are solid, air, and liquid waste. The liquid waste was known as palm oil mill effluent (POME).

POME is one of fermentation media that can be used as a substrate to produce some products, such as ethanol, total volatile acids. POME containing carbohydrate, protein, nitrogen compounds, oil and grease, highly mineral content [15], and other organic compounds, such as cellulose, hemicellulose, and starch. POME also containing 700-2.300 mg/L of starch, 1.500-3.000 mg/L of total sugar, 28.100-30.300 mg/L of carbohydrates, 4.200-6.000 mg/L of protein, 6.000-7.000 mg/L of oil and grease [23]. COD concentration in POME is 50.000 mg/L (Ma, 2000). According to Perez et al. [19], high level of organic contents in POME can be reduced by anaerobic treatment. The processing of wastewater with high organic concentration can be treated with the suspended growth in anaerobic reactors [24].

Anaerob fermentation process will produced some substances, such as ethanol and methane gas that can be alternative energy sources. Ethanol can be produced through the pathway in fermentation process involves at least the following stages : hydrolysis, fermentation, distillation, and dehydration. Ethanol formation under anaerobic condition was caused by the presence of

* Corresponding author: tiawidyap@students.itb.ac.id

 https://doi.org/10.1051/matecconf/201814704003

decarboxylase pyruvate and dehydrogenase alcohol enzyme in acidogenesis phase.

Anaerobic fermentation process involves cell metabolism influenced by micronutrients, such as cofactor metal ions for enzyme catalyst eventhough the amount of it is low concentration. As a biological process, fermentation is highly dependent on several factors, including pH, oxygen and temperature, which can greatly influence the specific rate of growth, and presence of enzyme modulator [9]. Metal ions can affect the rate of glycolysis phase and the conversion phase of pyruvic acid into ethanol and other organic acids [8]. Ethanol formation in the reactor was influenced by the presence of metal cofactors as trace elements especially on enzyme forming. This research will be carried out as a study of the presence of trace element, such as metals (Cu^{2+}, Mn^{2+}, and Mg^{2+}) that had the most influence on ethanol formation from palm oil mill effluent using two-way level factorial design.

2 Materials and Methods

2.1. Compositions of Wastewater

Substrate which used in this study was artificial POME, but the characterization of POME was also conducted. Characteristics of POME that analyzed was derived from PT Condong Garut, Garut Selatan, Kabupaten Garut, Jawa Barat. The characteristics of POME can be seen in **Table 1**.

Table 1. Characteristics of POME [14].

Parameters	Unit	Methods	Value
Dissolved COD	mg/L	SMEWW 5220-B	17,300
Total COD	mg/L	SMEWW 5220-B	24,500
BOD	mg/L	SMEWW 5210-B	12,450
Oil and grease	mg/L	SMEWW 5520-D	185
pH	-	SMEWW 4500-H$^+$-B	5.85
Total nitrogen	mg/L	SMEWW 4500-Norg-B	250
TSS	mg/L	SMEWW 2540-D	15,000

Artificial wastewater was used as a substitute for palm oil mill effluent. Artificial wastewater was made with such characteristic of the original wastewater. This wastewater can be used as a substrate or media in anaerobic process. The artificial wasterwater compositions can be seen in **Table 2**.

Table 2. Compositions of artificial wastewaster [1].

Components	Concentration (mg/L)	Components	Concentration (mg/L)
Pati	2,000	$CdCl_2$ $6H_2O$	0.00003
Gelatin	3,000	$NaHCO_3$	0.1571
Minyak-lemak	6,000	$MnCl_2$ $6H_2O$	0.0140
NH_4HCO_3	39.5	$ZnCl_2$ $6H_2O$	0.0056
K_2HPO_4	0.6743	$CuCl_2$ $6H_2O$	0.0047
$FeCl_3$ $6H_2O$	0.5661	$CoCl_2$ $6H_2O$	0.0020
$MgCl_2$ $6H_2O$	2	$CrCl_2$ $6H_2O$	0.00078
$CaCl_2$ $6H_2O$	17.21		

2.2 Seeding and Acclimatization

Seeding aims for duplicates the microorganism to be used in the treatment process. Biomass was used derived from biomass which had acclimatization phase in the previous study was conducted by Andrio [1]. Seeds in that process of seeding drawn from mixed cultures of microorganisms contained in cow's rumen and sludge from WWTP of palm oil industry at ratio 1:1. Seeding process is done by batch system in a volume reactor of 20 L with given glucose of 10 grams, nutrient, trace elements which consists of NH_4HCO_3, K_2HPO_4, $FeCl_3$, $MgCl_2$, $CaCl_2$, $NaHCO_3$, $CdCl_2$, $MnCl_2$, $ZnCl_2$, $CuCl_2$, $CoCl_2$, $CrCl_2$., N_2 *flushing* to remove oxygen.

Acclimatization is a step to adapts microorganism ability to treat POME into fermentation products. The substrate which used was artificial POME with COD concentration of 6,900 mg/L. Acclimatization was conducted in three phases in a batch. Ratio of glucose solution with artificial POME each phase was 30:70 for the first phase and 70:30 for the second phase. Acclimatization process is adding 2 L of biomass from seeding process and 8 L of a mixture glucose solution and aritificial wastewater.

2.3 Running Reactors

Circulating Bed Reactor (CBR) is made from flexyglass with diameter of 13.5 cm, a high of 43 cm and capacity

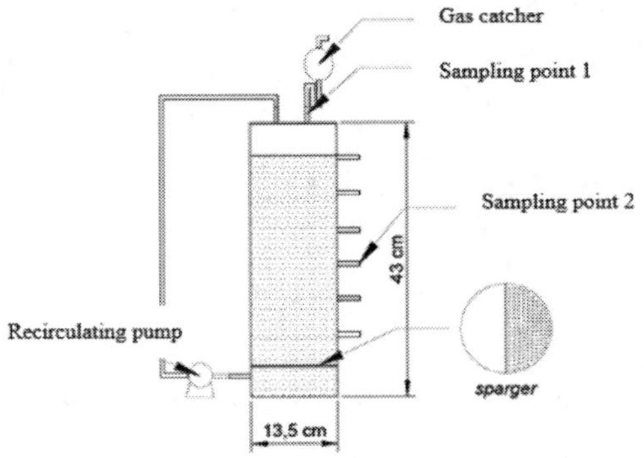

Fig. 1. Circulating Bed Reactor

of ± 6 L. Reactor equipped with a flexyglass disc (circular with a diameter equal to the diameter of the reactor) that given the holes in half the area of the circle area, aerator, sparger and silicon hoses. The pictures of the reactor CBR can be seen in **Fig. 1.**

This research used two Circulating Bed Reactors that has a volume operation of 3,5 L. Ratio of biomass and artificial wastewater at 1:4 (0,7 liters to 2,8 liters of biomass and artificial wastewater). A mixture of biomass and wastewater was added phosphate buffer solution to maintain neutral condition of pH. Each the time of operations, two reactors were operated at one variation of experimental design according to **Table 5**. The addition of metals in soluble form was done when mixed biomass with artificial wasterwater before it was operated in a batch reactor for 48 hours. Cu, Mn, and Mg metal ions were used in the study as $CuCl_2.H_2O$, $MnSO_4.H_2O$, and $MgCl_2.6H_2O$ with the concentration of those metals has been determined according to other researches. Concentration of metals can be seen in **Table 3.**

Table 3. Concentration of metals [21], [20].

Code	Metal	Concentration (mg/L)
A	Cu	0.1
B	Mn	3
C	Mg	0.05

The reactor was operated in a batch for 48 hours at laboratory conditions with a temperature of 25-27°C. Sampling for both reactors was conducted at the same time per 12 hours routinely. At the beginning of process, N_2 flushing was given for 15 minutes about 750 mL/min. The reactors will be connected to the aerator double port in order to air re-circulate inside the reactor with a flowrate about 1 L/min. This process is expected the substrate mixed completely.

2.4 Analytical methods

The measurement parameters in this study consists of COD, Volatile Suspended Solid (VSS), ethanol, pH, and Total Volatile Acids (TVA). **Table 4** contains the measurement parameters with their standard analysis method.

Table 4. Analytical methods.

No	Parameters	Analytical methods
1.	pH	pH meter (SM 4500 H^+)
2.	COD	Dichromate COD (SM 5220C)
3.	VSS	Gravimeter (SM 2540 E)
4.	TVA	Distillation (SM 5560 C)
5.	Etanol	Refraktometer

2.5 Experimental design

Two-way level factorial design was used to estimate main effects in the case of two level factorial experiments using very few runs. In addition, this factorial design can also estimate the interaction of several combined factors. For each variable, the presence and absence levels of variables were tested (all trials were performed in duplicate), resulting in 8 of independent experiments (**Table 5**). This study was used three factors which consists of Cu^{2+}, Mn^{2+}, and Mg^{2+}. These three factors had two levels, had siginificant effect (1) and had no significant effect (0). Concentration (mg/L) was used for Cu^{2+}, Mn^{2+}, and Mg^{2+} of 0.1; 3; and 0.05 respectively. The metal ions were screened to determine their effects on ethanol formation.

Table 5. Experimental design factorial 2^n.

No	Variation		A	B	C
1	1	1	0	0	0
2	A	Cu	1	0	0
3	B	Mn	0	1	0

No	Variation		A	B	C
4	AB	CuMn	1	1	0
5	C	Mg	0	0	1
6	AC	CuMg	1	0	1
7	BC	MnMg	0	1	1
8	ABC	CuMnMg	1	1	1

3 Result and Discussion

3.1 Characterization of artificial wastewater

Artificial wastewater was made from a mixture of starch as carbohydrates content, peptone as a source of protein, cooking oils as the content of oil and grease, and trace elements (**Table 2**). All of those substances is mixed in distilled water. Artificial wastewater was processed using Circulating Bed Reactor (CBR) with the addition of trace elements under anaerobic conditions. The result of the characteristics of artificial wastewater can be seen in **Table 6**.

Table 6. Characteristic of artificial wastewater.

Parameters	Unit	Value
COD	mg/L	6,818
Dissolved COD	mg/L	8,515
BOD	mg/L	4,750
Oil and grease	mg/L	4,210
pH	-	6.52
TVA	mg/L	150.4
TSS	mg/L	1,811
VSS	mg/L	1,630

3.2 Running Reactor

pH values, COD removal, and biomass growth, were analyzed as process control in reactor operation. Anaerobic digestion requires an interaction between fermentative and methanogenic organisms that are sensitive to the low pH values. Changes in the operating conditions of the reactor or the presence of toxic compounds can cause an imbalance in the process and the accumulation of total volatile acids. The pH value during operation has an important role to determine the composition of the acidogenesis effluent in the reactor. The optimum pH for acidogenesis phase is found between 6.0 to 6.5 to avoid the formation of excess propionic acid. However, when the pH values as low as 4.5, CO_2 and H_2 are also produced with butyric and acetic acid also ethanol. Shen, et al. [21] has studied that acidogenesis phase will reach a maximum at 48-72 hours. **Fig. 2** shows the pH values over 48 hours of reactor operation for each addition of the metal cofactors as trace element.

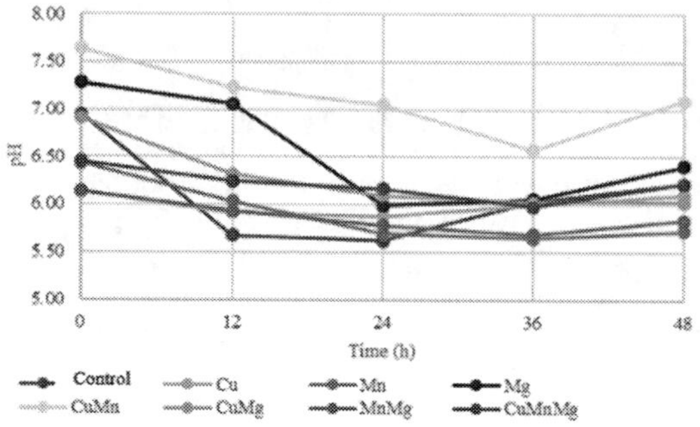

Fig. 2. pH values on operation of reactor

Based on **Fig. 2** the pH values decrease at 36 h and then begin to increase at 48 h. A decrease the pH values at 12 h showed acidogenesis phase that began to occur in the reactor, the pH range between 5.6 and 7.2. pH values still decreasing at 24 h, the pH range between 5.5 and 7.2. The increase pH value that began at 36 h nearly in all reactors indicate that domination acidogenesis phase in the reactor is slowly decreased and shifted to the next phase, methanogenesis.

Biomass growth was known approximately by volatile suspended solid measurement. VSS measurement to approximate the amount of microorganism which involves in running reactor process. VSS values in this study can be seen in **Fig. 3**.

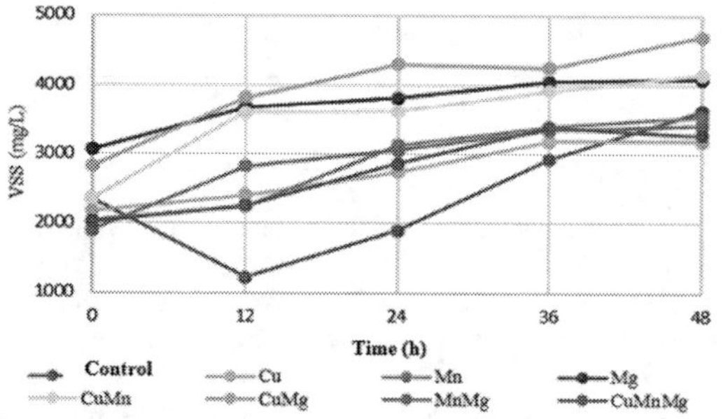

Fig. 3. VSS Values in running reactor

In the beginning of running reactor, The VSS values of 1,800-3,100 mg/L. VSS values tend to increase at each sampling time eventhough not too significant. The VSS values at 48 h of 3,200-4,700 mg/L.

COD concentration for each variation tend to decrease. In the beginning of running reactor, the COD concentration of 5,100-7,800 mg/L and in the final of running reactor (at 48 hours), the COD concentration of

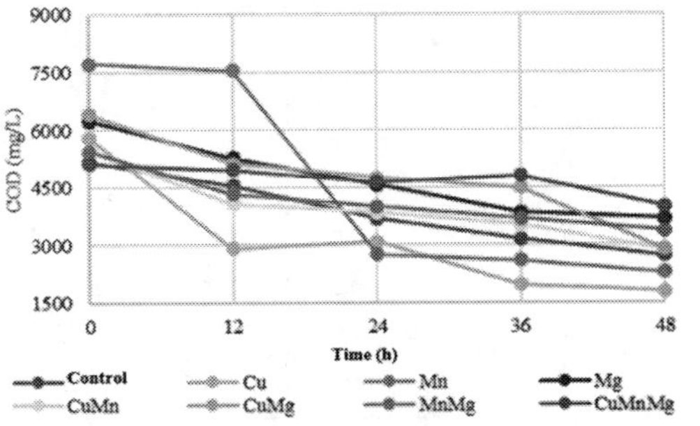

Fig. 4. COD concentration in running reactor

1,800-4,000 mg/L. The decreasing of COD concentration shows that organic content in substrate has converted into different substances, such as ethanol, TVA, and other gaseous products. COD concentration for each variation can be seen in **Fig. 4.**

3.3 Screening for significant effects of trace element on ethanol formation based on two-way level factorial design

Statistical analysis in this study using the statistical analysis of two-way level factorial design. This analysis aims to see the influence of each factor on every level, and estimate at the interaction between factors that affect the formation of ethanol. Statistical test is compared with the value of F_{table} at a significance level of 5% and 1% with the following hypotheses:

$$H_0: \mu 1 = \mu 2 = \mu 3$$

$$H_1: \mu 1 \neq \mu 2 \neq \mu 3$$

H_0 fail to reject if the significance of $F<F_{table}$ (5%) and $F<F_{table}$ (1%), which is the conclusion there is no significant effects between the mean of each population. It means that there is no significant effect on ethanol formation of various variables tested.

H_0 reject if the significance of $F>F_{table}$ (5%) and $F>F_{table}$ (1%), which is the conclusion there is a significant difference between the mean of each population. It means that there is a significant effect on ethanol formation of various variables tested.

Variations in this study correspond to the combination treatment is needed. The result of ethanol formation at the end of the process can be seen in **Table 7**. The results of calculations using statistical analysis 2^n method at various metal additions can be seen in **Table 8.**

Based on the result of two factors ANOVA test, H_0 on the C and AB variations were rejected, that meaning these factors at various levels had a significant effect on ethanol formation.

Table 7. Experimental values of ethanol concentration in variations addition of metals.

Factor			Ethanol concentration (mg/L)	
C (Mg)	B (Mn)	A (Cu)	R1	R2
0	0	0	3741.53	3741.53
		0.1	3741.53	1870.77
	3	0	7483.07	7483.07
		0.1	7483.07	5612.30
0.05	0	0	3741.53	1870.77
		0.1	5612.30	3741.53
	3	0	3741.53	3741.53
		0.1	7483.07	5612.30

Table 8. The results of calculations two-way level factorial design.

Variation	F	F_{table} 5%	F_{table} 1%	Result
1	784.47	5.59	12.25	H_0 reject
a	4.20	5.59	12.25	H_0 fail to reject
b	56.47	5.59	12.25	H_0 reject
ab	0.47	5.59	12.25	H_0 fail to reject
c	4.20	5.59	12.25	H_0 fail to reject
ac	22.87	5.59	12.25	H_0 reject
bc	11.67	5.59	12.25	H_0 fail to reject

abc	0.47	5.59	12.25	H_0 fail to reject
Error	-	-	-	-
Total	-	-	-	-

Variation 1 was a variation without the addition of metal ions. The hypothesis in this variation is reject, which concludes that factors had a significant effect on ethanol formation. In other words, the absence of any additional metal forming process of ethanol can occur naturally.
Mn^{2+} and Cu^{2+}; Mg^{2+} combination are screened as the significant variable which had positive effects on ethanol formation from POME at a significance level of 5% and 1%.

3.4 Screening for significant effects of trace element on TVA formation based on two-way level factorial design

Statistical analysis in this study using the statistical analysis of two-way level factorial design. This analysis aims to see the influence of each factor on every level, and estimate at the interaction between factors that affect the formation of TVA. Statistical test is compared with the value of F_{table} at a significance level of 5% and 1% with the following hypotheses.

H_0: $\mu 1 = \mu 2 = \mu 3$

H_1: $\mu 1 \neq \mu 2 \neq \mu 3$

H_0 fail to reject if the significance of $F<F_{table}$ (5%) and $F<F_{table}$ (1%), which is the conclusion there is no significant effects between the mean of each population. It means that there is no significant effect on TVA formation of various variables tested.
H_0 reject if the significance of $F>F_{table}$ (5%) and $F>F_{table}$ (1%), which is the conclusion there is a significant difference between the mean of each population. It means that there is a significant effect on TVA formation of various variables tested.
Variations in this study correspond to the combination treatment is needed. The result of TVA formation at the end of the process can be seen in **Table 9**. The results of calculations using statistical analysis 2^n method at various metal additions can be seen in **Table 10.**
Based on the result of two factors ANOVA test, H_0 on the b and c variations were rejected, that meaning these factors at various levels had a significant effect on TVA formation.
Volatile acid is the main product of organic material degradation at acidogenesis phase under anaerobic process. Volatile acids as acetic acid is the substrate that should be converted by methanogenic bacteria into methane and carbon dioxide. TVA concentration is closely related to the acidogenesis process because the increase of total volatile acids formation.

Table 9. Experimental values of TVA concentration in variations addition of metals.

Factor			**TVA concentration (mg/L)**	
C (Mg)	**B (Mn)**	**A (Cu)**	**R1**	**R2**
0	0	0	960	1056
		0.1	1056	1152
	3	0	1536	1440
		0.1	1152	1440
0.05	0	0	1536	1728
		0.1	1152	1536
	3	0	1536	1728
		0.1	1440	1536

Table 10. The results of calculations two-way level factorial design.

Variation	**F**	**F_{table} 5%**	**F_{table} 1%**	**Result**
1	2890.45	5.59	12.25	H_0 reject
a	6.67	5.59	12.25	H_0 fail to reject
b	15.93	5.59	12.25	H_0 reject
ab	0.50	5.59	12.25	H_0 fail to reject
c	34.45	5.59	12.25	H_0 reject
ac	2.70	5.59	12.25	H_0 fail to reject
bc	6.67	5.59	12.25	H_0 fail to reject

abc	4.46	5.59	12.25	H_0 fail to reject
Error	-	-	-	-
Total	-	-	-	-

Variation 1 was a variation without the addition of metal ions. The hypothesis in this variation is reject, which concludes that factors had a significant effect on ethanol formation. In other words, the absence of any additional metal forming process of ethanol can occur naturally.
Mn^{2+} and Cu^{2+}; Mg^{2+} combination are screened as the significant variable which had positive effects on ethanol formation from POME at a significance level of 5% and 1%.

3.5 The comparison of two-way level factorial design results on ethanol and TVA formation

The comparison of two-way level factorial design results on ethanol and TVA formation can be seen in **Table 11.** The difference of results may indicate predisposition of the metal ions to form a product. In addition, the switching material to other products may occur during the process.

Table 11. The comparison of two-way level factorial design results on ethanol and TVA formation.

Variations		TVA Formation	Ethanol Formation
1	Control	H_0 reject	H_0 reject
A	Cu	H_0 fail to reject	H_0 fail to reject
B	Mn	H_0 reject	H_0 reject
AB	CuMn	H_0 fail to reject	H_0 fail to reject
C	Mg	H_0 reject	H_0 fail to reject
AC	CuMg	H_0 fail to reject	H_0 reject
BC	MnMg	H_0 fail to reject	H_0 fail to reject
ABC	CuMnMg	H_0 fail to reject	H_0 fail to reject

Conclusions

High levels of organic contents in POME can be reduced by anaerobic treatment. Wastewater treatment with high concentration of organic can be treated by the suspended growth in anaerobic reactors. pH values, COD removal, and biomass growth, were analyzed as process control in running reactor. The pH values during operation has an important role to determine the composition of the acidogenesis effluent in the reactor. Acidogenesis phase in this study occur until the 36th hour. Based on screening experiments, Mn^{2+} and Cu^{2+}, Mg^{2+} metal ions combination are screened as the significant variable which had positive effects on ethanol formation and Mg^{2+} and Mn^{2+} metal ions were screened as the significant variable which had positive effects on TVA formation from POME at a significance level of 5% and 1%. The difference of results may indicate predisposition of the metal ions to form a product. In addition, the switching material to other products may occur during the process.

References

1. D. Andrio, *Mechanism and Strategy Control for Ethanol Formation as a High Organic Content Wastewater Recycle.* Doctoral Dissertation. Environmental Engineering, ITB. Bandung. (2014)
2. M. Azenha, M.T. Vasconcelos, and P. Moradas-Ferreira, *The influence of Cu concentration on ethanolic fermentation by Saccharomyces cerevisiae.* J. Biosci. Bioeng. 90, 163–167. (2000)
3. W. Cao, R. Liu, *Screening and optimization of trace elements supplement in sweet sorghum juice for ethanol production.* Biomass and bioenergy 50 (2013) 45-51. (2012)
4. C.A. Chernicaro, de Lemos. *Biological Wastewater Treatment Series Volume Four: Anaerobic Reactors.* IWA Publishing: London, UK. (1997)
5. D. Diviana, *Effect of Fe, Mn and Ni Metals in The Formation of Ethanol from Degradation Process of Waste of Cocoa Pulp in Anaerob Condition.* Undergraduate Thesis. Environmental Engineering, ITB, Bandung. (2014)
6. M.H. Gerardi, *Microbiology of Anaerobic Digesters.* John Wiley Sons, New Jersey. (2003)
7. G. Gottschalk, S. Peinemann, *The anaerobic way of life.* In: Balows A et al. (eds) The Prokaryotes, pp. 300–311. Berlin: SpringerVerlag. (1992)
8. A. Gumilar, *Effect of Flushing N_2 and Variations the Addition of Fe^{2+}, Zn^{2+} for Ethanol and Acidogenesis Products Formation in Degradation Process of High Organic Wastewater Under Anaerobic Condition.* Postgraduate Thesis. Environmental Engineering, ITB, Bandung. (2016)
9. M.A.B. Habib, F.M. S.M. Yusoff, K.J. Phang, S. Mohamed, *Nutritional values of chironomid larvae grown in palm oil mill effluent and algal culture.* Aquaculture, 158: 95-105. (1997)
10. A.N. Ma, *Environmental Management for the Oil Palm Industry.* Palm Oil Dev., 30: 1-10. (2000)
11. S. Onuki, J. A. Koziel, W.S. Jenks, J. van Leeuwen, D.A Greewell, *Ethanol production, purification, and*

analysis techniques a review. ASABE Annual International Meeting. Paper Number: 085136. USA. (2008)

12. D.H. Prakoso, *Effect of Flushing N_2 and Variations the Addition of Fe^{2+} for Ethanol Formation in Degradation Process of High Organic Wastewater Under Anaerobic Condition.* Undergraduate Thesis. Environmental Engineering, ITB, Bandung. (2015)
13. M. Perez, L.I. Romero, D. Sales, *Organic matter degradation kinetics in an anaerobic thermophilic fluidised bed bioreactor. Anaerobe*, 7: 25-35. (2001)
14. Pusat Data dan Informasi Departemen Perindustrian. *Gambaran Sekilas Industri Minyak Kelapa Sawit.* Jakarta: Departemen Perindustrian. (2007)
15. J. Saxena, R.S. Tanner, *Effect of Trace Metals on Ethanol Production from Synthesis Gas by the Ethanologenic Acetogen, Clostridium ragsdalei*. ASM Press, Washington, DC, pp 147–151. (2010)
16. P.K. Shin, M.H. Kim, J.H. Kim, *Biodegradability of degradable plastic exposed to anaerobic digested sludge and simulated landfill conditions*. J. Environ. Polym. 5, 1, 33-39. (2001)
17. S. Sinnaraprasa, P. Fongsatitkul, *Optimal Condition of Fenton's Reagent to Enhance the Alcohol Production from Palm Oil Mill Effluent (POME).* Environment Asia 4(2) (2011) 9-16. (2011)
18. G. Tchobanoglous, F. L. Burton, *Wastewater Engineering : Treatment, Disposal, and Reuse.* McGraw-Hill: Singapore. (1991)
19. World Growth. *Laporan World Growth: Manfaat Minyak Sawit bagi Perekonomian Indonesia*. (2011)
20. O. Deesuth, P. Laopaiboon, P. Jaisil, L. Laopaiboon, *Optimization of Nitrogen and Metal Ions Supplementation for Very High Gravity Bioethanol Fermentation from Sweet Sorghum Juice Using an Orthogonal Array Design*. J Energies 5, 3178-3197. (2012)
21. Y. Shen, H. Zhang, X. Zheng, X. Zhang, J. Guo, Y. Chen, *Very high gravity fermentation using sweet potato for fuel ethanol production*. Adv Mat Res. 2011:236 –238:59–62. (2011)

Kinetics of nutrient removal in an on-site domestic wastewater treatment facility

Va Vandith[1], *Ahmad* Soleh Setiyawan[1], *Prayatni* Soewondo[1], *Pom* Bophann[1] and *Hardjono*[1]

[1]Department of Environmental Engineering, Faculty of Civil and Environmental Engineering, Institut Teknologi Bandung, JL Ganesha 10 Bandung 40132, Indonesia

Abstract. Domestic wastewater from office building has not been maintained well especially in the areas where the sewerage system is unavailable. The aims of this research were to investigate the performance and kinetics of nutrient removal in an on-site domestic wastewater treatment facility which consists of anaerobic and aerobic systems for treating wastewater from office building. The experimental data obtained from the variations of COD:N:P ratio 250:28:2.5, 350:38:2.9, 450:47:3.3, and 600:60:3.7 with three different HRT 48h, 24h, and 12h. A One-way ANOVA was performed to investigate the effects of HRT and initial concentration of TN and TP on the performance of nutrient removal. In order to obtain the kinetic coefficients, First Order, Second Order and Stover-Kincannon Models were employed. The results showed that maximum TN and TP removal efficiency were 56% and 86%, respectively. The results of one-way ANOVA showed that HRT and initial concentration of TN and TP gave the significant effects on nutrient removal ($p < 0.05$). Second Order and Stover-Kincannon Models were found to be more appropriate models for prediction of TN removal in this facility. Controlling HRT and C: N: P ratio may keep good performance of nutrient removal in this facility.

1 Introduction

Presently, environmental destruction is becoming on a global scale, and people are showing a dramatic rise in concern for environmental protection. Water pollution is one of the major causes of the environmental pollution and destruction. As can be seen clearly, most of the developing countries including Indonesia are having a problem of water pollution due to an insufficient amount of wastewater treatment plant or sewerage system to treat wastewater. As a result, untreated domestic wastewater containing nitrogen and phosphorus is discharging into the environment without purifying action which will accelerate the water pollution such as eutrophication potential.

Subsequently, an on-site wastewater treatment facility is one of the preferable alternative technologies to treat nitrogen and phosphorus from domestic wastewater in low-density communities. The treatment facility is also introduced by USEPA in 1997 for decentralizing domestic wastewater treatment facilities where the sewerage system is unavailable. In Japan, Johkasou has been known as an advanced domestic wastewater treatment tank which has been widely used as a decentralized wastewater treatment system and it still continues to expand into other countries [1, 2].

Johkasou is an advanced wastewater treatment facility which contains anaerobic filter tanks, contact aeration tank, sedimentation tank, and disinfection tank [3]. It has been implemented as alternative treatment system in the rural areas of Japan, in addition from other areas that do not have wastewater treatment plants [4, 5]. The treatments in anaerobic filter tank are to separate solid matter, store sludge, and degrade organic matter by anaerobic microorganisms. Moreover, contact aeration tank is employed to degrade organic matter, oxidizes ammonia by aerobic microorganisms, and to return sludge with recirculation water to the anaerobic tank. Sedimentation tank can separate suspended solid from treated water through sedimentation, return sludge to contact aeration tank and send supernatant to disinfection tank. The last treatment unit is disinfection tank which supernatant is disinfected by chlorine tablets to get free from pathogenic microorganism. The effluent from Johkasou facilities can be discharged directly into the water body since the nitrogen and phosphorus concentration are in the permissible discharge criteria [6].

On-site domestic wastewater facility which consists of anaerobic and aerobic systems has been developed and become famous in developed countries and it also could be implemented in developing countries. However, it is hard to use this facility directly for individual house in Indonesia because of high cost and controlling. One of the strategies is to use this facility for communal use such as office building in Indonesia, but there is no information about the kinetics of nutrient removal of this facility for treating domestic wastewater from office building. Therefore, this research was to know the kinetics and performance of nutrient removal in an on-site domestic wastewater treatment facility which consists of anaerobic and aerobic systems for treating domestic wastewater from office building.

2 Method

The experiment was conducted in water quality laboratory of environmental engineering department of

* Corresponding author: vandith_va@yahoo.com

Institut Teknologi Bandung for a period of six months from August 2015 to February 2016.

2.1 Experimental Set up

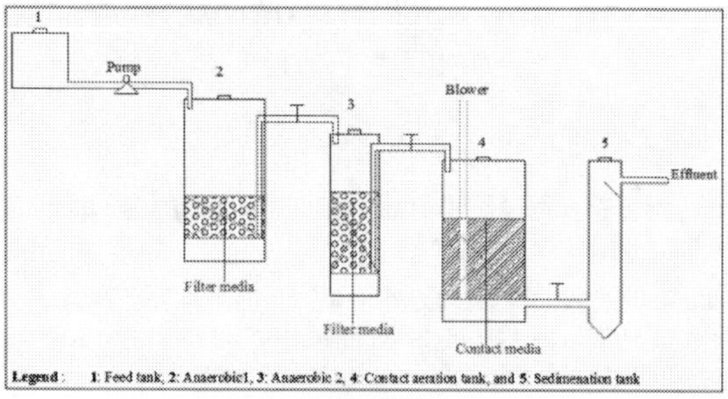

Fig. 1. Lab-scale model of on-site domestic wastewater treatment facility.

A lab-scale model of an on-site domestic wastewater treatment facility used in this research is shown in **Fig.1**.This facility consists of four compartments which are anaerobic 1, anaerobic 2, contact aeration tank, and sedimentation tank. In order to feed the wastewater to this facility, the feed tank was set in front of the facility and there was the water pump to pump it to anaerobic1 continuously with flow meter used to control the flow rate. The air flow rate was controlled by the air pump and placed diffuser at the bottom tank and at the middle of the contact aeration tank to supply the oxygen inside this compartment. The working volume of reactor is 100 L. In anaerobic 1, the volume of reactor is 32 L and bio-ball ball type was filled in order to let microbes attach it. The anaerobic 2, the volume is 19 L and bio-ball ball type were used as filter media in order to let microbes grow. Furthermore, the contact aeration tank, the volume is 34.5 L and bio-ball rambutan type was employed as contact media in this compartment. The final compartment is sedimentation tank, the volume is 14.5 L where the remaining suspended materials will settle before discharging to the environment.

2.2 Wastewater Preparation

Wastewater used in the experiment was the synthetic wastewater which had been characterized based on the characteristics of domestic wastewater from office building. The variations of CNP ratio were determined based on the characteristic of domestic wastewater from office building. The synthetic wastewater was used to control the variation of inlet concentration required in order to determine the kinetics of nutrient removal in this facility. The compositions of synthetic wastewater used were glucose ($C_6H_{12}O_6$), ammonium chloride (NH_4Cl), potassium nitrate (KNO_3) and sodium nitrite ($NaNO_2$), and potassium dihydrogen phosphate (KH_2PO_4) used as source for COD: N: P ratio, respectively [7].

2.3 Experimental Design and Reactor operation

The reactor was initially inoculated with domestic wastewater from dormitory and was carried out with a batch condition which was used for seeding process. The wastewater was filled into the reactor with the volume of 90% of total volume of the reactor. In the seeding process, glucose was used as a main source of the substrate because it was considered as a degradable source by microorganisms, and ammonium chloride (NH_4Cl) with potassium phosphate (KH_2PO_4) were used as nitrogen and phosphorus sources, respectively. The reactor started up involving daily feeding of synthetic wastewater until the microbes started to form on the media after three weeks, especially in the contact aeration tank in batch condition. The measurements of COD were made everyday to see how biomass grow against the elimination of organic matter. pH, DO, and Temperature were measured everyday to know the environmental conditions of biological wastewater treatment in this facility.

After the seeding process, the reactor was allowed to do the acclimatization process. The synthetic wastewater was pumped continuously to this facility. The wastewater used was synthetic wastewater for providing a continuous source of completely biodegradable organic matter and maintain the constant feed concentration into the reactor. The acclimatization process was finished when it reached the steady state condition at the variation of C: N: P ratio 250:2.8:2.5 and HRT 48h.

Based on **Table 1**, in the running process, this facility was supplied synthetic wastewater with the concentration varied from the lowest to the highest which flexible to characteristic of wastewater from office building fluctuation in real condition at the field with three different hydraulic retention times 48h 24h and 12h. The hydraulic retention time and C: N: P ratio were changed to another variation after it reached the steady-state condition.

In this experiment, sampling points were taken in five points namely: inlet, the effluent of anaerobic 1, the effluent of anaerobic 2, the effluent of contact aeration tank and effluent of sedimentation tank. TN and TP were taken and analyzed daily before and after it reached the steady state condition. pH, DO and Temperature were measured every compartment everyday to observe the environmental condition of biological wastewater treatment before and after it reached the steady-state condition. TSS and VSS were analyzed only when the running process was finished in order to know the average biomass inside the anaerobic tanks, contact aeration tank, and overall reactor. In order to know the biomass in the anaerobic tanks, aerobic tank, and overall reactor, the number of media is known in these compartments. The number of media is 512, 550, and 750 in anaerobic 1, anaerobic 2, and contact aeration tank, respectively. Three media of each tank were taken and measured to know the

Table 1. Variation of C: N: P ratio and HRT values in lab-scale model of on-site domestic wastewater treatment facility.

No	COD (mg/L)	TN (mg/L)	TP (mg/L)	Q (L/h)	V (L)	HRT (h)
I	250	28	2.5	2.08	100	48
	350	38	2.9	2.08	100	48
	450	47	3.3	2.08	100	48
	600	60	3.7	2.08	100	48
II	250	28	2.5	4.17	100	24
	350	38	2.9	4.17	100	24
	450	47	3.3	4.17	100	24

	600	60	3.7	4.17	100	24
III	250	28	2.5	8.33	100	12
	350	38	2.9	8.33	100	12
	450	47	3.3	8.33	100	12
	600	60	3.7	8.33	100	12

attached biomass. The average biomass is summed of attached biomass and suspended growth biomass.

2.4 Analytical Techniques

The following parameters were analyzed according to the standard method for the examination of water and wastewater: COD (SMEWW 5220-B), TSS (SMEWW 2540-D), VSS (SMEWW 2540-I), TN (SMEWW 4500-B), TP (SMEWW 4500-P-D), Temperature (SMEWW 2550), pH (SMEWW 4500H$^+$), and DO (SMEWW 4500-0-G). A one-way analysis of variance (ANOVA) was used to test the significant effect of HRT, initial TN and TP concentration on nutrient removal at the significant 0.05 level by using Microsoft Office Excel 2013.

2.5 Determination of Kinetics

There are three common kinetic models used to predict substrate removal as follows [8]:

2.5.1 First Order Substrate Removal Model

The rate of change in substrate concentration in the system with assuming the first order model for substrate removal could be expressed as follows [9].

$$-\frac{dS}{dt}=\frac{QS_o}{V}-\frac{QS}{V}-k_1S \quad (1)$$

Under the pseudo-steady state conditions, the rate of change in substrate concentration due to accumulation (-dS/dt) is negligible and the equation given above can be modified as:

$$\frac{S_o-S}{HRT}=k_1S \quad (2)$$

The value of the first-order kinetic constant can be obtained by plotting (So-S)/HRT versus S in (2). The value of k_1 is obtained from the slope of the straight line.

2.5.2 Second Order Substrate Removal Model

The general equation of a second-order model is given below [10].

$$-\frac{dS}{dt}=k_{2(s)}\cdot\left(\frac{S}{S_o}\right)^2 \quad (3)$$

If (3) is integrated and then linearized, (4) results:

$$\frac{S_o.HRT}{S_o-S}=HRT+\frac{S_o}{k_{2(s)}X} \quad (4)$$

If the second term of the right part of this equation is accepted as a constant, the equation will be modified as below:

$$\frac{S_o.HRT}{S_o-S}=a+bHRT \quad (5)$$

The value of a (per day) and b (without unit) will be drawn as y-intercept and the slope of the line, respectively. In the above-mentioned equation, HRT is per day, and pollutant removal speed constant or $k_{2(s)}$ is calculated by $a = S_0/(k_{2(s)} \times X)$. X is average biomass inside the reactor. If (So-S)/So expresses the substrate removal efficiency and is symbolized as *E*. Therefore, (6) can be written as follows:

$$\frac{HRT}{E}=a+bHRT \quad (6)$$

2.5.3 Stover Kincannon Model

In this model, the substrate utilization rate is expressed as a function of the organic loading rate by monomolecular kinetic for biofilm reactors such as rotating biological contactors and biological filters. The equation of the Stover-Kincannon model is as follows [11].

$$\frac{dS}{dt}=\frac{Q}{V}(S_o-S) \quad (7)$$

$$\frac{dS}{dt}=\frac{U_{max}(QS_o/V)}{K_B+(QS_o/V)} \quad (8)$$

If (-dS/dt) is taken as $V/[Q(S_0-S)]$, which is the inverse of the removed substrate loading rate and this is plotted against the inverse of the total loading rate V/QS_o, a straight line portion of intercept $1/U_{max}$ and a slope of K_B/U_{max} resulted.

$$\frac{V}{Q(S_o-S)}=\frac{K_B}{U_{max}}.\frac{V}{QS_o}+\frac{1}{U_{max}} \quad (9)$$

The substrate balance for the reactor at steady-state can be written as follows:

$$QS_O=QS+V\frac{dS}{dt} \quad (10)$$

Substituting of equation (8) and (10) gives

$$QS_O=QS+\frac{U_{max}(QS_O/V)}{K_B+(QS_O/V)}*V \quad (11)$$

This expression can be solved for either the effluent substrate concentration (12) or the removal efficiency of the reactor (13) by substituting kinetic constants U_{max} and K_B.

$$S=S_O-\frac{U_{max}S_O}{K_B+(QS_O/V)} \quad (12)$$

$$E=\frac{S_O-S}{S_O}=\frac{U_{max}}{K_B+(QS_O/V)} \quad (13)$$

3 Result and Discussion

3.1 Effect of HRT on TN and TP removal

After initial acclimatization of biofilm growth on support media, TN removal was obtained at hydraulic retention time 48 h when it reached the steady state condition. The synthetic wastewater which had been characterized based the domestic wastewater from office building was fed along with C: N: P ratio. The reactor was continuously operated at three different hydraulic retention times 48h, 24h, 12h and four different initial TN concentrations 27 mg/L, 38 mg/L, 47 mg/L and 60 mg/L, respectively. According to **Fig. 2**, the maximum TN removal efficiency can be achieved at C: N: P ratio 350:38:2.9 at 24h of HRT was 56% which was the optimum condition. In order to identify the effect of HRT on the performance of TN removal in facility, one-way ANOVA was performed to three principle components of HRT values.

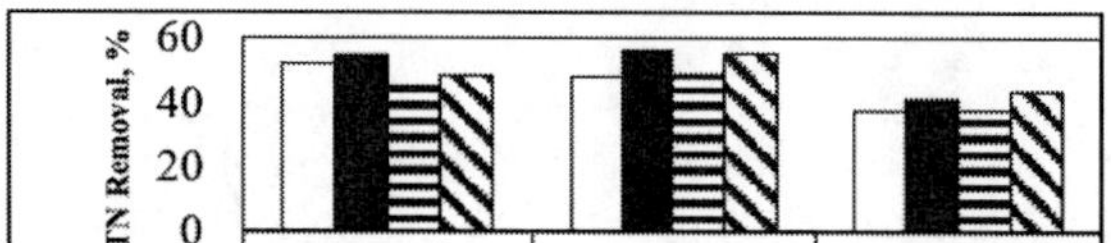

	HRT 48h	HRT 24h	HRT 12h
□27 mg/L	52	48	38
■38 mg/L	55	56	41
▤47 mg/L	45	49	38
▧59 mg/L	48	55	44

Fig. 2. Variation of TN removal at different HRT values in lab-scale model of an on-site domestic wastewater treatment facility.

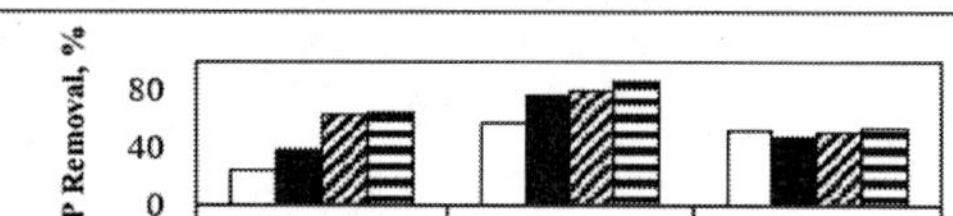

	HRT 48h	HRT 24h	HRT 12h
□2.5 mg/L	25	57	52
■2.9 mg/L	38	77	48
▨3.3 mg/L	63	80	51
▤3.7 mg/L	64	86	54

Fig. 3. Variation of TP removal at different HRT values in lab-scale model of an on-site domestic wastewater treatment facility.

The result of one-way ANOVA showed the effect of HRT on TN removal in the overall reactor at the 0.05 level.

TP removal was obtained at hydraulic retention time 48 h when it reached the steady state condition. The synthetic wastewater which had been characterized based on domestic wastewater from office building was fed along with C: N: P ratio. The facility was continuously operated at three different hydraulic retention times 48h, 24h, 12h and four different initial TP concentrations 2.5 mg/L, 2.9 mg/L, 3.3 mg/L, and 3.7 mg/L, respectively. **Fig. 3** showed that maximum TP removal efficiency can be achieved at C: N: P ratio 600:59:3.7 at 24h of HRT was 86 % which was the optimum condition. In order to identify the effect of HRT on the performance of TP removal in facility, one-way ANOVA was performed to three principle components of HRT values. The result of one-way ANOVA gave the significant effect of HRT on TP removal in the overall reactor of lab-scale model of an on-site domestic wastewater treatment facility at the 0.05 level.

3.2 Effect of initial TN and TP concentration on TN and TP removal

Based on **Fig. 4**, it showed the variation of TN removal at different initial TN concentrations in the facility. This facility was fed with four different initial concentrations, 28 mg/L, 38 mg/L, 47 mg/L and 60 mg/L and operated with three different hydraulic retention times 48h, 24h, and 12h. The One-way ANOVA was performed to four principle components of initial TN concentration. The effect of initial TN concentration was evaluated and all of the variations gave the significant effect on TN removal in lab-scale model of an on-site domestic wastewater treatment facility with the maximum TN removal can be

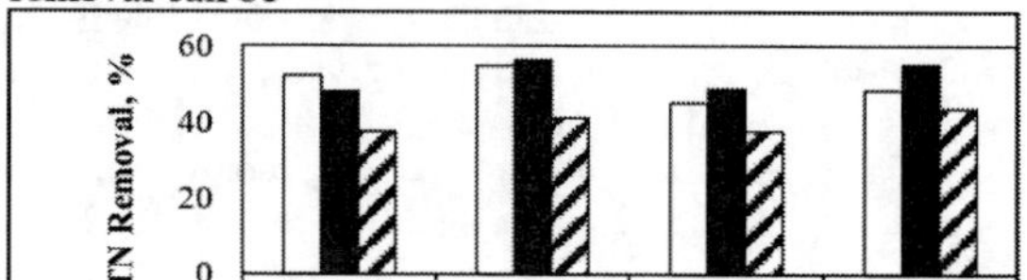

	28 mg/L	38 mg/L	47 mg/L	60 mg/L
□HRT 48h	52	55	45	48
■HRT 24h	48	56	49	55
▧HRT 12h	38	41	38	44

Fig. 4. Variation of TN removal at different initial TN concentration in lab-scale model of an on-site domestic wastewater treatment facility.

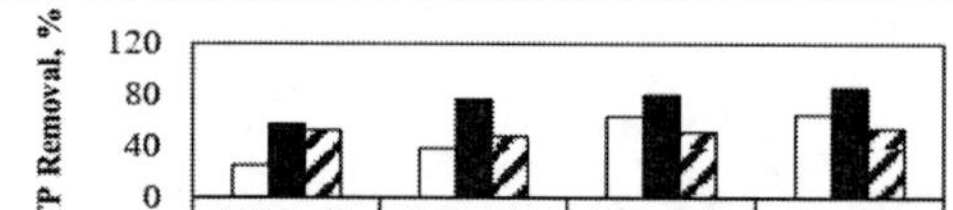

	2.5 mg/L	2.9 mg/L	3.3 mg/L	3.7 mg/L
□HRT 48h	25	38	63	64
■HRT 24h	57	77	80	86
▨HRT 12h	52	48	51	54

Fig. 5. Variation of TP removal at different initial TP concentration in lab-scale model of an on-site domestic wastewater treatment facility.

achieved at HRT 24h and C:N:P ratio 350:38:2.9 was 56%.

According to **Fig. 5**, it showed the variation of TP removal at different initial TP concentrations in the facility. This facility was fed with four different initial concentrations, 2.5 mg/L, 2.9 mg/L, 3.3 mg/L and 3.7 mg/L and operated with three different hydraulic retention times 48h, 24h, and 12h. A One-way ANOVA was performed to four principle components of initial TP concentration. The result of one-way ANOVA showed that all of the variations gave the significant effects of initial TP concentration on TP removal in the overall facility at the 0.05 level with the maximum TP removal can be obtained to be 86% at HRT 24h and C:N:P ratio 600:60:3.7.

3.3 Kinetics of nutrient removal

3.3.1 Kinetics of nitrogen removal

Based on **Fig. 6**, equation (2) was plotted and the result showed that k_1 was 1.01 d^{-1} with the correlation coefficient R^2 = 0.36. **Fig. 7** indicated a second-order kinetic model applied to lab-scale model of an on-site domestic wastewater facility. The equation (5) was plotted and the results showed that a and b were 0.19 d^{-1} and 1.94, respectively with correlation coefficient $R^2 = 0.97$. According to **Fig. 8**, The Umax and K_B values were calculated as 166.67 mg/L.d and, 316.33 mg/L.d, respectively with correlation coefficient $R^2 = 0.97$.

In this research, three kinetic models were applied to nitrogen removal. **Table 2** showed the kinetic parameters that were calculated by First Order, Second-order and Stover-Kincannon Models in different experimental conditions. k_1 is the first order kinetic model which is considered as the rate of substrate removed by the first order reaction. [12] Akhbari *et al.* (2012) found k_1 value

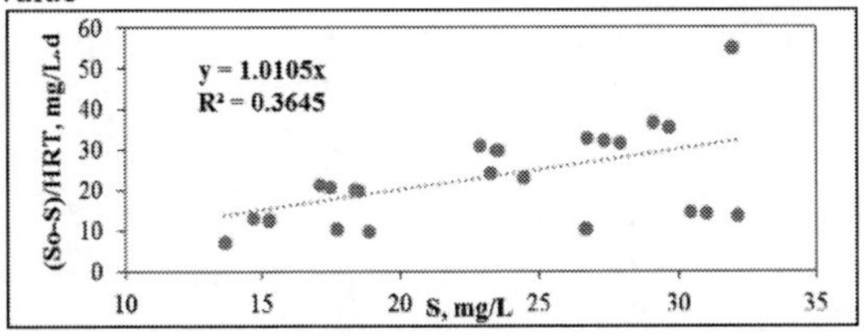

Fig. 6. First-order for TN removal in lab-scale model of an on-site domestic wastewater treatment facility.

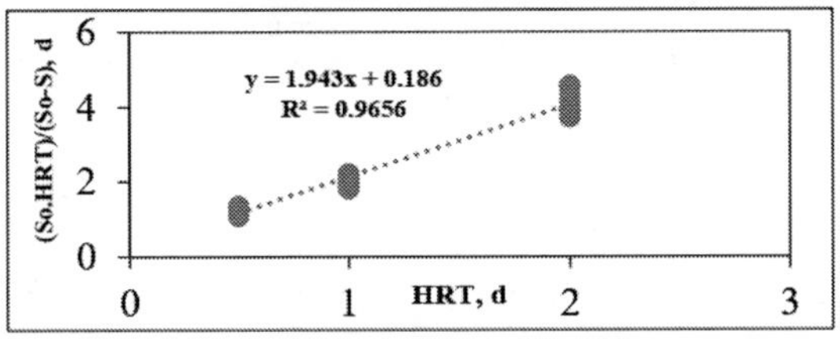

Fig. 7. Second-order for TN removal in lab-scale model of an on-site domestic wastewater treatment facility.

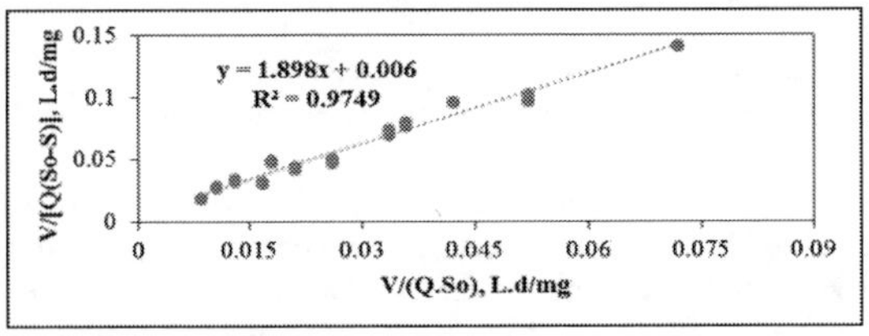

Fig. 8. Stover-Kincannon Model for TN removal in lab-scale model of an on-site domestic wastewater treatment facility.

to be 1.882 d^{-1} which was higher than this study. The most likely reason for the difference in k_1 value was the significant variation of inlet concentration, type of the reactor, the hydraulic retention time, wastewater characteristic and microorganisms in the facility. Moreover, it can be confirmed that TN removal process is so complex that can not be adequately described slowly by the first order reaction equation with a good degree of precision [13]. The saturation constant (K_B) and the maximum utilization rate (U_{max}), the rate constant in Stover-Kincannon model, were found to be 316.33 mg/L.d and 166.66 mg/L.d, respectively. Comparing with the results found by Akhbari *et al.* (2012) [12], in this study, K_B and U_{max} values was much lower, but the correlation coefficient was in the similar value. The correlation coefficient (R^2) in this model was 0.97 which consider as higher precision than the first-order model. In Second-order substrate removal rate, $k_{2(s)}$ is the speed constant at which the pollutant can be removed. Therefore, the higher the $k_{2(s)}$, the faster the pollutant can be removed. $k_{2(s)}$ (0.1522 d^{-1}) obtained in this study was lower than the one obtained by Akhbari *et al.* (2012) [11].The $k_{2(s)}$ will be increased as substrate removal rate increased depending on initial substrate concentration and microorganism inside the reactor.

3.3.2 *Kinetic of TP removal*

According to **Fig. 9**, equation (2) was plotted and the result showed that k_1 was 1.38 d^{-1} with the coefficient correlation $R^2 = -0.69$. **Fig. 10** indicated a second-order kinetic model applied to lab-scale model of an on-site domestic wastewater treatment facility. The equation (5) was plotted and the results showed that a and b were 0.84 d^{-1} and 2.84, respectively with correlation coefficient $R^2 = 0.60$. According to **Fig. 11**, The Umax and K_B values were calculated 1.94 mg/L.d and 6.93 mg/L.d, respectively after the equation (9) was plotted with correlation coefficient $R^2 = 0.73$.

In this research, two kinetic models were applied to phosphorus removal in facility. **Table 3** presents kinetic parameters for TP removal that were calculated by Second-order and Stover-Kincannon Models in different experimental conditions. The saturation constant (K_B) and the maximum utilization rate (U_{max}), the rate constant in Stover-Kincannon model, were found to be 6.93 mg/L.d and 1.94 mg/L.d, respectively. Comparing with the results found by Akhbari *et al.* (2012) [12], K_B and U_{max} values obtained in this study were much lower, but the correlation coefficient was in the similar value. The correlation coefficient in this model was 0.73 which consider as fairly good precision with the data obtained

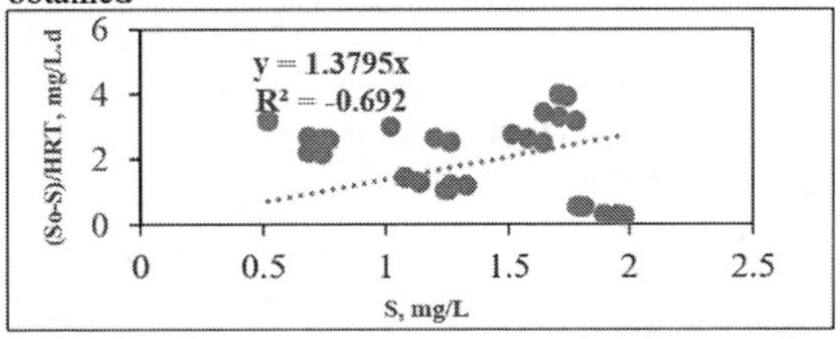

Fig. 9. First order for TP removal in lab-scale model of an on-site domestic wastewater treatment facility.

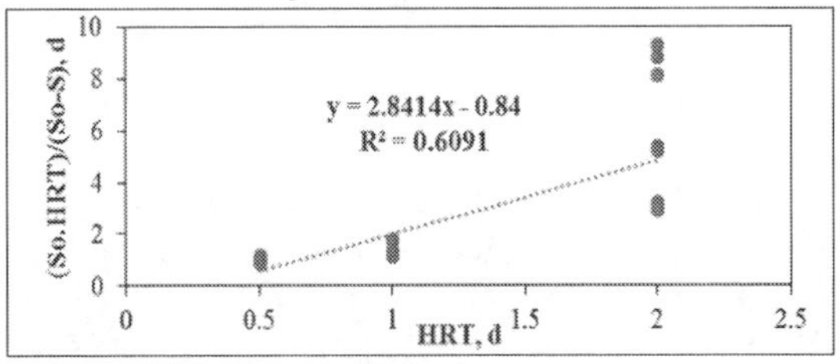

Fig. 10. Second order for TP removal in lab-scale of an on-site domestic wastewater treatment facility.

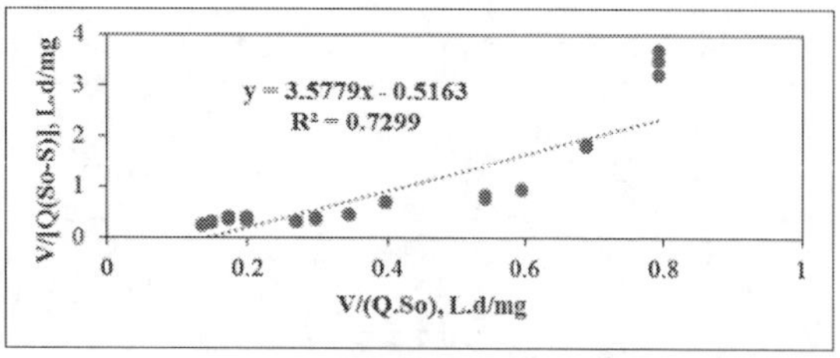

Fig. 11. Stover-Kincannon Model for TP removal in lab-scale model of an on-site domestic wastewater treatment facility.

Table 2. Kinetics of TN removal in lab-scale model of an on-site domestic wastewater treatment facility.

First-Order		Second-Order			Stover-Kincannon		
R^2	k_1 (d^{-1})	R^2	a (d^{-1})	b	R^2	Umax (mg/L)	KB (mg/L)
0.36	1.01	0.97	0.19	1.94	0.97	166.66	316.33

Note: k_1: First-order substrate removal rate constant, U_{max}: Maximum substrate removal rate constant, K_B: Saturation constant.

Table 3. Kinetics of TP removal in lab-scale model of an on-site domestic wastewater treatment facility.

Second-Order			Stover-Kincannon		
R^2	a (d^{-1})	b	R^2	Umax (mg/L.d)	KB (mg/L.d)
0.60	0.84	2.84	0.73	1.94	6.93

Note: k_1: First-order substrate removal rate constant, U_{max}: Maximum substrate removal rate constant, K_B: Saturation constant

from the experiment. $k_{2(s)}$ (0.0024 d^{-1}) obtained in this study was lower than the one obtained from another research conducted by Akhbari *et al.* (2012) [12].The $k_{2(s)}$ will be increased as substrate removal rate increased depending on initial substrate concentration and microorganism inside the facility. Simplified models are formed a few variables and can be used to determine the kinetic of the reactions. Three models were used to investigate the biological reactions, First Order, Second Order and Stover-Kincannon Models. After drawing regression diagrams for these models, second Order and Stover-Kincannon Models showed a very good and high adaptation in removing nitrogen in lab-scale model of an on-site domestic wastewater treatment facility as the determination coefficients were either the removal efficiency or effluent concentration at the field scale when it is operated in similar condition as lab-scale model of an on-site domestic wastewater treatment facility, accurately.

Moreover, the knowledge of kinetics of nutrient removal in this research provides a rational basis for the process analysis and control. It is possible to optimize reactor performance based on the kinetics study. It also gives the reaction rate of the system and provides a basis for reactor design and operation [14].

3.4 Evaluation of Facility Performance

Based on **Table 4**, it showed the predicted results of TN removal by using Second Order and Stover-Kincannon Models. The predicted result of maximum TN removal by using Second Order Substrate Removal Model was 57.87% based on the equation (6) and the predicted result of TN removal by employing Stover-Kincannon model was 50.4% according to equation (13). Comparing the maximum TN removal from experiment and maximum TN removal from kinetic models, they are similar. Based on **Table 5**, it showed the predicted results of TP removal by using Second Order and Stover-Kincannon Models and maximum TP removal by experiment. The predicted result of maximum TP removal by using Second Order Model was 30.66% based on the equation (6) and the predicted result of TP removal by employing Stover-Kincannon model was 23.64% according to equation (13).

Comparing with maximum TP removal from experiment, maximum TP removal from predicted models was lower. This is because of determination coefficient was lower than 0.75.

Table 4. Predicted results of TN removal by applying Second Order and Stover-Kincannon models and the results of TN removal from the experiment in this facility.

Model	a d^{-1}	b	Ep (%)	Ex (%)
Second Order	0.19	1.94	57.87	56%
Model	U_{max} **mg/l.d**	K_B **mg/L.d**	**Ep (%)**	**Ex %**
Stover-Kincannon	166.66	316.33	50.4	56%

Note: **Ep**: The predicted result of maximum TN removal from the kinetic models, **Ex**: Result of maximum TN removal from the experiment.

Table 5. Predicted results of TP removal by applying Second Order and Stover-Kincannon models and the results of TP removal from the experiment in this facility.

Model	a d^{-1}	b	Ep (%)	Ex (%)
Second Order	0.84	2.84	30.66	86%
Model	U_{max} **mg/l.d**	K_B **mg/L.d**	**Ep (%)**	**Ex %**
Stover-Kincannon	1.93	6.92	23.64	86%

Note: **Ep**: The predicted result of maximum TP removal from the kinetic models, **Ex**: Result of maximum TP removal from the experiment.

The result of maximum TN and TP removal from the experiment were 86% and 56%, respectively. Comparing with modified septic tank conducted by Kimleng and Effendi [15], the performance of TN and TP removal of that research was lower than this research. It indicates that an on-site domestic wastewater treatment facility had achieved good performance of nutrient removal.

4 Conclusion

Effect of HRT, initial TN and TP concentration on the performance of TN and TP removal were evaluated. All of the variations gave the significant effect of HRT, initial TN and TP concentration on the performance of nutrient removal. The optimum operating condition can be achieved at HRT of 24h and COD:N:P of 350:38:2.9 for TN removal was 56% and HRT of 24h and COD:N:P of 600:60:3.7 for TP removal was 86%.The kinetics of nutrient removal of lab-scale model of an on-site domestic wastewater facility for treating synthetic wastewater which had been characterized based characteristic of wastewater from office building was investigated using three different models such as First Order Substrate Removal, Second Order Substrate Removal, and Stover-Kincannon Models. Second Order and Stover-Kincannon Models were found to be more suitable than First Order Model for prediction of performance of TN removal in on-site domestic wastewater treatment facility as the correlation coefficients were 0.97.The maximum predicted results of TN removal efficiencies by applying kinetic models were ranging from 50.48 to 57.87 %. It showed that on-site domestic wastewater treatment facility has achieved the good performance of TN removal. Controlling HRT and initial TN and TP concentration may keep good performance of nutrient removal in an on-site domestic wastewater treatment facility.

This research was supported by P3MI KK ITB, Project of JICA ITB III, and Gifu Prefectural Environmental Management and Technology Center, Gifu, Japan.

Nomenclature

COD: Chemical Oxygen Demand, (mg/L)
HRT : Hydraulic Retention Time (h or d)
Q: Flow rate (L/d or L/h)
k_1:First-order kinetic constant rate (d^{-1})
$k_{2(s)}$: Second-order kinetic constant rate (d^{-1})
K_B: Saturation constant(mg/L.d)
So: Influent substrate concentration (mg/L)
S: Effluent substrate concentration (mg/L)
TN: Total Nitrogen (mg/L)
TP: Total Phosphate (mg/L)
U_{max}: Maximum substrate utilization rate (mg/L.d)
V: Working volume of the reactor (L)

References

1. Ichinari. T., Ohtsubo. A., Ozawa. T., Hasegawa. K., Teduka. K., Oguchi. T. and Kiso. Y., *Wastewater treatment performance and sludge reduction properties of a household wastewater treatment system combined with an aerobic sludge digestion unit*, Process Biochem, **43**, pp. 722-728, (2008).
2. JECES, *Japan Education Center of Environmental Sanitation*, (2009).
3. Ormori. D., *Nitrogen and Phosphorous Removal Technology in Johkasou System, Proceeding of Johkasou Session*, Kyoto, Japan, (2003).
4. Yang. X.M., Morita.A., Nakano, I., Kushida. Y. and Ogawa., H., *History and current situation of night soil treatment systems and decentralized wastewater treatment system in Japan*. Water Prac. Technol. **5**(4), (2010).
5. Fajri. J.A., Yamada. T., Setiyawan.A.S. and Fusheng. L., *Evaluation of water and sediment quality in open channels that received effluent from Johkasou facility,* J Water Environ Technol, **13**(3), pp. 2017-208, (2015).
6. Matsuo. T., *Japanese experienced in water pollution control and wastewater treatment technologies*, Water Sci. Technol, **42** (12), pp. 163-172, (2000).
7. Vandith. V., Setiyawan. A.S. and Soewondo. P., *Kinetics of organic removal in Johkasou as an-onsite domestic wastewater treatment system, The Fourth Joint Seminar of Japan and Indonesia: Environmental Sustainability and Disaster Prevention (ESDP)*, Bandung, Indonesia, (2016).
8. Reeves. E. G., *Kinetic Analysis of Kluyvermyces Maximum Yeast Strain Louisiana*, (2004).
9. Jin. R. and Zheng. P., *Kinetic of nitrogen removal in high rate anammox upflow filter*, Hazard. Mater, **170** (2), pp. 652-656, (2009).
10. Grau. P., Dohanyas. M. and Chudoba. J., *Kinetic of multicomponent substrate removal by activated sludge*, Water res, **9**, pp. 637-642, (1975).
11. Yu. H., Wilson. F. and Tay. J., *Kinetics analysis of an anaerobic filter treating soybean wastewater, Water Res*, **32**(11), pp. 3341-3352, (1998).
12. Akhbari. A., Zinatizadeh. A. L., Mohammadi. P., Mansouri. Y., Irandoust. M. and Isa. M. H., *Kinetic modeling of carbon and nutrients removal in integrated rotating biological contactor-activated sludge system,* Int.J.Environ. Sci. Technol, **9**, pp. 371-378, 2012.
13. Mansouri. A. M., Zinatizadeh. A. L. and Akabari. A., *Kinetic Evaluation of Simultaneous CNP Removal in an up-Flow Aerobic/Anoxic Sludge Fixed Film (UAASFF) Bioreactor,* IJEE, **5**(3), pp. 323-336, (2014).
14. Isik. R. and Sponza. T., *Substrate removal kinetics in an upflow anaerobic sludge blanket reactor decolorizing simulated textile wastewater*, Process Biochem, **40**, pp. 1189-1198, (2005).
15. Kimleng. T. and Effendi. A. J., *COD removal kinetics and microorganisms growth kinetics using lab-scale batch condition for toilet wastewater treatment in modified septic tank using plastic media, Master thesis*, Bandung, Indonesia, (2016).

https://doi.org/10.1051/matecconf/201814704005

Control of pH of retained water in the coastal waste disposal site

Ramrav Hem[1,*], *Tomohiro* Yamasaki[1], and *Hiroko* Sumida[1]

[1]Geoenvironment Laboratory, Technical Research Institute, Toyo Construction Co., Ltd.,
1-25-1, Naruohama, Nishinomiya City, Hyogo Prefecture, 663-8142, Japan

Abstract. After landfilling of wastes is completed, the stabilization of landfilled ground requires much time and cost. Therefore, this study aimed to control the pH of retained water in the coastal waste disposal sites during landfilling process, by conducting field surveys and laboratory experiments. In field surveys, we investigated the changes of retained water quality such as pH, salinity, and dissolved oxygen. The results show the pH of retained water has risen to about 10 when the volume of landfilled wastes reached about 25% of landfill capacity. In lowing the pH, we considered a low-cost method by pumping seawater from the adjacent sea into the landfill. The mechanism in this method is that, H^+ dissociated from HCO_3^- in the fresh seawater react with OH^- eluted from wastes would result in pH decrease. The laboratory experiments were conducted to verify the effect on pH change by adding fresh seawater to alkalized seawater. As a result, the effect of injecting fresh seawater into alkalized seawater with pH higher than 9 was confirmed. Therefore, this treatment method is suggested to enable the disposal sites to be used promptly after landfilling is completed, by adding fresh seawater to purify the retained water and waste at low cost during landfilling process.

1 Introduction

In Japan, especially in the metropolitan areas with narrow in-land spaces, waste disposal facilities are built at the coastal and sea areas. It is known that seawater has pH buffering capacity [1]. However, at some disposal sites where the landfilling is almost completed, the retained water and pore water within landfill ground become high alkaline with pH of 11 or more. Therefore, the landfilling may be completed, while the pH is still high [2]. In order to use these sites, e.g., for factory or industrial land-use, after the landfilling is completed, the quality of the leachate generated during rainfall must satisfy the standard, e.g., pH of 9.0 or less. However, the pH of pore water in the landfilled ground remains high at most of the landfill sites, where the long-term treatments are required. This is because the alkaline components and organic components contained in the landfilled wastes become the pollution sources and those components will continue to release into the pore water in the landfill ground for a long time.

The countermeasures applied to purify the landfilled ground after the landfilling is complete such as "active washout using rainfall" and "forced drainage using a catchment underdrain" [3] were reported. Since the landfill sites were landfilled with various kind of wastes, the landfill ground may have different water permeability and air permeability. For that reason, although the purification can be promoted in the landfill ground with large grain size that makes it easy to flow through and ventilate, but it is not effective for the ground with small grain size due to its low permeability.

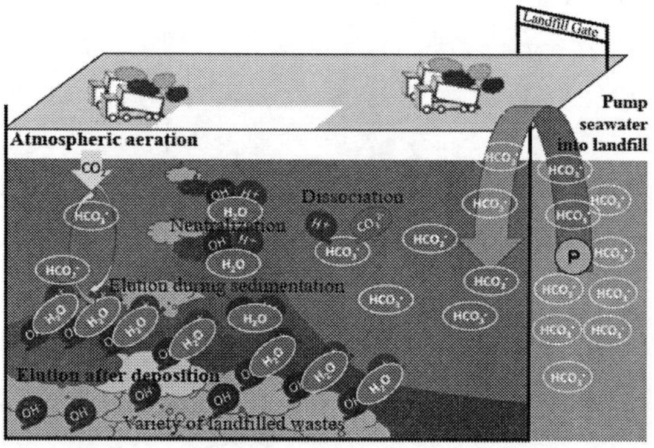

Fig. 1. Image of countermeasure and chemical reaction in retained water during landfill process

In such a case, in order to close the disposal sites, it is inevitable to choose uncertain or uneconomical countermeasures after the landfilling is completed.

As mentioned above, there is a need for a technology to quickly stabilize the leachate generated from the landfill ground, so that the disposal site can be closed early by implementing some countermeasures during the landfilling process, before the site is completed which makes it difficult to be purified. Therefore, in our research project, by focusing on the pH as the target of water quality, we assumed some prospective countermeasures that can be applied during the landfilling process. Those countermeasures are to maintain the buffering capacity of the retained water as much as possible before the

* Corresponding author: hem-ramrav@toyo-const.co.jp

landfilling is completed.

Among those countermeasures, in this paper, we present only the countermeasure with the pumping up of the fresh seawater into the retained water in the landfill site to lower the pH and to enhance the buffering capacity of the retained water (See Figure 1).

In this paper, we attempted to control the pH of retained water in the coastal waste disposal site during the landfilling process, by conducting field surveys and two separated laboratory experiments.

2 Field surveys

2.1 Method

2.1.1 Study site

The study site, namely, Port of Kinuura No. 3 Coastal Waste Disposal Site, is located in Taketoyo-cho, Chita-gun, Aichi prefecture, Japan (see Figure 2a). With the total capacity of 496 Mm^3, the site is divided into 2 compartments for stable waste (8.4 ha) and for controlled-type waste (34.4 ha) [4]. However, in this study, we focused on only the compartment for controlled-type waste which receive mainly fly ash. The landfill operation started to receive the waste in March 2011, and in March 2017, the volume of the landfilled waste was estimated to be about 25% of its total capacity.

2.1.2 Survey locations and dates

In order to investigate the changes of water quality of the retained water in this landfill site, four times of surveys had been conducted in summer and winter seasons from 2015 [5] to 2017. Figure 2b shows the survey locations, one location (Point C) is at the central part where water depth is relatively deep and the other four locations are at the East, West, South, and North parts of the site, where the water depths are relatively shallow.

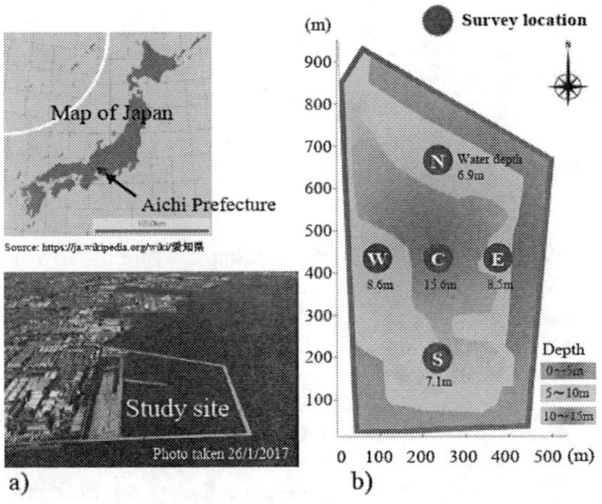

Fig. 2. a) Study site location and aerial view and b) survey locations and water depth

2.1.3 Survey equipment

In the surveys, the vertical distribution of pH, water temperature, DO, and salinity were observed by using a Multi-Parameter Water Quality Sonde (YSI 6600 V 2-4). In addition, we collected water samples basically at a 2 m interval from the water surface to conduct the laboratory analysis for pH, bicarbonate ion (HCO_3^-), dissolved metals (e.g., calcium, magnesium, iron, and aluminium), salinity, and chloride ion concentration.

2.2 Results and discussion

There was no difference between the observation values and the analysis values of the same depth in all the survey locations for all investigation days. From these results, in all the survey locations are assumed to be uniform in horizontal plan of the whole landfill site. Therefore, in this paper, only the data at point C is presented. Figure 3 shows the observation results at point C. From the survey results of 2017/1/31, the graphs show the pH increased with progress of landfilling, which were almost 10 from the top surface to middle layers and over 10 in the bottom layer. As for water temperature, the thermocline (steep vertical temperature gradient) occurred at the depth of 5~6 m in summer, but due to mixing due to high temperature in winter, it was uniform except the bottom layer. Salinity increased with the progress of landfill, and in the bottom layer it exceeded 40 ‰, and salinity halocline (steep vertical salinity gradient) developed at about 12 m in depth. Due to the influence of this halocline, water temperature, DO of water mass of 2017/1/31 was formed in the state that is not vertically mixed.

Figure 4 shows the results of samples analysis of point C. HCO_3^- decreased with the landfill progress, which was equivalent to less than a half of the value of general sea area (about 100 to 120 mg/L). Calcium as a whole was higher than general sea area (about 300 to 400 mg/L), and it was higher at the bottom layer. Magnesium was lower than the value of the general sea area (about 1000 to 1300 mg/L), and it was lower and lower at the bottom layer. The value also has decreased with the progress of landfill. According to the above results, it is considered that the chemical reactions as shown in the below chemical equations, modified from [1], occurred particularly in the vicinity of the bottom layer.

$$Ca(OH)_2 \Leftrightarrow Ca_2^{+} + 2OH^{-} \quad (1)$$

↑*Supplied from ash* ↑*Increase in seawater*

$$HCO_3^{-} \Leftrightarrow H^{+} + CO_3^{2-} \quad (2)$$

↑*Decrease in seawater*

$$OH^{-} + H^{+} \Leftrightarrow H_2O \quad (3)$$

$$Mg^{2+} + CO_3^{2-} \Leftrightarrow MgCO_3\downarrow \quad (4)$$

↑*Decrease in seawater* ↑*White deposit*

 https://doi.org/10.1051/matecconf/201814704005

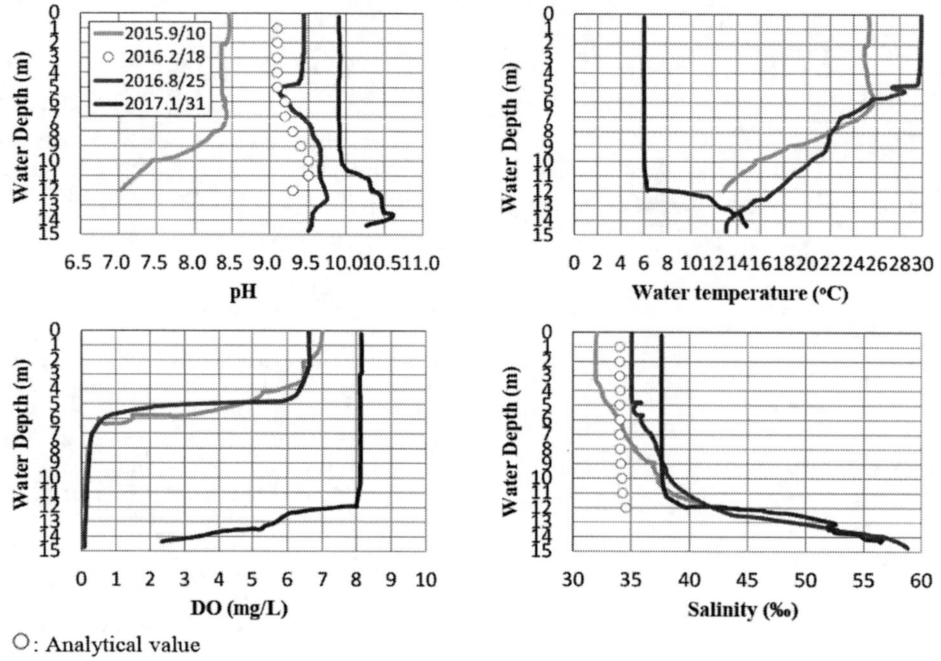

Fig. 3. Survey and analytical results of retained water quality (Point C)

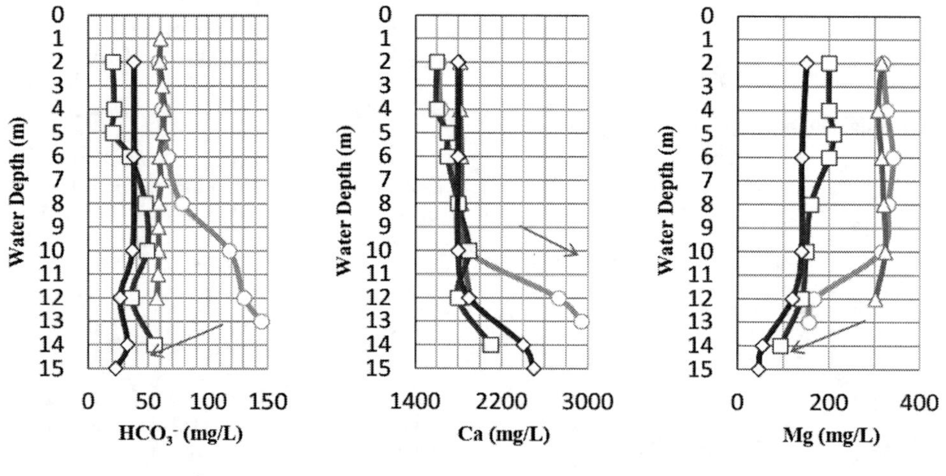

Fig. 4. Analytical results of retained water quality (Point C)

3 Laboratory experiments

3.1 Background

Two different laboratory experiments were carried out, separately. In both experiments, we used exactly the same fly ash taken from the site and artificial seawater made with same initial pH of about 8.2. Immediately after receiving the sample, we carried the sieve analysis for particle size distribution of fly ash. However, as show in Figure 5, the particle size of fly ash used in in the first experiment ranged from 0.85 to 2 mm, while particle size of less than 0.25 mm in the second experiment. With the smaller particle size of fly ash, we expected to have higher pH in the second experiment. Vertical pipes which assumed to be representative of the real landfill site in 1 dimensional vertically. However, the pipe length used in the first experiment is shorter than that of the second experiment.

By using short pipes, the experiment could be easily carried out to investigate the pH changes corresponding to different cases of water injection. However, with short pipes, the pH decrease due to not only from the effect of

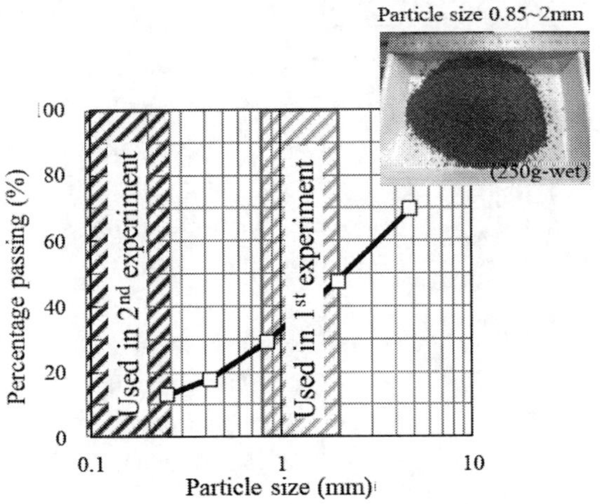

Fig. 5. Particle size of fly ash used in the experiments

the fresh seawater injected, but also from of CO_2 in atmosphere dissolved into alkaline seawater during experiment. Therefore, the second experiment was carried out with a longer pipe to investigate the depth of CO_2 influence and to clearly confirm the effect of fresh seawater injection.

The same pH meter (PH-230SDJ PE-21, Sato Corporation) has been used in both experiments. The pH meter can be used to continuously measure the pH at any depth by directly plunging into the seawater in the pipes.

3.2 Seawater injection experiment

3.2.1 Method

As shown in Figure 6, we used three acrylic pipes with an internal diameter of 10 cm and height of 1 m. At first, 5.0 L (height 63.7 cm) of artificial sea water with the initial pH of about 8.2 was input into the acrylic pipes, and the vertical pH was measured at an interval of 5 cm from the overlying water surface. Next, fly ash of 2.0 kg-wet was added little by little for 16 minutes. After that, it was left to stand for 24 hours, next, the vertical distribution of pH of overlying water was measured. And then, we started the injection of new artificial seawater using a small vertical pipe attached into the inside of each acrylic pipe. As shown in Table 1, the addition of new seawater, equivalent to 1% (50 mL) and 4% (200 mL) of the used seawater quantity, was carried out continuously every 24 hours for 6 days. The injection height of each case was set as shown in Table 1.

Table 1. Experimental cases for seawater injection

Seawater injection	Case 1	Case 2	Case 3
Ratio (%/day)	1	1	4
Volume (mL/day)	50	50	200
Injection height (cm)	70	40	40

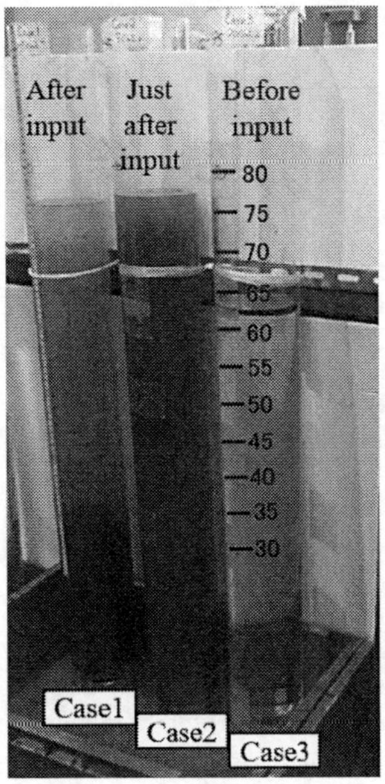

Fig. 6. Experiment conditions

3.2.2 Results and discussion

The changes of vertical distribution of pH before and after injection of new artificial seawater as a measure to lower the pH of alkaline seawater are shown in Figure 7. From the results, in Case 1, where the water injection height was 70 cm from the bottom surface of the acrylic pipe (about 5 cm below the overlying water surface), the pH from the injection position to the water surface greatly decreased. Similarly, in Case 2 and Case 3 with injection height of 40 cm from the bottom surface of the acrylic pipe (10 cm on the deposit surface), the pH below the injection height tended to decrease. The tendency was more remarkable in Case 3 with greater water injection volume. The pH of overlying water just above the sediment surface was suppressed small in Case 3, where much water injection volume and low water inject height compared to Case 2 and Case1.

The surface of the ash deposit, after 24 hours of injection, was covered with white precipitate and increased with increasing seawater injection in all cases. From the above results, it is considered that, the nearer to he deposited ash layer, the higher density of overlying water. Accordingly, when the water with a relatively low density injected, it diffuses towards the water surface, while neutralized with only the shallow part of overlying water. Therefore, this result suggests that a deep injection of new seawater is effective to neutralize the retained water as a whole.

After the injection was completed, it was left to stand. On the 49th day after injection, the overlying water and the deposited ash pore water of Case 3 were sampled and analyzed. Pore water was obtained by using a solid-liquid centrifuge separator. The analytical results of artificial sea

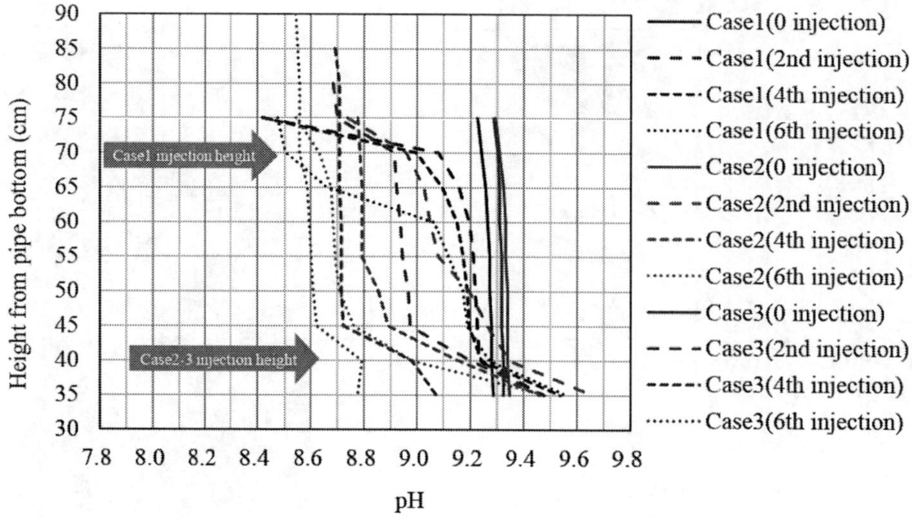

water before injection, overlying water, and pore water of Case 3 are shown in Table 2. Compared to the pH before being left to stand, the pH decreased as a result of atmospheric CO_2 dissolution from the overlying water surface, and the bottom layer was rising due to the influence of high alkaline pore water released from the ash deposit. The amount of HCO_3^- was decreased from the surface of overlying water to just above and within the ash deposit. Concentration of HCO_3^-, in general sea water with the maximum pH of 8.25, is about 90 % at pH 9 and about 50 % at pH 10 (see Figure 8). From a certain proportion of the dissolved HCO_3^-, the dissociated hydrogen ion H^+ (and carbonate ion CO_3^{2-}) is neutralized with OH^- derived from waste which can be considered to suppress the rising pH of seawater [1]. Compared to general sea water, the experimental value of HCO_3^- is small in the range of pH 8.2 to 9.5, which indicates that it was consumed by the neutralization reaction. At pH 10.5 it is considered that neutralization reaction does not occur because it is the same rate as general seawater. Therefore, the effect of injecting new seawater remarkably when the initial pH was 9.0 or more.

From the analytical results, the concentration of magnesium decreased downwards, and especially in the pore water. This is probably because the magnesium ion (Mg^{2+}) in the overlying water or pore water react with CO_3^{2-} dissociated from HCO_3^- and form as precipitate of magnesium carbonate $MgCO_3$, resulting in a decrease in dissolved concentration [1]. It is thought that the increased in the sediment at the surface layer of the sediment is caused by further neutralization and bonding reactions as the pH had increased.

3.3 Effect of atmospheric CO_2 and seawater injection experiment

3.3.1 Method

In order to make pH of the overlying water higher than the above experiment, smaller particle size of fly ash (under

Table 2. Analytical results of artificial seawater, retained water, and pore water (Case 3)

Items			pH	HCO_3^- (mg/L)	Mg (mg/L)	Ca (mg/L)
Artificial seawater (before input)			8.16	207.4	1300	640
Case 3	Height of the overlying water from the pipe bottom (cm)	85	8.29	116.1	1200	930
		80	N.D.	N.D.	N.D.	N.D.
		≀				
		55				
		50	8.72	123.0	1100	1000
		45	N.D.	N.D.	N.D.	N.D.
		40				
		35	9.31	100.0	N.D.	N.D.
	Pore water in deposited ash		10.53	54.6	110	2600

 https://doi.org/10.1051/matecconf/201814704005

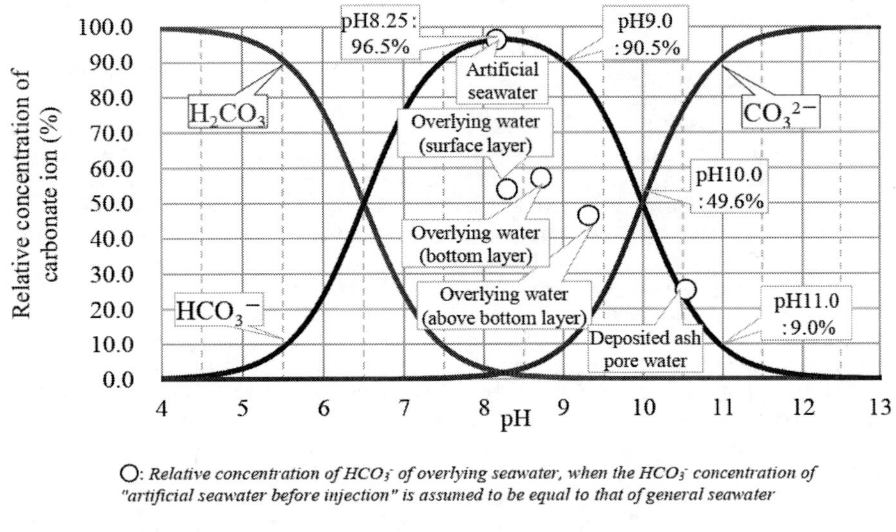

Fig. 8. Carbonate ion concentration of the general seawater and the experimental value

0.25 mm) was used. To clearly investigate the depth of influence of CO_2 dissolved into the overlying water and effect of the seawater addition, the use of a longer PVC pipe of 6 m in length and 75 mm in inner diameter was considered.

At first, as shown in Figure 9 the PVC pipe was firmly affixed to the wall of the experiment pit inside our research facility, where the environment is calm without strong wind. Then, about 24 L of artificial seawater with pH of 8.2 was injected into the PVC pipe. After that, 5 kg of fly ash with water content 12.36 % was input little by little. After 24, with an interval of 20 cm from the water surface, vertical distribution of pH of overlying water was measured.

As the injection method, a small PVC pipe was bended with 90 degree angle at the end so as not to allow the injected water drop below the exact depth, which was set to be about 2 m from the overlying water surface.

3.3.2 Results and Discussion

As a result, the average vertical distribution of pH of overlying water had been measured to be ranged from 11.0 to 11.5 for the first two months and a half, from the

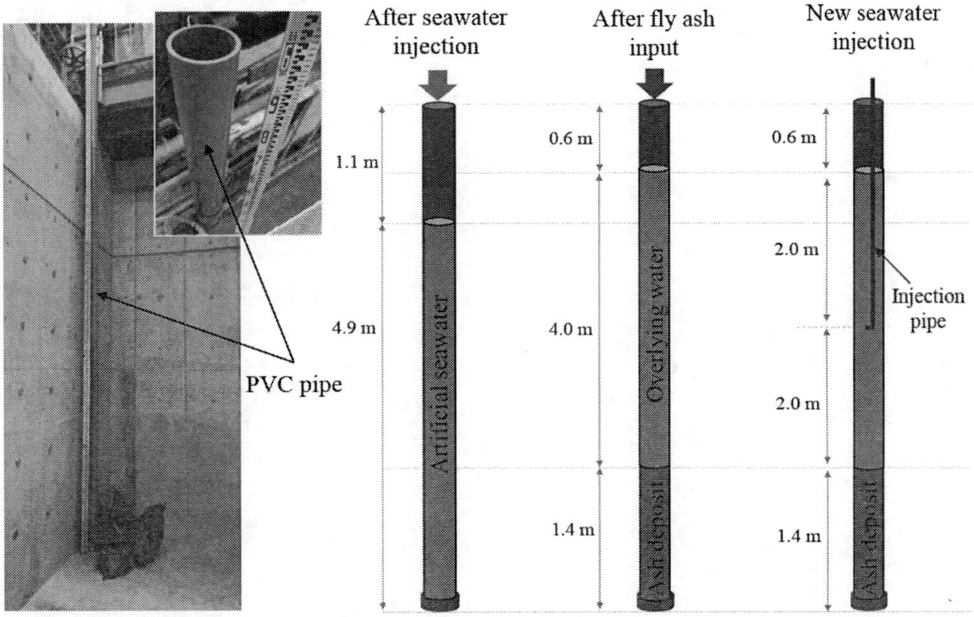

Fig. 9. Illustration of the second experimental condition

 https://doi.org/10.1051/matecconf/201814704005

beginning of the experiment. Based on the results during these period, the trend of vertical distribution of pH slightly decreased from the depth of 1 m downwards to the surface of overlying water. From this results, we could assumed that, under the effect of atmospheric aeration, CO_2 may dissolve into the overlying water during experiment until the depth of about 1 m from the overlying water surface.

After 3 months from the beginning of the experiment, the average pH of overlying water, measured from the surface to the injection depth of 2 m, was about 10.28. From then, we started to carry out the new artificial seawater injection. The first injection of about 220 mL of artificial seawater which was equivalent to about 1% of the total volume of the seawater initially used. After 24 hours, vertical distribution of pH was measured, and after that, immediately we carried out the second injection in the same way and the same amount of artificial seawater as the first injection. After 24 hours, the vertical distribution of pH was measured again. The results of vertical distribution of pH measured just before and after injection are shown in Figure 10.

We considered the average value of pH measured from the water surface to the injection depth of 2 with an interval of 20 cm. The average pH was about 10.28 just before the injection, had dropped to about 10.14 at 24 hours after the first injection, and then, dropped to 10.07 at 24 hours after the second injection. And, at 48 hours after the second injection the average pH very slightly decreased to just little below 10.07. Although the injection depth was at 2 m, however, the small effect of the seawater injection can be observed until the depth of 2.4 m. On the other hand, the pH distribution below the depth of 2.4 m was slightly increase. This may be due to the pore water with high alkalinity continued to release from the deposited ash layer as time pasted.

From these experimental results, especially, pH at 24 hours after the first injection, the effect of adding new artificial seawater in lowering pH of alkaline seawater was clearly confirmed.

4 Summary and concluding remarks

In Japan, coastal waste disposal sites are generally built with large scale and high construction cost. To use the sites after landfilling is completed for, e.g., industrial land-use, it requires much time and cost for stabilization of landfill ground (leachate treatment) to meet the standards. Therefore, our study aimed to control the pH of retained water in the coastal waste disposal sites during the landfilling process, by conducting field surveys and laboratory experiments.

Water quality surveys of the retained water were conducted for two consecutive years (in summer and winter) at the study coastal waste disposal site. In summer, a thermochemical layer and a halocline layer were confirmed, and in winter, the halocline of the bottom layer was confirmed. In such an environment that the halocline derived from the landfill waste is not eliminated throughout the full term, even if, for example, some of purification measures applied from the surface layer to the middle layer, because the purification effect may not reach the deep seawater layer with high density.

As the landfill progresses, the pH has risen to nearly 10, it is considered that the retained water in this study site reaches its buffering capacity limit [6] due to the decrease in HCO_3^- in the retained water, the increase in dissolved calcium concentration in the bottom layer, etc.

In order to control of pH of retained water of the study site, we consider a low-cost method by simply pumping seawater from the adjacent sea into the site. The added fresh seawater is rich in HCO_3^-. Therefore, the H^+, dissociated from HCO_3^-, react with OH^- eluted from landfilled waste to result in lowering the pH. With this mechanism the buffering capacity of the retained water can be maintained. From the results of the laboratory experiments, this measure was confirmed effective for the alkaline seawater with pH of 9 or over. Therefore, the study suggests this method as one of the effective and low-cost methods applied during landfill process, so the disposal site can used promptly after the landfilling is completed.

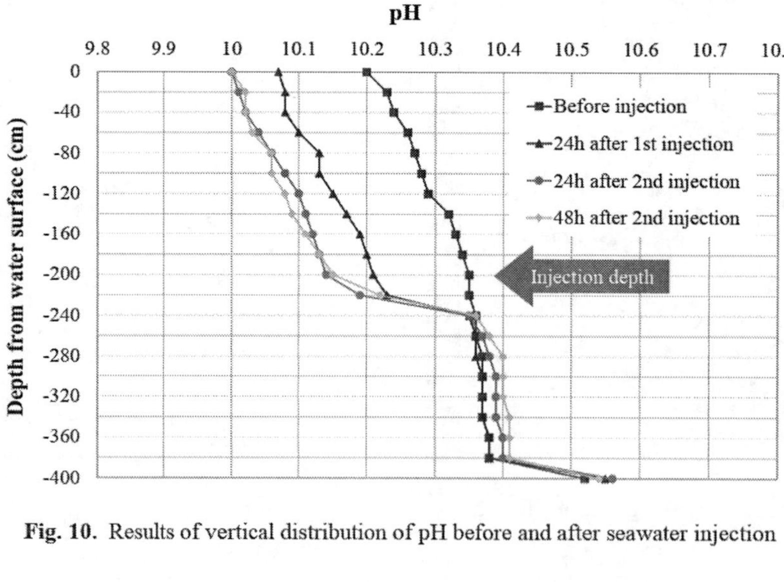

Fig. 10. Results of vertical distribution of pH before and after seawater injection

The authors would like to thank ASEC (Aichi Seaside Environment Center) for allowing us to carry out research on this coastal waste disposal site. We are also grateful for its partial data provision for this paper.

References

[1] H. Sakanakura, N. Nakagawa, N. Maeda, Y. Kadota, S. Mizutani, K. Endo, K. Miyawaki, *The 25rd Annual Conference of Japan Society of Material Cycles and Waste Management*, No. 25, pp. 399-400 (2014). (In Japanese)

[2] JESC (Japan Environment Sanitation Center), *Examination survey toward the technical standard of the sea area final disposal sites for closure or abolished*, pp. 18 (2005). (In Japanese, title is unofficially translated by authors)

[3] J. Tsukahara, T. Oshima, T. Sakaki, N. Maeda, *The Thirty-seventh Japan National Conference on Geotechnical Engineering*, **JGS37**, pp. 2437-2438 (2002A). (In Japanese)

[4] ASEC (Aichi Seaside Environment Center), Environmental Activities Report (Activity period: April 2016~March 2017), Issued 1 April 2017 (in Japanese). Link: http://www.asec.or.jp/images/profile/p_report.pdf

[5] T. Yamasaki and H. Sumida, *Proceedings of the 27th Annual Conference of Japanese Society of Material Cycles and Waste Management*, pp. 421-422 (2016). (In Japanese)

[6] T. Yamasaki, R. Hem, and H. Sumida, *Journal of Japanese Association for Coastal Zone Studies*, **28**, No. 3, pp. 41-49 (2015). (In Japanese)

Leachate Treatment from Sarimukti Landfill Using Ozone with Sludge from Water Treatment Plant as a Catalyst

Muhammad Yudha Ramdhani[1,*], *M. Rangga* Sururi[1], and *Siti* Ainun[1]

[1]Institut Teknologi Nasional, Environmental Engineering Department, 40124 Bandung, Indonesia

Abstract. Leachate is the liquid waste from anaerobic decomposition in a landfill. The ozonation process can be used for leachate treatment. Sludge from sedimentation in water treatment plant contains 5.96% of Al and 9.35% of Si which can affect of its cation exchange capacity and affects the active site in the catalyst. This study aims to determine the effectivity of sludge in the ozonation process to treat leachate. A 1,5 L semi-batch reactor containing 1 L sample was used in this experiment with the rate of oxygen supply was at 4 L/min taken from ambient air. Two groups of sludge weighing 1.5 grams, 3.0 grams and 4.5 grams were used and activate with physically and chemically activated. The best result was obtained by the physically activated sludge with mass of 4.5 gram O_3-L-4,5 AF. The differences of removal efficiency between O_3-L-4,5 AF with the control (O_3) for turbidity were respectively 13.02% and 7.81%, for EC were 10.57% and 8.29%, for COD were 49.44% and 37.50%, and for residual ozone concentration at the end of contact time were 7.6 mg/L and 9.7 mg/L. It can be concluded that activaed sludge and ozonation can be used as a catalyst in leachate treatment.

1 Introduction

Population growth will lead to an increase in the amount of waste disposed off to a landfill. The waste disposed to the landfill undergoes several simultaneous physical, chemical, and biological changes that produce a liquid called leachate [26,12]. Leachate is the liquid from solid waste decomposition in a landfill. Leachate will be produced, with a high chemical oxygen demand (COD), low biodegradibility, heavy metals, pathogens [25], organic materials (biodegradable and non-biodegradable), and ammonia-nitrogen [27].

The commonly used techniques to treat leachate in Indonesia is the use of stabilization ponds [5]. This technique has the disadvantages of which is the long detention time, and required a large area. Stabilization ponds also used at Sarimukti landfill, and based on measurement data from the west java provincial government through the department of settlements and housing (Diskimrum), 2016, the results of leachate treatment with stabilization pond technique has a BOD/COD ratio of 0.050 [29], and the effectiveness of COD removal was only 13,94% [20]. The effectiveness of biological leachate treatment in Sarimukti landfill is less effective, because leachate is a wastewater that is difficult to treat by biological [4], due to the complex nature of the COD component [2]. Therefore, chemical processing alternatives are required, i.e. using ozone.

Ozonation processes are attractive means for the treatment of landfill leachates due to the high oxidative power that ozone possesses [25]. Ozonation is a chemical process that can oxidize many organic and inorganic materials in water. Ozone is a very selective oxidant, and unstable in water because it can decompose quickly, however the unique feature of ozone is its decomposition into hydroxyl radicals ($OH^{\bullet}$) which are the strongest oxidants can react fast with many dissolved compounds and the water matrix [9]. According to Nuriana, 2015, leachate treatment with ozone has a removal efficiency for turbidity of 91.47%, electrical conductivity (EC) of 20.41%, and COD of 35% [14]. The ozonation process can be optimized by various advanced oxidation process (AOP) such as the addition of peroxide (H_2O_2), UV, and catalytic ozonation. According to Sururi et.al, 2014, the addition of 0.6 gr/L H_2O_2 in diluted the leachate sample taken from Sarimukti landfill has a removal efficiency COD of 28.9%, and with the use of UV lamps within 10 cm of the ozone contactor has efficiency of 23.4% [23,12].

In the last few years heterogeneous catalytic ozonation has been successfully used for the removal of organic compounds [9]. Heterogeneous catalytic ozonation is one of the oxidant techniques used to improve the efficiency of ozonation process [10]. The application of ozonation technique by using a catalyst is a promising alternative for leachate treatment. The sludge from water treatment plant can be used as a

* Corresponding author: ramdhani.yudha@outlook.co.id

catalyst, based on Muruganandham et.al, 2007, found that water treatment sludge powder was successfully used as an ozone decomposition catalyst in aqueous medium. Increase in the catalyst loading from 125 to 750 mg significantly promotes the percentage of ozone decomposition from 50.9% to 96.2% [13]. Thus, it is predicted that the use of sludge in the ozonization process of leachate treatment can accelerate the decomposition of ozone into $OH^{\bullet}$.

Because the leachate treatment at Sarimukti landfill which has a very low BOD/COD ratio, it needs chemical processing with ozone. Ozone is a powerful oxidant that can treat waste that is difficult to degrade such as leachate. In addition, this study aims to determine the effectiveness of the addition of sludge from drinking water treatment to treat leachate which is assumed to accelerate the decomposition of ozone to $OH^{\bullet}$.

2 Materials and Methods

2.1. Materials

Grab sample were taken from the Sarimukti landfill leachate collection points. The sludge that will be used as a catalyst was collected from sedimentation unit of Tirtawening water treatment plant in Bandung. The sludge was previously examined using X-ray Flourescence (XRF) which showed that the sludge consist of 9.35% Si, and 5.96% of Al, such that the Si/Al ratio is 1.56, this Si/Al value is dominated by silica content rather than aluminium, while Sururi et.al, 2016, says that aluminium acts as a catalyst. After that, the sludge as a catalyst was sieved based on ASTM D 421-85 to determine the coefficient of uniformity [22], which was 5.67, indicating a much more sand. Because the sludge was too fine sand, it is necessary to solidified, by compacting, to get a greater specific gravity than before. If the sludge is not solidified, it will cause the sludge to carried away during the ozonation process at the time of measurement. Egg whites were used for solidification as an adhesive, which serves to attached the sludge.

The catalyst is then physically and chemically activated. Physical activation was done by heating at 105°C for 24-36 hours [28]. For chemical activation, the solidified sludge was heated at 105°C for 24-36 hours, and impregnated with 24.5 ml of 3 mol/L per 10 gr of sample with $ZnCl_2$ for 24 hours at room temperature; after which the supernatant liquid was removed, and the sludge was dried at 105°C for 24 hours. For chemical activation, the sludge was than washed using 3 mol/L HCl to remove inorganic impurities, rinsed with de-ionized water and dried at 105°C for 24 hours [28]. A pH_{PZC} measurements was taken for the activated sludge, to know the pH where the catalyst is uncharged or the positive and negative charges are equal (isoelectric point) [7]. A pH_{PZC} in this study was measured to understand the mechanism of catalytic ozonation of the adsorption phenomenon from the catalyst when it is used in the ozonation process. Measurement of pH_{PZC} was done by titration [16]. Firstly the sludge as catalyst were washed with de-ionized water and dried in an oven at 108°C for 2 hours. Subsequently, sludge as much as 0.1 gr was added to 25 mL of 10^{-3}mol/L NaCl with continues stirring. The pH changed gradually and become constant at the point of zero charge (pH_{PZC}) [16]. Physically activated sludge has a pH_{PZC} value of 6.46-7.05 pH, while chemically activated with $ZnCl_2$ has a pH_{PZC} value of 4.32-4.42 pH.

2.2 Materials and Methods

This study was conducted in a semi-batch reactor with continuous ozone gas supply to the contactor filled with leachate sample. The equipment used in this studies consisted of an aerator, a flow meter, an ozone generator, and an ozone contactor that can be seen in Figure 1.

Fig. 1. Reactor Configuration.

The flow rate of oxygen supply was at 4 L/min taken from ambient air as connected to an ozone generator. The Ozone generator which changes the O_2 into O_3, has a spesification of OZF-1G, 200 W, and 220 V. The ozone generated was applied to an ozone contactor (1500 ml) which had previously been filled with 1000 mL of leachate sample. Sludge as a catalyst was then added to the ozone contactor with different activation and weight. Physically activated abbreviated as AF and chemically activated abbreviated as AK, have a weight variation of 1.5 gr (O_3-L-1,5 AF/AK); 3.0 gr (O_3-L-3,0 AF/AK); and 4.5 gr (O_3-L-4,5 AF/AK) was prepared. To make a homogeneous ozone condition through out the leachate sample, the contactor was equipped with magnetic stirrers. The ozonation process is carried out for 180 min, with samples taken ± 30 ml every 30 minutes. Parameter such as pH, dissolved oxygen (DO), residual ozone concentration (ROC), turbidity, electrical conductivity (EC), and chemical oxygen demand (COD) were measured every 30 minutes.

3 Results

3.1. Leachate Characteristics and Sludge as Catalyst

The results of measurement of the initial characteristics of leachate sample from Sarimukti landfill can be seen in Table 1.

Table 1. Leachate Characteristic.

No	Parameter	Unit	Concentration
1	pH	-	7.86-8.40
2	DO	mg/L	1.34-5.10
3	Turbidity	NTU	45.25-116.75
4	EC	mS/cm	4.31-29.57
5	COD	mg/L	4286-8378

Table 1 shows that the pH is neutral or slightly alkaline. This high pH value indicates that there are many hydroxide ions (OH^-) that can act to initiate the process of ozone decomposition [8]. The value of DO in the leachate sample were at 1.34-5.10 mg/L, indicating the amount of oxygen (O_2) available in the leachate. Oxygen fluctuations can affect to other chemical parameters, especially during oxygen-free conditions, which can lead to changes in the solubility properties of some chemical elements in the water [11, 6]. Sarimukti landfill and Jatibarang landfill is a landfill that has been operating for more than 10 years, and has similarities in treating lechate with stabilization pond. Sarimukti and Jatibarang landfills has a slightly similar pH were respectively 8.4 and 8.5, and for COD were 4286-8378 mg/L and 4000 mg/L [19].

The turbidity of the sample were 45.25-116.75 NTU indicate the amount of organic and inorganic content suspended in the leachate. The value of turbidity from Sarimukti landfill is lower than Jatibarang landfill of 300 NTU [19]. The EC of the sample were at 4.31-29.57 mS/cm indicating the inorganic content of the leachate sample. The value of EC is very high as compared to other leachates in other landfills in Indonesia which ranges from 1000-40000 µS/cm [24]. The concentration of COD were 4286-8378 mg/L indicating the number of organic compounds that were biodegradable and non-biodegradable. The value of COD is very high when compared with the effluent quality standard of leachate treatment plant in Indonesia, about the effluent leachate of 300 mg/L [17].

Sludge is predicted can used as a catalyst for ozone decomposition and also be utilized as an adsorbent. According to Ikhlaq et.al, 2012, the pH of a solution plays an important role in understanding the mechanisms of catalytic ozonation, since it is affected by ozone decomposition. At basic pH ($pH > pH_{PZC}$) the surface is negatively charged (no surface hydroxyl groups present), at acidic pH ($pH < pH_{PZC}$) its surface is positively charged (presence of hydroxyl groups), while at $pH = pH_{PZC}$ the surface is neutral [9]. Sludge as the catalyst used in this process has different pH_{PZC} between activation variations. pH_{PZC} on physically activated catalysts of 6.46-7.05 pH and pH_{PZC} on chemical activated catalysts of 4.32-4.42 pH. Because the pH of the solution is higher than pH_{PZC}, it is predicted that the sludge will act as adsorbent of O_3 and organic matter, so that the reaction of ozone and organics will occur at the surface of sludge, see Tabel 1.

3.2. Response to The Ozonization Process Based on pH and DO

pH is used as an indicator for hydrogen ion activity [21]. A high pH value indicates more hydroxide ions (OH^-) acting as an initiator in the decomposition of ozone [8]. In the ozonation system pH usually gives a positive effect on the removal rate of COD [3]. A high pH value is important because the hydroxide ions will start the ozone decomposition involving the following reactions [8]:

$$O_3 + OH^- \rightarrow O_2 + HO_2^- \quad (1)$$

$$O_3 + HO_2^- \rightarrow OH^\bullet + O_2^{\bullet -} + O_2 \quad (2)$$

if $pH \leq 8$:

$$O_3^{\bullet -} + H^+ \leftrightarrow HO_3^\bullet \quad (3)$$

$$HO_3^\bullet \rightarrow OH^\bullet + O_2 \quad (4)$$

if $pH \geq 8$:

$$O_3^{\bullet -} \leftrightarrow O^{\bullet -} + O_2 \quad (5)$$

$$O^{\bullet -} + H_2O \rightarrow OH^\bullet + OH^- \quad (6)$$

$$OH^\bullet + O_3 \rightarrow HO_2^\bullet + O_2 \quad (7)$$

These reactions will increase pH by time as can be seen in Figure 2. The pH of all variations increase with increasing contact time as indicated by the increasing of hydroxide ions. The physically activated catalysts resulted in a higher pH than the control condition, and the other variations. The more catalyst was added during the ozonation process will result in higher pH values. Based on the reaction (1) and (2) the initiation of ozone decomposition can be accelerated by increasing the pH or by the addition of hydrogen peroxide [1].

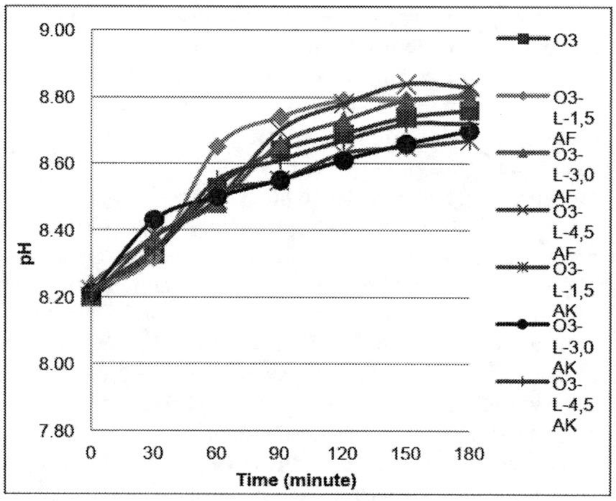

Figure 2. pH Change by Time.

Dissolved oxygen is one of the important parameters in leachate after going through ozonation process. Figure 3 shows that the DO fluctuates during the 180 minutes of the process. Increased DO occured because of the decomposition process of ozon based on reactions (1) and (2); and its decrease, occurred because oxygen acts as an oxidizing agent which react with the organic material in the leachate.

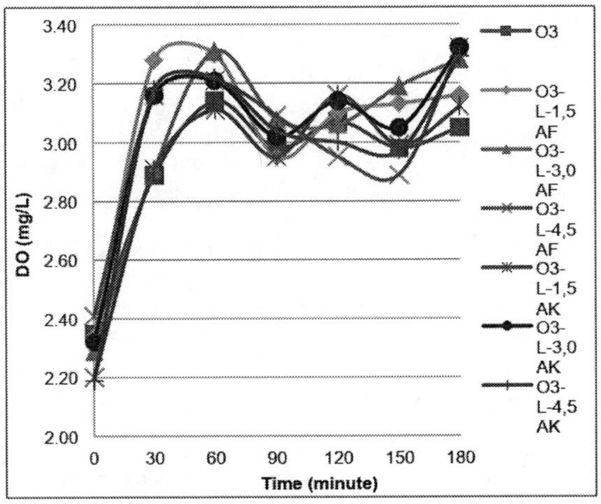

Figure 3. Dissolved Oxygen by Time.

3.3. Residual Ozone Concentration (ROC)

ROC measurements were conducted to find out the more dominant type of reaction (direct reaction and indirect reaction) occuring between ozone and the organic, and inorganic constituent in the sample. A direct reaction is a selective reaction that causes high ROC value due to the low ozone decomposition into $OH^{\bullet}$. The indirect reaction, however, is non-selective indicated by the low ROC value due to the high ozone decomposition into $OH^{\bullet}$. Measurements were made at the last 3 contact points because they were a critical points, and there has been decoloration on leachate sample [20].

Figure 4 shows a fluctuating ROC for all variations. The chemically activated catalysts resulted in higher ROC values than the physically activated, but it is lower than the control condition. The more catalyst was added, the faster the ozone decomposition takes place, which indicates an indirect reaction by $OH^{\bullet}$, evidenced by low ROC values due to the high ozone decomposition into the $OH^{\bullet}$. Based on the ROC data, it can be concluded, that the addition of sludge as catalyst can accelerate the decomposition of ozone into $OH^{\bullet}$.

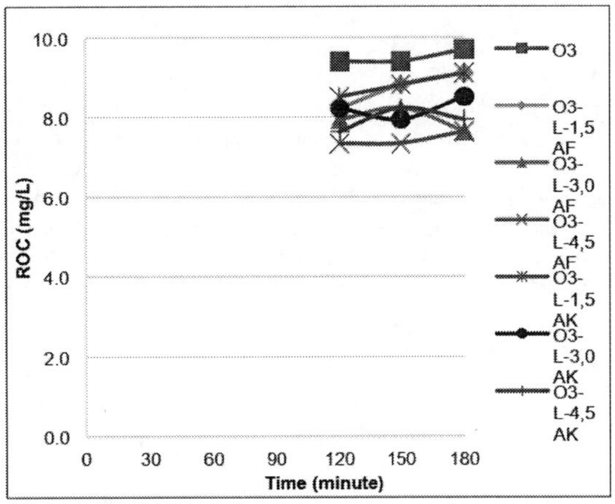

Figure 4. Residual Ozone Concentration.

3.4. Process Efficiency

3.4.1 Physical Parameter: Turbidity Removal

Turbidity, expressed in NTU (nephelometric turbidity units), is caused by material suspended from colloidal size to dispersed coarse material [21].

Figure 5 shows that turbidity can decreased by all variations; indicating an increase in the effluent discharge efficiency. The chemically activated catalysts removed turbidity the most, by 13.83-14.92% which were higher than the control of 7.81%, and other physically activated catalysts of 11.58-13.02%. This phenomenon occured because $ZnCl_2$ can bind water molecules contained in the sludge, which causes the pores of the sludge to increase its surface area [15], so that many pollutants can be adsorbed by the chemically activated catalysts.

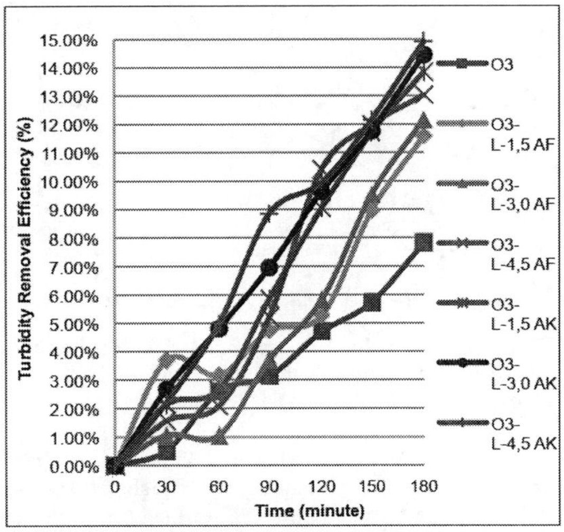

Figure 5. Turbidity Removal Efficiency.

Figure 5 also shows that the more catalyst was added, the higher the rate of turbidity removal. The highest level of removal was found for the O_3-L-4,5 AK, which decreases the turbidity from 45.25 NTU to 38.50 NTU or a removal efficiency of 14.92%. The decrease in turbidity occurs because ozone can degrade compounds that cause turbidity during the ozonation process. Continuous ozonization process create negatively charged colloidal particles which can be neutralized by ozone [18]. This phenomenon can be attributed to the high pH value of this variation as shown in Figure 2 and low ROC value as shown in Figure 4. The alkaline pH is indicated by the amount of OH^- ions, and the low ROC indicates the amount of ozone decomposed into $OH^\bullet$, causing the turbidity of the sample to decrease.

3.4.2 Physical Parameter: EC Removal

In this study the ion measured as electrical conductivity was estimate that to represent the inorganic content in the leachate sample.

Figure 6 shows that all catalyst variations increase the efficiency of removing the inorganic content of the leachate. The chemically activated catalysts removed EC by 7.82-8.77% which were higher than the control of 8.29%, but it is lower than the physically activated catalysts of 10.44-10.57%. The low removal efficiency for the chemically activated catalysts can be caused by saturation of it's surface area, as this variation was able to decrease turbidity more than the physically activated catalyst.

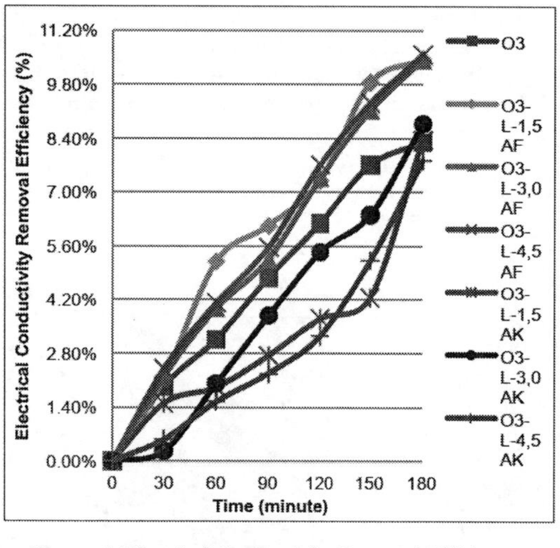

Figure 6. Electrical Conductivity Removal Efficiency.

The highest EC removal was found for of the O_3-L-4.5 AF which decreased the EC from 19.96 mS/cm to 17.85 mS/cm, or a removal efficiency of 10.57%. This phenomenon once again can be attributed to the high pH value of this variation as shown in Figure 2, and has a low ROC value as shown in Figure 4. High pH, and low ROC indicate the amount of $OH^\bullet$ formed that will react with the inorganic compound in the leachate sample. The low removal of inorganic compounds might be due to a material that is quite difficult to degrade, even using ozone which is a strong oxidizing agent.

3.4.3 Organic Parameter: COD Removal

COD removal was measured to find out how much organic compound can be eliminated, especially the non-biodegradables.

Figure 7 shows that each variation increases in COD removal by time. The chemically activated catalysts removed COD by 27.27-36.36% which were higher than the control of 30.0 0%, but the physically activated catalysts has a removal efficiency for COD of 27.27-40.00% higher than the other variations.

The highest COD removal was found for of the O_3-L-4.5 AF which decreased the COD from 7616 mg/L to 4570 mg/L, or a removal efficiency of 40.00%. This phenomenon can also be attributed to the high pH value of this variation as shown in Figure 2. In addition, this variation has a low ROC value as shown in Figure 4, the low ROC indicates that ozone decomposes to $OH^\bullet$. The decrease in COD concentration is caused by a non-biodegradable organic substance which is oxidised either directly with ozone which is selective or indirect by $OH^\bullet$ which has a non-selective property and converts it into an inorganic substance such as CO_2.

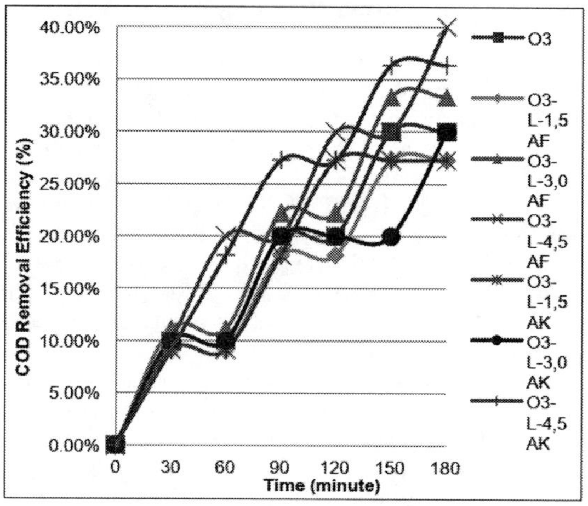

Figure 7. COD Removal Efficiency.

4 Conclusion

Several important conclusion can be drawn as follows:

1. The addition of sludge as a catalyst in the ozonation process can increase the ozone decomposition into $OH^\bullet$, thereby increasing the efficiency for removing organic and inorganic materials in leachate.
2. Sludge as a physically and chemically activated catalyst, in this study, showed a slightly different role in removal of pollutants. The physically activated catalysts were best to remove EC and COD. The chemically activated catalysts were best to remove turbidity.
3. The difference in sludge weight used during the research had an effect on the rate of removal, with more catalyst was added resulting in the higher the rate of removal, and the faster the decomposition of ozone.

4. Catalyst efficiency is an important parameter for commercial use. Further research is needed to determine the effect of catalyst if it is used more than once, and for optimum weight variation in leachate processing. Eventually the efficiency of a catalyst determine its practical applications.

References

1. Acero, J.L., & Gunten, U.V., Influence of Carbonate on the Ozone/Hydrogen Peroxide Based Advanced Oxidation Process for Drinking Water Treatment, Ozone: Science & Engineering: The Journal of the International Ozone Association, **22**:3, 305-328, (2000).
2. Barratt, P.A., Baumgartl, A., Hannay, N., Vetter, M., & Xiong, F., Chemox^{TM}: Advanced Waste Water Treatment With The Impinging Zone Reactor, Published by Elsevier Science Ltd, Wat. Sci. Tech, **35**, No. 4, pp. 347-352, (1997).
3. Beltran, F.J., *Ozone Reaction Kinetics for Water and Wastewater Systems*, Lewis Publishers : Boca Raton, (2004).
4. Cortez, S., Teixera, P., Oliveira, R., & Mota, M., Evaluation of Fenton and Ozone-Based Advanced Oxidation Processes as Mature Landfill Leachate Pre-treatments, Journal of Environmental Management **92**, 749-755, (2011).
5. Damanhuri, E., & Padmi, T., *Pengelolaan Sampah*, Diktat kuliah TL 3104, Institut Teknologi Bandung, Bandung, (2010).
6. Effendi, H., *Telaah Kualitas Air Bagi Pengelolaan Sumber Daya dan Lingkungan Perairan.*, Kanisius. Yogyakarta, (2003).
7. Gulicovski, J.J., Čerović, L.S., & Milonjić, S.K., Point of Zero Charge and Isoelectric Point of Alumina, Journal of Materials and Manufacturing Processes, **23**:6, 615-619, (2008).
8. Gunten, U.V., Ozonitation of Drinking Water : Part I. Oxidation Kinetics and Products Formation, Water Reasearch, **37**, 1443-1467, (2003).
9. Ikhlaq, A., Brown, R. David., & Kasprzyk-Horden, B., Mechanisms of Catalytic Ozonation on Alumina and Zeolit in Water: Formation of Hydroxyl Radicals, Applied Catalysis B: Environmental 123-124, 94-106, (2012).
10. Ikhlaq, A., Brown, R. David., & Kasprzyk-Horden, B., Catalytic Ozonation for The Removal of Organic Contaminants in Water on Alumina, Applied Catalysis B: Environmental **165**, 408-418, (2015).
11. Jeffries, M., & Mills, D., *Freshwater Ecology, Principles and Management, Development in Hydrology*, Pergamon Press, Oxford, UK, **14** (1996).
12. Krisnawati, A., Sururi, M.R., & Ainun, S., Pengaruh Karakteristik Lindi Terhadap Ozonisasi Konvensional dan Advanced Oxidation Process (AOP), Jurnal Rekayasa Lingkungan, 2, **2** (2014).
13. Muruganandham, M., Chen, S.H., & Wu, J.J., Evaluation of Water Treatment Sludge as a Catalyst for Aqueous Ozon Decomposition. Catalysis Communications **8**, 1609-1614, (2007).
14. Nuriana, W., Pengolahan Lindi TPA Sarimukti dengan Menggunakan Ozonisasi Konvensional dan Advanced Oxidation Process (AOP) Tipe Ozon/H_2O_2, Institut Teknologi Nasional, Bandung, (2015).
15. Prasetyo, Y., & Nasrudin, H., Penentuan Konsentrasi $ZnCl_2$ Pada Proses Pembuatan Karbon Aktif Tongkol Jagung dan Penurunan Konsentrasi Surfaktan Linier Alkyl Benzene Sulphonate (LAS). UNESA Journal of Chemistry, **2** (2013).
16. Preočanin, T., & Kallay, N., Točka nul-naboja i površinska gustoća naboja TiO2 u vodenim otopinama elektrolita dobivene potenciometrijskom masenom titracijom. Croatica chemica acta; **79**: 95-106 (2006).
17. Republic of Indonesia's Government. Regulation of The Minister of Environment and Forestry Number 59 of 2016 Regarding The Quality Standards of Leachate for Business and/or Activities Place of Final Waste Processing, (2016).
18. Rezagama, A., & Notodarmojo, S., Kinetika Transfer Ozon dan Tren Kekeruhan Dalam Air Lindi Dengan Pengolahan Ozonisasi, Institut Teknologi Bandung, Bandung, (2012).
19. Rezagama, A., Hadiwidodo, M., Purwono, P., Ramadhani, N.F., & Yustika, M., Penyisihan Limbah Organik Air Lindi TPA Jatibarang Menggunakan Koagulasi-Flokulasi Kimia, Jurnal Teknik, 37, **2**, 78-83 (2016).
20. Safria, P., Pengolahan Lindi Dengan Menggunakan Advanced Oxidation Process (AOP) Berbasiskan Ozon (O_3/H_2O_2) Dengan Penambahan Zeolit Pada Kontaktor. Institut Teknologi Nasional, Bandung, (2016).
21. Sawyer, C.N., McCarty, P.L., & Parkin, G.F., *Chemistry for Environmental Engineering and Science*, 5th ed., McGraw-Hill. New York, (2003).
22. Standard Practice for Dry Preparation of Soil Samples for Particle-Size Analysis and Determination of Soil Constants, ASTM D421-85, (1998).
23. Sururi, M.R., Ainun, Si., & Krisna, A., Pengolahan Lindi Dengan Proses Oksidasi Lanjut Berbasis Ozon, Jurnal Reaktor, **15**, 20-26 (2014).
24. Sururi, M.R., Siti, S.A., & Safria, P.O, Leachate Treatment Using Advanced Oxidation Process with Zeolite as a Catalyst, 3rd International Postgraduate Conference on Biotechnology, ISBN 978-602-73103-1-5 (2016).
25. Tizaoui, C., Bouselmi, L., Mansouri, L., & Ghrabi, A., Landfill Leacheate With Ozone and Ozone/Hydrogen Peroxide System, Journal of Hazardous Materials **140**, pp. 316-324 (2007).
26. Tchobanoglous, G., Burton, F.L., & Stensel, H.D, *Water and Wastewater Engineering*, Metcalf & Eddy, Inc, (2013).
27. Wang, Z.P., Z. Zhang., Lin, Y.J., Deng, N.S., Tao, T & Zhou, K., Landfill Leachate Treatment by Coagulation-Photooxidation Process, J. Haz. mater., B95, 153-159 (2002).
28. Wen, G., Pan, Z.H., Ma, J., Liu, Z.Q., Zhao, L., & Li, J.J., Reuse of Sewage Sludge as a Catalyst in

Ozonation – Efficiency for The Removal of Oxalic Acid and The Control of Bromate Formation, Journal of Hazardous Material 239-240. 381-388 (2012).
29. West Java Provincial Government, Department of Settlement and Housing, Quality Testing Center for Construction and Environment, A.2.BPSR.24-2-3.51.11.16 (2016).

Removal of Copper (II) Ions in Aqueous Solutions by Sorption onto Alkali Activated Fly Ash

Lita Darmayanti[1,2,*], *Suprihanto* Notodarmodjo[1], *Enri* Damanhuri[1], and *Rino* R Mukti[3]

[1] Environmental Engineering Department, Institut Teknologi Bandung, Jalan Ganesha No. 10 Bandung 40132, Indonesia
[2] Civil Engineering Department, Universitas Riau, Kampus Bina Widya Simpang Baru, Pekanbaru Riau 28293, Indonesia
[3] Chemistry Department, Institut Teknologi Bandung, Jalan Ganesha No. 10 Bandung 40132, Indonesia

Abstract. Fly ash is a particulate material produced from coal combustion power plants with major components are silica, alumina, iron oxide, calcium oxide, magnesium oxide, and carbon which are ideal for metal adsorbents. The potential use of fly ash in the wastewater treatment process is obvious because it can be obtained cheaply in large quatities and it can be used as an adsorbent. However, fly ash still shows lower adsorption capacity unless it is activated. In this study, fly ash activated by NaOH 14 M and KOH 14 M solutions. The batch experiments were carried out to study the sorption of copper ions from aqueous on alkali activated fly ash. The influence of initial concentration and contact time were examined at constant pH and dose of adsorbent. The sorption capacity of copper ions increased with the initial concentration and contact time. The sorption capacities followed the order Na1>Ka1>FA. The adsorption isotherm model exhibited that the Langmuir model is very suitable with copper ions adsorption onto fly ash and alkali activated fly ash. Kinetic study shows that adsorption of copper ions onto FA, Na1, and Ka1 follows the pseudo second-order kinetics.

1 Introduction

Rapid industrilization and economic development cause heavy metals to be introduced to environment continuously through point and non-point sourcers. Heavy metals are generated as a rest of metal refinishing products. Therefore, heavy metals pollution is still an environmental problem nowadays in whole of countries. Heavy metals are becoming a public health problem since these elements do not degrade and tend to accumulate in living organism, causing various diseases and disorders. Some of metals like Hg, Pb, and Cr can become very toxic to organism, others like Fe, Mn, Cu, Zn, Ni and Co though essential for living beings, when present in excess certain limits can be very harmful. Copper is highly toxic because it is bioaccumulative, persistent which does not break down in the environment and carcinogenic [1-3].

There are various treatment processes available for treatment metal-contaminated water, such as, chemical precipitation, coagulation, oxidation with ozone/hydrogen peroxide, ion exchange, photo catalytic degradation, and adsorption [4]. Among of these methods, adsorption appears to be the most widely used for the removal of heavy metals [5]. There are many low-cost material have been investigated for removal of heavy metals from water such as hardwood and corn straw [6], oil shale ash [7], carrot residues [8], *Lonicera japonica* flower biomass [9] and lignin [10,11].

Fly ash is a particulate material produced from coal combustion power plants that requires ultimate disposal. The major components of fly ash are silica, alumina, iron oxide, calcium oxide, magnesium oxide, and carbon which are ideal for metal adsorbents. The potential use of fly ash in the wastewater treatment process is obvious because it can be obtained cheaply in large quatities and it can be used as an adsorbent and as a neutralization agent [12]. Several research reported that significant amounts of heavy metals were removed from solution by adsorption onto fly ash. Sharma [13] studied the removal of chromium from aqueous solutions and wastewaters, Polowczyk [14] examined the efficiency of fly ash agglomerates (particles size 1.0-1.6 mm) for removal of boron from aqueous solution.

However, fly ash still shows lower adsorption capacity unless it is modified. Alkali activation is a technology that can enlarge utilization of fly ash. The alkali activation or geopolymerization formed of mixing the fly ash with alkaline activators and then curing the paste at a certain temperature to result a solid material [15]. Alkali activated fly ash have been more widely used for cement replacement. Nowadays, alkali activated fly ash also used as adsorbent for removal heavy metals with higher adsorption capacity [1, 17-18]. The objective of this study is to investigate the effectiveness of alkali activated fly ash for removal copper ion from aqueous solution using batch experiment to determine adsorption kinetic and isotherm.

* Corresponding author: litadarmayanti@eng.unri.ac.id

https://doi.org/10.1051/matecconf/201814704007

2 Experimental

2.1 Materials

A sample of the raw fly ash was collected from a coal-firing power plant of PT Trisula Textile, in Bandung city, Indonesia. This material sieved to a particle sizes less than 0,074 mm (sieve no. 200). All chemical were analytical grade reagents supplied by Merck. Solutions containing copper ions were prepared by dissolving the appropriate weight of $CuSO_4.5H_2O$ in distilled water.

2.2 Alkali activation of fly ash

Treatment method involved mixed the certain amounts of fly ash with two different alkaline activator, NaOH 14 M and KOH 14 M. Liquid/solid ratio was 0.6. After adding all the component, the mixture was mixed for 15-20 min and poured into polypropylene bottles. Bottles were closed to prevent moisture evaporation. The fresh pastes were heat at 85^{0}C for 24 h in an oven. After that, geopolymer were removed from oven and kept at room temperature for 3 days. Activated fly ash bodies were washed with acetone to remove the excess of alkaline activator and then crushed, sieved to obtain particle diameter less than 0.074 mm, and stored in a desiccator prior to use.

2.3 Sorption experiments

The sorption of copper ions on raw fly ash and activated fly ash adsorbent were studied by shaking 0.1 L of varying Cu concentration from 5-60 mg/l at temperature of 25^{0}C with a constant amount of adsorbent (0.15 g) in batch reactors. The mixtures were shaking with horizontal shaker at 250 rpm for 60 min. The concentration of the copper ions in the filtered sample was determinate by atomic absorption spectrophotometer. Duplicate batch were applied for all experiment and mean values were used.

3 Results and Discussion

3.1 Characterisation of adsorbents

The BET surface areas and pore volumes of fly ash and activated fly ash were obtained from N_2 adsorption and given in Table 1. The fly ash has the lowest surface area and pore volume. Activated fly ash with alkaline activator produces materials with higher surface area and pore volume.

Table 1 Textural properties of fly ash and activated fly ash

Sample	S_{BET} (m^2/g)	V (cm^3/g)
Fly ash	9.5	0.022
FA-NaOH	21.9	0.052
FA-KOH	23.0	0.055

Fly ash is an essentially vitreous material (halo recorded at 2□=15-35^0) that also contains a series of minority crystalline phase like Q, M, and H. Activation of fly ash with different solutions change the diffraction pattern. It has to be highlighted the shift in the position of the halo attributed to the vitreous Phase in the initial ash to sligthly higher angular values (2θ=20-40^0). The changes indicate the formation of an alkaline aluminosilicate gel (Fig. 1).

The original fly ash peaks of quartz, mullite, and hematite were also found in the samples of activated fly ash. However, these peaks tended to lose their intensity. This finding suggest that reactions disorganized the structure of the crystalline of the fly ash particles (Q, M, H) produced by the alkaline activator. There was also an increase in the number of crystalline compounds which demonstrated on the formation of new crystalline compounds (albite and nahcolite).

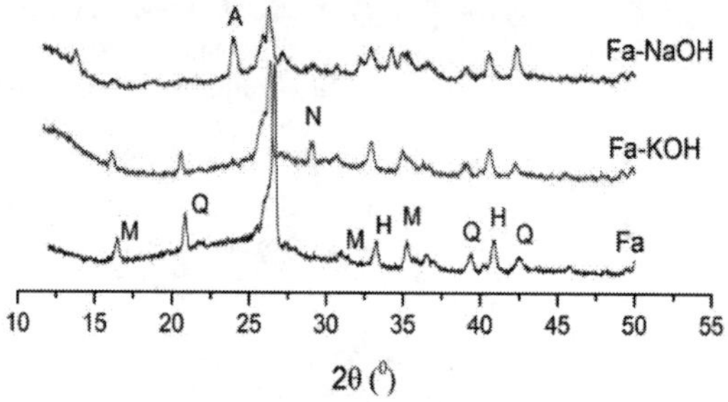

Fig. 1 XRD patterns of fly ash and alkali activated fly ash

The change in morphology of the original surface of the FA particles, can be explained by the dissolution of the vitreous aluminosilicate phase in the raw FA and the crystallization of zeolitic materials onto the external surface of FA particles as pseudomorphs, taking into account that the zeolitic aggregates keep the original shape of the FA spheres after alkaline activation (Fig. 2).

Fig. 2 SEM images of fly ash and synthesis products

3.2 Factors affecting Cu^{2+} adsorption

The removal of Cu onto fly ash activated by NaOH (Na1), KOH (Ka1), and raw fly ash (FA) as a function of the their concentrations was studied under the following conditions: pH: 4; adsorbent dose: 1,5 g/l, temperature: 25^{0}C, equilibrium time: 60 min, and initial concentrations ranging from 5-60 mg/l. Percentage of Cu^{2+} removal were compared (Fig. 3).

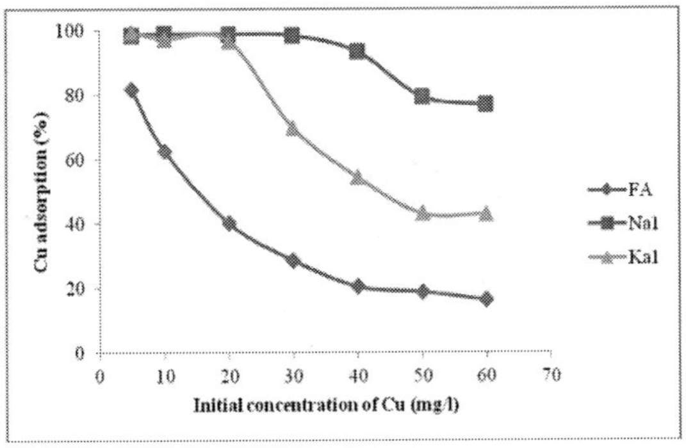

Fig. 3 Effect of initial concentration on Cu removal efficiency

Percentage of adsorption decrases with increasing initial concentration in aqueous solutions. The removal of Cu decreases from 81 to 16%, from 99 to 77%, and from 99 to 43% by adsorption onto FA, Na1, and Ka1 respectively by increasing concentration from 5 to 60 mg/l. Percentage removal of Cu was increased when fly ash activated by NaOH and KOH. This results indicate that adsorption percentage of Na1 is higher than Ka1.

The Cu uptake may be attributed to different mechanisms of ion exchange process as well as to adsorption process. During the ion exchange process, metal ions move through the pores of the alkali activated fly ash mass and channels of the lattice and then they replace exchangeable cations (Na and K). Diffusion was faster through the pores and was retarded when it moved through the smaller diameter channels. In this study, the Cu uptake could mainly be attributed to ion exchange in microporous of alkali activated fly ash.

The effect of contact time on Cu^{2+} ions removal was conducted at initial concentration 50 mg/l, adsorbent dose 1,5 g/l, value of pH 4, temperature 25^{0}C, and contact time varying from 5 to 120 min. the results are shown in Fig. 4.

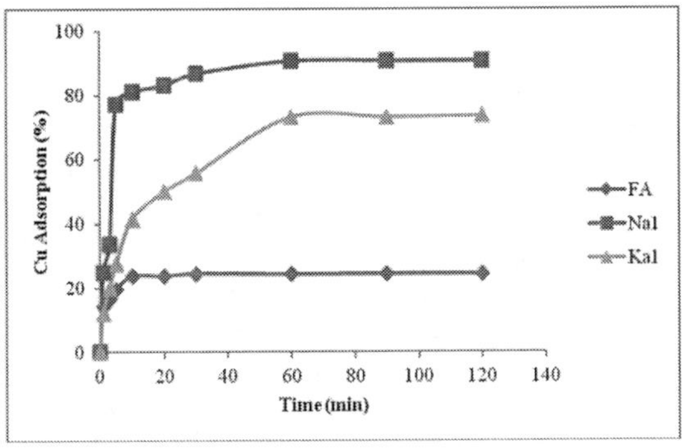

Fig. 4 Effect of contact time on Cu removal efficiency

The sorption of Cu^{2+} ions onto alkali activated flyash is depended on time. A significant removal of Cu^{2+} occurs after 10 min contact time (23%, 81%, and 42% for Fa, Na1, and Ka1 respectively) and only change slightly in removal efficiency occurred after this period of time. The results ensured that the removal of Cu^{2+} from the aqueous solution had occurred under equilibrium conditions. As expected, the using of alkali activated pretreatment will increase adsorption performance of Cu^{2+} onto fly ash. Adsorption of Cu^{2+} by sorbent that developed in this study is relatively fast and 60 min contact time is suffient to reach the adsorption equilibrium and this contact time was applied in following test.

3.2 Adsorption isotherm

Adsorption isotherms are essential to explore the nature interaction between metal ion with adsorbent. The adsorption data have been analysed with two adsorption isotherm methods which are Langmuir and Freundlich equations [19].

3.2.1 Langmuir isotherm

The Langmuir adsorption model is based on the asumption that sorption occurs at specific homogeneous sorption sites within the sorbent and intermolecular forces decrease rapidly with the distance from the sorption surface. The model is also based on the asumption that all the sorption sites are energetically idential and sorption occurs on a structurally homogeneous sorbent [19-21]. The equilibrium data for each initial concentration of Cu^{2+} range from 5 to 60 mg/l have been fitted with the Langmuir model (Eq. 1):

$$\frac{C_e}{q_e} = \frac{1}{q_m K_L} + \frac{C_e}{q_m} \quad (1)$$

where Ce is equilibrium concentration of Cu(II) in solution (mg/l); q_e is the amount of Cu(II) adsorbed onto adsobents at equilibrium (mg/g); K_L is the Langmuir equilibrium constant related to sorption energy; and q_m is the maximum sorption capacity (mg/g).

A linear plot is gained when Ce/qe is plotted againts Ce for the entire of concentration range of Cu^{2+} studied (Fig. 5). The value of q_m and K_L were calculated from the slopes and the intercepts of lines. The isotherm model will effectively and significantly described the sorption data if R^2 values ≥ 0.95.

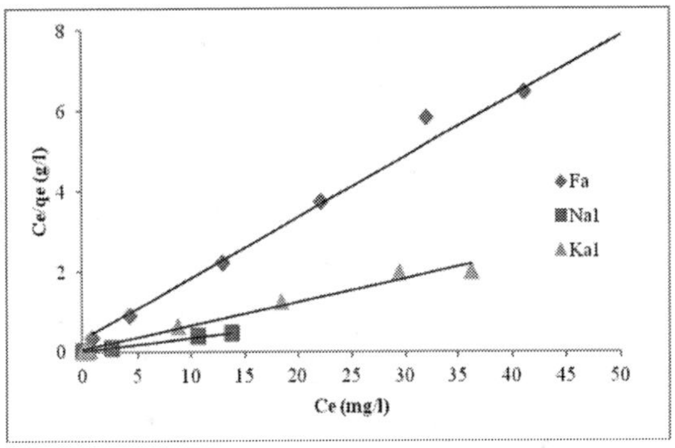

Fig. 5 Langmuir plots for Cu^{2+} ions adsorption onto fly ash and activated fly ash

The adsorption data of Cu^{2+} was found to fit well with the Langmuir isotherm (Table 2). It shows the maximum sorption capacity (q_m) which reflects monolayer coverage of sorbent with sorbate and K_L reflects enthalpy of sorption.

Table 2 Langmuir parameters for Cu^{2+} adsorption onto fly ashand alkali activated fly ash

	Langmuir parameters		R^2
	Q_m (mg/g)	K_L (l/mg)	
FA	6.61	0.45	0.988
NaI	30.21	2.30	0.996
KaI	16.89	0.99	0.979

Raw fly ash usually has low adsorption capacity. Fly ash can be effective adsorbent depending on the composition and treatment if comparison with other adsorbent. Activation by alkali treatment would increasing the adsorption capacity thus increasing the value for application. Activation by NaOH more effective from by KOH. Regression coefficient (R^2) of the plotted isotherm curve show that the Langmuir isotherm found to be suitable for all of the fly ashes. Therefore, Cu^{2+} ion adsorption of all of the adsorbent that used in this study was estimated as adsorption on a monolayer and sorption occurs on a structurally homogeneous sorbent.

Langmuir isotherm may be reflected by a dimensionless parameter called separation factor. The dimensionless separation factor (R_L) was calculated from the Langmuir isotherm using the equation:

$$R_L = \frac{1}{1 + K_L C_0} \quad (2)$$

Where K_L is the Lagmuir adsorption constant and C_0 is the initial concentration. The R_L values that were in the range of $0<R_L<1$ for fly ash and activated fly ash (Table 3). Separation factor indicate that isotherm will be according to the following adsorption characteristics: $R_L > 1$ unfavorable; $R_L = 1$ corresponds to linear; $0 < R_L < 1$ is favorable and $R_L = 0$ is irreversible. The results show that the separation factor less than 1 suggests a favorable adsorption [21].

Table 3 R_L values calculated at different initial concentrations (mg/l)

Adsorbent	R_L at initial concentration						
	5	10	20	30	40	50	60
FA	0.30	0.16	0.09	0.07	0.05	0.04	0.03
NaI	0.07	0.04	0.02	0.02	0.01	0.01	0.01
KaI	0.15	0.09	0.05	0.03	0.02	0.02	0.02

3.2.2 Freundlich isotherm

This isotherm developed by Freundlich (1926), describes the equilibrium on heterogeneous surface and does not assume monolayer capacity [7, 22]. The isotherm provides an expression involving the surface heterogenity, exponential distribution of active sites, and their energies. The Freundlich isotherm can be represented as:

$$\log q_e = \log K_F + \frac{1}{n} \log C_e \quad (3)$$

Where q_e is the amount of Cu(II) ions adsorbed at equilibrium per unit weight of adsorbent (mg/g); C_e is the equilibrium concentration of Cu^{2+} ions in solution (mg/l); and K_F and n are Freundlich constants corresponding to adsorption capacity and intensity respectively. A linear plot is gained when log q_e is plotted againts log C_e and the vaues of K_F and n can be calculated from intercept and slope of the straight line respectively (Fig. 6). The isotherm parameters are summarized at Table 4.

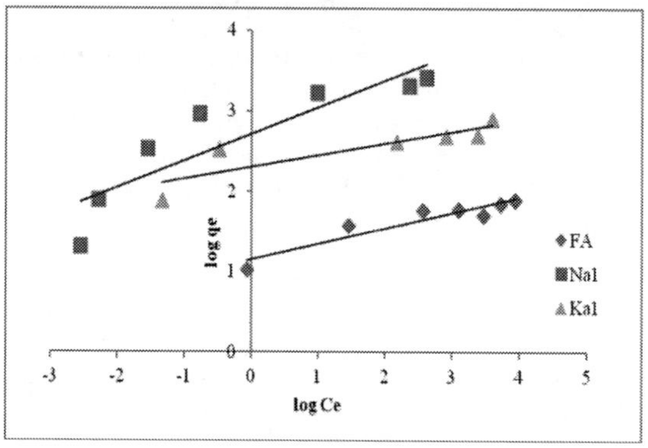

Fig. 6 Freundlich plots for Cu^{2+} ions adsorption onto fly ash and activated fly ash

Table 4 Freundlich parameters for Cu^{2+} adsorption onto fly ashand alkali activated fly ash

	Freundlich parameters		R^2
	K_F (mg/g)	n	
FA	3.16	5.16	0.886
NaI	15.18	3,02	0.798
KaI	8.91	5.18	0.874

It was found that the Freundlich isotherm fits well with Cu^{2+} adsorption by raw fly ash. Furthermore, the adsorption intensity n is > 1 suggesting the existence of a strong bond between adsorbent and adsorbate. Generally, if the exponent lies between $1<n<10$, it represents advantageous adsorption.

As indicated above, Langmuir model could elucidate the adsorption of Cu^{2+} on the fly ash and fly ash that activated by NaOH as well as KOH since value of R^2 Langmuir's are larger than Freundlich's for all of fly ashes. Maximum adsorption capacity of alkali activated by NaOH (30.21 mg/g) is higher than activated by KOH (16.89 mg/g). This result suggests that the adsorption of Cu^{2+} onto fly ash and activated fly ash are monolayer. Similar results were also obtained by others [19, 20, 23].

3.3 Adsorption kinetics

Kinetic models were used to analyze the experimental data regarding the adsorption mechanism and speed control steps. The adsorption process could be controlled with different kind of mechanisms like mass transfer, diffusion control, chemical reaction, and particle diffusion. Several adsorption models were employed to assess the experimental data. The Lagergren's pseudo first-order and pseudo second-order kinetic models were utilized and fitted with the experimental data for evaluate the adsorption process [24]. Fig. 7 and 8 present the plot of the pseudo first-order and pseudo second-order models for adsorption of Cu^{2+} onto raw fly ash and activated fly ash. Experimental and theoretically calculated adsorption capacities at equilibrium values and coefficients that related to kinetic plots are shown in Table 5.

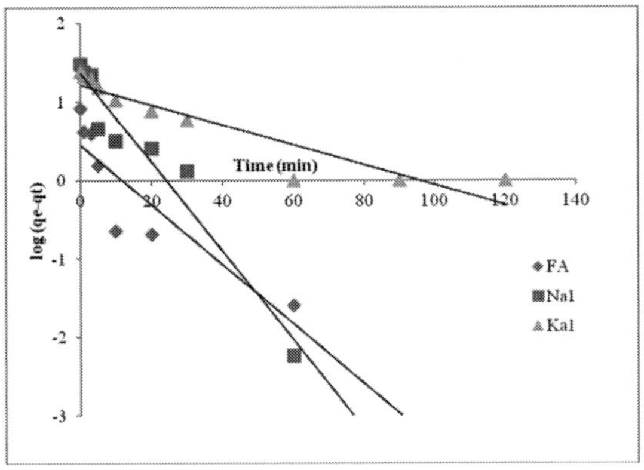

Fig. 7 Fitness of pseudo first-order kinetic

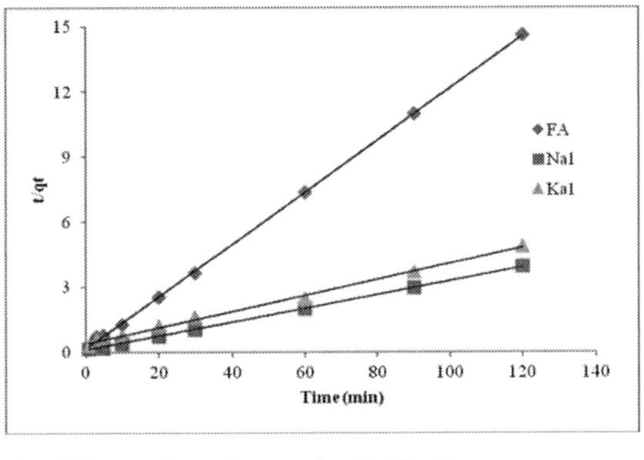

Fig. 8 Fitness of pseudo second-order kinetic

Table 5 Adsorption kinetic model rate constant

	Pseudo first-order			Pseudo second-order			q exp
	qe (mg/g)	k1 (min^{-1})	R^2	qe (mg/g)	k2 (g/mg/min)	R^2	(mg/g)
Fa	2.8	0,087	0,79	8.31	0.104	0.99	8.2
NaI	23.5	0.131	0.94	31.45	0.009	0.99	30.4
KaI	21.1	0.047	0.94	27.10	0.004	0.99	24.5

As can be seen from Table 5, the linear correlation coefficient for pseudo first-order is not good when compared to the pseudo second-order model. These results indicate that the adsorption of Cu^{2+} ions onto fly ash and activated fly ash is not regarding to first-order reaction. It is obvious from the results of the correlation coefficient of pseudo second-order model is very high. Also, the theoretical and experimental qe values are very suitable. This model based on the assumption that the rate limiting step may be a chemical adsorption involving the valence forces through exchange of electrons between the adsorbent and the adsorbate [24, 25].

4 Conclusion

Fly ash is a cheap and effective adsorbent for removal Cu^{2+} ions from aqueous solutions. Changing the nature of fly ash by alkali activated can increase its ability to adsorb Cu^{2+} ions significantly. The increasing of adsorption capacity of fly ash that activated by NaOH was greater than fly ash that activated by KOH. The adsorption of Cu^{2+} ions onto alkali activated fly ash fit the Langmuir isotherm very well as well as the adsorption of Cu^{2+} ions onto fly ash. Maximum adsorption capacities were calculated from the Langmuir isotherm as 30.21 mg/g and 16.89 mg/g for the fly ash that activated by NaOH and KOH respectively. The adsorption of Cu^{2+} ions onto fly ash and alkali activated fly ash follow the pseudo second-order rate kinetics due to being more suitable with experimental data. The adsorption mechanism is a physical process since it contains electrostatic interaction between Cu^{2+} ions and adsorbent.

References

1. T.C. Hsu, C.C. Yu, C.M.Yah, *Fuel*, **87**, 1355–1359 (2008)
2. K.S. Hui, C.Y.H. Chao, S.C. Kot, *Journal of Hazardous Materials*, **B127**, 89-101 (2005)
3. F.A. Lopez, C. Perz, *Journal of Materials Science Letters*, **15**, 1310-13 (1996)
4. M. Ahmaruzzaman, *Colloid and Interface Science*, **166**, 36–59 (2011)
5. K.G. Bhattcharyya, S.S. Gupta, *Advances in Colloid and Interface Science*, **140**, 114-131 (2008)
6. X. Chen, G. Chen, L. Chen, Y. Chen, J. Lehmann, M.B. McBride, A.G. Hay, *Bioresource Technology*, **102**, 8877–8884 (2011)
7. R. Shawabkeh, Al-Harahsheh, Al-Otoom, *Separation and Purification Technology*, **40**, 251-257 (2004)
8. B. Nasernejad, T.E. Zadeh, B.B.Pour, M.E. Bygi, A. Zamani, *Process Biochemistry*, **40**, 1319-1322, (2005)
9. L.P. Lingamdinne, J-K. Yang, Y.Y. Chang, J.R. Koduru, *Hydrometallurgy*, xx, xxx–xxx (2015)
10. Y. Wu, S. Zhang, *Bioresource Technology*, **99** (16), 7709-7715 (2008)
11. X. Guo, S. Zhang, X. Shan, *Journal of Hazardous Materials*, **151** (1), 134-142 (2008)

12. C.H. Weng, C.P. Huang, *Colloids and Surfaces A: Physicochemichal Engineering. Aspects*, **247**, 137–143 (2004)
13. Y.C. Sharma, Uma, S.N. Upadhyay, C.H. Weng, *Colloids and Surfaces A: Physicochem. Engineering Aspects*, **317**, 222–228 (2008)
14. I. Polowczyk, J. Ulatowska, T. Kozlecki, A. Bastrzyk, W. Sawifiski, *Desalination*, **310**, 93–101 (2013)
15. A. Fernández-Jiménez, A. Palomo, *Cement and Concrete Research,* **35**, 1984–1992 (2005)
16. L. Li, S. Wang, Z. Zhu, *Journal of Colloid and Interface Science,* **300**, 52-59 (2006)
17. M. Visa, *Applied Surface Science*, **263**, 753–762 (2012)
18. M. Kobya, *Bioresource Technology*, **91**, 317–321 (2004)
19. A. Papandreou, C.J. Stournaras, D. Panias, *Journal of Hazardous Materials*, **148**, 538-547 (2007)
20. M.S. Al-Harahsheh, K. Al-Zboon, L. Al-Makhadmeh, M. Hararah, M. Mahasneh, *Journal of Environmental Chemical Engineering*, **3**, 1669-1677 (2015)
21. S. Mohan, R. Gandhimathi, *Journal of Hazardous Materials*, **169**, 351–359 (2009)
22. S. Wang, L, Li, Z. Zhu, *Journal of Hazardous Materials*, **B139**, 254-259 (2007)
23. H. Xiyili, S. Cetintas, D. Bingol, *Process Safety and Environmental Protection*, **109**, 288-300 (2017)
24. Y.S. Ho, G. McKay, *Process Biochemistry*, **34**, 451–465 (1999)

Effect of Aeration Rates on Removals of Organic Carbon and Nitrogen in Small Onsite Wastewater Treatment System (*Johkasou*)

Joni Aldilla Fajri[1,*], *Tomonari* Fujisawa[2], *Yenni* Trianda[3], *Yasushi* Ishiguro[4], *Guangyu* Cui[2], *Fusheng* Li[4], and *Toshiro* Yamada[5]

[1]Faculty of Civil Engineering and Planning, Universitas Islam Indonesia, Kaliurang 55584, Indonesia
[2]Graduate school of Engineering, Gifu University, Yanagido 501-1193, Japan
[3]Department of Chemical Engineering of Natural Product, Polytechnic ATI Padang, Adinegoro Bungo Pasang 25586, Indonesia
[4]River Basin Research Center, Gifu University, Yanagido 501-1193, Japan
[5]Faculty of Engineering, Gifu University, Yanagido 501-1193, Japan

Abstract. Onsite application of oxygen supply in domestic wastewater system may be influenced by several factors that can inhibit the oxidation and nitrification processes. In this study, the influence of aeration rate on the *Johkasou* performance was focused using two *Johkasou* facilities serving up to five persons household. In the *Johkasou* A (JO-A) system, we increased the aeration rate from 30 to 63 $L.min^{-1}$ whereas, in the *Johkasou* B (JO-B), it was decreased from 59 to 34 $L.min^{-1}$. Water and sludge samples were collected from the anaerobic-anoxic-oxic zones before and after adjustment of the aeration rate measured for organic matters and nitrogen parameters. Increasing the aeration rate in JO-A resulted in a high removal of organic matter (82.5%) and nitrogen (60.3%) compared to decreasing of aeration rate in JO-B (52.0% and 33.0%, respectively). Simultaneous nitrification and denitrification (SND) exhibited a maximum percentage when the aeration rate was increased compared to decreasing of aeration rate. These results indicate that application of a high aeration rate increases removal of organic matter and nitrogen and enhances ammonia transformation. It is therefore recommended to apply high aeration rates in Johkasou system.

1 Introduction

Small-scale onsite domestic wastewater treatment systems named *Johkasou* have been widely used in Japan`s rural areas and are still developing the performance. *Johkasou* systems treat for both black and gray waters from household wastewater. The effluent of *Johkasous* is generally discharged into a local environmental receiving body, which the biochemical oxygen demand (BOD) and total nitrogen (TN) should be less than 20 m/L [1]. *Johkasou* system is considered to be an effective means for wastewater treatment in rural areas, which are designed, located and maintained satisfactorily. *Johkasou* system consists of an anaerobic stage as a primary treatment unit (a primary settling tank) and a secondary anaerobic (anoxic) stage that both are equipped filter bed. And the aerobic (oxic) stage is a biological treatment unit typically like; activated sludge process, aerobic submerged biofilter, fluidized bed reactor, or biofilm filtration reactor. A disinfection chamber is also installed at end of the process to inactivate microorganisms and fecal indicator bacteria in the effluent before disposing to the environment.

During the last decade, biological nitrogen removal (BNR) has been promisingly method used both in domestic and industrial wastewater treatment [2]. A biological system is mostly eliminated the nitrogen compounds through combination processes of nitrification and denitrification. This system consists of two successive steps: 1) autotrophic nitrification in which NH_4^+-N is oxidized to nitrite (NO_2^--N) and then nitrate (NO_3^--N) under aerobic condition; and 2) heterotrophic denitrification in which NO_3^--N is reduced to nitrogen gas (N_2) by heterotrophic denitrifiers under anoxic conditions with organic carbon as the electron donor [3].

Several studies reported the factors that inhibited nitrification-denitrification in biological system. These factors included higher free ammonia [4]–[6], hydraulic retention time (HRT) [7], [8], dissolve oxygen (DO) concentration and inhibitor [9]. However, the efficiency of biological nitrogen removal can be improved by maintaining and monitoring the ORP, pH and DO in the nitrification-denitrification process [10]. Peng *et al.*[11] conducted the real-time aeration control using pH and DO in sequence batch reactor, and found that the end of nitrification could be identified by the binding the point in pH and increased in DO profile.

Recently, studies of BNR process mainly used DO as a limiting factor to achieve the nitrification and

* Corresponding author: joni.af@uii.ac.id

denitrification [7], [11], [12]. Dissolved oxygen is also economically feasible control the parameters particularly at low concentration that can save the aeration consumption, while in some cases the efficiency of BNR can be achieved at high DO concentration. Therefore, there is no definitive explanation regarding an appropriate DO concentration to achieve complete nitrification. Guo *et al.* [7] conducted the partial nitrification by comparing the high and low DO concentration (2.5 and 0.5 mg/L, respectively) on long-term aeration time. The complete nitrification was reached at both conditions, which corresponded increase nitrite concentration and kept constant or nearly negligible the nitrate concentration. Under limited oxygen condition, Jianglong and Ning [13] reported that significantly higher ammonia oxidation rate was reached when increased the DO concentration from 0.5 to 1.5 mg/L. Another study found that temporal nitrite accumulation was obtained at medium of DO at 1.7 mg/L, however, the excess of DO concentration had no influence on nitrite accumulation, while at low DO concentration decreased the nitrite accumulation and ammonia consumption [14].

The successful of DO concentration in BNR process can also be measured by simultaneous nitrification and denitrification. The efficiency of SND mainly occurs as consequence of DO concentration gradients within microbial flocs of biofilms due to diffusional limitation. This indicates the appropriate application of DO for degrading organic matter and producing less sludge. In batch scale reactor, many researchers pointed that high efficiency of SND was achieved during low DO concentration [7], [15], [16], which remained organic compounds used by denitrifying bacteria. Contrary, Zhu *et al.* [8] noted that the SND could be obtained even at high DO concentration under variances of alkalinity concentration.

The studies on nutrient and organic removal using biological system were mostly conducted in a batch reactor that is easy to control and less complexity in the system. The application of BNR in the real system or in-situ study faces many factors such as air supply clogging, low oxygen availability and out of usage of a blower. These factors may reduce the aeration rate and then decrease organic and nutrient removal in wastewater treatment system. It will result in poor effluent quality discharged to the local water and can deteriorate downstream water environment. Moreover, little research was conducted in *Johkasou* system to study the effect of aeration rate and there is no authentic regulation on setting the aeration rate by third parties. Therefore, the objective of this study was to know the effect of aeration rate in the biological process involved in the removal of organic and nitrogen in *Johkasou* system during onsite research and to find the appropriate aeration rate that can apply in this system.

2. Material and methods

2.1 Description of *Johkasou* system

We studied the effect of aeration rate in two *Johkasou* facilities. The experiments were carried out in two installed *Johkasou* facilities (CE-5, Fuji clean, Japan) used up to five family members (**Fig. 1**). *Johkasou* system is a household-size unit with a capacity of 1-2 m^3 per day. It composes of a primary treatment unit consisting of an anaerobic settling tank as sludge store and a down-flow anaerobic filter packed media as an anoxic process, the aerobic reactor coupled with a fixed filtration bed, and then a final settling tank and a disinfection chamber [1]. The filter packed media in the anaerobic zone and the fixed filtration bed in the oxic zone is made of polystyrene material. A blower was utilized to provide the air for sufficient gas-liquid mixing in aerobic treatment. Adjustable opening valves in the aeration tank were used to control the air supply and recirculate the settled sludge from the bottom of the aeration tank to the anaerobic settling tank. The maintenance of the system is generally conducted once in four months, while the cleaning and sludge dragged are scheduled once a year.

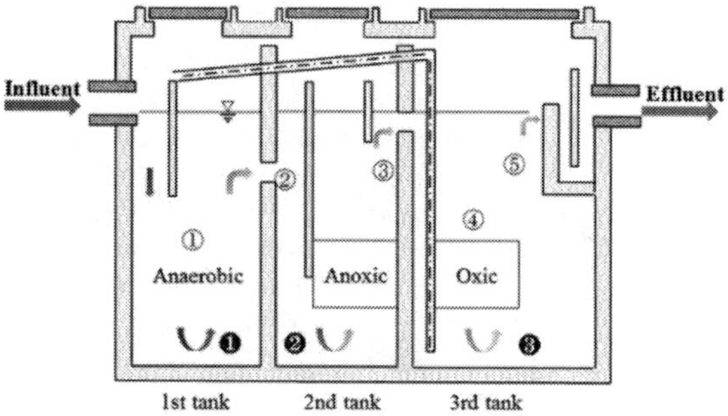

Fig. 1. Structure of *Johkasou* used in the study and sampling points. Open numbers from 1 to 5 are water sampling in each zone; influent in the anaerobic, effluents of anaerobic and anoxic, oxic and settling tank, respectively. Bold numbers from 1 to 3 indicate sludge samplings in each zone.

2.2 Experimental operation and sampling

The two *Johkasou* systems, *Johkasou* A (JO-A) and *Johkasou* B (JO-B), were operated under different aeration supplies. The JO-A was utilized at high aeration rate, which an initial rate was 30 L/min started from October 15th, 2015. After two weeks of operation and collecting sample prior adjustment, the aeration rate was increased to 63 L/min and then maintained till end of the operation (December 17, 2015). We assumed that the steady state of the system in JO-A was achieved within two weeks after increasing aeration and the sample collections were started on this date and continued for bi-weekly sampling. The total operation time of JO-A was 71 days and mean influent loading was 0.58 m^3/d.

The JO-B was operated as low aeration rate with an initial rate of 59 L/min started from September 1st, 2015. After ten days operation and collecting samples prior

Table 1. Initial influent characteristics in both *Johkasou*.

	Influent flow rate (m^3/d)	A_r[a] before adjustment (L/min)	A_r After adjustment (L/min)	BOD (mg/L)	DOC (mg/L)	TN (mg/L)	TP (mg/L)	NH_4^+-N (mg/L)
Increase aeration (JO-A)	0.52 ± 0.08	30	63	258.0 ± 21.7	65.4 ± 24.0	50.2 ± 4.1	5.9 ± 0.4	38.4 ± 2.9
Decrease aeration (JO-B)	0.38 ± 0.06	59	34	225.0 ± 54.5	40.0 ± 12.4	51.8 ± 8.3	5.0 ± 0.9	38.0 ± 8.8

[a] A_r indicates aeration rate
Values are mean ± standard deviation.

adjustment, the aeration rate was decreased to 34 L/min and maintained up to end of the operation (November 5th, 2015). We assumed that the steady state of the system in JO-B was achieved within a month operation after decreasing the aeration and then the samples were collected and continued for bi-weekly samplings here after. The total operation time of JO-B was 63 days and the mean influent loading was 0.38 m^3/d.

2.3 Wastewater and sludge characteristics

Each of *Johkasou* system received continuous influent (black and gray waters) from single house. The influent characteristics of both *Johkasou* systems were described in **Table 1**. During the operation, there was no sludge recycled either before or after adjustment of aeration. The mixed-liquor suspended solid (MLSS) and mixed-liquor volatile suspended solid (MLVSS) of sludge were measured and the result is displayed in **Table 2**.

2.4 Sample collection and analytical methods

We collected water samples from several sampling points in order to study the differences of effluent quality for each stage before and after adjustment of aeration rate. Water samples were obtained from the influent of anaerobic zone, effluents from anaerobic, anoxic and oxic zone, and settling zone. From each sampling points, water samples were taken once before adjustment as initial condition and bi-weekly after it. And, we also grabbed sludge samples at the bottom of each zone. The samples of water and sludge phases were transferred into one litter polystyrene plastic bottle and stored in a cooling box before transporting to the laboratory. Then, the samples were analyzed soon on the same day for organic matter and nutrients.

We measured the physico-chemical parameters to identify the basic effluent quality at each zone. The physical parameters such as oxidation reduction potential (ORP), pH, DO, and water temperature (T_w) were measured onsite once a week. The measurements of ORP, pH, DO, and Tw were done using the corresponding portable probe meters (DKK-TOA, Tokyo, Japan). The organic matter and nutrients constituents including biochemical oxygen demand for 5 days at 25°C (BOD), dissolved organic carbon (DOC), turbidity, suspended solid (SS), volatile suspended solid (VSS), total nitrogen (TN), total phosphorous (TP), ammonia-N (NH_4^+-N), nitrite-N (NO_2^--N) and nitrate-N (NO_3^--N) in the water samples; and MLSS, MLVSS and sludge volume index (SVI) in the sludge samples were analyzed according to standard methods [17].

2.5 Statistical analyses

Statistical analysis was calculated to find the differences among means and influence of aeration rates on the wastewater quality variables. One-way ANOVA with Post-Hoc Honest tests, Tukey HSD was used to obtain the variance data between the *Johkasou* systems.

3. Results and discussion

3.1 Physical parameters

The mean values of physical parameters at each zone are displayed in **Table 2**. The ORP values showed the difference before and after adjustment of aeration rate for both JO-A and JO-B. At anoxic and oxic zones, the mean of ORP ranged from -110.0 to -250.5 mV before adjustment and from -24.1 to 93.4 mV after adjustment for JO-A, and from 102 to 116 mV before adjustment and from -101.4 to 49.5 mV after adjustment for JO-B. These results implied that the ORP increased by increasing oxygen supply and dropped when the oxygen supply was shortened. The pH in both *Johkasou*s (ranged from 6.7 to 7) was sufficient for nitrification (pH > 6.5) and demonstrated a slight drop during nitrifying aeration phase as according to Canziani et al. [18]. The water temperature was similar before and after adjustment of aeration rate for both JO-A and JO-B, which indicated that seasonal effect was not the main factor reflecting the water temperature.

Difference concentrations in DO were recognized between before and after adjustment of aeration rate. In the anaerobic stage, concentration of DO was recorded at zero and ORP was lower than -250 mV for both *Johkasou*s. In the anoxic zone, significance in DO concentrations was showed after adjustment of aeration rate. It increased from 0.2 to 2.7 mg/L after increasing DO for JO-A and decreased from 5.1 to 0.9 mg/L after decreasing DO for JO-B, which indicated that the limited oxygen supply could be transferred from the oxic zone through upflow system. High DO concentrations in the anoxic zone might be caused the excess oxygen supply in the oxic zone whereas the mean values of ORP were -

Table 2. Parameter indexes measured in this study.

	Before adjustment					After adjustment				
	Inffluent	1st Stage Anaerobic	2nd Stage Anaerobic	Aerobic	Settling	Inffluent	1st Stage Anaerobic	2nd Stage Anaerobic	Aerobic	Settling
Increasing aeration (JO-A)										
ORP (mV)	-250.5 (2)	-	-236 (2)	-110.5 (2)	-250.5 (2)	-249.1 ± 24.0 (8)	-	-24.1 ± 50.3 (8)	93.4 ± 25.1 (8)	44.3 ± 63.6 (8)
pH	6.1 (2)	6.2 (2)	7.3(2)	7.3 (2)	7.2 (2)	6.2 ± 0.1 (8)	6.2 ± 0.1 (8)	6.7 ± 0.2 (8)	6.7 ± 0.3 (8)	6.6 ± 0.2 (8)
DO (mg/L)	0 (2)	0 (2)	0.2 (2).	1 (2)	0 (2)	0 ± 0.0 (8)	0 ± 0.0 (8)	2.7 ± 0.9 (8)	5 ± 0.9 (8)	1 ± 0.7 (8)
Tw (°C)	24.6 (2)	-	24.6 (2)	-	24.3 (2)	24.0 ± 1.8 (8)	-	21.3 ± 1.8 (8)	-	21.0 ± 2.0 (8)
Turbidity (NTU)	55 (1)	80 (1)	23 (1)	23 (1)	21 (1)	113.3 ± 15.1 (4)	117.8 ± 166.7 (4)	9.1 ± 2.7 (4)	4.5 ± 0.3 (4)	3.9 ± 1.0 (4)
TSS (mg/L)	63 (1)	97 (1)	12 (1)	13 (1)	14 (1)	67.0 ± 12.5 (4)	78.5 ± 21.3 (4)	12.6 ± 4.6 (4)	9.6 ± 3.1 (4)	5.1 ± 34 (4)
VSS (mg/L)	57 (1)	95 (1)	9.4 (1)	11 (1)	13 (1)	58.8 ± 12.8 (4)	70.0 ± 19.1 (4)	11.5 ± 4.7 (4)	8.2 ± 2.4 (4)	4.1 ± 2.2 (4)
MLSS (mg/L)	16,435	–	11,173	4,028	4,000	16,690.0 ± 7322.4	–	9,745.6 ± 725.5	4,337.8 ± 408.3	5,275.6 ± 2,386.4
MLVSS (mg/L)	16,310	–	10,283	3,970	3,928	15,910.0 ± 7087.4	–	8,966.3 ± 831.0	4,174.4 ± 459.2	4,960.0 ± 2,066.1
Deacreasing aeration (JO-B)										
ORP (mV)	-343 (2)	-	104.5 (2)	116 (2)	102 (2)	-265.0 ± 104.3 (8)	-	-101.4 ± 75.5 (8)	49.5 ± 40.1 (8)	- 97.4 ± 147.8 (8)
pH	6.7 (2)	6.7 (2)	6.5 (2)	6.4 (2)	6.5 (2)	6.3 ± 0.3 (8)	6.5 ± 0.2 (8)	7.1 ± 0.2 (8)	7.1 ± 0.2 (8)	7.0 ± 0.2 (8)
DO (mg/L)	0 (2)	0 (2)	5.1 (2)	7 (2)	2.8 (2)	0.0 ± 0.0 (8)	0.0 ± 0.0 (8)	0.9 ± 0.8 (8)	2.3 ± 0.8 (8)	0.1 ± 0.2 (8)
Tw (°C)	26 (2)	-	26.0 (2)	-	26.0 (2)	23.2 ± 1.6 (8)	-	23.3 ± 1.9 (8)	-	22.8 ± 1.6 (8)
Turbidity (NTU)	34 (1)	35 (1)	6.9 (1)	6 (1)	6.9 (1)	63.7 ± 18.9 (3)	66.7 ± 14.6 (3)	13.2 ± 6 .0 (3)	7.2 ± 0.6 (3)	5.7 ± 1.2 (3)
TSS (mg/L)	40 (1)	18 (1)	9.4 (1)	12 (1)	9 (1)	60.3 ± 23.4 (3)	70.0 ± 27.9 (3)	21.0 ± 1.0 (3)	19.0 ± 3.6 (3)	13.0 ± 2.0 (3)
VSS (mg/L)	36 (1)	16 (1)	7.5 (1)	10 (1)	8.5 (1)	55.7 ± 25.7 (3)	63.0 ± 21.7 (3)	18.3 ± 0.6 (3)	15.7 ± 3.5 (3)	10.8 ± 2.9 (3)
MLSS (mg/L)	20,740	–	10,130	5,750	2,685	14,467.5 ± 5203.6	–	8,266.3 ± 2,198.1	5,697.1 ± 715.5	3,603.8 ± 2,174.5
MLVSS (mg/L)	19,970	–	9,705	5,573	2,400	13,442.5 ± 5568.9	–	7,595.0 ± 2,186.7	5,295.0 ± 881.9	3,280.0 ± 2,153.4

Values before adjustment are means/raw data and parentheses indicate number of samples.
Values after adjustment are mean ± standard deviation and parentheses indicate number of samples.

24.5 mV and -101.4 mV for JO-A and JO-B respectively, which these zones indicated as the facultative aerobic state. In the oxic zone, the mean concentrations of DO were 5 and 2.3 mg/L for JO-A and JO-B, respectively, and the mean ORP was +49.5 and +93.4 mV for JO-A and JO-B, respectively, which indicated both *Johkasou* were well supplied with oxygen (**Table 2**).

3.2 Suspended solid and biomass properties

The concentrations of TSS and VSS in water samples are shown in **Table. 2**. Concentration of TSS and VSS before adjustment for JO-A and JO-B showed similar values in the anoxic and oxic zones. After adjustment, however, concentrations of TSS and VSS were nearly two times lower in the JO-A then in the JO-B. The ratio of VSS/TSS in JO-A turned into less than JO-B with a value from 85.0 to 87.5 % and 82.7 to 91%, respectively. These results indicated that increasing of aeration rate could increase conversion of ammonia and other soluble organic into biomass.

Initial concentrations of MLSS and MLVSS in the JO-A and JO-B were operated in different sludge concentrations; whereas, in the oxic zone, concentrations of MLSS and MLVSS pointed a slightly lower at high aeration rate rather than low aeration rate (**Table 2**). These results were found at same circumstance after adjustment of aeration rates. This indicated that high aeration rate could be an important factor to degrade organic carbon and increase the number of biomass. The ratios of MLVSS/MLSS for the whole operating condition were 91.6% and 94.4% respectively in JO-A and JO-B, represented a high proportion of biomass in the sludge.

3.3 Removal of organic carbon and nitrogen by increasing aeration rate in JO-A

Concentrations of organic compound and nutrient in JO-A before and after adjustment of aeration rate are shown in **Fig. 2**. The removals of DOC and BOD were used as representative of a part of total organic matter and biodegradable organic matter, respectively. Removal percentages of BOD and DOC after adjustment of aeration rate were not significantly different.

Increasing DO concentration showed an improvement of organic carbon removal in the *Johkasou* system. When the aeration rate increased from 30 to 63 L/min, the DO concentration lifted from 0.19 to 2.7 mg/L and from 1 to 5 mg/L in the anoxic and the oxic zones, respectively. Before adjustment, at low aeration rate (30 L/min), total removals of BOD and COD were 81.7% and 40%, respectively. However, total removals of organic carbon increased sharply onto 98.1% and 91.1% for BOD and DOC, respectively, when the aeration rate was increased into 63 L/min. Proportions of organic carbon removal were detected mainly occur in the anoxic zone. This indicated that high DO improved organic carbon oxidation in the anoxic zone and remained organic carbon was utilized by autotrophic bacteria in nitrification process in the oxic zone. Effect of increasing DO on improvement of organic removal was also reported by several studies [16], [19]–[22].

Concentrations of TN and NH_4^+-N in JO-A before and after adjustment aeration rate are displayed in the **Fig. 2b**. Before adjustment of aeration rate, at low DO of 0.19 L/min, removals of TN and NH_4^+-N could not be found in both the anoxic and the oxic zones. It reveals that the operation of *Johkasou* at low DO is inappropriate for nitrifier bacteria to produce nitrite and nitrate from oxidized ammonia as shown in **Fig. 4.**

 https://doi.org/10.1051/matecconf/201814704008

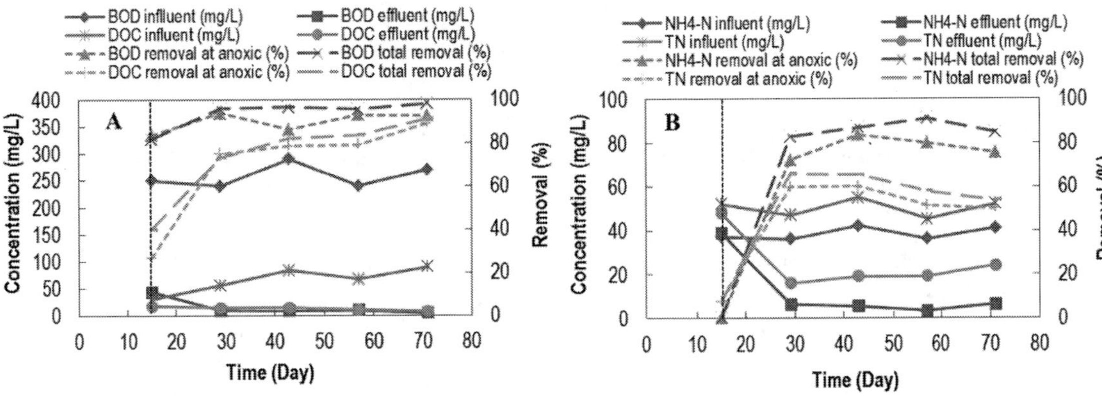

Fig. 2. Concentrations of a) organic carbon and b) nitrogen in JO-A during operation. Concentrations at dash line indicate the operation before adjustment with an aeration rate of 30 L/min and then, the aeration rate was increased to 63 L/min (after adjustment).

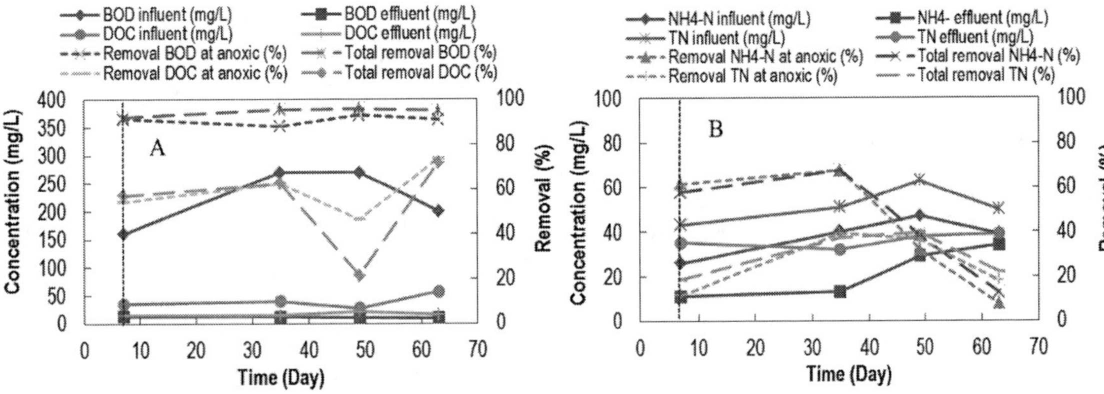

Fig. 3. Concentrations of organic matter and nitrogen in JO-B during operations. Concentrations at dash line indicate the operation condition before adjustment with an aeration rate of 59 L/min and then, the aeration rate was decreased to 34 L/min (after adjustment).

Conversely, TN and NH_4^+-N removals increased sharply when DO was increased to be 2.7 mg/L in the anoxic zone and the removals of these compound ranged from 50 to 60 % and from 72 to 83 %, respectively. The excess DO in the oxic zone could only remove the TN up to 65% while NH_4^+-N removal reached to 91%, respectively. These results indicate that high DO is an appropriate condition for *Johkasou* system to oxidize TN and transform the ammonia into NO_2^--N and NO_3^+-N forms.

3.4 Removal of organic carbon and nitrogen by decreasing aeration rate in JO-B

Concentrations of organic carbon and nitrogen during low aeration condition in the JO-B system are displayed in **Fig. 3**. The removals of BOD and DOC were statistically not difference between before and after adjustment. Before adjustment with a high aeration rate (59 L/min), DO concentration in the anoxic and oxic zones was in very rich condition on the level of 6.1 mg/L and 7.5 mg/L, respectively. At this condition, BOD and DOC removal could be obtained with mean percentages of 95.5% and 71.9%, respectively, and most removals of these organic carbon occurred in the anoxic zone (91.9% and 62.5%, respectively). The removal of NH_4^+-N was in moderate level around 57%, however, TN removal could not be obtained. These results suggested that excess DO concentration in the oxic zone changed the state of anoxic condition into rich DO that was not sufficient for reducing TN and ammonia.

Contradictory results were obtained when the aeration rate was decreased. When aeration rate decreased from 59 to 34 L/min, DO concentration dropped from 5.1 to 0.9 mg/L and from 7 to 2.3 mg/L both in anoxic and oxic zones, respectively. Decreasing DO concentration leads to slightly declined DOC removal from 73.7% to 71.9% and sharp declined of

 https://doi.org/10.1051/matecconf/201814704008

NH_4^+-N from 57.7 to 12.8%. However, total removal of TN showed increased trend to 22%. This condition suggests that the organic consumption was not properly achieved during nitrification in which nitrifier growth much slower than heterotrophic bacteria, and caused competition of oxygen consumption between both groups of organisms.

These results indicate that high DO concentration is a favor condition for niftrifier bacteria to oxidize the ammonium and improves the removal of TN and NH_4^+-N. However, decreasing aeration rate caused high dissolved organic carbon concentrations, which inhibited nitrification both in the anoxic and oxic zones and thereby increased the concentration of nitrogen in the effluent. Similar results were reported by [3], [15], [16]. Therefore, it is suggested that *Johkasou* system is recommended to applied continuous oxygen supply at high rate to achieve simultaneously nitrification.

3.5 Effect of aeration rate on simultaneous nitrification and denitrification

Adjustment of aeration rates affected the transformation of ammonia into nitrite-nitrogen and nitrate-nitrogen forms during nitrification process (**Fig. 4**). The amount of oxidized ammonia nitrogen via simultaneous nitrification and denitrification (SND) is calculated from the difference between total consumption of ammonia nitrogen and NH_4^+-N assimilated into heterotrophic biomass (Eq. 1) [15]. While, nitrate accumulation rate (NAR) based on the Eq. 2) [22].

$$\% \text{ SND} = \left(1 - \frac{NO_x^- - N}{NH_4^+ - N_{oxidized}}\right) \times 100 \qquad 1)$$

$$\% \text{ NAR} = \frac{NO_2^- - N}{NO_x^- - N} \times 100 \qquad 2)$$

Where NO_x^--N is the sum of nitrite and nitrate produced at the moment when ammonia was depleted, and $NH_4^+ - N_{\text{oxidized}}$ is the amount of ammonia nitrogen oxidized during the aerobic circumstance. NAR during nitrification stage more than 50% is indicated as nitrification via nitrite.

In the JO-A system, before adjustment of aeration rate, concentrations of ammonia were not significant differences and also NO_x-N formations exposed limited to a negligible amount in each process (**Fig. 4a**). It indicated that insufficient DO level inhibited the nitrifier bacteria to convert ammonia into NO_2^--N and NO_3^-N and caused the SND and NAR could not be measured as well. Contrast when the aeration was raised into 63% L/min, sufficient oxygen supply into the system could enhance nitrification with the NH_4^+-N was depleted and NO_x^--N productions increased. The production of NO_2^--N detected very less because rapid conversion of NO_2^--N into NO_3^--N by Nitrobacter bacteria, which resulted percentages of NAR rose to 74.5 % and 60.9%, in the anoxic and oxic zones respectively. At this condition, the SND exhibited increased values in average 68.9% in the anoxic and 71.8% in the oxic. The maximum of SND value (84.6%) was found at the beginning after increasing DO concentration in the oxic zone and the SND percentage at the last sampling was slowly declining to 64.9% during long term aeration.

Decreasing aeration rate lead to in opposition results in the JO-B system particularly after decreasing aeration rate (**Fig. 4b**). Before adjustment, aeration rate was set upped at high rate around 59 L/min. The oxidized ammonium nitrogen was confirmed by detection of nitrite and nitrate with percentages of SND and NAR in the anoxic zone were 94% and 93.2 %, respectively. At this condition, the nitrification process had been achieved due to appropriate oxygen supply used by nitrifier bacteria, while nitrification effect in the oxic zone not differed with in the anoxic zone.

After adjustment, DO concentration declined linearly with decreasing aeration rate to 34 L/min. This circumstance affected DO changed into very low concentration, increased ammonium concentration and depleted NO_x^--N productions. Particular in day of 50, the sharp dropped of nitrite and nitrate could be detected at the end of experiment with percentages of SND and NAR fell onto zero level. This circumstance resulted low accumulations of nitrite and nitrate due to limitation of oxygen for nitrification bacteria. These results indicate

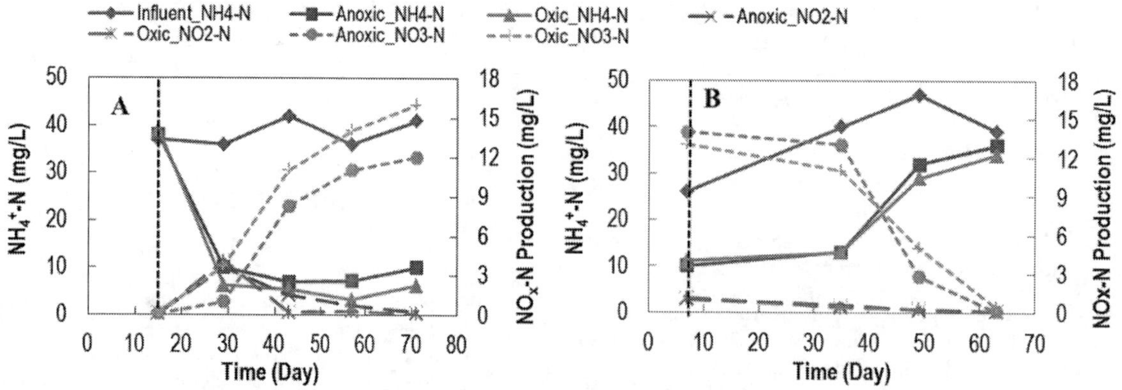

Fig. 4. Oxidized ammonia nitrogen into nitrite and nitrate affected by aeration rates in the a) JO-A and b) JO-B.

that high DO concentration favors conditions for niftrifier bacteria to oxidize the ammonium and improve the removal percentages of TN and NH_4^+-N. However, decreasing aeration rate caused high dissolved organic carbon concentrations, which inhibited nitrification both in the anoxic and oxic zones and thereby increased the concentration of nitrogen in the effluent. Similar results were reported by [3], [22], [23].

Limited DO concentration had no influence on the ammonia and nitrite oxidization, meanwhile the continuous supply of high DO concentration possibly inhabited oxidation of NO_3^--N into Nitrogen by denitrification reaction. This pointed that DO concentration plays an important role for nitrifier bacteria to oxidize the ammonia and the appropriate DO concentration or intermittent aeration supply may has a necessity to retain the oxidization, which can also save the cost of operation. The results in this study were also comparable with report by several researchers. Ruiz *et al.* ([14]) studied the behavior of the nitrification system during consecutive changes in DO values in high ammonia concentration. At a DO of 1.7 mg/L there was a temporal accumulation of nitrite and at DO of 1.4 and 0.7 mg/L accumulation increased with the same ammonia consumption. At a DO of 0.5 mg/L, both nitrite accumulation and ammonia consumption decreased. Similar results were also obtained by Bae *et al.* [24] that the maximum specific oxidation of ammonia (k_a) and nitrite (k_n) were similar under DO = 0.5 mg/L and increased as DO increased.

4. Conclusion

Aeration rate is an important variable in the biological wastewater treatment process and affects removals of organic matter and nitrogen. Effect of aeration rate on performance of two *Johkasous* was studied to know the degree oxidation of organic matter and nitrogen. High removals of BOD, DOC, NH_4^+-N and TN were achieved when the aeration was increased. The percentages of BOD and DOC removals could rich 98% and 91% respectively. And oxidization of ammonia nitrogen into nitrite and nitrate was obtained, which total mean percentages of SND and NAR were 71.8% and 60.9%. Contradictory results were attained when the aeration sharply decreased and followed declining removals of organic matter and nitrogen, and increase NH_4^+-N concentration in the effluent. Therefore, it is suggested that *Johkasou* system is recommended to applied continuous oxygen supply at high rate to achieve simultaneously nitrification.

5. References

[1] T. Ichinari, A. Ohtsubo, T. Ozawa, K. Hasegawa, K. Teduka, T. Oguchi, Y. Kiso. *Process Biochem.* **43**, 7 (2008).

[2] P. H. Chen, K. C. Leung, J. T. Wang, **26**, no. 1–2, pp. 63–68 (2000)

[3] M. Zhang, P. G. Lawlor, G. Wu, B. Lynch, X. Zhan. Bioproses Biosysts Eng **34**, 9, pp. 1049–1056 (2011).

[4] S. Park and W. Bae. *Process Biochem.* **44**, 6, pp. 631–640 (2009).

[5] Y. Peng, S. Zhang, W. Zeng, S. Zheng, T. Mino, H. Satoh. Water Res **42**, no. 4–5, pp. 883–892, (2008).

[6] V. M. Vadivelu, J. Keller, Z. Yuan. Water Sci. Technol.**56**, 7, pp. 89–97 (2007).

[7] J. Guo, Y. Peng, S. Wang, Y. Zheng, H. Huang, Z. Wang. Bioresour. Technol **100**, 11, pp. 2796–2802 (2009).

[8] G. bing ZHU, Y. zhen PENG, S. yun WU, S. ying WANG, S. Wei XU. J. Environ. Sci **19**, 9, pp. 1043–1048 (2007).

[9] G. Ruiz, D. Jeison, O. Rubilar, G. Ciudad, R. Chamy. Bioresour. Technol **97**, 2, pp. 330–335, (2006).

[10] N. Kishida, J.-H. Kim, M. Chen, H. Sasaki, R. Sudo. J. Biosci. Bioeng **96**, 3, pp. 285–290, (2003).

[11] Y. Z. Peng, Y. Chen, C. Y. Peng, M. Liu, S. Y. Wang, X. Q. Song, Y. W. Cui. Water Sci. Technol **50**, 10 (2004).

[12] H. Yoo, K. H. Ahn, H. J. Lee, K. H. Lee, Y. J. Kwak, K. G. Song. Water Res **33**, 1, pp. 145–154, 1999.

[13] W. Jianlong and Y. Ning. Process Biochem **39**, 10, pp. 1223–1229 (2004).

[14] W. Yang, J. Vollertsen, T. Hvitved-Jacobsen. Water Sci. Technol **48**, 3, pp. 135–141 (2003).

[15] R. Naseer, S. Abualhail, L. Xiwu. Saudi J. Biol. Sci **20**, 1, pp. 11–21 (2013).

[16] W. yi Dong, H. jie Wang, W. guang Li, W. chao Ying, G. Hua Gan, Y. Yang. J. Memb. Sci **344**, 1–2, pp. 219–224, **2009**.

[17] APHA., AWWA., WEF. *Standard method for the examination of water and wastewater. 21st Ed.* New York: (APHA, 2005).

[18] R. Canziani, V. Emondi, M. Garavaglia, F. Malpei, E. Pasinetti, G. Buttiglieri. J. Memb. Sci **286**, 1–2, pp. 202–212 (2006).

[19] J. Pan, H. Fei, S. Song, F. Yuan, L. Yu. Bioresour. Technol. **191**, pp. 327–331 (2015).

[20] M. Kumari and B. D. Tripathi. Ecol. Eng **62**, no. 3, pp. 48–53 (2014).

[21] H. Dong, Z. Qiang, T. Li, H. Jin, W. Chen. J. Environ. Sci.**24**, no. 4, pp. 596–601(2012).

[22] X. Yuan and D. Gao. J. Hazard. Mater **178**, 1–3, pp. 1041–1045(2010).

[23] Y. Rahimi, A. Torabian, N. Mehrdadi, B. Shahmoradi. J. Hazard. Mater **185**, 2–3, pp. 852–857 (2011).

[24] W. Bae, S. Baek, J. Chung, Y. Lee. Biodegradation **12**, 5, pp. 359–366 (2001).

Study on e-waste (CRT TVs/monitors and washing machines) generation in Bandung

Nadya Humaira[1], and *Emenda* Sembiring[2,*]

[1,2]Environmental Engineering Study Program, Faculty of Civil and Environmental Engineering, Bandung Institute of Technology, Jl. Ganesha No. 10, Bandung, West Java, Indonesia

Abstract. Modern day's rapid development in technology has forced a shift in trends and popularity of electronic products. This causes early obsolescence of former technologies such as cathode ray tubes (CRT), leading to massive disposal in a short amount of time. To be able to predict this newly developing waste stream, a study on the generation of such electronic waste products is needed. In a case study focusing on Bandung, questionnaires to primary sources of e-waste such as households, offices, schools and laundromats were conducted to determine not only the number of e-products that is used and discarded, but also how e-waste is treated firsthand when it becomes unwanted. The e-waste generation predicted is about 65,000 units of CRT TVs per year, 19,000 CRT monitors per year and 48,000 washing machines per year. Survey results show that when proper waste collection and recycling is implemented, CRT TVs and monitors will have been eliminated within a decade (2028 being the marking year) meanwhile the trend for washing machine consumption continues to incline up until today.

1 Introduction

This study focuses on specific e-products, that is CRT TVs, CRT monitors, and washing machines. The history of how such electronic products entered Bandung is not precisely known, but it can be predicted that CRT TVs became popular in the 60s when TV stations were still monopolized, meanwhile CRT monitors probably became popular in the 90s when internet and computer networking was first developed by local students. Washing machines were first put into international market around 1957, meanwhile in Indonesia, it can be assumed that the product boomed between 1980-1990 when household appliance producers started to arise.

According to Widyarsana [4], sales of e-products in general continues to incline up to the year 2000, though sales and use of desktop computers began to diminish after 2005. Widyarsana [4] stated through statistics of his 2008 research findings that the average number of e-products used in Bandung (new and secondhand) per 100 households is around 100 TVs (CRTs and flat screens), 50 computers and 60 washing machines.

Early obsolescence of cathode ray tube products is caused by the rising popularity of new technologies such as LCD, LED and plasma screens. This shift in technology and trend eventually causes consumers to have their old electronic products disposed of, even though they may still be in good shape. Though these "old-fashioned" products have become a dying industry, they still contain materials that are valuable and can be recycled. Meanwhile, CRT products are estimated to reach the peak of disposal between 2015-2020 according to Gregory [1].

2 Materials and method

The study of e-waste generation in Bandung is done primarily through questionnaires sent out to (randomly picked) households, schools, offices and laundromats, and through literature or secondary data. The number of survey respondents according to their categories are given in Table 1.

The survey is done with two types of media, the first being an online questionnaire with a total number of respondents making up 75% of the number of samples needed to represent the domestic sector, and the second being door-to-door manual interviews to the remaining samples needed for households, schools, offices and laundromats. The main objective of the survey is to determine the number of electronic products (namely CRT TVs, CRT monitors and washing machines) that are and have been of use throughout Bandung, if and when such products have become unwanted (nonworking or obsolete) then discarded, and the behavior of society in handling unwanted electronics. From such data, the lifespan of e-products can be determined, and e-waste generation can be predicted using a life cycle analysis method.

* Corresponding author: emenda@ftsl.itb.ac.id

Table 1. Number of survey respondents.

Category of Respondent	Population in Bandung	Confidence Level	Number of Samples	Description
Domestic				
Household (HH)	657,769	93%	200	36 HH of low economy (17.84%) 62 HH of mid-class economy (30.83%) 102 HH of high-class economy (51.32%)
Nondomestic				
Schools	2,113	80%	25	1 public kindergarten and 5 private kindergartens (24.8%) 10 public elementary schools and 2 private elementary schools (46.81%) 1 public junior high school and 2 private junior high schools (10.13%) 1 public senior high school and 1 private senior high school (6.39%) 1 public vocational school and 1 private vocational school (6.25%) 1 public university and 1 private university (5.63%)
Offices	6,310	80%	25	Government office Private office Government-owned enterprise
Laundromats	750	80%	25	Population is predicted, no specific categorization in samples
TOTAL			**275**	

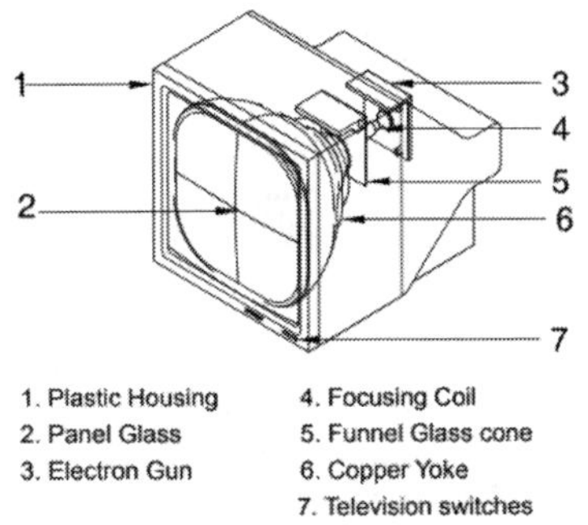

Fig. 1. Components of a CRT monitor.

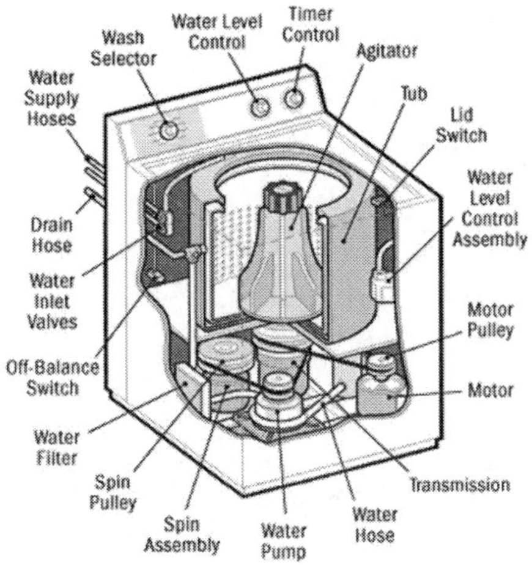

Fig. 2. Components of a washing machine.

3 Results

3.1 Survey derived data

Survey results hint that the use of CRT TVs and CRT monitors show a rise in the 90s era, and reached its peak in the early 21st century. Between the period of 2007-2010, the trend for those e-products have drastically declined (mass disposal) and continues to diminish, especially in nondomestic sectors (schools and offices) due to the introduction of new and improved technology that is flat screens. Although the nondomestic units tend to follow the technology trend, many households still own or use CRT TVs and CRT monitors, especially in low and middle-class economy households. The trend for CRT TV and CRT monitor sales are heading towards zero. Even most offices have already gotten rid of such products so that the e-waste generation also declines within these past couple of years.

The only difference in consumption trends are found in that of washing machines, specifically from households and laundromats. Unlike CRT products, washing machines have not been replaced with any new technology so that the use of washing machines within our society is still normal, if not increasing in demand. The prediction of waste CRT products cannot be done with a period of time that is too long because consumption is declining rapidly (until it may reach zero altogether), meanwhile the prediction of waste washing machines has to take into account the trend of sales and higher levels of consumption. Table 2 summarizes the average e-waste generated in Bandung.

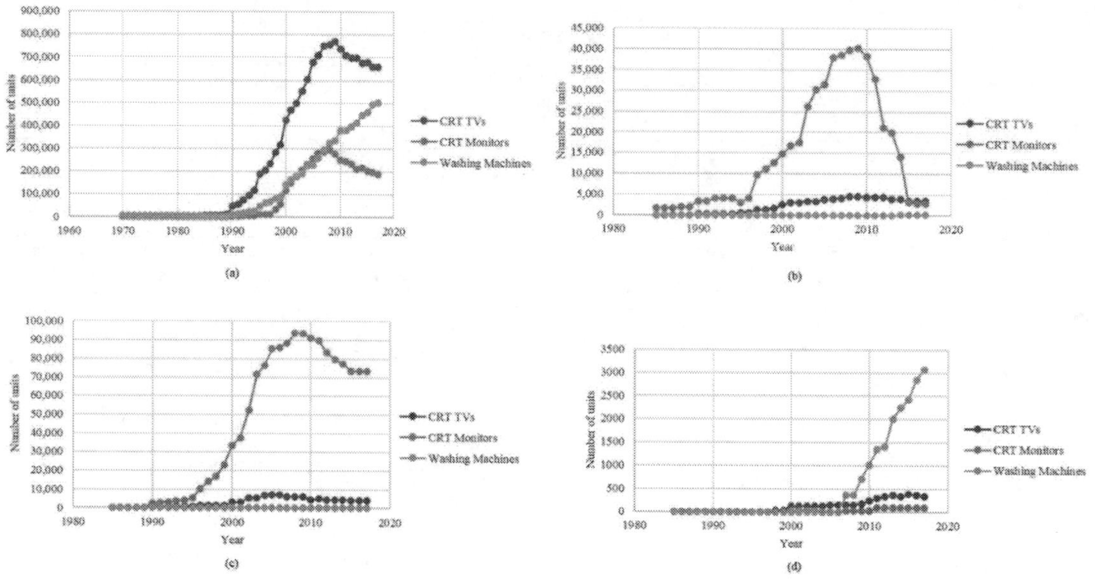

Fig. 3. Trend of CRT TVs/monitors and washing machines possessions in Bandung; (a) households, (b) schools, (c) offices, (d) laundromats.

Table 1. Average amount of e-waste generated annually in Bandung.

Source of E-waste	Average e-waste generated per year (units/year)		
	CRT TVs	**CRT Monitors**	**Washing Machines**
Household (n=200; N=657,769)	22,827	16,615	15,260
Schools (n=25; N=2,113)	261	3,978.4	~
Offices (n=25; N=6,310)	662.6	3,821.1	~
Laundromats (n=25; N=750)	~	~	80

The generation of waste CRT products from laundromats can be ignored because consumption of such products is very insignificant compared to the consumption in other sources of e-waste. The same applies to waste washing machines from schools and offices.

Because the prediction of e-waste generation depends on a life-cycle analysis, then another objective of the survey is to determine the lifespan and shelf life (period of storage) of CRT TVs, CRT monitors and washing machines. Lifespan is defined by the period of time in which the electronic product is still in use, which is calculated from the time of purchase until the time of defection or the time it becomes obsolete. Shelf life is defined as the period of time during which time the product has already passed its lifespan, but is still kept in storage and not yet disposed of. The following equations will better define such terms;

$$L_u = t_o - t_p \quad (1)$$

$$L_s = t_d - t_o \quad (2)$$

whereas

L_u = lifespan of product (years),
L_s = shelf or storage life of product (years),
t_p = year of product purchase,
t_o = year of defection or obsolescence, and
t_d = year of disposal.

Survey results show that the life span and shelf or storage life varies with each product. The data obtained also gave a wide range of variety in such numbers within the same product category. This is mainly due to the fact that the condition of the electronic product, whether new or secondhand, upon purchasing was not considered. Table 2 and Table 3 summarizes the average life span and storage life of CRT TVs, CRT monitors, and washing machines that were obtained from survey data.

The life span of electronic products given in Table 3 shows numbers smaller in households of lower class economy, meanwhile the life span of e-products in households of middle and high-class economy are of approximately the same value. This statement follows the results of Widyarsana's [4] research of e-waste generation in Bandung, 2005-2006, which shows that residents of lower-class economy households tend to buy secondhand products, whose life span is shorter than new products. On the other hand, new products are more favorable to higher economy households, which is why the life span is longer. According to Widyarsana [4], an approximate 40% of low economy households prefer to buy secondhand electronics, meanwhile only 10%-20% of high-class economy households use secondhand electronics.

Table 2. Average life span of CRT TVs, CRT monitors and washing machines according to survey results.

Source of E-waste	Life Span (years)		
	CRT TVs	CRT Monitors	Washing Machines
Domestic			
HH of high-class economy	9.57	6.08	8.27
HH of middle-class economy	9.53	5.48	9.11
HH of lower-class economy	8.03	4.13	4.78
Nondomestic			
Schools	9.3	8	~
Offices	7.2	8.8	~
Laundromats	~	~	4.1

Table 3. Average storage life of CRT TVs, CRT monitors and washing machines according to survey results.

Source of E-waste	Storage Life (years)		
	CRT TVs	CRT Monitors	Washing Machines
Domestic			
HH of high-class economy	2.52	2.37	0.66
HH of middle-class economy	0.86	2.39	1.15
HH of lower-class economy	0.86	0.88	1.12
Nondomestic			
Schools	1.95	2.14	~
Offices	2.33	5.23	~
Laundromats	~	~	1

Data on storage life also corresponds to Widyarsana's [4] research, in that electronics purchased as new products are kept in storage for longer periods of time than secondhand electronics. According to survey results, with the assumption that a larger portion of low economy households use secondhand electronics, the storage life is then shorter than that of the electronics purchased in new conditions by households of higher class economy.

3.2 E-waste generation prediction

Data on current e-waste generation given by survey results need to be further analyzed to predict the e-waste generation in the upcoming years. The e-waste generation cannot be directly predicted from the previous trend data because the trend for e-waste disposal does not show a general pattern, but a quite random trend.

From the survey, we can conclude that the purchase of CRT TVs and CRT monitors have nearly come to zero since 2015 in households and schools, and further since 2012 in offices due to the shift to flat screen displays. For unprofessional use, CRT TVs and CRT monitors are still popular in households, either as old appliances or stored and waiting to be disposed of. Nevertheless, purchase of new CRT products are still declining sharply. Therefore, the prediction of waste CRT products generation will simulate the diminishing of said products until it reaches complete elimination throughout Bandung. With the assumption that there will be no more CRT TVs and CRT monitors purchased in the coming years, the future e-waste generation is predicted from the existing number of electronic products that are already in use in Bandung, and disposal is calculated according to life span and storage life already given in Table 3 and Table 4.

As stated before, the prediction of waste washing machine generation cannot depend only on the existing number already used in Bandung, but rather with the consideration of future sales trends. Sales trends are projected from the existing trend of sales from mid 1980s to present day. Though the trend fluctuates, the trend for washing machine sales can be summarized generally to show a positive inclination. These trends (of washing machine possessions and sales) can be graphically presented by Figure 4 and Figure 5.

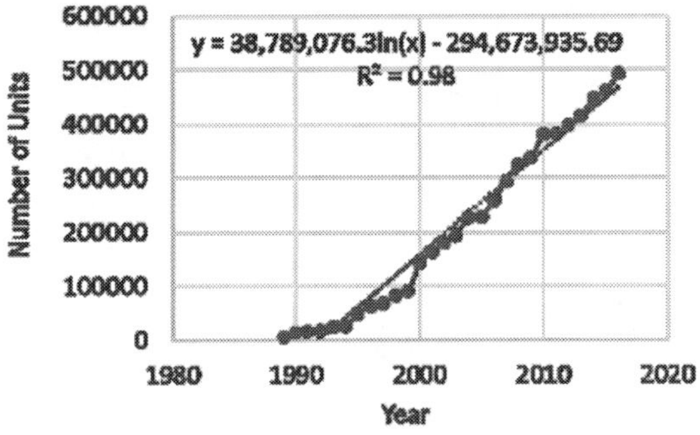

(a)

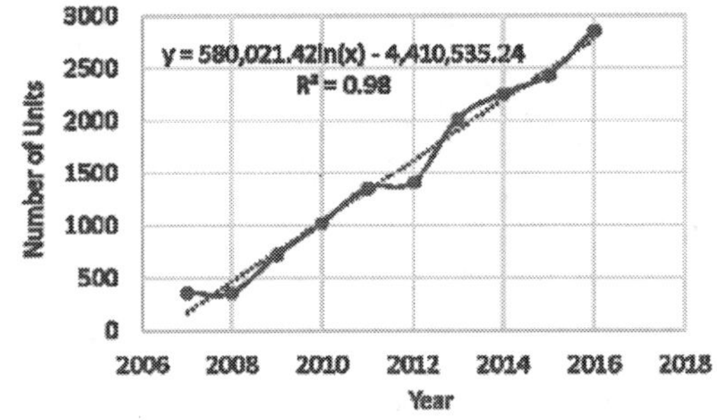

(b)

Fig. 4. Trend of washing machine possessions in Bandung; (a) households, (b) laundromats.

A logarithmic trend is used because electronic products follow a 4-step life cycle, from introduction of product, to rapid growth, a stagnant trend and ending with a rapid decline. Derived from the graphics in Figure 4, the number of washing machines accumulated is given by Equation 3 and Equation 4 respectively for households and laundromats.

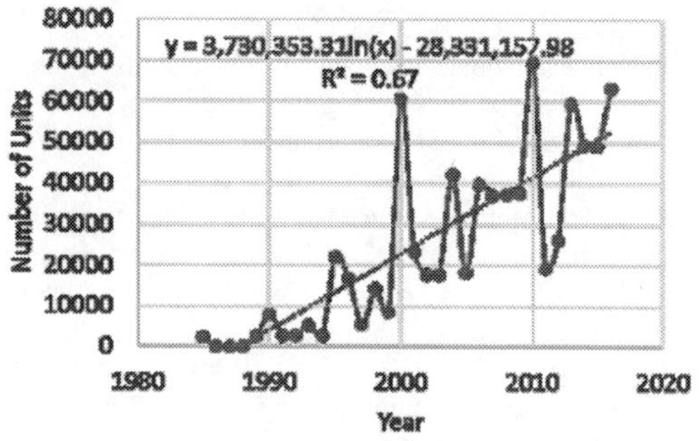

(a)

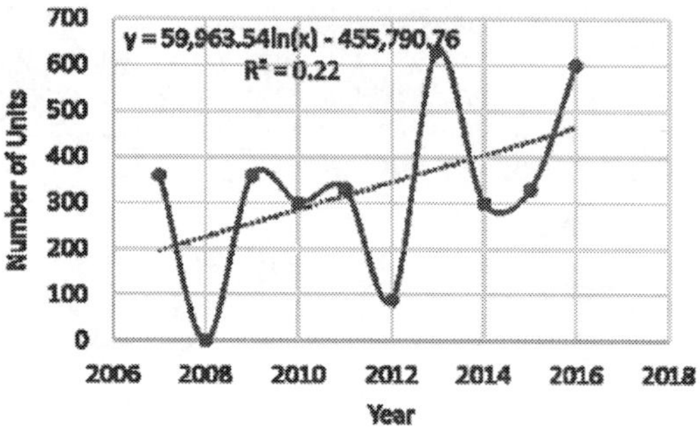

(b)

Fig. 5. Trend of washing machine sales in Bandung; (a) households, (b) laundromats.

$$y_{aHH} = (38,789,076.3 \times \ln(x)) - 29,673,935.69 \quad (3)$$

$$y_{aL} = (580,021.42 \times \ln(x)) - 4,410,535.24 \quad (4)$$

whereas

y_{aHH} = accumulation of washing machines owned in households in year x,

y_{aL} = accumulation of washing machines owned in laundromats in year x, and

x = year of study.

The following equations are used to determine the number of washing machine sales, derived from Figure 5.

$$y_{sHH} = (3,730,353.31 \times \ln(x)) - 28,331,157.98 \quad (5)$$

$$y_{sL} = (59,963.54 \times \ln(x)) - 455,790.76 \quad (6)$$

whereas

y_{sHH} = number of sales/purchase of washing machines from households in year x,

y_{sL} = number of sales/purchase of washing machines from laundromats in year x, and

x = year of study.

Meanwhile, the number of discarded washing machines is determined with the following equation.

$$D_n = AC_{n-1} + SP_n - AC_n \quad (7)$$

whereas

D = disposal of waste washing machine,

AC = accumulation of washing machines owned,

SP = sales/purchase of washing machines, and

n = year of study.

The prediction of e-waste generation is done per year, which is then used to determine the annual average as follows:

- Waste CRT TVs as many as 65,195 units/year
- Waste CRT monitors as many as 19,191 units/year
- Waste washing machines as many as 47,987 units/year.

For a number of households of approximately 650,000 units, the prediction shows that within 10 years, almost every household will discard a single CRT TV and washing machine, and as many as 0.3 units of CRT monitors. This is a relatively high number, considering a desktop possession coefficient of 0.57 in Bandung in 2008 as stated by Widyarsana [4]. Widyarsana [4] also predicted that by 2015, as much as 2,000 tons/year worth of computer waste will be generated, which can be estimated to 100,000 units (for an average weight of 20 kg as stated by Huisman [2]). The difference found in Widyarsana's [4] studies are probably due to the different survey methods --Widyarsana [4] not only studied the waste generation from domestic and nondomestic units, but also the informal sectors that play different roles in e-waste collection and management-- and that this prediction applies for both CRT and flat screen computers. Referencing to other Asian countries such as China, with a household population of about 450 million in 2012, Singh [3] stated that by 2012, it is estimated that around 190 million PCs and 74 million TVs will become obsolete, with CRTs making up of around 80% of the number. In a study by Yoshida [5] the annual amount of e-waste generated in Indonesia, the Philippines and Vietnam (all of which are categorized as middle-income countries) ranged from 2 to 3 kg per capita. Assuming one household holds 4-5 people, then the household generation is estimated to 15 kg per year. Yoshida's [5] study result gives a rather small number in comparison to the numbers given by this study.

4 Conclusion

Based on the survey study done, it can be concluded that waste CRT products will diminish within the next decade, if an integrated waste collection and recycling system is implemented in Bandung. The number of waste CRT products can be summed (from TVs and monitors) to 74,000 units/year, meanwhile the number of waste washing machines generated is estimated to be around 48,000 units/year.

References

1. J.R. Gregory, M.C. Nadeau, R.E. Kirchain, Env. Sci. & Tech., **43(24)**, 9245-9251 (2009)
2. J. Huisman, F. Magalini, R. Kuehr, R., C. Maurer, C. Delgado, E. Artim, J. Szlezak, S. Oglivie, J. Poll, A. Stebels, *2008 Review of directive 2002/96 on waste electrical and electronic equipment (WEEE) final report* (United Nations University, Germany, 2007)
3. N. Singh, J. Wang, J. Li, Procedia Env. Sciences, **31(2016)**, 465-474 (2016)

4. I.M.W. Widyarsana, I.M.W. *Pengembangan metode proyeksi timbulan limbah "e-waste" berdasarkan masa pakai (end-of-life) barang "e-product" sebagai dasar dalam prediksi material flow analysis (MFA)*, Disertation. Environmental Engineering Program, Faculty of Civil and Environmental Engineering, Bandung Institute of Technology, Bandung (2008)
5. A. Yoshida, A. Terazono, F.C Jr. Ballesteros, D. Nguyen, S. Sukandar, M. Kojima, S. Sakata, Res. Cons. and Rec., **106(2016)**, 48-58 (2015)

Study on the leaching performance of chromium (Cr) and cadmium (Cd) from the utilization of solidified nickel slag as concrete floors

Maryam Dewiandratika[1,*], *Sukandar*[1], and *M. Tafdhila* El Akmam[1]

[1] Department of Environmental Engineering, Faculty of Civil and Environmental Engineering, Institut Teknologi Bandung, Indonesia

Abstract. With the increasing demands on more metals extraction, the hazardous waste resulted from mining industry has also raised. In this paper, the treatment process of nickel slag containing hazardous metals is described. The treatment involved a solidification process, where nickel slag is used as fine aggregates substituting agent. This study aimed to determine the optimum dosage of nickel slag that met specification in the construction of concrete floors. A series of technical feasibility test were carried out by replacing the weight proportions of sand with nickel slag gradually from 0% to 100% to examine the compressive strength, absorption rate, durability and leaching performance. The results indicated that even though all mixture variations possessed high value of absorption rate indicating high permeability concrete mix, all mixtures, however, had good durability. Furthermore, based on the compressive strength test, a mixture with 60% substitution of nickel slag is considered as a viable replacement of sand in the concrete floors construction in conformance to American Concrete Institute. The leaching performance test showed that leachability of heavy metals content, particularly Cr and Cd from mortar specimens were below the standard according to Indonesian Government Regulation No 101 of 2014, therefore, in terms of environmental point of view, the utilization of solidified nickel is considered safe to be used as concrete floors.

1 Introduction

Mining is one of the potential sectors that contribute to the national income, due to wide range of uses of metals and growing of their demand. However, the operation of mining industry could not be separated with the waste or by-products resulting from its process, of which hazardous waste is included. Rapid growth of mining industry has obviously led to steep increase of hazardous waste that should be properly managed. Otherwise, serious negative impacts on human and environment may occur.

Slag is the most by-product left over in the mining industry, which resulted from a process separation between metals with its raw ore through heating process. Originally, the slag is produced in liquid form, but then as the temperature cools down it is naturally transformed to be a solid. In various mining industry, slag can be found in the form of steel slags, phosphorus slag, zinc slag, nickel slag, etc, depending on the metals that the industry produces.

One of the nickel mining industry in Indonesia, generates approximately 50 ton of nickel slag to produce one ton of pure nickel, this means an annual production of 40,000-60,000 ton of pure nickel leading to annual slag production equal to 2-3 million ton [1]. Moreover, slag is categorized as hazardous waste with code B403 due to its heavy metal content, in conformance to Indonesian Government Regulation No 101 of 2014. Therefore, the need of having treatment alternatives for slag becomes crucial, given that the huge amount of hazardous waste produced annually.

Solidification and stabilization process is an alternative treatment that may be able to convert slag to any others harmless product by way of immobilizing contaminants or waste within any other constituent materials in concrete production, hence, the migration of contaminant to human and environment is prevented. The contaminant immobilization can be achieved through a mixing process between a binding material and the contaminated material [2]. In addition, the incorporation slag into concrete volume would benefit construction industry as slag could be utilized as a replacement of constituent material like fine aggregates or cement [3].

This current study focuses on nickel slag generated from nickel mining industries. The objective of the study reported here was to investigate the possibility of nickel slag to substitute fine aggregates material in the

* Corresponding author: md2115.tl.itb@gmail.com

production of concrete floors. Additionally, this study aims to determine the optimum of compositional variation between nickel slag and binder with respect to the required products specifications and lastly, to analyze the leaching performance on Cr and Cd from the solidified nickel slag.

2 Experimental Methods

2.1 Materials

Nickel slag sample taken from one of the largest nickel mining industry in eastern part of Indonesia was expected to be used as fine aggregates replacement in the construction of concrete floors. An Ordinary Portland Cement (OPC) was used as a binder in the solidification process. Fine aggregates used was a typical of natural sand found in Indonesia. Nickel slag, OPC and fine aggregates have specific gravity of 2.73, 3.15, and 2.53, respectively [4].

2.2 Analysis

Analysis on nickel slag, OPC, fine aggregates and water used in the experiment were carried out in order to determine the physics-chemical characteristics of each material. The analysis including water content, dry matter, absorption rate, ash content, volatile content, metals oxide and heavy metals were undertaken based on methods available in Indonesian National Standard (SNI) and American Standard Testing and Material (ASTM) [5-8]. The sieve analysis was also performed on the nickel slag samples and the sand in conformance to ASTM C136:2012 [9]. The main elements of nickel slag and the OPC used are shown in Table 1 and the heavy metals content of the nickel slag are shown in Table 2.

Table 1. Main elements of nickel slag and OPC.

	LOI	Main elements (% w/w)					
		CaO	SiO_2	Al_2O_3	Fe_2O_3	MgO	Na_2O+K_2O
Nickel slag	16.7	10.34	48	1.17	10.19	9.75	0.88
OPC		60-65	17-25	3-8	0.5-6	0.5-4	0.5-1

Table 2. Heavy metals content in nickel slag, OPC and fine aggregates.

Metal content (mg/kg)	As	Ba	Cd	Cr	Co	Cu	Pb
Nickel slag	0.432	0.1	0.25	25.4	4.1	1.25	0.025
Sand	1.25	279.75	8.5	1.25	3.25	50.75	31.75
OPC	3.75	251	5.5	51.25	3	46	65.25
Metal content (mg/kg)	**Hg**	**Mo**	**Ni**	**Sn**	**Se**	**Ag**	**Zn**
Nickel slag	0.012	0.025	310.5	0.025	0.205	0.025	2.45
Sand	0.007	<0.001	0.074	4.75	ND*	0.005	ND*
OPC	0.008	<0.001	0.182	8.75	ND*	0.004	109.5

* Not detected

2.3 Preparation of solidification solids

Based on the results of sieve analysis from the previous stage and the data of specific gravity, it is obtained that nickel slag has met the requirements of fine aggregates, as it has fineness modulus of 3 compared to the standard value for fine aggregates according to ASTM C 33-03, which is lied within the range of 2.3-3.1. Therefore, in this current study, a total of 10 concrete mixtures for each variation were prepared by varying the proportions of nickel slag and the sand, whereas the OPC was maintained constant. Here, where nickel slag was used to substitute fine aggregates in concrete floors construction, the required weight ratio between OPC and fine aggregates is within the range of 1:4 – 1:6 [1]. The proportion variations of nickel slag and sand are shown in Table 3.

Table 3. The mix variations of concrete specimens in percent weight.

Weight ratio of OPC:sand	NS:S*	NS:S	NS:S	NS:S	NS:S
1:4	0:100	55:45	65:35	75:25	100:0
1:5	0:100	35:65	45:55	55:45	100:0
1:6	0:100	15:85	25:75	35:65	100:0

*NS: nickel slag; S: sand

The amount of added water to each of the sample variations corresponded to the flow table test based on ASTM C1437-07, resulting in water-to-cement ratio within the range 0.7-1. According to the proportional variations, the mix concrete paste was placed into cubical molds (5 cm x 5 cm x 5 cm) and compacted. After 24 hours and naturally dried, the concrete specimens then demolded. To estimate the duration needed by the concrete specimens to completely dry, an analysis on setting time based on ASTM C191-04-a [10] was performed by dropping a Vicat needle to the mortar paste. Dried samples were ready to undergo a further test after being cured in water for certain period of curing days.

2.4 Technical feasibility test

To evaluate the mechanical properties of the concrete specimens, a series of testing in conformance to ASTM were carried out. The technical feasibility test includes compressive strength test following ASTM C 109-01 by using Universal Testing Machine (UTM) [11]. The test was undertaken to the mortar specimens that have been cured in water during period of 3, 7, 14, and 28 days. The absorption test was carried out on all the variation of 28 days old mortar specimens [12]. The durability test was done on similar number of specimens as in the absorption test, but following the procedure on ASTM D 4843-88 R04 for 13 cycles, which one cycle equals to 48 hours [13].

* Corresponding author: md2115.tl.itb@gmail.com

2.5 Leaching test

Leaching test includes the Toxicity Characteristics Leaching Procedure (TCLP) test, static and dynamic test were carried out on both nickel slag and concrete specimens. All the concrete specimens that met criteria required by previous technical feasibility tests were included in the leaching test. In the TCLP test, acetate acid was used according to US EPA SW-846 Test Method 1311 [14], where the selected mortar specimens were agitated in acid solution during ±18 hours.

The static and dynamic leaching test used a modified method based on ASTM D3987-12 [15]. In the present study, the observation of leaching test is focused on Cr and Cd since the predominant metals in nickel slag, such as Ni and Co have been examined in another study. In the static test, with regards to the procedure in USEPA Test Method 1311 [13], concrete specimens were contained and H_2SO_4 and CH_3COOH with pH 4.92 ± 0.05 were added to soak the specimens. The liquid-to-solid ratio (L/S) of 20:1 was used to quantify the volume of acid solution and the amount of solid sample. Cr and Cd contaminant in the leachate samples were observed during curing period of 1, 3, 6, and 9 days using Atomic Absorption Spectroscopy (AAS).

On the other hand, carbonate acid was used in the dynamic test to represent the condition of acid rain, in which samples of concrete specimens and nickel slag were placed in glass cylinders and acid solution was flown into the cylinder for 3 hours. The leachate produced then was underwent a heavy metals analysis [8]. The test was continued by drying the specimen samples and the nickel slag that have been flown by acid solution for 24 hours in the temperature of 60°C. The dried specimen samples and the nickel slag then for the second time were flown by the carbonate acid, while produced leachate were taken hourly to conduct the heavy metals analysis using AAS.

3 Results and Discussion

3.1 Compressive strength test

A concrete compressive strength is a maximum strength that concrete can hold for an area functional unit, of which it is one of the most necessary parameter in the building construction materials. Concrete compressive strength is highly affected by several factors like the water-to-cement ratio, aggregates type and characteristics, workability, maintenance and period of curing days. Figure 1 presents the results of compressive strength of concrete specimens for various composition as sand substitution agent at 7, 14, and 28 days curing periods.

In general, the results show a decreasing trend as more nickel slag in the concrete mixtures. This is probably due to the high value of water to cement ratio in the mortar specimens with nickel slag. According to the SNI 03-6468-2000 [16], the required water-to-cement ratio for high strength concrete should be within the range of 0.2-0.5, whereas the ratio value in this experiment reached 0.7-1. The high ratio of water to cement may be due to low water content in the nickel slag, which is 1.812%, that is quite lower than most aggregates value, which is lied between 3-8% [17].

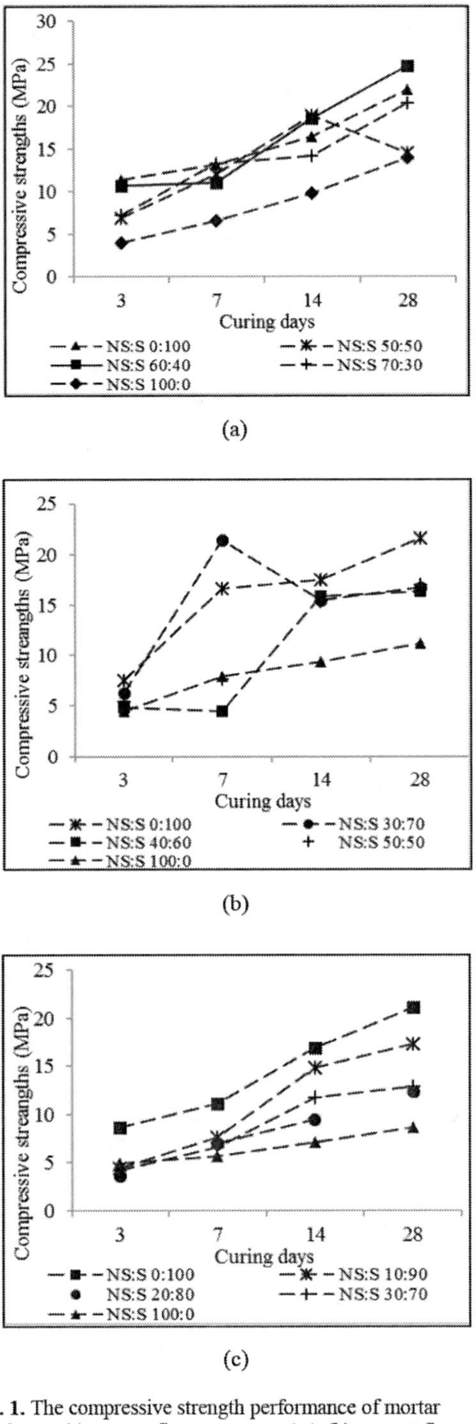

Fig. 1. The compressive strength performance of mortar specimens (a) cement:fine aggregates 1:4, (b) cement:fine aggregates 1:5. (c) cement:fine aggregates 1:6.

Concrete mix with cement-to-aggregates ratio of 1:4 presents the highest value of compressive strengths, whereas between 2 others ratio, the difference of compressive strength on the control specimen is pretty small. The value of compressive strength obtained was far below the value when copper slag is used as substituting agent for aggregates in building construction material, even for the control specimen as reported by Al-Jabri, et al [18]. In accordance with American Concrete Institute [19], only the mixture with 60% substitution of nickel slag that met the concrete strength requirements as concrete floors level 1,2, and 3 that require minimum strength of 21 MPa.

The compressive strength raises as the curing period was extended due to longer hydration duration allowing a more complete bond forms between concrete constituents and at the same time preventing the presence of excess water that may form internal voids in the concrete mix. The existence of porosity in the concrete mix is unexpected, as it may weaken the bond between the concrete components, which is one of determining factors for the strength of concrete [20].

The fineness modulus in the nickel slag is beyond the range of most aggregates commonly used in the construction. According to Wu et al. [20], this condition is beneficial as finer particles of nickel slag leading to the less porosity that may form in the concrete mix, but on the other hand, the property is naturally denser than most aggregates, hence to fill the same volume of space that sand occupies, more nickel slag is needed. Considering this fact, the concrete strength with more nickel slag is more likely weaker than those with less nickel slag due to larger surface area of nickel slag per unit volume exposed, resulting in inadequate cement content for binding the concrete matrix together. Furthermore, as the fineness level of aggregates increases, the shrinkage possibility increases as well. In some cases, the shrinkage that cause deep cracks does not appear in surface of concrete mix, but inside the concrete [21].

Based on the metal oxides and heavy metals estimation, it was obtained that nickel slag constitutes a relatively high content of ferric oxide and chromium. This may be another reason that as nickel slag content is higher, the compressive strength of concrete mix is lower. Previous study showed that the duration of cement hydration goes longer when heavy metals present. The delay of cement hydration becomes more serious with the addition of nickel slag that leaves more excess water in the concrete and forms more flaws or weakest links harm to the concrete structure [20].

3.2 Absorption test

Absorption test aims to determine the ability of concrete mix in absorbing water. The maximum limit of absorption rate of concrete mix in conformance with ARTO Brick and California Pavers is 12%. The absorption rate of concrete specimens was presented in Table 4. The results showed that all specimens have met the requirement of desired concrete mix with regards to absorption rate.

Table 4. Value of absorption rate of all mortar specimens.

Cement aggregates ratio	Proportional variations (% w/w)	Absorption rate (%)
1:4	NS:S 0:100	8.54
	NS:S 50:50	9.51
	NS:S 60:40	9.38
	NS:S 70:30	9.8
	NS:S 100:0	11.6
1:5	NS:S 0:100	11.3
	NS:S 30:70	9.73
	NS:S 40:60	8.75
	NS:S 50:50	10.9
	NS:S 100:0	12.2
1:6	NS:S 0:100	9.18
	NS:S 10:90	8.42
	NS:S 20:80	8.87
	NS:S 30:70	9.6
	NS:S 100:0	11.82

It can be seen from the results that there was an inverse relationship between compressive strength and absorption rate, of which this result is supported by Zhang [3], the specimens that show highest strength will have lowest absorption rate. However, the absorption rate that is evaluated in current study only represented a surface water absorption and did not take into account an internal water absorption, which is also one of determining factor of concrete strength.

Based on study done by Zhang [3], a linear correlation exists between absorption and permeability, where permeability is one of the main characteristics affecting the durability of concrete. It can be seen from the results that higher nickel slag content possessed higher absorption rate indicates higher permeability. This is primarily due to higher water cement ratio in the mortar specimen that comprise more nickel slag than that in the control concrete. The high level of water cement ratio can cause volume change, that eventually results in the development of micro-cracks and disintegration [22].

3.3 Durability test

The durability of concrete was evaluated to understand the effect of harsh environments that can result in failure of structures. Table 5 presents the results of durability test on all variations of concrete specimens. In conformance to ASTM D 4843-88 R 04 [13], the maximum percentage of mass loss after durability test for 13 cycles is 30%. The smallest mass loss from present study was shown by a mixture with 60% substitution of nickel slag. This value is even lower than those for three values of control mortar. This result reveals that even though the absorption rate on all mixtures are high, the mass loss percentage, however, still met the relevant criteria.

Table 5. The results of durability test of all compositional variations.

Cement aggregates ratio	Proportional variations	Durability (%)
1:4	NS:S 0:100	7.7
	NS:S 50:50	6.34
	NS:S 60:40	6.01
	NS:S 70:30	7.32
	NS:S 100:0	8.52
1:5	NS:S 0:100	9.36
	NS:S 30:70	7.54
	NS:S 40:60	6.2
	NS:S 50:50	8.24
	NS:S 100:0	10.97
1:6	NS:S 0:100	7.6
	NS:S 10:90	7.33
	NS:S 20:80	7.08
	NS:S 30:70	7.78
	NS:S 100:0	8.08

3.4 TCLP test

In current study, TCLP test was conducted to identify the effect of pozzolan characteristics in stabilizing heavy metals and to determine the heavy metals presence in the leachate. As it was expected, most of heavy metals level in all mix variations decrease after solidification was carried out (Figure 2). The pozzolan physical nature is very effective in the metals immobilization leading to reduced leachability, primarily because most of the compound is reacted into undissolved hydroxide [23].

Even though most of the metals level decline after solidification took place, several metals concentration like Ba, Cu, Ag and Zn raise. This is probably due to the contribution of heavy metal content in the OPC and sand themselves, in which Ba, Cu, and Zn content in OPC and sand were far greater than those in nickel slag.

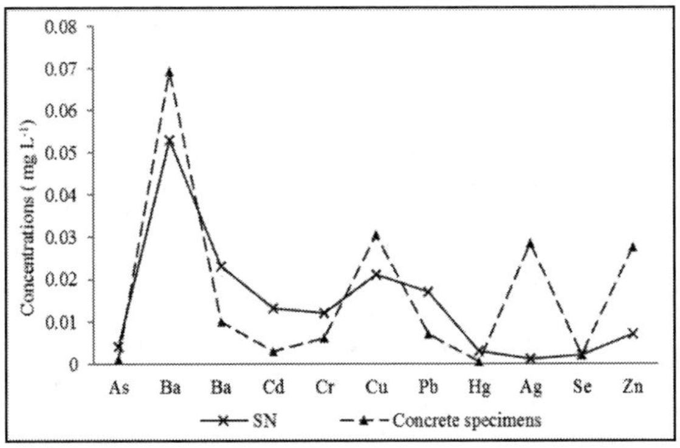

Fig. 2. Comparison of TCLP analysis of concrete specimen and nickel slag.

3.5 Static and dynamic leaching test

Within this test, the leaching behavior of Cr and Cd that present in the concrete mix was measured. This test differs from previous TCLP test, in which within this study, TCLP is a used method to estimate the leachability of heavy metals that consider severe environment condition that involve agitation within the procedure. On the other hand, both static and dynamic test were used as leaching test that represent natural environment condition, where agitation is not necessary.

Both static and dynamic test were carried out on the nickel slag and concrete mix with the highest compressive strength value and 28 days curing period, to compare the leaching performance prior and after the solidification took place. The result of leaching performance test showed that the concentration of Cr and Cd for both nickel slag and mortar specimens in all conditions were not detected in the AAS. This means that the Cr and Cd concentration was extremely low, even below the detection limit in the used AAS.

3 Conclusions

The results of this study have demonstrated the effectiveness of solidification process in immobilizing heavy metals in nickel slag Therefore, the cement-based solidification technique is proven can be an option to treat nickel slag when it is intended to be utilized as concrete floors. Through this study it is identified that a dosage of 60% nickel slag with cement aggregates ratio of 1:4 met the requirements of concrete floors construction in terms of compressive strength, absorption rate, and durability. The TCLP test also indicates that heavy metals content were below the standard and in particular, leaching behavior of Cr and Cd in mortar specimens are also found still within tolerable concentration criteria in accordance with Government Regulation No 101 of 2014.

4 References

[1] Sugiri. *Penggunaan Terak Nikel sebagai Agregat dan Campuran Semen untuk Beton Mutu Tinggi.* Thesis Program Magister, Program Studi Teknik Sipil, Institut Teknologi Bandung, (2005).

[2] Wilk, C. Solidification/Stabilization Treatment and Examples of Use at Port Facilities. (2004) .https://doi.org/ 10.1061/40727(2004)92

[3] Zhang, S. P., & Zong, L. Evaluation of Relationship between Water Absorption and Durability of Concrete Materials, (2014). https://doi.org/10.1155/2014/650373

[4] El Akmam, M.T. *Leachability of Chromium and Cadmium Heavy Metals from Nickel Slag Utilization as Concrete Floors Aggregate Constituent with Solidification Method.* Final Project for Bachelor Degree, Environmental Engineering Study Program, Institut Teknologi Bandung, (2017)

[5] ASTM C128-97, Standard Test Method for Specific Gravity and Absorption of Fine Aggregate, ASTM International, West Conshohocken, PA, (1997), www.astm.org

[6] ASTM C114-07, Standard Test Methods for Chemical Analysis of Hydraulic Cement, ASTM International, West Conshohocken, PA, (2007), www.astm.org

[7] Badan Standardisasi Nasional. SNI 13-6793-2002 Metode Pengujian Kadar Air, Kadar Abu dan Bahan Organik dari Tanah Gambut dan Tanah Organik Lainnya. (2002)

[8] Badan Standardisasi Nasional. SNI 6-2464-1991 Metode Ekstraksi Logam Total. (1991)

[9] ASTM C128-97, Standard Test Method Sieve Analysis of Fine and Coarse Aggregates, ASTM International, West Conshohocken, PA, (1997), www.astm.org

[10] ASTM C191-04-a, Standard Test Method for Time of Setting of Hydraulic Cement by Vicat Needle, ASTM International, West Conshohocken, PA, (2004), www.astm.org

[11] ASTM C 642-06, Standard Test Method for Compressive Strength of Hydraulic Cement Mortars Cube Specimens, ASTM International, West Conshohocken, PA, (2007), www.astm.org

[12] ASTM C 642-06, Standard Test Method for Density, Absorption, and Voids in Hardened Concrete, ASTM International, West Conshohocken, PA, (2006), www.astm.org

[13] ASTM D 4843-88 R09, Standard Test Method for Wetting and Drying Test of Solid Waste, ASTM International, West Conshohocken, PA, (2009), www.astm.org

[14] US EPA SW-846 Test Method 1311, Toxicity Characteristic Leaching Procedur. (1992). https://www.epa.gov/hw-sw846/sw-846-test-method-1311-toxicity-characteristic-leaching-procedure

[15] ASTM D3987-12, Standard Practice for Shake Extraction of Solid Waste with Water, ASTM International, West Conshohocken, PA, (2012), www.astm.org

[16] Badan Standardisasi Nasional. SNI 03-6468-2000 Tata Cara Perencanaan Campuran Beton Berkekuatan Tinggi dengan Semen Portland dan Abu Terbang, (2000)

[17] H.Kosmatka, S., & L.wilson, M. *Design and Control of Concrete Mixtures. Construction*. Retrieved from http://www.cement.org/bookstore/supporting/cd100/EB001Frt.pdf. (2011)

[18] Al-Jabri, K. S., Hisada, M., Al-Oraimi, S. K., & Al-Saidy, A. H. Copper slag as sand replacement for high performance concrete. Cement and Concrete Composites, 31(7), 483–488, (2009) https://doi.org/10.1016/j.cemconcomp.2009.04.007

[19] ACI Committee 318. Guide for Concrete Floors and Slab Construction: (ACI 302.1R-96). Farmington Hills, MI: American Concrete Institute, (1995).

[20] Wu, W., Zhang,W., Ma, G. Optimum content of copper slag as a fine aggregate in high strength concrete. *Journal of Material and Design*, 2878-2883.(2010)

[21] Pacheco-Torgal, F., Jalali, S., Labrincha, J., & John, V. *Eco-efficient concrete*. Cambridge: Wodhead Publishing Limited.(2013)

[22] Tandon, G. Durability and Permeability of Concrete. India.(2005)

[23] Wiles, C. A review of solidification/stabilization technology. *Journal of Hazardous Materials* , 5-21.(1987)

The Treatment of Used Reagent Glass Bottles Using Solidification Technology

Anggun Farida[1,*], *Iwan* Juwana[1], and *Hazairin*[2]

[1]Environmental Engineering Department, Institut Teknologi Nasional (Itenas), 40123 Bandung, Indonesia
[2]Civil Engineering Department, Institut Teknologi Nasional (Itenas), 40123 Bandung, Indonesia

Abstract. Used reagent glass bottles are solid waste that people rarely pay attention to in Indonesia. In many cases, the bottles are disposed off without proper handling and treatment, due to the lack of research and information. One potential treatment for such waste is to transform the glass bottles into raw materials for concretes. The objective of the research is to determine the influence of crushed glass particles to the quality of the concrete. The potential reduction of glass bottles will also be calculated.The glass bottles used in this research are the amber ones. The concrete was designed to meet the K225 type requirements. Experiments were done using 0%, 25%, 50%, and 75% of crushed glass particles. In the concrete production, these crushed glass particles will replace the coarse aggregates. This research measured compressive strength and Toxicity Characteristic Leaching Procedures (TCLP) tests in accordance to SNI and TCLP Extraction Procedures from BAPEDAL. At the end, it was found that the concretes using all variations meet the requirements. As for the TCLP, it showed that concentrations of Cr^{6+} for the concretes using all variations were all under the threshold, according to Hazardous National Regulation 101/2014. The potential reduction is 600 – 2100 waste bottles.

1 Introduction

Waste is the residue of a human and animal activity and/or production process in solid, liquid, or gasses form [8]. Hazardous waste is the waste from a work and/or activity that contains hazardous materials. The reduction of such waste can be done through waste treatment and/or reuse [2,8].

People are rarely concerned about solid hazardous waste from laboratory activity including used reagent glass bottles. If such waste are disposed off without proper handling and treatment, they will damage the environment and endanger human health. Thus, hazardous waste management is required for reducing such wastes. According to Republic of Indonesia's Government Regulation Number 101 of 2014 Regarding Hazardous and Toxic Waste Management, hazardous waste treatment is the processes undertaken to reduce the toxicity of waste. Stabilization and solidification are the alternatives of hazardous waste treatments that can be done to reduce the toxicity of waste [8]. Solidification using cement are the most common process. It is a process that physically encapsulates the contaminant by using portland cement [7,14], which will be done in this research. Coarse aggregates, as raw materials to produce concrete, will be replaced by crushed brown glass particles with different concentration. Then, the concrete will undergo two tests: the compressive strength and TCLP tests. The materials in this research were used reagent glass bottles because of its abundance in every laboratory and have not been recycled

Up to this stage, research on concrete and concrete materials has been widely reported, and the effects of raw materials subtitution to the produced concrete have been described [1, 4, 10-15]. The use of waste glass in concrete materials has attracted wide attention. The waste glass addition is believed to have effect on the concrete compressive strength. Suhartini (2014) showned that 2,5% of crushed green glass bottles used to subtitute fine aggregate has increased concrete compressive strength. However,the subtitution of fine aggregates by 5% - 10% has decreased its compressive strength. Hadi's research also showed that 25% - 100% of coarse aggregates weight substituted by crushed green glass bottles has decreased its concrete compressive strength [4,12].

Compressive strength is the capacity of a material or structure to withstand loads and causes the sample to be destroyed when loaded with a certain force [6]. Compressive strength is affected by the age of the concrete. Usually, concretes were tested after 28 days of castings [3]

This research aims to determine the effects of glass bottles waste addition on its compressive strength, TCLP quality, and the number of bottles that can be reduced due to these activities.

* Corresponding author: anggunfrd@gmail.com

2 Materials and Methods

2.1 Glass Bottles

Samples in this experiment were used reagent glass bottles. These glass particles will substitute the coarse aggregate in concrete production. The bottles were washed and crushed to meet the same size of coarse aggregates. The size of crushed glass particles were in the range of 4,75 and 19 mm. As many as 25 brown glass bottles were obtained from X Water Laboratory, also known as "amber glass". The detailed characteristics of glass bottles used can be seen in Table 1 and 2, as well as in Figure 1.

Table 1. Chemical substances on glass bottles and the number of glass bottles.

No.	Chemical Subtances	The Number of Glass Bottles Used	Volume for 1 Bottle (ml)
1	G Solution (Reagent for COD Meters)	3	500
2	$KMnO_4$ 0,05 N	1	2500
3	$K_2Cr_rO_7$ 0,025 N	1	1000
4	Potassium Dichromate Solution	3	500
5	Chloroform	3	2500
6	Nitric Acid	2	2500
7	Sulphuric Acid	8	2500
8	Hydrochloric Acid	3	2500
9	Ethanol	1	1000

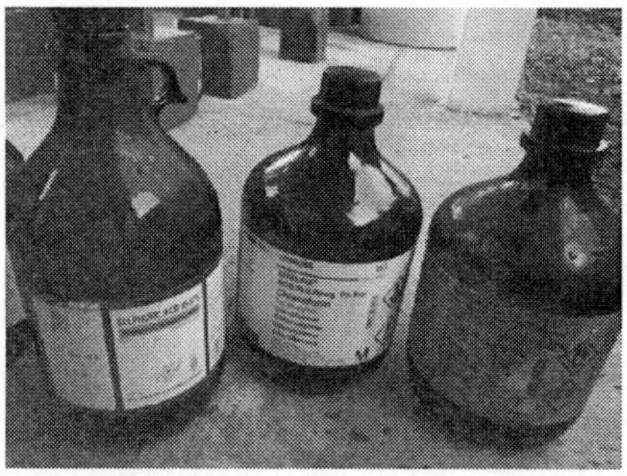

Figure 1. Some glass bottles were used for this research.

Table 2. The weight of glass bottles.

No.	Volume (ml)	Glass Bottles Weight (kg)
1	500	0,26
2	2500 (round glass bottles)	1,17
3	2500	1,01
4	1000	1,2

The parameters to be tested by TCLP were obtained from the chemical substances/reagents data and required parameters of relevant regulations. Then, the weight of glass bottles data in Table 2 was used to calculate the potential reduction of glass bottles waste in this particular laboratory.

2.2 Experimental Methods

Weight of each materials were obtained from concrete mix design based on SNI 03 – 2834 – 2000. For this research, 8 cylinder concretes were created for each variations of mix. A total of 32 cylinder concretes for 4 variations were created as samples for this study.

For mix design result, materials weight for normal concrete is given in Table 3. And the materials for each variation were obtained from mix design. Besides that, cylinder concrete diameter and height in this research were 10 cm and 20 cm. Density of normal concrete is 2200 – 2600 kg/m^3. Weight of concrete can be obtained from volume of concrete and density of concrete. The result is given also in Table 3.

Table 3. Quantity of Concrete

Fine aggregates weight (kg/m^3)	Coarse aggregates weight (kg/m^3)	Water weight (kg/m^3)	Cement weight (kg/m^3)	Weight of Concrete (kg)
401,79	769,21	978,99	225	3768

This experiment were carried out in Itenas Concrete and Pavement of Roads Laboratory and X Water Laboratory. The design of experiment is given in Table 4.

Table 4. Design of experiment and materials weight for each variation

Variation of glass particles added (%)	0%	25%	50%	75%
Fine aggregates weight (kg)	10,62	10,62	10,62	10,62
Coarse aggregates weight (kg)	13,53	10,15	6,77	3,38

Variation of glass particles added (%)	0%	25%	50%	75%
Glass weight (kg)	0	3,38	6,77	10,15
Water weight (kg)	3,11	3,11	3,11	3,11
Cement weight (kg)	5,55	5,55	5,55	5,55
TOTAL (kg)	42,51	33,83	20,3	22,2

3 Results and Discussion

3.1 Slump Test

Slump test measures the consistency of fresh concrete. The test is carried out using Abrams cone filled with fresh concrete in three stages. The slump test are illustrated in Figure 2 and the results of slump test are provided in Table 5.

Figure 2. Slump test.

Table 5. The result of slump test.

Variation of Crushed Glass Particles Added	Date		Concrete Age (days)	Slump (cm)
	Production	Test		
0%	May 24, 2016	June 21, 2016	28	6,4
25%	May 25, 2016	June 22, 2016	28	5
50%	May 26, 2016	June 23, 2016	28	4,5
75%	May 26, 2016	June 23, 2016	28	3

It can be seen that the use of waste glass particles in the production of concrete decreases the slump value [11]. The decrease were due to the diameter of the crushed glass particles which were smaller than those of the coarse aggregates. Hence, more water is needed in the concrete production, because crushed bottle particle's surface area were larger than that of the coarse aggregate. In consequence, the slump value decreases with the increasing amount of crushed bottles particles

The slump concrete will take various shapes, in accordance to concrete mix. A collapse shape generally means that the mix is too wet because of the high slump value. High workability mix can influence the concrete production. If the concrete mixes were too dry, it will be difficult to work with.

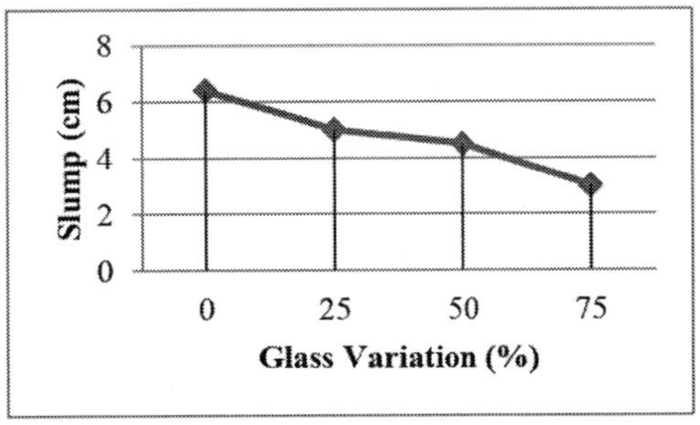

Figure 3. Influence of glass variation added on slump value.

3.2 Compressive Strength

All cylinder concretes were tested using a Universal Testing Machine (UTM). The concretes were tested after 28 days of castings, allowing the concretes to reach 99% of its strength. The concrete compressive strength test is illustrated in Figure 3 and the results are shown in Table 5.

Figure 4. Compressive test.

Table 6. The result of compressive strength test.

Variation of Crushed Glass Particles Added	Date		Compressive Strength (kg/cm^2)	Compressive Strength (MPa)
	Production	Test		
0%	May 24, 2016	June 21, 2016	278,9	27,89
			245,6	24,56
			314	31,4
			273,2	27,32
			305,5	30,55

https://doi.org/10.1051/matecconf/201814704011

Variation of Crushed Glass Particles Added	Date		Compressive Strength (kg/cm^2)	Compressive Strength (MPa)
	Production	Test		
Average of Compressive Strength			**283,44**	**28,34**
25%	May 25, 2016	June 22, 2016	213,8	21,38
			227,9	22,79
			238	23,8
			213	21,3
			224	22,4
Average of Compressive Strength			**223,3**	**22,33**
50%	May 26, 2016	June 23, 2016	189,3	18,93
			201,4	20,14
			199	19,9
			197,8	19,78
			202,3	20,23
Average of Compressive Strength			**197,9**	**19,79**
75%	May 26, 2016	June 23, 2016	225,3	22,53
			193,3	19,33
			183,5	18,35
			199	19,9
			204,1	20,41
Average of Compressive Strength			**201**	**20,1**

Table 6 shows that high concentration of crushed glass particles in concrete production has decreased its compressive strength due to many factors, such as interlocking, bonding, and the aggregates compressive strength itself. Aggregate's surface characteristic affects the bond between cement paste and aggregate. Rough aggregate's surface causes a good bond while the crushed bottles has a poor bond because of the slippery surfaces.

It can be concluded that waste glass addition in concrete affected its compressive strength due to its slippery surfaces. Figure 5 shows the influence of glass variation added on concrete compressive strength.

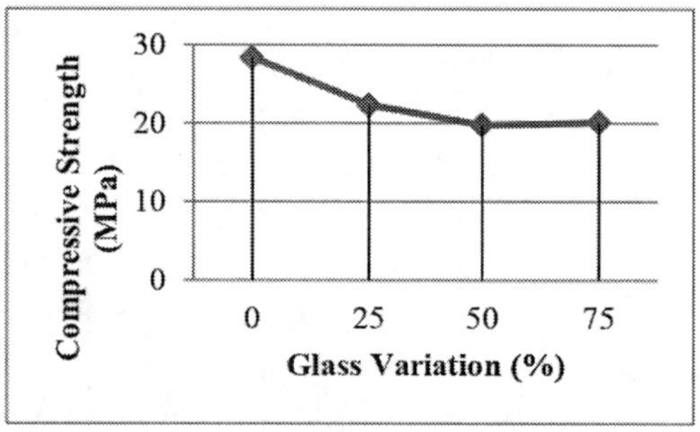

Figure 5. Influence of glass variation on concrete compressive strength.

As mentioned earlier, the concrete was designed to meet the K225 type requirements. The value of concrete compressive strength planned was225 kg/cm^2 or 19 MPa. The concretes using all variations of glass concentrations did meet the requirements because it has more than 19 MPa for the all compressive strengths, namely 28,34 MPa, 22,33 MPa, 19,71 MPa, and 20,1 MPa for 0%, 25%, 50% and 75% of crushed brown glass bottles substitution respectively.

According to SNI-2847-2013 on Structural Concrete Requirements for Buildings, the requirements of structural concretes should not be less than 17 MPa [5]. Structural concretes are concretesusedto withstand heavy loads, such as a building. The result of this research shows that the concretes using all variations of glass concentrations meet the requirements for structural concretes.

Table 7. Percent reduction of concrete compressive strength.

Glass Variation Added	Compressive Strength (kg/cm^2)	Reduction of Compressive Strength (kg/cm^2)	Reduction of Compressive Strength (%)
0%	283,44	-	-
25%	223,34	60,1	21,2
50%	197,96	85,48	30,16
75%	201	82,44	29,09

Even if the reduction of concrete compressive strength from crushed glass particles substitution was between 20 – 30%, the resulting concrete can still be used as building materials. However, if it is desired to get the same quality of a normal concrete, then it can be done by increasing the concrete compressive strength during the mix design process

3.3 Toxicity Characteristic Leaching Procedures (TCLP)

TCLP (Toxicity Characteristic Leaching Procedure) test was carried out in X Water Laboratory. TCLP test is used to determine leaching potential from concretes. This test was done according to TCLP Extraction Procedures for Hazardous Waste (from BAPEDAL's draft in 1992)".

TCLP test parameters can be obtained from comparing the chemical substances data with the parameters on the regulations. K2Cr2O7 were detected in one of glass bottles waste. The Appendix of Republic of Indonesia's Government Regulation Number 101 of 2014 Regarding Hazardous and Toxic Waste Management informs, Cr^{6+} is one of the heavy metal parameters that should be examined. The results of TCLP test can be seen in Table 8 below.

Table 8. The results of TCLP test.

No.	Sample	Cr^{6+}Level (mg/lt)	Average of Cr^{6+}Level (mg/lt)
1	Concrete with 0% crushed glass particles	0,6	0,6115
		0,623	
2	Concrete with 25% crushed glass particles	0,717	0,762
		0,807	
3	Concrete with 50% crushed glass particles	0,691	0,6785
		0,666	
4	Concrete with 75% crushed glass particles	0,547	0,6185
		0,69	

Table 8 shows the result of TCLP test for Cr^{6+} levels in the concretes using all variations of glass concentrations. It shows that the concrete with 25% glass particles has the highest concentration of Cr^{6+} (0,762 mg/l). It is possibly caused by Cr^{6+} found within the crushed glass particles added.

The Cr^{6+} level, however, was decreasing with the increasing concentration of glass particles added,. which might be due to the reaction between Ca^{2+} of cement paste with CrO_4^{2-} from the chemical substances in glass bottles waste.Cr^{6+} immobilization can be done with the formation of CaCrO4.

There were 2 quality TCLP standards used in this research. The first standard is taken from the Appendix III of Republic of Indonesia's Government Regulation Number 101 of 2014 Regarding Hazardous and Toxic Waste Management. This regulation is used to identify the category of hazardous waste. The quality standard for TCLP B requires Cr^{6+} concentration of less than 2,5 mg/l and for TCLP A, the Cr^{6+} concentration should not be over the 15 mg/l. As for the results of this study, it showed that concentrations of Cr^{6+} from the concretes using all variations of glass concentrations were all under the 2,5 mg/l. So, it can be considered as not having toxic characteristic from Cr^{6+} concentration.

The second standards is the Appendix IV of Republic of Indonesia's Government Regulation Number 101 of 2014 on Hazardous and Toxic Waste Management. This regulation is used to establish the hazardous waste treatment before placing in a dumping location. The limit of quality standard is 2,5 mg/l. In this research, TCLP test showed that concentrations of Cr^{6+} from the concretes using all variations of glass concentrations were all under the threshold of 2,5 mg/l. In accordance to these quality standards, the glass particles concretes can be safely placed in any dumping location. But, a Paint Filter Test needs to be done beforehand to finally determine whether this solidification of waste can be placed on dumping sites.

3.4 Potential Reduction of Glass Bottles Waste

Potential reduction of glass bottles waste was determined by counting the weight differences of coarse aggregate's and crushed glass particles. This study shows that a 75% glass particles mix used as coarse aggregates substitute can still have a good quality in terms of concrete compressive strength and TCLP test. Therefore, the concrete using 75% glass concentrations was chosen for the calculation of reduction glass bottles waste.

Table 9. Potential reduction of glass bottles waste.

Coarse Aggregate Needed (kg/m^3)	Glass Needed as Coarse Aggregate – 75% (kg/m^3)	Bottles Size (ml)	Bottles Weight (kg)	Potential Reduction of Glass Bottles Waste (Bottles)
978,99	734,24	500	0,26	2824,01
		2500	1,01	726,97
		2500	1,17	627,56
		1000	1,2	611,87

Table 9 shows that the weight of coarse aggregate needed was obtained from mix design of this research. Potential reduction of glass bottles waste is considered to be significant. The K225-type concrete can reduce 2.825 bottles of 500 ml bottles size. If the 2.500 ml bottles size were used, then it can reduce up to 700 bottles, whilst as many as 612 bottles can be reduced if the 1.000 ml bottles size are used. This research shows that subtitution of aggregates of concrete can be used to reduce used reagent glass bottles in a laboratory.

4 Conclusions

Several important conclusion can be taken as follows:

1. Waste glass can be used to subtitute coarse aggregates by 0%, 25%, 50% and 75% in concrete production.
2. Concretes produced using waste glass meet the concrete compressive strength of K225-type requirements, hence they can be safely used for construction.
3. The TCLP test results meet the requirements of Republic of Indonesia's Government Regulation Number 101 of 2014 Regarding Hazardous and Toxic Waste Management, indicating the concretes to be environmentally friendly.
4. The subtitution of coarse aggregates in concrete potentially reduces glass bottles waste in laboratories. The reduction of glass bottles can be up to 2.100 bottles with any kind of size and weight.
5. The reduction of these waste bottles will also reduce the cost of disposing them into a landfill.

References

1. Aswad, N. H., & Soeparyanto, T. S.,Glass Bottles Fractions Used as Coarse Aggregates on Concretes Mixes,Stabilita, **2**(1)., (2014)
2. Damanhuri, E., & Padmi, T.,*Solid Waste Management.Environmental Engineering Book*, Institut Teknologi Bandung, (2010)
3. Dipohusodo, I.,*Reinforced Concrete Structures*, PT Gramedia Pustaka Utama, Jakarta, (1994)
4. Hadi, G. Utilization of Glass Waste as a Substitute of Coarse Aggregates in Eco-Friendly Concretes, Thesis, Civil Engineering Department, Institut Teknologi Nasional, Bandung, (2009)
5. Indonesian National Standard,Structural Concrete Requirements for Buildings,SNI-2847-2013, (2013)
6. Indonesian National Standard, Compressive Strength Test Method, SNI 03 - 1974 - 1990, (1990)
7. LaGrega, M. D., Buckingham, P. L., & Evans, J. C.,*Hazardous Waste Management*, **2nd** ed.,McGraw-Hill, Singapore, (2001)
8. Republic of Indonesia's Government. Government Regulation Number 101 of 2014 Regarding Hazardous and Toxic Waste Management, (2014)
9. Sagel, R., Kole, P., & Kusuma, G., *Guideline of Concrete Constraction*, Penerbit Erlangga, Jakarta, (1997)
10. Setiawan, B.,The Effect of Glass Aggregates Used on Concretes Reviewed from Concrete Strength and Shrinkage,Thesis, Petra Christian University, (2006)
11. Shi, C., & Zheng, K., A Review on the Use of Waste Glasses in the Production of Cement and Concrete. Elsevier: Resources, Conservation and Recycling 52, pp 234-237, (2007)
12. Suhartini, A., Srie Gunarti, A. S., & Hasan, A., The Effect of Crushed Glass Added as Fine Aggregates Subtitutes to Compressive Strength and Flexural Strength, BENTANG Journal, (2014)
13. Topcu, I. B., & Canbaz, M., Properties of Concrete Containing Waste Glass, Cement and Concrete Research 34, pp 267-274, (2004)
14. Utomo, M. P., & Laksono, E. W., Study of Solidification/Stabilization of Heavy Metal Process with Portland Cement, (2007)
15. Wang, S., & Vipulanandan, C., Solidification/stabilization of Cr (VI) with Cement Leachability and XRD Analysis. Cement and Concrete Research 30, pp 385-389, (2000)

Analysis of ocean wave characteristic in Western Indonesian Seas using wave spectrum model

Rima Rachmayani[1], *Nining Sari* Ningsih[1], *Hani* Ramadhan[1], *Suliskania* Nurfitri[1]

[1]Department of Oceanography, Faculty of Earth Science and Technology, Bandung Institute of Technology, Indonesia

Abstract. Understanding the characteristics of the ocean wave in Indonesian Seas particularly in western Indonesian Seas is crucial to establish secured marine activities in addition to construct well-built marine infrastructures. Three-years-data (July 1996 - 1999) simulated from Simulating Waves Nearshore (SWAN) model were used to analyze the ocean wave characteristics and variabilities in eastern Indian Ocean, Java Sea, and South China Sea. The interannual or seasonal variability of the significant wave height is affected by the alteration of wind speed and direction. Interactions between Indian Ocean Dipole Mode (IODM), El Niño Southern Oscillation (ENSO) and monsoon result in interannual ocean wave variability in the study areas. Empirical Orthogonal Functions (EOF) analysis produces 6 modes represents 95% of total variance that influence the wave height variability in the entire model domain. Mode 1 was dominated by annual monsoon and has spatial dominant contribution in South China Sea effected by ENSO and Indian Ocean influenced by IODM. Java Sea was influenced by Mode 2 which is controlled by semi-annual monsoon and IODM. A positive (negative) IODM strengthens (weakens) the winds speed in Java Sea during the East (West) season and hence contributes to Mode 2 in increasing (decreasing) the significant wave in Java Sea.

1 Introduction

Various activities on seas, either for sea transportation activities, fisheries, marine resources exploration, and also development in marine sector are very sensitive to weather and sea condition changes. Two hundred sixty cases of ship accidents were due to natural factor increased with years [1]. Among various natural factors, one that greatly influences marine activities is wave, and therefore, in the marine meteorological services, in addition to wind information, wave information is the most important part that should exist in every kind of marine weather information [1]. The existence of extreme high tides could threaten the safety on sea and may result in great losses.

The wave conditions in the waters are strongly influenced by wind variability as the dominant force of ocean waves [2]. Indonesia is an area affected by the monsoon wind cycle. The monsoon interacts with an interannual phenomenon such as El Niño-Southern Oscillation (ENSO) and Indian Ocean Dipole Mode (IODM); affects regional and local wind circulation patterns in Indonesian waters [3]. Characteristics of waves and the influence of interannual phenomenon such as ENSO and IODM to wave variability in the waters can be determined from long-term wave measurements. Some sources of data buoy measurement, Voluntary Observing Ship (VOS), and altimeter satellite measurement can be utilized as a basis for determining wave climate. In addition, the wave hindcasting data using model is also can be used as an alternative data.

* Corresponding author: rrachmayani@oceanography.itb.ac.id

Characteristics and variability of waves in the waters of western Indonesia, including the South China Sea, Indian Ocean, and Java Sea are addressed in this study. The wind field in the mentioned three waters is strongly influenced by the relative position of the sun against the earth and it is influenced by the IODM system. This study addresses two main issues: (1) how the dominant variability pattern can influence the variability of the wave, (2) how it interacts with the interannual phenomenon of ENSO and IODM. This paper is organized as follows: the study area is introduced in Sect. 2, and the details of data used and the methodology are presented in Sect. 3. Section 4 presents the results and discussion of the study, and the conclusions are given in Sect. 5.

2 Study area

This study focusses on the territorial waters of western Indonesia between 20°N – 23°S and 95°E – 120°E. The boundary covers the eastern Indian Ocean, South China Sea, and Java Sea. Domains are further extended towards the Indian Ocean to accommodate the perfect wave formation state in the region. Figure 1 shows the model domains and bathymetry profiles. The study area is strongly influenced by the monsoon wind. There is a trough in southern Java and Sumatra marked by a very steep bathymetry changes. Meanwhile, the Java Sea is a shallow sea with a depth ranging from 50 m. The northern part of the South China Sea reaches a depth of about 4500 m while the southern part is around 100 m. Seamount in the South China Sea can cause a significant wave energy reduction either through the process of dissipation due to basic friction and dissipation due to breaking waves.

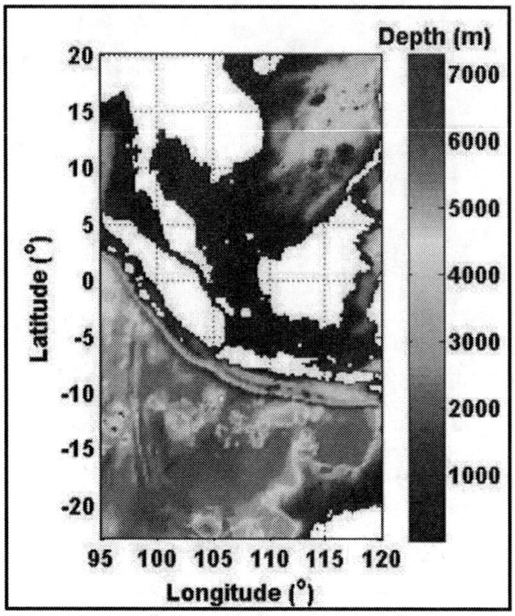

Fig. 1. Domain model and bathymetry profiles

3 Methodology

3.1. Data

In this study, wind and bathymetry data are applied as model input. The wind field is taken from 10 m above sea level called CCMP (Cross Calibrated Multi Purposed) with a gradient resolution of 0.25° x 0.25° and intervals every 6 hours. Meanwhile, bathymetry data from GEBCO (Global Bathymetric Chart of Oceans) produced with 1' x 1' resolution [4]. The influence of tides and changes in depth due to sedimentation is negligible, so the depth value is considered constant during the hindcasting period (July 1996 - 1999).

3.2 Model

Simulating Wave Nearshore (SWAN) is a third-generation wave model for obtaining realistic estimates of wave parameters in coastal areas, lakes and estuaries from given wind, bottom and current conditions. However, SWAN can be used on any scale relevant for wind-generated surface gravity waves. The model is based on the wave action balance equation with sources and sinks [5, 6].

In this study, hindcasting wave was performed by using model of SWAN with nonstationary mode where the input of wind field always change every step of time. The physical processes activated in this simulation are development of wind wave, quadruplet nonlinear interaction, dissipation due to white-capping, basic friction, and breaking waves. In this study, the initial spectrum used in the model is the initial spectrum of Joint North Sea Wave Project (JONSWAP). Meanwhile, the discrete wave propagation in the geographic space uses the BSBT (backward in space and backward in time) scheme. The formulation of the wind source/development of wind wave and white-capping function employed in the model is described in [7, 8] and the basic friction function performed in the study followed the formulation in [8, 9]. Here are the SWAN model configurations applied to this study:

Resolution	:	1/6° x 1/6°
Frequency	:	Frequency with lowest frequency 0.04 Hz and highest 1 Hz
Space- θ	:	180 mesh in-space θ for a full circle 360°
Domain	:	20°N – 23°S and 95°E – 120°E with resolution 1/6° x 1/6°

The model runs for 3 years (July 1, 1996, 00 00 UTC to June 30, 1999, 18 00 UTC), with a spin-up period (from calm state) for the first 7 days. Model output is the significant wave height and direction of propagation.

3.2. Emphirical Ortogonal Function

Empirical Orthogonal Function (EOFs) technique aims at finding a new set of variables that capture most of the observed variance from the data through a linear combination of the original variables. EOFs have been introduced in atmospheric science since the early 50's [10, 11]. EOF techniques are deeply rooted in statistics and EOFs was introduced as principal component analysis (PCA) [12] and it is analyzed details [13, 14].

To obtain the main patterns of wave fields variability in this current study, EOFs was performed in determining the variation of significant wave height (H_s) values in the model domain either temporally or spatially. The time series of the EOFs principal components (PCs) are normalized to have unit variance, so that the corresponding EOFs represent the typical variability of the data in their original units.

4 Results and Discussion

4.1 Wind and Significant Waves Height (H_s)

The model results show a linier relationship between the higher of wind speed and the higher of H_s. The direction of wave propagation is strongly influenced by the wind direction. Waves generally move in the direction of the wind, except in areas with considerable refraction effects such as waters near the coast. The seafloor indicated by significant depth as found in the South China Sea (SCS) causes significant wave heights in the area to be relatively smaller as the wave energy is dissipated by increasingly larger friction and steep bathymetry changes can cause breaks.

The significant wind direction and wave height of the model in the west season (DJF) is shown in Figure 2. The significant monthly wind and wave height is the result of the reduction of the wind and H_s for each month for 3 years. The x and y direction components of the wind field and H_s are averaged separately to determine the mean direction of wind movement and significant wave propagation. In December, maximum winds are located at South China Seas (latitude 10°N - 20°N) and in January and February months winds in the South China Seas weaken. The direction of wave propagation is also in accordance with the direction of the wind. In narrow and closed waters, H_s tends to be relatively lower than H_s in wide and open waters, as large and open waters support perfect waveforms and the inclusion of wave energy from other places through wave propagation phenomena.

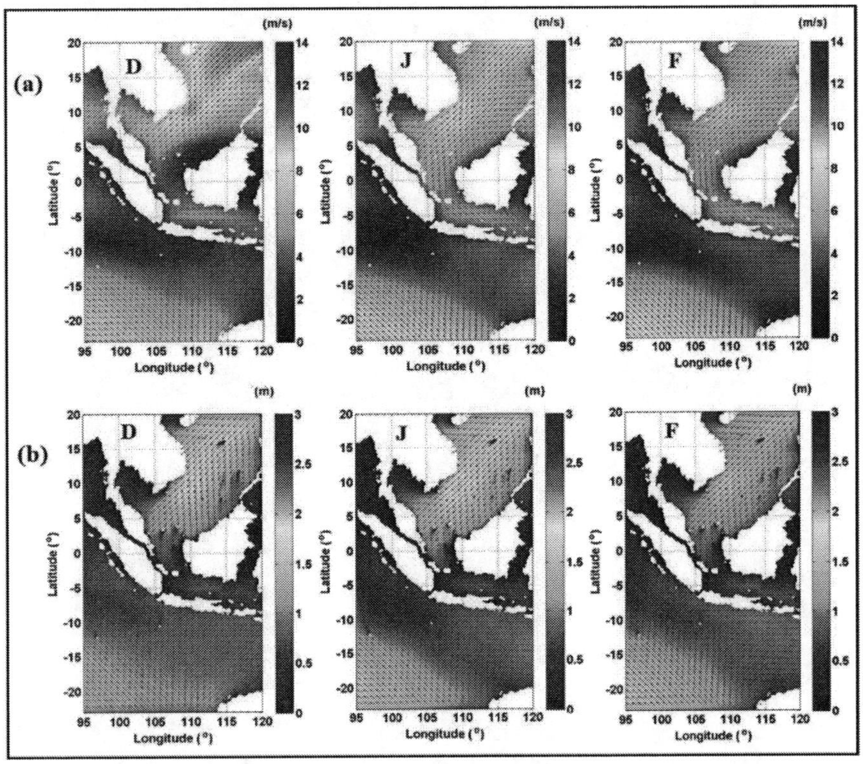

Fig. 2. Averaged of (a) the wind field (m/s) and (b) the significant wave height (m) in December-January-February (DJF)

Meanwhile, the wind field and H_s in the east season (JJA) are shown in Figure 3. During east season (JJA), the winds in South China Sea strengthen again and move toward northeast. Similarly, the winds in Java Sea (JS) also increase and move from the east and reach its maximum in August. The wave simulation results show the similar seasonal pattern with the wind field. Moreover, during JJA the wind field increases strongly in Indian Ocean (IO) between 10°S – 20°S. In general, the wind field in Indian Ocean near the equator is smaller than any other place in Indian Ocean. However, the Hs in Indian Ocean near the equator is a result of traveled wave energy from other regions in Indian Ocean.

 https://doi.org/10.1051/mateconf/201814705001

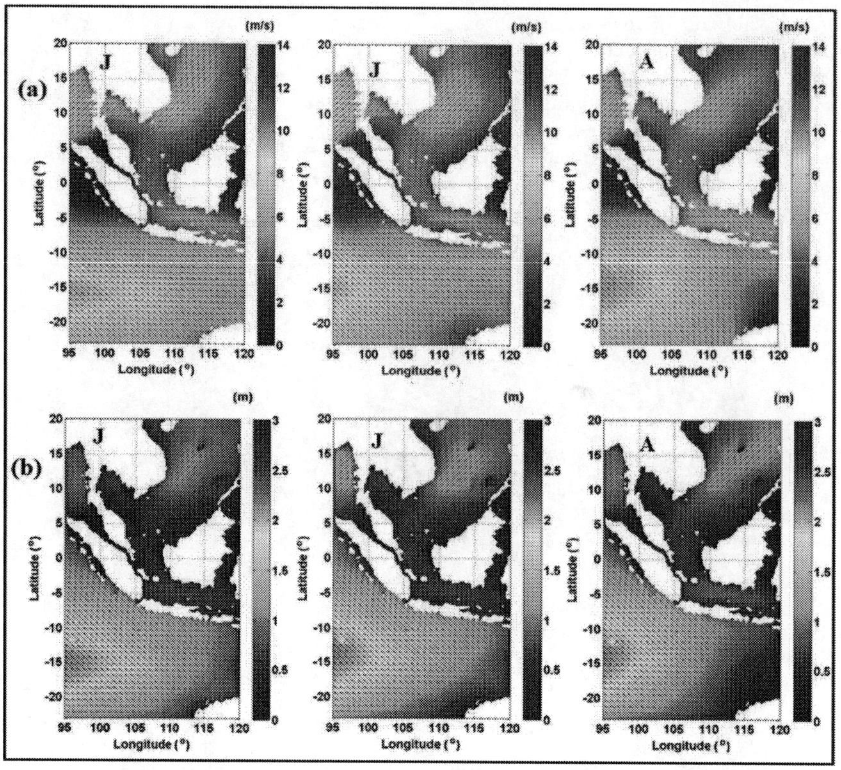

Fig. 3. As in Fig. 2 but for June-July-August (JJA)

4.2 Results of EOF Analysis

The result of significant wave height simulation affirms significant variation of H_s to time at the three waters. Figure 4 shows the mean significant wave heights in the model domains. The highest average H_s up to 1.75 m is in Indian Ocean, followed by South China Sea up to 1.3 m. The average H_s value in Java Sea is lower than 1 m, smaller than in South China Sea and Indian Ocean due to a narrower and closed water in Java Sea. Moreover, wind speed passing through Java Sea is not as big as in South China Sea and Indian Ocean.

The EOF analysis in this study uses the monthly average of H_s value and yields 6 modes representing 95% of the total variance as shown in Figure 5. Mode 1 represents 52.97%, Mode 2 represents 27.02%, and Mode 3 to Mode 6 represents only 15.10% of the total variance. Significant wave height variability in the model domains is dominated by the effects of Mode 1 and Mode 2. EOF analysis accommodates temporal/amplitude patterns of modes.

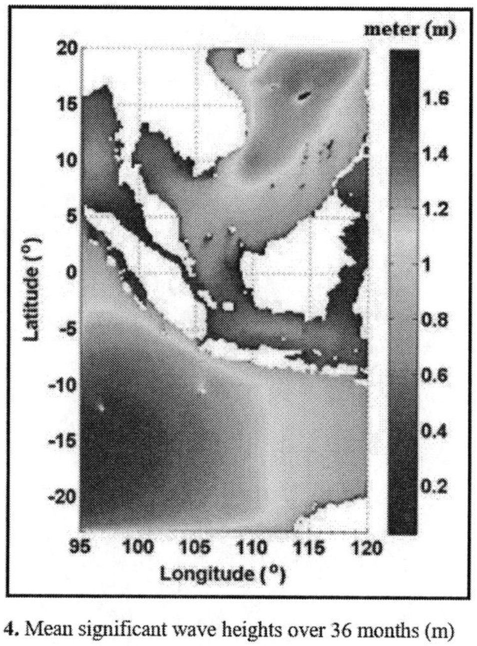

Fig. 4. Mean significant wave heights over 36 months (m)

* Corresponding author: rrachmayani@oceanography.itb.ac.id

https://doi.org/10.1051/matecconf/201814705001

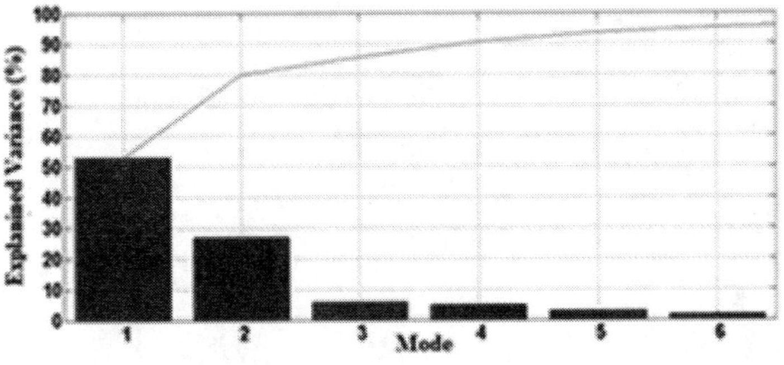

Mode	Explained Variance (%)	Explained Variance (%)
1	52,97	79,99
2	27,02	
3	5,75	15,10
4	4,79	
5	2,91	
6	1,65	
Total	95,09	95,09

Fig. 5. Six modes represents of 95% total variance

The spatial pattern of Mode 1 is exhibited in Figure 6a. South China Sea is dominated by two positive poles located at latitude 10°N – 17°N. The position of the two poles coincide with the maximum wind in the South China Sea during the west season (DJF). Meanwhile, Indian Ocean is controlled by a negative spatial pattern around latitude 12°S. This indicates a reversed temporal variability phase between South China Sea and Indian. Temporal pattern of Mode 1 is displayed in Figure 6c. In general, the high (low)-peak temporal pattern of Mode 1 occurs in the west (east) season. During west (east) season, South China Sea experiences a higher (lower) H_s and Indian Ocean captures a lower (higher) H_s value than its average. Temporal patterns Mode 1 is a composite of various signals with different periods. To find out the various signals, a power spectrum analysis is performed on the temporal pattern of Mode 1. The result of power spectrum analysis for the six PCs is affirmed in Figure 7. Annual monsoon plays a role in Mode 1 (Figure 7a).

* Corresponding author: rrachmayani@oceanography.itb.ac.id

 https://doi.org/10.1051/matecconf/201814705001

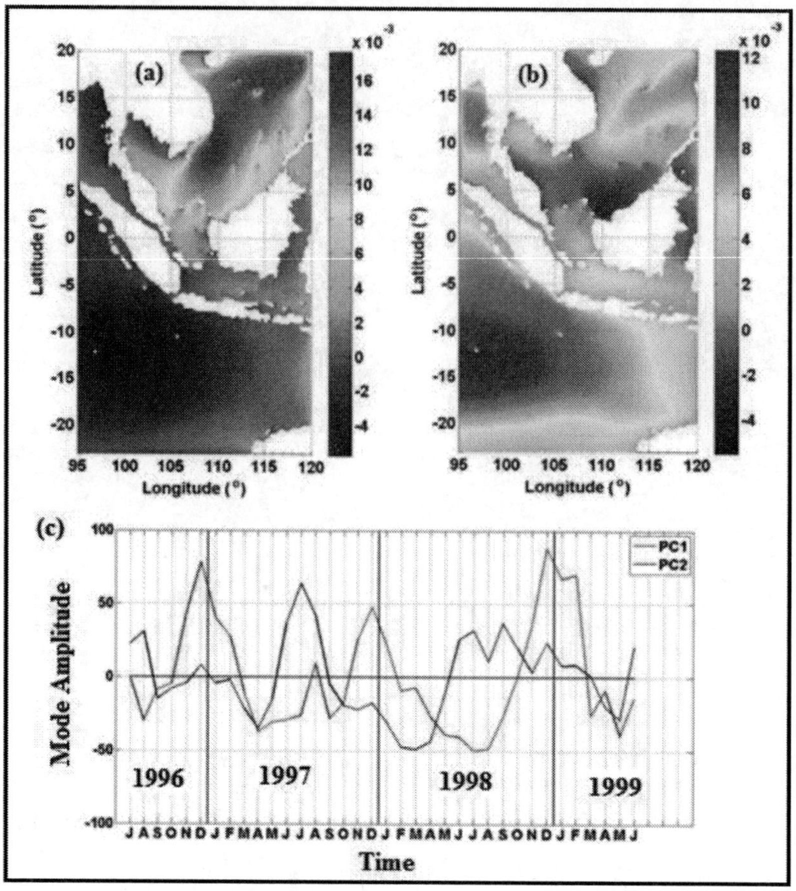

Figure 6. EOF analysis results: (a) Spatial pattern of Mode 1, (b) Spatial pattern of Mode 2, and (c) mode amplitude/temporal pattern of Mode 1 and Mode 2

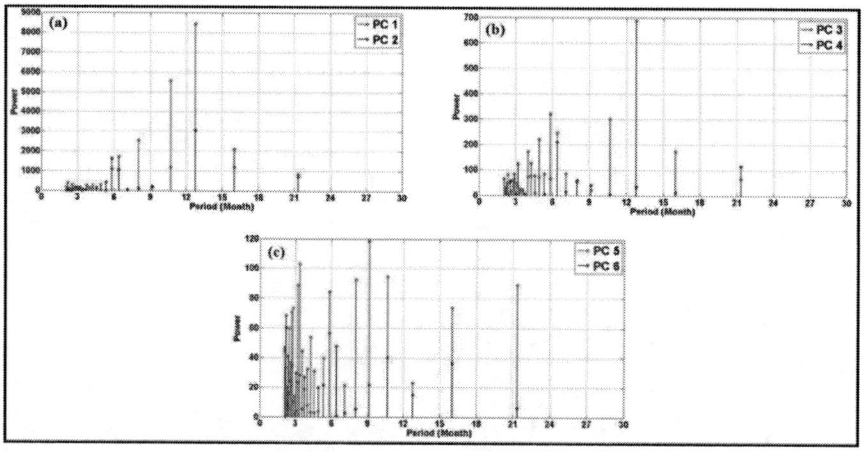

Fig. 7. The spectrum energy of mode amplitude/temporal pattern: (a) Mode 1 and Mode 2, (b) Mode 3 and Mode 4, and (c) Mode 4 and Mode 5

* Corresponding author: rrachmayani@oceanography.itb.ac.id

Spatial pattern of Mode 2 is presented in Figure 6b. Almost all domains have a positive spatial pattern. This shows a similar phase of variability with the maximum value in Indian Ocean. The value of spatial patterns in Java Sea and Atlantic Ocean (AO) is quite large; indicates the huge effect of Mode 2 in both waters. Meanwhile, some areas in the South China Sea respond as negative and positive spatial pattern values; indicates differences phase variability by Mode 2 in South China Sea. Temporal pattern/mode amplitude of Mode 2 is exposed in Figure 6c shows the presence of two peaks in one year (i.e., in the west and east seasons). Spectrum analysis for temporal pattern of Mode 2 is shown in Figure 7. Based on the spectrum analysis, Mode 2 is determined by the semi-annual monsoon.

The spatial patterns of Mode 3 to Mode 6 are respectively displayed in Figures 8 and 9. Mode 3 to Mode 6 contributes only 15% to total variance. Mode 3 illustrates two areas with different spatial pattern in Indian Ocean, as well as in South China Sea. Based on spectrum analysis, Mode 3 is still controlled by 12 monthly signals but modulated with smaller-than-6-months signals. Intraseasonal signals play a role in Mode 4, Mode 5, and Mode 6.

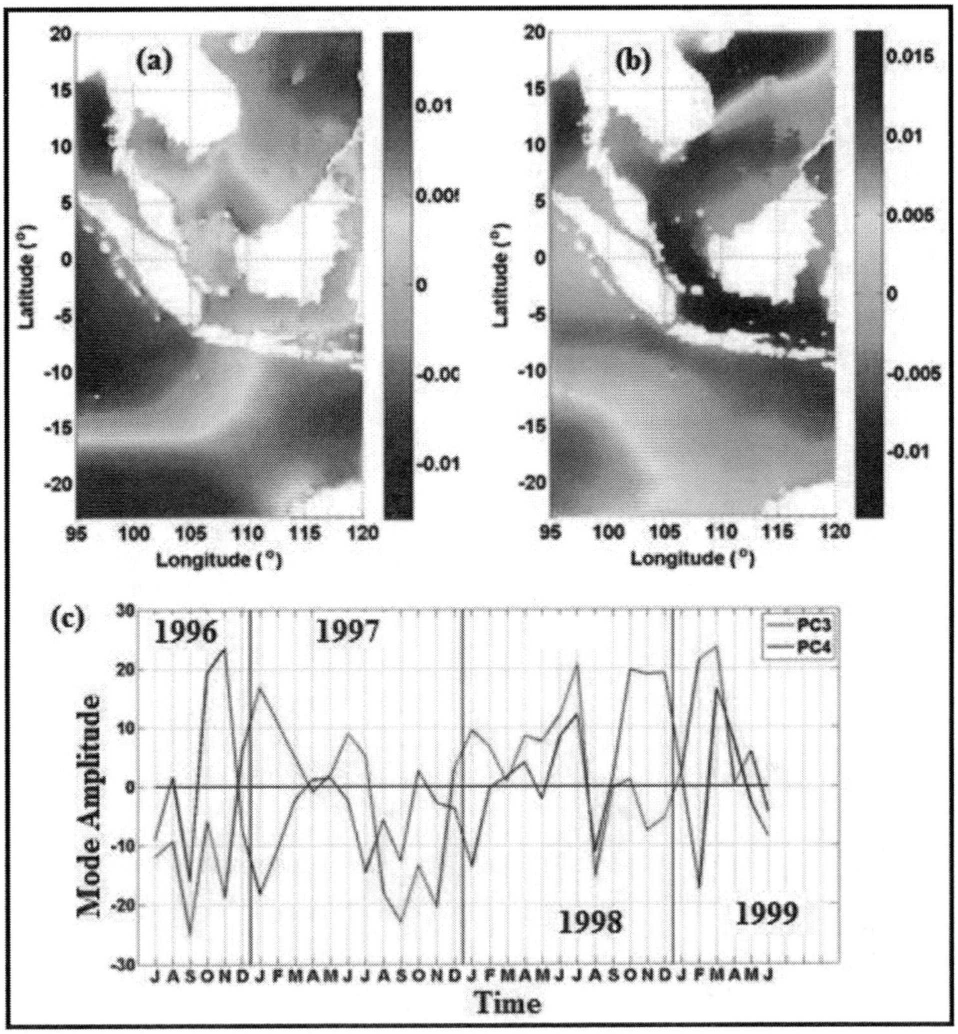

Fig. 8. As in Fig. 6 but for Mode 3 and Mode 4

* Corresponding author: rrachmayani@oceanography.itb.ac.id

https://doi.org/10.1051/matecconf/201814705001

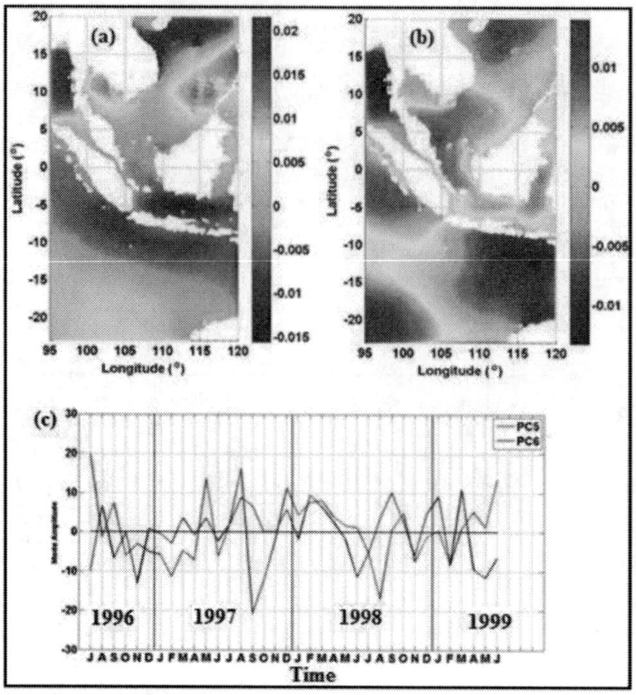

Fig. 9. As in Fig. 6 but for Mode 5 and Mode 6

4.3 Results of EOF Analysis Test

The EOF analysis produces modes that affect significant wave height variability in the model area. The EOF analysis test has been done by comparing the number of multiplication products between the spatial pattern value and the temporal pattern with the input data (not shown). Figure 10 displays the five test points of the EOF analysis results.

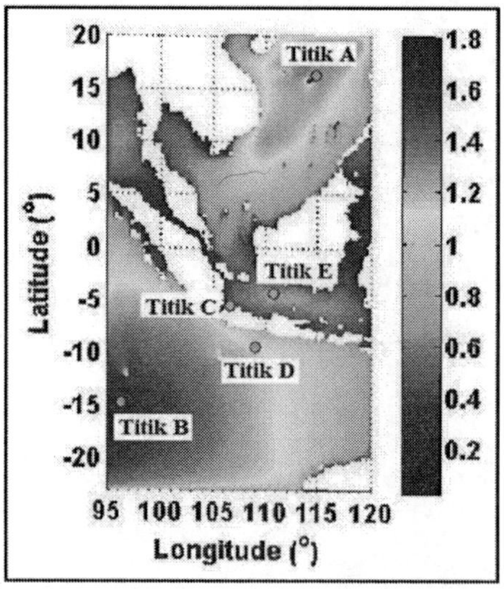

Fig. 10. Five test points of the EOF analysis results

Based on the multiplication of the spatial pattern (u) and the temporal pattern (a) for the six modes as well as the sum of the spatial (u) and temporal (a) at six PCs at point A to point E exhibits that wave variability from Mode 1 to Mode 6 at Point A is strongly dominated by the effect of Mode 1. The wave variability from Mode 1 to Mode 6 at Point B is dominated by the effects of Mode 1 and Mode 2. At Point C, D and E, the effect of Mode 1 is minor, while Mode 2 contributes in influencing wave variability. Mode 1 becomes the most dominant mode of the whole domain in general due to the largest temporal pattern or amplitude mode.

4.4 Seasonal Wind and Significant Wave Height

Furthermore, based on the results of EOF analysis, in this study, domain model can be divided into 6 areas that are considered to have the characteristics of wave variability. The divided area is illustrated in Figure 11. The wind field and the average H_s for each month for 3 years at the six consecutive areas are shown in Figure 12 and Figure 13. The wind moves from the southeast in Area 1 throughout the year. The maximum wind occurs during east season (JJA) with a great average speed ranging from 6.46 m/s to 7.90 m/s. This corresponds to the maximum average value of simulated H_s results. The minimum average H_s in Area 1 occurs during west season with values between 1.17 m to 1.26 m and corresponds to the lowest wind velocity conditions in Area 1. There is a difference in the wind field pattern between Area 1 and 2. In Area 2, the wind moves from the southwest during west season with an average H_s value of 1.0 m.

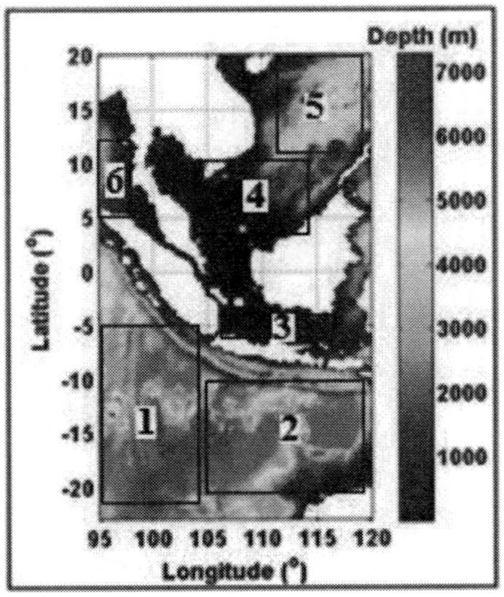

Fig. 11. Six divided area

Maximum waves occur during east season with a range of 1.1 m to 1.4 m. Area 3 represents Java Sea with a wind field pattern that changes direction almost 180° in one year. The wind moves from the southeast with a maximum speed of 4.54 m/s to 5.61 m/s during east season and a wave height of 0.6 m to 0.76 m. In Area 4, the wind moves from the northeast during west season with a significant wave height of 1.20 m to 1.60 m. During east season, the wind moves from the southwest with a significant wave height of 0.57 m to 0.66 m. The maximum wind speed and maximum wave height in Area 5 occurs twice in a year (west and east season).

The wave height is significant up to 1.30 m to 1.90 m during west season, whereas during east season the significant wave height reaches 0.90 m to 1.0 m. In Area 6, the dominant wind moves from the northwest from May to September with a significant wave height of 0.56 m to 0.87 m. During west season, the wind moves from the northeast with a significant wave height of 0.34 m to 0.44 m.

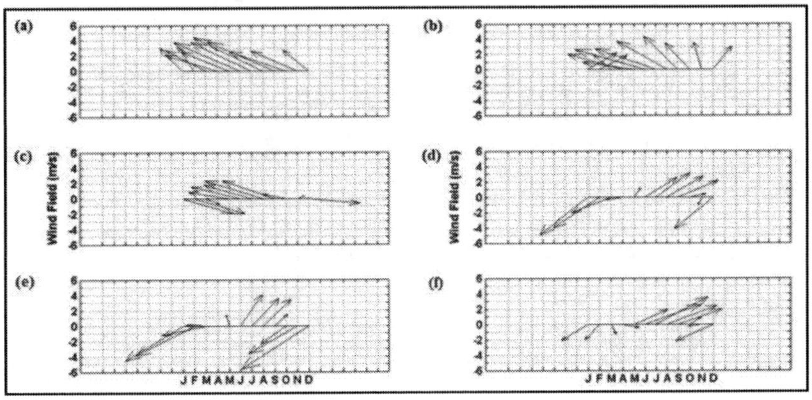

Fig. 12. Average monthly wind field at: (a) Area 1, (b) Area 2, (c) Area 3, (d) Area (4), (e) Area 5, and (f) Area 6

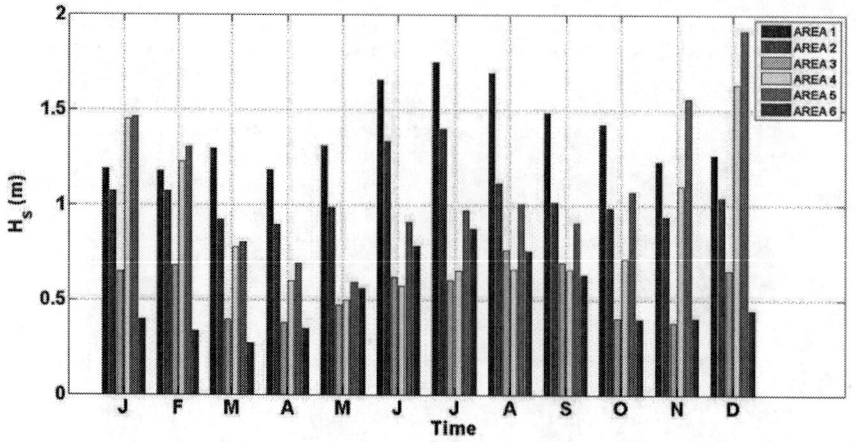

Fig. 13. The averaged significant wave height at all six areas of study

4.5 Analysis of Annual Variations

Based on the results of EOF analysis, it is known that wave variability in domain model is dominantly influenced by Mode 1 and Mode 2. In the previous section also has been explained that a monsoon with a period of 12 months plays a role in Mode 1 and a semi-annual monsoon plays a role in Mode 2. However, there is a discrepancy in the temporal pattern at certain intervals and it is associated with the influence of the interannual phenomenon of IODM and ENSO.

Mode 1 gives the dominant influence in the Indian Ocean and South China Sea with different phase variability. Thus, the analysis was performed separately for both waters and selected a point that has the largest spatial pattern value in the Indian Ocean and South China Sea. In addition, the average monthly wind vector supports the analysis. From Figure 11, point A (South China Sea) and point B (Indian Ocean) are chosen to illustrate interaction between IODM and ENSO to Mode 1 and point E represents Java Sea is picked to capture the interaction of the interannual phenomena to Mode 2.

Figure 14 - 16 expose the plot of Ocean Nino Index (ONI), Dipole Mode Index (DMI), spatial and temporal pattern and the averaged monthly wind field at point A, B, and C, respectively. Interval 1 indicates a normal IODM and ENSO states, index 2 (interval 2) reveals the time interval of positive El-Niño and IODM events, and index 3 (interval 3) affirms the time interval of La-Niña events.

During west season, the mean value of significant wave height at point A ($h_{1,A}$) (South China Sea) at interval 2 is smaller compared at interval 1. This is due to the weakening of monsoon winds during the El-Niño incident. At interval 3, the value of $h_{1,A}$ is greater due to the strengthening of the wind during La Niña event. To conclude, ENSO phenomenon contributes in Mode 1 at point A (South China Sea).

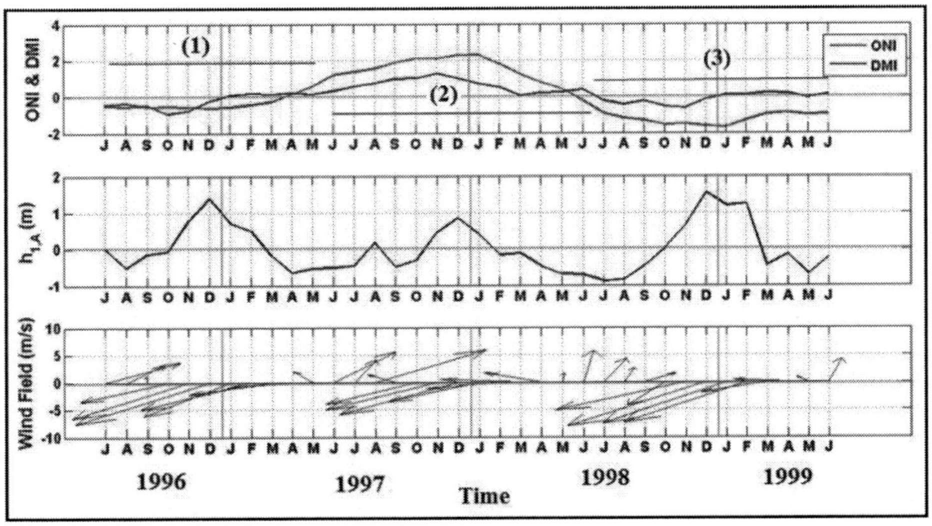

Fig. 14. ONI and DMI (top panel), spatial and temporal pattern (middle panel) and the average monthly wind field (bottom panel) at point A (South China Sea)

The maximum (minimum) wave height occurs during east (west) season at point B in Indian Ocean. This is supported by Mode 1 at interval 2 during west season where Mode 1 has a smaller effect in reducing wave heights in the Indian Ocean than its normal state. Positive IODM at interval 2 makes the winds speed larger and move westward; indicates that the IODM affects Mode 1 in Indian Ocean.

As mentioned above, point E is selected to represent Java Sea where contribution of Mode 2 is dominant in affecting wave variability rather than Mode 1. In July to September (JAS) at interval 2, winds in the Java Sea move westward and it is reinforced by a positive IODM which drive stronger winds westward around Java Sea. Meanwhile, in December to February (DJF) at interval 2, the wave at point E becomes smaller than at interval 1. This is associated with a movement of winds eastward in Java Sea which is reduced by the positive IODM.

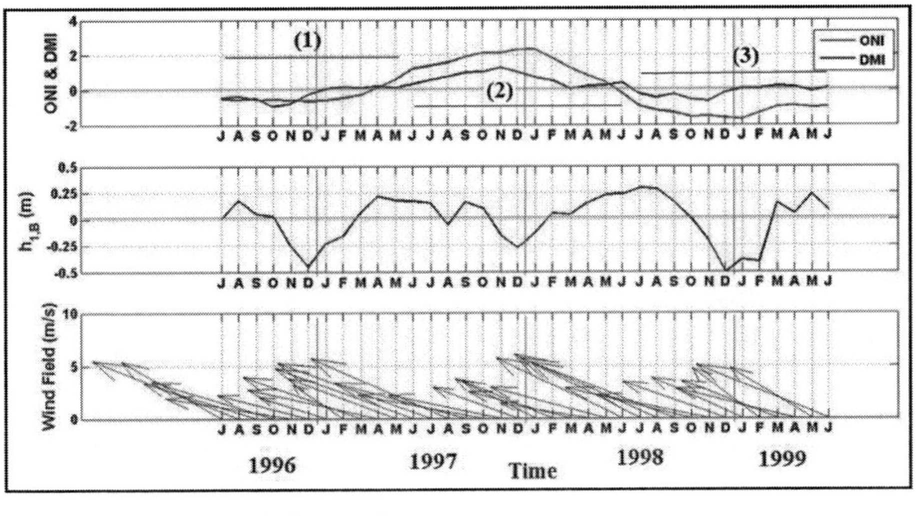

Fig. 15. As in Fig, 14 but for point B (Indian Ocean)

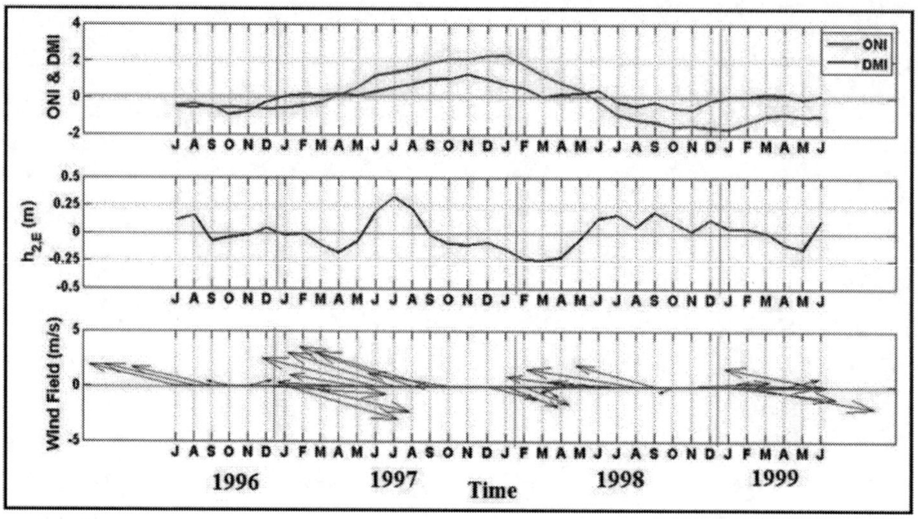

Fig 16. As in Fig, 14 but for point E (Java Sea)

5 Conclusion

Empirical Orthogonal Functions (EOF) analysis produces 6 modes represents 95% of total variance that influence the wave height variability in the entire model domain. Annual monsoon plays a role in Mode and has spatial dominant contribution in South China Sea and Indian Ocean. Java Sea was influenced by Mode 2 which is controlled by semi-annual monsoon. ENSO effect was identified in South China Sea and IODM effect was captured at Indian Ocean in Mode 1 and at Java Sea in Mode 2. A positive (negative) IODM strengthens (weakens) the winds speed in Java Sea during the East (West) season and hence contributes to Mode 2 in increasing (decreasing) the significant wave in Java Sea

References

1. R. Kurniawan and M. K. Khotimah, *Ocean Wave Characteristics in Indonesian Waters for Sea Transportation Safety and Planning*, IPTEK, J. Tech. and Sci., **26**(1), pp. 19-27, (2015).
2. R. A. Arinaga and K. F. Cheung, *Atlas of global wave energy from 10 years of reanalysis and hindcast data*, Renewable Energy, **39**(1), pp. 49-64, (2012).
3. A. Ramdhani, *Variabilitas Iklim Gelombang di Perairan Indonesia (Validasi Model Gelombang Wavewatch III)*, Laporan Kemajuan 1, Program Doktor Sains Kebumian, ITB, Bandung, (2013).
4. IOC, IHO dan BODC, *Centenary Edition of the GEBCO Digital Atlas*, Intergovernmental Oceanographic Commission dan the International Hydrographic Organization sebagai bagian dari General Bathymetric Chart of the Oceans; British Oceanographic Data Centre, Liverpool, (2003).
5. I. R. Young, *Seasonal variability of the global ocean wind and wave climate*, Int. J. Clim., **19**(9), pp. 931 – 950, (1999).
6. N. Booij, N., R. C. Ris, and L.H. Holthuijsen, *A third-generation wave model for coastal regions. Part I: Model description and validation*, J. Geophys. Res., **104**, 7649 – 7666, (1999).
7. G. J. Komen, S. Hasselmann, and K. Hasselmann, *On the existence of a fully developed wind-sea spectrum*, J. Phys. Oceanography, 14, 1271 – 1285, (1984).
8. The SWAN Team, *SWAN Scientific and Technical Documentation*, Delft University of Technology, Netherlands, (2013).
9. O. S. Madsen, Y. -K. Poon, and H. C. Graber, *Spectral wave attenuation by bottom friction: Theory. Proc.*, 21st Coastal Engineering Conf., ASCE, (1988).
10. A. M. Obukhov, *Statistically homogeneous fields on atmosphere*, Usp. Mat. Navk., **2**, pp. 196 – 198, (1947).
11. A. M. Obukhov, *The statistically orthogonal expansion of empirical functions*, Bull. Acad. Sci, USSR Geophys. Ser. (English Transl.), pp. 288 – 291, (1960).
12. H. Hotelling, *Analysis of a complex of statistical variables into principal components*, J. Educ. Psych, **24**, pp. 417 – 520, (1933).
13. D. S. Wilks, *Statistical Methods in the Atmospheric Sciences*, Academic Press, San Diego, (1995).

Development of Risk-Realiability Based Underwater Inspection for Fixed Offshore Platforms in Indonesia

Ricky L. Tawekal[1,*], *Faisal D.* Purnawarman[1], *Yati* Muliati[2]

[1]Bandung Institute of Technology, Ocean Engineering, Jl. Ganesha 10, Bandung, Indonesia

[2]Institut Teknologi Nasional, Civil Engineering, Jl. P.H.H. Mustafa 23, Bandung, Indonesia

Abstract. In RBUI method, platform with higher risk level will need inspection done more intensively than those with lower risk level. However, the probability of failure (PoF) evaluation in RBUI method is usually carried out in semi quantitative way by comparing failure parameters associated with the same damage mechanism between a group of platforms located in the same area. Therefore, RBUI will not be effective for platforms spread in distant areas where failure parameter associated with the same damage mechanism may not be the same. The existing standard, American Petroleum Institute, Recommended Practice for Structural Integrity Management of Fixed Offshore Structures (API RP-2SIM), is limited on the general instructions in determining the risk value of a platform, yet it does not provide a detail instruction on how determining the Probability of Failure (PoF) of platform. In this paper, the PoF is determined quantitatively by calculating structural reliability index based on structural collapse failure mode, thus the method in determining the inspection schedule is called Risk-Reliability Based Underwater Inspection (RReBUI). Models of 3-legs jacket fixed offshore platform in Java Sea and 4-legs jacket fixed offshore platform in Natuna Sea are used to study the implementation of RReBUI.

1 Introduction

Risk Based Underwater Inspection (RBUI) method has been developed and implemented in offshore platform inspection in Indonesia [1]. Previously, the inspection of offshore platform used Time Based Inspection (TBI) system in accordance to Indonesian government regulations since 1977. TBI is conducted regularly based on the types of inspection. Minor inspection is carried out annually, while major inspection was carried out once for 2 years, and complete inspection was carried out once for four years [2]. Due to the development of technology and demand in efficiency, Indonesian government encourages application of RBUI method as an alternative of TBI in determining inspection interval. By RBUI method, platform with higher risk level will need more often inspection than those with lower risk level.

Existing standard API RP-2SIM has not discussed the detail of risk determination, such as determining Probability of Failure (PoF), and Consequence of Failure (CoF). In previous RBUI, PoF evaluation is usually conducted in semi quantitative way by comparing failure parameters associated with the same failure mechanism between some platforms in the same area. However, RBUI method will not be efficient to be conducted for platforms spread in distant areas where failure parameters associated with the same damage mechanism may not be the same. Therefore, the risk estimation should be conducted quantitatively based on the same failure mode.

In this paper, Risk-Reliability Based Underwater Inspection (RReBUI) is developed, where the PoF is determined quantitatively by calculating structure reliability index based on collapse failure mechanism in which uncertainty of wave load and structural resistance are considered. Meanwhile, CoF factor calculated quantitatively in accordance to API RP-2SIM. By using RReBUI method, risk on every platform can be determined quantitatively without comparing failure parameter based on expert justification.

For case studies, 2 (two) offshore platform models installed in Java Sea (Indramayu offshore, West Java Province) and West Natuna sea (Anambas Islands, Riau Islands Province) are used as case study. Figure 1 shows the models of offshore platform.

2 Methodology

RReBUI analysis is started by collecting structure data in form of drawing and basis design. This structure data is then used to create model in SACS software for further analysis. Wave data is also collected, which has significant wave height parameter (H_s), wave period (T_n), and also maximum wave height (H_{max}).

* Corresponding author: ricky@ocean.itb.ac.id

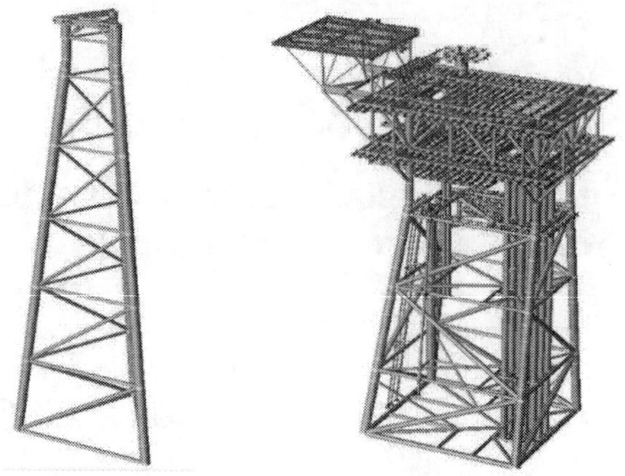

Fig. 1. Offshore platform 1: Java Sea (left) and Offshore platform 2: Natuna Sea (right)

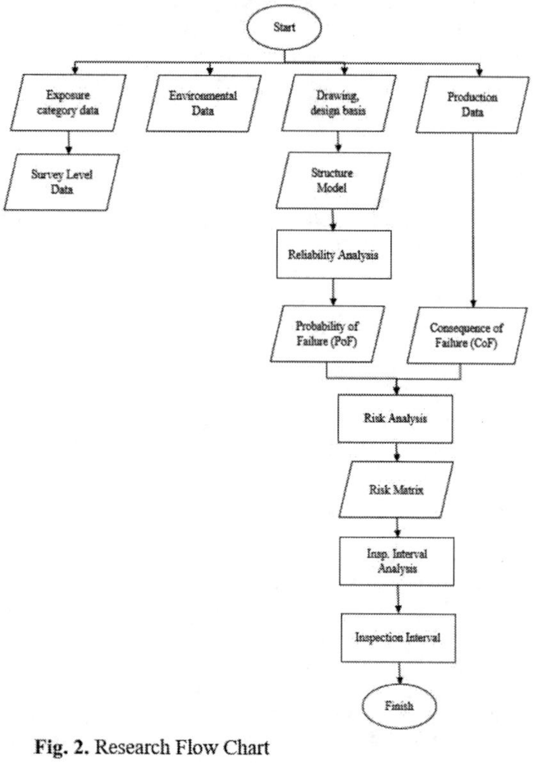

Fig. 2. Research Flow Chart

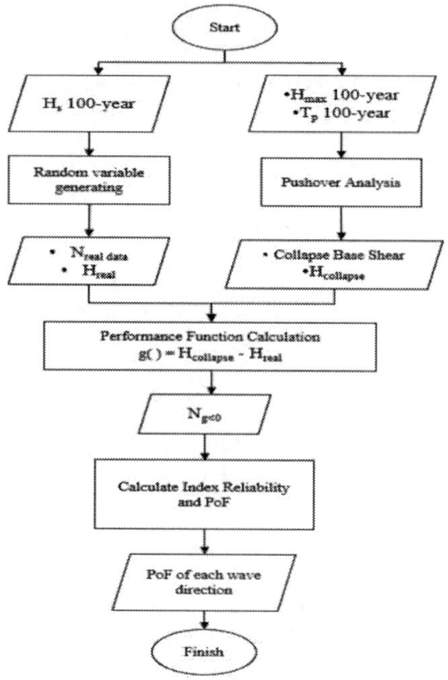

Fig. 3. Flow chart of index reliability and PoF calculation

Figure 2 and Figure 3 show research and probability of failure (PoF) calculation, respectively.In this research, the reliability analysis is divided into two stages; the random wave data analysis and collapse analysis

2.1 Random wave data analysis

One of loads to be concerned in structure analysis is load affected by the environment, in which an offshore structure is operated. Environmental load applied in this design is wind, current, and wave load.

The loading in structure is predicted as the function of some variables, such as wave height, wave period, wind speed, wind direction, and current [3].

The data of wave height and period can be obtained from these following results:

a. Direct observation by using wave recorder in the observed location
b. Visual observation of vessel passing certain coordinate
c. Observation of water surface using altimetry system
d. Prediction of wave and wind data in certain location

* Corresponding author: ricky@ocean.itb.ac.id

The wind data can be obtained from Indonesian Agency for Meteorology, Climatology, and Geophysics, and National Oceanic and Atmospheric Administration, while the current data at platform locations can be calculated by using numerical modeling. Generally, wave load is the most affecting load to the structures.

For structural reliability calculation based on collapse failure mode, we need wave data for return periods 1 year, 10 years and 100 years. Those data can be calculated using long term wave data analysis. The wave distribution in long term period can be approached using continuous distribution from Weibull. The linear formula of the probability density function (PDF) is given in this following formula:

$$\ln\left[\ln\left\{\frac{1}{1-P(H)}\right\}\right]=\xi\ln(x)-\xi\ln(\lambda) \quad (1)$$

P(H) = probability of wave existed

– = form parameter with common value is between 0,75 to 2,0; Whereas for ocean wave, the value is ranged between 0,9 t o. 1,1 (Naess: 0,7 to 1,3) 18

λ = scale parameter whose value is depend on extreme value of variable x; while for the ocean wave, the high extreme value occurs once in some period of time (m)

x = object/intensity parameter observed, for instance height wave, hence $x = H$

Weibull distribution can be approximated with linear curve if variable x in the right side of the above equation is replaced with $(H - a)$. Variable here is threshold wave height, which is the smallest wave height in the ocean. For closed water, a may have very small value (≈0), while for open water, it may have value between 0,5 to 2,0 m. Weibull distribution curve will be linear if it is drawn on the graphic correlating *ln{ln[1/1-P(H)]}* as ordinat and *ln(H –a)* as abcissa. The wave height used is significant wave height (H_S), if the wave known is maximum wave height (H_{max}), so H_S can be defined as:

$$H_S = H_{max}/1.86 \quad (2)$$

Beside Weibull, random wave distribution can be approached using Rayleigh distribution,

$$f_x(x)=\frac{2x}{R}\exp\left(-\frac{x^2}{R}\right),\ x\geq 0 \text{ with } R=\frac{1}{n}\sum_{i=1}^{n}x_i^2 \quad (3)$$

If the lower limit of one per three of the biggest height is noted as x*, the probability value of wave height occurrence exceed x* is 1/3 as seen in this equation:

$$\Pr\{X\geq x_*\}=\int_{x_*}^{\infty}\frac{2x}{R}\exp\left(-\frac{x^2}{R}\right)dx=\frac{1}{3} \quad (4)$$

From equation (4), the lower limit value, x_* can be a deritative into function of parameter in this equation below:

$$x_*=\ln(3)\sqrt{R}=1.048\sqrt{R} \quad (5)$$

H_s in PDF can be observed in Figure 4. Thus we can see that Hs is the central point of area under the PDF curve with lower limit of x_*.

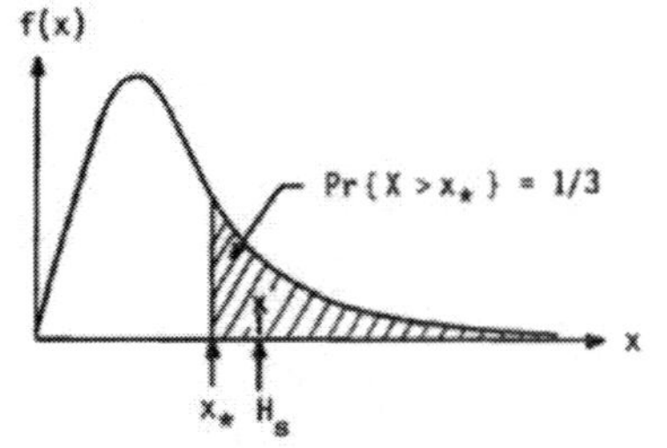

Fig. 4. Definition of significant wave height (H_s) in a PDF curve (Ochi, 1998)

In order to find out the central points of area under PDF curve, we can calculate it using this following formula.

$$H_s=x_{1/3}=\frac{\int_{x_*}^{\infty}x\cdot f_x(x)dx}{\int_{x_*}^{\infty}f_x(x)dx}=\frac{x_*\cdot\exp\left(-\frac{x_*^2}{R}\right)+\sqrt{\pi R}\left\{1-\Phi\left(\sqrt{\frac{2}{R}}x_*\right)\right\}}{1/3} \quad (6)$$

With substitution of 2 (two) above equations, the value of H_s can be calculated as follows.

$$H_s=\left[\sqrt{\ln(3)}+3\sqrt{\pi}\left\{1-\Phi\left(\sqrt{2\ln(3)}\right)\right\}\right]\sqrt{R} \quad (7)$$

$$H_s \sim 1.42\sqrt{R} \quad (8)$$

2.2 Reliability analysis

In engineering design, uncertainties in structural analysis may exist in loading, material properties, geometry, and surface condition. These uncertainties must be taken carefully to ensure that the design can perform its intended function with desired confidence that failure will be avoided. The traditional way of dealing with uncertainties is by using safety factors which set conservative limits for the design values of factors such as stress that are derived from uncertain design parameters.

Reliability analysis of structure is used to evaluate failure of structure related to force-load system probability. The information of average and uncertainty variable distribution defines structure failure probability. Structure reliability design model can define load and resistance as random variable probability. Figure 5 shows that reliability equations depend on the intersecting area of probability density function between load and resistance. As in Figure 5, it can also be stated that there are no area which has no risk.

In general, failure function is defined as

M = $g(R,S) = R - S$ (9)

M = $g(R,S)$ = Performance function

R = resistance function

S = Load Functions

Probability of failure can be defined as

$$P_f = P(R < S) \quad (10)$$

The reliability index can be calculated as

$$\beta=\frac{\mu_M}{\sigma_M} \quad (11)$$

μ_M = average M (safety margin)

σ_M = standard deviations M (safety margin)

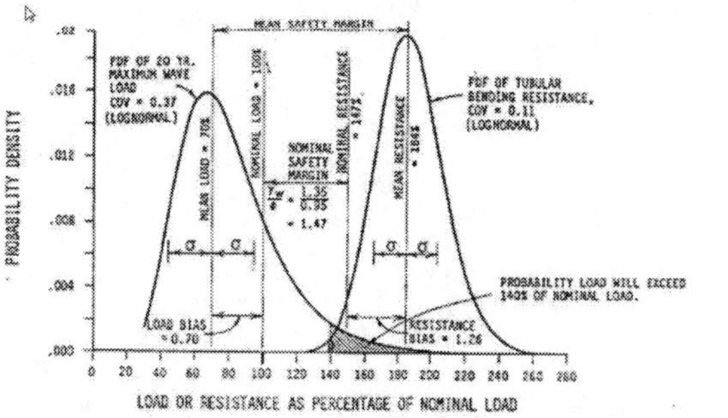

Fig. 5. Probability density function of structure failure [4]

Reliability analysis can be started by determining performance function that defines the parameter of structure failure mode. Structure failure mode can be defined based on pushover [5], joint fatigue [6], fracture mechanics [7] and structural member failure.

The reliability analysis is started by determining the performance function which defines the failure parameter of the structure. Random wave data analysis is an analysis of individual wave data to generate random waves with return period of 1000 years in order to obtain random wave data which is predicted to be structure load. This data-generating produces 100 million data of random wave height that will be used as wave data input to calculate the forces acting on the structure. Pushover analysis is carried out using SACS software, by increasing the wave load incrementally. The simulation is done until the structure collapses and then the relation between the wave height and the load that makes the structure collapses is obtained. These both analyses (random wave analysis and pushover analysis) are combined so the numbers of wave data concurrence that make the structure collapse is obtained. This number is compared to the number of total wave that occurrences, so the value of probability of failure (PoF) is obtained. This PoF value will be used later in RReBUI analysis.

RReBUI analysis is started by calculating the platform risk matrix by x- axis parameter as PoF and y-axis as CoF. The PoF value has been defined by using reliability analysis. Meanwhile, CoF analysis is determined qualitatively based on API RP-2SIM standard, thus we can obtain point in range 1-5 based on the level of failure consequence severity. The combination of PoF and CoF value result to platform risk value in risk matrix. API RP-2SIM provides conversion method and risk matrix that are obtained into an inspection interval. This conversion method uses exposure category factor and inspection level classification. The result of this inspection interval is then classified into inspection interval for each inspection level that ranges from 1 to 4. Each of this inspection level has different detail examination structure, with different interval schedule.

3 Results and discussions

Reliabity analysis is started in determining performance function which is used as the parameter of structure failure. Parameter of structure failure used is as follows.

$$RSR > RSRmin$$

$$\frac{ultimatelateralload}{100yearswaveload} > RSRmin$$

$$\frac{collapsebaseshear}{initialbaseshear} > RSRmin$$

$$collapsebaseshear > (RSRmin)initialbaseshear$$

$$H_{(collapse)} > (RSRmin)H_{(initial)}$$

$$(RSRmin)H_{(initial)} - H_{(collapse)} < 0$$

In addition, the performance function is,

$$g = Resistance?\,Load$$

$$g = maxallowablebaseshear - realbaseshear$$

$$g = collapsebaseshear - baseshearH_{(real)}$$

$$g = H_{(collapse)} - (RSRmin)H_{(real)}$$

The analysis of random wave results to the value of wave maximum height with return period 1000 years. In addition, pushover analysis results to the value of wave height affecting toward the collapse structure. The results of these both analyses are described in Figure 6 below.

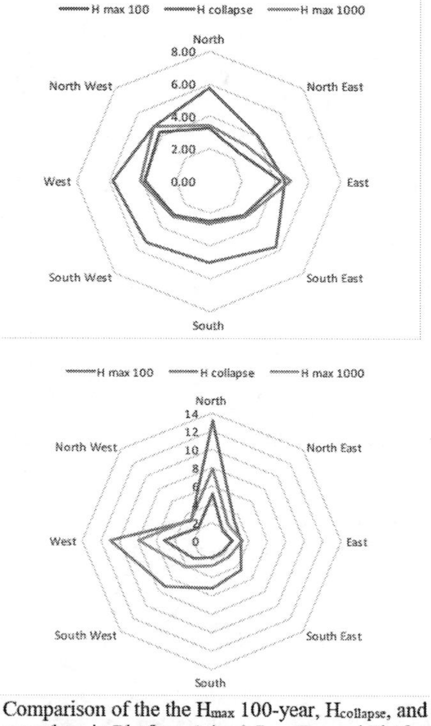

Fig. 6. Comparison of the the H_{max} 100-year, $H_{collapse}$, and H_{max} 1000-year values in Platform 1 (top) Java Sea and Platform 2 (bottom) Natuna Sea

The Probability of Failure is then calculated based on the data of 100 milion results from the analysis value above by using Matlab software and by comparing each score of Hcollapse and Hmax 1000 year based on the performance function. Negative performance function means that the structure is failed. Then, the total of negative value is calculated and compared to 100 million wave data. This value then is called as the Probability of Failure (PoF) presented in Table 1 below.

Table 1. Calculation of PoF value

Direction	Platform 1: Java Sea		Platform 2: Natuna Sea	
	N fail	PoF	N fail	PoF
Omni	16130	1.61.E-04	580434	5.80.E-03
North	0	0.00.E+00	26364	2.64.E-04
North East	0	0.00.E+00	616570	6.17.E-03
East	34816	3.48.E-04	746055	7.46.E-03
South East	0	0.00.E+00	4293	4.29.E-05
South	0	0.00.E+00	1581	1.58.E-05
South West	0	0.00.E+00	13452	1.35.E-04
West	8	8.00.E-08	300665	3.01.E-03
North West	16282	1.63.E-04	531329	5.31.E-03
Maximum	**34816**	**3.48.E-04**	**746055**	**7.46.E-03**

The PoF value obtained is then converted into reliability index value [8] as presented in Table 2a and Table 2b

Table 2a. Conversion of PoF value into reliability index

Platform	Direction	PoF	Reliability(β)
Java Sea	East	3.48.E-04	3.26
Natuna Sea	East	7.46.E-03	2.87

Table 2b. Conversion of PoF value into reliability index

Safety Class	Target (PoF)	Reliability Index (β)
Low	0.01	2.32
Normal	0.001	3.09
High	0.0001	3.72

For CoF calculation, additional information is needed related to the determination of failure consequence in accordance to API RP 2-SIM [9] which is the parameter from the safety, environment, and business point of view. Table of additional information is shown on this Table 3 below.

Table 3. Data and information needed in RReBUI analysis

	Java Sea	Natuna Sea
Reliability Index	3.2636	2.8728
Platform Crew	Unmanned	Manned
Oil Production (bpd)	4200	1050
Gas Production (mmscfd)	55.8	105.2
Platform Price (million USD)	<100	100 - 300

The calculated CoF and PoF parameters are then used to analyse the severity level in range 1-5. Table of parameter CoF and PoF categories, and the classification of risk matrix severity level is shown in Table 4 and Table 5.

Based on API RP 2SIM, the consequence criteria were divided into three levels as described in Table 6. These levels determined the inspection interval as stated in API RP2A-WSD and API RP 2SIM.

Table 4. Parameter of PoF, CoF, and classification of severity level

Parameter	Factor	Score 1	2	3	4	5
CoF	Safety	Unmanned	-	-	-	Manned
	Bussiness (oil and gas production)	Production < 2500 BOEPD	Production 2501 – 5000 BOEPD	Production 5000 – 7500 BOEPD	Production 7501 – 10000 BOEPD	Production > 10000 BOEPD
		Platform Price <$100 MM	-	Platform Price $100 -300 MM		Platform Price <$100 MM
	Environment (oil production)	Production < 2500 BOPD	Production 2501 – 5000 BOPD	Production 5000 - 7500 BOPD	Production 7501 – 10000 BOPD	Production > 10000 BOPD
PoF	**Reliability Index**	**>4.44**	**3.72-4.44**	**3.09-3.72**	**2.32-3.09**	**<2.32**

Table 5. Calculation result of PoF and CoF

Platform	Data (β)	Platform Crew	Oil Production (bpd)	Gas Production (mmcfd)	Platform Price	CoF Safety	Business	Environment	CoF Value	PoF Value
Java Sea	3.26	Unmanned	4200	55.8	< 100	1	2	2	2	3
Natuna Sea	2.87	Manned	1050	105.2	100 - 300	5	3	1	5	4

Table 6. Exposure level-based inspection program recommended by API

Exposure Category Level	Survey Level (Years)			
	I	II	III	IV
L-1	1	3–5	6–10	
L-2	1	5–10	11–15	Case by Case only
L-3	1	5–10	Case by Case only	

The calculated CoF and PoF are then used to determined platform's exposure category and platform positions in risk matrix as shown in Table 7 and Table 8 below

Table 7. Exposure Categories of Platforms

No	Platform Location	PoF	CoF	Risk	Exposure Category
1	Java Sea	3	2	Medium	L3
2	Natuna Sea	4	5	High	L1

Table 8. Platform positions in risk matrix

Consequence of Failure	1	2	3	4	5
5				Platform 2	
4					
3					
2			Platform 1		
1					
	Probability of Failure				

Inspection intervals are then determined in accordance to API RP-2SIM and the result is shown in Table 9 below.

Based on Table 9, it is clearly seen that the value difference resulted by each analysis of different platforms. Factor of platform characteristic, wave force, and information related to safety, environment, and business become the important points of the risk and inspection interval difference.

4 Conclusion

Reliability index value can be used for Probability of Failure (PoF) calculation in RReBUI analysis. By using RReBUI method, risk comparison of 2 (two) platforms located in distance areas can be determined quantitatively without comparing failure parameter based on expert justification. The results of RReBUI analysis for 2 (two) fixed offshore platform models is an inspection plan that accentuated safety, but also had a longer interval when compared with the previous time-based methods. It means, the efficiency objectives of inspection operation could be accomplished.

Table 9. Calculation result of inspection interval

Platform	PoF	CoF	Risk	Exposure Category	Inspection Interval			Last Inspection		Next Inspection		
					Level 1	Level 2	Level 3	Level 1	Level 2	Level 1	Level 2	Level 3
Java Sea	3	2	Medium	L3	1	9	*	2016	2015	2017	2024	*
Natuna Sea	4	5	High	L1	1	3	6	2016	2015	2017	2018	2021

* Corresponding author: ricky@ocean.itb.ac.id

References

1. Tawekal, R.L., et al., *Risk Based Underwater Inspection (RBUI) For Existing Fixed Platforms In Indonesia.* International Journal of Research in Engineering and Science (IJRES), 2017. **5**(7).
2. Indonesian General Directorate of Oil and Gas, *Inspection Guidelines for Construction of Oil and Gas Platform*, in *No. 21K/38/DJM/1999.* 1999, Indonesian General Directorate of Oil and Gas: Jakarta.
3. Zachary, S., *Multivariance Extrapolation in the Offshore Environment.* Journal of Applied Ocean Research, 1998. **20**(5): p. 273-295.
4. American Petroleum Institute, *Recommended Practice for Planning, Designinf, and Constructing Fixed Offshore Platform - Load Resistance Factor Design*, in *API Recommended Practive 2A-WSD.* 2000, American Petroleum Institute: Washington D. C.
5. Tawekal, R.L. and R. Heriana, *Reliability of Offshore Jacket Fixed Platform based on Pushover Loading.* Media Komunikasi Teknik Sipil, 2007. **14**(2).
6. Tawekal, R.L., *Reliability Assessment of Aging Offshore Platforms.* Journal of Infrastructure and Built Environment, 2008. **IV**(2).
7. Tawekal, R.L. and M.A. Iqbal, *Fatigue Reliability Index of Jacket Offshore Platform based on Fracture Mechanics*, in *The Eleventh East Asia-Pacific Conference on Structural Engineering and Construction.* 2008: Taipei.
8. Bai, Y., *Marince Structural Design.* 2003, Oxford: Elsevier.
9. American Petroleum Institute, *Recommended Practive for Structural Integrity Management*, in *API Recommended Practice 2SIM.* 2014, American Petroleum Institute: Washington D. C.

Dynamic Response of Breasting Dolphin Moored with 40,000 DWT Ship due to Parallel Passing Ship Phenomenon

Heri Setiawan[1,*], *Muslim* Muin[2]

[1,2] Coastal Engineering Research Group, the Faculty of Civil and Environmental Engineering, Institut Teknologi Bandung, Ganesha Street No. 10, Bandung, West Java, Indonesia

Abstract. When a ship is moving through another ship moored nearby, hydrodynamic interactions between these ships result in movements of the moored vessel. The movement may occur as surge, sway, and/or yaw. When a ship is passing a moored vessel parallelly, this effect will give a dominant lateral force on the moored ship and response from this phenomenon will appear in a certain time. Only dynamic response due to sway force is considered in this study, the sway force shall be absorb by the breasting dolphin. 40,000 DWT shall be moored to the breasting dolphin. Three passing ships size are considered, the breasting dolphin shall be modeled as a single degree of freedom model. This model will be subjected to a force caused by parallel passing ship. The model is assumed to be in a state of quiet water, this assumption is taken so that the fluid does not provide additional force on the model. The SDOF system shall be analyzed using a computer program designed to solve an ordinary differential equation.

1 Introduction

Muga and Fang [1] stated that if a ship is moving through another ship moored nearby, hydrodynamic interactions between these ships result in movements of the moored vessel. The movement may occur as surge, sway, and yaw. When a ship sail, this ship shall develop waves that carries energies. When two ships passed as presented in Figure 1, the moored vessel shall withstands force and moment due to this phenomenon. This interaction shall be very important in mooring system design.

Fig. 1. Parallel passing ship laboratory model. [3]

If the ships are closely spaced or having minimum under keel clearance and the passing ship passes through with high speed, the load due to this phenomenon can be very high. Seelig [2] documented a numbers of serious accidents due to this interaction. Various researches have been performed to estimate the force that is caused by this phenomenon. However, there are few researches that have been published where the structure response due to this force is analyzed.

This paper analyzed the effect of passing ship velocity to the breasting dolphin moored by a 40,000 DWT ship. The peak force that is considered is taken from Kriebel [3].

2 Study on Parallel Passing Ship

Today there are several methods to predict the hydrodynamic interaction due to the parallel passing ship phenomenon. There are two methods commonly used, the first one is numerical method that simulates fluid velocity and pressure due the passing ship and then integrate the pressure around the vessel to determine the force. Another one is a simple engineering technique that provides a direct force estimation to the moored vessel through a simple equation or graph. It has discovered that both methods were performed in the same laboratory and with similar scale. Hence, the applicability of the results outside the model are questioned. Based on the above findings, Kriebel [3] developed a model to obtain a non-

* Corresponding author: heri.setiawa@ocean.itb.ac.id

dimensional equations that predicts the maximum force values as presented in Figure 2. The equations are:

$$F_X = 0.5\ C_X \rho\ \underline{D}\ LOA\ V^2 \quad (1)$$

$$F_Y = 0.5\ C_Y \rho\ \underline{D}\ LOA\ V^2 \quad (2)$$

$$M = 0.5\ C_{MO} \rho\ \underline{D}\ LOA^2\ V^2 \quad (3)$$

$$C_X = 0.0074\Delta_R\, e^{+2.6(D/h)}\, e_{-}^{1.5(S/LOA)} \quad (4)$$

$$C_Y = 0.0126\Delta_R\, e^{+3.6(D/h)}\, e^{-2.0(S/LOA)} \quad (5)$$

$$C_{MO} = 0.0044\Delta_R\, e^{+3.2(D/h)}\, e^{-3.4(S/LOA)} \quad (6)$$

where
F_X = Maximum surge force.
F_Y = Maximum sway force.
M = Maximum yaw moment.
$\underline{D}$ $\overline{D}$ = Moored ship draft.
LOA = Moored ship length overall.
V = Passing ship velocity.
CX = Peak surge force coefficient.
CY = Peak sway force coefficient.
CMO= Peak yaw moment coefficient.
Δ_R = Displacement ratio of passing ship to moored ship.
h = Water depth.
S = Distance between ship

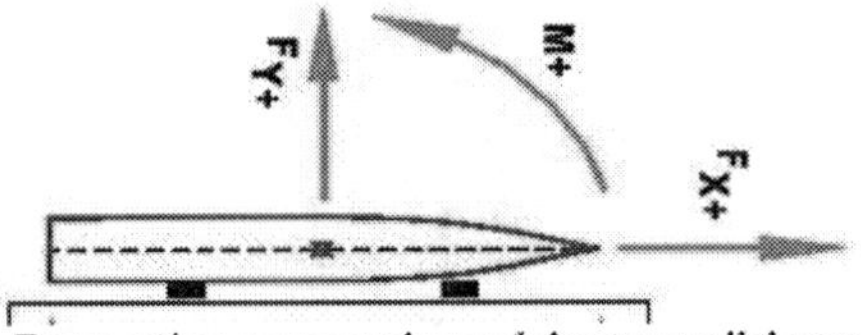

Fig. 2. Force acting on moored vessel due to parallel passing ship. [3]

Other than the peak forces, Kriebel [3] also presented a non-dimensional graph that shows a relationship between force and distance between ships. The graph is presented in Figure 3.

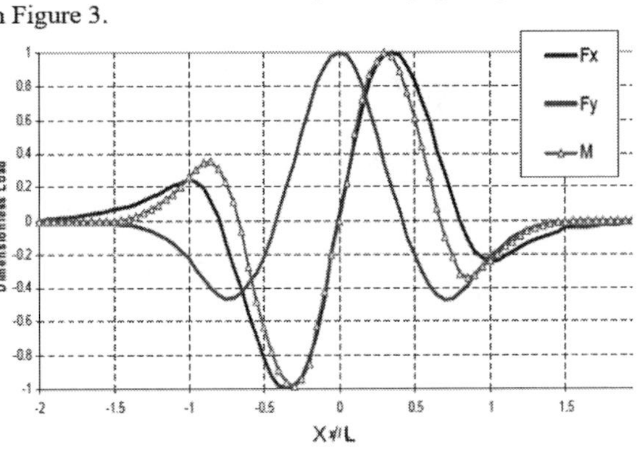

Fig. 3. Comparison of normalized load to distance over ship length. [3]

According to Marine Oil Terminal Engineering and Maintenance Standards (MOTEMS) [4] the parallel passing ship effect shall be considered when following criteria are met,

- the passing ship is larger than 25,000 DWT;
- distance between ships is less than 700 feet or 213.36 meter;
- the passing ship velocity is higher than the critical velocity, that is calculated as:

$$v_{cr}(\text{Knot}) = 1.5 + \left(\frac{S-2B}{700-2B}4.5\right) \quad (7)$$

where B is the moored ship breadth.

3 Research Methodology, Data and Assumption

3.1 Research Methodology

The mooring system is represented as a massless single pole, which is assumed as a single degree of freedom system. The moored vessel is assumed to be attached to the mooring dolphin and it moves as a unit with the dolphin. Hence, the vessel is considered as an additional mass to the system. Beside the additional vessel mass, the model obtains additional mass from the displaced fluids. This model is exerted by force extracted from Kriebel [3].

In single degree of freedom model mass, stiffness and damping are required to solve the system. The mass is obtained directly from engineering drawings and specification. The stiffness on the other hand is obtained by assuming the system as fixed cantilever. The stiffness value is obtained by its relations with the modulus elasticity and the cross section properties. The model damping ratio is assumed to be 5 %.

The fluids that surrounds the dolphin is assumed to be in static condition, hence it will not impose additional loads to the model. Computational program built code shall be used to solve this model.

3.2 Vessel Characteristics

3.2.1 Moored Ship Characteristics

A moored vessel with 40,000 DWT shall be considered in this analysis. The vessel dimensions are presented in Table 1.

3.2.2Passing Ship Characteristics

As presented in Section 2, the effect of parallel passing ship shall be considered if the passing ship is larger than 25,000 DWT. The passing ship that is considered in this research is presented in Table 2. Additional 1,000 DWT ship is considered to see the effect of ship with fast speed to the system.

Table 1. Moored Ship Characteristics.

Description	Value
Mass	40,000 DWT
Displacement	50,000 tonne
LOA	195 m
Breadth	29 m
Draft	11.5 m
Mass	40,000 DWT

Table 2. Passing Ship Characteristics.

DWT (tonne)	Displacement (tonne)	Typical Maximum Ship Speed (m/s)	Passing Ship Speed (m/s)
200,000	236,000	8	3, 5 and 8
40,000	50,000	8	3,5 and 8
1,000	1,400	10	3, 5, 10 and 20
200,000	236,000	8	3, 5 and 8
40,000	50,000	8	3,5 and 8
1,000	1,400	10	3, 5, 10 and 20

3.3 Distance between Ship

As suggested by MOTEM [4] parallel passing effect shall be considered if the distance between ships is less than 700 ft (213.36 m). In this research the distance is set to 200 m.

3.4 Breasting Dolphin Characteristic

In this study, the system to be considered is a coal offloading dolphin to accommodate a 200,000 DWT ship. This port has an overall length of 315 m, 11 m width, deck elevation of + 4.9 m and located at 20.9 m water depth. Therefore, the system is modeled as a 25.8 m long beam.

This dolphin is constructed with a series of deck on pile breasting and mooring dolphin. The breasting dolphin is designed to absorb the berthing force, meanwhile the mooring dolphin is designed to moor the vessel during docking. The force due to parallel passing ship shall be absorbed by breasting dolphin. Hence, only the breasting dolphin is considered in this study.

The breasting dolphin is constructed of a deck that is sitting on 36 piles, all piles fixity point is located at 7 m below mudline. The breasting dolphin sketch is presented in Figure 4 and Figure 5.

The breasting dolphin properties are presented in Table 3.

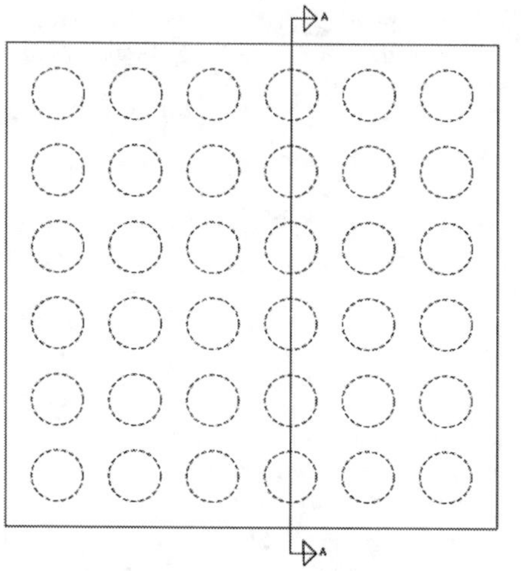

Fig. 4. Breasting dolphin sketch

Table 3 Breasting Dolphin Characteristics.

Description	Value
Concrete compression strength (*fc'*)	35 MPa
Reinforce concrete yield strength (*fy*)	400 MPa
Deck Size	17 m x 17 m x 2 m
Steel yield strength (*fys*)	240 MPa
Steel ultimate strength (*fu*)	370 MPa
Steel pile diameter (*DO*)	1.524 m
Steel pile thickness (*WT*)	25.4 mm
Steel pile length (i)	32.8 m

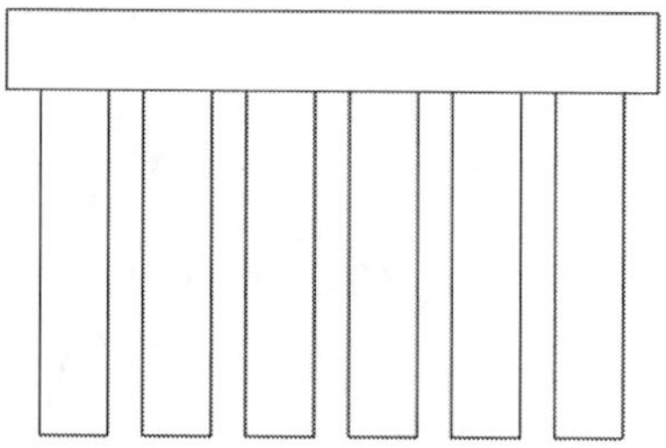

Fig. 5. Breasting dolphin cross section.

The dolphin model is assumed as the summation of the deck and 1/3rd of the pile mass and the dolphin mass (M_{BD}) is calculated by following equation,

$$M_{BD} = \rho_c V_c + \rho_s V_s \quad (8)$$

where ρ_c and ρ_s is the concrete and steel density respectively, and V_c and V_s is their volume.
Then the model total mass (M_{TOT}) is calculated as,

$$M_{TOT} = M_{BD} + M_{ship} + M_{add} \quad (9)$$

where *Mship* is the ship mass and *Madd* is additional mass due to the fluid displaced by the dolphin and vessel movements. From equation (8) and (9), the model total mass is obtained 26.34 tonne.

In this study the breasting dolphin is modelled as a cantilever. The cantilever stiffness (K) can be calculated by following equation,

$$K = \frac{3EI}{l^3} \quad (10)$$

where E is the section elasticity modulus and l is the section length. I is the section moment of inertia calculated by following equation,

$$I = \frac{\pi}{64}\left(D_0{}^4 - D_i{}^4\right) \quad (11)$$

where D_O and D_i is the pile outer and inner diameter. From equation (10) and (11), the model stiffness in lateral direction is obtained 10,539 kN/m. By assuming a 5% damping ratio, the model damping is obtained 1,666 kNs/m.

3.5 Force due to Parallel Passing Ship

With the above mention parameters, the peak forces of each case shall be calculated based on Equation (1) to (3). The governing load for berthing dolphin design comes from sway force. Therefore, only sway force is considered in this study. As presented in Figure 4, 4 berthing dolphins are used to absorb the sway force from the vessel movement. Therefore, it is assumed that the maximum sway force is divided equally to each dolphin. The maximum sway force for each case is presented in Table 4.

Figure 7 presents the time series estimation of the sway force acts on the moored ship. The time series is extracted from Kriebel [3]. The vertical axis represents the ratio of the sway force (Fsway) compared to the maximum sway force (FY) obtained from Equation (2). The horizontal axis represents the distance between passing ship (x) to the moored vessel LOA. As presented in Figure 6, there will not be any sway force at the moored ship until the passing ships moves closer to the moored ship. This force gradually increases and reaches its maximum value when the ships are in the same x location. After that, the force gradually reduces when the ship moves away.

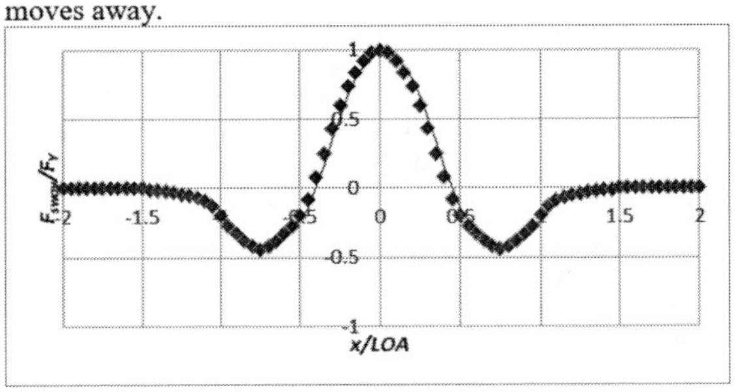

Fig. 6. Idealized variation of loads with overtaking distance. [3]

Table 4. Maximum Sway Force for Each Ship.

Passing Ship Displacement (tonne)	Passing Ship Velocity (m/s)	Maximum Sway Force (kN)
236,000	3	143.33
	5	398.14
	8	1,019.24
50,000	3	30.37
	5	84.35
	8	215.94
1,400	3	0.85
	5	2.36
	10	9.45
	20	37.79

To obtain the sway force time series, a number of points are estimated from Figure 7. To obtain the time step (t), following equation is used,

$$t = \frac{x}{LOA}\frac{LOA}{V} \quad (11)$$

By using Equation (2), (5) and (12), the estimated time series for a 1,000DWT passing vessel with distance between vessel of 200 m and velocity of 10 m/s is presented in Figure 7. To obtain smaller time discretization, polynomial fitting is performed. 20th order polynomial fitting time series is presented in Figure 8. This time series is applied to the model to obtain the berthing dolphin displacement.

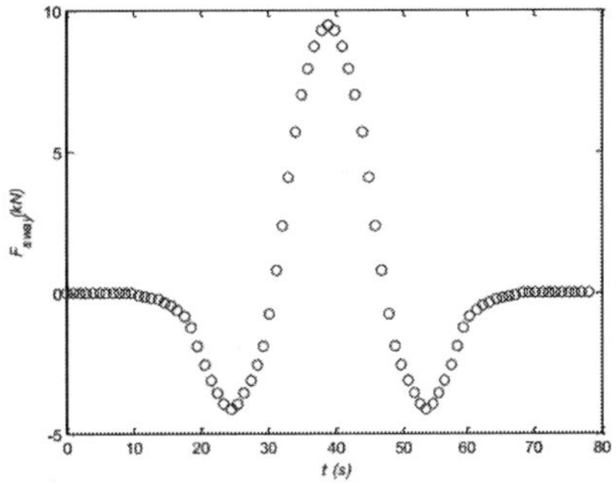

Fig. 7. Fs_{way} time series estimation.

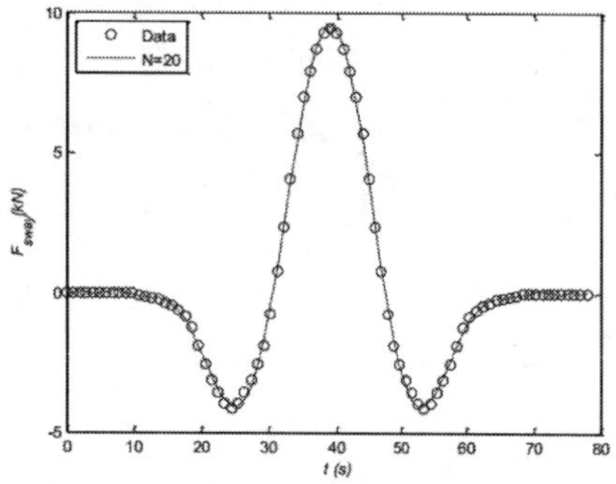

Fig. 8. Fs_{way} time series with 20th order polynomial fitting.

4 Dynamic Analysis

Before the structure motion can be defined, a representation of the structure along with the force and boundary condition is defined with a mathematical model. Based on the free body diagram, the equation of motion can be defined based on Newton's second law. The equation of motion can be modeled as a single or multi degree of freedom. The system is modeled as a damped spring mass that is exerted by a certain load.

The free body diagram of a single degree of freedom stucture is illustrated in Figure 10. Where *m* is the model mass, *k* is model stiffness, *c* is damping, *F(t)* is force acting on the system and *y* is the model displacement.

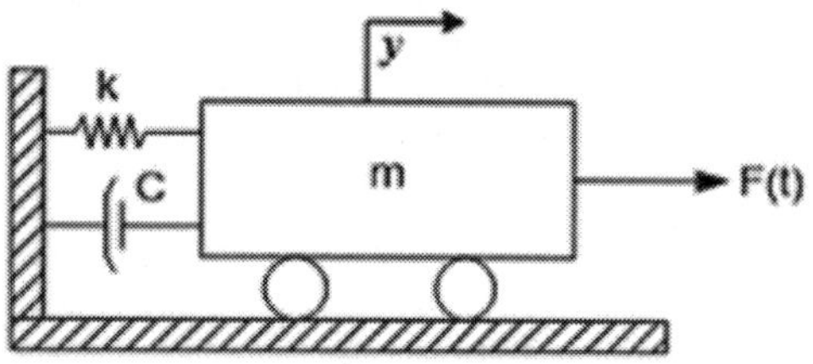

Fig. 9. Mathematical model for single degree of freedom system.

As presented in Figure 10, it is assumed that each system elements represents one particular characteristic. The structure mass (*m*) only defines its property of inertia. The (*k*) spring stiffness only represents its elasticity not its inertia or energy loss. On the other hand, the damping element (*c*) only represents the energy loss.

Based on Newton's second law, the force equilibrium presented at Figure 9 can be written as,

$$m\ddot{y} + c\dot{y} + ky = F_0 e^{i\omega_0 t} \tag{13}$$

where ω_0 is the force period.

The displacement from Equation (13) for a harmonic exerted system can be written as,

$$y = e^{-\xi\omega_n t}\left(A\cos(\omega_D t) + B\sin(\omega_D t)\right) + \frac{y_s \sin(\omega_0 t - \alpha)}{\sqrt{(1-r^2)^2 + (2\xi r)^2}} \tag{14}$$

Where $y^{\cdot}$ and $y^{\cdot\cdot}$ is the model speed and acceleration respectively, ω_D undamped natural frequency, ω_n natural frequency, ξ damping ratio, A and B are constants that is evaluated through the initial condition.

To solve the SDOF model, several programs are developed. The programs are:

- Force-Plot and Motion

This program is used to calculate the force time series that represents the parallel passing ship phenomena. The program executes a polynomial fitting to the estimated value presented in section 3.5.

- ODE

This program utilize a Runge-Kutta method and it is used to solve an ordinary differential equation.

The dynamic analysis program that are used to obtain the structure response is presented in Fig. 10.

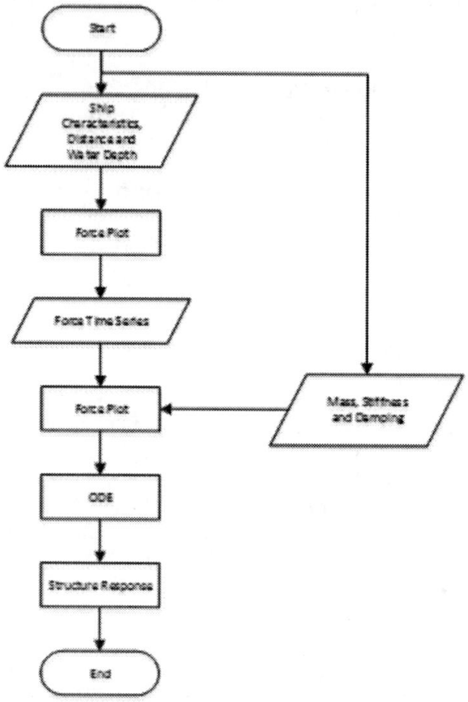

Fig. 10. Modelling flowchart.

To solve the single degree of freedom model, it is assumed that the structure is in a static condition prior to the loading application.

4.1 Program Validation

To assess the program accuracy, the program is validated by comparing the structure displacement with two cases of harmonic loads and half sine loads. The result accuracy shall be measured by the deviation, which is derived from following equation,

$$\delta = \sqrt{\frac{\sum(\widehat{x_i} - x_i)^2}{N-1}}$$

4.1.1 Validation with Harmonic Loads

To validate the model a model with 44.36 tonne mass, 1,751.27 kN/m stiffness, 1 s natural period and 5% damping ratio is imposed by a harmonic load of 10 sin ($\pi t/0.6$) kN. The structure is assumed to be in static condition.

The displacement comparison for both analytic and the program is presented in Fig. 11. The deviation between the two solution is obtained 0.00105%.

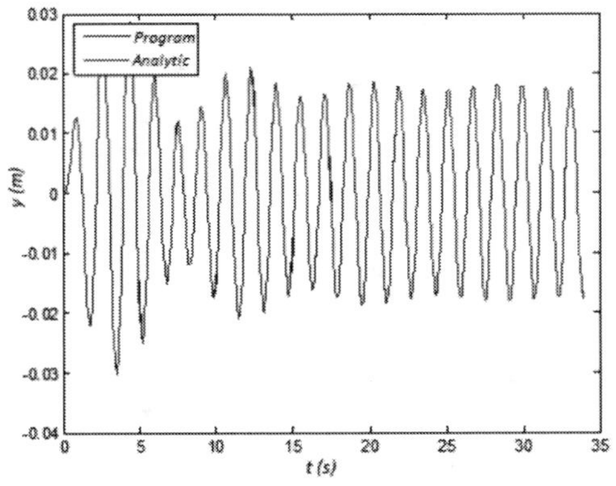

Fig. 11. Model result comparison with harmonic loads.

4.1.2Validation with Half Sine Loads

The displacement comparison for both analytic and MATLAB program is presented in Fig. 12. The deviation between the two solution is obtained 8.12%.

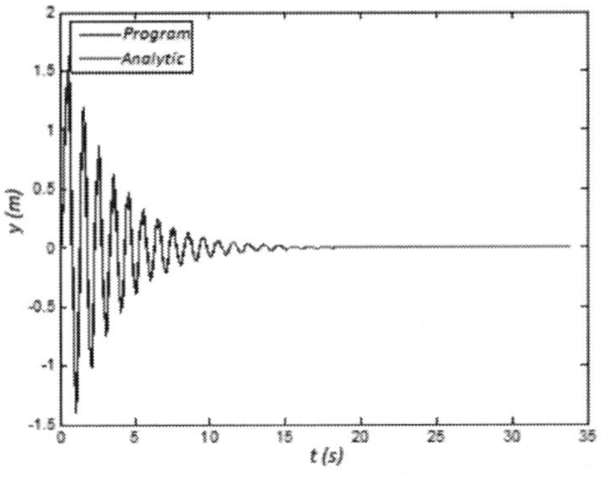

Fig. 12. Model result comparison with half sine loads.

As presented above, the program has a decent accuracy to solve the SDOF dynamic model.

5 Result and Discussion

The berthing dolphin displacement due to 200,000 DWT passing ship is presented in Figure 13. As presented in Figure 13, the structure response and loads trends are almost the same, it has one crest and two trough. The maximum deflection for all cases occurred once, the maximum deflection occurs when both ships center lines are inline. Due to different ship speed, the maximum deflection occurs at different time steps. The dolphin maximum displacement for each case is (a) 13.89 mm, (b) 39.32 mm and (c) 104.35 mm.

On the other hand, the berthing dolphin displacement due to 40,000 DWT passing ship is presented in Figure 14. The displacement trend for this case have similar trends with the 200,000 DWT passing ships, it has one crest and two trough. The difference for this case is its maximum displacement, where the dolphin maximum displacement for each case is (a) 2.94 mm, (b) 8.33 mm and (c) 22.11 mm.

The last case in this study is the berthing dolphin displacement due to 1,000 DWT passing ship. The time series of the displacement is presented in Figure 15. There are two displacement trends observed on this case. The displacement trends for the case with passing speed smaller than 10 m/s are similar with the 200,000 and 40,000 DWT passing ship. Meanwhile, the displacement with 20 m/s passing speed possess a different trend. Unlike other cases, the displacement trend do not resemble the exerting force. It possesses a maximum displacement that occurs quickly. Then unlike other cases, it do not reduce significantly but reduce slightly until it stabilize at static condition. The maximum displacement for each case is (a) 0.09 mm, (b) 0.24 mm, (c) 1.04 mm and (d) 6.84 mm.

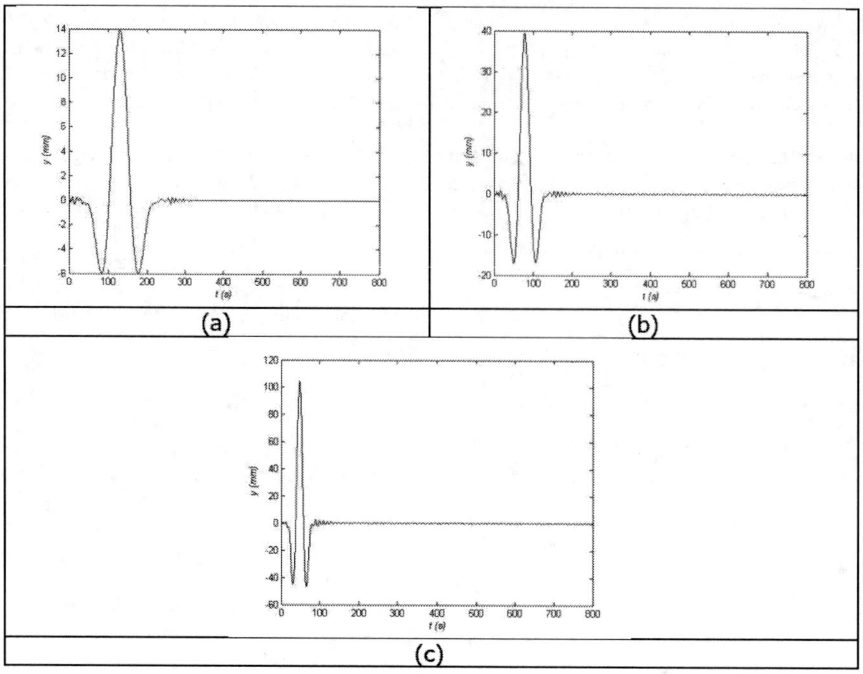

Fig. 13. Time series of dolphin deflection (y) due to 200,000 DWT passing ship with velocity of (a) 3 m/s, (b) 5 m/s and (c) 8 m/s.

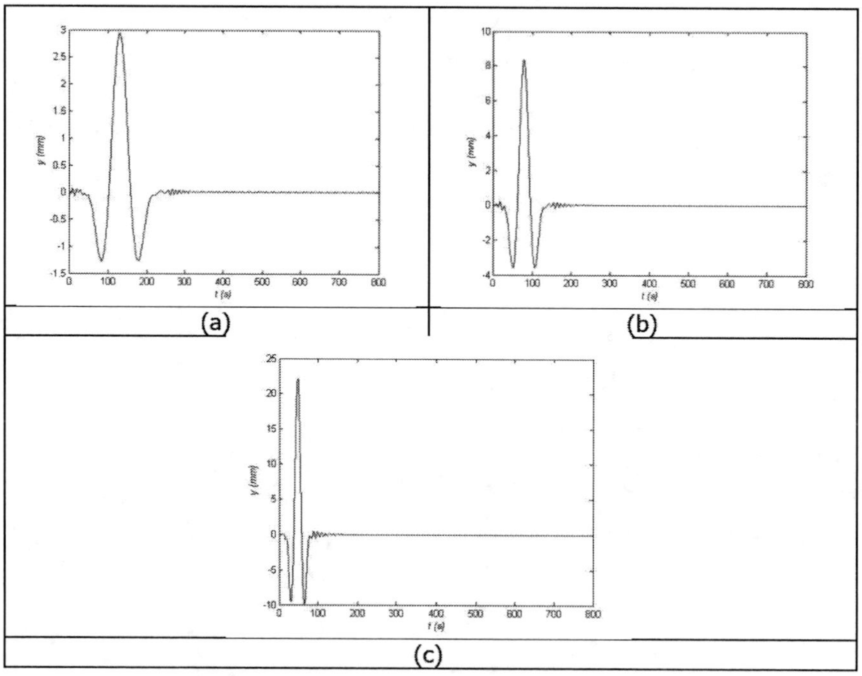

Fig. 14. Time series of dolphin deflection (y) due to 40,000 DWT passing ship with velocity of (a) 3 m/s, (b) 5 m/s and (c) 8 m/s.

* Corresponding author: heri.setiawa@ocean.itb.ac.id

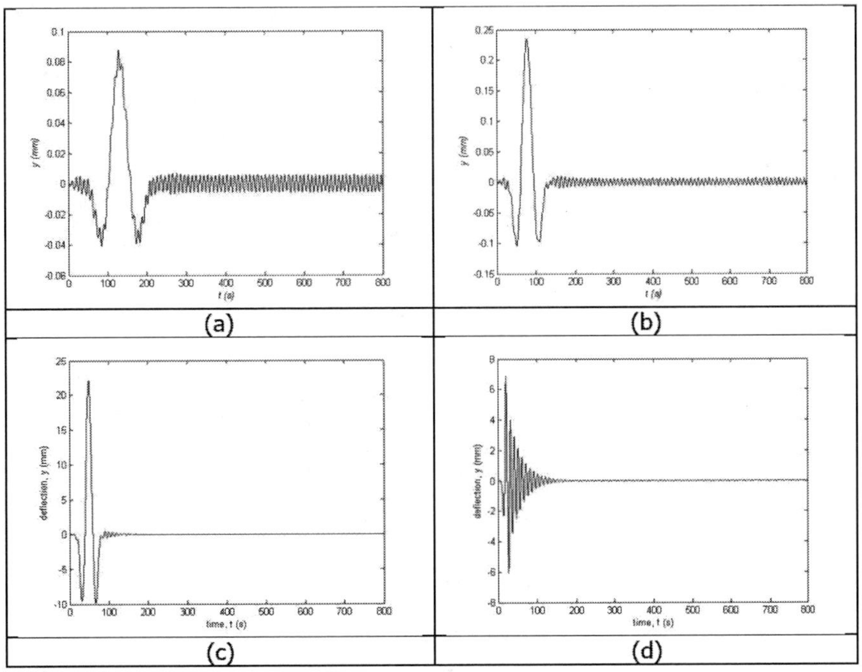

Fig. 15. Time series of dolphin deflection (y) due to 1,000 DWT passing ship with velocity of (a) 3 m/s, (b) 5 m/s, (c) 10 m/s and 20 m/s.

Summary of the maximum displacement and its comparison to the maximum force is presented in Table 5. Meanwhile the comparison between maximum displacements to its static deflection is presented in Table 6.

As presented in Table 5 and Table 6 the ratio of maximum displacement over maximum sway force and ratio of maximum displacement over static displacement are the same for all cases except for the case where the passing ship speed is 20 m/s. The ratio of this cases is almost doubled compared to other cases. It can be concluded that there are amplification to the displacement when the passing ship speed is 20 m/s. This phenomenon happens because the structure natural periods are closer to the force period compared to other cases.

6 Conclusions

Based on the results, it is observed that not only big vessel can cause high displacement to the structure but also smaller vessel with high speed. Therefore, careful consideration also shall be given to small ship with high passing speed, since due to the amplification, the displacement may equal with displacement caused by bigger passing vessel with slower speed.

Further study shall be made for other moored vessel size, passing vessel size and speed to confirm this amplification phenomenon. With enough datas a speed limit can be establish to make sure that the amplification phenomenon did not occur. The data shall be used to develop a non-dimensional relations between moored ship, passing ship characteristics and the maximum displacement due to passing ship effect.

https://doi.org/10.1051/matecconf/201814705003

Table 5 Maximum Displacement Summary

Passing Ship Displacement (DWT)	Passing Speed (m/s)	Maximum Sway Force (F_Y) (kN)	Maximum Displacement (y_{max}) (mm)	y_{max} / F_Y (kN/mm)
236,000	3	143.33	13.88	0.10
	5	398.14	39.32	0.10
	8	1,019.24	104.34	0.10
50,000	3	30.37	2.94	0.10
	5	84.35	8.33	0.10
	8	215.94	22.11	0.10
1,400	3	0.85	0.09	0.11
	5	2.36	0.24	0.10
	10	9.45	1.04	0.11
	20	37.79	6.84	0.18

Table 6 Dolphin Displacement Comparison

Passing Ship Displacement (DWT)	Passing Speed (m/s)	Maximum Sway Force (F_Y) (kN)	Maximum Displacement (y_{max}) (mm)	y_{max} / F_Y (kN/mm)
236,000	3	13.60	13.88	1.02
	5	37.78	39.32	1.04
	8	96.72	104.34	1.08
50,000	3	2.88	2.94	1.02
	5	8.00	8.33	1.04
	8	20.49	22.11	1.08
1,400	3	0.08	0.09	1.12
	5	0.22	0.24	1.07
	10	0.90	1.04	1.16
	20	3.59	6.84	1.91

* Corresponding author: heri.setiawa@ocean.itb.ac.id

References

1. Muga, B. and Fang, S., "Passing Ship Effects – From Theory and Experiment," Paper OTC 2368, Offshore Technology Conference, Houston, pp. 319-338 (1975).
2. Seelig, W., "Passing Ship Effects on Moored Ships," Technical Memorandum TM- 6027-OCN, Naval Facilities Engineering Service Center, (2001).
3. Kriebel, D., "Mooring Loads Due to Parallel Passing Ships" Technical Report TR- 6056-OCN, Naval Facilities Engineering Service Center, (2005).
4. Marine Oil Terminal Engineering and Maintenance Standards (MOTEM), California State Land Commission, (2008).
5. Cohen, S. and Beck, R., "Experimental and Theoretical Hydrodynamic Forces on a Mathematical Model in Confined Waters," Journal of Ship Research, Vol. 27, No. 2, pp. 75-89, (1983).
6. Flory, J.F., "The Effect of Passing Ships on Moored Ships," Prevention First 2002 Symposium, California State Lands Commission, Long Beach, CA, 10-22, (2002).
7. Pinkster, J., "The Influence of Passing Ships on Ships Moored in Restricted waters," Paper OTC 16719, Offshore Technology Conference, Houston, (2004).
8. Remery, G., "Mooring Forces Induced by Passing Ships," Paper OTC 2006, Offshore Technology Conferences, Houston, pp. 349-363, (1974).

* Corresponding author: heri.setiawa@ocean.itb.ac.id

Efficiency Comparison between Conventional and Modern Port Operation System for Small-Scale Dry Bulk Cargo

Aulia Tiara[1,*] and *Julfikhsan Ahmad* Mukhti[2]

[1,2]Institut Teknologi Bandung, Jalan Ganesha No. 10 Bandung, Indonesia

Abstract. Since the launching of Sea Toll Road Program in 2015, the improvement in ports' operation systems has become Indonesia's foremost necessity. This improvement commonly leads to equipment modernization, while realistically, modern equipment does not always amount to a productive performance, especially in the context of small-scale ports. Instead, it is prone to creating wasteful capital and maintenance cost as well as making the planning time ineffective. This study compares two options of port operation systems in a small port, which is conventional and technologically-advanced method for dry bulk cargo. It results in thin gaps between each option's financial assessment variables, which are Internal Rate of Return, Benefit/Cost Ratio and Payback Period, regardless of a stark difference between each option's equipment cost. This study concludes that with the right approach, the conventional operation system is still the most efficient option compared to its contemporary opposite.

1 Introduction

In 2015, Indonesia started a program to set the route to connect existing main ports throughout the nation through a fixed shipping schedules. The program, known as "Tol Laut" (Sea Toll Route) aims to improve the efficiency of logistics and distribution by directing smaller ports, which consists of collector and feeder ports, to ship the cargo from or to the main ports connected in the Sea Toll Route.

The Sea Toll Route is predicted to increase the overall logistics distribution throughout the nation as it may stimulate the current economic development in Eastern Indonesia which is relatively low compared to the market in the Middle and Western Indonesia [1]. Part of the Sea Toll Route plan is to improve the infrastructure and technology on the existing national ports. As an example, PT Pelabuhan Indonesia III launched Teluk Lamong Terminal in 2016, which is a container and dry bulk terminal with modern unloading equipment such as Grab Ship Unloader and Ship-To-Ship Crane.

The technological improvement that is undergoing on larger scale ports owned by national port operators (PT Pelabuhan Indonesia I to IV), however, should not be adapted for smaller-scale ports without careful planning. The improvement method is critically suitable for some ports to achieve a beneficial return, especially for smaller ports in Indonesia in which facility, human resource, and logistics throughput disparity between port scales are relatively large.

One example of reverted approach is in the Sunda Kelapa Port, which is a collector port owned by PT Pelabuhan Indonesia II. A multi-purpose terminal with harbor mobile cranes was planned to be built in the Post VI of Sunda Kelapa in 2016. However, the plan eventually canceled due to the low projection of incoming sand bulk cargo, which was set as the main handled goods in the planned terminal, and a new plan to make the Sunda Kelapa Port as a heritage port in the future with completely different infrastructures. Should the planning approach started from the most basic improvement in form of deck-on-pile terminal with excavator and truck for unloading operation, the terminal may be built before the new plan is settled and generate valuable income for the upcoming port improvements.

As to Sunda Kelapa Post VI Jetty, this approach will have the same possibility to be applied in other smaller ports. However, the approach is not only risky for small infrastructures with barriers in operational use but also burdening regarding budget and planning time due to added maintenance and preparation scheme planning. As a nation with a high number of population, a simpler method that utilizes Indonesia's human resource should be prioritized over advanced, complicated technology.

In this study, two approaches are compared for port operation improvement based on these philosophies: human resource utilization and advanced technology. Sunda Kelapa Post VI Terminal development is chosen as the sample case. The first approach uses Excavator for cargo unloading, representing the human resource utilization approach and will be referred as "Conventional System" in this study. The second approach uses Vertical Screw Conveyor for cargo unloading, representing the advanced technology approach and will be referred as "Modern System" in this study. The terminal berth structure for both approaches, however, is designed to serve Harbour Mobile Crane, which is heavier than

* Corresponding author: auliauliatiara@gmail.com

Excavator and Vertical Screw Conveyor mentioned before. Finally, a financial analysis is conducted to compare the efficiency and effectiveness for each approach which uses parameters such as Benefit/Cost Ratio (BCR), Internal Rate of Return (IRR) and Payback Period (PP).

2 Operational System for Dry Bulks

The study is based on a survey, investigation, and design (SID) project of Post VI Jetty in Indonesia Port Company (IPC) branch in Sunda Kelapa, Jakarta [1]. Jakarta. It has an elevation of +3.5 m from the low water surface (LWS), a length of 100 m and is expected to receive a single barge of 5,000 ton of sand bulk cargo.

Due to the change of level of the barge towards the jetty in the unloading process, the operational scheme in all the options is divided into two separated phases. The elevation change is shown in **Fig 1**. The first phase, named Phase One, is the unloading of the barge's cargo from full to 1/3 part unloaded. After the 1/3 part of the cargo is unloaded, the barge reaches the same level as the jetty. The second phase, named Phase Two, is the unloading of the barge's cargo from 1/3 part unloaded to fully unloaded. In a fully-unloaded cargo condition, the barge is on a higher level than the jetty, which requires all unloading methods presented in this study to adjust mid-operation.

2.1 Conventional System (Excavator – Dump Truck

Taking an operational reference on PT Pelabuhan Indonesia II's Muarajati Coal Wharf in Cirebon, this study simulates an operational scheme for sand bulk using two units of Excavators as unloading equipment and Dump Truck as transfer equipment as shown in **Fig 2**. This operating system is widely used in unloading dry bulks in Indonesian ports, thus making it a 'conventional' type of system in this study.

In Phase One condition, 2 Excavators unload the barge's cargo into 2 Dump Trucks per activity. However, Excavator has a limited arm reach, so to fully unload the barge into the end of Phase Two condition, the Excavators and Dump Trucks need to get inside the barge. Therefore, in between Phase One and Phase Two condition, a preparation for loading the equipment to the inside of the barge is conducted. This preparation includes the opening of the cargo hold, the moving and locking of ramp door, maneuvering of Excavators and preparation time to make ways for Dump Trucks to move into the barge. This whole preparation process is named the Intermediate Phase.

After the Intermediate Phase is done, Phase Two condition can be conducted. In this phase, each of the Dump Trucks enters the barge via one-way ramp door. In the barge, those 2 Excavators perform the unloading process in parallel until all the sand bulk cargo is unloaded into Dump Trucks.

For this condition, the study refers to the survey conducted by PT LAPI ITB on Excavator's duration to fill a 25 Ton Dump Truck [2].

2.2 Modern System (Vertical Screw Conveyor – Conveyor Belt)

Taking an operational reference on Talin Power Plant in Taipei, Taiwan, this option studies operational scheme of a Vertical Screw Conveyor ship unloader which moves the sand bulk cargo into a Conveyor Belt system as shown in **Fig 3**. The Screw Conveyor Belt then transfers the sand bulk cargo to a dedicated cargo bulk terminal.

In Phase One condition, the Screw Conveyor unloads the 1/3 part of the barge's cargo to a Conveyor Belt system. The Conveyor Belt system of the equipment then transfers the cargo to the terminal.

After the 1/3 part of the cargo is loaded, the Vertical Screw Conveyor needs another equipment to push the cargo into the Screw Conveyor's suction direction. For that reason, a single Bulldozer needs to be loaded into the barge. The condition marks the beginning part of the Intermediate Phase condition. During this condition, a preparation for Bulldozer loading into the barge is conducted, which includes the opening of the cargo hold, the moving and locking of ramp door, and lastly, the maneuvering of Bulldozer into the inside of the barge. After Intermediate Phase is finished, the process will be proceeded with the Phase Two condition, in which the unloading of cargo is continued with the help of Bulldozer.

3 Productivity Measurement

PT Pelabuhan Indonesia II [3] provides an approach for berth time measurement as shown in **Fig 4**, which consists of Berth Working Time and Not Operating Time. Berth Working Time consists of actual operational time, stated as Effective Time and Idle Time. Using the approach as reference, this study divides each Effective Time of the two systems into three comparative phases, which are Phase One, Intermediate Phase and Phase Two.

Based on **Fig 4,** Effective Time (ET) follows Equation (1) below.

$$\boldsymbol{ET = Phase\ One + Intermediate\ Step + Phase\ Two} \quad (1)$$

In this study, Idle Time (IT) is assumed to follow Equation (2), while Berth Working Time (BWT) automatically follows Equation (3).

$$IT = 20\%\ X\ (ET) \quad (2)$$

$$BWT = ET + I \quad (3)$$

For both systems, Not Operating Time (NOT) is assumed to occur once for every work shift. One work shift is 8 hours long, and Not Operating Time is assumed to be one-hour long. That way, this variable follows Equation (4).

* Corresponding author: auliauliatiara@gmail.com

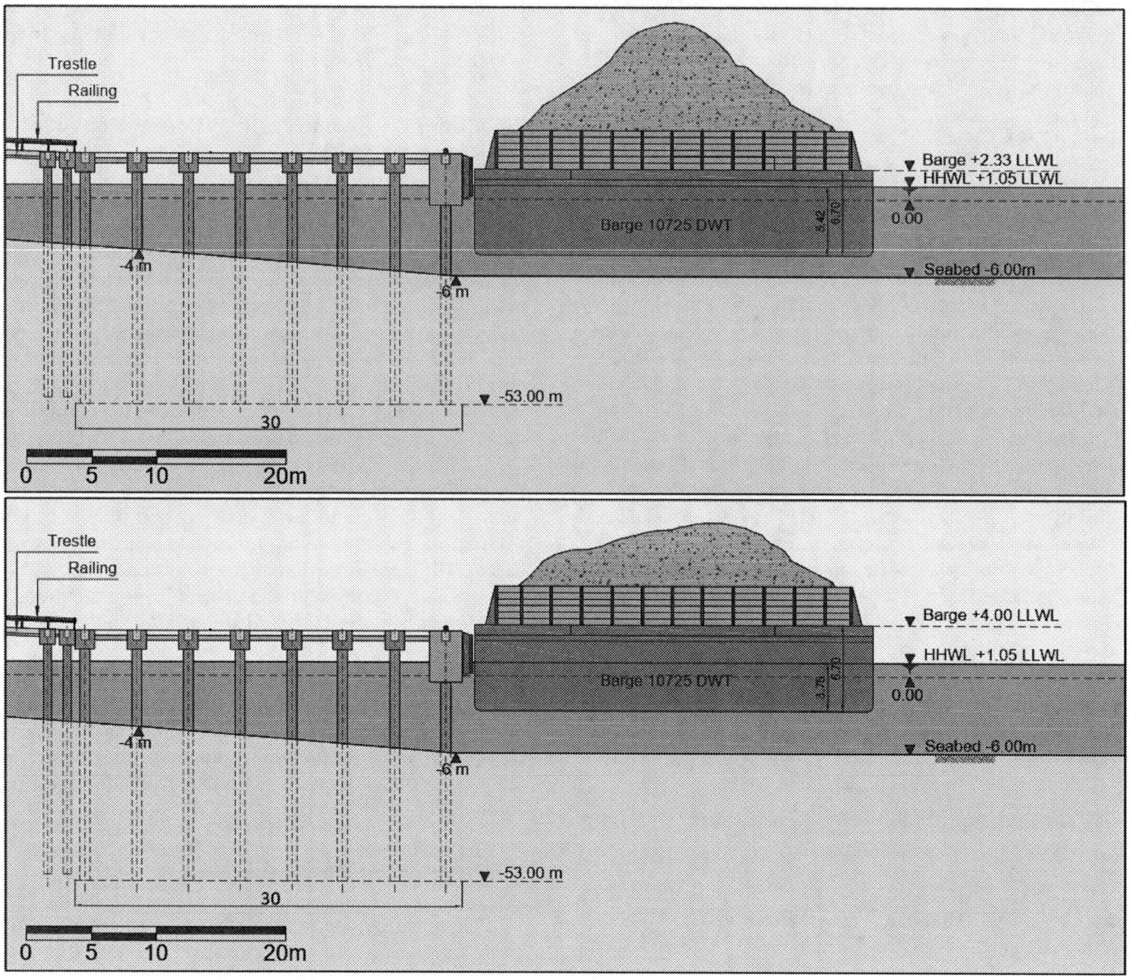

Fig 1 Illustration of barge and jetty level during (a) Phase One and (b) Phase Two.
Source: PT LAPI ITB [2]

$$NOT = \frac{BWT}{7} \quad (4)$$

Finally, Berth Time (BT) is measured using Equation 5 below.

$$BT = BWT + NOT \quad (5)$$

After berth time is measured, measurement of annual cargo throughput is conducted to estimate the value of each system's annual productivity further. The annual throughput of each system follows Equation (6) below.

$$Throughput = \frac{YOT}{BT} \; x \; C \; x \; BOR \quad (6)$$

While,

YOT = Yearly Operational Time, or the amount of port's operational hours in a year. In this study, these variable equals to 24 hours x 365 days or 8760 hours.

C = The amount of cargo unloaded from a barge, which equals to 5000 tons.

BOR = Abbreviation for Berth Occupancy Ratio. For this variable, the value of 80% is chosen based on Sunda Kelapa Port's BOR value (70%) [4]. This study chooses a higher BOR to represent the operational readiness of one jetty, which should be greater than of one port that consists of more than one jetty.

3.1 Conventional System (Excavator – Dump Truck

Berth Time measurement for this system follows the approach given in **Fig 4**. Analysis based on the system's operational condition regarded as Effective Time is described thoroughly in the subchapters below.

* Corresponding author: auliauliatiara@gmail.com

Fig 2 Excavator and Dump Truck coal unloading in Muarajati Wharf, Port of Cirebon. Source: PT LAPI ITB [2]

Fig 3 Vertical Screw Conveyor in Talin Power Plant, Taiwan. Source: Company Profile Video Siwertell Ship Unloader

3.1.1 Phase One

Phase One of this system consists of the unloading process of bulk cargo from the barge using Excavator into a Dump Truck, and the duration of Dump Truck maneuver to and from the unloading location as shown in **Fig 5.** The time length of this condition is calculated using the formula Equation (7) below.

$$Phase\ One_C = \left(\frac{5000\ tons}{P_{Excavator}}\ x \frac{1}{3} x\ OF\right) + (M1_{DT}\ x\ N\ x\ OF) \quad (7)$$

Where,

$Phase\ One_C$ = The duration of Phase One condition for conventional operational system

$P_{Excavator}$ = Productivity of Excavator (ton/hour)

$M1_{DT}$ = The duration of Dump Truck's single maneuver

N = The amount of Dump Truck's maneuvers, which are: (a) Dump Truck movement from the dumping place to the unloading place, and; (b) Dump Truck movement from unloading place to dumping place

OF = Operational factor of the equipment (taken as 1.2)

3.1.2 Intermediate Phase

Intermediate Phase consists of activities in preparation for Excavator and Dump Truck's movement into the barge. First, the duration of cargo hold unlocking to open a channel for equipment to move inside the barge (presented as Ul_{CH}) is measured. The duration of Ramp Door movement to the channel location by Mobile Crane (presented as M_{Ramp}) is calculated, continued by the measurement of the duration of Ramp Door preparation (P_{Ramp}) and Ramp Door locking (L_{Ramp}). Finally, the duration of Excavator unloading into the barge and equipment's setting ($D_{Excavator}$) is calculated. The whole calculation is briefly summarized in Equation (8) below.

$$\boldsymbol{Intermediate\ Step_C = Ul_{CH} + M_{Ramp} + P_{Ramp} +}$$

* Corresponding author: auliauliatiara@gmail.com

$$L_{Ramp} + D_{Excavator} \quad (8)$$

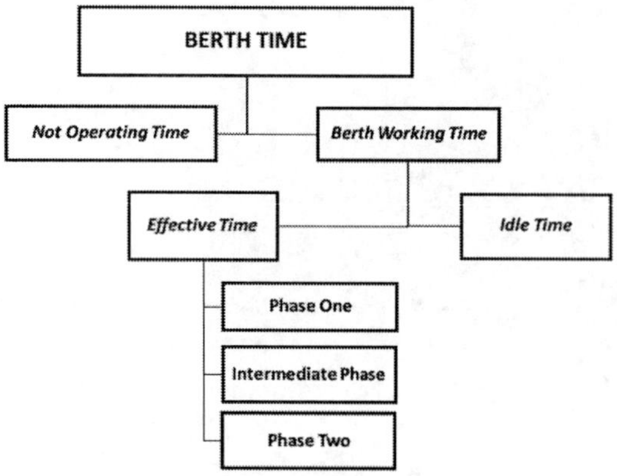

Fig 4 Berth time measurement hierarchy.
Source: Sutanto, *et al.* [3]

Due to the specification of a conventional Ramp Door, it can only be passed by one vehicle at a time. For that reason, the value $D_{Excavator}$ is the maneuver duration of 2 Excavators, plus the duration of equipment setting.

$$D_{Excavator} = S_{Excavator} + (M_{Excavator}\ x\ OF\ x\ N) \quad (9)$$

Where:

$D_{Excavator}$ = Duration of Excavator maneuvering and setting

$S_{Excavator}$ = Duration of Excavator setting

$M_{Excavator}$ = Duration of Excavator maneuver into the barge

N = Number of Excavators

3.1.3 Phase Two

Truck which located inside the cargo barge as shown in **Fig 6**. The duration of this condition follows Equation (10).

$$Phase\ Two_C = (U2_{Excavator} x\ OF) + (M1_{DT}\ x\ N\ x\ OF) \quad (10)$$

Where:

$Phase\ Two_C$ = The duration of Phase Two condition for conventional operational system

$U2_{Excavator}$ = The unloading duration of 2/3 part of bulk cargo by Excavator inside the cargo barge

$M1_{DT}$ = The duration of Dump Truck's single maneuver into the cargo barge.

N = The amount of Dump Truck's maneuvers, which are: (a) Dump Truck movement from the dumping place to the unloading place inside the barge, and; (b) Dump Truck movement from unloading place inside the barge to dumping area.

OF = Operational factor of the equipment, regarded with a value of 1.2.

A conventional Ramp Door can only be passed by one vehicle at a time. Thus, the process of unloading by 2 Excavators will not be in exact parallel. Instead, two unloading processes will work in parallel with the interval duration of Dump Truck maneuver ($M1_{Dump\ Truck}$) to the inside of the barge. Then, the duration of Phase Two's unloading process ($U2_{Excavator}$) will be a summation of the first Dump Truck's maneuver into the barge, Excavator's unloading process ($U1_{Excavator}$), the maneuver of the second Dump Truck into the barge, and the maneuver of the first Dump Truck out of the barge. This process is described in Equation (11) below.

$$U2_{Excavator} = \left(\frac{5000\ tons}{P_{Excavator}}\ x \frac{2}{3}\right) + (3\ x\ M1_{DT}) \quad (11)$$

After the variables in the subchapters are measured, ET can be concluded by following Equation (1). Then, following Equation (2), Equation (3), Equation (4) to Equation (5) respectively, *IT*, *BWT*, *NOT* and finally, *BT* can be measured. Then, annual throughput measurement can be done using Equation (6).

3.2 Modern System (Vertical Screw Conveyor – Conveyor Belt)

Similar to the Conventional System, the Berth Time measurement for Modern Operational System also follows the approach given in **Fig 4**. Analysis based on the system's operational condition regarded as Effective Time is described thoroughly in the subchapters below.

3.2.1 Phase One

Phase One of this system consists of the unloading process of bulk cargo from the barge using Vertical Screw Conveyor as shown in **Fig 7**. The duration is calculated using the formula Equation (12) below.

$$S1_M = (U1_{VSC} x\ OF) \quad (12)$$

$$U1_{VSC} = \left(\frac{5000\ tons}{P_{VSC}} x \frac{1}{3}\right) \quad (13)$$

Where:

$S1_M$ = The duration of Phase One condition for Modern Operational System

$U1_{VSC}$ = The unloading duration of 1/3 part of bulk cargo by Vertical Screw Conveyor

* Corresponding author: auliauliatiara@gmail.com

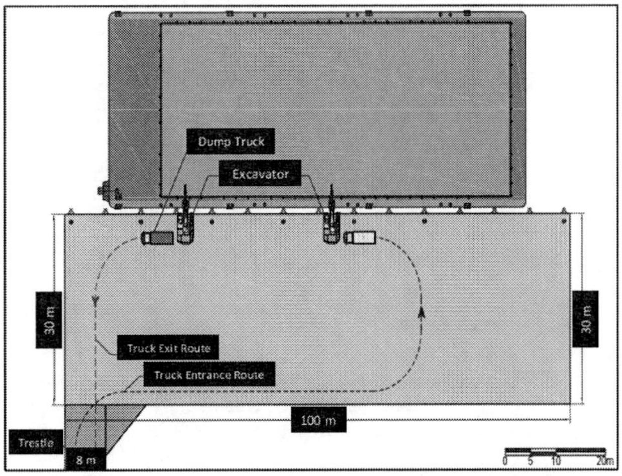

Fig 5 Conventional Operational System during Phase One.

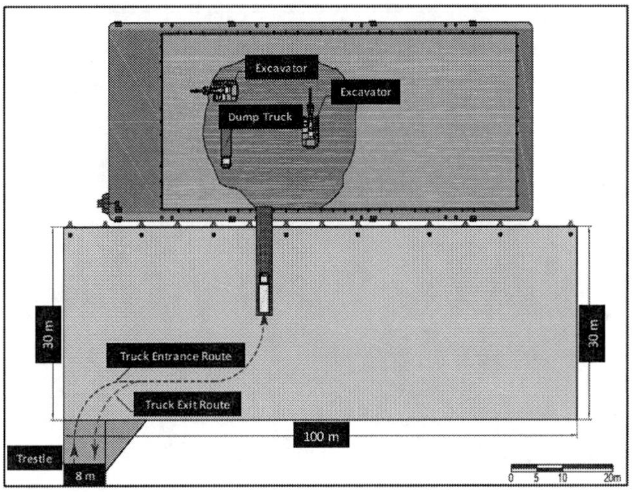

Fig 6 Conventional Operational System during Phase Two.

P_{VSC} = Productivity of Vertical Screw Conveyor (ton/hour)

3.2.2 Intermediate Phase

Intermediate Phase for this system consists of activities in preparation for Bulldozer movement into the barge to conduct Phase Two as shown in **Fig 8.** The events that occur in this condition follows Equation (14) below:

$$Int\ Phase_M = Ul_{CH} + M_{Ramp} + P_{Ramp} + L_{Ramp} + M_{Bulldozer} \quad (14)$$

Where:

$M_{Bulldozer}$ = The duration of Bulldozer's maneuver into the cargo barge

3.2.3 Phase Two

Phase One condition of this system consists of the unloading process of bulk cargo from the barge using Vertical Screw Conveyor. The duration of this condition is calculated using Equation (15) below.

$$\boldsymbol{Step\ Two_M = (U2_{VSC} x\ OF)} \quad (15)$$

$$\boldsymbol{U2_{VSC} = (\frac{5000\ tons}{P_{VSC}} x \frac{2}{3})} \quad (16)$$

Where:

$Phase\ Two_M$ = The duration of Phase Two condition for Modern Operational System

$U2_{VSC}$ = The unloading duration of 2/3 part of bulk cargo by Vertical Screw Conveyor

* Corresponding author: auliauliatiara@gmail.com

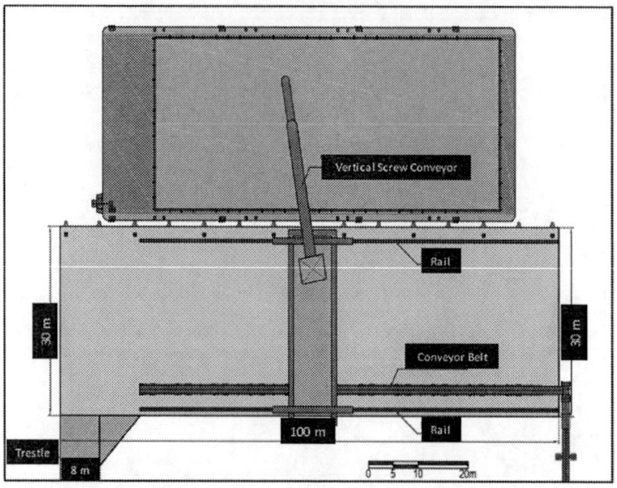

Fig 7 Modern Operational System during Phase One.

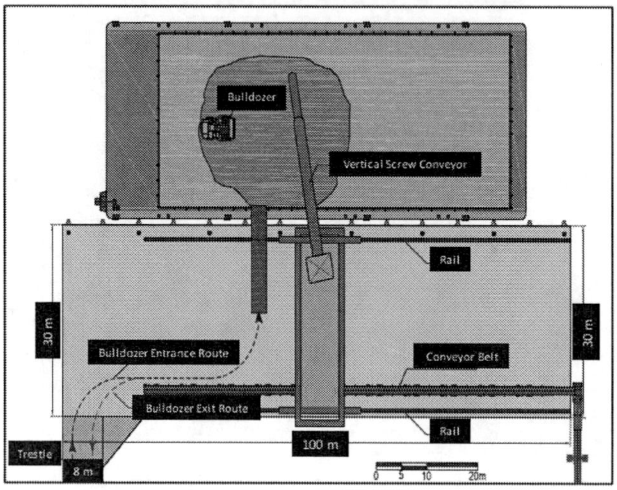

Fig 8 Modern Operational System during Phase Two.

P_{VSC} = Productivity of Vertical Screw Conveyor (ton/hour)

After the variables in the subchapters are measured, ET can be concluded by following Equation (1). Then, by following Equation (2), Equation (3), Equation (4) to Equation (5) respectively, IT, BWT, NOT and finally BT can be measured. Then, annual throughput measurement can be done using Equation (6).

3.3 Summary and Discussion

The summary of productivity measurement is shown in **Table 1** for the Conventional System and **Table 2** for the Modern System. The comparison between the systems is shown in **Fig 9**. The result shows that regarding unloading productivity, the Modern System is almost three times faster than the Conventional System. The most time-consuming phase of both system is Phase Two, due to the large amount of cargo unloaded. Despite the Conventional System's long berthing time (26 hours for 5,000-ton dry bulk unloading), two days of berthing time is still considered as acceptable based on the observation in Port of Cirebon conducted by PT LAPI ITB [2].

4 Financial Simulation and Analysis

Financial simulation and analysis of each operating system is crucial to measure the feasibility of the investment for port operators, and as an approach to choose the best operating system with higher profits and less capital cost. In simulating the cash flow of the operating system and infrastructure, this study follows a set of scenario, income, and expenditure components. After the financial simulation is conducted, financial analysis is done using three kinds of feasibility measurement, which are BCR, IRR, and PP. These

* Corresponding author: auliauliatiara@gmail.com

financial assessment variables will show the best operational scheme for investment.

Table 1 Productivity Measurement of Conventional Operational System

Variables	Value	Source
$P_{Excavator}$	357 ton/hour	Survey on Cargo Unloading Mechanism in IPC Cirebon, December 2016 [1]
OF	1.2	Assumption
$M1_{DT}$	1 minute	Assumption
N	2 units	Survey on Cargo Unloading Mechanism in IPC Cirebon, December 2016 [1]
$Phase\ One_C$	4.7 hours	
Ul_{CH}	20 minutes	Sasono, Herman Budi (2012) [5]
M_{Ramp}	1 minute	Assumption
P_{Ramp}	5 minutes	Assumption
L_{Ramp}	5 minutes	Assumption
$S_{Excavator}$	10 minutes	Assumption
$M_{Excavator}$	1 minute	Assumption
N	2	Survey on Cargo Unloading Mechanism in IPC Cirebon, December 2016 [1]
$Intermediate\ P$	0.72 hour	
$Phase\ Two_C$	16.7 hours	

Table 2 Productivity Measurement of Modern Operational System

Variables	Value	Source
P_{VSC}	1000 ton/hour	Siwertell Ship Unloader's productivity [6]
OF	1.2	Assumption
$U1_{VSC}$	1.67 hours	
$Phase\ One_M$	2 hours	
Ul_{CH}	20 minutes	Sasono (2012) [5]
M_{Ramp}	1 minute	Assumption
P_{Ramp}	5 minutes	Assumption
L_{Ramp}	5 minutes	Assumption
$S_{Excavator}$	10 minutes	Assumption
$M_{Bulldozer}$	1.2 minute	Assumption
$Intermediate\ Phase_M$	0.5 hour	
$Phase\ Two_M$	4 hours	

4.1 Financial Simulation Scenarios

List of scenarios used in this financial simulation are listed below:

a. The construction year of Post VI Jetty and its operating system is assumed to be 2017.

b. The first operational year of Post VI Jetty is assumed to be 2018 with the project concession period of 50 years.

c. The cargo throughput of the first operational year follows Equation 17 below.

$$Throughput_{1st} = Throughput_{2015}\, x\, (1+i)^n \quad (17)$$

In which,

$Throughput_{1st}$ = Cargo throughput of Post VI jetty's first operational year.

$Throughput_{2015}$ = Sand bulk cargo throughput of Sunda Kelapa Port in 2015 for one jetty

i = Throughput growth percentage per year (assumed to have a value of 10%)

n = The year of the first operational throughput – the year of the data

= 2018 – 2015

d. After the first operational year, the throughput percentages of the following years grow each year with the value of 10%, as stated in Equation (17) until finally being constant in throughput percentage of 100%.

e. 30% part of the capital investment is funded with bank loan with the installment plan of 50 years and 10% bank interest.

f. 70% part of the capital investment is funded with equity.

g. Increase in port service rates is assumed to occur every three years with the amount of 10%.

h. Discount Rate or the rate value for accounting the future value factor in the project's cash flow is assumed to be 10%.

i. The amount of cargo stored in the port's terminal is assumed to be 70% of its annual throughput.

4.2 Income Components

Income components included in this study refers to port service rates for port users. Details on income components are described in **Table 3**.

4.3 Expenditure Components

The components of expenditure in this financial analysis consist of Post VI Jetty construction cost, the capital cost of each operational systems, operational tariffs, insurance, and taxation, which is presented in detail in **Table 4**.

* Corresponding author: auliauliatiara@gmail.com

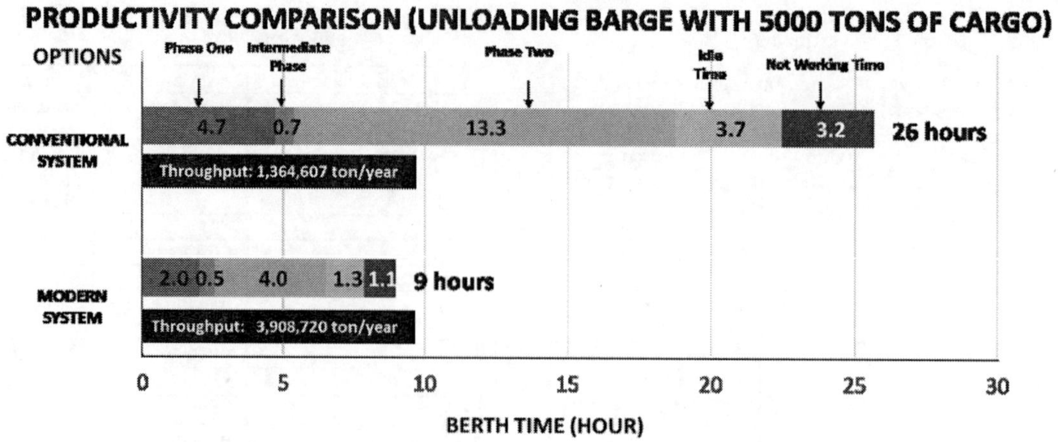

Fig 9 Productivity Comparison between Conventional and Modern Operational System.

4.4 Financial Feasibility Components

4.4.1 Benefit/Cost Ratio (BCR)

This component is a ratio between income components, presented as the value of benefit, and expense components as the value of cost in the cash flow. This ratio follows the formula below:

$$\boldsymbol{B/C} = \frac{\boldsymbol{Present\ Value\ Benefits}}{\boldsymbol{Present\ Value\ Cost}} \tag{18}$$

The decision guideline for this feasibility measurement component, as stated by Blank and Tarquin [10] is given below:

If $B/C \geq 1.0$, accept the project as economically acceptable for the estimates and discount rate applied.

If $B/C < 1.0$, the project is not economically acceptable.

4.4.2 Internal Rate of Return (IRR)

Internal rate of return is a discount rate that makes the net present value (NPV) of the project's cash flow equals zero. IRR follows the formula below:

$$NPV = \sum_{t=1}^{T} \frac{C_t}{(1+r)^t} - C_o = 0 \tag{19}$$

Where:

C_t = Net cash inflow during the period t

C_O = Total initial investment cost

r = The value of IRR

t = The year value of the cash flow

Table 3 Income Components

No.	Income Components	Value	Source
1	Vessel Anchoring Tariff	IDR 100 per GT (Gross Tonnage)	Port of Indonesia's decree [7]
2	Vessel Mooring Tariff	IDR 100 per GT (Ramp Door setting included)	Port of Indonesia's decree [7]
3	Jetty Facility Tariff	IDR 2,250 per cargo tonnage	Feasibility Study of Port of Sunda Kelapa [8]
4	Stevedoring Tariff	IDR 22,000 per m^3 of cargo	Port of Indonesia's decree [7]
5	Bulk Terminal Tariff	IDR 250 per cargo tonnage	Port of Indonesia's decree [7]
6	Water Supply Tariff		Feasibility Study of Port of Sunda Kelapa [8]
	Bulk Water Tariff	IDR 40,000 per ton	
	PH Meter Maintenance	IDR 20,000 per unit vessel	
	Other Maintenance	IDR 7500 per unit vessel	
7	Cargo Pass Tariff	IDR 8,182 per 25 tons of cargo	Feasibility Study of Port of Sunda Kelapa [8]

The project can be decided as economically acceptable if the value of IRR (r) follows this condition:

$$r > \boldsymbol{Discount\ Rate} \tag{20}$$

* Corresponding author: auliauliatiara@gmail.com

4.4.3 Payback Period (PP)

Payback period (PP) method is used to quantify the length of time required to recover the cost of an investment. The payback period of an investment is an important determinant of whether to commence the project, as longer payback periods are not desirable for investment positions. Thus, for two comparable investment options with the same rate of return value and risk, this feasibility measurement is used to select an option with the quickest investment recovery.

Table 4 Expenditure Components

No.	Expenditure Components	Value	Source
1	Post VI Jetty Construction Cost	IDR 72,128,902,102	Post VI Jetty Engineering Report [1]
2	Operational System Capital Cost	Depends on each system	Port of Indonesia's decree [7]
3	Bulk Terminal Operational Cost	70% of Bulk Terminal income	Feasibility Study of Port of Sunda Kelapa [8]
4	Water Supply Operational Cost	30% of Water Supply income	Port of Indonesia's decree [7]
5	Cargo Pass Operational Cost	30% of Cargo Pass income	Port of Indonesia's decree [7]
6	Vessel Handling Cost	30% of Vessel Handling income	
7	Conveyor Belt Maintenance Cost	25% of Conveyor Belt's capital cost	Assumption
8	Indirect Operational Cost	0.1% of the total income	Assumption
9	Other Operational Cost	0.08% of the total income	Assumption
10	Insurance Cost	0.05% of the construction cost	Assumption
11	Tax	10% of the total income	Assumption based on Bank of Indonesia (BI) Rate in 10 years [9]

4.5 Financial Simulation Result

After conducting the simulation of two investment options by calculating all income and expenditure components detailed in subchapter **4.1** to **4.4**, the financial feasibility assessment based on subchapter **4.5** is measured. The result of the process is shown on **Table 5**.

4.6 Discussion

Table 5 shows that the Conventional System has a resembling quantity of financial performance to its opposite. With the capital cost of almost the half the Modern System, Conventional System will result to 13-year payback period, only a year later than the Modern System. That way, by choosing the Conventional System, port operator will save 50% of capital cost to have only one added year of investment recovery.

Table 5 Financial Simulation Result

Options	BCR	IRR	PP	Capital Cost (IDR)
Conventional System	1.69	14.60%	13 years	75,178,382,000
Modern System	2.12	15.78%	12 years	145,178,634,000

5 Conclusion

This simulation concludes that with the exact circumstances, these operational systems, with two times difference in capital cost, result in similar financial feasibility values. That way, conventional operating system will be the most economically-suitable option for Post VI Jetty of Sunda Kelapa Port. This phenomenon happens because, in a small infrastructure where the operational capability is limited, a modern operating system will not work on its maximum capability. With a limited amount of operational and financial productiveness, the cash flow of the project will take just the similar amount of time to the conventional operating system to recover the capital investment. Hence, the use of a traditional operating system will bring more advantages to the project.

References

1. Prihatono, Bambang, *Pengembangan Tol Laut Dalam RPJMN 2015-2019 dan Implementasi 2015,* State Ministry of National Development Planning (BAPPENAS) (2015)
2. PT LAPI ITB, *Laporan Akhir Perencanaan Pembangunan Dermaga/Jetty Curah Kering di Apron Pos VI Muara Baru Cabang Pelabuhan Sunda Kelapa*, PT Pelabuhan Indonesia II Cabang Sunda Kelapa (2016)
3. Sutanto, Harry et al., *Referensi Kepelabuhanan Seri 06 Edisi II: Pengoperasian Pelabuhan*, PT Pelabuhan Indonesia (Persero), p 27-50 (2010)
4. Directorate General of Sea Transportation, *Peraturan Direktur Jenderal Perhubungan Laut Nomor: HK.103/2/18/DJPL-16 tentang Standar Kinerja Pelayanan Operasional Pelabuhan pada Pelabuhan yang Diusahakan Secara Komersial*, Ministry of Transportation (2016)
5. Sasono, Herman Budi, *Manajemen Pelabuhan dan Realisasi Ekspor Impor*, Andi Offset, p 135 (2012)

* Corresponding author: auliauliatiara@gmail.com

6. Siwertell, *High Capacity Ship Unloader,* Cargotec Sweden AB Bulk Handling, https://cargotec.picturepark.com/Website/?Action=downloadAsset&AssetId=33001 (2016, December 16)

7. Board of Directors of PT Pelabuhan Indonesia II Cabang Sunda Kelapa, *Surat Keputusan Direksi PT Pelabuhan Indonesia II (Persero) tentang Tarif Jasa Kapal dan Barang Pelabuhan Sunda Kelapa,* PT Pelabuhan Indonesia II Cabang Sunda Kelapa, (2012)

8. PT LAPI ITB, Feasibility Study Penyiapan Lahan Untuk Penataan Pelabuhan Sunda Kelapa, PT Pelabuhan Indonesia II Cabang Sunda Kelapa (2014)

9. Bank of Indonesia, *Data BI Rate,* Bank of Indonesia, http://www.bi.go.id/en/moneter/bi-rate/data/Default.aspx (2017, February 2)

10. Blank, Leland & Tarquin, Anthony, *Basics of Engineering Economy,* McGraw-Hill, p 166 – 168. (2008)

* Corresponding author: auliauliatiara@gmail.com

MATEC Web of Conferences **147**, 05005 (2018) https://doi.org/10.1051/matecconf/201814705005
SIBE 2017

Application MuTsunami in Mentawai Island Indonesia

Muslim Muin[1], *Izqi Yustina Ammylia* Yusuf[1,*], and *Hendra* Achiari[1]

[1]Coastal Engineering Research Group, Faculty of Civil and Environmental Engineering, Institut Teknologi Bandung, Indonesia

Abstract. Software MuTsunami as research tool in Coastal Research Group Institut Teknologi Bandung, was applied to simulate propagation of tsunami and sediment transport in Mentawai. The Tsunami occurred in 25 October 2010, 21:42:22 (GMT +7). The hydrodynamic model was based on Non-Orthogonal curvilinear coordinate in spherical coordinate. The results of simulation were compared with observational data, which was collected by group of researcher from Institut Teknologi Bandung, Waseda University, and Yokohama National University. The agreement between the model and observational data are very good.

1 Introduction

Three dimensional ocean hydrodynamics model using non-orthogonal curvilinear coordinate technique, also known as boundary fitted technique, has been well developed [1-3], and was world widely applied to simulate ocean circulation, for example in New York Harbor [4], Cone Bay Western Australia [5], Bay of Fundy Canada [6], etc. Funded by the National Research Council Indonesia, Muin [7] further developed a model of Tsunami propagation using two-dimensional version of the non-orthogonal boundary fitted hydrodynamics model [1]. The propagation model is integrated with tsunami generation model in a Geographic Information System (GIS), and developed in windows system, named MuTsunami as in-house models in Coastal Engineering Research Group (CERG), Institut Teknologi Bandung (ITB). MuTsunami had been successfully applied for Aceh Tsunami [8].

The main advantage of the MuTsunami is that it can be run using affordable notebook. Since the model use semi-implicit technique, the time step is no longer restricted by courant number. Therefore the results of simulation can be obtained in few minute. Once the magnitude and position of earthquake is known, the results of simulation can be run and posted on internet in 15 minute.

The 25 October 2010 Mentawai tsunami earthquake (Mw 7.8) ruptured the shallow portion of the Sunda megathrust seaward of the Mentawai Islands, offshore of Sumatra, Indonesia, generating a strong tsunami that took 509 lives [9]. They iteratively adjusted the data weighting, rupture velocity, spacing, and lateral extent of the finite-fault model grid in the joint inversion of the hr-GPS and teleseismic signals to reproduce the tsunami observations through modeling of nonlinear and dispersive ocean wave processes. The results of their simulation matched the observations. However this iterative approach is not practical in Mitigation which quick response is required.

This paper presents the results of simulation of MuTsunami for the case of Mentawai Tsunami on October 25, 2010. The results of model were compared with available observation [10].

2 Methodology

MuTsunami consist of four models:

1. Wave Generation (Near Field)
   a. Simple energy approach
   b. Specified the initial condition
2. Wave Propagation (Far Field)
3. Wave Run-up
4. Sediment Transport Model

The propagation and sediment transport model use the non-orthogonal curvilinear coordinate technique in spherical coordinate. MuTsunami utilize the state of the art of three-dimensional sediment transport model MuSed3D [11] to simulate sediment movement by integrating over depth the basic equation to vertically averaged 2D sediment transport model.

2.1. Tsunami generating model

2.1.1 Simple energy approach

A large tsunami was caused by the movement of the sea floor. The best way to simulate the generation of Tsunami is to go back to the basic equations of conservation of mass and momentum (Navier-Stokes). Grilli and Watts [12] developed numerical model of Tsunami generation by submarine mass failure. This requires a dense grid systems and accurate data of seabed. Big rise in the sea floor is very difficult to predict accurately because it is highly dependent on mechanical or geotechnical characteristics of the seafloor in which data are very minimal. For these reasons and quick response and early warning data, MuTsunami use an empirical approach. The

* Corresponding author: izqiyustina@gmail.com

empirical approach is to consider the potential energy of the water received from the energy generated by the quake would be a Tsunami Energy [13].

2.1.2 Initial condition is specified

Initially, the model was developed with a simple waveform generation models where the tsunami generation is derived by a simple empirical formula based on the energy of earthquake as described above. The latest version was developed to allow user use particular initial wave condition. However, the first approach is useful, for quick response.

2.2 Wave propagation

The wave propagation model (far field) requires the input of wave generation models (near field). The result of the wave generation model is the initial conditions of the wave propagation models. In this computational domain, a mathematical model was derived using vertically averaged long wave's equation in spherical coordinate. Non-orthogonal curvilinear coordinate technique [1-3, 14] was employed by neglecting the density gradient effect. The simulation results of the far field model are the input to Wave Run-up model.

Non-orthogonal Curvilinear Coordinate or also known as Boundary-Fitted technique is excellent model in areas with complex geometry. The advantage of Non-Orthogonal Boundary Fitted Model is that the orthogonal grid system is not required. It is impossible to generate orthogonal grid in complex area. In Boundary-Fitted Techniques, a grid system is no longer square, the grid can be fitted to the shoreline. Use of the model with square grid will produce an error for the region with a complex geometry. Of course by using a smaller simulation grid size would minimize this problem, but this will require huge computer memory and CPU time.

The model use a spherical coordinate system as the basis of physical domain, meaning this model is much more accurate than the models using cartesian coordinate system given that our earth is not flat. Figure 1 illustrates the physical domain in spherical coordinate (ϕ,θ) and computational domain in curvilinear coordinate (ξ,η).

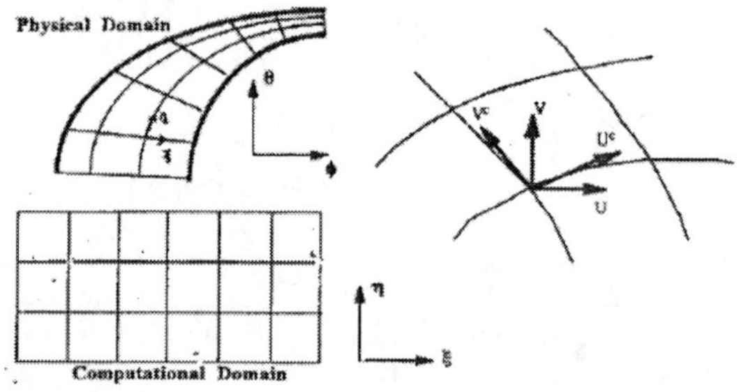

Fig. 1. Physical and Computational Domain of Non-Orthogonal Curvilinear Coordinate System [1]

Following Muin [1], after neglecting the density gradient effect, the governing equations in curvilinear coordinate system (ξ,η), can be written as follows:

Conservation of Mass

$$JR\cos\theta\frac{\partial\zeta}{\partial t}+\frac{\partial}{\partial\zeta}(\cos\theta\, JU^cD)+\frac{\partial}{\partial\eta}(\cos\theta\, JU^cD)=0 \quad (1)$$

Conservation of Momentum

ξ - Direction

$$\begin{aligned}\frac{\partial U^cD}{\partial t}=&-\frac{\theta_\eta\theta_\eta+\cos^2\theta\phi_\eta\phi_\eta}{J^2R\cos^2\theta}Dg\frac{\partial\zeta}{\partial\xi}+\frac{\theta_\xi\theta_\eta+\cos^2\theta\phi_\xi\phi_\eta}{J^2\rho_0R\cos^2\theta}Dg\frac{\partial\zeta}{\partial\eta}\\&-\frac{\theta_\eta}{J^2\cos^2\theta}\left\{\left[\frac{\partial}{\partial\xi}\left(\phi_\xi\cos^2\theta JU^cU^cD+\phi_\eta\cos^2\theta JU^cV^cD\right)\right]\right.\\&\left.+\left[\frac{\partial}{\partial\eta}\left(\phi_\xi\cos^2\theta JU^cV^cD+\phi_\eta\cos^2\theta JV^cV^cD\right)\right]\right\}\\&+\frac{\phi_\eta}{J^2R\cos^2\theta}\left\{\left[\frac{\partial}{\partial x}\left(\theta_\xi\cos^2\theta JU^cU^cD+\theta_\eta\cos^2\theta JU^cV^cD\right)\right]\right.\\&\left.+\left[\frac{\partial}{\partial h}\left(\phi_\xi\cos^2\phi JU^cV^cD+\theta_\eta\cos^2\theta JV^cV^cD\right)\right]\right\}\\&+\frac{fD}{J\cos\theta}\left[\left(\theta_\xi\theta_\eta+\cos^2\theta\phi_\xi\phi_\eta\right)U^c+\left(\theta_\eta\theta_\eta+\cos^2\theta\phi_\eta\phi_\eta\right)V^c\right]\\&+\frac{\tau_{s\xi}-\tau_{b\xi}}{\rho_o}\end{aligned} \quad (2)$$

η- Direction

$$\begin{aligned}\frac{\partial V^cD}{\partial t}=&-\frac{\theta_\xi\theta_\xi+\cos^2\theta\phi_\xi\phi_\xi}{J^2R\cos^2\theta}Dg\frac{\partial\zeta}{\partial\eta}+\frac{\theta_\xi\theta_\xi+\cos^2\theta\phi_\xi\phi_\xi}{J^2R\cos^2\theta}Dg\frac{\partial\zeta}{\partial\eta}\\&+\frac{\theta_\eta}{J^2\cos^2\theta}\left\{\left[\frac{\partial}{\partial\xi}\left(\phi_\xi\cos^2\theta JU^cU^cD+\phi_\eta\cos^2\theta JU^cV^cD\right)\right]\right.\\&\left.+\left[\frac{\partial}{\partial\eta}\left(\phi_\xi\cos^2\theta JU^cV^cD+\phi_\eta\cos^2\theta JV^cV^cD\right)\right]\right\}\\&-\frac{\phi_\xi}{J^2\cos^2\theta}\left\{\left[\frac{\partial}{\partial\xi}\left(\phi_\xi\cos^2\theta JU^cU^cD+\theta_\eta\cos^2\theta JU^cV^cD\right)\right]\right.\\&\left.+\left[\frac{\partial}{\partial\eta}\left(\phi_\xi\cos^2\theta JU^cV^cD+\phi_\eta\cos^2\theta JV^cV^cD\right)\right]\right\}\\&-\frac{fD}{J\cos\theta}\left[\left(\theta_\xi\theta_\xi+\cos^2\theta\phi_\xi\phi_\xi\right)U^c+\left(\theta_\xi\theta_\eta+\cos^2\theta\phi_\xi\phi_\eta\right)V^c\right]\\&+\frac{\tau_{s\eta}-\tau_{b\eta}}{\rho_0}\end{aligned} \quad (3)$$

ξ,η	=	curvilinear coordinates
ϕ,θ	=	longitude, latitude (degree)
t	=	time (s)
f	=	Coriolis parameter
g	=	gravitational acceleration (m/s^2)
ζ	=	water surface deviation from mean level (m)
h	=	water depth (m)
D	=	$h+\zeta$ (m)
D_h	=	horizontal eddy diffusivity (m^2/s)
R	=	earth's radius (m)
τ_s	=	surface shear stress (Pa)
τ_b	=	bottom shear stress (Pa)
J	=	Jacobian = $\phi_\xi\theta_\eta-\phi_\eta\theta_\xi$

The above equation is written using incompressible flow assumption. Vertical acceleration is ignored (hydrostatic approximation).

As a boundary condition, the shoreline is considered as an impermeable layer where the water velocity in the direction perpendicular to the beach is zero. Radiation boundary is used along open boundary. On the surface of the water and the seabed, shear forces is determined by the quadratic shear stress formula.

The transformed vertically averaged equations are solved by semi-implicit technique, in which the water elevation in the wave equation is solved implicitly and the other explicit. Time step is no longer limited by the speed of wave propagation, so that the can be performed in more quickly and economically.

The computational is performed in a staggered grid system. Three-level time differencing is used to discretize the time domain to avoid artificial damping. The model had been tested to simulate standing wave in closed basin. The damping is negligible [7, 8]. A more detailed description of the three-dimensional version of this model is presented in Muslim Muin [1]. The ocean hydrodynamics model has been applied in worldwide.

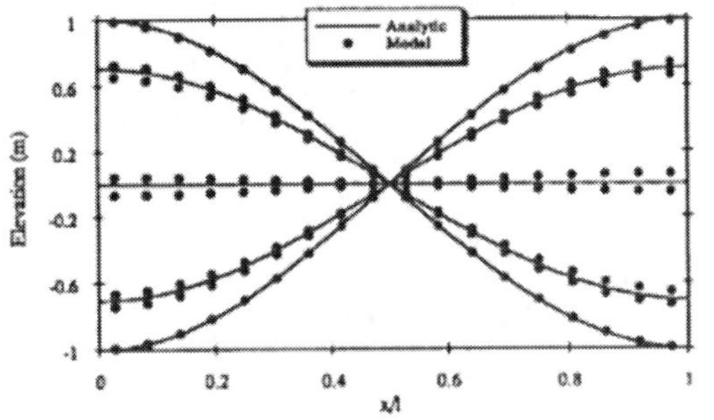

Fig. 2. Comparison of Model Predicted Surface Elevation with Analytic Solution for Standing Wave in Closed Basin at 1/8 Period Increment, l=23 km, h=10m, Δt=58s

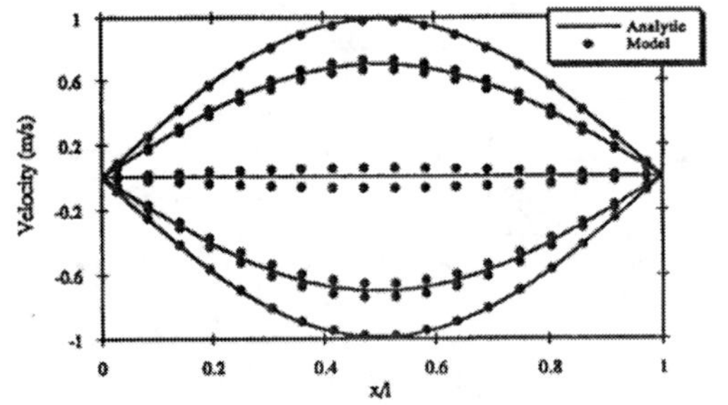

Fig. 3. Comparison of Model Predicted Velocities with Analytic Solution for Standing Wave in Closed Basin at 1/8 Period Increment, l=23 km, h=10m, Δt=58s

Figure 2 and Figure 3 present a comparison between the model prediction and analytic solution in the closed basin, with length of 23 km, 4 x 18 grid, water depth 10 m. It shows that there is no numerical damping in the model which is very important in Initial Value Problem.

3 Model Application for Mentawai Tsunami

MuTsunami is applied to simulate for Mentawai Tsunami. The grid system is presented in Figure 4. It is finer in the interest area (West Sumatera Province) and coarser in others area. The grid size ranges from 5 km to 30 km.

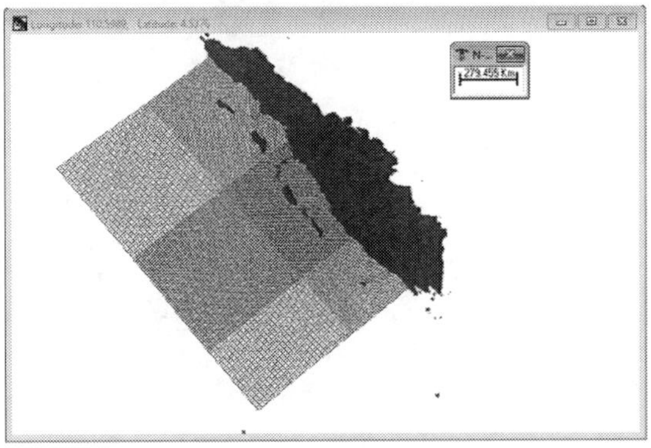

Fig. 4. Non-Orthogonal Curvilinear Coordinate Grid System in the Study Area

Figure 5 shows the water depth in the grid system. Bathymetric data was obtained from Geodas with one minute resolution. The water depth in the water cell is an averaged value of available data set.

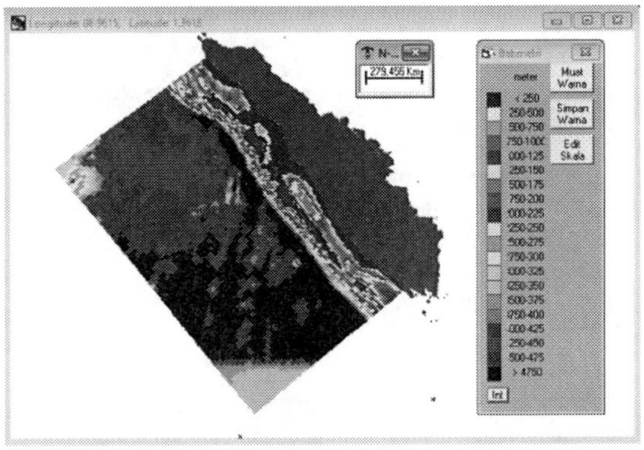

Fig. 5. Water depth data in the grid system

The model was run using; (1) simple energy approach for quick response mode, (2) specified Tsunami generation for better accuracy which is based on simulation by Han et. al, [9]. The computation was performed in affordable notebook computer, equipped with processor I5 2.7 GHz, 16GB Ram. The time step is 5 seconds. The bottom friction was set 0.005. The required CPU Time is less than one minute to complete 4 days simulation.

https://doi.org/10.1051/matecconf/201814705005

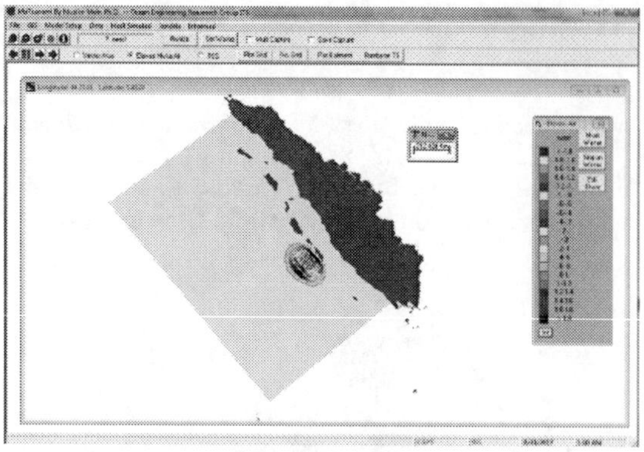

Fig. 6. Water elevation at 7 minute after earthquake, using Specified Tsunami Generation

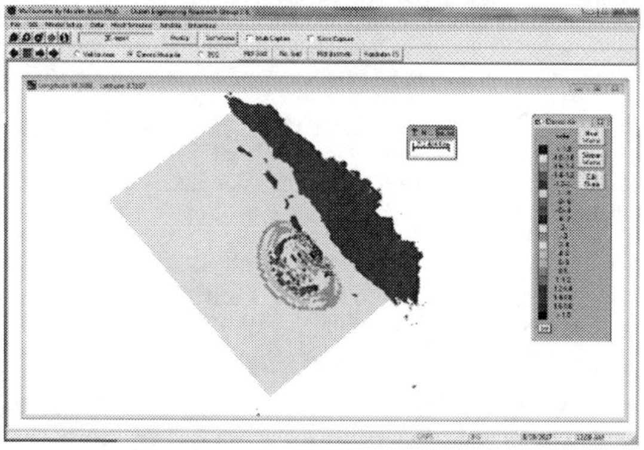

Fig. 7. Water elevation at 20 minute earthquake, using Specified Tsunami Generation

Figure 6 shows the predicted water elevation at 7 minute after the earthquake or when the big Tsunami hit Mentawai. Figure 7 presents the predicted water elevation at 20 minute. The Tsunami also propagate to Indian Ocean because of with small amplitude, less than one meter. The modelling results indicate the effectiveness of Mentawai Islands to protect West Sumatera from Tsunami.

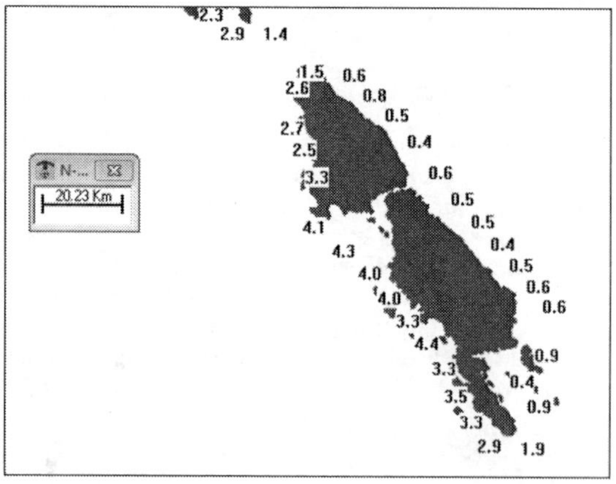

Fig. 8. Water elevation along the coast of Pagai Selatan Island, using Specified Tsunami Generation

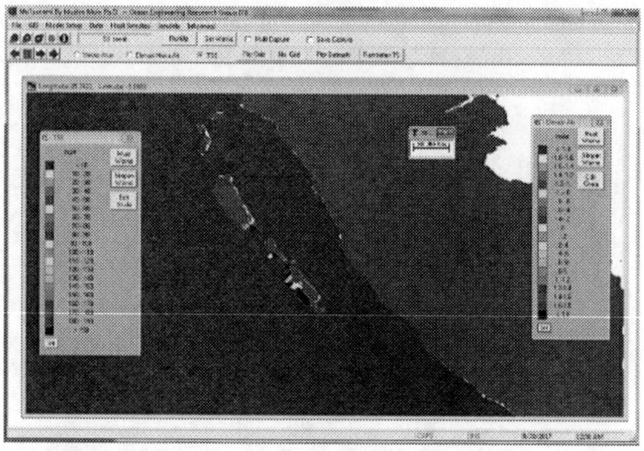

Fig. 9. Sediment Plume at 40 minute after earthquake

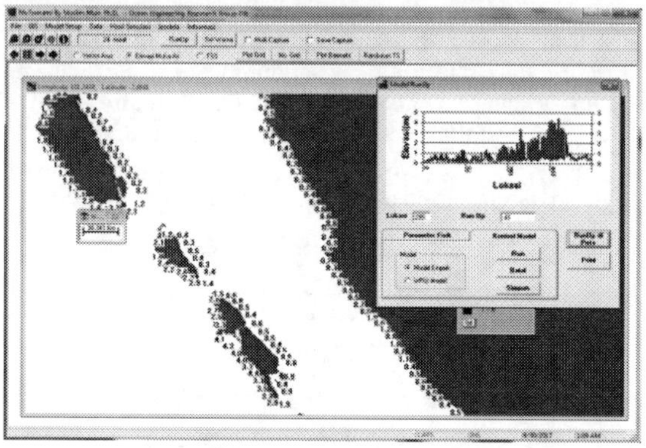

Fig. 10. Water elevation along the coast of West Sumatera Province and Mentawai Islands, the model was run using specified well known Tsunami Generation

Figure 8 presents the predicted maximum water elevation along the coast of Pagai Selatan Islands. Despite the simulation is conducted in relatively coarse grid system, the results of simulation is not much different from very high grid simulation by well known others model. The simulation can be completed in very short time and used as early warning system using affordable computer.

The model was also run to simulate one class of cohesionless sediment. The Erosion rate coefficient is specified 0.05 mm/hour, Critical Shear Stress for Erosion = 0.2 Pa, and Critical Shear Stress for Deposition = 0.1 Pa. The settling velocity = 20 mm/sec. The results of simulation are presented in Figure 9. The model indicates that the seabed erosion in the area of Mentawai. It is because of the water current of Tsunami exceed the critical shear stress for erosion. The model also indicate that Tsunami will not erode the West Sumatera Coast.

Figure 10 shows the wave height along Mentawai Islands and along West Sumatera Coast. The results of simulation are comparable with field observation. The model clearly indicates that the coast of West Sumatera are well protected by Mentawai Islands.

Table 1 present the comparison between simulated wave height and observed wave height by Mikami [10].

Table 1. Comparison between results of simulation and observation.

Location	Latitude	Longitude	Simulation (m)	Observation (m)
Bosua	-2.3664	99.8027	4.50	4.69
Old-Gobik	-2.3687	99.8212	4.80	5.69
Masokut	-2.3462	99.7883	4.40	4.42
Bere-Berilou	-2.3323	99.7289	4.30	3.18

The main problem in the specified initial condition that the model is not linked into sophisticate model Tsunami Generation which requires higher resolution. Therefore, the use of simple energy approach to estimate the initial condition is not avoidable to get quick response.

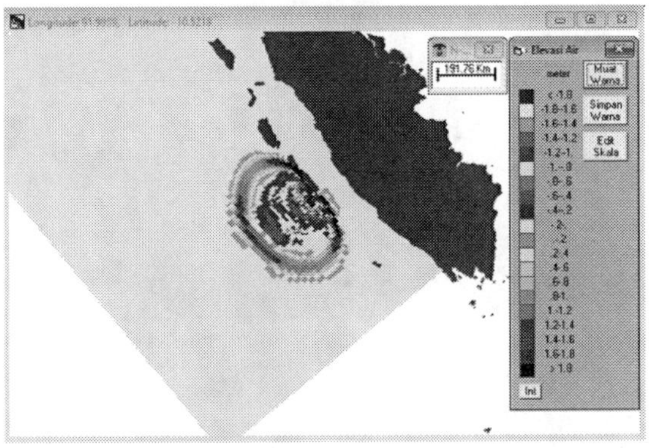

Fig. 11. Water elevation at 16 minute after earthquake, the model was run using Simple Energy Approach

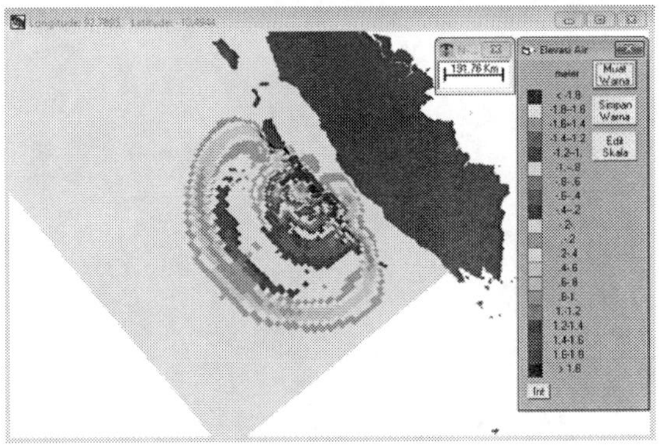

Fig. 12. Water elevation at 28 minute after earthquake, the model was run using Simple Energy Approach

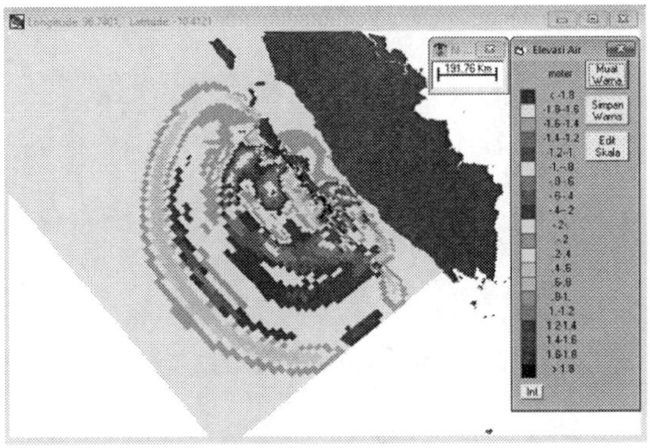

Fig. 13. Water elevation at 39 minute after earthquake, the model was run using Simple Energy Approach

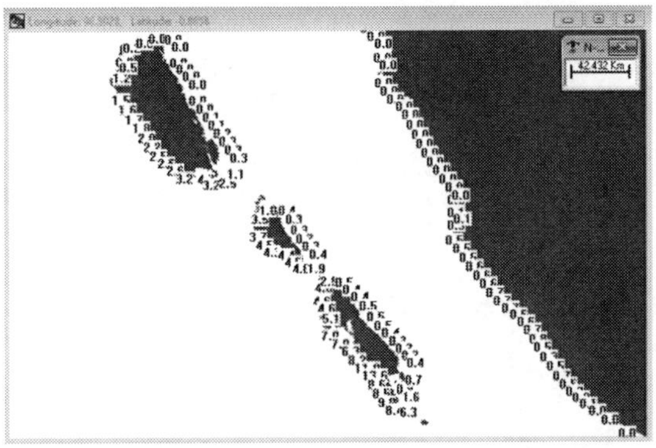

Fig. 14. Water elevation along West Sumatra Province coastal water, the model was run using Simple Energy Approach

Figure 11 to Figure 13 shows the results of simulation in quick response mode using simple energy approach at 16 minutes, 28 minutes, and 39 minutes after the initial condition respectively. The wave height along Aceh coastal area is shown in Figure 14. Clearly, the results of simulation are less accurate compare to previous simulation. However, the simulated wave is a worth information for quick response. The model does show correct impacted area.

4 Conclusion

Software MuTsunami, model of Tsunami propagation and sediment transport model using non-orthogonal curvilinear-coordinate technique in spherical coordinate as in house research tool of Tsunami simulation in CERG ITB had been applied for Mentawai Tsunami. The comparison between the modeling results and observation are very good.

Like in application of model in Aceh, the simulation results also shows that Simple Energy Approach for Tsunami Generation can be used as a preliminary simulation or for quick response as early warning although it will be less accurate. The impacted area is consistent with field observation.

References

1. M. Muin, *Ph.D. Dissertation, Department of Ocean Engineering, University of Rhode Island* (1993)
2. M. Muin, M.L. Spaulding, J. Hydraulic Engineering **123**, 13 (1997)
3. M. Muin, M. L. Spaulding, J. Hydraulic Engineering **123**, 2 (1997)
4. S. Sankaranarayanan, Continental Shelf Research **25**, 2233 (2005)
5. S. Zigic, O. Makarynsky, S. Langtry, G. Westbrook, *Proceedings of the International Conference on Estuarine and Coastal Modeling* **388**, 428 (2010)
6. S. Sankaranarayanan, D.F. McCay, J. of Waterways, Port, Coastal, and Ocean Engineering **129**, 114 (2003)
7. M. Muin, Dewan Riset Nasional Indonesia (1998)
8. M. Muin, A. Wurjanto, *Proceedings of the Twenty-third International Offshore and Polar Engineering* (2013)
9. H. Yue, T. Lay, L. Rivera, Y. Bai, Y. Yamazaki, K.F. Cheung, E.M. Hill, K. Sieh, W. Kongko, A. Muhari, J. of Geophysical Research: Solid Earth **119**, 5574 (2014)
10. T. Mikami, T. Shibayama, M. Esteban, K. Ohira, J. Sasaki, T. Suzuki, H. Achiari, T. Widodo, Natural Hazards **71**, 851 (2014)
11. M. Muin, *Manual MuSed3D and Technical Guidance, Ocean Engineerng, Institut Teknologi Bandung* (2001)
12. S.T. Grilli, P. Watts, J. of Waterway, Port, Coastal, and Ocean Engineering **131**, 283 (2005)
13. F.E. Camfield, *Chapter 10 Handbook of Ocean Engineering* (1992)
14. M. Muin, M.L. Spaulding, J. Hydraulic Engineering **122**, 512 (1996)
15. M. Esteban, V. Tsimopoulou, T. Mikami, N.Y. Yun, A. Suppasri, T. Shibayama, International Journal of Disaster Risk Reduction **5**, 84 (2013)

Adaptive Port Planning under Disruptive Trends with focus on the case of The Port of Kuala Tanjung, Indonesia

Andre Prakoso[1,*], *Poonam* Taneja[2], and *Tiedo* Velinga[2]

[1]Delft University of Technology, Master program in Hydraulic Engineering, Delft, The Netherlands [2]Delft University of Technology, Hydraulic Engineering Department, Delft, The Netherlands

As a national strategic project, the Port of Kuala Tanjung draws significant attention at national and international level. Considering the semi-greenfield nature of the port, the diverse set of stakeholders, and the prevailing disruptive trends in the world port business, a robust first-phase port layout is required to kick-start the project and guarantee the overall sustainability of the port development. The objective of this research is to identify any uncertain or disruptive trends, both present and future, and access their implications towards the Port of Kuala Tanjung. Adaptive Port Planning (APP) framework[a] will be used as the main methodology in this research. A combination of a literature review and interviews with experts are used to both identify the sources of the uncertain and disruptive trends mentioned above and also to propose adaptation strategies. Based on our qualitative study and interviews with experts, we have concluded that the consolidation of major shipping lines, multi-nationality partnerships, Indonesian regulation, and ship breaking regulations are currently the four most relevant contributions to uncertain and disruptive trends towards the port. To mitigate potential risk and seize opportunity provided by these trends, an industrial port complex concept might become the most promising alternative.

1 Introduction

1.1 Background

As a strategic national project, the Port of Kuala Tanjung in Indonesia draws significant attention at regional, national, and international level. Currently, Pelindo 1, the Port of Rotterdam, Indonesian ministries, regional government, private sector, and local communities are involved in the planning of the port. Considering the many challenges facing the project, an adaptive port masterplan is required to satisfy these challenges. These challenges include the semi-greenfield nature of the Port of Kuala Tanjung, along with the diverse set of stakeholders, international agreement in fulfilling the United Nations' goals and Paris climate agreement towards a more sustainable development, and finally the prevailing uncertain and disruptive trends in world port business aided by globalization, internet, and digitalization. This study is conducted on the Port of Rotterdam and as a part of thesis research to obtain a Master of Science in Hydraulic engineering at TU Delft. The author would like to clearly state that while this study is conducted during the author's internship at the Port of Rotterdam, the views expressed in this paper are those of the author alone and not the Port of Rotterdam Authority.

1.2 Project Description

The Port of Kuala Tanjung is an approximately 3000-ha-wide integrated port industrial complex development which is located approximately 140 km south-east of the Port of Belawan in North Sumatera, Indonesia [1]. It is strategically situated in the Malaka Strait which is the second-busiest shipping route in the world [2]. In the surrounding hinterland lies the Sei Mangkei Special Economic Zone and Medan, the second largest economic city on the island of Sumatra. A population of 13 million people living within a 250 km radius will be directly benefited by the presence of this industrial complex port.

Through several years, the Port of Kuala Tanjung status has gone through some changes. Beginning as a tertiary port in 2008, it was intended to support the Port of Belawan in terms of short-sea shipping [3]. In 2015, the Indonesian government changed the status of the Port of Kuala Tanjung into an International Hub Port and placed as a national strategic project [4]. There are plans to make it the largest port in Indonesia to support the Indonesian government's "World Maritime Axis" agenda. It also serves as one of the backbones of "Tol Laut", an

[a] Taneja P, *The Flexible Port*, 2013

* Corresponding author: author@e-mail.org

Indonesian national, primary shipping line route [5]. Therefore, the port received significant attention from both a national and international audience ever since. In November 2016, the Port of Rotterdam Authority signed a JV agreement with Pelindo 1 and committed to develop a world-class port industrial complex in Kuala Tanjung.

Nevertheless, the importance of the Port of Kuala Tanjung's existence towards Indonesian connectivity and the North Sumatra region's development cannot be represented simply by its status in a national document [1]. A world-class port industrial complex located in North Sumatra will surely introduce positive influence towards the region's economy and its people's livelihood if developed and managed in efficient and sustainable way. However, like every massive infrastructure complex development, the Port of Kuala Tanjung needs to be built in phases. Prior to the JV agreement between Pelindo 1 and the Port of Rotterdam Authority, there is already a multi-purpose terminal construction which might be regarded as the first phase of the Port of Kuala Tanjung in most literature.

However, in this study we will **not** refer the multi-purpose terminal (MPT) as the first phase of the Port of Kuala Tanjung's development. By first phase, we mean the first set of infrastructure which will be built by Pelindo 1 and the Port of Rotterdam JV. Since the main idea of the Port developed by the JV is an integrated port industrial complex, and the MPT simply do not compatible to work with the other port tenants. The focus of the Port of Kuala Tanjung JV project organization in this period would be to develop a robust, self-sustainable, first-phase development which would provide catalyzing effect towards the next phases of development.

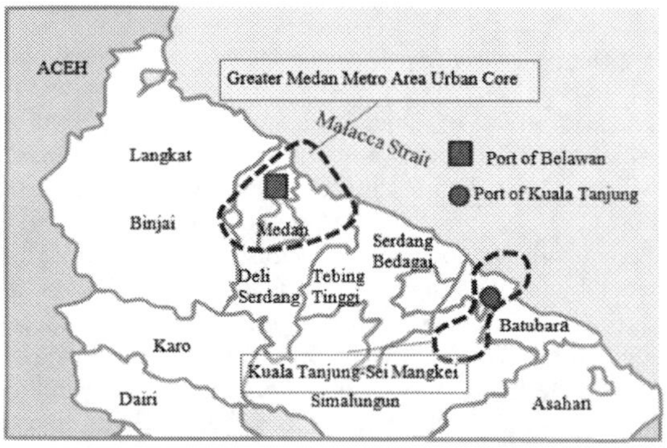

Fig. 1. Port of Kuala Tanjung site map.

1.3 Objective of the study

A study is being carried out in order to develop a robust, first-phase layout for the Port of Kuala Tanjung, which is self-sustainable and provides catalyzing effect towards the next stages of port development by taking into account port disruptive trends during the planning process. The result of this study will provide additional insight for decision makers and other interested parties in how to optimize the potential of the Port of Kuala Tanjung and avoid less outcome than intended.

Robustness implies that the layout and design of the port is such that the port is able to function in most plausible future scenarios. Self-sustainable indicates the ability of the Port of Kuala Tanjung's first-phase layout to gain more revenue than operating expenditure without any dependence on the next stage of development while adding significant added value to the neighboring region. A catalyzing effect applies to both a commercial context and a context based on infrastructure. In a context based on infrastructure, it implies that the first-phase port infrastructure either creates the environment of the next phase's infrastructure or avoids obstructing future construction developments, thereby enabling faster and cheaper construction work and enables clusters to develop at the next phase. In a commercial context, a catalyzing effect means that the first phase of the Port of Kuala Tanjung will consistently generate a good financial portfolio to attract potential investors and future tenants in the industrial port complex. APP framework will be used as the main framework in this research [6]. This paper presents the first results of the study, namely the analysis of the current and future uncertain or disruptive trends and their implications towards the Port of Kuala Tanjung. This is Step II in the Adaptive Port Planning Framework [6]. The completed study will include an adaptive port masterplan, robust first-phase layout, and a business case.

1.4 Research methodology

Port developments cover broad multi-disciplinary subjects in real practice. In order to systematically identify the implication of each ongoing trend towards the future of the Port of Kuala Tanjung development, we used port research themes to pinpoint in which topic of port development might a trend affect the Port of Kuala Tanjung. These port research themes are a result of extensive literature reviews conducted by Taneja, in which it represents every possible aspect of a port system that could be scientifically observed [6].

First, we conduct a literature review to gain a list of relevant ongoing trends which are relevant towards the Port of Kuala Tanjung project. These trends should possess both a significant disruptive effect and level of uncertainty embedded in its nature, although its actual level cannot be measured and decided through the author's qualitative judgement. We will then conduct interviews with five experts at the Port of Rotterdam who have backgrounds in port development (technical), finance, stakeholder management, economics, and commercial points of view. The focus of these interviews was to acquire expert insights in ranking the most relevant trends towards the Port of Kuala Tanjung, while identifying which port research would be affected by each trend and predict the implication of those trends towards the port's design process.

1.4.1 Underlying principle

We believe that sustainable port developments should always introduce significant added value for the surrounding community while still being profitable financially and respecting nature and environment at the same time. In order to realize that, efforts in optimizing environmental and social benefit should be the main goal right from the beginning of the planning process and not just as auxiliary item for corporate social responsibility. We refer to the United Nations Sustainable Development Goals (UN-SDG) in this regard as our guiding principles [7]. It should be noted that due to its large construction and activity scale, all 17 aspects of the UN-SDG will be relevant to the Port of Kuala Tanjung's development, with number 8 (decent work and economic growth), 9 (industry, innovation and infrastructure) and 11 (sustainable cities and communities) being the top three most relevant goals in the development of the Port of Kuala Tanjung.

Fig. 2. United Nations Sustainable Development Goals [8].

2 Uncertainty analysis

2.1. Properties of uncertainty

Uncertainty is defined as the existence of unknown information regarding the cause and effect of a particular or series of events towards a system of interest's performance that originated either from exogenous or endogenous sectors of the system itself. In this case, the port environment will be treated as a system, or an engineering system to be exact. An engineering system is defined as a man-made system designed to meet specific objectives and functional needs that fulfill pre-determined requirements [9]. The port environment is defined as a combination of technological, political, social and economic factors directly affecting the port itself [6]. It should be noted that our intention is to minimize and manage uncertainty rather than eliminating it, which is impossible. The topology of uncertainty itself is best described by Walker who categorizes it into location, level, and nature, all of which is summarized in **Error! Reference source not found.** below [10]. An important concept in the nature of uncertainty which needs to be further elaborated is epistemic uncertainty and inherent uncertainty. Epistemic uncertainty originates from the lack of knowledge of a system, which may be reduced through scientific research, optimization, or knowledge exchange. On the other hand, inherent uncertainty originates from the random nature of a variable itself. For example, in a steel mill, the probability that a produced steel will be up to standard may be increased through decreasing the factory's epistemic uncertainty. This means it may be achieved through better technology, production processes, or other forms of optimization. However, due to the inherent uncertainty contained in the production process, the probability that a produced steel will be up to standard will never be 100%. This inherent uncertainty may be located in the variability of the iron ore's quality as the main steel mill's feedstock or in the performance variability of the steel production equipment itself.

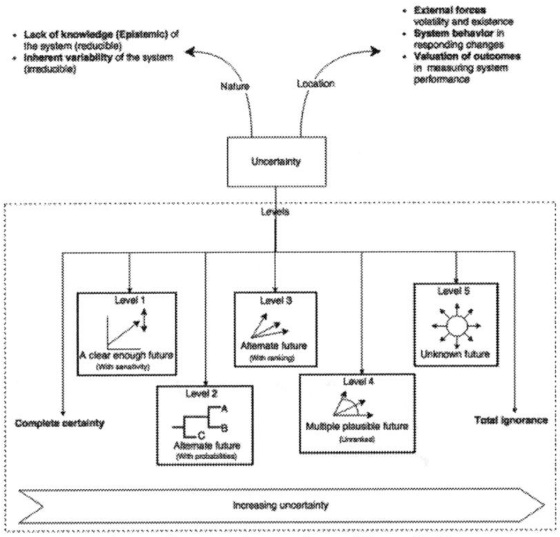

Fig. 3. Topology of uncertainty, adopted from [10].

2.2 Properties of disruptive trends

Daneels wrote: "Disruptive technology can be thought of as a technology that changes the essence of competition among firms by transforming the performance metrics" [11]. Based on this explanation and the uncertainty topology of Walker, we would like to propose a more elaborate definition of disruptive trends. Disruptive trends are defined as trends which add volatility into a relatively stable, long-term trend, while changing entirely how the performance of a specific topic in the system of interest is measured and eventually causing long-term trend reversal after a certain period of time. The emphasis on our definition is that disruptive trends should be time-bounded. Each disruptive trend should be identified and assessed sequentially by taking into account the contribution of the previous trends and knowing that each of the trends are unique for their specific era and field of interest. However, the argument on the period, or the start and finish, of each trend depends heavily on the observer's point of view and the field of interest. Therefore, in categorizing disruptive trends it is not advisable to use time as a reference frame. The study of the effect of previous, ongoing, and potential disruptive trends on a system's performance have been conducted on various fields, primarily with intention to maintain the competitive edge of a company ahead of its competitors. This edge involves having the knowledge to assign appropriate resources in a way that increases the chance of benefitting or mitigating the potential vulnerabilities from the existence of disruptive trends [12].

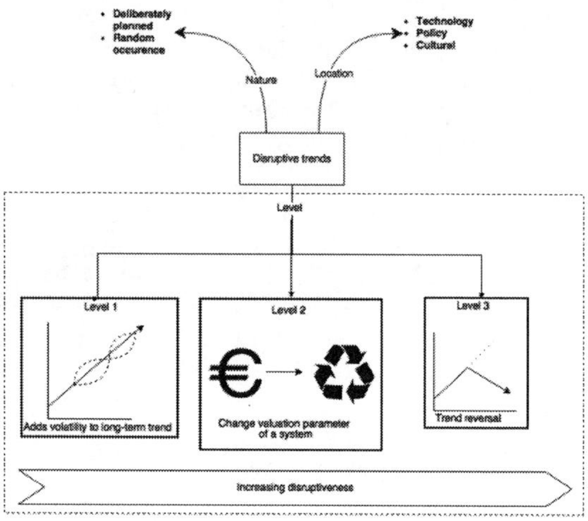

Fig. 4. Topology of disruptiveness

2.2.1 Handling uncertainties and disruptive trends

Uncertainties and disruptive trends can jointly be perceived as a blessing or curse in disguise depending on one's system point of view. The nature of their interaction can be visualized as in **Fig. 4** with some degree of independency between each other but still somehow related. Each trend started as being "uncertainty dominated" and grew overtime towards being "disruptive-trend dominated". The uncertainty mentioned in this section only applies to epistemic uncertainty which arises from technical and scientific uncertainty, which is therefore may be reduced.

When a trend starts to occur, there is still very little data available regarding its effect towards the imposed system. Therefore, there is a huge amount of uncertainty towards this trend's performance and its profile as a disruptive trend has also not been proved. As time during the development progresses, the way both properties interact is assumed to be independent with a tendency of lowering epistemic uncertainty and growing more disruptive with time. This assumption is based on additional data and knowledge gathered overtime showing the uncertain element of the system can be reduced and the true extent of the disruptive trend effect can be discovered.

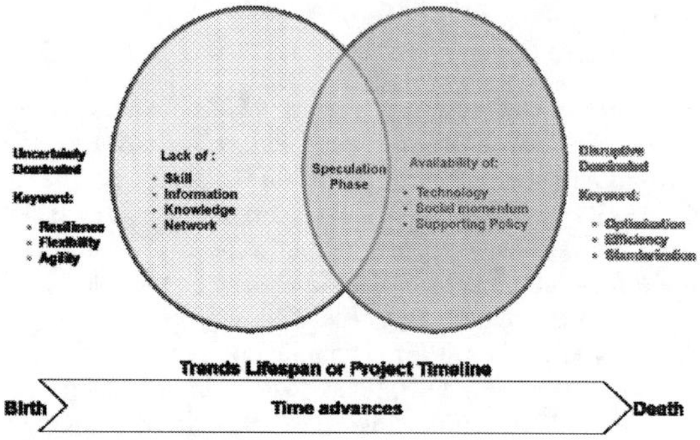

Fig. 3. Relationship between uncertainty and disruptive properties of a trend

We argue that the relationship between uncertainty and disruptive trends in **Fig. 5** is also applicable for port projects. In the **green-field stage,** ports are dominated by uncertainty in their development. There are simply too many elements in the port construction, management, and commercial services that are still not known, even with a state-of-the-art forecasting method. The challenge for this stage of development is that ports need to invest in a significant amount upfront for large infrastructure without having a guaranteed revenue stream in the process. This condition places greenfield ports in huge financial risk. Therefore, the focus of these ports should be on **resilience, flexibility, and agility**.

While for a matured, **brownfield port**, epistemic uncertainties should be near diminished due to prolonged exposure towards the port operation as well as experienced workers and a well-established port organization. These ports are dominated by ongoing disruptive trends. The scale of the disruptive trends differ depending on the scale of the ports. For instance, gateway ports are affected by global disruptive trends while small, feeder ports are only affected by regional disruptive trends. In order to prevent stagnantization and maintain its competitive edge compared to other port competitors sharing the same hinterland, brownfield ports should focus on **optimization, efficiency, and standardization**.

Having illustrated the complexity surrounding uncertainty and disruptive trends in port industry, proper identification and relevant strategies are required to manage it optimally. Nonetheless, only those who are cautious, agile, knowledgeable, and visionary enough can gain real benefit and grow from it. Uncertainties and disruptive trends are part of the eternal cycle of the world and will stay forever. Ports will be forced to either take advantage and prosper from it or suffer defeat and lose from it.

2.3 Major trends relevant to The Port of Kuala Tanjung

2.3.1 Consolidation of shipping lines

Fair competition has always been the nature of shipping line industries. In order to attract customers, large number of shipping lines constantly compete with each other by reducing costs while maintaining their safety level to ensure their cargo arrives at destinations. As a result of this competition, a trend of larger ships has started to occur in maritime industries to further lower the cost of logistics. Larger ships mean more,cargos to transport in one trip which leads to better efficiency. However, as the trend progresses, the world shipping line industry will eventually reach overcapacity meaning there are more ships available than the cargo which has to be transported. To solve the problem of overcapacity, shipping companies started making alliances in order to optimize the utilization of large ships in their fleet [13]. This action proved to be a good one as it benefits all shipping line companies within the alliances. Currently, there are four major alliances recognised in the world as presented in Table 1. The trend of alliances is expected to continue in dynamic fashion as the power balance between each shipping line company always fluctuates. This trend will then lead towards a consolidation of shipping lines in the world, resulting in fewer calls with same or even more throughput for all the ports in the world.

Table 1. Current shipping alliances in operation [14]

Alliance	Members	Est.	Market Share in the World	Fleet Size
CKYHE	COSCO, K-line, Yangmig, Hanjin, Evergreen	2002	25 %	3.3 M TEUs
G6 Alliance	APL, Hapag-Lyod, HMM, MOL, NYK Lines, OOCL	2012	28 %	3.5 M TEUs
2M Alliance	Maersk, MSC	2015	27 %	5.7 M TEUs
Ocean Three Alliance	CMA-CGM, China Shipping, United arab shipping co	2015	15 %	3.0 M TEUs

Since major shipping line companies are highly competitive, they only call to terminals which offer the best service in terms of time and cost. Relying on a transshipment business model introduces significant risk to the Port of Kuala Tanjung considering there are already exists the well-established Port Klang and the Port of Singapore which are both placed as the top 15 container ports in the world. Confronting direct competition with these ports through transshipment activity would not be a wise move, since they could simply reduce their price or increase their productivity which then would hurt the Port of Kuala Tanjung financially. Moreover, transshipment ports introduce minimum added value towards the neighboring region since the cargo flow which is generated directly by these activities would not be retrieved by the hinterland. This is often referred to as "lose feet" cargo.

2.3.2 Multi-national company partnerships

There is a considerable difference between the Port of Rotterdam and the Pelindo 1 Port's governance model. Before 2008, Indonesia adopted the service port model, where the role of Terminal Operator and Port Authority was both conducted by four state-owned companies called Perusahaan Pelabuhan Indonesia (Pelindo). According to state regulation, No. 21, Year 1992 about shipping, these four pelindos were given the authority to regulate ports in all cities within their territory as illustrated in **Fig.6**.

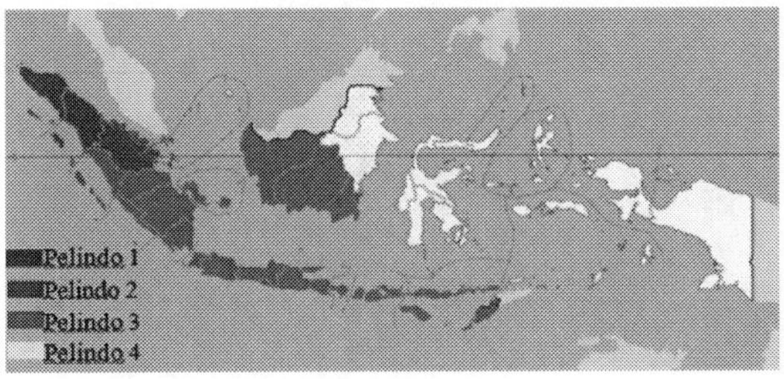

Fig. 4. Pelindos designated territories

In a newly amended shipping regulation in 2008 (state regulation No. 17), the government made significant changes to the shipping laws. Pelindo, which previously held total control of the planning, construction, regulation, and operation of all ports in the region, now only acts as the port operator as a result of the amendment. The Port Authority role will be given to a new government institution, Otoritas Pelabuhan, which acts under the Ministry of Transportation. This indicates the Indonesian Government's intention to change the service port governance model into a land-lord port governance model. It should be noted here that Pelindo does not only act as the terminal operator in their capacity as the port operator. They also have the power to control a range of operations within the port, ranging from berthing navigation, cargo handling, and hinterland activities. Therefore, their role extends beyond a terminal operator. The new entity, Otoritas Pelabuhan will assume the role of port master planner and will act as a regulating body for all port related activity. For instance, if Pelindo wishes to expand their terminal, they would need to first obtain approval from Otoritas Pelabuhan.

However, due to unclear separation of duties between Pelindo and Otoritas Pelabuhan and the absence of a clear, time-bound transition plan in 2008 regulation No. 17, changes to the Indonesia port governance model have been minimal since then. Typically, in a landlord port model, the Port Authority represents a form of publicly owned enterprise. One such example would be the Port of Rotterdam which owns the land under the municipality jurisdiction. They have the authority to regulate the lands given to them by the government provided that they abide by certain city rules and regulations. They obtain their main revenue stream from a land-lease fee and port dues. They do not handle daily port operations directly as this task is handled by terminal operators which rent the land from the Port of Rotterdam. In Indonesia however, the majority of land ownership still belongs to Pelindo and not to the newly established Otoritas Pelabuhan. There are even some pieces of land inside port areas that belong to private companies or individuals. Otoritas Pelabuhan is only acting as a government regulatory agency without any commercial activity in the institution's agenda. A summary of the difference in responsibility allocation between the Port of Rotterdam and Pelindo 1 is presented in **Table 2.**

Table 2. Responsibility allocation in Port of Rotterdam and Pelindo 1.

Responsibility	Port of Rotterdam	Pelindo 1	Otoritas Pelabuhan
Commercial	■	■	
Planning	■		■
Infrastructure	■	■	■
Nautical safety & security	■	■	■
Regulation	■		■
Stakeholder management	■		
Marine service & utilities		■	
Supra-structure		■	
Logistics		■	

Another source of uncertainty and disruptive trends related to the JV culture is the differences between Indonesia and The Netherlands with regards to communication style and the way in which projects are approached. Based on his working experience, Lewis has conducted a study on how cultural background might affect someone's communication and behavior [15]. He then measured each country in terms of three general criteria and plotted them into **Fig.7**. This is commonly referred to as the "Lewis model".

Indonesia and The Netherlands are assigned to opposite sides of the figure in the Lewis model, indicating significant difference in communication style and behavior influenced by each country's respective cultural-background differences. Indonesia is said to be culturally reactive, while The Netherlands is assumed to be linear-active. Countries with reactive cultural values place courtesy and politeness in high regard. They prefer to look at the general picture, conceal their feelings, and most notable, they prefer to listen. On the other side of the spectrum, The Netherlands tends to prefer a more direct approach. They plan ahead, step-by-step in detail, and culturally have no difficulty in expressing their feelings.

The approach on problem-solving and the way Indonesians and Dutchs people treat their business partner is indeed very much different. These differences might become the source of conflict which put the partnership at stake or become the source of strength that gives the Port of Rotterdam and Pelindo 1 JV a unique advantage against their competitors through collective knowledge and diverse yet comprehensive approach on port related problems. We believe that there is even more potential than risks that could be realized through this partnership. Dedicated studies and discussions in deciding how Pelindo 1 and Port of Rotterdam should cooperate with each other is required in order to truly complement each other's strengths, and making Pelindo 1 and Port of Rotterdam the best port managers in their respective region.

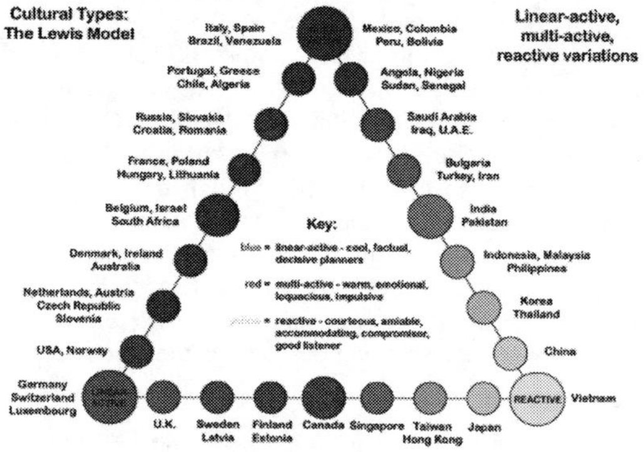

Figure 5: Lewis Model [15].

2.3.3 Ship breaking regulations

Even after eight years since Hong kong convention on ship breaking in 2009, most of ship breaking activity still take place in substandard ship breaking yard facilities located in Bangladesh, Pakistan, and India which violates international safety and environmental standard [16]. There could be several reason for this trend, one of them arguably is the lack of law environment for ship owners and ship breaking yards which violates the international regulations [17]. Ship owners could simply choose the cheapest, non regulated ship breaking yard available in the world without having any real consequence in doing so. Numerous worker causalities and massive environmental damage in the coast happened where non regulated ship breaking activities take place.

However, recently global pressure towards a more green ship breaking facility have increased. European commission have published a list of registered ship recycling facilities on December 2016, as a shown commitment from the EU in attempt towards a more green ship breaking industry [18]. If we take into account the raised global awareness aided by internet, global common goal provided by UN-SDGs, we believe that sooner or later ship breaking industries will transformed into a more sustainable industry. If we could eliminate the negative effect of ship breaking activity by having the state of the art ship breaking facilities, a potentially huge benefit could be introduced into the region through: steel production, massive low education-employment, generated economic activity and the compatibility of a ship breaking facility to integrate with industrial cluster.

Table 3. Total dismantled ships worldwide vs ships dismantled in substandard ship breaking facilities [16].

Year	Total dismantled ships worldwide	Ships dismantled in substandard beaching facilities in India, Pakistan, and Bangladesh
2012	~1213	850 (70 %)
2013	1213	645 (53 %)
2014	1026	641 (62 %)
2015	768	469 (61 %)
2016	862	668 (77 %)

2.3.4 Regulation and legal permits in Indonesia

Based on interviews with experts at the Port of Rotterdam, it was concluded that the legal frame work is not yet ready in Indonesia in order to apply a landlord port model. In order for a landlord port model to work optimally, a single entity which takes the role of the port manager is required. This usually takes the form of a port authority which regulates, manages, and plans all related port activities and developments. By giving full authority to the port manager, full integration of port shared facilities and sustainable development can be achieved. In contrast, Indonesian ports see various industrial activities under the authority of many different entities. Ports are, in essence, an established set of required logistics infrastructure which allow cargo movement between sea and land. They are not considered to be industrial complexes. Hence, in Indonesia, the ports lay under the authority of the Ministry of Transportation, while industrial complexes fall under the authority of the Ministry of Industry. Moreover, the division of the regulatory and management role of the port manager between Otoritas Pelabuhan (regulatory) and Pelindo 1 (managerial) make the application of a landlord port model difficult. Therefore, customized regulatory efforts are required in order to simplify but not undermine the regulatory framework related to the construction of the Port of Kuala Tanjung.

2.4 Implications for The Port of Kuala Tanjung

Considering the various challenges posed by ongoing and forecasted future disruptions, the Port of Kuala Tanjung needs to develop adaptive plans to truly become a world-class port as envisioned by the Indonesian government. Relying on a static and rigid port masterplan would not be advised as the nature of these disruptive trends are uncertain. Many parameters in the masterplan which are being used for the current design criteria may change in the future. For this reason, the first crucial step in determining the future of the Port of Kuala Tanjung would be affirming what kind of port it should be in order to become a world-class port. This will support the Indonesian "World Maritime Axis" and the "Tol Laut" agenda. By comparing common characteristics between the top 15 busiest container ports in the world, we have identified key elements required to become a world-class port. There should be an optimal combination of: populous urban hinterland, significant industrial activity inside the port area, and a frequent stream of dedicated shipping line traffic. The management of each port in Table 4 has focused on at least one of the identified key elements resulting in the port maximizing its performance.

Table 4: Top 15 largest container ports in the world [19].

Rank	Port	Terminal Operator
1	Shanghai, China	APMT, Hutchison,
2	Singapore	PSA and Jurong
3	Shenzhen, China	CMHI and COSCO
4	Ningbo –Zhousan,	Hutchison, TIL, etc.
5	Hong Kong, S.A.R,	DP World, PSA, etc.
6	Busan, South Korea	Hutchison
7	Qingdao, China	DP World and
8	Guangzhou, Harbour,	PSA and COSCO
9	Jebel Ali, Dubai,	DP World
10	Tianjin, China	APMT, DP World,
11	Rotterdam, Netherlands	APMT, Hutchison,
12	Port Klang, Malaysia	APMT
13	Kaohsiung, Taiwan,	DP World
14	Antwerp, Belgium	PSA, TIL, and DP
15	Dalian, China	APMT, PSA, etc.

In the case of the Port of Kuala Tanjung, we will first describe the implications of the trends mentioned in previous sections and then draw conclusions on how to optimize the port. First, the implication of the consolidation of major shipping lines is considered. Bigger and less frequent vessel traffic would result in higher peak cargo flow in the respective port. This trend would put additional stress on not only the port infrastructure, but also its hinterland connections. Wider reaching cranes, stronger quays, and wider roads among other things are required in order to avoid possible congestion at the moment of cargo discharge from large vessels. The Port Authority and the city municipality need to cooperate fully to mitigate this risk. The focus in this regard should not be on the vessel size itself, but on the maximum cargo flow capacity that the whole port system and the cities in the surrounding hinterland area could manage at any one time. On the other hand, the birth of major shipping alliances offers a significant opportunity. Large, steady numbers of yearly cargo flow are expected as a result of these new alliances. The existence of a world-class terminal operator in this region is crucial to attracting shipping traffic from these major shipping line alliances. Even when operating individually, shipping lines tend to be attracted to the most efficient terminal in the region so that they maintain their competitive position. The consolidation of major shipping alliances would surely increase their demand for more efficient cargo handling in a terminal. The best course of action for the Port of Kuala Tanjung manager (Pelindo 1 and the Port of Rotterdam JV) in this case would be to maintain a neutral position when appointing the terminal operator and perform objective periodical assessment on the terminal operator's performance. By doing this, the Port of Kuala Tanjung can maintain a competitive position over Port Klang and the Port of Singapore in terms of its container terminal handling facility and efficiency.

Next, we would like to elaborate on the implications due to the Multi-nationality partnerships and Indonesian governmental regulations. Both aspects will directly affect the management and day to day operation of the Port of Kuala Tanjung. A major cause of debate resulting from the multi-nationality partnerships was authority designation for each company department or subsidiary. This was due to differences between the Pelindo 1 and the Port of Rotterdam port governance models. It should be noted that each port governance model is not a fixed rule. There are no perfect port governance models and each of them are developed according to the objectives and needs of the port in which they are applied. Therefore, in the Port of Kuala Tanjung, it is possible that a hybrid port governance model will be used, which in essence would involve a combination of a service and landlord port model. The main idea would be to adjust the landlord port governance model in such a way that it follows Indonesia's current regulatory framework. However, it should be emphasized that since the ambition of the Port of Kuala Tanjung is to become a world-class port, the port managers should stay neutral when it comes to the selecting terminal operator. The focus of the JV between the Port of Rotterdam and the Port of Kuala Tanjung should be on cooperation and harmony between both companies in order to facilitate knowledge transfer and the sharing of expertise. One solution would be to establish remote Project Organization (PO) in the Port of Kuala Tanjung with the Port of Rotterdam and Pelindo 1 JV as its shareholder. The main idea is to separate the detailed day to day operation task and the long-term political or strategic planning responsibility. The PO would be in charge of day-to-day operations, port development, and port master planning while the JV will have the authority to conduct long-term strategic planning and negotiate political terms with relevant stakeholders outside the Port of Kuala Tanjung system. A significant benefit of this solution would be the simplification of the decision-making process in the Port of Kuala Tanjung. This increases the port's responsiveness and efficiency and allows it to compete with Port Klang and the Port of Singapore. If done correctly, the advantages of the partnership with the Port of Rotterdam can be fully realised. Its experience in developing the largest port industrial complex in Europe combined with the expertise of Pelindo 1 in conducting stakeholder management with relevant Indonesian state-owned enterprises and government bodies would surely give rise to another world-class port facility.

Finally, the implication of ship breaking regulations is considered with respect to the Port of Kuala Tanjung. This trend offers a great long-term opportunity for port of Kuala Tanjung industrial complex. It is expected that the shipping industry will continue to grow to facilitate world economic growth. It follows that the demand for ship breaking will steadily increase over the coming years. The existence of a state-of-the-art ship breaking yard which fully complies with existing international regulations (i.e. IMO & EU) in the Malaka strait region will give the Port of Kuala Tanjung a competitive edge against its competitors, further securing Indonesia's position as the "World Maritime Axis". However, it is not our intention so simply advise the Port Authority to construct ship breaking yard in the Port of Kuala Tanjung. This would not fully achieve the desired benefits.The true strength of a port industrial complex resides in its seamless integration between port infrastructures and industrial cluster thereby enhancing its overall system efficiency and self-sustainability. By doing so, a significant amount of jobs are created when compared to a regular gateway or transshipment port. Through a combination of numerous cargo handling activities in the port and industrial activity in the industrial cluster stimulus, significant regional economic growth can be achieved. Moreover, due to its high efficiency, environmental risk could be kept to a minimum. For instance, a cluster of a ship breaking yard, steel plant, aluminium plant, coal stacking area, coal-powered electricity plant, and dry bulk terminal would provide a complete industrial cycle in the Port of Kuala Tanjung. The key objective in this case would be to find industrial activity opportunities which are profitable and promotes significant added value to the region. This must be done without compromising the profitability of planned and pre-existing industries. This concept put into practice may involve the recycled steel from the ship breaking facility being used in a nearby steel plant before finally being shipped outside the region through the Port or distributed to hinterland areas in the region. This industrial cluster model would enhance the robustness and self-sustainability of the Port of Kuala Tanjung, minimizing its dependency on cargo flow from the shipping line companies.

3 Conclusions

In order to establish an adaptive port plan for the Port of Kuala Tanjung, relevant industry trends and the implications of such trends have been analysed. Based on qualitative studies and interviews with industry experts, we have concluded that the consolidation of major shipping lines, JV culture, Indonesian governmental regulations, and ship breaking regulations are four critical areas of uncertainty and disruptive trends for the Port of Kuala Tanjung. It should be noted that each of these areas have inherent uncertainty which is not reducible with further research. The true extent of these uncertainties and the resulting effects are difficult to quantify.

One solution posed to mitigate the negative effects of uncertainty while taking advantage of available opportunities is the establishment of a port industrial complex. The advantage of this concept being that the port area is that it produces steady cargo traffic due to its own industrial activity as compared to gateway or transshipment ports which have a huge dependency on shipping lines for cargo. Moreover, this concept focuses on the integration of port facilities and activities which cultivates a more efficient and environmentally friendly port ecosystem. Another advantage of this concept would be the significant added value such a development brings to the industrial port complex and to the neighboring region in the form of jobs and stimulus for regional economic growth while presenting minimum environmental risk due to its high efficiency.

However, an industrial port complex requires a landlord port model to be implemented. Due to the existing Indonesian governmental regulatory framework, it might be a challenge to directly apply a landlord port model in exactly the same manner as in the Port of Rotterdam. The port governance model itself should be adapted to meet Indonesian regulatory framework. A hybrid port model might be a solution in this case. As the port model is implemented, it could be adapted accordingly provided there is a single port manager entity which maintains a neutral position in the regulation, management, and planning of the Port of Kuala Tanjung area.

Acknowledgements

The authors would like to express our gratitude toward the Indonesian Ministry of Finance for funding this study through the LPDP program. Also acknowledged is an appreciation of the Port of Rotterdam International Department for their huge support in providing the required data and in coordinating interviews with industry experts.

References

[1] Santoso, P., *Direktur Pemberitaan Media Indonesia Berkunjung ke Pelindo 1*, Majalah Gema Pelabuhan, April 2016 (4), http://www.pelindo1.co.id/wps/wcm/connect/316a3513-f677-4dc8-a323-6687f0d00496/Gema+April+2016.pdf?MOD=AJPERES, 2016

[2] Hirst, T., *The world's most important trade route?*, World Economic Forum, https://www.weforum.org/agenda/2014/05/world-most-important-trade-route/, 2014

[3] Pemerintah Provinsi Sumatera Utara, *Rencana Pembangunan Jangka Panjang Daerah (RPJPD) Provinsi Sumatera Utara, Tahun 2005 – 2025*, Peraturan Daerah Provinsi Sumatera Utara, 2008

[4] Presiden Republik Indonesia, *Percepatan pelaksanaan proyek strategis nasional*, PP 3 th 2016, 2016

[5] Direktorat Transportasi., *Laporan Implementasi Konsep Tol Laut 2015*, Bappenas Indonesia, http://nusantarainitiative.com/wp-content/uploads/2016/02/150915-Buku-Tol-Laut-bappenas.pdf, 2015

[6] Taneja, P., *The flexible port. Proefschrift*. With assistance of H. Ligteringen, W. E. Walker. Delft: Next Generation Infrastructures Foundation (NGInfra PhD thesis series on infrastructures, 59), 2013

[7] Osborn, D., Ullah, F., *Universal Sustainable Development Goals, Understanding the Transformational Challenge for Developed Countries*, Stakeholder Forum, https://sustainabledevelopment.un.org/content/documents/1684SF_-_SDG_Universality_Report_-_May_2015.pdf, 2015

[8] United Nations Department of Public Information, *Sustainable Development Goals. Guidelines for the use of the SDG Logo, Including the colour wheel, and 17 icons*, United Nations Department of Public Information, 2016

[9] Blanchard, B., Fabrycky, W.J., *System engineering and analysis*. NJ, USA: Prentice Hall, 2011

[10] Walker, W., Harremos, P., Rotmans, J., van der Sluijs, J., van Asselt, M., Jannssen, P., and von Krauss, M., *Defining uncertainty: A conceptual basis for uncertainty management in model-based decision support*. Integrated Assessment, 4(1):5-17, 2003

[11] Danneels, E., *Disruptive Technology Reconsidered, A Critique and Research Agenda*. Journal of Product Innovation Management 21 (4), pp. 246-258. DOI: 10.1111/j.0737-6782.2004.00076.x, 2004

[12] Dotsika, F., Watkins, A. *Identifying potentially disruptive trends by means of keyword network analysis*. Technological Forecasting and Social Change. *DOI:* 10.1016/j.techfore.2017.03.020, 2017

[13] Wright, R., *Container shipping lines mired in crisis*. Financial times. New york. Available online at https://www.ft.com/content/1e98963c-1853-11e6-bb7d-ee563a5a1cc1?mhq5j=e1, 2016

[14] OECD, *Competition issues in liner shipping*. Working Party No. 2 on Competition and Regulation. Edited by Ana Rodrigues. OECD. Genova, 2015

[15] Lewis, R.D., *When Cultures Collide. Leading Across Cultures*. Boston: Nicholas Brealey International, 2006

[16] NGO Shipbreaking Platform, *Global Dumpers List*. 4 volumes: NGO Shipbreaking Platform, http://www.shipbreakingplatform.org/annual-lists-of-scrapped-ships/, 2013-2016

[17] mrlee, *Off the Beach*, NGO Shipbreaking Platform, http://www.offthebeach.org/, 2013

[18] European Commission, *European list of ship recycling facilities*, http://ec.europa.eu/environment/waste/ships/pdf/list_ship_recycling_facilities.pdf, 2016

[19] World Shipping Council, *Top 50 World Container Ports*, http://www.worldshipping.org/about-the-industry/global-trade/top-50-world-container-ports, 2015

Structural Development for Well Service Activity in Limited Capacity Platform

Marciela Esa Dewa[1,*], *Muhammad* Arifin[1], and *Eki* Primudi[1]

[1]Facility Engineering & Integrity (FEI), PT PHE West Madura Offshore (WMO), Jakarta 12520, Indonesia

Abstract. The idea is to replace drilling rig, which is utilized for work over/well services activities, with working barge. The challenge comes from the operation and structural platform capacity since the majority of PHE WMO wellhead platform has limited capacity issue and braced monopod type. To minimize the risk of drilling equipment placed on the platform, engineering study is proposed to ensure the reliability of the platform. The process and method of operation during well service activity will also be developed. The proposed mitigation includes: Global Structural Check for Platform Integrity, Local Structural Check for Strengthening, Localized load for Additional Equipment, and Weather-Time-Frame Constraint. Using the proposed methods, PHE WMO can ensure the platform integrity and ensure SAFETY during well service activities.

1 Introduction

Pertamina PHE WMO as a Cooperation Contract Sharing (K3S) of SKK MIGAS operates in the West Madura Offshore block. Currently WMO operates 8 well-head platforms, and most of these platforms are brace monopod platforms. This type of platform has some advantages, especially from the installation side. However it has redundancy of only one main support.

Well services activity, which has been programmed periodically, is required to maintain the well integrity and production demand. Drilling rig has been much preferred to be utilized since the minimum impact to the platform, as the equipment are mostly placed on the rig itself.

Current oil industry situation demands efficiency. The use of Drilling Rig needs to be reviewed and the idea to replace it with Working Barge is proposed.

Replacing drilling Rig with working Barge will save the operational activities of well service thus saving operation cost. However, the safety and security of workers and facilities shall still be addressed. The equipment and the operation that was once placed on the Rig during well service must take place on the platform. As a result the platform capacity shall be revisited.

2 Data and Method

Working barge utilization for the well-service activity requires all related equipment to be placed on the platform. This is the main concern since the capacity of the platform is limited.

Several previous structural studies show the platform limitation, such as:

Table 1. WMO Structural Analysis Result

Document & Title	Remarks
PHEWMO-S-TEN-0012, "Engineering Technical Note - Plan of Additional 30in conductors at KE-30"	Executive Summary: "it is not recommended to add any additional equipment loaded on deck more than 5.0 MT (due to pile safety factor in storm condition is 1.5 nearest to the minimum required as per API RP2A)".
PHEWMO-PHE38A-S-RPT- 0001 Rev.1, "In Place Analysis Report for PHE-38A Platform"	Executive Summary: "Based on in-place analysis result, it is found that the PHE-38A platform is non-compliance to the code. However, the platform is still in marginal integrity for operation …"

After identifying the condition and determining the objective, two solutions are then proposed:

- Well Service Method & Operation
- Global and Localized Platform

This is also including barge positioning, in regards to the pipeline and weather constraint.

* Corresponding author: marciela.dewa@gmail.com ; marciela.dewa@pertamina.com

https://doi.org/10.1051/matecconf/201814705007

3 Result and Discussion

3.1. Well Service Method & Operation

The Well Service Method & Operation relate with the necessary mitigation for well service activity. These include:

1. Well Service Equipment
2. Barge Type
3. Barge Positioning

Determining the well service equipment is needed to accommodate the platform limitation, both capacity and space. Criteria for the equipment shall be:

- Well Perform
- Lighter (than conventional equipment)
- Space efficient type

The proposed equipment is as follow,

Table 2. Proposed Well Service Equipment Weight and Properties

Type	Footprint (Ft.)	Weight (tonne)
Slickline & E-Mast	19,7 x 1,5 x 1,5	4,5
Coil Tubing Jacking Frame	10 x 12 x 16,25	10
Hydraulic Workover Unit	17 x 12 x 54	20

Barge type shall consider the necessary capacity carrying the well service equipment and safety of operation. Barge shall have minimum capacity of 3200tonne and clear deck space of 45 x 15 m2. Minimum 4 (four) points mooring are also necessary in accordance with PHE WMO marine regulation.

The barge positioning shall also consider wind prevailing direction, crane's platform, pipeline, and conductor location for safety operation. The proposed locations are as follow:

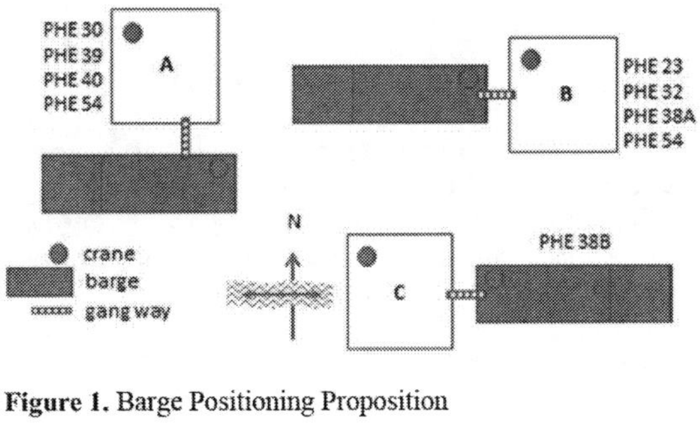

Figure 1. Barge Positioning Proposition

3.2 Platform Capacity

Platform capacity is based on structural analysis using standard industry code (API, AISC). It depends on many factors including: equipment load, environmental condition, material (member, pile) capacity (yield strength), etc.

The proposed assessment is outlined on the following chart,

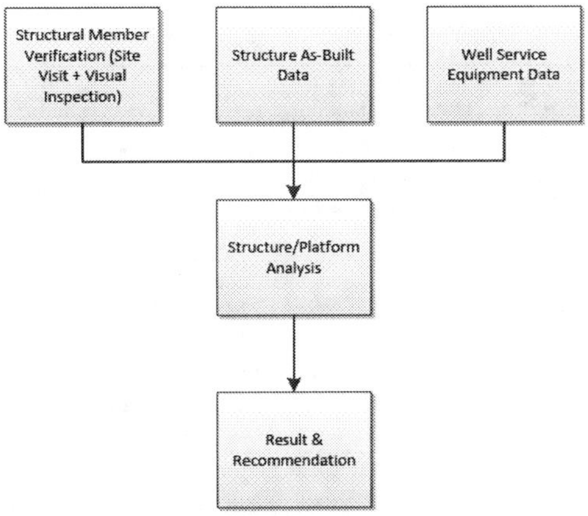

Figure 2. Flow Chart of Platform Analysis

Platform analysis using SACS software divided into two: (1) Global Platform Analysis and (2) Local Platform Analysis.

- Global platform analysis determines the overall platform capacity. Jacket and Pile capacity are the main indicators of this analysis.
- Local platform analysis provides structural capacity at specific locations. This analysis does not affect the integrity of the entire platform, but this damage still poses hazard/ safety risk. By knowing the local capacity of member, mitigation can be done to minimize the risk hazard. For example by strengthening the member or reduce the load on a particular part of the platform.

Structural analysis was performed for all active well-head platform. However, this essay takes one example that represents the most, which is The Structural Review of PHE-38A Platform.

Figure 3. PHE-38A SACS Model

PHE-38A platform has the most wells than any other wellhead platform. There is also a tilting issue (1.08° based on 2014 measurement). PHE-38A platform details are as follow,

Table 3. PHE-38A Platform Specification

Item	*Remarks*
Water Depth	+/- 195 ft.
Platform Type	Braced Monopod (1 caisson with 2 batterpile)
Jacket Diameter	61"
Pile Diameter	56"
Jacket Elevation	EL.(-) 79' – Bracing EL.(-) 130' – Bracing EL. (-) 195' – Mudmat
Topside Elevation	EL. (+) 54' – Main Deck EL. (+) 42' – Mezzanine Deck EL. (+) 32' – Cellar Deck EL. (+) 21' – Sub-Cellar Deck

The result of the structural analysis shows that some members have member check exceeding 1.00 (the ratio between the applied load versus maximum allowed load). Although the capacity has been exceeded, the structure is still operational. This is due to the contingency of loading and material.

To assure the structural capacity, several mitigations proposed:

- Directional Wave Analysis
- Localized Loading
- Time-Framing Environmental Condition
- Local Structural Strengthening

Wave load plays significant effect to the offshore structure. The current analysis uses the Omni Directional Wave/Current assumption. This is true for the new platform design despite conservative result, since it takes the maximum value of each load. Yet, for Brownfield development the Actual Directional Wave is used.

The result of the analysis using Directional Wave shows significant decrease in member check,

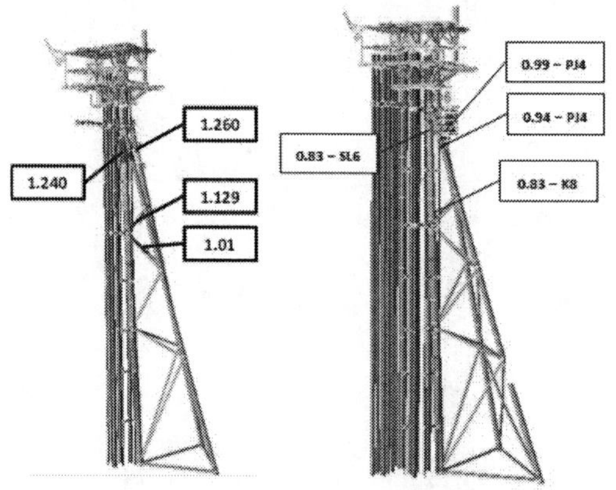

Figure 4. Analysis Result (Before – After Wave Directional)

Localized loading for well service equipment also need to be considered. This localized load assures the member integrity to avoid structural failure. Structural analysis is also performed to localize area, which is suitable for equipment laydown.

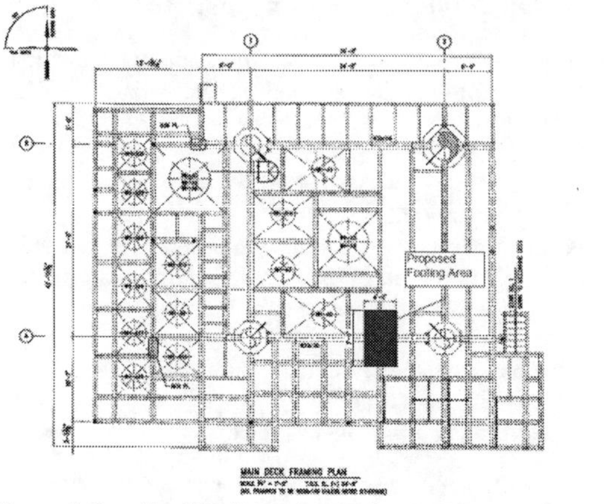

Figure 5. Localized Well Service Equipment Placement

Time-Framing Environmental Condition is based on the result of structural analysis. Offshore analysis considers two conditions, which are normal-operating and storm environmental conditions.

The result of structural analysis shows that the platform is able to accept loads with wave conditions below 2.5 meters (or non-monsoon conditions). This condition limits well service operation for monsoon condition, since the structural failure risk is greater (in addition to the barge mooring condition).

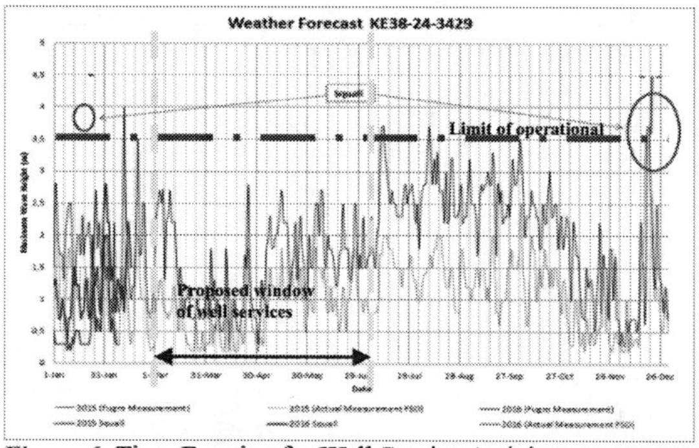

Figure 6. Time-Framing for Well Service Activity

Local Structural Strengthening is the least mitigation for local structural failure. Proposed strengthening is adding temporary columns or support during well service operation. Additional support will distribute excessive well service load to the main members.

https://doi.org/10.1051/matecconf/201814705007

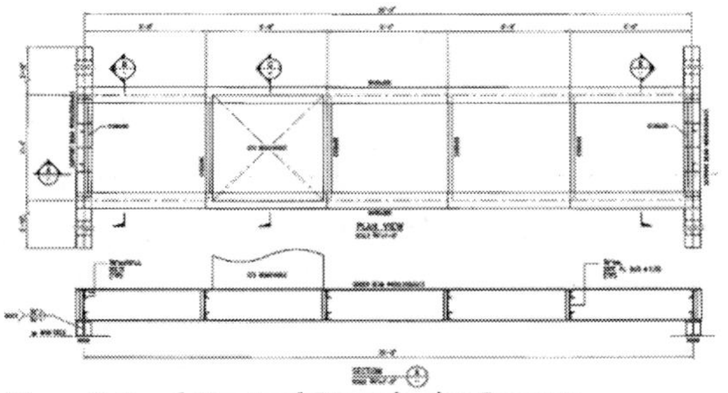

Figure 7. Local Structural Strengthening Supports

4 Conclusion

It can be concluded that the barge utilization within platform structural limitation is applicable for Well Service Operation under several conditions. These conditions including engineering assessment and operation method shall be followed to minimize the risk of well service activity to the platform and ensure safety of operation. This mitigation shall be supported with the reliability of inspection to assure mitigation and control.

Implementation of working barge utilization is expected to be executed in early 2018. The cost efficiency is expected to be achieved with the rental cost of working barge is lower than drilling rig.

References

1. API RP 2A-WSD, Recommended Practice for Planning, Designing and Constructing Fixed Offshore Platform – Working Stress Design, 21st edition. (Dec 2000)
2. API RP 2A-WSD, Errata and Supplement 3, (October 2007)
3. Facility Engineering & Integrity PHE WMO, PHEWMO-FEI-S-TEN-0002: Engineering Technical Notes, Wellhead Platform Capacity due to Work Over/Well Service Activity, (2016)
4. Project Department PHE WMO, PHEWMO-S-TEN-0012: Engineering Technical Notes, Plan of Additional 30in conductors at KE-30, (2015)
5. Facility Engineering & Integrity PHE WMO, PHEWMO-PHE38A-S-RPT- 0001 Rev.1: In Place Analysis Report for PHE-38A Platform, (2014)

Renewable Energy Potentials along the Bay of Bengal due to Tidal Water Level Variation

Myisha Ahmad[1,*], *G. M. Jahid* Hasan[1], *H. M. Mohaymen Billah* Chy[1], and *Gazi Masud Md.* Mahenoor[1]

[1] Department of Civil Engineering, Military Institute of Science and Technology (MIST), Dhaka, Bangladesh

Abstract. The projected increase in energy demand coupled with concerns regarding present reliance on fossil fuel and associated environmental concerns had led to increased interest in exploiting renewable energy sources. Among different renewable energy sources, tidal energy is unique and most suitable because of its predictable nature and capability to ensure supply security. Tide consists of both kinetic and potential energy which can be converted to electricity using well-proven technology. The potential energy of tides - the principal focus of the study, is stored due to rise and fall of the sea level. Head difference created due to tidal variation between basin side and sea side of a barrage stores potential energy which is converted into fast-moving water that rotates turbine and generates electricity. Bangladesh with its long coastline has promising prospects of tidal energy resource development. The study focuses on tidal energy resource exploration and exploitation along several competent locations of the Bengal coastline. Tidal records of flood and ebb tide of these locations are analyzed to calculate the potential energy. Finally, available potential techniques of energy extraction are evaluated for annually generated energy estimation. This study investigates the prospect and utilization of tidal energy concept and reviews the possibilities and opportunities of employment of the technology for sustainable development and climate change mitigation in context of Bangladesh.

1 Introduction

Energy is the pinnacle of human civilization. It is an indispensable input of survival for human. Currently, the consequence of the depletion and price fluctuation of the traditional energy resources are at the forefront of the global energy debate. Energy needs, their associated security of supply, coupled with environmental concerns and the impact of climate change requires constructive attempts to exploit non-polluting renewable energy sources. There is now an urgency to ensure sustainable development that will meet the demand of the present without reducing the ability to meet future demand. As per sustainable development goals 2030, the reliance on clean renewable energy resources has to be increased. The improved energy consumption efficiency and conversion measures have to be adopted to minimize the loss of primary resources without risking the climate. This has resulted in search of new energy generation methods with renewable sources which are promoted globally and therefore offer a window of opportunity for expanding utilization of renewable energy technologies through research and development activities.

Over the last thirty years, harnessing tides as a source of electric power have been a part of world trend that marks the introduction of renewable energy sources and the tapping of local potential for exploitation of these sources. There has been a concerted effort worldwide to make harnessing of renewable sources, particularly tidal power compatible with energy demands and environmental concerns. Tidal energy is unique because of its ability to provide predictable, clean power ensuring supply security. The tide and the resultant power is predictable and dependable, which makes the energy attractive in a future renewable weighted energy market when compared to wind and wave. The concept of conversion of tidal power for human use is very ancient, there being instances from Roman epoch describing its use for milling grain. Tide mills have also been used by the Spanish, French and British, date back to 787 A.D. In the Middle Ages tide mills have been used in Europe and Northern America [1]. The most recent endeavors at harnessing the tide have focussed on converting the hydraulic power of tides into electricity.

Currently, there are three major options for generating power from the tides: a barrage, a fence, and a tidal farm. The foremost method, which is the main focus of this study, is an efficient approach to power generation by placing a barrage across a tidal inlet. The barrage requires building a dam across an enclosed tidal basin. Sluice gates are employed along the structure that assists in filling the basin with the rising water level. The gates are then closed to develop a head difference and power is thereby generated through turbines with the falling of tides.

According to Rathod and Gaikwad [1], the first commercial-scale modern-era tidal power plant was built, near St. Marlo, France. In 1966, the first tidal barrage was completed in France with commercial level electricity generated by La Rance power station. Between 1966 and 2011 a number of small tidal plants were constructed in countries with an abundance of tidal energy resources namely, Canada, China, Iran, Russia, Korea, the UK, Mexico, Brazil, and India [2]. The largest and the newest tidal barrage in the world is the Sihwa dam in north-eastern South Korea, built in 2011 and operational from 2012. It is a multi-functional tidal barrage which prioritizes ecological water quality improvement.

In Bangladesh, Karnaphuly hydropower station is the only hydropower plant located at Kaptai, Chittagong having a capacity of 230 MW where potential energy of stored water is used. But with an existing long coastline, coastal energy potentials have not yet been explored which can be a great source of sustainable renewable energy in order to meet future demand. However, fewer studies have been carried out to explore the potential of electricity generation from tides along the Bay of Bengal. Hossain et al. [3] studied the potentiality at Sandwip Channel with an average tidal range of 4.6 m, 23 sluices and a barrage having the capacity to generate 16.49 MW of electricity daily. However, none of these studies have been implemented practically till date at least to the author's knowledge.

In this paper, the main subject is to take a closer look at the potential of tidal energy resource along the Bay of Bengal. In addition, the work establishes a constructive approach that could be assisted for technology and project development aspects, through analysis of results based on experimental and field observations. The aim here is to highlight the future prospects of energy production utilizing the tidal resources, and also to project the approaches needed in the development phases of the technology; because tidal energy might be a vital, viable and reliable option of the future energy market.

2 Tidal Energy

2.1 Origin of Tide

Tides are caused by the gravitational pull of sun and moon on earth's oceans, coupled with the relative position of sun and moon and the rotational force of earth. The rise and fall of tides create potential energy while the flows due to flood and ebb currents create kinetic energy. Both of the energy forms can be harvested by tidal energy conversion technologies. This paper studies the former category that deals with the tidal range concept.

Tide cycles are calculated and can be predicted with high accuracy. It is possible to predict the tide levels of a particular location. The basic harmonic changes of tide level are also eminently predictable. Tide changes twice daily in the region where semi-diurnal governs. The difference between the high and low tide is the tidal head available for each tide cycle. On a global scale, some tidal head is as small as 0.5 meter to some are as large as 12 meters [4]. Broad mouthed estuaries create larger tidal range than long straight coastlines. Tidal power is directly proportional to the square of the tidal range and hence higher tidal ranges are the most attractive location for tidal power generation.

2.2 Energy Extraction Techniques due to Tide

Tidal barrage technologies harvest the potential energy created by the difference in head between ebb tide and flood tide. Tidal barrages are built across suitable bay or estuary, are designed to extract this potential energy using turbines located inside the barrage. The potential energy caused by the water level difference across the barrage is converted into kinetic energy in the form of fast-moving water passing through the turbines. This, in turn, converted into rotational kinetic energy by the blades of the turbine. The spinning turbine drives a generator to produce electricity. The highest output is achieved by operating these turbines at the highest head. The highest head is available at extreme low tide or extreme high tide. The period of the high and low tide is roughly two hours but the half an hour before and after each high and low tide the water level does not change significantly. Thus a total generation period of three hours is achieved in each cycle in semidiurnal areas. Using the combination of flow direction through the turbine and arrangements of the pool commonly five different methods of power generation are employed, explained by Zabihian and Fung [4]. Brief of each of these methods are summarized as follows:

2.2.1 Single Pool Ebb Type Generation

This is a one-way generation method. Water flows from basin-side to sea-side. Power is generated during the low tide or ebb tide. The sluice side of the barrage is kept open during flood tide and the basin is allowed to fill. At the beginning, both the seawater and basin water will rise up to M as depicted in Figure 1. When the basin and seawater reaches the same level, both the sluice side and turbine portion of the barrage is kept close, till the

 https://doi.org/10.1051/matecconf/201814705008

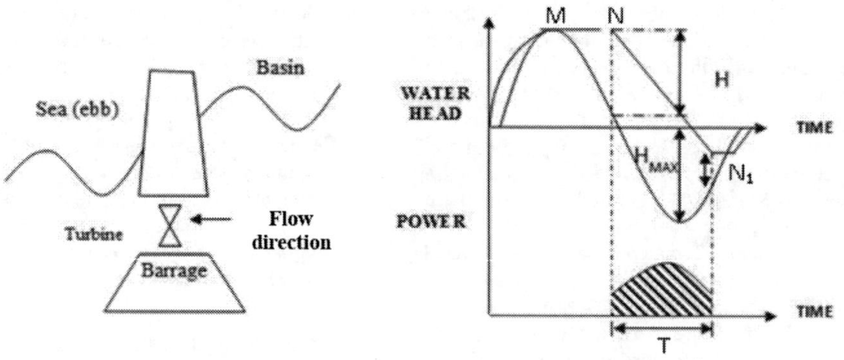

Fig. 1. Schematic diagram of a single pool ebb system.

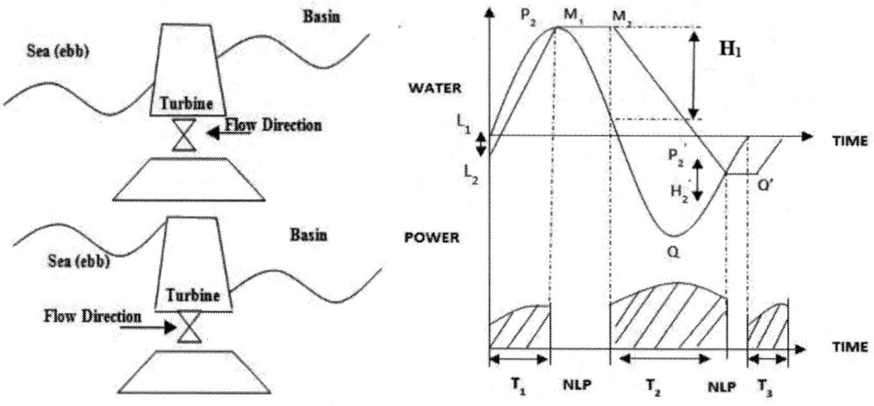

Fig. 2. Schematic diagram of a single pool two way (flood and ebb) system

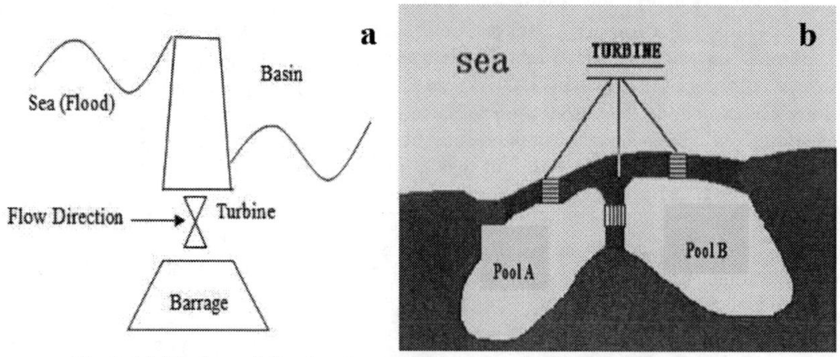

Fig. 3. (a) Single pool flood system and (b) two pools ebb and flood system.

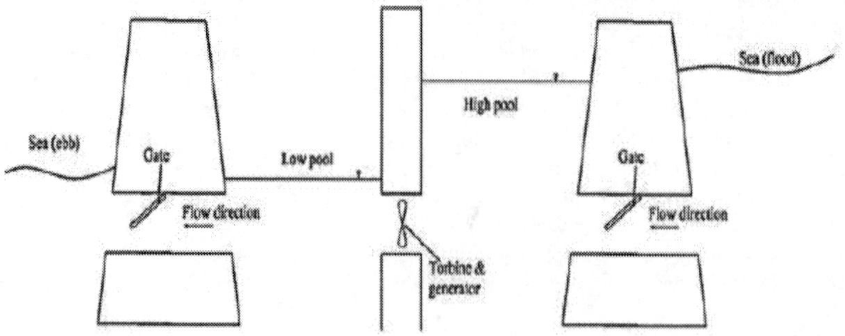

Fig. 4. Schematic diagram of a two-pools one-way system [4]

seaside water level fall due to ebb tide and head H is created. When sufficient head is achieved the gates are opened at point N. The water is allowed to pass through the turbines for the duration of T till the basin water level reaches point N_1. At this point, both turbine and sluice are kept close, till the water levels on both sides are equal. The next cycle starts again when the water level is same on both sides of the barrage. This type of plant operates for five hours and remains off for seven hours in one cycle of the tide.

2.2.2 Single Pool Flood Type Generation

The single pool flood generation system is similar to the first one. The only difference in this mechanism is that the power is generated when the basin is filling. During the flood tide, the sluice gates and the turbines are kept close until a considerable head is formed. The sluice gates are then opened and water flows from the sea to basin (Figure 3a). Though the time period of the generation is almost same of the first method, the head difference is much less. As a result, power generation is lower than the first method.

2.2.3 Single Pool Two-Way (both Flood and Ebb) System

The third method uses the advantage of the bi-directional turbine. Electricity can be generated both during the flood tide while the basin is filling and during the ebb tide by emptying the basin. During flood, tidewater will pass through turbine and therefore there should be a head difference which is between point L_1 and L_2 (Figure 2). Eventually basin water level will level up with sea water level. At this point, no electricity will be generated. This time period is termed as No Load Period (NLP). This period will continue until sufficient head is built up due to falling of sea water level during ebb tide. At point M_2, when sufficient head has built up and the gates of turbines are opened in the opposite direction and basin water will fall. The generation will last until point P_2', when the head difference is not sufficient to generate electricity. Turbines will be shut but sluice will remain open till point Q. Sluice gates will be closed at point Q' when the water level will be same both at sea and basin side. After the next head difference is built, sluice gates open and next cycle begins. From the Figure it is obvious that total generation time is $T_1 + T_2$, thus the total generation is higher than the previous two methods. Sometimes water is pumped from the sea to basin side keeping the sluice closed to increase the head and thereby the generation.

2.2.4 Two Pool Ebb and Flood System

This method is a combination of first two methods. Where one of the two pools operates based on the single pool flood type system and another operates based on the single pool ebb type system. Usually, turbines are places between pool A and pool B to generate additional power while transferring water from A to B as shown in Figure 3b. This type of generation produces more power than single pool flood or ebb type system. But this requires large basin area and a minimum three barrage needs to be constructed across the channel.

2.2.5 Two Pools One-Way System (high and low pools)

The fifth method needs two separate physically adjacent pools i.e. high and low pools. The high pool is filled during the flood tide. The low pool is emptied on the other side during ebb tide. Power is generated during the transfer of water from the high pool to low pool through the turbine (Figure 4). This type of generation plant can generate power almost continuously, and the NLP becomes very short. It requires estuary or bays with very high tidal range. Sites with such requirements are very limited around the world.

3 Study Location

Bangladesh has a long coastline of 710 km with an average tidal variation of 2m to 5m [5]. A number of potential locations along the coast are suitable to be used as basin areas. Generally, sites having the high tidal range and large basin with a narrow mouth are preferred for barrage construction. For this study, total seven tide stations were selected along the coastal area stretching from Saint Martin to Hiron Point. Locations are displayed in Figure 5. The stations Hiron Point, Khepupara and Charchanga are surrounded by several tidal rivers, creeks and are located near the Sundarban region. The location of Hiron Point is near the mouth of Pussur River. Khepupara is located in the shallow region near west of Dimer Char and Pokhkhir Char. Charchanga is at the mouth of Meghna River on the western side of Hatiya Island and surrounded by a group of small islands. Jahaizzer Char is a newly developed island located near Swandip and has two distinct channel in its east and west side. The tide station at Cox's Bazar is nearly 10 km offshore from Mahes-Khali channel, and Saint Martin is located on the southern tip of Bangladesh. In these locations, the tidal range varies from 5 m in monsoon to 2 m in dry season on an average scale. All these selected seven sites have deep channels with a moderate basin area and narrow mouth for easier construction of tidal barrages.

4 Generation of Energy

4.1 Energy Estimation

Harnessing energy through construction of a barrage and creating a reservoir has a principle very similar to conventional hydroelectric project installed in any river. According to Tousif and Taslim [6], mathematically, average potential energy E due to the height difference of flood and ebb tide in each tidal cycle for a single pool ebb type or flood type generation can be calculated using the following equation:

$$E = ½\rho g A H^2 \quad (1)$$

where ρ is the density of the sea water in Kg/m^3, g is the acceleration due to gravity in m/s^2, A is the basin surface area in m^2, and H is the water level difference between flood and ebb tide in m.

4.2 Data Collection and Analysis

Tidal records of the aforementioned seven stations for the year 2014 were collected from different organizations that maintain and operate tidal gauges at those locations, i.e. Bangladesh Water Development Board (BWDB), Port Authority Chittagong (PAC), Bangladesh Navy Hydro-graphic and Oceanographic Centre (BNHOC). Hourly tidal data was first filtered into daily high tide and low tide. Difference between the highest high tide and lowest low tide is the tidal range of a particular site for a particular day. The annual mean of these daily data was finally considered as the tidal range of a particular location. Moreover, based on the criteria of wider basin area and narrow channel mouth, the selected sites were analyzed through satellite images using Arc GIS and thereby, the probable basin area and barrage length was estimated. This area of the reservoir was used in energy calculation using Equation [1]. The basin areas are highlighted in Figure 6. The maximum area, thereby maximum generated power was obtained for Jahazzer Char West among the seven study locations. Figure 7 projects a graphical representation of the change of generated energy with a daily variation of tidal range, for Jahazzer Char West.

The tidal range data of the selected locations are used to estimate energy generation following one-way generation method. The main reason for selecting this method is that it provides a more effective method of generation with simpler equipment and easier installation. It is the most widely experimented method and most commonly used technology among the existing plants. Moreover, suitability of the method in the selected locations also influenced the preference.

5 Results and Discussion

As previously mentioned, in order to estimate the average power output of potential locations of the Bay of Bengal, one-way power generation method has been used. The annual mean difference of daily high and low tide is calculated to find the tidal range H. The basin area A is calculated with the help of Arc GIS and these values are then incorporated into Equation [1] to estimate the energy generated due to water level variation from the tide. This energy per lunar day (24 hr or 86400 sec) measures the generated power P_{actual} for one-way power generation method using the following equation.

$$P_{max} = \frac{2 \times ½ \times \rho \times g \times A \times H^2}{86400}$$

$$P_{actual} = \eta \times P_{max} \quad (2)$$

where P_{max} is the maximum power generated by the tidal turbine for one-way power generation method and η is the coefficient of turbine efficiency which has been considered as 75% for this study. Table 1 illustrates the annual average power generation for the year 2014 in the selected seven sites.

It is observed that the tidal range reduces gradually from eastern part to the western part of the coast. Saint Martin and Cox's Bazar has a high tidal range of 2.22 m and 2.54 m respectively, but the

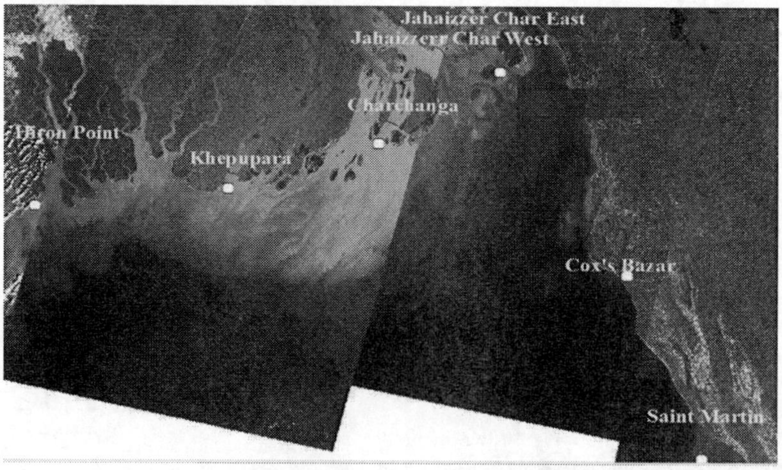

Fig. 5. Study locations along the coastline of Bangladesh considered for energy estimation.

 https://doi.org/10.1051/matecconf/201814705008

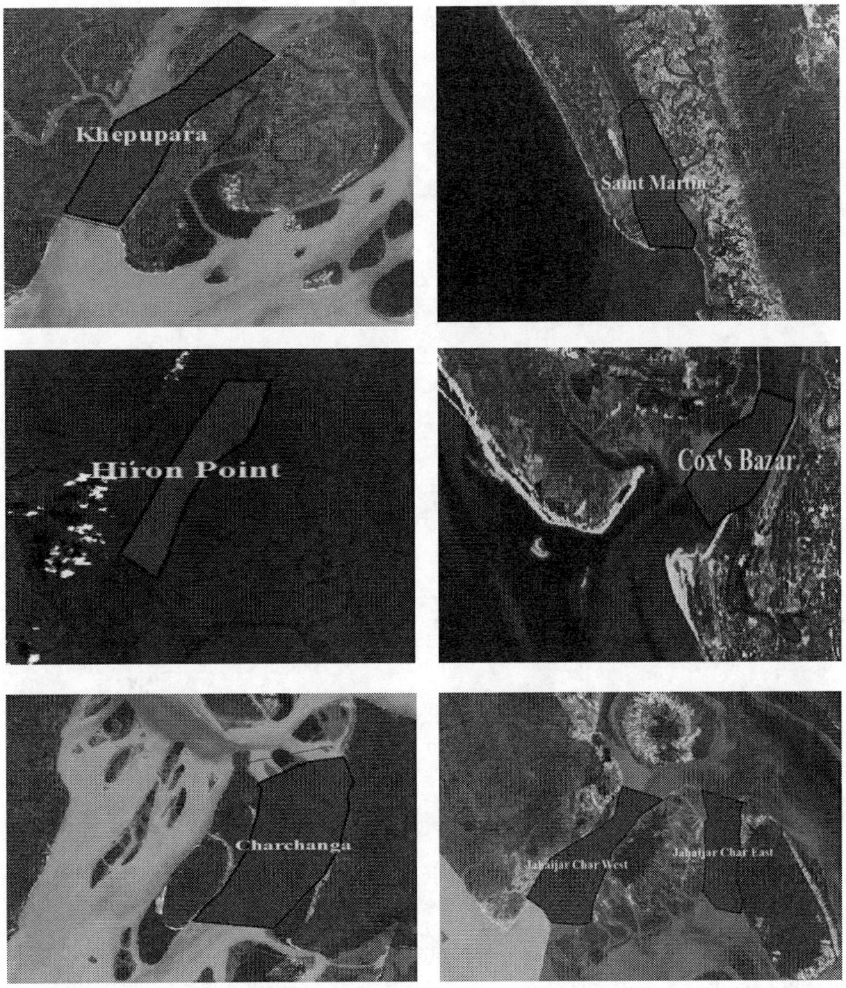

Fig. 6. Basin areas of all study sites based on the satellite image of November 2014.

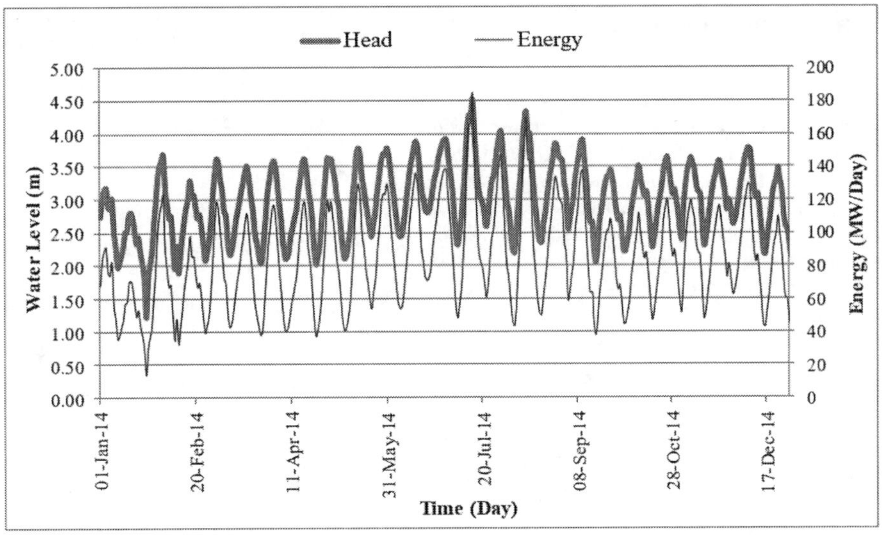

Fig. 7. Variation of daily tidal water level and estimated energy of Jahazzer Char East for the year 2014.

https://doi.org/10.1051/matecconf/201814705008

Table.1. Summary of the generated annual energy at the study sites.

Serial No	Station Name	Latitude	Longitude	Tidal Range, H (m)	Basin Area, A (km^2)	Annual Average Power Generation (MW)
L1	Saint Martin	20.59583	92.331388	2.22	38.13	18.1
L2	Cox's Bazar	21.439464	92.007732	2.54	90.26	55.04
L3	Jahaizzer Char West	22.37436	91.449405	2.94	103.66	80.73
L4	Jahaizzer Char East	22.37436	91.449405	2.94	60.5	47.06
L5	Charchanga	22.03662	90.92496	2.41	159.2	84.65
L6	Khepupara	21.81791	90.279263	2.05	68.3	28.2
L7	Hiron Point	21.72046	82.432934	2.04	35.8	13.78

locations are remote and grid connectivity will be difficult. Charchanga has a large basin area and moderate tidal range of 2.41 m. The tidal range of Hiron Point and Khepupara is low (2.04 m) and not suitable for an effective generation. Jahaizzer Char shows greater potentiality with a large basin area of 103.66 km^2 on the west part and a highest tidal range compared to the other stations. A maximum generation of 80.73 MW is obtained for this location considering an efficiency of 75%. The newly formed island has two deep channels in its east and west side. The channel has a narrow mouth and is suitable for barrage construction.

6 Concluding Remarks

In response to the growing concern of reduction in worldwide fossil fuel availability and impacts of climate change due to greenhouse effects and carbon emission, it is time to stimulate the society to alter energy policy towards renewable energy sources. Energy generation from renewable sources is a clean and environmentally efficient method. Though the generation capacity is not as high as other generation methods, the main advantage is that the resources are infinite.

Tide is one of the few clean energy resources which is predictable with high accuracy and thus allows a finite amount of daily supply to be firmly contracted. This paper briefly outlined the status of tidal range technology potentials along the Bay of Bengal. The analysis of tidal records of seven suitable locations along the coast shows that there are several sites available along the coastline with a tidal range of 2-3 m. Considering easier installation, cost-effectiveness, suitability in study area, one directional generation method is used for energy generation. Few sites like Charchanga and Jahaizzerchar has an annual average output up to 80 MW. It is observed that the tidal range is greater on the south-east part of the coast and gradually reduced to the western part of the coast. The tidal range of Saint Martin and Cox's Bazar is considerable but the absence of suitable basin location, also the locations being far away from grid connections, will increase the cost of connectivity with the grid. Charchanga has a larger basin area available and Khepupara and Hiron Point has narrow channel mouth but the tidal range varies very little in these locations. Jahaizzer char possess greater potentiality (80.73 MW) in terms of higher tidal range, large basin area, and narrow channel mouth. These locations show better potentiality of electricity generation from tide and thereby secure optimistic opportunities for marine renewable resources.

Nomenclature

A = Basin surface area (m^2)

E = Energy generated from tidal turbines (Joule)

H = H is the water level difference between flood and ebb tide (m)

P_{actual} = Power generated by tidal turbine for one-way generation method (Watt)

g = Acceleration due to gravity (m/s^2)

η = Turbine efficiency

ρ = Sea water density (1025 kg/m^3)

References

[1] G. J. Rathod, and G. R. Gaikwad, "Tidal Power Energy- A renewable energy in future", *ARD IJEET*, **3(3),** (2015).

[2] L. Hammar, J. Ehnberg, A. Mavume, B. C. Cuamba, and S. Molander, "Renewable Ocean Energy in the Western Indian Ocean", *J. Renewable and Sustainable Energy Rev.*, **16(7),** Pp 4938–4950, (2012).

[3] Md. A. Hossain, Md. Z. Hossain, and Md. A. Rahman, "Perspective and Challenge of Tidal Power in Bangladesh", *International Journal of Electrical, Computer, Energetic, Electronic and Communication Engineering,* **8(7),** Pp 1140–1143, (2014).

[4] F. Zabihian, and A. S. Fung, *J. Renewable and Sustainable Energy Rev.,* **15(5),** Pp 2461–2474, (2011).

[5] L. Rose, and P. K.Bhaskaran, "The Role of Environmental Forcing on Tidal Dynamics along Complex near-Shore Waters off Bangladesh", *J. Ocean Eng.,* **116,** Pp 68–81, (2016).

[6] S. Md. R. Tousif, and S. Md. B. Taslim, " Tidal Power- an Effective Method of Generating Power", *IJSER,* **2(5),** (2011).

https://doi.org/10.1051/matecconf/201814705009

Hydrodynamics Modeling of Khung Krabaen Lagoon, Chanthaburi Province, Thailand

Pokavanich Tanuspong[1,*], *Buranapratheprat* Anukul[2], and *Aryuthaka* Chittima[1]

[1]Department of Marine Science, Faculty of Fisheries, Kasetsart University, Bangkok, Thailand
[2]2Department of Aquatic Science, Faculty of Science, Burapha University, Chon Buri, Thailand

Abstract. Khung Krabaen Lagoon (KKBL) is a small low-inflow water body. There are vast areas of tidal flat occupied nearly 60% of the lagoon that host some of the most productive seagrass habitats in the region. The lagoon is surrounded by mangrove forest and intensive shrimp farms behind it. The KKBL was used as an intake and recipient water for the farms. However due some shrimp disease epidemics and possibly deteriorated water quality, the farms are now taking the intake water from the outer sea through very expensive (to construct and to maintain) irrigation system. Objective of this study is to investigate the KKBL's hydrodynamics using a numerical simulation model validated with measured data. The simulation model was setup two-dimensionally based on the Delft3D model. Results suggested that water currents inside, at the mouth and at the outer sea of the lagoon are mainly governed by tide and wind. Offshore of the lagoon, there are strong tidal currents flowing along northwest and southeast direction. The tidal currents flow into the lagoon through its mouth before dispersion rapidly inside the lagoon. Mean circulation largely varied seasonally and had direct correlations outer sea seasonal mean currents and the monsoons.

1 Introduction

Khung Krabaen Lagoon (KKBL), also known as Kung Krabaen Bay, is a small low-inflow water body situated at the eastern part of Thailand. According to the Kung Krabaen Bay Royal Development Study Center. The KKBL was formed naturally by mountain erosion and coastal currents generated Tombolo and gradually drop of the sea level. This form of water body origin is typical for a lagoon. Therefore the Kung Krabaen is called a lagoon rather than a bay in this study. Coverage area of the KKBL about 6.7 km2 (Fig. 1). Shape of the lagoon resembles shape of "Krabaen" meaning sting ray in Thai. The lagoon has narrow mouth and has its body oriented along northwest and southeast axis. Dimension of the lagoon along the axis is approximately 3.6 x 2.5 km. It has approximately 11 km long coastline which is surrounded by mangrove forest and vast area of tidal flats. The tidal flat occupied nearly 60% of the lagoon that host some of the most productive seagrass habitats in the region. The KKBL is a shallow water body and has no major freshwater source discharged into it.

The KKBL was used as an intake and recipient water for some intensive shrimp aquaculture complex in the country since 1986 (Sangrungruang et al., [1]). The complex is located right behind the mangrove forest. However due some shrimp disease epidemics and possibly deteriorated water quality, the complex is now taking the intake water from the outer sea through very expensive (to construct and to maintain) irrigation system. It has been reported also that there has been a decline of coastal fish stocks at the KKBL. Sangrungruang et al. [1] also reported that activities of the shrimp farming deteriorate sediment quality and adversely affected the lagoon's sea-shore communities. It is clear that human developments at the KKBL is not sustainable and to meet the challenges, not only proper planning, design, implementation of infrastructure, but also clear policy and strategy direction of the infrastructure are required through better understanding of the water system. Objectives of this study is to update and provide some fundamental understandings of the hydrodynamics of the KKBL in which described with insufficient details by Sasaki and Inoue [2].

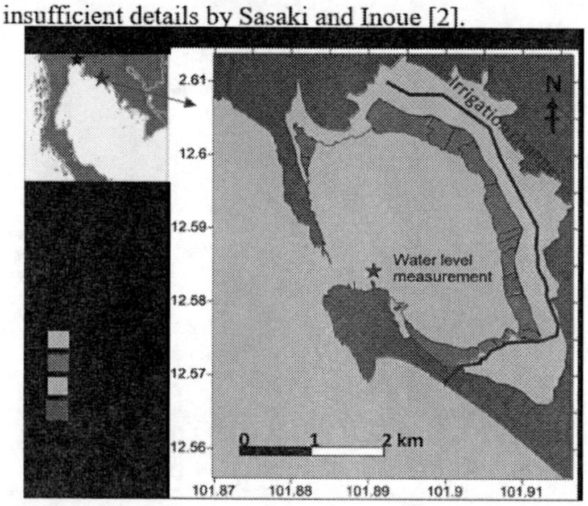

Fig. 1. Location maps of study site and measurement.

* Corresponding author: ffistop@ku.ac.th

2 Field Observation

There are very limited field observations for the physical oceanographic characteristics of the KKBL. In this study, two preliminary hydrographic survey were carried out between 24th April to 13th May 2017 to obtain spatial variation of water depth and temporal variation of water level. The water depth survey utilized an echo-sounder (HDS-9, LOWRANCE-USA) for the day and night survey to cover the whole area of the lagoon. Fig. 2 shows a photo of the model, model setting up on a boat and measurement track. The water level survey measured water depth temporal variation for 15-days using an automatic pressure logging sensor (HOBO Water Level Logger, ONSET-USA) at 30 minute interval. The sensor was placed in a PVC tube tighten to a stable concrete column at a fish cage located at 12° 34.96'N and 101° 53.42'E. Fig. 3 shows details of the pressure logger deployment. The recorded pressure data was convert to water depth by assuming constant water density and constant atmospheric pressure. The water depth were subtracted by its 15-day averaged value to produce the water level (from Mean-Sea-Level, MSL). Measured data from the depth survey was corrected to the MSL depth using measured water level in this study.

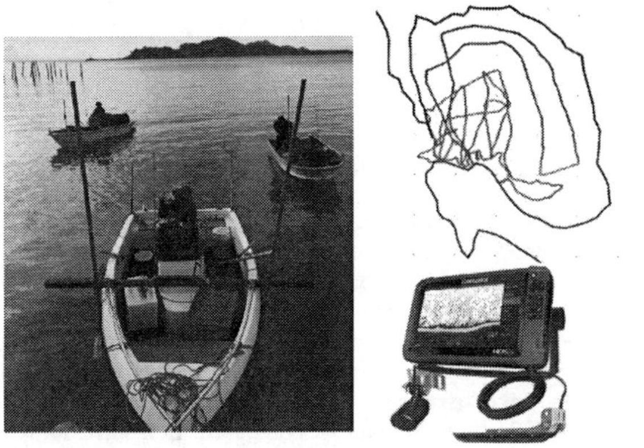

Fig. 2. LOWRANCE HDS-9 echo sounder, its field setup and measuring route of the water depth survey conducted on 24th April 2017

Fig. 3. HOBO water level logger and its deployment methodology for the water level variation survey between 24th April and 13th May 2017

3 Setting Up Numerical Model

In this study, two-dimensional (depth averaged) hydrodynamic modeling of the KKBL applied a well-established program named Delft3D-FLOW (Deltares, [3]). Two hydrodynamic models were set up, i.e., a coarse resolution model that covers whole area of the Gulf of Thailand (GOT) called GOT Model and a fine resolution model that covers whole coastal waters of the Chanthaburi Province including the KKBL area called Chathaburi model. Fig.4 show the overall coverage area and calculation mesh of both models. Fig. 5 shows interpolated depth data used for the coarse model and at the KKBL. Bathymetric data applied here were derived from a newly measured water depth conducted in this study, digitized navigational charts from the Hydrographic Department, Royal Thai Navy adjusted to the mean sea level and from the ETOPO1 1/60 data (http://www.ngdc.noaa.gov/mgg/global/global.html).

The GOT model were forced at offshore boundaries by a set of tidal constituents derived from time-series data from TPXO6.2: Global Inverse Tide Model (Egbert and Erofeeva, [4]). Time-series of simulated water level at five locations situated along the offshore boundaries of the Chanthaburi Model were forced that derived from the GOT model. The simulation for year 2011 applied spatial and temporal varying wind data at 6 hourly interval with 2.5 degree resolution from NCEP/NCAR Reanalysis 1 (Kalnayet al., [5]). The validating simulation for year 2017, applied 3 hourly measured wind data from Chathaburi Meteorological Station (12° 37.00N, 102° 06.80E) from Thailand Meteorological Department. Bed roughness coefficient was uniformly applied. General setup of the GOT and Chanthaburi model are given in Table 1.

Table 1. General setup of the hydrodynamic model for the KKBL

Item	GOT & Chanthaburi
Simulation period	1 January 2011- 31 December 2011
Validation period	1 January 2017- 31 May 2017
Mesh	Curvilinear grid in spherical coordinate
Time step	3 minutes & 1 minute
Initial condition	Uniform rest water
Bottom roughness	Chezy 100
Horizontal eddy viscosity	10 m2/s & 1 m2/s

https://doi.org/10.1051/matecconf/201814705009

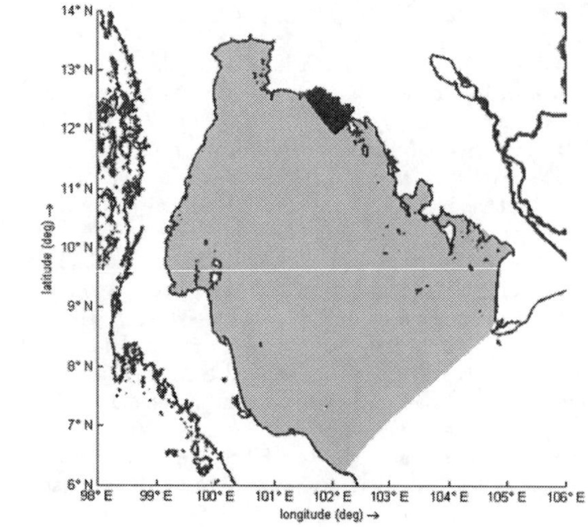

Fig. 4. Calculation mesh of the GoT model (grey) and Chathanburi model (blue)

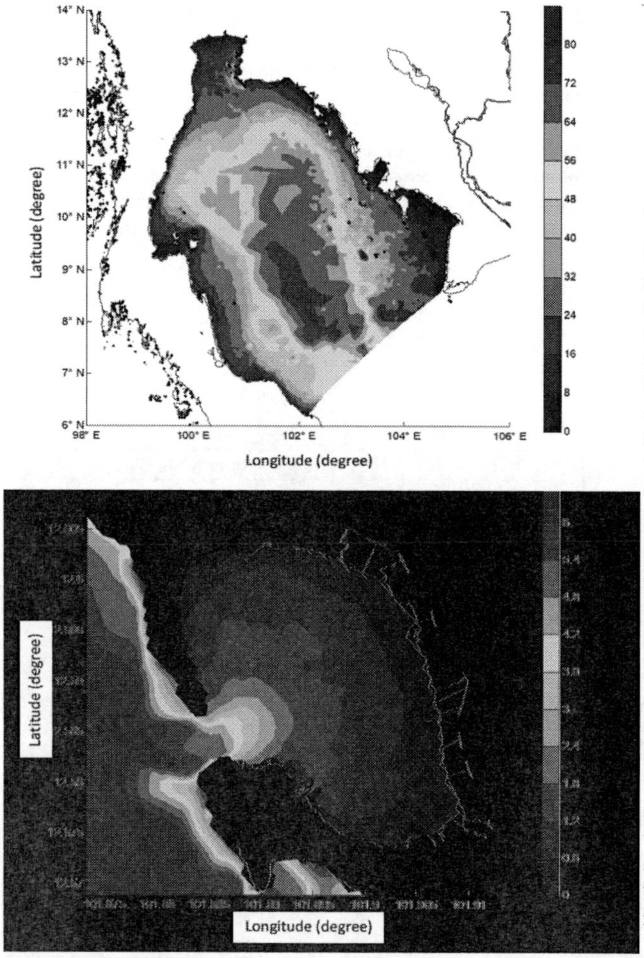

Fig. 5. Interpolated water depth data for the GoT and the KKBL area. Unit in meter.

4 Results and Discussion

4.1 Model validation

Hydrodynamic models were validated with water level data measured at a location inside the KKBL shown in Fig. 6. Results suggested that the Chathaburi Model can reproduce tidal water level variations very well in terms of amplitude and phase. This suggests that the model can reproduce the barothopic tidal currents and wind-driven current with acceptable accuracy.

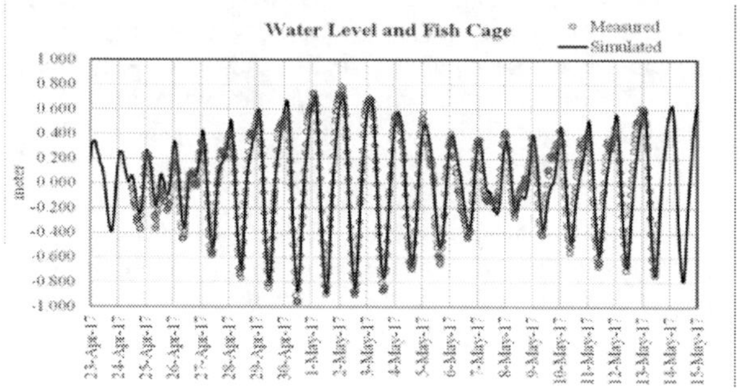

Fig. 6 Comparison between measured and simulated water level variation inside the KKBL for 15 days

4.2 Hydrodynamic characteristics of the KKBL

Results of this study, during the water level measurement period, tidal ranges during spring and neap tide are 1.6 m and 0.6 m, respectively. Detailed calculation in this study suggested that the lagoon has averaged depth is 0.77 m-MSL. Horizontal area of the lagoon is approximately 6.69 km2 and calculated volume of the lagoon at MSL is 5.14 km3. Due to the fact that the KKBL is very shallow compared to its tidal ranges. Tidal current plays dominant role in pushing and drawing water and material in the water in and out of the KKBL. Fig. 7 and Fig. 8 show current characteristics at different tidal phases at spring and neap tide, respectively. Note that there are vast areas dry up at low and ebb tide inside the bay (areas in brown color).

Tide at the KKBL is mainly diurnal that raise and fall generally once a day. Tidal currents at the mouth of the lagoon can reach the speed of 0.42 m/s during the spring tide and are around 0.15 m/s during neap tide. The currents gradually reduce their speed once enter the lagoon. At the middle of the lagoon, the current's speeds are 0.18 m/s during spring tide and 0.08 m/s during neap tide. General directions of the tidal current are northwest and southeast at the outer sea, northeast and southwest at the mouth and toward and from the coastline at the region inside the lagoon.

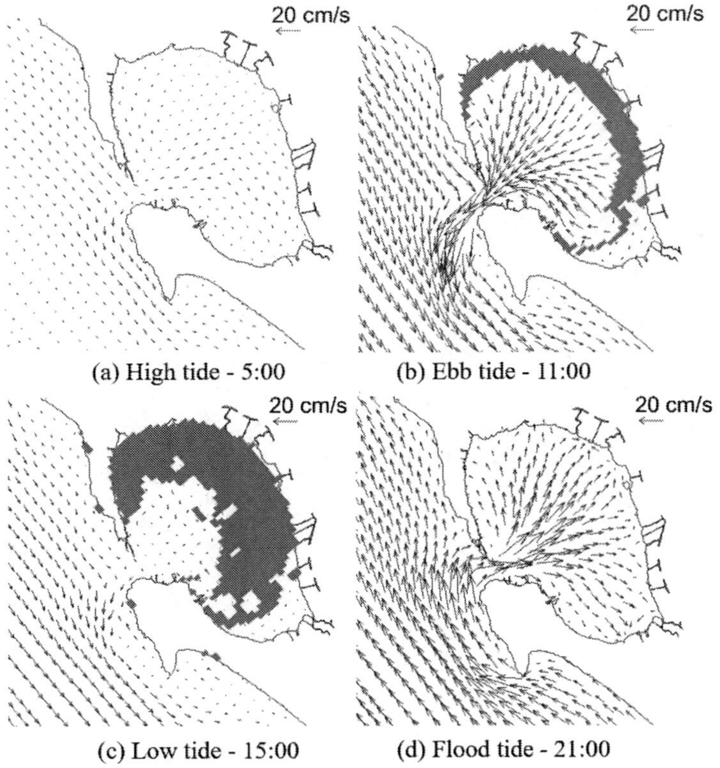

Fig. 7. Flow velocity patterns (blue arrows) and dry-up tidal flats (brown areas) at different tidal phase during spring tide on 13th January 2017

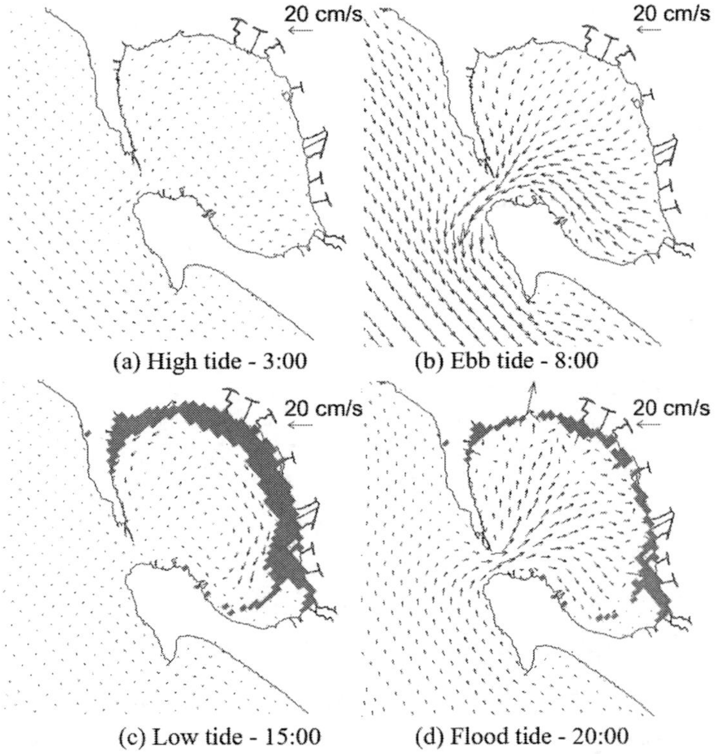

Fig. 8. Flow velocity patterns (blue arrows) and dry-up tidal flats (brown areas) at different tidal phase during neap tide on 20th January 2017

4.3 Mean water circulation affected by topography and monsoonal wind

Apart of investigating the short-term hydrodynamic features of the KKBL by tide, this study also studied long-term hydrodynamic characters of the lagoon affected by monsoonal wind by means of the monthly-mean flow velocity. Fig. 9 shows monthly residual flow velocity fields or monthly-averaged velocity fields of the lagoon at different monsoonal time in Thailand. According to a report prepared by Thailand Meteorological Department (https://www.tmd.go.th), climate in Thailand is dominated by Northeast (NE) Monsoon between middle of October to middle of February and by Southwest (SW) Monsoon between middle of May to middle of October. From April to middle of May are considered as a transition period between the two monsoons.

As can be seen in Fig. 9, there are distinct differences features of the mean flow velocity fields (mean circulation) at different monsoonal period. During the NE monsoon, offshore mean current flow toward northwest direction and generate clockwise gyre with diameter of around 1 km located in front of the mouth of the lagoon. There is a strong inflow current straight from the mouth flowed through the middle of the lagoon. The current deviated to right and left side once reaching the coastline and splitted the lagoon into two circulation cells, i.e. counterclockwise in the cell at the northwest and clockwise in the cell at the southeast. During the transition between NE monsoon and SW monsoon, the feature of the offshore mean current drastically changes. The strong northwest flowing current and the clockwise gyre at the front of the lagoon`s mouth relax. The strong inflow current from the mouth to the inner part of the bay disappeared and a counterclockwise gyre with the same scale of the mouth was generated. The mean current in the lagoon is sluggish and has not visible trend. During the SW monsoon, offshore mean currents rapidly flow toward the southeast direction. The gyres at the mouth disappeared. There is a strong out-going current flowing straight toward the mouth of the lagoon but deviate to northwest and southeast direction before leaving the lagoon. The deviated currents form two gyres inside the lagoon, i.e., the clockwise and counterclockwise gyres at the northwestern and southeastern part of the lagoon.

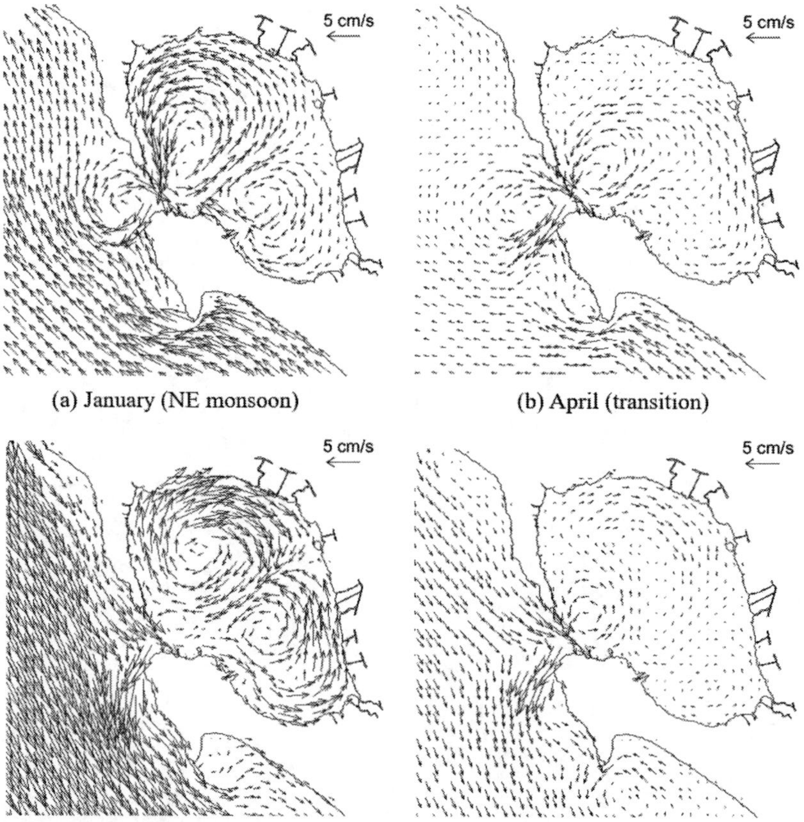

(a) January (NE monsoon) (b) April (transition)

(c) August (SW monsoon) (d) October (transition)

Fig. 9. Monthly residual flow velocity (averaged flow velocity) fields at different months in year 2011

5 Concluding Remarks

Present study reports about a hydrodynamic characteristics of a small shallow seawater lagoon with low-inflow. When compare the lagoon averaged depth and its tidal features, we speculated that the lagoon water column should be well-mixed especially during the spring tide. Two-dimensional hydrodynamics model of the KKBL was developed and the model perform very well when compared with the measured water level data at a station inside the lagoon. Results of the model suggested that tide and wind dominate overall hydrodynamics of the lagoon. While tides dictate the diurnal fluctuating current, the monsoon winds dominate overall water and material circulation patterns at the outer sea and inside the lagoon. Findings of this study also imply that there are important seasonal variations in residence time of water inside the lagoon and carrying capacities of the lagoon. To legitimately confirm this, future study is needed for the model validation against the flow velocity and other physical parameters. Three-dimensional model to investigate of effects of water temperature and seawater salinity variations to the hydrodynamics of the lagoon is needed. Details circulation patterns, the characteristics of mixing and transport pathway should be further investigated in the near future. They will be useful for the water resource management of the lagoon.

Acknowledgement

The authors would like to convey their sincerely thanks to the Southeast Asian Fisheries Development Center (SEAFDEC) for providing the echo-sounder for the bathymetry survey especially to Mr.Isara Chanrachkij and Dr.Nopporn Manajit for their full supports, to the Kung Krabaen Bay Royal Development Study Center for permission to install instrument and all assistance for the water level measurement and to the Thailand Meteorological Department for providing the wind data. This study is a self-funded research project under Kasetsart University Research and Development Institute.

References

[1] C. Sangrungruang, W. Sakares, M. Anongponyoskun, B. Damrak, KKBRDSC, (1999)
[2] T. Sasaki, H. Inoue, In: Mangrove Estuarine Ecology in Thailand, Thai-Japanese Cooperative Research Project on Mangrove Productivity and Development, (1985)
[3] Deltares, Delft3D-FLOW User Manual, Hydro Morphodynamics, (2011)
[4] G.D., Egbert, S.Y. Erofeeva, J. of Atm. and Ocean. Tech. 19(2) (2002).
[5] E. Kalnay, and Coauthors, Bul. of Amer. Met. Soc. 77, (1996).

MATEC Web of Conferences **147**, 06001 (2018) https://doi.org/10.1051/matecconf/201814706001
SIBE 2017

A Global Review of Public Private Partnerships Trends and Challenges for Social Infrastructure

Adrianto Oktavianus[1,*], Iris Mahani[1], and Meifrinaldi[1]

[1]Construction and Management Research Group, Faculty of Civil and Environmental Engineering, Institut Teknologi Bandung, Indonesia

Abstract. In developing countries, the government which has limited budget for public infrastructure development should choose which infrastructure should be developed. Most countries decided to build more economic infrastructure than social infrastructure because former have direct economic impact for society. The involvement of private sector in public infrastructure financing has been accomplished for decades in the form of Public Private Partnership (PPP). However, the implementation is also more often for economic infrastructure, but some countries have started to implement PPP for social infrastructure (education, healthcare, care of the elderly, etc.) when they think to add human capital and improve quality of life. This study attempts to review a set of public private partnership implementation models relevant for social infrastructure development in some countries. Moreover, this study also more explores to the challenges and issues in different areas of social infrastructure. The outcome is to show a trend public-private partnership for social infrastructure in some successful projects from different countries. The challenges and issues about implementation public-private partnership for social infrastructure also be a part of the results from this study. Finally, the study has a valuable input for implementation of PPP on social infrastructure in Indonesia.

1 Introduction

Since 1980s, public private partnership (PPP) is gaining global popularity as a viable alternative to public funding for building and financing infrastructure projects (Mital, 2016) [1]. PPP is a means to effectively deliver projects in the public sector because it emerges against the backdrop of financial constraints and management capacity in the public sector to support largescale infrastructure projects (Ahmed, 2013) [2]. PPP not merely means a way to accomplish financial problems by bringing private capital, but it also aims to bring private sector efficiency and best practices in delivering infrastructure. Although, focus of the current debates is primarily on economic infrastructure, but investment on social infrastructure has received surprisingly little attention and it is predicted to continue growing (Inderst, 2015) [3].

Social infrastructure investment is important to the society, so actually it is also being a considerable factor in the economy. Inderst (2015) describes that PPP is about twice as important in social as in economic sector, even though the PPP on social infrastructure in Europe runs only 6-7% and is still concentrated in a few countries [3]. Meanwhile, Preqin (2017) states that social infrastructure emerges with accounted for a promising proportion (6%) of deals completed in 2016, and represent around one tenth of deals completed since 2009 [4].

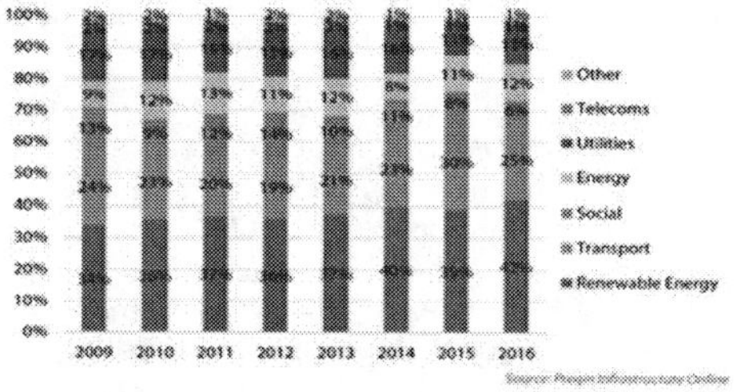

Fig. 1. Infrastructure Deals by Industry, 2009-2016.

Moreover, Inderst (2015) presents that Preqin (2014) also has reported approximately 6,000 infrastructure deals. Of these, around 13% of total deals equate to social infrastructure deals, with an estimated aggregate deal value of $150bn [3]. In terms of geographic scope, data from Preqin (2014) shows that Europe is the dominant region in terms of total deals completed, accounting for 78% of social infrastructure deals. North America accounts for 10% of deals, Asia 3%, and countries outside these three regions represent 9% of completed transactions Inderst (2015) [3]. The largest number of deals took place in the United Kingdom, representing 64% of all social infrastructure deals completed between 2008 and 2014. So, it indicates that

* Corresponding author: oktavianus.adri.@gmail.com

social infrastructure spending tends to be more significant in a country's economic development.

Indeed, in general, PPP implementation in a country is different from other countries, depending on many factors such as social, economic, and political factors. For example, Germany has a challenge about institutional task force through arrange responsibilities and coordination amongst the parties, meanwhile Columbia which begin to apply PPP model to social infrastructure face an interesting problem about risk allocation because the uncertainty associated to each risk. In other side, so many countries face other important issues related to policy and government support for implementing the PPP in social infrastructure. However, the number of problems indicates that the implementation of PPP in social infrastructure is actually happening and becoming a growing topic.

The variety of problems and challenges faced by each country actually enriches the understanding of this partnership model. Moreover, if it is associated with social infrastructure, this partnership model has an interesting side where there is no contribution of income from users. Besides, the size and form of cooperation used by each model of partnership between public and private does not have one standard form. Therefore, this study focuses on a review of PPP trends undertaken by several countries that have applied this PPP model to social infrastructure and analysed the challenges faced as well as emerging issues. Furthermore, this study provides lessons learned for countries that want to implement the partnership in social infrastructure, such as Indonesia.

2 Analytical Framework

This research strategy is divided into three parts. First, this study begins with a literature review on concept and characteristics of PPP and social infrastructure. Understanding about those concepts brings out the basis to know way to implement PPP in social infrastructure beyond the diverse conditions in each country. The notion of PPP in social infrastructure that explained by PPP theory and the function of social infrastructure are the detailed elaboration in this part.

The second part is exploration about international experiences by comparing PPP in different sectors and countries. The investigation highlights some significant factors concerning the critical factors, problems, and challenges. The review of the PPP system some countries from different continents. The detail in this part aims to see how the application of PPP on social infrastructure. This section explains the international best practices in using the PPP to build social infrastructure that adjusted with the different characteristics of each country. A deeper analysis of the PPP implementation allows getting lesson learned of delivered PPP projects. The findings of the analysis are associated with theoretical explanations.

Finally, the analysis concludes the important factors that should be consider implementing the PPP and the challenges that the private and public must face. This conclusion can be developed in the form of a model of cooperation that can be done by the public and private sector, and their positive influences in delivering social infrastructure in Indonesia.

3 Concept and Characteristics of PPP and Social Infrastructure

3.1 Public Private Partnership

Public Private Partnerships (PPP) can be depicted as a long-term contractual agreement and relationship between the public authorities and private sector companies. PPP is a delivery method that aim at financing, designing, implementing and operating public sector facilities and services (Beck, 2010) [5]. In other word, PPP which defined as private-sector investment in public infrastructure have the following key characteristics (Yescombe,2007) [6]:

- The authors, in the form: initials of the first names followed by last name (only the first letter capitalized with full stops after the initials),
- a long-term contract (a 'PPP Contract') between a public-sector party and a private-sector party;
- for the design, construction, financing, and operation of public infrastructure (the 'Facility') by the private-sector party;
- with payments over the life of the PPP Contract to the private-sector party for the use of the Facility, made either by the public-sector party or by the public as users of the Facility; and
- with the Facility remaining in public-sector ownership, or reverting to public-sector ownership at the end of the PPP Contract.

This partnership model provides benefits for both parties. For the public sector, PPP improve value to the public against conventional projects in terms of reducing risk, cost effectiveness, higher level of service, efficiently and quicker completion. In addition, PPP reduces public capital investment because PPP joins public and private capital in the hand of the private partner but under monitoring of the public partner.

Meanwhile, the private sector also has great benefit in PPP model. They have the possibility to implement better solutions, new technology and improvements during the building process. PPP also gives the private sector access to secure, long-term investment opportunities, so they can generate business with the relative certainty and security of a government contract. Besides, private sector get advantage from PPP by achieving efficiencies based on their managerial, technical, financial and innovation capabilities.

PPP actually does not have a standard defined form. PPP includes a range of structures and concepts which involves the allocation of risks and responsibilities between the public and private sectors so that many variants of the model of cooperation. Based on the nature of the service and risk transfer inherent in the PPP Contract., there are two categories, namely (Yescombe,2007) [6]:

• User payment, traditional payment systems under the concession model, characterized by a direct relationship between private partners and end users. The private sector provides services to the public and obtained concession rights can charge service fees to users of the service.

• Availability based payment, or Private Finance Initiative (PFI) model, the private partner to provide and manage the infrastructure for public authorities. In this model, the remuneration for the private partner does not take the form of fees paid by users of the service, but the periodic payments by the public side based on the level of service provided.

These two categories are commonly associated with two types of infrastructure: economic and social infrastructure. Economic infrastructure uses user payments as income for private sector. Meanwhile, social infrastructure, where there is no user payment, uses an availability-based payment model with a support from the public sector. However, few infrastructure projects using PPP models also refer to this understanding.

The PPP model can be much more diverse and tends to require closer management, e.g. Through central government PPP Units, PPP laws or standardized forms of PPP Contract, as discussed below. Some projects provide an important lesson in the success of other projects in the term of the implementation of PPP.

3.2 Social Infrastructure

Public infrastructure is one of the drivers of sustainable growth and act as an enhancer of a nation's competitiveness. Public infrastructure can be defined as facilities which are necessary for the functioning of the economy and society (Yescombe, 2007) [6]. These are thus not an end in themselves, but a means of supporting a nation's economic and social activity, and include facilities which are ancillary to these functions, such as public-sector offices or accommodation.

World Economic Forum (2012) defines three types of infrastructure, i.e. economic, social, and soft infrastructure [7]. The economic and social infrastructure can be categorized as hard infrastructure which refers to the large physical networks necessary for the functioning of a modern industrial nation. The definition of social infrastructure is the assets to support the provision of public services (World Economic Forum, 2012) [7]. Examples include social housing, healthcare facilities, educational institutions and green infrastructure (green multifunctional space in urban areas, such as parks, gardens and green corridors, which increases social livelihoods and encourage biodiversity). Meanwhile, Economic infrastructure defines as projects that generate economic growth and enable society to function which include transport facilities (air, sea and land), utilities (water, gas and electricity), flood defences, waste management and telecommunications networks.

New Zealand Social Infrastructure Fund (NZSIF) in 2009 has similar definition which mentions that social infrastructure is a subset of the infrastructure sector and typically includes assets that accommodate social services [8]. It is defined more clearly using example that social infrastructure assets include schools, universities, hospitals, prisons and community housing which show in the following table for detailed further.

Table 1. List of Social Infrastructure Assets Examples.

No	Sector	Examples
1	Health	• Medical facilities • Ancillary infrastructure (e.g. offices, carparks, training facilities)
2	Education	• Schools (primary and secondary) • Tertiary facilities • Residential student accommodation
3	Housing	• State or Council housing • Defence force housing
4	Civic and Utilities	• Community & sports facilities • Local government facilities • Water and wastewater treatment
5	Transport	• Bus stations • Park and rides • Availability-based roading (excluding demand-risk toll roads)
6	Corrections and Justice	• Prisons • Court houses

Source: NZSIF, 2009 [8]

According to Blanc-Brude (2012), one of its main characteristics is that social infrastructure projects provide public infrastructure assets and services in exchange for revenue streams paid directly by the public sector [9]. This is contrary to economic infrastructure, such as toll roads, which collect revenue from end users. That means that social infrastructure is not generating revenue from users but is paid by the public sector in the long term.

Some other characteristics that occur in some social infrastructure projects using PPP are that most projects in the field of health, education and other social logistics are relatively small in terms of project size. In addition, in terms of risk, PPP contracts for social infrastructure are often considered relatively "low risk" by investors because their projected returns are considered relatively stable and more certain. Although there is always an element of political risk, due to regulatory changes and unusual renegotiations in this area, or "social risk" and "reputation risk" if the project is opposed by some community groups.

4 International Experiences for PPP in Social Infrastructure

4.1 PPP in Canada

One of the problems faced by Canada is the need for housing that is safe, adequate and affordable for all Canadians. Ministry of Finance of Canada found their affordable housing a large influence on the whole aspects - from raising healthy children to pursue education, jobs and opportunities - that will enhance Canada's economic well in the end. Investment is necessary to meet those needs. By making new investments in social infrastructure - in such things as affordable housing and early learning and childcare, it will encourage inclusive growth for Canadians and elevate more Canadians - including children and the elderly - due to poverty.

These investments fall into the three investment priorities of the Government of Canada that are expected to lay the groundwork for future growth. The investment is planned for public transportation, green infrastructure and social infrastructure for 10 years starting 2016. As stage 1 of the infrastructure plan, the Government proposes to provide $ 11.9 billion over five years allocated for [10]:

• improve and improve the public transit system in Canada, for $ 3.4 billion over three years;
• investments in water, sewage and green infrastructure projects across Canada, for $ 5.0 billion over five years; and
• social infrastructure, including affordable housing, early learning and childcare, cultural and recreational infrastructure, and safe public health care facilities; For $ 3.4 billion over five years.

In addition, from the new funding, $ 1.2 billion will be invested specifically for indigenous peoples in the northern region. In more detail, the Budget to support social infrastructure in budget detail as follows [10]:

• $ 270.2 million over five years to expand and improve health facilities in the First Nations community;

• $ 732.0 million over two years to meet the housing needs of the reserves and in the Inuit and northern communities;

• $ 129.4 million over two years for the improvement and renovation of original early childhood care and care facilities and to support the establishment of an Early Learning and Childcare Framework;

• $ 76.9 million over two years to support the development of a combined cultural and recreational community; and

• $ 10.4 million over three years for renovation and construction of a new shelter for victims of family violence in the First Nations community.

As one of the strategies used in running the program, most of the $ 2.3 billion worth of investments is allocated to provinces and regions, which can identify communities where the greatest demand for affordable housing is. Then, to ensure that investment reflects the needs of the Canadian community, the Canadian Government also continuously consulted with stakeholders to determine where future social infrastructure investments should be undertaken.

In addition to these funds, the Government of Canada also supports community infrastructure priorities by providing approximately $ 3 billion annually in special funding for municipal infrastructure projects such as maintaining national parks and urban infrastructure. Partnerships with private parties must be a priority so that the Government diverts responsibility for PPP Canada Inc. to the Minister of Infrastructure and Society, so that the Government institutions are aligned to support the infrastructure innovation as well as possible. Cooperation with provincial, territorial and municipal partners can speed up spending and investment programs through PPP for existing infrastructure.

4.2 PPP in Chile

According to Third Global Infrastructure Investment Index (2016), Chile is the most attractive investment infrastructure market in Latin America [11]. Indeed, Chile's experience in infrastructure development is similar to that of many other developing countries around the world. It began in 1990 when the Chilean government faced worsening infrastructure and decided to seek other sources of financing. The government triggered a PPP concession program that was originally used to rebuild damaged infrastructure.

In fact, there are three phases of the PPP concessions that started from 1991 to 1994 where Chile expanded its highway network. Then, from 1995 to 2002, the Government focused on the construction of urban highways and airports. And finally, from 2003 to 2010, Chile concentrated on building social infrastructure such as hospitals, prisons, and public buildings.

Much of Chile's success in the PPP project is due to a regulatory framework that is directly linked to two things: concessions and 'investors and investments'. The government created a strong concession law that must touch many key areas to encourage investor confidence and encourage investment. The law should handle all phases of the PPP project, starting from the proposal until the completion of the asset is completed by the end of the concession agreement (Hill, 2011) [12]. The concession law aims to create "an environment that supports PPP and serves as a possible marketing Tool for investors." For more than twenty years, even though the system has changed, there are three key elements that are key to the implementation of the partnership.

First, Chile establishes a concession process in order to remain clear, transparent and fair so that the private sector can know the criteria in the evaluation of the offer. Offer details and open contracts for public access; as well as a fair process, without bilateral or reverse transaction negotiations.

Second, Chile's regulatory framework remains stable and predictable to ensure private sector investment because of the low risk of government expropriation and compensation for any unilateral changes clearly stated in the contract. Stability allows accurate pricing and project valuation. In the event of a dispute, foreign investors are convinced that they get their way through an arbitration or court, just like a national investor. In addition, policies in the investment and trade tax are also able to prevent foreign capital investment leaving the country.

Finally, the Chilean government and the public believe that the terms of the agreement will be respected because the contract with the private sector clearly determines the level of service and expectation throughout the concession (Hill, 2011) [12]. Chile's law allows termination of contracts for serious offenses by private partners and encourages compliance at every stage of the deal.

4.3 PPP in UK

United Kingdom (UK) is a prominent country in terms of private capital investment in infrastructure over the last few decades. The country has developed public-private partnerships in PPP models since the 1990s, they are developing a form of PFI for social infrastructure. In the UK, HM Treasury (2013) describes that there are more than 600 PPPs are in the form of a $ 100 billion Private Finance Initiative (PFI) arrangement for hospitals, schools, prisons, bridges, roads and military equipment (Roehrich et al, 2014) [8]. Inderst (2017) argues that infrastructure needs in the UK in the future are quite high, both economic and social infrastructure [14]. The need for economic infrastructure is expected to increase to 3-3.5% of GDP or £ 55-65 billion per year, while social infrastructure becomes 1-1.5% of GDP or £ 20-25 billion per year.

UK is also known as one of the countries that have expertise in developing PPP for social infrastructure. However, in fact, the UK still has many problems in the development of the PPP model. Roehrich et al (2014) states that PFIs in the UK are too expensive, too opaque, too slow and too inflexible [8]. This has caused the private sector to have a more pronounced advantage while the transfer of risk and future liabilities to the public sector is unclear. Later, it encouraged the reform of the PFI model. PFIs with new models (often called PF2) try to make changes in public equity shares, faster procurement times, easier renegotiation and greater transparency.

The country has a friendly investment environment and open with clear property rights, the judicial system that works, and political frameworks that are relatively stable; so, it is seen as one of the most attractive investment targets. Their political system and a stable macroeconomic stability has a major influence on investment. Policies on investment is also supported by the presence of a solid institutional and legal. On the other hand, the amount of investment into forming a capital market that continues to grow rapidly with a strong investor base of diverse backgrounds.

Policy investment in infrastructure is also seen as relatively stable and strong. The balanced attention between economic and social infrastructure makes this country stronger both in economic growth and in building people. Infrastructure policy is complemented by a robust and comprehensive infrastructure plan so that the development of social infrastructure with PPP model also fits the needs and development plan. The development of social infrastructure using PPP with availability-based payment is also one of the strengths of this country so that the level of investment trust is also increasing.

However, the current UK condition in which the political system is a little dynamic with the Brexit decision raises a little concern about the future of investment, market access openness and changes in policy professionals who have been centered in this country.

4.4 PPP in South Korea

South Korea began developing private infrastructure investment in early 1994 with launched "The Private Capital Inducement Act" (PPIAF, 2009) [15]. At that time, PPP in the country aimed at creating infrastructure facilities by the private sector to reduce public spending and to take advantage of private sector efficiency. The first legal framework that specifically promoted PPIs was the "Private Capital Laws Legislation" which was officially launched in 1994. In 1999 when there was an economic crisis in Asia, the PPP Law was changed to promote the PPP market in order to further encourage the private sector to Developing projects with a clearer distribution of risk. And in 2005, the PPP Law was changed by introducing Build-Transfer-Lease (BTL) schemes for social infrastructure.

Based on data from Public & Private Infrastructure Investment Management Center (PIMAC) in 2012, there are 400 social infrastructure projects from a total of 600 projects in South Korea (Hyeon, 2012) [16]. That represents two-thirds of the total infrastructure projects with a value of 21.7 billion USD. Value is indeed only a quarter of the value of infrastructure projects because of the characteristics of social infrastructure that is relatively less costly than the economic infrastructure. This indicates the number of social infrastructure development in South Korea using PPP model.

Table 2. Signed PPP Projects in Korea (as of December 2011).

		TOTAL (# of projects) Project Cost (USD)	TOTAL (# of projects) Project Cost (%)	Under Operation	Under Construction	Under Preparation
TOTAL		(600) 79.6 bn	(100%) 100%	(417) 46.1 bn	(143) 22.1 bn	(40) 11.3 bn
Competent Authority	Central Government	(151) 46.6 bn	(25.2%) 58.4%	(79) 26.2 bn	(52) 10.9 bn	(20) 9.6 bn
	Local Government	(449) 33 bn	(75.2%) 41.6%	(338) 19.8 bn	(91) 11.2 bn	(20) 1.8 bn
Project Type	BTO	(200) 57.9 bn	(33.3%) 72.9%	(144) 34.9 bn	(34) 12.7 bn	(22) 10.2 bn
	BTL	(400) 21.7 bn	(66.6%) 27.3%	(273) 11.1 bn	(109) 9.4 bn	(18) 1.1 bn

Source: Hyeon, 2012 [16]

In general, it can be seen that this country has enacted comprehensive laws to be regarded as a political commitment to encourage PPIs. South Korea has clear rules and criteria established at each stage of the PPI process and specific procedures have been developed to evaluate proposals submitted for PPP projects. In addition, to complete this framework, the Government has also established an effective organization to promote and coordinate PPIs.

One of the highlights of PPP development in this country is the development of a regulatory framework where initial arrangements may not be ideal until a more transparent and effective PPP framework is established. By establishing clear responsibilities between different authorities and by incorporating various forms of government support supported by a special law in 1999 to speed up the PPI. In addition, monitoring of PPP programs is also important and flexibility to adapt according to the program.

To date, South Korea has become an attractive country for private investment in the infrastructure sector. The country is one of the pioneers in PPP that contribute to meeting the increasing demand for infrastructure in the Asia Pacific Region. The efficient development of private sector which is closely related to the construction industry is an important point that South Korea should develop. And much more important is that Korea should start to adopt a more efficient allocation process for its PPI project especially regarding the effective capacity of its institutions that govern the bidding process.

4.5 PPP in Australia

Australia is one of the most advanced countries in adopting Public Private Partnerships (PPP) to deliver social infrastructure projects. The use of PPP in Australia has grown rapidly since its inception in the 1980s and is now used to provide services in various economic sectors (roads, railways, airports etc.) and social infrastructure (schools, hospitals and prisons etc.). Jetro (2010, p.9) states that the Australian market now has a reputation for a clear and transparent procurement process used to realize infrastructure projects in many sectors [17].

The Australia Infrastructure Partnership (2015) says that more than 55% of investors seek social infrastructure projects in the country. PPP on social infrastructure in Australia focuses on addressing the growing demand for services in the areas of: health; education; Social housing; correction; and justice [18]. In general, Australia has developed this model of PPP and has well-developed institutional and regulatory tools and is supported by highly developed research. The widespread use of PPP throughout Australia is currently being facilitated by the policies and guidelines of the National PPP developed by Infrastructure Australia on behalf of the Federal Government, and supported at the State and Territory level by the Infrastructure Department with their own PPP policies and guidelines and special PPP units for Overseeing the delivery of state-owned infrastructure projects.

Key elements of implementing PPP for social infrastructure in Australia Due to the following (Jetro, 2010, p.9) [17]:

• Legislative framework that allows private sector ownership, management and operation of public infrastructure in various infrastructure assets;

• Clear and transparent procurement process for large-scale infrastructure projects;

• A Federal Minister for Infrastructure dedicated to a set of centralized project and policy guidelines overseen by Australia's Infrastructure;

• KPS support networks and well supported Ministries of Infrastructure throughout the State and Territories with well-understood procurement policy methodologies and guidelines.

Meanwhile, Ernst and Young Australia (2008) also noted that PPP developments in Australia concentrate on three major concepts, namely value for money, collaborative and comprehensive planning, and strong relationships [19]. First, the PPP conveys value for money to the public sector with a focus on end-users and government perspectives. The value of money is one of the key factors for governments pursuing the PPP model to realize social infrastructure. It requires time and cost efficiency, great assurance of income, innovation, and the effort to release public sector resources to deliver its core services. Furthermore, they also apply high standards to this partnership to provide attention to core services. The core services carried out using PPP transfer the responsibility to the private sector in terms of delivery services, maintenance and infrastructure support.

Secondly, the success of the PPP project requires collaborative and comprehensive planning which means that it requires time and effort investing in the beginning in the framework of the partnership process. With good planning, it can determine more clearly the needs of the country and the purpose of the project. Ultimately, it will affect the formulation of appropriate specifications for a project to increase added value in efficient project operations and deliverables and ultimately will make a development project successful. In addition, the existence of a mature planning will have an impact on the opening of space for private parties to innovate.

Finally, PPP projects also require effective partnerships built between government and the private sector in a strong and balanced contract. The relationship between the public and private sectors really does not have to sacrifice commercial value or end user time. However, government agencies do need to adapt to different working environments, especially in terms of balancing the demands of becoming an effective commercial partner for the private sector. The demand for change is that the public sector focuses on developing policies and procedures that strengthen the transfer of risk to private operators, so that the spirit of partnership can continue to be established.

5 Global Challenges and Lesson Learned

5.1 PPP Global Challenges in Social Infrastructure

The implementation of PPP on social infrastructure that has been done by many countries makes the development more interesting because of the dynamic conditions in terms of politics and investment. The main challenge for the global PPP market can be summarized in the following discussion.

Arrangements on sectoral and regional opportunities. The role of public institutions is key to the implementation of PPP. Indeed, investment opportunities by private parties are abundant but the interest in social infrastructure must be generated by the Government. The existence of market arrangements on a sectoral basis will make it easier for private parties to identify risks. In addition, regional market arrangements will also provide a significant impact on investment opportunities because the expected social impacts of a region can also be measured more clearly. On the other hand, the arrangements between national and local governments are also highlighted in this market share of investment. The local level is expected to play an important role in the channelling of long-term investment funds for social infrastructure.

The need for financial innovation and access to investment opportunities. In most countries, there is often a lack of consistency and regulatory stability resulting in an impact on access to the infrastructure market. Therefore, there is a need for a definite policy framework that allows the private sector to have a broad opportunity to invest in public infrastructure with measurable risks in accordance with government needs, including through innovation of private investment. The innovation in question is, for example, the opening of private opportunities as new investors to form or join a consortium to bid in a project. In addition, the development of information access to social infrastructure projects is also indispensable to attract the interest of private parties, especially for new entrants in order to conduct competitive bidding.

Maximizing the social impact of social infrastructures. The development of social infrastructure using PPP models should have a positive social impact on community development. Efforts to maximize social impacts need to be a big consideration in making investments both in the process of formulating needs, planning, offering, and institutionalizing. Furthermore, investment policy also needs to consider social impacts locally as well as in a social complexity, and its implications are microeconomic.

So, market segmentation in financial scaling could be a solution for making regional be a player in PPP project. Meanwhile, prioritizing particular sector which have low or middle risk for PPP pilot project are supposed to make regional have more opportunities to involve in this kind of partnership. In addition, opening access to municipal and regional masterplan data and asset management could be a way to attract the private sector to participate in PPP projects. The private sector with experience and technology can have ideas to create partnerships with governments if they know the projects they can follow. Finally, the development of social infrastructure should involve the community. They can contribute in planning by giving ideas or gathering community needs; and in its implementation through cooperation in the construction phase of PPP projects, such as partnerships between government and the public as a private sector.

5.2 Lessons learned from implementing PPP in social infrastructure for Indonesia

The need to open social infrastructure investments to the private sector through PPP can be due to external and internal factors. Their incoming foreign investment and initiative to make an investment, or the internal needs of a country that wants to build infrastructure but will take the role of the private sector. South Korea is a good example of how they are moving to take the initiative to realize the PPP in social infrastructure, while Canada provides a smooth process in achieving development in accordance with the needs of their communities.

All countries in this study who are experienced in the implementation of PPP models have a clear political will and legal framework. This indicates that the political will be a very important factor in the effort to build social infrastructure using PPP models. The support of the legal framework is also needed in order to provide clear rules to private parties. Australia and Chile proved that with strong legal protection they can guarantee investment in the country. Political will and regulation framework should be run with transparency and fairness.

Public sector need to provide long-term capital-related facilities that can attract investors. Comprehensive and comprehensive planning (masterplan) is needed so that the private sector can identify and measure risk. In addition, with good planning, both parties can estimate how the project life cycle of a project and establish the scope and size of the projects to be offered in the PPP.

Subsequently, Chile and UK have shown that a procurement process requirement and a clear and transparent strongly supports the success of the PPP project. That makes it easy for both parties to get the desired benefits through the PPP model. The government can get the private sector which is professional and has expertise, while the private sector is also able to compete with innovation. Although, on the other hand, Government also needs to ensure a clear and definite revenue stream for the private sector. In addition, Government also needs to establish a steady institutional and stakeholder engagement as undertaken by Canada and UK.

Indonesia has a strong potential in terms of PPP project implementation initiatives by the government. Private can play a role when the government opens opportunities related to the project and which sectors can use the PPP model. However, it is necessary to be covered by clear rules and procedures. The main strength of Indonesia is the presence of strong political will so that the rules must be structured to support the will. The steps taken by the Government of Indonesia (2017) to trim the tortuous investment process can also encourage the private sector to run this model. However, a transparent procurement process becomes the key to be able to run this partnership with private parties that really have added value.

6 Conclusion

An important message of social infrastructure is the clear need that development is able to provide benefits to the community. Public and private stakeholders have different needs and goals, but they must have the same goal to provide social infrastructure in order to resolve the problems of society by using partnerships.

The experience of countries like the United Kingdom, Canada, Chile, and Australia have been relatively stable in implementing PPP for social infrastructure provide a deeper understanding of how the process is done. The development of South Korea also provides another positive perspective in studying the application of PPP to social infrastructure. The other interesting point is how Chile, which are socially similar to Indonesia, has a solid strategy and a quality service in providing social infrastructure for residents.

Governments have a central role in policies and regulations to provide a transparent and fair 'investment field', as well as providing payments to private parties. In investing, the private sector actually requires stability, predictability and a framework that supports their investment in social infrastructure. Meanwhile, the private sector has a role to provide expertise in building social infrastructure efficiently and has added value in innovation.

References

1. Mital, K. M., and Vivek Mital. "Public Private Partnership and Social Infrastructure." Computer Society of India: Mumbai, India, 2016. https://www.csi-sigegov.org/1/13_353.pdf.
2. Ahmed, Mansoor. "Partnerships Aim to Unite Economically Divided Region." Colliners International, April 2013, 130.
3. Inderst, Georg. "Social Infrastructure Investment: Private Finance and Institutional Investors," 2015. https://papers.ssrn.com/sol3/papers.cfm?abstract_id=2696421.
4. Preqin Infrastructure Onli. "Infrastructure Deal Flow," February 2017. https://www.preqin.com/docs/newsletters/ra/Preqin-Real-Assets-Spotlight-February-2017.pdf.
5. Beck, Benedikt. "PPP in Sweden and Germany." KTH Architecture and Built Environment, 2010.
6. Yescombe, E. R. Public-Private Partnerships: Principles of Policy and Finance. Elsevier Finance. Amsterdam ; Boston: Elsevier, 2007.
7. World Economic Forum. Strategic Infrastructure Steps to Prioritize and Deliver Infrastructure Effectively and Efficiently. Geneva, 2012.
8. NZSIF. "New Zealand Social Infrastructure Fund (NZSIF)." NZSIF, 2009. http://www.nzsif.co.nz/.
9. Blanc-Brude, Frédéric. "Pension Fund Investment in Social Infrastructure." ADHEC-Risk Institute, February 2012. https://www.edhec.edu/sites/www.edhec-portail.pprod.net/files/publications/.
10. Government of Canada, Department of Finance. "Budget 2016: Chapter 2 - Growth for the Middle Class," March 22, 2016. http://www.budget.gc.ca/2016/docs/plan/ch2-en.html.
11. Arcadis. "Third Global Infrastructure Investment Index 2016," n.d. www.arcadis.com.
12. Hill, Andrew. "Foreign Infrastructure Investment in Chile: The Success of Public-Private Partnerships through Concessions Contracts." Nw. J. Int'l L. & Bus. 32 (2011): 165.

13. Roehrich, Jens, M. K. Lewis, and Gerard George. "Are Public-Private Partnerships a Healthy Option? A Systematic Literature Review of 'Constructive' Partnerships between Public and Private Actors," 2014. https://spiral.imperial.ac.uk/handle/10044/1/13799.
14. Inderst, Georg. "UK Infrastructure Investment and Finance from a European and Global Perspective," 2017. https://papers.ssrn.com/Sol3/papers.cfm?abstract_id=2972704.
15. PPIAF. "Module 6: Tools." In Toolkit for Public-Private Partnerships in Roads & Highways. Public – Private Infrastructure Advisory Facility (PPIAF), 2009. https://ppiaf.org/sites/ppiaf.org/files/documents/toolkits/highwaystoolkit/6/toolkit_files/index.html.
16. Hyeon, Park. "Government Support for PPP Projects in Korea." 2012. http://www.unescap.org/sites/default/files/4b-Govt-support-PPPs-in-ROK.pdf.
17. JETRO. "Public Private Partnerships in Australia and Japan." Japan External Trade Organization, 2010.
18. Infrastructure Partnership Australia. "Australian Infrastructure Investment Report 2015," 2015.
19. Ernst & Young. The Journey Continues PPPs in Social Infrastructure. Ernst & Young Australia, 2008. www.ey.com/au.

Initial Study on Building Information Modeling Adoption Urgency for Architecture Engineering and Construction Industry in Indonesia

Fauzan Alfi Agirachman[1*], *Ilham Fajar Putra*[2] *& Adam Angkawijaya*[2]

[1]School of Architecture, Planning and Policy Development, Institut Teknologi Bandung, Indonesia
[2]BicaraBIM Research Team, Bandung, Indonesia

Abstract. This paper presents our initial study on Building Information Modeling (BIM) adoption urgency for architecture, engineering and construction (AEC) industry in Indonesia. Currently, BIM is being adopted by many countries around the world because of its' efficiency and other benefits. Meanwhile, most of AEC industries in Indonesia still using the conventional method and there are no regulations from Indonesian AEC authority for adopting BIM. With that situation, a study of BIM adoption in Indonesian AEC industries is important. This study uses a qualitative approach with explorative type. Input from the survey is evaluated qualitatively using content analysis, distribution analysis and correspondence analysis method. Based on analysis result, it shows that BIM as a mean to encourage a more sustainable approach in AEC industry is still in its development phase but it shows great potentials and it gives stakeholders a better way to achieve sustainable built environment. Current lack of awareness and understanding of BIM in Indonesia, particularly in the education sector, is a key factor that impedes BIM adoption and one that can be addressed by integrating BIM into AEC curriculum. Government and practitioners alike need to develop a strategic roadmap to pave way for successful BIM implementation.

1 Introduction

Building Information Modeling (BIM) has been used both as a modeling tool and design-construction process, around the world. Eastman et.al. [1] mentioned that Building Information Modelling is one of the promising developments in the industry, especially in architecture, engineering and construction (AEC) industry. Stakeholders are able to create a more accurate model to will-be-constructed building and see any problems long before the construction happens. At the same time, BIM is also a whole new approach which requires implementation of new policies, contracts and relationships among project stakeholders [2].

Currently, BIM is being adopted by many countries around the world because of its efficiency and other benefits. United States of America (USA) is one of the countries that's pioneering the adoption of BIM nationwide. In the United States, National 3D-4D BIM program was already launched in 2003 by General Services Administration (GSA) [3]. Wong et.al. [3] mentioned that half of the industry in the USA was using BIM or BIM-related tools in 2009 alone. It represents 75% of usage compared to the level in 2007. Also, US National Building Information Modeling Standard was released in 2006 to support the adoption. GSA also released a BIM Guide that helps designers to develop final concept designs that require them to submit a BIM Industry Foundation Classes (IFC) model defined by the buildingSMART organization.

In Europe, BIM also adopted by a lot of countries such as Finland and Denmark. In Finland, BIM adoption is led by Senate Properties which is the public owner running pilot projects using BIM and IFC [3]. BIM also being researched in Finland by Helsinki University of Technology and Tampere University of Techno. Those institutions are performing research and development in BIM for educational context. Meanwhile, in Denmark, BIM was initiated by at least three public owners: The Palaces and Properties Agency, The Danish University and Property Agency and Defence Construction Service. They have a significant impact due to IFC model requirements.

In Asia itself, Singapore is known as one of first nation adopting and implementing BIM nationwide specifically in public sector. BIM development and implementation in Singapore is started by Construction and Real Estate Network (CORENET), an initiative launched by the government in 1995. Currently, most of development and implementation is led by the Building Construction Authority (BCA) aiming 80% of construction industry adopting BIM by 2015. This intention was to improve industry productivity up to 25% over the next decade. In Hong Kong, where under the Works Branch of the Development Bureau (DevB),

* Corresponding author: fauzan.alfi@gmail.com

BIM was introduced into the existing CAD standard for Works Projects in 2007. With that phenomenon, a study of BIM adoption in Indonesian AEC industries is important.

Currently, in Indonesia, most of AEC industries in Indonesia are still using conventional methods, such as 3D modelling and CAD, and there is no regulation from Indonesian authority for adopting BIM as part of building design and construction process. In this paper, we would like to present our initial study on BIM adoption urgency in Indonesia and discuss how AEC industries are able to implement BIM in Indonesia as a sustainable approach for themselves as industries and also for the environment itself.

2 Research Method

This research is using qualitative [4] and explorative [5] research method. Explorative qualitative research is done to get data with several information possibilities from respondents and also to get factors from the gathered information. Study input based on survey data that had been taken from Indonesian BIM Implementation and Protocol Seminar held by Study Program of Architecture ITB in 2016. The survey was a questionnaire form that answered by 106 respondents from various stakeholders in AEC industries in Indonesia.

Data analysis is done qualitatively using content analysis, distribution analysis and correspondent analysis method. The content analysis method is used first to enrich information about how urgent the implementation of BIM in AEC industry for Indonesia. Then, distribution analysis is done to get the frequency of dominant and non-dominant answers from respondent related with BIM adoption urgency. In the end, correspondent analysis is done to see the correspondent between respondent's profession and BIM urgency to see the needs of each profession.

3 Analysis & Interpretation

Content analysis from questionnaire result is done to create a base statement about the importance of BIM adoption in Indonesia. Answers which can be analyzed in the content analysis must directly correlated or relevant to the issues. The first step in content analysis method is open coding. Open coding is a phase to identify all keywords of respondents' statement in survey answers. Here's one for the example,

> *"BIM is important to be implemented because of the viability of design process until construction. From the base information that can be collected, a design team can explore and choose the best design decision in the process. Besides of that advantage, BIM help designer to deliver information to other engineer consultants using visual presentation. Information that modelled using BIM in construction phase will become an accurate database for construction (material, dimension, method, etc.)" (One of the respondent's answer)*

Based on that answer, there are several keywords like "viability on design process", "make/help better coordination" & "accurate building information" are categorized as "BIM adoption advantages". After finishing the open coding phase, axial coding is done to categorize keywords into specific categories. Categorizing is done by doing several discussions with researchers to prevent bias result.

Several categories that come up from axial coding process described in Table 1 below,

Table 1. Partial Axial Coding Result from Questionnaire

No	Main Category	Category	Sub Category
1	Urgent	Efficiency	Time
			Cost
			Design Process
			Production Process
			Human Resource
			Tools
		Project Control	-
		Coordination and Collaboration	-
		External Demands	-
		Information Processing	-
		Future Needs	-
		Accuracy	-
		Large Scale Project	-
2	Not Urgent	Lack of Regulation	-
		Lack of Knowledge	-
		Adaptation Problem	-

After processing keywords into several categories, next phase is Distribution Analysis that intended to count frequency that shows up from respondents' answers. Distribution analysis reveals how important of each category for this BIM urgency factor in Indonesia. From this process, sub-category "design process" from "efficiency" category is known as most frequently answered by correspondent in BIM urgency question. More results are shown in Table 2.

Table 2. Analysis Distribution Results

No	Main Category	Category	Sub Category	Freq.
1	Urgent	Efficiency	Time	8
			Cost	3
			Design Process	18
			Production Process	11
			Human Resource	1
			Tools	1
		Project Control	-	8
		Coordination and Collaboration	-	12
		External Demands	-	10
		Information Processing	-	4
		Future Needs	-	5
		Accuracy	-	7
		Large Scale Project	-	4
2	Not Urgent	Lack of Regulation	-	1
		Lack of Knowledge	-	7
		Adaptation Problem	-	1

Last step to analyze base information about BIM urgencies in Indonesia is selective coding phase by doing correspondent analysis. This phase identifies the correlation of correspondent profession with their perspective of BIM urgency. This analysis described by ward hierarchical clustering in Figure 1 & 2.

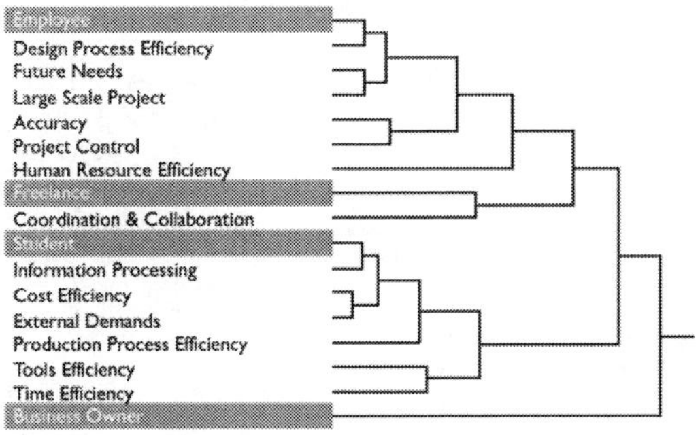

Fig. 1. Relationship Between Respondent's Work Sector and Categories of BIM Adoption Urgency

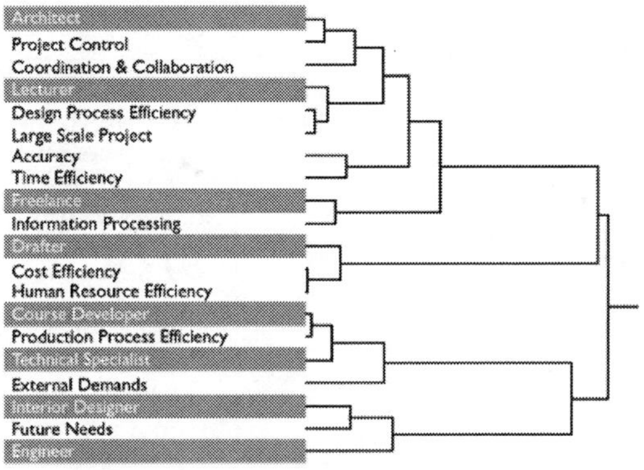

Fig. 2. Relationship Between Respondent's Occupation and Categories of BIM Adoption Urgency

Figure 1 describes the correlation between urgencies categories and specific profession that closely related to it. Meanwhile in Figure 2, it describes the similar correlation but profession view in general. Based on that figure, "design process efficiency" category is highly correlating with architect and lecturer's perspective about BIM. Other important categories for BIM urgency based on Table 2 are closely correlated with architect, course trainer and technical specialist perspective or expectation when using BIM methods for work. Comprehensive analysis about correlation between important factor and job sector is described in Table 3. This analysis result becomes our basic information to seek further strategies on BIM adoption urgencies in Indonesia.

Table 3. Correspondent Analysis Table on BIM Adaption Urgency Issues in Indonesia

No	Category/ Sub Category	Work Sector/ Occupation	Summary
1	Efficiency/ Time	Education/ Lecturer	Time efficiency become main concern in educational institution
2	Efficiency/ Design	Professional /	Lecturer & Architect are

	Process	Lecturer & Architect	important sector that most concern at design process efficiency in BIM
3	Efficiency/ Production Process	Education/ Course Trainer & Technical Specialist	Production process efficiency become main concern in BIM training sector both inside corporation or independent training facilities
4	Project Control	Professional /Architect	Architects that work in a company are concern for BIM method abilities to help project control aspect
5	Coordination & Collaboration	Freelancer/ Architect	Freelancer Architect is sector that most concern about coordination & collaboration aspect in BIM
6	External Demands	Education/ Technical Specialist	External demands (trends & competition) is main concern in companies BIM training sector
7	Accuracy	Professional /Architect	BIM abilities to provide accurate building information also become architect concern
8	Not Urgent/ Lack of Knowledge	Professional /Architect	Some architects that work in companies concern that BIM is not urgent because lack of BIM education

4 Discussion

Based on survey answers, there are several expected benefits from BIM usage by AEC actors in Indonesia. Further studies will be needed to evaluate BIM usage development and user expectation.

4.1 Design Process Efficiency

From several expected BIM benefits, design process is the most frequently used answer. In design process, sustainability & energy efficiency become fundamental factor in making design decision. It really depends on project owner's perspective because there is still some project owner who sees the project just in economical aspect which sometimes leads to much energy waste or not sustainable method way of work. Implementation of BIM on design method is not simultaneously matching up with way of work of AEC industries in Indonesia.

BIM software that shows up with the pretty expensive price also becomes a problem for small firm and consultant. Since BIM design methodology is based on database and information management, only large architecture & engineering consultant that can initiate comprehensive studies in BIM implementation. Implementation of BIM methodologies in whole design process also has different phase compared with conventional design methodologies that have been implemented in Indonesia right now (CAD methods). This difference between each method shown in Figure 3.

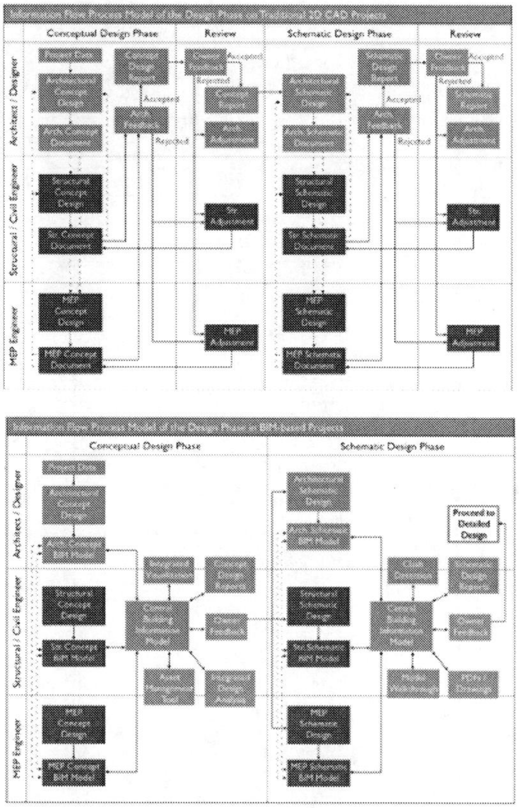

Fig. 3. Design Process Difference Between Traditional Method and BIM Method [9]

4.2 Coordination & Collaboration

Good & efficient coordination between AEC sectors while doing productive collaboration in project lifecycle is another BIM implementation advantage that mentioned by many respondents. Though, in Indonesia, work management & project phases are regulated by guideline from several institutes that regulating public infrastructure development. So that is why BIM method implementation is easier to implement in private sector. Because BIM method depends on information exchange & management into building model, it is hard to not share confidential company information when doing BIM collaboration. Without support from a government institution to control & provide BIM information right & law regulations,

BIM collaboration will be hard to implement both in public & private sector.

4.3 Production Process Efficiency

Document integration, organized clash detection, construction progress simulation and many production processes support that provided by BIM implementation are also the main reason why AEC actors & industries in Indonesia need to start BIM adoption. Indonesia has a regulation that describes documentation drawing in infrastructure project, construction operational standards, but it still lacks detailed guideline & information that control product quality issues. Even though, this standard & guideline issue still need to be fixed immediately. In 2015, Government of Jakarta launched CAD layering guideline & standard for project procurement in public sector. This kind of guideline will help settlement process for creating future BIM protocol & guideline in Indonesia.

Also, the maturity of BIM & stakeholder capacity plays a great role in BIM implementation. Common value about BIM among stakeholders for shaping the proactive market needs to be created [6]. Better knowledge about this BIM maturity for AEC actors is important to do efficient production process.

4.4 External Demands

More valuable deliverables in business competition and keep up with technological trends also the value that many AEC actors seek in implementing BIM methods. BIM methods initially implemented for the public infrastructure project in the USA at 2003 [7]. After years of study and development, BIM implementation turns out increasing efficiency and many aspects of AEC industries. Since around 2010 until nowadays, many AEC industries and academic institutions in Indonesia still doing research & improvement in BIM implementation on infrastructure & business development. Recently in 2017, Institut BIM Indonesia (iBIMi) was officially launched to gather Indonesian AEC actors for creating future strategies, protocol and many other developments that will be needed for BIM implementation in Indonesia.

4.5 Time Efficiency

Accomplishing project lifecycle in less amount of time is another BIM beneficial factor that respondent seeks in BIM implementation. This advantage comes from several integrated procedures that facilitated in BIM tools and workflow. Even further, efficiency potential lies in coordination and collaboration facilities that provide clash detection & revision distribution in real time. Also without investing effort in BIM education, proper training & technologies, BIM implementation in Indonesia will turn out to become "silver bullet syndrome" in time management. By that condition, rather than make time efficiency by implementing BIM method, it will take much time to finish project compared to conventional method.

4.6 Project Control and Accuracy

Project control & accuracy are another BIM implementation advantage that helps BIM, coordinator/manager, to calculate, record & simulate future probability in a project then make a decision based on it. To have fully implemented BIM project in Indonesia, the same type of issue that been explained before about lack of technological and human resource education in BIM method are big obstacle to meet this condition.

4.7 Lack of BIM Knowledge

Based on several discussions above, lack of BIM knowledge for AEC actors about good BIM implementation in project lifecycle become the main reason why BIM implementation is still not an urgent condition for Indonesia right now. Taylor and Bernstein [8] explored four paradigms of BIM implementation at the firm level: visualization, coordination, analysis and supply chain integration. With firm's increasing project experience, these paradigms evolve along a trajectory from visualization, to coordination, to analysis and finally to supply chain integration [7]. So, better knowledge about BIM implementation won't be achieved either if AEC actor or firm in Indonesia never try to implement it properly.

5 Conclusion

BIM as a mean to encourage more sustainable approach in AEC industry in Indonesia is still in its development phase but it shows great potentials and it gives stakeholders a better way to achieve sustainable built environment. This initial study for BIM urgencies in Indonesia leads up to 6 important aspects of consideration for AEC stakeholders for adopting BIM methods in their project. These aspects are designed to process efficiency, coordination and collaboration, production process efficiency, external demands, time efficiency, and project control and accuracy.

Current lack of awareness and understanding of BIM in Indonesia, particularly in the education sector, is a key factor that impedes BIM adoption and one that can be addressed by integrating BIM into AEC curriculum. Government and practitioners alike need to develop a strategic roadmap to pave way for successful BIM implementation.

Greater adoption of BIM will be possible if stakeholders have better understanding of BIM and its usage through education, regulation and protocols and standardization, it also takes time and collaborative effort from AEC stakeholders to reach its maximum potential, therefore early adoption and collaboration is imperative so that users will be able to achieve a better built environment sooner, with Building Information Modeling.

References

1. Eastman, C., Teicholz, P., Sacks, R. & Liston, K., BIM Handbook: A Guide to Building Information Modeling for Owners, Managers, Designers, Engineers, and Contractors, Second Edition, John Wiley & Sons, Inc., 1-2, 2011.
2. Aranda-Mena, G., Crawford, J., Chevez, A. and Froese, T., Building information modelling demystified: does it make business sense to adopt BIM?, International Journal of Managing Projects in Business, 8(3), pp. 419-420, 2009.
3. Wong, A.K.D., Wong, F.K.W. and Nadeem, A., Attributes of Building Information Modelling Implementations in Various Countries, Architectural Engineering and Design Management, Vol. 6, pp. 288-302, 2010.
4. Creswell, J.W., Research Design: Qualitative, Quantitative, and Mixed Methods Approaches, Third Edition, SAGE Publication, 173-190, 2009.
5. Groat, L.N. & Wang, D., Architectural Research Methods, Second Edition, John and Wiley Inc., 215-258, 2013.
6. Zhang, D., Wei, L. & Rowlinson, S., Exploring BIM Implementation: A Case Study in Hong Kong, pp. 5-8, 2010.
7. Wong, A.K.D., Wong, F.K.W. & Nadeem, A., Government Roles in Implementing Building Information Modelling System, pp. 4, 2010.
8. Taylor, J.E. & Bernstein, P.G., Paradigm Trajectories of Building Information Modeling Practice in Project Networks. Journal of Management in Engineering, pp. 25,69, 2009.

Development of Work Breakdown Structure (WBS) Standard for Producing the Risk Based Structural Work Safety Plan Of Building

Vania Elsye [1], vania_htj@yahoo.com, *Yusuf* Latief [1] *and Leni* Sagita [1]

1 *Civil Engineering Department, Engineering Faculty, University of Indonesia, Depok, West Java, 16424, Indonesia*

Abstract. A construction project is inseparable from a series of tasks in the form of activities. Activities that take place on a project can be disrupted due to various reasons, one of the causes of the disruption of project activities are accidents on construction projects. The risk of workplace accidents can be prevented with early identification and analysis of the potential danger that exist in every activity contained in the project's WBS. The need for a standardized WBS in preventing the risk of workplace accidents is very important because it would present a risk assessment, impact and frequency arising from workplace accidents. The aim of this study is to develop a risk-based WBS standard particularly for structural work, by using a qualitative approach. The results of this research are WBS standard for building, source of potentially dangerous risk at structural work,and the development of safety plan using a risk-based WBS that has been standardized, as a form of prevention, reduction or even nullify the risk of workplace accidents (to obtain zero accidents) during construction project implementation.

1 Introduction

Construction industry is a very unique industry and more dangerous than other industries. The construction sites are constantly changing and temporary. Each construction sites involve of many sub contractors and they perform different types of work in close proximity to each other. The safety of workplace is an essential component of efficiency and productivity[6]. Construction workers, both staff and field workers are protected assets in order to work properly and productively until construction is completed without any workplace accident (zero accident). Workers need to be protected against the threat of accidents and health in work. Occupational safety measures are indispensable, both to workers and to corporations and to their production. Workers must follow safety-related regulations, in this case the OSH program, as the OSH program is closely related to the safety of the workers. A comprehensive OSH program and self-awareness of workers is essential to reduce the number of accidents [2].

Work accidents occurring on a project can actually be minimized, if activities on the project are well defined and each job can be placed at a level and level corresponding to a standard WBS, WBS constructing used during the project phase with a basic scope that has been identified. Initially, WBS was created with a restricted formation of scope, however WBS will be reviewed in addition to scope information developed or developed through complete project-to-project analysis. To complete the description of WBS and its application refers based on standard practical guidance according to the principle used so that will get high quality WBS [8], that is easy to determine the amount of activity in project.

Each defined WBS level brings WBS to a more complex level of activity, each activity that has been grouped in a standard WBS, will be easier to identify and mitigate its potential risks, so it is important to develop a standard Work Breakdown Structure (WBS) for the manufacture of safety based risk [11] to obtain output that can be used as a guideline for the implementation of construction.

2 Research Objectives

The objectives of this research are :

1. To develop of building standardized work breakdown structure.
2. To identify a potentially dangerous source of risk for structure work.
3. To develop safety plan using a risk-based WBS that has been standardized.

3 Literature Review

3.1 Work Breakdown Structure

Work Breakdown Structure is a results-oriented analysis of the work covered in a project called the total scope of the project. WBS is a fundamental document in project management as it provides the basis for planning and managing the schedules, costs and changes that occur. WBS serves as a deliverable oriented tool [4]. WBS formulation enables the project to be better in its definition, both in terms of resources and the estimated time required to complete the project to be further enhanced. However, it is important to note that WBS projects change over time, depending on the needs and constraints that the project is experiencing [10].

WBS forms the basis for planning, estimating, scheduling, monitoring, management and control of all project activities. The proposed methods for developing a deliverable are well defined and comprehensive, and the importance of increasing the probability of project success by ensuring that the best resources applied to the project [12]. A standardized WBS for the project is used to facilitate the preparation of project financing and a standardized WBS is also used to define project activities [1].

The best method of constructing WBS is the hierarchically displayed decomposition of the entire scope that the project team will undertake to complete the project objectives and produce the intended work output. In PMBOK 5th Edition, 2013 The planned work at the bottom of the WBS called work package. (deliverables) and not the activities.

3.2 Risk Management

Risk management is an important part of the decision-making process in construction project management, especially regarding integration, scope, time, cost, quality, human resources, communications and project procurement. Creating hierarchical risk management with RBS can improve future project prospects for identifying uncertainty and probabilities [13]. Risk is considered a negative term, but in the engineering construction industry in managing risks that arise is necessary and implemented in a structured way, knowledge of risk management, can nullify and minimize risks occurring in construction projects [7].

3.3 Performance Concepts of Occupational Safety and Health

Accidents occurring at construction sites may result in serious physical injuries and accidents such as people falling from altitudes and crashing material or equipment, the cause of occupational accidents may occur due to worker negligence and also undisciplined workers in the project environment, preventive measures against accidents should Implemented and implemented, so that workers and top management will raise awareness in preventing The workplace hazard handling strategy for construction may introduce potential hazards [5] , further investigation of frequent hazards and may result in safety plans in construction projects.

Safety Plan is a plan document containing practical safety that can assist companies in avoiding potential hazards and can control them in the best way when in these hazardous conditions, the importance of safety planning training during the planning, design and implementation phase of a project because it is able to minimize the risk of construction projects and Can improve workers' safety.

4 Methodology

This research is using qualitative approach to achieve formulation of standardised WBS for building construction. A survey and deep interview was also conducted by means of a structured questionnaire to contractor's experts who have had more than 20 years experiences in highrise building construction projects. Here in Figure 1. The Research Flow, there is a sequence of implementation of this study.

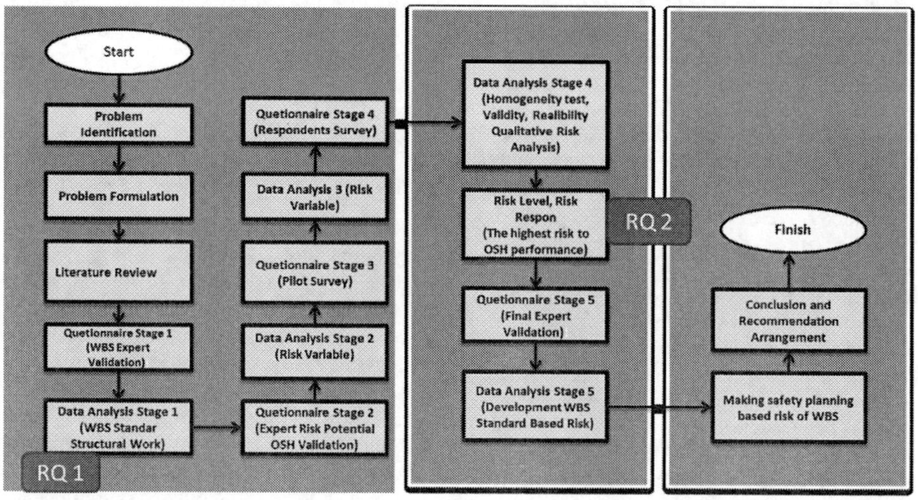

Fig. 1. The Research Flow

4.1 Questionnaire

In this study there are 4 (four) questionnaires to be made: questionnaires in stage I, II, questionnaires in stage III, questionnaire at stage IV, and questionnaire at stage V.

- Questionnaire in phase I is with expert form validation related to standard WBS level will be done variable validation by expert first. There is also the number of experts / experts are 3-5 people with a minimum of 10 years experiences in the field of high-rise construction and education S1. The validation process contains expert responses to agree / disagree that variable X is an activity in high-rise building.

- Questionnaires in phase II will be validated variables by experts in advance related risk factors that affect OSH. There is also the number of experts / experts are 3-5 people with a minimum of 10 years experiences in the field of high-rise construction and education S1.Proses validation contains expert responses upon agree / disagree that the variable X effect on the risk of high building construction OSH.
- After obtained a valid X variable by expert / expert, conducted phase III that is pilot survey to know whether questionnaire made easy understood responder. Pilot survey is in the form of questionnaires to be distributed to 5-10 respondents. Here is an example of a questionnaire in stage III.
- Questionnaires in stage IV are used to obtain the highest risk factors that affect the quality performance of high construction work. This questionnaire will be distributed to respondents who are / have worked in the process of building high-rise project and handling high building construction work with experience of at least 5 years. Here is an example of a stage IV questionnaire.
- Questionnaire at stage V is expert validation aimed at soliciting expert responses related to the results of risk analysis while discussing the causes, preventive actions, impacts, and corrective actions of the highest risk on the process of implementing high-rise building construction work. Here is an example of a stage V questionnaire.

4.2 Respondent

A total of 35 respondents, the profile of respondent is defined in Table 1. Profile of Respondents as follows:

Table 1. Profile of Respondents

No	Description	Total
1	**Position**	
	Staff/Engineering	28
	Manager	6
	> Manager	1
2	**Work Experience**	
	< 5 years	19
	5 - 10 years	7
	> 10 years	9
3	**Education**	
	D3	2
	S1	31
	S2	2

In homogenity test with Krusskal Wallis test, according to position, work experience, and education, the result is ho >0,05.

WBS research variables the standard derived from the archive analysis of the WBS work package from 5 high construction project data from the structural work down to the top structure in the last 5 years there are 31 work packages, and based on expert validation of the potential risks that affect the performance of OSH 173 risk variables. Then the variable is defined in Table 2. Research Variables as follows:

Table 2. Research Variables

WBS LEVEL 6 Activities		RISK VARIABLES INFLUENCING OSH	
Work Package: Land Clearing Method: Conventional (Land Work)			
Clearing	X1	Workers attacked wild animals in the area around land clearing	
	X2	Feet affected by hoe at the time of land clearing	
	X3	Hands exposed to sickle	
Land Alignment	X4	Feet affected by hoe at the time of land clearing	
Work Package: Land Clearing Method: Mechanical			
Clearing	X5	Workers sprinkled or doused with herbicide solution	
	X6	Tools crashing the workers / facilities around (backhoe) at the time of land clearing	
	X7	Slip due to the condition of the steep soil	
Land Alignment	X8	Accidents during machine mobilization	
	X9	Tool crashing into nearby worker / facility (bulldozer	

WBS LEVEL 6 Activities		RISK VARIABLES INFLUENCING OSH	
		and combine harvester) at ground level	
Work Package: Dewatering			
Channel Preparation	X10	Tool hit the worker / facility around(excavator)	
	X11	Struck by a stone as a shade	
	X12	Workers are dropped due to slippery area conditions in the channel preparation	
	X13	Injured by equipment	
Excavation	X14	Excavation crashing the worker / facility around (backhoe)	
Excavation	X15	Worker fell into the excavation	
	X16	Landslide excavation	
Installation of submeasible pumps	X17	Workers are electrocuted from the pumping machine wires	
	X18	Workers fall / slips due to slippery conditions on pump installation	
Making Well Point	X19	Workers exposed to water runoff during the making of well points	
Making Well Test	X20	Workers fall due to slippery areas around the well	
Channel Preparation	X21	Transfer of existing channel	
	X22	Stricken casing	
	X23	Tools hit the workers / facilities around (excavator)	
	X24	Injured because of equipment	
	X25	Affected crane maneuver	
	X26	Hit the mobile crane	
	X27	Material regardless of the hook / sling mobile crane is broken	
	X28	Tools crashing workers / facilities around	
	X29	Workers fall into the excavation	
	X30	Landslide excavation	
Work Packages: Digging and Landing			
Slope Reinforcement	X31	Landslides/side wall collapse	
	X32	Workers fall due to slope reinforcement activity	
Excavation	X33	Tools crashing workers / facilities around	
	X34	Workers fall into the excavation	
	X35	Landslide excavation	
Landscapes	X36	Workers fell / stumbled while throwing away the soil	
Slope Reinforcement	X37	Landslides / collapse of side walls	
Excavation	X38	Tools crashing workers / facilities around	
	X39	Workers fall into the excavation	
	X40	Landslide excavation	
Landscapes	X41	Workers hit by heavy equipment (backhoe)	
	X42	The leg is exposed to the hoe while doing soil discharges	
	X43	Exposed to dump truck maneuvers	
Work Packages: Stockpiling and Compacting			
Stockpiling and Compacting	X44	Workers fell / clogged due to soil conditions	
	X45	Exposed swing excavator	
	X46	Exposed maneuver dump truck	
Work Package: Anti Termite			

https://doi.org/10.1051/matecconf/201814706003

WBS LEVEL 6 Activities		RISK VARIABLES INFLUENCING OSH	
Procurement and Installation of Anti Termite	X47	Sprayed or doused with anti-termite solution	
Work Package: Deepth Foundation (Lower Structure)			
Erection	X48	Worker struck by a stake	
	X49	Worker picked up sling crane	
	X50	Crushed pile during lifting	
	X51	The effect of noise on the surrounding environment	
Pile Linking	X52	Fire due to leaking tube	
	X53	Respiratory disorders due to exposure to welding fumes	
	X54	Workers suffered burns due to burst tubes	
	X55	Irritation to the eye due to smoke (Plunged into the hole during pole connection)	
Making Test Well	X56	Workers fall into the excavation	
	X57	Tools crashing workers / facilities around	
Drilling (Wash Drilling / Dry Drilling)	X58	Workers fall while doing lubrication on the drilling tool	
	X59	Workers fall while refueling the drilling machine	
	X60	Drilling tools crashing into workers / facilities	
	X61	Road dirty / slippery due to spilled soil	
Installation of Casing	X62	Worker struck by reinforcement	
	X63	Worker stricken casing	
Drilling uses mud fluid	X64	Splashed or flushed by fluid	
Cleaning of drill holes	X65	Skin dermatitis due to dust and smoke	
	X66	Eye irritation due to smoke	
Installation of Bored Pile Repetition	X67	Workers puntured iron	
	X68	Worker hands are exposed to an iron	

WBS LEVEL 6 Activities		RISK VARIABLES INFLUENCING OSH	
		cutting machine (bar cutter or bar bender)	
Rebuild with a vacuum	X69	Skin dermatitis due to dust and smoke	
Installation of tremie pipes for casting	X70	Tools crashing into nearby workers / facilities	
	X71	Pipe off / fall into the boring hole casting drill	
Casting drill pole	X72	Traffic congestion due to the truck mixer queue on the drill biting work	
	X73	Tremie pipes jammed so concrete setting with casing	
Withdrawal of the casing	X74	Worker stricken casing when making withdrawal	
Work Package: Test Pile			
Installation of cross beam	X75	Injured by equipment	
Loading	X76	Overwritten material	
Drilling pole	X77	Accident due to boring tool collapsed and overturned on drilling work	
Installation of PDA tools Test	X78	Wounded by equipment	
Work Package: Diaphragm Wall			
Grab soil	X79	Tools crashing workers / facilities around	
	X80	Workers fall into the excavation	
	X81	Landslide excavation	
Landscapes	X82	Workers fall into the pit	
Pouring Slurry Fluid	X83	Eyes splashed or watered solution	
Precast Panel Installation	X84	Overwritten material precast at lifting material	
Spawning	X85	Enlargement Worker hands are exposed to a cut-iron machine (bar cutter or bar bender) during fabrication	
	X86	Working legs	

https://doi.org/10.1051/matecconf/201814706003

WBS LEVEL 6 Activities		RISK VARIABLES INFLUENCING OSH	
		wedged in iron	
	X87	The hands of the workers is stabbed with wire	
	X88	The hands of workers exposed pliers when tying the ring	
	X89	Stumbled iron	
Formwork	X90	Overwritten Formwork	
	X91	Workers are hammered	
	X92	Workers fall down due to porous wood	
	X93	Workers are stricken with material	
	X94	Punctured legs of messy material	
	X95	Legs scratched cluttered material	
Concretion	X96	Sprayed or sprayed with concrete	
	X97	The concrete bucket fell off the TC due to the sling severed so that it spilled over the worker underneath	
	X98	Traffic congestion due to truck mixer queue at casting job diapraghma wall	
	X99	Ready mix from concrete pump spilled on workers	
Work Packages: Soldier Pile, Contiguous Pile, Secant Pile			
Drilling for concrete pile on soldier pile	X100	Accident due to boring tool collapsed and overturned on work Soldier Pile, Contiguous Pile, Secant Pile	
Drilling bentonite pile on contiguous pile	X101	Holes filled with bentonite fluid left open not closed sheeting tarpaulin	
Drilling for boneless concrete on secant pile	X102	The drilling mud disposal is not considered	
Work Package: *Ground Anchor*			
Soil drilling	X103	Workers fall while performing	

WBS LEVEL 6 Activities		RISK VARIABLES INFLUENCING OSH	
		lubrication on the drilling tool	
Drilling	X104	Workers fall while refueling the drilling machine	
	X105	Accident due to boring tool collapsed and overturned on ground anchor work	
Work Packages: Raft Foundation, Pile Cap, Tie Beam, Retaining Wall, Ramp, GWT/RWT			
Spawning	X106	The iron scaffolding is not solid	
	X107	Iron material falls when lifting material with TC	
	X108	Slip on the way up the column	
	X109	Worker hands stabbed wire	
	X110	Workers hands hit by pliers while tying the ring	
	X111	Triple iron	
Formwork	X112	Overwritten Formwork	
	X113	Workers are hammered	
	X114	Workers fall down due to porous wood	
	X115	Workers are stricken with material	
	X116	Legs exposed nails dismantling formwork form work	
	X117	Legs scratched cluttered material	
Concretion	X118	Pipe concrete pump clogged then broke and exposed workers	
	X119	Overloaded concrete bucket drops from TC due to slashed sling so it spills over the worker underneath	
	X120	Traffic congestion	
	X121	Due to truck mixer queue at foundry	

WBS LEVEL 6 Activities		**RISK VARIABLES INFLUENCING OSH**	
		work raft foundation, pile cap, tie beam, retaining wall, ramp, gwt / rwt	
Procurement & Installation of	X122	Ready mix from concrete pump spilled on workers	
Precast Concrete	X123	Hook / sling TC disconnected when transporting precast raft foundation, pile cap, tie beam, retaining wall, ramp, gwt / rwt	
	X124	Form wrok collapse on the installation of precast raft foundation, pile cap, tie beam, retaining wall, ramp, gwt / rwt	
	X125	Workers fall from altitude on the installation of precast raft foundations, piles, tie beam, retaining walls, ramps, gwt / rwt	
Work Packages: Columns, Beams, Plates, Ladders, Core wall / Shear wall (Upper Structure)			
Spawning	X126	Worker hands are exposed to an iron cutting machine (bar cutter or bar bender) when manufacturing iron for upper structure	
	X127	Worker legs wedged in iron	
	X128	Slip on the way up the column	
		Worker hands stabbed wire	
	X129		
	X130	Workers hands hit by tweezers while tying the ring	
	X131	Triple iron	
Formwork	X132	Overwritten Formwork	
	X133	Workers are hammered	
	X134	Workers fall down due to porous wood	
	X135	Workers are stricken with material	
	X136	Feet punctured spikes former dismantling formwork top structures	
	X137	Dropped while installing formwork	
	X138	Hook / sling TC disconnected when transporting material formwork top structure	
	X139	Form work collapse	
	X140	Workers fall from altitudes when block work, columns, plates, and upper structure ladders	
	X141	Formwork / scaffolding falls and affects workers / facilities	
Concretion	X142	Scrubbed or poured concrete	
	X143	Overloaded concrete bucket drops from TC due to slashed sling so it spills over the worker underneath	
	X144	Traffic congestion due to queue mixer queue on top structure foundry work	
	X145	Ready mix of concrete pump spilled on workers	
Procurement & Installation of Precast Concrete	X146	Overwritten precast concrete material	

WBS LEVEL 6 Activities		RISK VARIABLES INFLUENCING OSH	
Work Packages: Beams, Plates, Core wall / Shear wall (Roof Structures)			
Spawning	X147	Worker hands are exposed to an iron cutting machine (bar cutter or bar bender) when fabricating iron for roofing structures	
	X148	Worker legs wedged in iron	
	X149	Slipped when corewall firing	
	X150	Worker hands stabbed wire	
	X151	Workers hands hit by pliers while tying the ring	
	X152	Triple iron	
Formwork	X153	Overwritten Formwork	
	X154	Workers are hammered	
	X155	Workers fall down due to the wood formwork porous	
	X156	Workers are stricken with material	
	X157	Feet punctured nail used dismantled formwork roof structure	
	X158	Legs scratched cluttered material	
	X159	Hook / sling TC disconnected while transporting material	
	X160	Form work collapse	
	X161	Workers fall from the height when the work of the beam, and core wall roof structure	
	X162	Formwork / scaffolding falls and affects workers / facilities	

WBS LEVEL 6 Activities		RISK VARIABLES INFLUENCING OSH	
Concretion	X163	Sprayed or sprayed with concrete	
	X164	Overloaded concrete bucket drops from TC due to slashed sling so it spills over the worker underneath	
	X165	Traffic congestion	
	X166	Due to the truck mixer queue at the foundry work of the roof structure	
Procurement & Installation of Precast Concrete	X167	Overwritten precast concrete material	
	X168	Hook / sling TC disconnects when transporting precast roofing material	
	X169	Form work collapse	
	X170	Workers fell from the height during the installation of precast roof structure	
Installation of roof covers	X171	Workers exposed to sunlight while working on the roof structure	
	X172	Workers fall from height during erection of roof steel	
	X173	Workers fall from altitude when helipad plates work	

Data collection was carried out using questionnaire survey to understand the perception of the practitioners to the risk factors. Once the probabilities and impacts are determined, the risk score can be calculated with following :

$$R = P \times I \quad (1)$$

Where R = risk factor, P = probability and I = impact.

The probability and impact matrix or risk level matrix (Table 3 and Table 4) illustrates a risk rating assignment for risk factors.

This analysis is done on the results of questionnaires given to respondents by using Likert measurement scale. The probability value of the occurrence of risk and the impact generated multiplied to produce a number that can be made ranking of risk factors. The risk matrix shows the combination of impact and probability as shown below:

 https://doi.org/10.1051/matecconf/201814706003

Table 3. Probability Matrix

Grade	1	2	3	4	5
Probability Criteria	Very Low	Low	Moderate	High	Very High
Threats	0,10	0,30	0,50	0,70	0,90

Table 4. Impact Matrix

Grade	1	2	3	4	5
Impact Criteria	No Influential	Less Influential	Quite Influential	Influential	Very Influential
Threats	0,05	0,10	0,20	0,40	0,80

5 Result and Discussion

The results of this study are as follows:

1. To answer RQ 1 found WBS Building Standard Building in Figure 2. WBS Standard For High Rise Building a

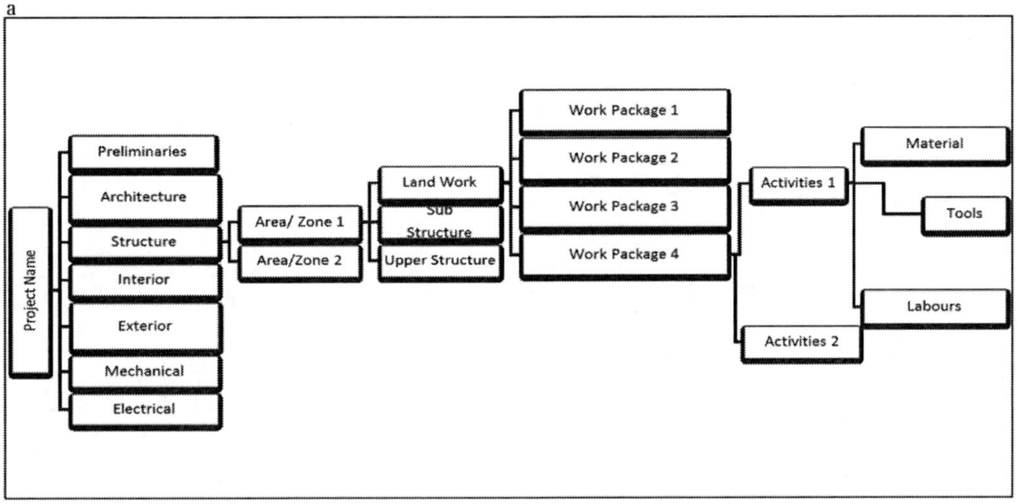

Fig 2. WBS Standard For High Rise Building

Level 1 is for the name of project, level 2 is work sections, preliminaries, structure, architecture, external work, interior, mechanical and electrical works. Level 3 is for area, level 4 are for sub work sections among others soil work, sub structure work, upper structure work, and roof work. Level 5 is for work package, level 6 is for activities, and level 7 is for resources.

2. To answer RQ 2 there is a potentially hazardous source of risk and affect the performance of OSH on Structural Work in Table 5. Dominant Risk as follows:

Table 5. Dominant Risk

No	Risk Variables Influencing OSH		Rank	Level
1	X9	Tool crashing into nearby worker / facility (bulldozer and combine harvester) at ground level	23	H
2	X15	Worker fell into the excavation	13	H
3	X16	Landslide excavation	5	H
4	X27	Material regardless of the hook / sling mobile crane is broken	16	H
5	X123	Hook / sling TC disconnected when transporting precast raft foundation, pile cap, tie beam, retaining wall,	21	H

No		Risk Variables Influencing OSH	Rank	Level
		ramp, gwt / rwt		
6	X124	Form work collapse on the installation of precast raft foundation, pile cap, tie beam, retaining wall, ramp, gwt / rwt	15	H
7	X125	Workers fall from altitude on the installation of precast raft foundations, piles, tie beam, retaining walls, ramps, gwt / rwt	4	H
8	X135	Workers are stricken with material	24	H
9	X139	Form work collapse in uppers structure work	18	H
10	X140	Workers fall from altitudes when block work, columns, plates, and upper structure ladders	1	H
11	X141	Formwork / scaffolding falls and affects workers / facilities	11	H
12	X143	Overloaded concrete bucket drops from TC due to slashed sling so it spills over the worker underneath	19	H
13	X146	Overwritten precast concrete material	17	H
14	X159	Hook / sling TC disconnected while transporting material	9	H
15	X160	Form work collapse	14	H
16	X161	Workers fall from the height when the work of the beam, and core wall roof structure	6	H
17	X162	Formwork / scaffolding falls and affects workers / facilities	3	H
18	X164	Overloaded concrete bucket drops from TC due to slashed sling so it spills over the worker underneath	12	H
19	X167	Overwritten precast concrete material	20	H
20	X168	Hook / sling TC disconnects when transporting precast roofing material	10	H
21	X169	Form work collapse	22	H
22	X170	Workers fell from the height during the installation of precast roof structure	8	H
23	X172	Workers fall from height during erection of roof steel	7	H
24	X173	Workers fall from altitude when helipad plates work	2	H

In the next discussion, we will answer the formulation of the second problem in this study that is the source of any risks that are potentially dangerous on the work of building structures. Based on 173 risk factors, there are 24 dominant risk factors in every activity step which have an effect on OSH performance in construction project especially high rise building construction project. Where from 24 indicator variables belonging to high level of risk dominant happened at Upper Structure and Roof Structure that can decrease OSH performance in similar project, while 111 indicator of variable of medium risk level, and 38 indicator of variable belong to low category. From some of the above dominant risk events with a high level of risk level if related to the literature that has been obtained the author is as follows:

a. The variable risk events X125, X140, X161, X170, X172, and X173, are risk factors for accidents falling from a height or mired in a perforated place. The opinion by Expert who thinks that workers are still less aware of the importance of occupational safety when working at high altitude and there are also workers who are still violating the rules to wear PPE (Personal Protective Equipment) such as helmets, glasses, safety shoes , full body safety harness.
b. The variable risk events X141 and X162 are risk factors formwork / scaffolding fall and override workers / facilities on the work of the roof structure. The following analysis of possible causes of collapse from the use of scaffolding is the inability to accept the load to obtain maximum

results in accordance with the designed, then the use of raw materials with good quality becomes absolutely necessary.

c. The risk event X 135 on the installation of the hazard formwork fell from a height according to Expert, the control measures undertaken were SOPs to work at altitudes, then creating a solid platform for workers, using rubber boots for workers To avoid slipping while climbing the formwork and monitoring the workers working at altitude.

d. In the work involving Tower Crane on X27, X123, X143, X159, X164, and X168 the dangers of material falling according to Expert shall be made of operational SOP tower crane, inspection SOP, use, and maintenance of sling. the riggers provide precise directions to the TC operator to lower the material appropriately.

3. To answer RQ 3 Preparation of Safety Plan using WBS standard is as follows, based on ROSHK in attachment of Ministerial Regulation 05 / PRT / M / 2014, and has been added with WBS level taken sample at some risk with high potential value in Table 6 Work Safety Plan Structure and taken 3 highest risk for example as follow:

Table 6. Safety Plan For Structural Work

NO	WBS Level 5 (WP)	WBS Level 6 (Activities)	Hazard Identification	Probability	Impact	Risk Level	Safety Risk Control
X140	Upper Structure Works (Column, beam, plate, core wall, stairs)	Formwork on Upper Structure	Workers fall from altitude when block work, column, plates, and upper structure ladders.	0,35	0,55	0,288 High	Complete PPE for each worker Installing OSH signs Make SOP pairs formwork upper structure Socializing OSH Program
X173	Roof Structure Works Beams, Plates, Core wall / Shear wall	Roof Structure	Workers fall from the heights when the helipad plates work	0,297	0, 605	0,264 High	Complete safety equipment for each worker Installing OSH signs
X162	Roof Structure Works Beams, Plates, Core wall / Shear wall	Roof Structure Works Beams, Plates, Core wall / Shear wall	Formwork / scaffolding falls and affects workers / facilities on roof structure work	0,314	0,25	0,25 High	Check the installation of formwork Make SOP pairs of roof structure

6 Conclusion

Based on the results of testing and analysis that has been done then it can be concluded as follows:

a. Creating WBS building standards divided into 7 levels, Level 1: Project Name, Level 2: Work Section, Level 3: Zoning Area / Location, Level 4: Sub Work Section, Level 5: Work Packages, Level 6: Activity, and Level 7: Resources. WBS structural work standards can be established according to the classified level after collecting and analyzing data / archives based on BOQ and RKS projects and have been validated by related experts.

b. Identify potentially hazardous sources of risk on structural work. There are 24 high risk risk event variables (X9, X15, X16, X27, X123, X124, X125, X135, X139, X149, X141, X161, X162, X164 and X167, X168, X169 , X170, X172, X173) are identified to be the dominant risk in every stage of the high rise building development process. Having obtained the risk event variable is also equipped with the risk response of each variable so that the potential danger can be prevented / mitigated.

c. Preparation of safety plan by using risk based WBS standard is completed and developed according to document of safety planning / ROSHK Government Regulation PU 05 / PRT / M / 2014 at occupational safety planning section for building construction work which can be used either as assessment material of service provider auction process and also as guide for contractors in the preparation of safety planning.

Acknowledgements

The authors would like to express gratitude to University of Indonesia for granting a support through PITTA no 862/UN2.R3.1/HKP.05.00/2017 scheme in order to assist the researchers in completing this research.

7 References

[1] Aishah Momoh, Rajkumar Roy and Essam Shehab. *A Work Breakdown Structure for Implementing and Costing ERP Project.* (Communications of the IBIMA,2008) **94-103**

[2] Al-Anbari.S, Khalina, A. Safety and Health Risk. Ass.at.Oman. Build. Const. Pro. Oman. IJRET: International Journal of Research in Engineering and Technology eISSN: **2319-1163 Volume: 02.571-578** (2013).

[3] Alex, Albert. Emerging Strategies for Construction Safety & Health Hazard Recognition. Journal of Safety, Health & Environmental Research **152-160**, (2014).

[4] Hans, Robert. Wor.Bre.Stru: A Tool for Soft Pro. Sco. Ver. International Journal of Software Engineering & Applications (IJSEA), **Vol.4, No.4. 19-25**, (2013).

[5] Kai Chen Goh1,a, Hui Hwang. Accidents Preventive Practice for High-Rise Construction. (EDP Sciences, MATEC Conference, 2016). **04004-p.1-04004-p.6**

[6] Kamar, Mohd- N.S. Lop, N. Mat. Contractor's Awareness on Occupational Safety and Health(OSH) Management Systems in Construction Industry, (EDP Science Confrences, 2014) **01019-p.1-01019-p.6,**

[7] Mhetre, K; Konnur, B.A; B, Amarsinh. Risk Man.t in. Con. Ind. International Journal of Engineering Research **Volume No.5. 153-155**, (2016)

[8] PMI, *Practice Standard for Work Breakdown Structures Second Edition*,Pennsylvania: Project Management Institute, (2006).

[9] Satish Kumar. *Current State of Construction Safety Planning Practice*. Proc. of Int. Conf. on (Advances in Civil Engineering, 2011), **47-50**

[10] Shlomo Globerson. Imp. Of. Var. WBS on Pro. Concept. (International Journal of Project Management, 1994), **Vol. 12. 165-171**

[11] Su,Lei. *WBS-based Risk Identification for the Whole Process of Real Estate Project and Countermeasure*. (National Conference on Information Technology and Computer Science, 2012).**780-783**

[12] Van Tonder, J.C. and Bekker, M.C. (2001). *Analysis of A Methodology to Obtain a Work Breakdown Structure Built Up From Interdependent Key Project Deliverable Packages*. (African Rhythm Project Management Conference, 2002),**ISBN Number: 0-620-28853-1**

[13] Zid Chaher, Ali Raza Soomro. Fac. Risk.Man in Cons. Pro by using Hie. RBS. (International Journal of Scientific and Research Publications, 2016).**Volume 6, Issue 7. 703-709**

The influence of wage, age and experience to labor productivity in construction works in Kota Langsa, Aceh

Ipak Neneng Mardiah Bukit[1,*], *Yulina* Ismida[1], *Rizcy* Maulana[1], and *Muhammad* Nasir[1]

[1]Civil Engineering Department, Engineering Faculty, Samudra University, Meurandeh Kota Langsa, Indonesia

Abstract. Construction work, as well as other production process depends on how people work towards it. The success of the work is determined by labor productivity. Labor productivity is affected by many factors such as human factors, environment, labor condition, leadership, type of work, level of difficulty etc. This article will discuss how these aspects influence the value of labor productivity. We limit our research on wage, age and experience from 15 to 30 workers in the brick work. The research is held in 4 construction sites in Kota Langsa, Aceh, Indonesia in late 2016 until early 2017. We observe the effective time, time of contribution, ineffective time work and the work volume. Productivity is obtained by dividing the work volume and the effective time work. The result shows that the amount of wage influences the labor productivity. Nonetheless, we found that age and experience do not affect labor productivity. Interestingly, the experience workers in any ages will increase the value of labor productivity.

1 Introduction

Kerzner [1] states that the real failure of project management is when the outputs or results are not what we expected. He also studied that in 1990's a failure is attributed to poor morale, poor motivation, poor human resources, no employee commitment and poor productivity. Borchering and Liaou in Gundecha [2] state that productivity is related with inputs and outputs of work. That is the expression of what people provide in the end of the work depends on how they do their work. Productivity is also understood as how many hours people need to deliver a unit of activity which will determine how long the work could be finished. Planning the productivity, especially labor productivity is believed will give effect to project planning and schedule. People who work in construction industry tend to optimize their ability according to the project schedule, because time is very delicate in the industry. PMI [3] guides that resource productivity has an effect on the activity duration, thus it is difficult to be estimated in project planning phase. Furthermore, planning productivity means project management is right on track.

Labor is not a simple resource to be organized. Numerous factors give influence in these resources such as environment, local tradition, experience, age, leadership, wages, etc. Kaming (1997) in Ervianto [4] classifies the influential factors of labor productivity in 4 (four) categories, i.e. 1). Method and technology; 2). Project management in site; 3). Work environment; and 4). Human factors (wage rate, work satisfaction, profit share, worker-supervisor relationship, worker-workerrelationship and discipline). These factors could give positive or negative consequences to labor productivity.

The past studies try to measure or calculate the impact of these factors to productivity; however it is hard to standardize this impact [2]. This article discusses how the factors influence the labor productivity positively or negatively by measuring labor productivity in site. We limit this research in three factors i.e. wages, ages and experiences, since they are considered as sensitive aspects to human factors. The sample is 15 workers for wages and 30 workers for ages and experiences. The observation is held on brick work activity. The result found that the higher of wages give a better productivity. Interestingly as people getting older, their productivity is associated with their working experiences. An older worker with good experiences provides a better productivity. A younger worker with good experiences gives a better productivity as well. But an older one with lack of experience will be less productive as well as a non-experience young worker.

2 Labor Productivity Definitions

Productivity is an important aspect on project schedule. It is related to worker capability that will determine the

* Corresponding author: ipakbukit@unsam.ac.id

project execution time. Productivity also encourages the project performs well. Datta, Guthrie & Wright [5] from Delery & Shaw (2001) state labor productivity is the key indicator of workforce performance. This means the labor productivity confirms project schedule process result as plan, or the people work well and the schedule runs properly.

Calculating the project schedule is unique for every project planning. The good labor productivity cannot be repeated in different place and different situation. Koch & McGrath [6] study that superior productivity cannot be imitated, cannot be replicated easily, uniquely valuable in given firm setting and may add unique value to strategic inputs factors. This argument gives reason to emphasize the project productivity in order to create a firm's competitive position and to reach its project goals.

3 Research Goal(s)

Labor productivity as other project performance indicators is affected by various factors that could bring the project performance into positive or negative impacts. Wolgemuth et al [7] found that poor health and nutritional status of employee may reduce labor productivity. Labor health condition related to how much wages they earn to supply good nutrition. Therefore we put wages as one of the influential factors on labor productivity.

Other factor is age, because human grows older. Generally speaking, worker's energy decreases by time of age. Hence, it is a significant factor in labor productivity. We want to see whether this factor will affect their productivity or not.

Koch & McGrath [6] state there are three resources as predictor project performances; they are physical resources, organizational resources, and human resources. Human resources are representing the know-how and skills related working in an organization. Skills are related to worker's experiences. Worker with experiences may have more skills than others. However, we need to clarify whether experience influence labor productivity or not.

Based on the background, subsequently this research aims to figure out how wages, ages and experiences affect the productivity in construction works. The research is held on 4 (four) construction project sites in Kota Langsa, Aceh during late 2016 until the early 2017.

4 Methodologies

Research methodology is descriptive, observation and interview. Wages observation is held in 4 (four) sites, while ages and experiences observation is held in 2 (two) project sites altogether. Data collection is done by three steps; 1. Preliminary survey, in order to get a fit location; 2. Interview with the laborers; and 3. Observation of the actual time work.

Data collected is: number of worker in sites, effective time work (ET), ineffective time (IT), and contribution time (CT). The observation is done in 5 to 6 days, depends on the number of data needed.

Work volume is obtained by observing the worker's brick work achievement per day in sites. The volume is measured for every ET. ET is calculated from the start time of the activity until it stops for resting, and continues until the end of work day during 5 to 6 days. It is measured using measuring tool for every ET. The result is an average volume in m^2.

Productivity of this work is calculated using the formula below [8]:

$$Pr = Volume/n.ET \tag{1}$$

Where
Pr = productivity (m^2/hour)
n = number of worker
ET = effective time work (minutes)

5 Results and Discussion

5.1 Results

The results of the research can be seen in tables and figures below.

Table 1. Summary of ET, CT, IT, volume & productivity by wages

Workers	Amount of wages (Rp.)	ET (minutes)	CT (minutes)	IT (minutes)	Average Volume (m^2)	Average Productivity (m^2/hr)
1	100,000	330	46	125	8.88	1.615
2	80,000	328	40	116	8.55	1.564
3	80,000	255	67	158	7.54	1.774
4	80,000	250	48	182	5.84	1.402
5	80,000	244	55	181	7.24	1.780
6	80,000	271	47	173	8.61	1.906
7	80,000	223	41	231	5.95	1.601
8	80,000	236	68	176	6.91	1.757
9	80,000	278	79	123	8.52	1.839
10	120,000	295	73	117	8.7	1.769
11	120,000	304	66	113	9.03	1.782
12	120,000	233	89	160	11.9	3.064
13	120,000	280	62	141	5.78	1.239
14	120,000	202	88	206	6.27	1.862
15	120,000	265	91	133	6.07	1.374

The data collected for wages was taken from 4 (four) projects. However, the wages range does not vary in the area; therefore it is unnecessary to obtain more respondents in this category. The result of the observation in wages can be seen in figure 1.

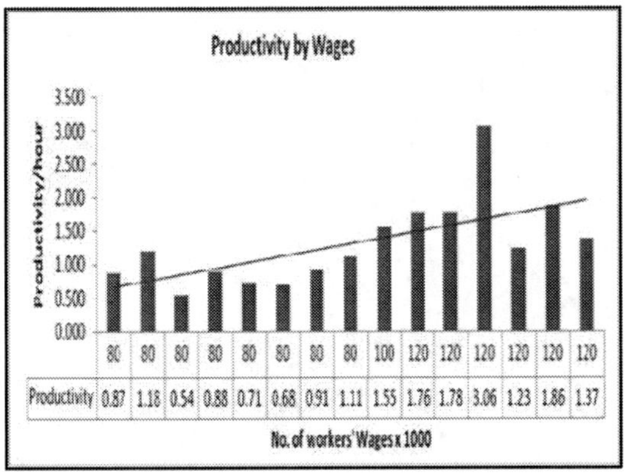

Fig. 1. Productivity vs. wages

This Fig. 1 shows that wages influence the productivity value. Worker with higher wage will work faster compare to the one who earns smaller wage. The trend describes the value of productivity significantly increases. It tells us that people will work as how he is rewarded. Subsequently, wage influences the value of labor productivity.

Table 2. Summary of ET, CT, IT, volume & productivity by Age and Experiences

Workers	Age	Years of experience	ET (minutes)	CT (minutes)	IT (minutes)	Ave. Volume (square meter)	Ave. Productivity (m^2/hr)
1	23	6	236	176	68	6.91	1.757
2	26	7	223	231	41	5.95	1.601
3	30	12	283	218	44	5.60	1.187
4	31	10	258	165	55	5.54	1.288
5	32	7	250	182	48	5.84	1.402
6	32	7	283	157	35	5.80	1.230
7	32	11	175	205	65	7.54	2.585
8	32	-	240	175	40	5.23	1.308
9	33	8	255	158	67	7.54	1.774
10	33	11	248	192	45	4.06	0.982
11	34	7	252	183	50	3.87	0.921
12	36	9	273	210	63	6.01	1.321
13	36	14	260	175	45	6.84	1.578
14	37	10	244	181	55	7.24	1.780
15	38	10	271	173	47	8.61	1.906
16	38	12	240	175	40	6.13	1.533
17	38	15	232	197	41	6.53	1.689
18	40	19	217	222	48	5.71	1.579
19	40	9	255	195	28	7.64	1.798
20	40	14	236	210	62	7.08	1.800
21	41	10	215	205	35	4.93	1.376
22	42	18	328	116	40	8.55	1.564
23	42	13	298	150	40	8.46	1.703
24	46	13	275	140	65	6.58	1.436
25	46	15	265	175	45	6.66	1.508
26	48	16	278	123	79	6.06	1.308
27	51	18	265	185	30	8.36	1.893
28	52	20	227	206	48	8.52	2.252
29	54	18	277	166	37	7.78	1.685
30	62	23	252	175	60	6.79	1.617

The data collected based on ages and experiences was taken from 2 (two) projects, however we have enough reliable amount of workers. The workers' ages are varying from the youngest, 23 years old and the oldest, 62 years old, while their experiences are from 6 to 23 years. The trend of these categories can be seen in figures 2 and 3 as follow.

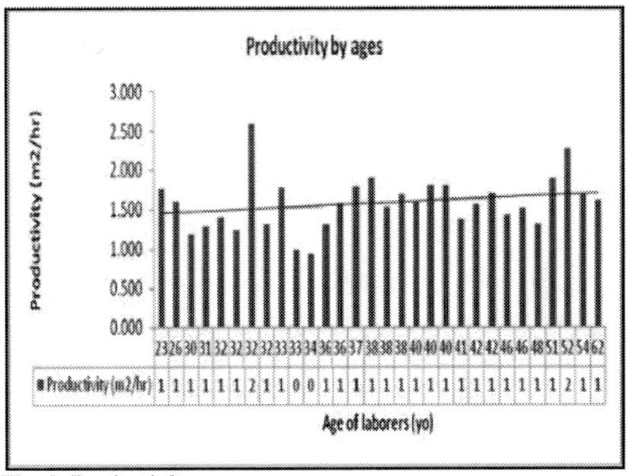

Fig. 2. Productivity vs. ages

Fig.2. shows that getting older does not cause workers to be more productive. It tells us that age does not significantly influence the value of labor productivity. The trend just slightly increases.

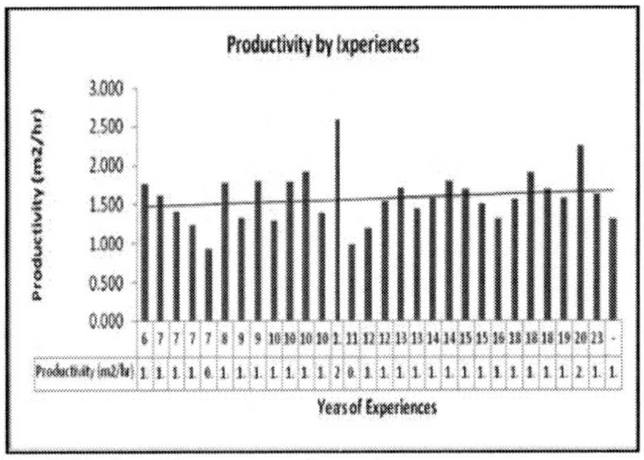

Fig. 3. Productivity vs. year of experiences

Fig.3 shows that the level of productivity on experience is fluctuated. The longest experience worker is not the most productive worker.

However, if we combine both figures we can see in fig. 4 below.

https://doi.org/10.1051/matecconf/201814706004

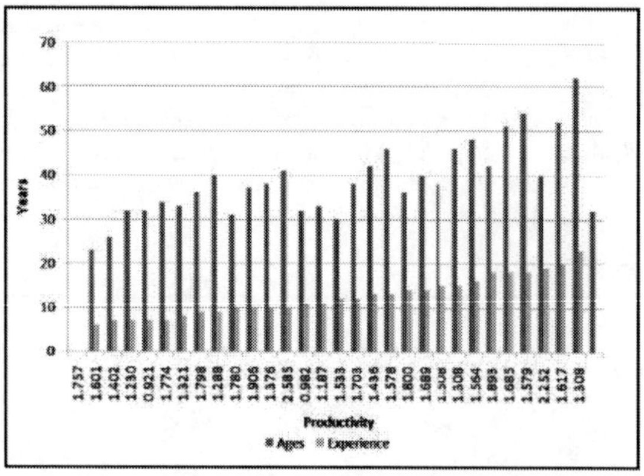

Fig.4. Productivity vs. year of age and experiences

Fig.4. shows that old worker with good experience apparently has a better productivity compares to the one who doesn't. It also applies to younger worker. It shows that experience in any ages will affect the level of productivity.

5.2 Discussion

The trend of wages to productivity is quite influential. Even though there is discrepancy between workers with Rp,120,000,- per day rate, the trend convinces that the more wages the worker earns, the more productive he is. This is a sign that worker will work faster if they are well paid.

In contrary, worker's ages and experiences' trend just slightly increase the productivity. The older is a bit more productive than the younger one, but some younger workers with good experience have better productivity.

This is related with skills and work ability. Young worker has a good skill because he has good experience. While the older labor gains working skills from their long experience. Experience determines whether workers have enough skills in construction or not, therefore they have high productivity.

The best ages of worker in construction are in their 30's. In this age, stable productivity is reached. On the other hand, old and young workers with less experience are usually less productive. Obviously, experience gives influence to labor productivity, both at the young and the old age.

6 Conclusions

In conclusion, the wages of worker is influential to labor productivity. The trend shows that the significant increases in labor productivity level caused by increasing wage. The age doesn't give significant influence to labor productivity, neither does experience. However, if both ages and experiences combined, they show interesting fact that workers with experience is more productive, even though they are still at the young age. Thus we believe that worker who lacks experience is less productive. Hence, experience at any ages is influential to labor productivity.

In this paper, we would like to express our appreciation and thankfulness to our colleagues at Civil Department, Engineering Faculty, Samudra University. We also want to convey highest gratitude to the SIBE Conference 2017 ITB - Bandung for giving us the remarkable opportunity. Additionally, we would like to appreciate the entire project's stakeholders in project sites which with their cooperation who allow us to collect the data smoothly.

References

1. H. Kerzner, *Project Management A System Approach to Planning, Schedulling and Controlling,* Eight Edition, Jhon Wiley & Sons, (2003).
2. M.M. Gundecha, S*tudy of Factors Affecting Labor Productivity at a Building Construction Project in the USA: Web Survey*, Master Thesis, the Faculty of Agriculture and Applied Science, North Dakota University, Fargo, North Dakota, (2012).
3. PMI, *Project Management Book of Knowledge,* Third Edition, Project Management Institute Inc., (2004).
4. W.L. Ervianto, Peng. Prod. Kel. PB dlm Proy Konst, Jurnal Teknik Sipil, **9(1)**, pp.31-42, (2008).
5. D.K. Datta, J.P. Guthrie, & P.M. Wright, HRM and Lab. Prod.: Does Ind. Matter, Academy of Management Journal, **48(1)**, pp.135-145, (2005).
6. M.J. Koch, & R.G. McGrath, Imp. Lab. Prod.: HRM Policies Do Matter, Strategic Management Journal, **17**, pp.335-354, (1996).
7. J.C. Wolgemuth, M.C. Latham, A. Chesher, & D.W.T. Crompton, Wkr Prod. and Nutrition. Stat. of Kenyan Road Constr. Lab., The American Journal of Clinical Nutrition, **36**, pp.68-78, (1982).
8. Hansen, *Menghitung Produktifitas Pekerja,* Wordpress.com, retrieved in August 2017, from https://hansenkammer.wordpress.com/2009/08/21/menghitung-produktivitas-pekerja/, (August 21st 2009).

Value engineering application for conceptual design of seawater desalination plant in Jakarta

Rian Mantasa Salve Prastica[1,*] *Almasul* Auzan[1], *Niswatin* Farika[1], *Nonik* Eko Wahyuning Tiyas[1], *Pratomo* Cahyo Nugroho[1], and *Riko* Apriatresnayanto[1]

[1]Graduate Student, Civil Engineering Department, Faculty of Engineering, Universitas Indonesia, Kampus UI Depok, 16424, West Java, Indonesia

Abstract. Jakarta has 10.2 million inhabitants. It has thoughtful problem about clean water availability. Today, only 54% of clean water availability could be supplied to citizens in Jakarta. The urgent need of new clean water resources is highly demanding for Jakarta future life. By employing abundant sea water in the northern part of Jakarta, desalination is a highly effective way that has been proven through value engineering. A conceptual design of desalination project is completed through FAST diagram and benchmarking method. According to life cycle cost analysis, it results IRR 13%, net product value Rp3.782 trillion, payback period of 13.39 years, and benefit cost ratio of 3.00. To make this project complete and reasonable, the public-private partnership should be conducted for modal sharing. This research piloted four analyzed scenarios. The best alternative scenario makes the government budget 60% for initial cost, spend half for operational and maintenance cost, and obtain 20% from the whole project revenue.

1 Introduction

Jakarta, the capital of Indonesia with approximately 10.2 million people, faces excessive problem of water availability [1]. Jakarta, the city with area of 661 km^2, should fulfill water demand from outside of the city. Its water supply reaches 98% from the outside of the city. Ciliwung River, once able to meet the water needs of Jakarta population. However, its water quality is significantly decreasing. Along with population growth, Ciliwung River is no longer able to meet all the water demand of the city. Today, two operators of water services in Jakarta, PT. PAM Lyonnaise Jaya and PT. Aetra Air Jakarta, are only able to meet about 54% of the clean water demand for Jakarta inhabitants. 46% of water supply to the residents are obtained from groundwater resources. In addition, the quality of water resources from Jatiluhur reservoir through Banjir Kanal Barat is declining. In 2010, the level of ammonia of water was about 2.9 mg/liter and it doubled to 4.8 mg/liter from January to November 2011. The number level is far above the maximum threshold set, i.e. 1.0 mg/liter. But, it could be denied that over 80% of Jakarta water demand are met from this resource.

Jakarta water demand problem could be solved through benchmarking of sophisticated technology from another countries. For example, China. According to Zheng et al. [2], China has water requirement which has been increased significantly. 22% of the world's inhabitants stays in China which has only 8% of the world's water resources. The supply and demand imbalance in Chinese coastal areas is even worse, compared with Chinese inland areas. It is happened because of the essential to encourage more population and industry with less water resources. There are three factors that make water supply and demand imbalance, they are economic development, population growth, and urbanization. China struggles with its imbalance of population growth and water resources due to the three factors. One of the most challenging issues facing water resources professionals in China is how to address the increasing imbalance between water supply and demand. Therefore, China is the most potential country for desalination plant project.

The struggling issue of China is as same as Jakarta. Water demand in Jakarta will be dramatically increasing due to those three factors. The planning for another water resource today is an essential objective for Jakarta. The availability of clean water in Jakarta will be a big obstacle in the future if it is not projected today. The location of Jakarta as a coastal city provides benefits to access the abundant of sea water that has not been used optimally. Therefore, this research aims to produce conceptual design of sea water desalination to meet water supply and demand balance in Jakarta.

Furthermore, several researchers conducted more comprehensive sight about desalination. In the process of seawater desalination, a large amount of water would be generated [3, 4] and would damage the sea-shore environment [5]. Nie and Tao [6] simulated the effect of

[*] Corresponding author: rianmantasasp@gmail.com, rian.mantasa@ui.ac.id

desalination system in 100,000 metric tons/day capacity on Bohai water quality. They found that the salinity would increase four times in 10-day continuous discharge, increase seawater temperature because of waste heat discharge, and decrease algae population. The using of reverse osmosis technology would decrease these impacts. Many researchers proposed to use concentrated seawater by combining the process of seawater desalination and salt production [7]. The lack of application for zero-discharge technology is limited due to high cost [8].

This research uses alternative approach to produce value-added projects through Value Engineering (VE) method. Berawi et al. [9] conducted a research to produce conceptual design of Sunda Strait Bridge (SSB) and Soekarna-Hatta International Rail Link (SHIARL) in order to obtain optimum result and to generate added values to the projects. The research employed a combination of quantitative and qualitative methods through questionnaire survey that distributed to the related stakeholders of the projects and focus group discussion (FGD).

This research proposes conceptual design of sea water desalination plant in Jakarta through extensive literature review. Value Engineering (VE) method is expected to produce creative or novel ideas and innovative methods in promoting various projects and obtaining added value and optimum benefits to attract private sectors to contribute in the infrastructure development.

2 Research Method

Creating added value in the proposed desalination project was done by benchmarking. The literature study sources are derived from various journals, trusted study reports, research reports, government regulations, and other related publications. The using of literature study aims to obtain benchmarking of certain implementation systems that have been done by several countries. So, it is expected to be able to create reliable innovation and capable ideas for increasing the added value of clean water infrastructure development. This research uses quantitative method by analyzing secondary data. The lack of primary data in this research would be the weakness of this research. However, it could become recommendation for future research.

Value engineering is the main concern method in this research. Miles [10] defined that value engineering is as same as value analysis to obtain to make various functions to meet customer needs and requires the lowest cost. Zimmerman [11] defined that value engineering could be the best practice to get the most balance cost, reliability, and performance of a project. The same definition was clarified by Connaught and Green [12]. Standing [13] added that value engineering could make the effective cost with efficient identification and elimination of unnecessary cost. But, value engineering concerns in safety, reliability, and efficiently product. Hence, it is suitable to seek the appropriate design for proposed sea water desalination to balance water demand and supply in Jakarta.

Value engineering would generate creative ideas and innovation [14, 15]. Innovation is idea generation in the creativity step [16, 17], formalization processes [18], and the successful implementation of product concept [19].

2.1 Data Collection

Secondary data are obtained from several official sources that have been validated by their institutions. The population of Jakarta in 2015, obtained based on the projected population census results in 2010, reaches approximately 10,177,924 inhabitants. It has 1.02 percent population growth rate per annum. The following years are calculated with the same steps. The projection data result could be observed in the following table 1.

The other data that needed by this research are (1) the number of customers and total water that channeled by Jakarta water supply company (PAM Jaya), (2) the customer distribution map of PT. Palyza and PT. Aetra, (3) tariff data of clean water from PAM Jaya, and (4) the quality of river and reservoir water in Jakarta province. From obtained data, further analysis of population and water demand (domestic and non-domestic) projections are conducted from 2012 to 2030. The same time interval would be the basic time of proposed desalination project. The end year of projection corresponds to the construction and installation of desalination project‘s design life year, 2030.

2.2 Data Analysis

Data analysis for this research consists of several stages. Firstly, the analysis of desalination project is using a model diagram, called Task/Customer Function Analysis System Technique (FAST) model diagram. It is expected to ensure that all determined functions in the model are being ready for analysis, revealing problems with simple way, being used to define, simplifying and clarifying the problem, so that it could support communication activities, study the cost allocation, help the process of creativity, and define the scope of the project. Secondly, financial analysis was employed. Financial planning involves of pre-construction phase (including the cost of land acquisition, if it is required), construction phase, operation and maintenance phase, and revenue in accordance with the year of age planning. Next, the method of Value Engineering (VE) was expected to create creative ideas and innovative in promoting various projects and could create added value and optimum benefits that would attract the private sector to contribute in the development of infrastructure. Life cycle cost includes the components of the initial cost, operating and maintenance cost, and revenue that would be generated. All of cost estimation data are obtained from journals, research reports, and brochures that relate to the proposed infrastructure cost system. Finally, several scenarios were simulated to meet the optimum value of net present value (NPV), and the amount of fees that could be tolerated by the government. From the initial cost, operation and maintenance cost, and revenue simulate cash flows during the operation and the concession period between the government and private sectors. Next, simulation is done by dividing each components according to the portion of cost sharing between government and private sectors.

Table 1. The analysis of water demand and water balance in Jakarta year 2012-2030

No	Planning Parameter	Unit	Year				
			2012	2015	2020	2025	2030
1	Total Population	People	9,862,705	10,245,541	10,800,000	11,800,000	12,500,000
2	Services Coverage Target	%	60.59	82.6	92.9	98.0	98.0
3	Natural Resources Water	%	42.38	33.32	22	17	14
4	Total Clean Water Demand	L/s	23,192	28,414	24,608	26,667	27,263
5	Total Raw Water Demand	L/s	24,352	29,835	28,839	28,000	28,626
6	Deficiency of Clean Water	L/s	5,317	810,389	6,358	8,417	9,013

3 Result and Discussion

3.1 Conceptual Design of Seawater Desalination Plant

Initially, innovative ideas for desalination project is listed in table 2. Then, the innovative ideas are schemed through FAST diagram. Figure 1 depicts FAST model diagram of proposed seawater treatment plant. The main concept design construction of water infrastructure in Jakarta is water desalination plant according to value engineering studies with several functions development. Three functions integrate desalination concept. They are sea water reverse osmosis (SWRO) function, energy function, and water brine processing function. Figure 2 depicts the design plan of seawater desalination plant.

According to Figure 1 and 2, the whole proposed concept is depicted clearly. Seawater enters the intake to be treated in the post treatment. Firstly, it travels to sediment and trash screening. The solid waste then distributed to the landfill. Next, seawater particles settle in the particle treatment to gain residual to be treated. The solid residual treatment product next is recycled. The next stage, seawater enters sand filters and diatomaceous earth filters to gain remained sand. Then, it would be pumped through cartridge filters to reach the first reverse osmosis process. The product from this process is divided into two types. The first type continues to the second reverses osmosis to be treated in the post treatment 1 for domestic and non-domestic consumption. The second type is used for energy function. This type of seawater could not be used as domestic and non-domestic consumption purposes. The water then is distributed to the post treatment 2 to be processed. Crystal salt and water usage for waterpark are the proposed result from this stage. Furthermore, the rest of seawater then turn back to the sea.

This research states that the analysis of the condition of Jakarta that would allow wind power and solar power effectiveness are still less comprehensive. However, they still have probability for energy generation. The wind data and sunlight availability could be explored from Meteorological, Climatological, and Geophysical Agency of Indonesia. For further research, hydrological and climatological analysis should be analysed to make scenario for energy generation from wind power and solar cell.

Table 2. Innovative ideas for desalination project

Reference	Innovative Ideas
Limited water resources	Desalination
Renewable energy	Construction of solar cell
Wind energy potential	Wind turbine energy production
Abundant of residual salt	Water park business
Increase regional economy	Salt production
Needs to protect environment	Sea water reverse osmosis project

 https://doi.org/10.1051/matecconf/201814706005

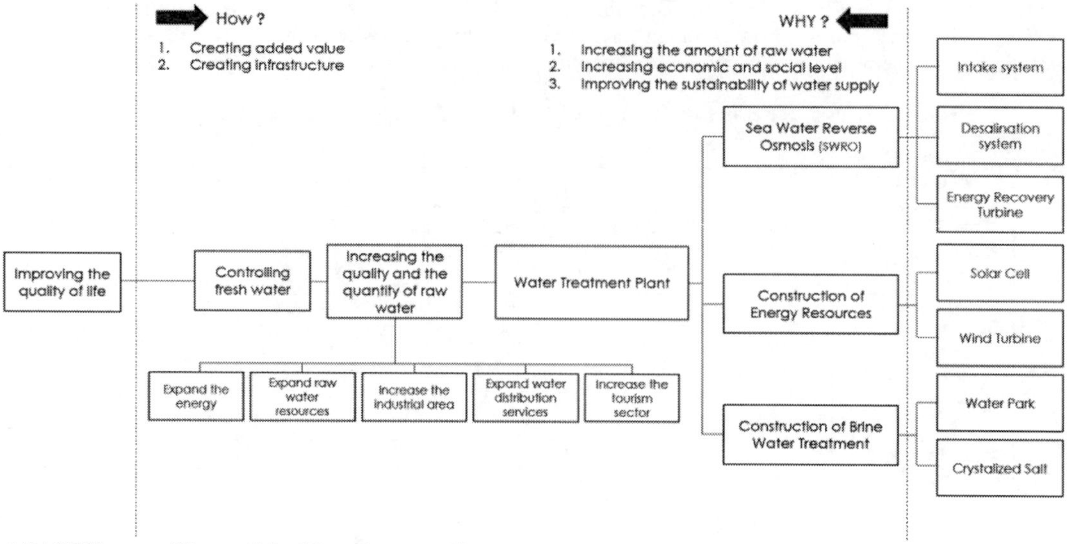

Fig. 1 FAST Diagram of Proposed Sea Water Treatment Plant

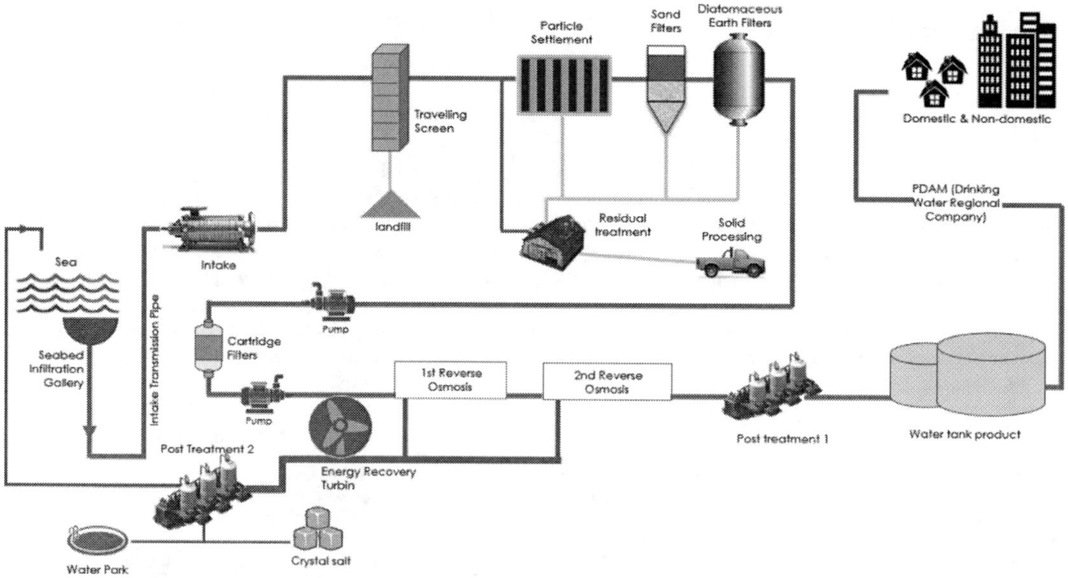

Fig. 2 Design Concept of Sea Water Desalination

3.2 Financial Engineering: Life Cycle Cost (LCC) Analysis

To sum up, the assumption of design life is 18 years, from 2012 to 2030. The design capacity of desalination is projected according to the number of population that shown in table 1. MARR of 12% is used [20].

3.2.1 The Sea Water Reverse Osmosis Installation Function Cost

SWRO function, as seen in table 3, is benchmarked with the application in Australia, Emirates, and Singapore by interpolating the value of cost and capacity. The production capacity of SWRO is expected to reach 39,318,576 m^3/year. SWRO is estimated to be built in Cilincing, North Jakarta City. Jakarta has 16 hectares land area. The value of land acquisition in the taxable value of *Nilai Jual Obyek Pajak* (NJOP) is Rp1,675,779 [21].

Table 3. Seawater Reverse Osmosis Function Installation Cost

Components of LCC	Total Amount (million)
Initial Cost	Rp12,500,000
Operating and Maintenance Cost	Rp250,000
Revenue	Rp1,275,000

* Corresponding author: rianmantasasp@gmail.com, rian.mantasa@ui.ac.id

3.2.2The Energy Function Cost

The energy is performed from solar photovoltaic and wind turbines. For an approximate price of energy, benchmarking functions are derived from the application in Abu Dhabi and China. Solar Photovoltaic installation is laid on the roof of the desalination system building. The projection of initial cost and operational and maintenance costs of solar photovoltaic (PV), are benchmarked from solar photovoltaic system at Masdar and Abu Dhabi by conducting production capacity and cost data interpolation. Initial cost and operational and maintenance costs of wind turbines, benchmarking against the installation of wind turbines in China is analyzed. Revenue is derived from the number of generated photovoltaic solar energy production by reducing energy requirements of SWRO system, multiplied by the price of electricity tariffs in 2016, that is Rp1,040. Revenue of wind turbine would be earned from the multiplication of wind turbine energy capacity, the operational multiplied turbine/year (5-hour operation per day), and the price of electricity tariffs in 2016. The energy function cost is estimated in table 4.

Table 4 Energy Function Cost

Components of LCC	Total Amount
Initial Cost	Rp151,473,200,000
Operating and Maintenance Cost	Rp2,180,464,000
Revenue	Rp4,278,766,303

3.2.3The Brine Water Installation Function Cost

The brine water processing result gives benefit for floating swimming pool. This function releases as water park concept. The estimation cost is analyzed by benchmarking to Sugar World Waterpark in Australia. Water Park is planned on the north side of Jakarta with five hectares area. Then, the value of land acquisition is multiplied by the taxable value of NJOP, approximately Rp1,675,779 [21]. Furthermore, benchmarking in San Diego gives cost estimation for salt crystals from brine water process result. Table 5 explains the brine water installation function's cost.

Table 5 Brine Water Installation Function

Components of LCC	Total Amount
Initial Cost	Rp419,076,424,756
Operating and Maintenance Cost	Rp36,057,920,714
Revenue	Rp87,181,254,417

Finally, life cycle cost for each function is conducted and it is listed in table 6. The following table depicts the detailed result of internal rate return (IRR) and net present value (NPV).

Table 6 Life Cycle Cost of Functions

Function	IRR	NPV (in million)
All functions combined	13.00%	Rp3,782,350.44
Sea Water Reverse Osmosis Installation	12.95%	Rp3,298,780.49
Energy	0.420%	Rp159,573.30
Brine Water Treatment	17.62%	Rp331,640.78

4 Discussion

The main design construction concept of water infrastructure in Jakarta is water desalination plant according to value engineering studies with several functions development. Three functions integrate desalination concept. They are sea water reverse osmosis (SWRO) function, energy function, and water brine processing function. This research goals to produce conceptual design of sea water desalination to meet water demand in Jakarta. As a proposed project, according to Berawi [20], to make the conceptual design feasible financially, public-private partnership funding should be done.

Based on to the result in table 5, it could be seen that with the addition of functions, the IRR would be higher. Therefore, this project is financially feasible. But it looks a very significant difference between the IRR among functions. So, modal sharing of one function to another is analyzed to reach the optimum IRR. The following table 7 result depicts the difference of previous and afterwards analysis by using modal sharing.

Table 7 Modal Sharing Analysis

No	Function	Modal Sharing	Previous Initial Cost (in million)	Afterwards Initial Cost (in million)	Previous IRR	Afterwards IRR
1	Sea Water Reverse Osmosis (SWRO) Installation	1.69%	Rp12,500,000	Rp12,405,077	12.95%	13.02%
2	Energy	0.96%	Rp151,473	Rp25,473	0.42%	12.99%
3	Brine Water Treatment	0.73%	Rp419,077	Rp640,000	17.62%	12.68%

4.1 Public-Private Partnership

4.1.1Funding Pattern

Financial scheme of the proposed project in this conceptual design is done by scenarios between the government and private sectors for initial cost, operational and maintenance cost, and revenue. In the following steps, several scenarios are simulated to meet the optimum value and net present value (NPV). The amount of funding should be tolerated by the government. Government funding comes from the government, either in the form of sum cost, soft loan or foreign loan. The private sector could be either in the form of the borrower, the project sponsor, or banking [20].

4.1.2 Cost Components Sharing Analysis for Funding Pattern

The life cycle cost analysis before the simulation produces internal rate of return of 13% for all functions and hits Rp3.782 billion net present value, benefit cost ratio of 3, and payback period as long as 13.39 years. The next simulation of scenarios in each components divides initial cost, operation and maintenance cost, and revenue according to public-private partnership scenarios portion.

4.1.3 Initial Cost Sharing

To determine the increasing or decreasing value of IRR by sharing portion, simulation of construction costs takes three scenarios: (1) 40% government and 60% private, (2) 50% government and 50% private, and (3) 60% government and 40% private, respectively. The following table 8 figures that the more the government involve in the initial cost sharing, the more IRR and NPV of the project would be.

Table 8 Initial Cost Sharing

Scenario	Government	Private	IRR
1	60%	40%	26%
2	50%	50%	22%
3	40%	60%	19%

4.1.4 Operational and Maintenance (O&M) Cost Sharing

The calculation step is similar with the previous initial cost sharing. From table 9, it could be analyzed that operational and maintenance cost sharing would not be impacted significantly to the proposed project IRR and NPV.

Table 9 Operation and Maintenance Cost Sharing

Scenario	Government	Private	IRR
1	60%	40%	14.01%
2	50%	50%	14.03%
3	40%	60%	14.04%

4.1.5 Revenue Sharing

Revenue sharing scheme runs for income during 30 years. This scheme will decline the rate of return (IRR) and reducing the net benefits (NPV). This is due to the private annual profits earned will be reduced because of the division of revenue to the government. In the conventional theory, infrastructure development for the construction of public facilities. It means that the government does not expect revenue after the project is done. But, with the concept of revenue sharing, the government would earn revenue each year after providing viability gap funding (VGF). The simulation is conducted to determine the level of reduction in IRR as shown in the table 10.

Table 10 Revenue Sharing

Scenario	Private	Government	IRR
1	60%	40%	7%
2	50%	50%	5%
3	40%	60%	2%

4.1.6 Ideal Funding Scheme Scenario of Public-private Partnership

The analysis runs for 4 scenarios of costs sharing. Then, the best scenario would be chosen as the public-private partnership best scenario choice. Table 11 figures the analysis of four cost sharing scenarios.

Table 11 Cost Sharing Scenarios

Scenario	Cost Components	Cost Sharing		IRR	NPV (million)
		Government	Private		
A	Initial Cost	60%	40%	26%	Rp1,685,882
	O&M Cost	0%	100%		
	Revenue	0%	100%		
B	Initial Cost	45%	55%	23%	Rp777,745
	O&M Cost	60%	40%		
	Revenue	0%	100%		
C	Initial Cost	60%	40%	21%	Rp2,160,415
	O&M Cost	0%	100%		
	Revenue	20%	80%		
D	Initial Cost	60%	40%	24%	Rp938,056
	O&M Cost	50%	50%		
	Revenue	20%	80%		

* Corresponding author: rianmantasasp@gmail.com, rian.mantasa@ui.ac.id

All scenarios have positive net present value (NPV) and the IRR is above the assumed MARR (12%). Scenario D is recommended as an alternative project funding scheme. Although the government spend up to 60% in the initial cost, but they will receive revenue each year through revenue sharing scheme which can be used as the initial capital to build another infrastructure for the society. Figure 3 depicts the proposed institutional scheme.

Figure 3 depicts institutional scheme according scenario D. It means that the institutional scheme is conducted based on initial cost, O&M, and revenue sharing between the government and private sectors. In this scheme, it is formed a Joint Venture of 4 sectors with different disciplines. Company A runs SWRO, Company B runs energy sector, Company C runs industrial business, and Company D is responsible for tourism development. All of the sectors work together to form main developer company with the government to manage sea water desalination plant in Jakarta.

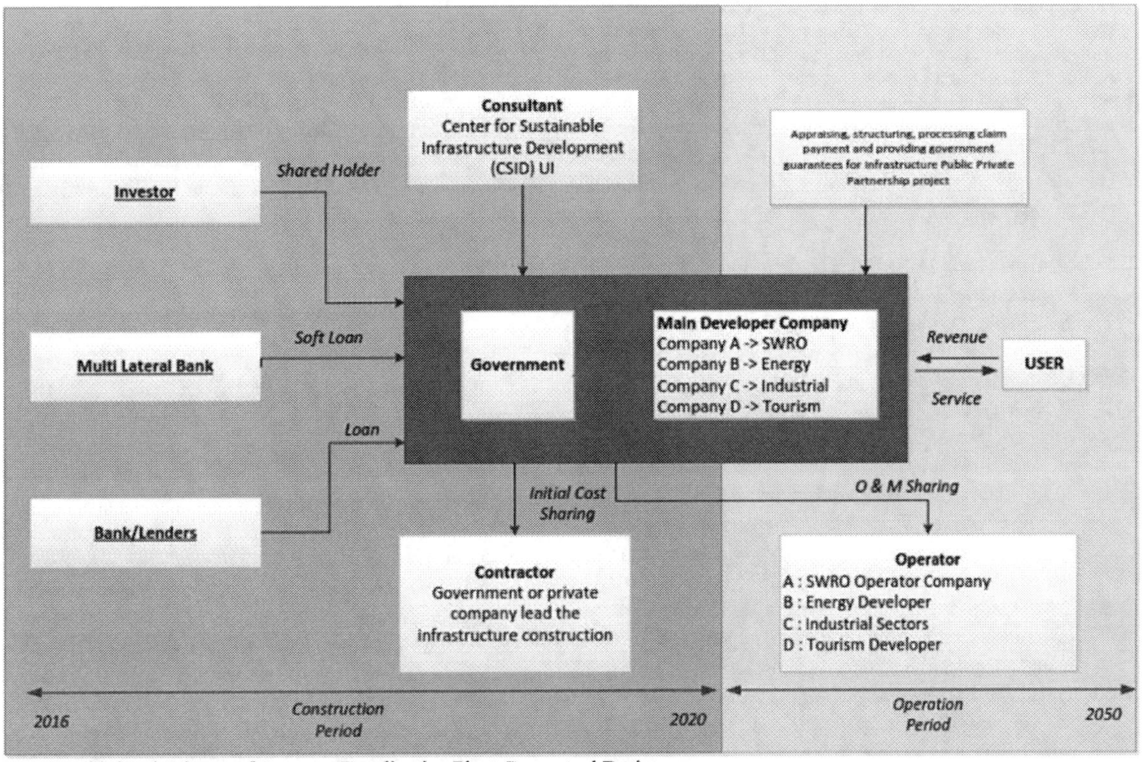

Fig. 3 Institutional Scheme of Seawater Desalination Plant Conceptual Design

4.2 Suggestion

This research has lack of primary data. Berawi et al. (2014) finished a research to produce conceptual design of Sunda Strait Bridge (SSB) and Soekarna-Hatta International Rail Link (SHIARL) which employed a combination of quantitative and qualitative methods through questionnaire survey distributed to the related stakeholders of the projects and focus group discussion (FGD). However, this research used extensive literature review for secondary data. But, Berawi et al. [9] discussed about existing project, while this paper proposes desalination plant project for clean water sustainability in Jakarta. This value engineering analyses presented here are performed for a representative research example problem that should be sufficient to demonstrate real-world applicability of conceptual design of sea water desalination to meet water demand in Jakarta. For further research, or this concept would be applied in the future for real-world case studies, primary data and more comprehensive research should become the primary concern.

5 Conclusion

The application of value engineering method on the development of conceptual design of sea water treatment plant gives the added functions, from the previous limited function of standard water supply function for Jakarta, into several innovative functions, they are (1) the development of renewable energy-based electricity generation through solar photovoltaic and wind turbine, and (2) the utilization of waste results from desalination process for water supply of water park floating pool and the production of salt crystallization. Besides innovation, value engineering method could increase value for money that is shown by the increasing of IRR from 12.95% to 13.02%. Finally, it hits 23.94% after public-private partnership funding scheme. To conclude, value engineering method is proven to give optimum added value and to be feasible both technically and financially for conceptual design of seawater desalination plant.

The present work was performed as part of Value Engineering research project in Civil Engineering Department of Universitas Indonesia, Indonesia, supported by research grants. The author appreciate for Center for Sustainable Infrastructure Development (CSID) Universitas Indonesia for their insightful comments and suggestions. The author wish to thank to Mohammed Ali Berawi, M.Eng.Sc., Ph.D for mentoring this research. The first author wishes to thank Indonesia Endowment Fund for Education (LPDP) scholarship which legislated by Ministry of Finance of Indonesia who give me occasion as a awardee.

6 References

1. Statistics of DKI Jakarta Province, Jakarta in figures, (2016).
2. X. Zheng, D. Chen, Q. Wang, Z. Zhang, *Seawater desalination in China: retrospect and prospect,* Chemical engineering journal **242**, 404-413 (2014).
3. R.X. Yu, Y. Wang, S.C. Wang, *A review in brine disposal from desalination plants*, Technol. Water Treat **6**, 1-3 (2005).
4. Y.L. Zhang, H.J. Ni, A.G. Chen, Z.H. Jiang, D. Yuan, H. Zhang, *Progress in the research on effect of desalinated seawater on environment and human health*, J. Environ Occup. Med. **27**, 317–318 (2010).
5. Z.W. Gao, Z.Q. Lin, D. Wang, C.J. Gao, *Seawater utilization and impact on environment in China*, Marine Environ. Sci. **6**, 671–676 (2008).
6. H.T. Nie, J.H. Tao, *Impact of coastal exploration on the eco-environment of Bohai Bay*, Ocean Eng. **26**, 44–50 (2008).
7. H.J. Feng, C.G. Xie, *Status and prospect of Chinese seawater desalination technology*, Chem. Ind. Eng. **27**, 103–109 (2010).
8. X.H. Ma, Z. Lan, S.F. Wang, L. Lu, *Impact of discharge in seawater desalination on marine environment and progress of zero liquid discharge*, Chem. Ind. Eng. Prog. **30**, 233–242 (2011).
9. M. A. Berawi, B. Susantono, P. Miraj, A. R. B. Berawi, H. Z. Rahman, Gunawan, A. Husin. *Enhancing value for money of mega infrastructure projects development using value engineering method*, Procedia technology **16**, 1037-1056 (2014).
10. L. D. Miles, *Techniques of Value Analysis and Engineering 2nd ed*, New York: McGraw-Hill, (1972).
11. L. W. Zimmerman, *Value Engineering: A practical approach for owners, designers and contractors*, New York; London: Van Nostrand Reinhold, (1982)
12. J. N. Connaughton, S. D. Green, *Value management in construction: A client's guide*, London: Construction Industry Research and Information Association, (1996).
13. A. Standing, *Value management incentive programme*, London: Thomas Telford, (2001)
14. State Ministry of National Development Planning, *Public private partnership: infrastructure projects in Indonesia*, Jakarta, (2011).
15. M. A. Berawi, *Roles, challenges and prospects of construction of services in Indonesia*, National seminar of GAPEKSINDO 12 October 2010, Jakarta, (2010)
16. Coordinating Ministry for Economic Affair, *Masterplan for accelereation and expansion of Indonesia*, Jakarta, (2011).
17. P. S. Fong, Q. P. Shen, *Is the Hong Kong construction industry ready for value management?*, International journal of project management, **18** (5), 317-26, (2000).
18. R. Woodhead, M. A. Berawi, *An alternative theory of idea generation*, International journal of management practice, **3**(1), 1-19, (2008).
19. M. Zack, J. McKeen, S. Singh, *Knowledge management and organizational performance: an exploratory analysis.* Journal of knowledge management, **13** (6), 392-409, (2009).
20. J. Van der Geer, J. A. J. Hanraads, R. A. Lupton, *The art of writing a scientific article*, J Sci Commun **163**, 51-9, (2000).
21. M. A. Berawi, *Quality revolution: leading the innovation and competitive advantages*, International journal of quality and reliability management, **21**(4), 425-38, (2004).
22. M. A. Berawi, *Innovation engineering of infrastructure mega project: study case in sunda strait bridge*, Depok: UI Press, (2015).
23. Jakarta Governor Regulation No. 265 year 2014.

Improving Role of Construction Industry for More Effective Post-Disaster Emergency Response To Road Infrastructure in Indonesia

Krishna S. Pribadi[1,*], Eliza Puri[1], Eliya Hanafi[1], and Tri Hadinata[1]

[1]FTSL Institut Teknologi Bandung, Department of Civil Engineering, Jl. Ganesha 10 Bandung 40181, Indonesia

Abstract. Geo- and hydro-meteorological disasters typically caused disruptive impact to road networks due to damaged road infrastructure, which in turn disconnect access to and isolate the disaster affected areas. Road clearing work and emergency road recovery operation are considered a priority to reconnect the access during post-disaster emergency response. However, the operation is not always smooth and in many cases delayed due to various problems. An investigation is conducted to understand the current practice of post-disaster emergency road recovery operation in Indonesia and to study possible participation of construction industry in order to improve its effectiveness. In-depth interviews with Local Disaster Management Agencies (BPBDs) and local road agencies in West Java Province were conducted to understand current practices in emergency road recovery operation and to view perspectives on local contractor participation. The surveys showed supports from the local governments for contractor involvement as long as it is still under guidance of related agencies (Ministry of Public Works and Housing) despite some possible obstacles from the current regulation that may hamper contractors' participation, which indicate that there is a potential role of construction industry for more effective post-disaster emergency response, provided that contractor associations are involved and existing procurement regulation is improved.

1 Introduction

Indonesia is vulnerable to many different types of hazards. According to National Disaster Management Agency (BNPB), there are at least 20 categories of natural and man made hazard in Indonesia. Related to road networks, landslide is an important secondary hazard in Indonesia and is closely associated with flooding, earthquake, and volcanic activity (CFE-DMDA, 2015). These hazards have been known for the disruptive effect to the road service due to damages and debris caused by the impact of the hazards, and in some cases causes disruption of access to the disaster affected locations.

Transport/access road disruption into and out of the affected area have been considered as a vital constraint in providing efficient response in disaster and post disaster reconstruction activity [1]. Not only does a functioning access road may save lives through enabling access for evacuations, but also help a speedy distribution of goods and helps into the affected area [2]. Thus functioning road networks or access road have been proven to have a major and important role [2]. However, some cases have shown that fast and propre emergency response to recover road functions was often delayed by several factors, such as lack of heavy equipment and skilled operators, unavailability of standby emergency fund, and delayed response decisions due to regulation constraints. The pre-arranged participation of private sector or construction industry may be considered as a solution to improve the process, as [3] suggested that construction companies already in the affected areas may contribute better labour, materials and equipments supported by good networks and supply chain for immediate relief which also include debris removal and emergency repair to put back the road function.

This paper tries to identify factors including scheme and obstacles during emergency response and provide recommendation on how to improve construction industry participation as a solution for an effective and timely post-disaster emergency response.

2 Post Disaster Emergency Response Mechanism in Indonesia

2.1. Role Sharing and Coordination

The Law No. 24/2007 concerning Disaster Management was enacted in 2007 and the National Disaster Management Agency or Badan Nasional Penanggulangan Bencana (BNPB) was established in

* Corresponding author: kspribadi@gmail.com ; ksuryanto@si.itb.ac.id

2008. As the national agency, BNPB is responsible to coordinate and implement disaster management activities at the national level, while at the local level, Local Disaster Management Agencies or Badan Penanggulangan Bencana Daerah (BPBDs) are established in the provinces, cities and regencies, with the similar responsibilities with BNPB at their own jurisdictions. Fig.1 shows the coordination mechanism between different organizations and agencies for disaster management. The Law 24/2007 establishes three stages of disaster management, i.e. pre-disaster, emergency response, and post-disaster. This study focus on the emergency response stage. During this stage, rapid and immediate respond is needed to prevent further victim and infrastructure losses. Activities in this stage consist of rapid assessment of location affected by the disaster, damage and losses, and needed resources, decide on disaster emergency status, rescue and evacuation of disaster affected community, provision of basic necessities, protect vulnerable groups and immediate recovery of essential facilities and infrastructure.

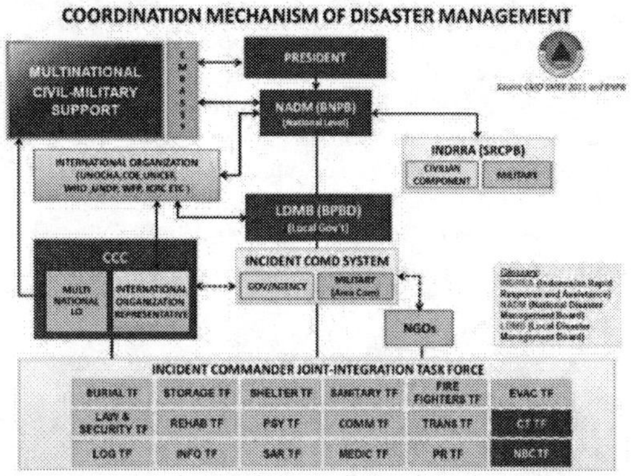

Fig. 1. Coordination Mechanism of Disaster Management (Source: [4]).

Immediatelly after a disaster strikes and is reported, Rapid Response and Assistance Team from Local Disaster Management Agency (BPBD) will be dispatched to conduct rapid assessment related to the disaster. Their report and recomendation will be collected and submitted to Governor or Regent/Mayor as a basis for declaring disaster emergency status. Next, all local agencies representatives will join the Rapid Response and Assistance Team (Tim Reaksi Cepat) dispatched by the local BPBD to the impacted area. They will coordinate with BPBD on the issues of rescues, evacuation and delivery of basic needs aid.

In particular, related to emergency restoration of road access, the agency responsible for the operation is the road administration agency, depending on the level of the road hierarchy, e.g. for the national road the authority is the Directorate General of Road Development under the Ministry of Public Works and Housing (DGRD-MoPWH), for the provincial roads, the authority is under the Provincial Public Works Agency and in some cases the Provincial Road Agency, depending on the provincial government structure. For the City/Regency Road, the authority is under the Public Works or Road Agency of the municipality/regency. These agencies are in close coordination with the Local Disaster Management Agency (BPBD) for conducting fast response related to road and other infrastructure recovery in the post-disaster emergency situation. In the post-disaster emergency response, emergency access road restoration should be conducted rapidly to support other activities (rescues, evacuation, aid distribution). While to reconnect the disconnected access, road clearing is usually the first task for the post-disaster process [5].

2.2 Post-Disaster Emergency Funding

Disaster management fund sources and fund use, financing, supervising, reporting and accountability guidelines are provided in Presidential Regulation No.22/2008 concerning Disaster Aid Financing and Management. In the implementation, the state budget (APBN), the local budget (APBD) and social/community funding are stipulated as the main fund sources. However, during disaster emergency response, fund source may also utilize work unit budget, and in this case of road clearing and temporary repair, national or local Road Authority or Public Works Department (PWD) unit. The provision of on-call budget for emergency response is regulated in the Head of BNPB Regulation (PERKA) No.6A/2011 concerning Emergency Fund Guideline. Basically the Regulation divides emergency response situation into three statuses, i.e. emergency alert, emergency response and emergency transition to recovery. Activities related to debris removal and emergency temporary repair of road infrastructure in order that the road can be functioned again in the emergency situation are considered as in the emergency response status within the emergency relief category. Table 1 explains the use of the on-call funding.

Table. 1. Use of on-call budget for road clearing and repair (Source : PerkaBNPB6A/2011 concerning emergency fund guideline)

Emergency Status	Activity	Use of on-call fund
Emergency response status	Emergency relief	Procurement of good and services required to clear debris and landslide, repair of dikes, repair/build emergency road/bridge/ quay and other relevant equipment/plant (Clause II.B.2.b.2)

The Minister of Home Affaires (MoHA) regulation No.13/2006 about guidance for local government allocation states that Government can use Unexpected Expense Fund or Belanja Tidak Terduga (BTT) during post-disaster emergency response to prevent further damage and social distructive effect, based on the Mayor/Head of Local Government statement of disaster emergency situation. In emergency situation, BTT can be used and the expenses and/or reimbursement should be reported. The treasurer of the office/agency in charge will be responsible for the reporting. The decision of the Mayor/Head of Local Government to declare emergency statement should be based on the recommendation of the relevant agency (BPBD). Most of the local governments have their on local regulation stipulating the procedure and responsibility of using BTT for emergency response.

A typical procedure for using BTT from a local government usually requires the following documents:

- A rapid and proper assessment to identify location, victims, damage and loss of infrastructure
- Declaration of disaster emergency status by the Head of Local Government
- A brief action plan stating activities during the emergency response with the detailed allocation and expense.

Based on the above documents, the Head of Local Government will issue the permit to use BTT fund stating the amount of BTT fund. More detailed procedures are usually stated in the SOPs of each local government.

3 Common Practices in Other Countries

It is very common to have agreements between local government and business companies or private sector in Japan, which may concern electricity recovery, medication, logistics and foods companies. Support of the private sector includes companies in the construction, transportation, and retail sectors. They played a key role in the response as they brought in expertise and assets that benefited the response, although in some cases it was improvised and unanticipated, as suggested by Taniguchi et al. in [5].

Contractors in Japan enter into agreement with local governments to do some road clearing without waiting for the instruction in case a natural disaster occurs. The scheme of the framework agreement in Japanese is called 'Kyoutei'. This practice is very useful to speed up the response action to put back the affected roads in operation [6]

In general, if an emergency process is used for selecting a contractor, it is still important that proper procurement procedure be followed and that the afore mentioned considerations in establishing a pre-disaster plan be applied [7]. Specifically for debris removal, it was suggested that bids and/or quotes from contractors should contain the following information:

1. The approximate quantity of debris
2. The specific disposal location
3. The unit price per cubic yard or ton
4. Mobilization time and cost
5. References and proof of insurance

4 Post Disaster Emergency Response Works in Indonesia

4.1 Civil Works in Post Disaster Emergency Response

To prevent further damage and losses to the affected areas caused by delayed response operation as well as to prevent work accident, road clearing and emergency repair works should be conducted by professional personnel, which may include local experienced contractors with adequate resources. However, in actuality, there are cases where the locals with lack of skill and knowledge are participating in emergency operation, such as in Banyumas Regency landslide in February 2017 [8] where road clearing operation was under the command of Banyumas BPBD, involving military personnel and volunteers from local community with no special skill on debris removal under hazardous situation.

Civil works recovery on disaster affected areas, which include road clearing, debris removal and temporary road structure works are needed to put back the function of the road infrastructure, either temporarily or permanently. The works may require the mobilization of heavy equipment for earthwork, such as dozers, excavators, shovel, and compactors, to rapidly recover the access. Hazardous site condition and constraints on logistics as well as pressure to reopen the road back to traffic as soon as possible will increase the complexity of the work on site, thus competent field engineers and supervisors, skilled operators and workers are needed for optimum results.

4.2 Contractor Involvement During Post Disaster Emergency Response

Road clearing and emergency access road recovery become critical activities when disaster impacts road which serve as the only access to deliver logistics to the affected areas. Since road clearing and emergency access road recovery are in the critical path of post-disaster emergency response activities, the construction industry is well placed to participate in and contribute to the process. Another benefit of involving construction industry or private sector is that when they prepare their own Business Continuity Plan (BCP), they can take advantage of their involvement in the emergency operation by supporting a wider region contingency and survival-ability plan enabling resource mobilization from private for the public [5], which in turn will help in the quick recovery of businesses in the disaster affected region. In addition, the participation of local contractors in the recovery process will also help the local economy recovery, by providing livelihood to the local people and by promoting the local economic activity through the provision of good and services in the region, thus providing opportunity for local suppliers (materials and equipment) to be engaged again in activity.

In accordance with Presidential Regulation No.70/2012 on Procurement of Goods and Services for

Government, the private sector can be directly appointed by local agency considering the need of immediate works to prevent further losses caused by the disaster. The experience of the local contractor can become an added value to complete the task immediately and reliably. The presence of a local contractor will cut the local agencies coordination, and make the road clearing task more efficient.

4.3 Views of Local Government Agencies Related to Contractor Involvement During Post Disaster Emergency Response

During this research, several interviews had been conducted with local agencies related to contractor involvement during post-disaster emergency phase. The local agencies such as military and police organization (TNI/Polri), transportation agency, Indonesian Red Cross and Social agency are not included in this research since the research is more concerned with Local Disaster Management Agency (BPBD) and Road Agency or Public Works Department (PWD) as the main actors in the reopening of impacted road and bridges and they are considered as having better understanding on the technical operation. Table 2 and Table 3 show the result of all local agency perspective about contractor involvement.

Four local agencies seem to agree with the idea of contractor involvement. They consider that the works would be much easier to be conducted. The procurement process would be under PWD because of their mandate and the technical expertise.

Table 2. Interview Response Related to Contractor Involvement and Procurement

Local Agencies	Response related to contractor involvement	Procurement Process
BPBD of Bandung District	Does not agree with contractor involvement	Response managed by PWD of Province/district
BPBD of Sumedang District	Agree with contractor involvement	Managed by PWD Province/district
BPBD of West Java Province	Agree with contractor involvement	Managed by PWD Prov/district
PWD of Bandung District	Agree with contractor involvement	Direct Appointment
Bina Marga West Java Province	Agree with contractor involvement with amendment of regulation especially for procurement process	- Direct Appointment - Tendering method can not be implemented as it will take time and can not follow the procedure stipulated in Pre.Reg No 70/2012

Table 3. Interview Response Related to Funding Source and Obstacle

Local agencies	Source of Post-Disaster Emergency Fund	Perceived Obstacles
BPBD Bandung District	Emergency/On-call Fund (Dana Siap Pakai/DSP)	Non trustable Contractor, will be difficult to manage
BPBD Sumedang District	Emergency/On-call Fund (Dana Siap Pakai/DSP)	Lack of experience to respond to disaster in Sumedang Area. BPBD Sumedang District was established in 2015
BPBD West Java Province	Emergency/On-call Fund (Dana Siap Pakai/DSP)	BPBD doesn't have supporting facility like the presence of technical experts
PWD Bandung District	Bina Marga's Maintenance Allocation first, then request reimbursement using Emergency/On-call Fund (Dana Siap Pakai) to BPBD's	-The lack of availability of heavy equipment on disaster site -The availability of Emergency Fund

4.4 Constraints to Road Clearing and Temporary Repair Operation in Post-disaster Emergency Response

4.4.1 Availability of Heavy Equipment in Local Government

Delays in the emergency road clearing and repair operation have been experienced in many post-disaster cases, due to the unavailability of adequate heavy equipment in the vicinity of the affected roads. Fig.2 shows the unbalanced distribution of heavy equipment registered at the MoPWH, belonging to various construction and heavy equipment firms as well as to various MoPWH local offices. The figure was based on an equipment register database maintained by MoPWH, which is far from being complete, totalling only about 4800 units. The equipment relevant to road clearing and temporary repair are about 50% of the registered equipment. Actually, there are more heavy equipment available belonging to other organizations/agencies, such as mining companies, military engineering divisions, aggregates and asphalt/cement mix producers, etc. The figure indicates that provinces with high vulnerability are not always equipped with sufficient number of heavy equipment, while DKI Jakarta (the capital) with low landslide hazard is well equipped.

4.4.2 Lack of Skilled Human Resources

During post-disaster emergency operations, the availability of suitably skilled labour may be limited due to a combination of small local capacities and high demand [9]. Those are also experienced in Indonesia, as an interview with one of local PWD officers shows that

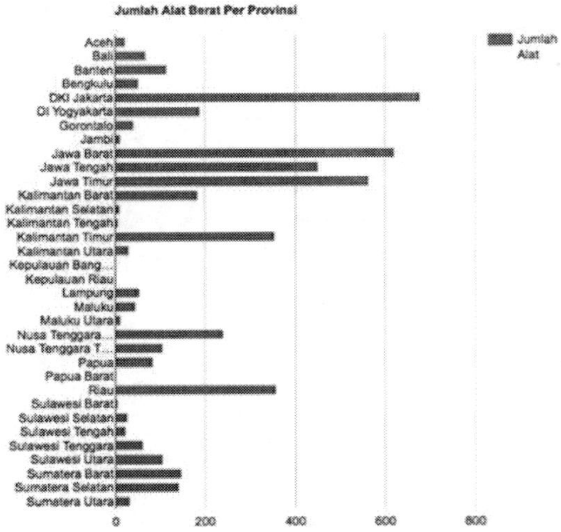

Fig. 2. Heavy Equipment Distribution by Province in Indonesia. (http://mpk.binakonstruksi.pu.go.id/summary/prov, downloaded on 20 Sept, 2017).

no special response unit with adequate skilled personnel to operate is available in the region. People on disaster sites are not always qualified to operate heavy equipment or conduct civil works. A lot of locals may want to help, but it can be dangerous for them because they have no or little experience and knowledge and skill to do the works properly. Therefore, it is essential to assess the construction skills of the local population and capacity of the local contractors.

4.4.3 Availability of Liquid Fund during Post-Disaster Emergency Response

According to the interviews, most of post-disaster emergency works including road clearing and emergency repair are hampered by the lack of availability of liquid fund on time. It is not rare for related agency, local PWD, to use their own source and budget in order that task can be done, despite the provision stipulated in the government No 22/2008 on reporting and accountability procedure. The Provincial Road Authority in West Java Province stated that most of road works fund for post-disaster emergency response utilised operational budget allocation to temporarily cover their works. However, there were issues of unreimbursed expenses when they tried to propose to BPBD to get reimbursement. In addition, it is a common situation on post-disaster emergency situation that fuel, equipment and operators cost much more than in the normal situation and will affect internal operational budget of the agency. Delayed or unapproved reimbursement usually will impact their internal budget for the current maintenance works and hamper the realization of their initial plan target.

4.4.4 Restrictions in Procurement Schemes

Under the local agencies, Bina Marga Provincial Government or Bina Marga Regency/City Government, contractors may be engaged to do road post-disaster clearing work. The Presidential Regulation No 70/2012 on Procurement of Goods and Services for Government stipulates a scheme for direct appointment for emergency post-disaster works, but government agencies are not allowed to do so if there is no declaration of disaster by the head of the local (or national) authority. However, during a post-disaster emergency situation, where timely response is really needed, selecting and appointing a contractor to do emergency works may take some time to process, initiated by inviting recommended (by local contractor associations) contractors, assessing their capacity (financial, skill and availability of equipment), negotiating the contract and then appointing the selected contractor.

4.5 Multi Stakeholder Forum Group Discussion

A Forum Group Disccussion where representatives from BPBDs, PWDs and Contractor Association from different cities/regencies in West Java has been conducted in November 2017 in order to get a better understanding of stakeholders' view related to contractor involvement in emergency response,

It was agreed during the discussion that participation of local contractors / private sector sides in disaster emergency response operations was an implementable and useful solution. However there were areas that need to be considered and elaborated more in detail, such as preparation, technical implementation, cost, coordination, and in particular supporting regulations.

Other factors to be considered when involving contractors in emergency response that were agreed during the discussion include :

1. Response time:
   - Response time is the main variable, prioritize speed and readiness in solving the problem
   - Good coordination
   - Placement of facilities and resources in the proximity of disaster prone areas.
   - Consideration of existing rules and regulation.
   - Equipment types and distance to impacted areas
   - Priority based on the level of importance of the service to be recovered, for example to prioritize between national roads or local roads
2. Approach to implementation:
   - Improvement / revision of existing regulations related to disaster management and preparation of relevant SOP/guidelines for good emergency response to avoid obstacles during the implementation process.
   - Prepare good MoU with the contractor association
   - Local political will to involve the private sector in disaster management (disaster response)
   - Involvement of local legislators, community, local government leaders before contractor involvement

- Commitment from all parties involved in disaster management, including local leaders, for humanitarian purpose.
- Ensure budget allocation for disaster management supported by legal rule.

5 Discussion

5.1 Improving The Contractor Association Role

Participation of contractor association is needed to improve the preparedness and readiness of contractors, in order that the contractor can be safely involved in the emergency response with less common obstacles such as regulation restriction, uncertainty of estimated basic price for activities, and avoiding non trustable contractor in the field.

Cahill in [7] argues that taking the time to formulate a detailed management plan before a disaster strikes helps to identify and resolve potential issues that might arise ahead of time, without the pressure and time constraints that would be faced in the immediate aftermath of a disaster. Furthermore, if a disaster does occur, having a careful plan to follow minimizes response times and maximizes recovery efforts. Considering this perspective, contractor association can be involved in the preparation of the emergency response management plan before disaster strikes, and by assisting and preparing the contractor to be ready when disaster strikes in certain area.

Bellow are some recommended roles for contractor association to enhance contractor involvement during post disaster emergency response:

- Assists selected contractors to get earthwork specialty. This may help contractors to get required certification, which will justify his/her direct appointment for clearing works.
- Recommends selected contractors to obtain special emergency response construction permit based on their qualification and classification, which will help MoPWH in ensuring qualified contractors mobilized for post disaster emergency operation.
- Mapping and recommending the selected contractors to be based within a specific disaster prone area to enhance the effectiveness and immediate response during disaster emergency situation.
- Building capacity of the contractors to cope with disaster emergency situation so that sufficient resources can be ready.
- Requesting an amendment of Presidential Regulation No.70/2012 Concerning Procurement of Goods and Services for Government that will enable contractors to be selected before disaster happens in order to permit them to be mobilized by direct appointment once there is a disaster..
- Assessing contractor works after emergency situation for next recommendation.

5.2 Improving Regulation

As mentioned in [6], Kyoutei or agreement between local government and local contractor in Japan is effective and useful to speed up the response during disaster emergency situation. Unfortunately, in the current situation, the scenario like Kyoutei can not be implemented in Indonesia, as it is unsuitable in Indonesia according to the Presidential Regulation No.70/2012. Indonesian government agencies are not allowed to procure contractor services if the payment item is not clear and still uncertain. The scope of works of the contractor, the coordination line, the quantity of works, location of works and amount of contract need to be ensured before a contract is signed. Without clear definition of the items, road clearing and emergency works during disaster emergency can raise the issue of improper conduct of the responsible officer and can be considered as conducting corruption crime.

In practice, the available procurement scheme for post disaster response by direct appointment of contractors by the local agencies is still felt as far from effective and timely, in particular when the local agencies are also affected by the disaster and need to evacuate, while at the same time they have also the responsibility to conduct the evacuation of the survivors.

Another issue to consider also is the trust between government agency and the contractors, as in the eye of many government officers, contractors are not trustable and their participation in the emergency response is just for profit only, so they will maximize the profit during the uncertain emergency situation. This view was observed during the interview with local agencies, whereas many government officers felt the best way to respond to a disaster is by force-account or self-managed work, as they can not trust the contractor to provide satisfactory work, and in some cases even enter into dispute. Nevertheless, it is thought that under the responsibility of trustable contractor, emergency operation works can be implemented in a timely manner, without to much hassle from bureaucratic administrative procedure, at the same time as the local government agencies are occupied with the priority of saving people's life.

The procurement scheme for direct appointment is valid only for work under Rp. 200 million (approx..USD 150.000), low risk with simple work method/technology, and can be awarded only to small business entity, and the scheme can only be implemented when a disaster has been declared. According to our interview, usually post-disaster emergency status is only declared for a week and the longest time to respond is two weeks. In this case there will not be enough time for local agency to make the direct appointment process while responding to other activities of the post-disaster emergency. For them, it is easier and sometime faster if they respond by their own resources without any third party, provided that they have the require resources to do so.

To improve construction industry participation in post-disaster emergency response, there is a need to improve the regulations. The basic principle to apply is that it would be better to be prepared before the disaster happens, and contractor participation should be well arranged before disaster strikes. First, the Road Agency should conduct the process of contractor selection for future emergency response. Ideally, in one disaster prone area there would be one contractor selected to respond in post-disaster emergency but in case no sufficient budget allocation for that, one contractor can be assigned for one district. The criteria for selected contractor among others include the availability of qualified personnel and heavy equipment with experienced operators, good financial situation, high integrity, trustable in all services, recommendation from construction association, and have a good attitude for humanitarian operation..

Regarding the supply of construction materials, which experience shows that in post disaster situation, they are typically subject to high inflation, and the quality deteriorates as production processes becomes overstretched in order to meet the large scale demand [8], consideration should also be given to accommodate the fluctuation during the process. It is desirable that a robust post-disaster reconstruction planning is conducted prior to the occurrence of disaster. Systematic mapping of resource availability should be made available to the disaster prone areas to provide for effective emergency road repair. Over the emergency response and recovery period, to reduce potential resource cost fluctuations and their impacts on recovery, systematic mapping and monitoring of various construction resources should be conducted [10]. The use of E-catalogue will be useful for this process, but as construction costs are often fluctuating in post-disaster situation, a separation between normal situation E-catalogue and crisis or post-disaster situation E-catalogue need to be exercised.

5.3 Standard Operational Procedure (SOP)

Based on the FGD result, there are four main stakeholders which are to be involved in the emergency response preparation, namely BPBDs, PWDs, contractor association and the contractors. It is understood that the emergency works will be implemented in the emergency response stage, however the procurement process has to be completed during the preparedness stage.

If the Presidential Regulation No.70/2012 has been amended, BPBD may initiate the procurement process and coordinate with PWD who will conduct all the procurement process. PWD will have to determine the scope of works, price policy (e-catalogue) and contractor criteria/contractor qualification. The works can be confined to non-permanent works such as road clearing or temporary road access works. Unit price contract may be suitable for this kind of situation, because the total amount of the contract could not be determined in the beginning.

For emergency stage, contractor should mobilize their resources (personnel and heavy equipment) under 24 hours after disaster strikes as the policy is to reopen the damaged road within 24 hours. At this stage, communication become the key to the success of the operation and BPBD and PWD need to monitor the contractor works to mitigate new risk of another landslide during the evacuation or construction process.

In the transition stage, contractor together with the PWD Officers have to measure and verify the works that have been done. Later, approved request for payment based on the join measurement will be delivered to BPBD for the reimbursement process by using emergency fund or BTT. Fig.3 shows a scheme depicting the SOP for the implementation of the mechanism by all the involved parties during preparedness and emergency response stages.

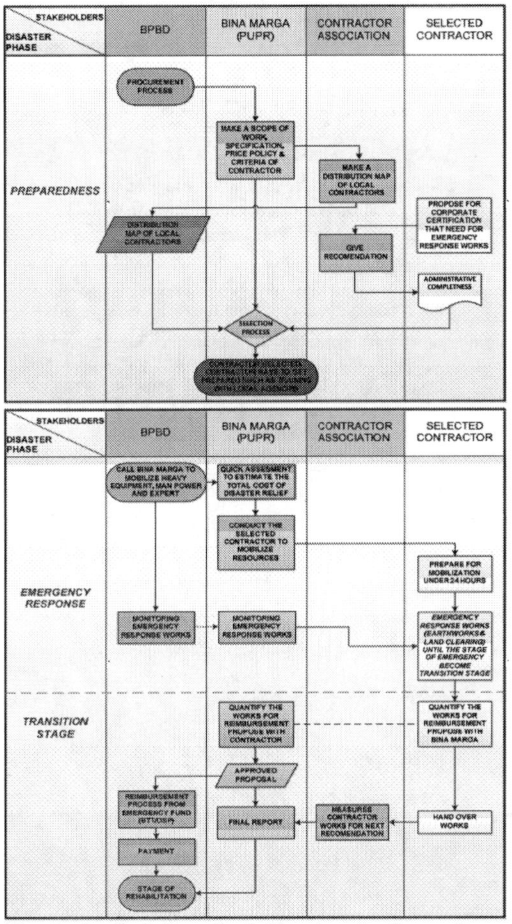

Fig. 3. Standard Operational Procedure for Preparedness, Emergency Response and Transition Stages

6 Conclusions

Findings revealed some problems that occurred during post-disaster emergency response include lack of heavy equipment, lack of skilled human resource, the availability of emergency fund and tight policy of procurement scheme. Those barriers can be overcome by improving contractor roles for road clearing and emergency/temporary repair works during post disaster emergency situation.

The survey also concluded that Local Disaster Management Agency (BPBD) as a substantial stakeholder in the disaster emergency response does not give any objection for contractor involvement as long as it is conducted under the guidance of related department (Road Agency or PWD or MoPWH) despite the possibility of several obstacles, particularly regarding the current regulations.

In conclusion, some suggestions for improving the role of construction industry are proposed:

1. To improve contractor association involvement in role sharing among local contractors for more effective post-disaster emergency response.
2. To amend regulation the current government regulation on Procurement of Goods and Services for Government. The amendment should involves changes that allow procurement process to be initiated before the occurrence of disaster in order to promote disaster mitigation measure, as it is believed that preparation during normal time will produce better result rather than arrangement during post-disaster crisis. Both direct appointment and tender schemes with unit price based contract are to be implemented before the disaster.
3. Road Agency or PWD should maintain an e-Catalogue special for post-disaster emergency response purposes to facilitate the procurement process and control the fluctuation of price due to scarcity of resources.

The authors wish to extend their gratitude and appreciation to the Institute of Research and Community Outreach of ITB who provided the grant for this research under the P3MI Program.

References

1. F Grünewald, A Binder, Y Georges, Inter Agency Real Time Evaluation in Haiti. Final Report. Global Public Policy Institute, 2010
2. E. Hayat, RDG. Amaratunga, Road reconstruction in post - disaster recovery; the challenges and obstacles (University of Salford Manchester, 2011) http://usir.salford.ac.uk/23372/1/23372.pdf
3. P Stringfellow, Construction contractors involvement in disaster management planning, Australasian Journal of Constr. Econ. & Build., **14 (2)** 120-132 (2014)
4. ASEAN Disaster Management Reference Handbook , CFE-DMHA, 2015
5. M.B.F. Bisri, Choose Your Partner Correctly in Emergency Response: Partnership Innovation Learned from the Great East Japan Earthquake and Tsunami 2011 – Final Report, UNOCHA 2013 Humanitarian Research and Innovation Grant Program, New York (2014).
6. M. Onishi, K.S. Pribadi, The Role of Construction Industry in Post Disaster Recovery (Proc.of 3rd International Conference on Earthquake Engineering and Disaster Mitigation - ICEEDM, Bali, 2016)
7. K. Cahill, - Proper Debris Management —Key to Disaster Recovery and FEMA Assistance, Disaster Recovery Today **13** (Adjusters International Inc. New York, 2011)
8. https://www.gatra.com/nusantara/jawa/287430-bencana-tanah-longsor-landa-4-desa-di-kabupaten-banyumas
9. J de Silva, Lessons from Aceh - Key Considerations in Post-Disaster Reconstruction, Practical Action Publ., 2010
10. Y. Chang, S. Wilkinson, E. Seville, R. Potangaroa, Changes in resource need for post-disaster reconstruction: a longitudinal study in China (Journal of Build. Res.& Inf.**40(3)** 327-336 (2012)

Grouting design for slope stability of kedung uling earthfill dam

Najib [1,*], *Agus* Setyawan[2], and *Dwiyanto* Joko Suprapto[1]

[1]Geological Engineering Department, Engineering Faculty, Diponegoro University, Indonesia
[2]Physics Department, Mathematics and Science Faculty, Diponegoro University, Indonesia

Abstract. Kedung Uling earthfill dam locates at Wonogiri Regency, Central Java, Indonesia. The dam encountered sliding and settlement at the embankment wall. To minimize sliding and settlement and to optimize the dam, both field investigation and laboratory tests have been proceeded for slope stability analysis and remedial embankment wall. Soil and rock investigation around the dam, which is followed by 10 core drillings, have been conducted. Laboratory tests such as direct shear and index properties have also been carried on. The results were further used for dam slope stability model using slide 6.0 and were used to analyzed factor of safety (FS) of Kedunguling dam. 10 conditions of dam were simulated and strengthening body of dam with grouting was designed. The results showed two conditions, which are condition of maximum water level with and without earthquake at downstream, were unsatisfy Indonesia National Standard (SNI) for building and infrastructure. These conditions can be managed by using grouting for increasing stabilization of embankment wall. By setting up grouting, factor of safety increases and meet the SNI standard requirement.

1 Introduction

Kedung Uling earthfill dam is one of many old earthfill dams in Indonesia. The dam which was established on 1917, locates at Wonogiri Regency, Central Java, Indonesia. The dam lies on between Jatirejo and Tempuran river. Kedung Uling dam which has main function for irigation has 479.000 m^3 capacity and distributing irigation for 596 hectare rice fields. Other functions of this dam are for tourism and fishery fresh water. In 2013 to 2014, the dam was encountered sliding and settlement at the embankment wall. These such problems affect function of dam such as releasing water in big quantity, disturbance risk people or infrastructures at the downstream [1].Therefore, It is strongly required to analyze the slope stability of Kedung Uling dam for overcoming the problems. Besides, water capacity of the dam can be increased by optimizing the dam for fulfilling water needs.

Dam Optimazing can be done by increasing height of top of the dam. This solution can be done by adding filling material at the body of dam. However, increasing the top of dam with impervious filling materials without good engineering solution cause the problems such as sliding, settlement or piping as happened in Kedunguling dam.

One of solutions for strenghtening filling soil in earthfill dam is by using grouting method [2,3]. Grouting has more benefit than another method such as grouting can be installed in the finite area, more efective and no need formidable tools and does not change slope geometry. Curtain grouting has already implemented in overcoming dam seepage with the bedrock of dam is limestone, eventhough piping still continue due to its porous litology [4]. Therefore, it needs to calculate the effective of composition for implementing of curtain grouting especially for dam foundation rehabilitation [5]. It is expected that by appropriate composition, grouting will increase value of cohesion and decreasing internal friction angle of soil [6].

In this research, we will design the grouting based on field data and laboratory tests so that the grouting is fulfill the requirement standard. In the implementation in earthfill dam, it is expected that strengthening of soil after grouting is not changing the materials to be concrete so that the soil can move plastically when the water flooding the dam.

2 Methods

2.1 surface and Subsurface Observation

Surface observation is figure out soil and rock identification around Kedung Uling dam. Meanwhile, core drilling is conducted to get soil samples either undisturbed sample or disturbed sample. Core drilling is followed by SPT test by 1 meter interval and undisturbed sample has interval 5 meter/ depth for mechanical soil test. Core drilling is carry out for 4 points at sliding location and 4 points at body of dam with the depth around 10 to 15 meter.

* Corresponding author: najib@ft.undip.ac.id

2.2 Laboratory Tests

Laboratory tests from undisturbed samples are set for 30 samples i.e direct Shear test and Index properties test. The results of tests will be used for slope stability analysis at the embankment wall of dam and for grouting design. The analysis has been conducted at soil mechanics laboratory (ISO 9001: 2008) at Selimut Bumi Adhi Cipta Company.

2.3 Slope Stability Analysis

Slide 6.0 is used for simulation of dam slope stability. In this simulation, cross section of dam model was made based on litological units from drilling discription and laboratory analytical results i.e unit weight, cohesion and internal friction angle. Before slope stability of dam is simulated, analytical of earthquake in Kedung Uling is determined based on Indonesia earthquake zonation map (fig. 1). Afterward, dam risk class (Risk factor) based on the classification as shown in Table 1 [7].

Class of Building and earthquake loading (FR_{tot}) can be determined by using equation 1[7]:

$$FRtot = FRk + FRt + FRe + FRh \quad (1)$$

Which FRk is capacity of dam, FRt is height of dam, FRe is evacuation need and FRh is downstream damage level. Afterward, risk class can be determined by using table 3. Earthquake loading criteria for dam design can be seen in table 2.

Kedung Uling dam will be analyzed for 100 years (T =100 years); Factor of safety (FS) agree with the criterion. For T = 5000 years, Factor of safety >1, if dinamic analysis is not fulfill. Earthquake velocity coeficient (K) has given by equation 3:

$$Ad = Ag.V \quad (2)$$

$$K = \frac{Ad}{g} \quad (3)$$

Where Ad is corrected ground velocity (cm/s^2), Ag is Peak soil velocity (gal), V is type of soil corrected factor, g is gravitation velocity (cm/s^2). Z (earthquake zone coeficient) can be determined by plotting research location on earthquake zonation map based on its zonation and contour. Ac can be determined from repeat period (T) where the value depends on type, class and how long the dam will be used. Ground maximum velocity (ag) is calculated from Z multiply by Ac. Wave velocity based on NSPT of soil. V is calculated from equation 4:

$$Ts = 1.25 \sum_{i=1}^{n} \frac{4Hi}{Vsi} \quad (4)$$

(1)

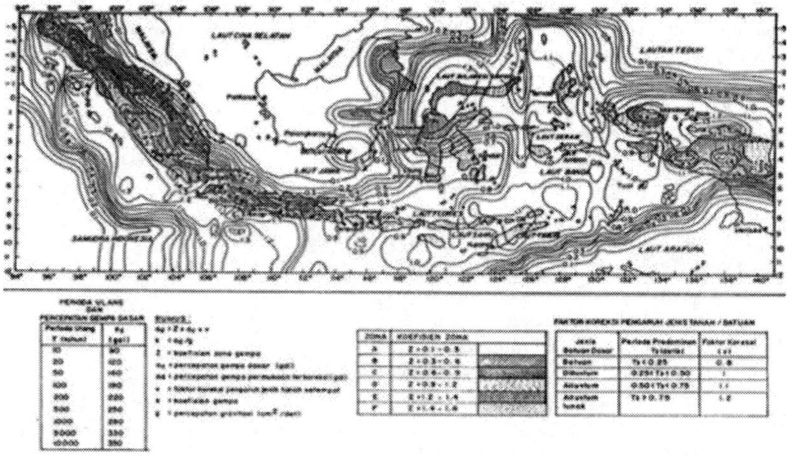

Fig.1. Indonesia earthquake zonation map [7]

Table 1. Dam Risk Classification [7]

	weighting value in brackets				
Risk Factor	Ekstreme	High		Moderate	Low
Capacity ($10^8 m^3$) (FRk)	>100 (6)	100-1,25 (4)		100-0,25 (2)	<0,125 (0)
Height (m) (FRt)	>45 (6)	45-30 (4)		30-15 (2)	<15 (0)
Evacuation need (people number)(Fre)	> 1000 (12)	1000-100 (8)		100-1 (4)	0 (0)
Downstream damage level (FRh)	Very High (12)	High (10)	Moderately high (8)	Moderate (4)	None (0)

Table 2. Earthquake loading criteria for dam design [7].

Risk class based on time use	Requirement without devastaging	Requirement allowing devastating without fail		
	T (year)	Analytical method	T (year)	Analytical method
IV	100-200	Earthquake coeficient	10000	Earthquake coeficient or dynamic*
N=50-100	ad ≥ 0,1 g		(MDE)	
III	50 - 100	Earthquake coeficient	5000	Earthquake coeficient or dynamic*
N=50-100	ad ≥ 0,1 g		(MDE)	
II	50 - 100	Earthquake coeficient	3000	Earthquake coeficient or dynamic *
N=50-100	ad ≥ 0,1 g		(MDE)	
I	50 - 100	Earthquake coeficient	1000	Earthquake coeficient or dynamic*
N=50-100	ad ≥ 0,1 g		(MDE)	

Table 3. Risk class dam and water infrastructure [7].

Total risk factor	Risk class
(0-6)	I (Low)
(7-18)	II (Moderate)
(19-30)	III(High)
(31-36)	IV(Extreme)

Slope stability analysis of body dam based on some condition for slope stability analysis. Ten conditions were set for Kedung Uling dam stability analysis:

1. Empty dam without earthquake loading at upstream condition.
2. Empty dam with earthquake loading at uppstream condition.
3. Empty dam without earthquake loading at downstream condition.
4. Empty dam with earthquake loading at downstream condition.
5. Maximum water level dam without earthquake loading at upstream condition.
6. Maximum water level dam with earthquake loading at upstream condition.
7. Maximum water level dam without earthquake loading at downstream condition.
8. Maximum water level dam wit earthquake loading at downstream condition.
9. Sudden water level drop without earthquake at upstream condition.
10. Sudden water level drop without earthquake at downstream condition.

Dam of stability analysis model use some parameters from soil mechanic tests such as unit weight, cohesion and internal friction angle. For earthquake analysis, it uses earthquake zone map[6].

2.3 Grouting Design

After slope stability of dam is being analyzed, grouting design will be simulated by using slide 6.0.

3 Results and Discussion

3.1 Surface Investigation

Kedung Uling earthfill dam is basically underlain by Baturetno formation (Qb), which consists of grey to black clay (Surono et al, 1992). Based on surface investigation, grey to black color clay is dominant, intersected with white yellowish sand and calcareous pebble. This intersection is only found in some limited area in the dam area.

Some cracks and sliding were found at the top of the dam and embankment walls (see fig. 2). Those cracks and sliding possibly occured due to poor foundation, poor filling materials or when constructed the filling materials was not same with the procedure.

3.2 Core Drilling

To investigate subsurface of Kedung Uling dam, core drilling was set up in ten points and was represent dam subsurface condition. The Locations of drilling points can be seen in fig. 3. The depth of drilling was 10 m, 12 m and 15 meter. Besides, SPT and permeability tests have been assembled. The results show that soil dominantly by grey to black stiff-firm clay with N-SPT 4-20.

3.3 Laboratory Tests

The mechanical soil tests i.e direct shear test, index properties and atterberg limit have been done. The numbers of samples are 30 samples. The results of soil tests, as shown in table 4, were used for slope stability of

https://doi.org/10.1051/matecconf/201814707001

Kedung Uling dam namely unit weight, cohesion and internal friction angle.

Fig. 2. Sliding (a) and cracks (b) at the top of Kedung Uling Earthfill dam.

Table 4 Mechanical soils characteristics of Dam body

Material	Unit weight (kN/m³)	Cohesion (kN/m²)	Internal Friction angle (°)
Filling impervious clay material	18.3	15.7	18.3
Filling old impervious black clay material	22.2	15.2	11.92
Filling old impervious grey clay material	20.17	14.42	14.97
Bedrock	20	7.1	30.59

3.4 Slope Stability Analysis of Dam

Based on Indonesia earthquake zonation map [7] (Fig. 2) and equation (1-4), Kedung Uling Earthfill dam can be categorized as shown on table 5.

Table 5 Determination of building class and earthquake loading.

Dam condition	Information	Factor
Dam capacity	366.646 m³	FRk=2
Dam Height	9 m	Frt=0
Evacuation need	0-200 people	Fre=4
Upstream danger level	Moderately high	RFh=8
		Frtotal=14

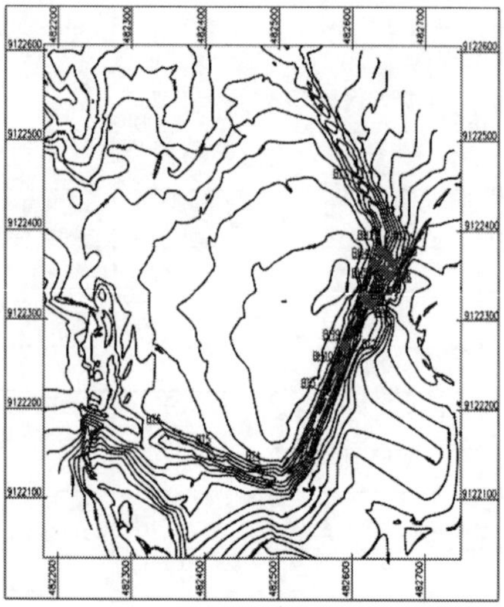

Fig.3 Core drilling location (dots are core drilling location).

Based on table 5, the total risk factor, Kedung Uling dam is categorized in class II (moderate). Table 6 shows the results of calculation of earthquake coeficient based on equation 1-4.

Table 6 Earthquake coeficient calculation

Parameter	Information
Top soil velocity (Ac)	T = 100 year Ac = 190 gal
Earthquake zone coeficient (z)	z = 0.2 (average)
Ground maximum velocity (ag)	ag = 38 g
Wave disperse velocity (Vs)	N_{SPT} = 12 Vs = 337.5 m/s
Soil type correction (Ts)	Ts = 0.444 Detik Foundation :Dilluvium V = 1 with 0.25 <Ts <0.50
earthquake velocity plan(ad)	Ad = 38 gal
Earthquake velocity coeficient(K)	K = 0,0387 g

Based on the 10 conditions of Kedung Uling dam, there are two conditions of Kedung Uling dam which are not fulfill SNI standard (table 7).

The condition of dam with the maximum water level with and without earthquake load are not satisfy value of safety factor of SNI standard (table 7). This means Kedunguling has sliding potential. In fact, kedung uling dam has settlement so that dam body falled down (see fig. 2). Cracks were found in several parts top of dam where the materials are black clay with high plasticity and low permeability.

3.5 Analytical slope stability after grouting in sliding location

https://doi.org/10.1051/matecconf/201814707001

To understand value of factor of safety, it has been conducted simulation of slope stability after grouting and compared cohesion value with its factor of safety(fig. 4). Cohesian should be at least ≥17 kN/m^2 to get minimum SNI standard requirement of the body of dam. The model can be seen in fig. 5.

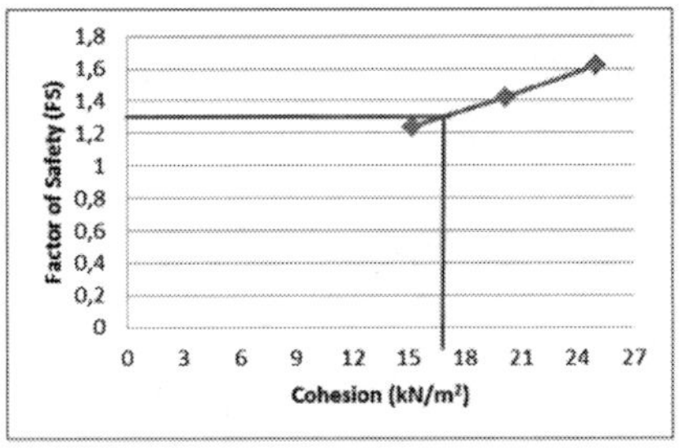

Fig.4. Comparison between FS and Cohesion .

3.6 Grouting Simulation

Grouting was simulated at three locations i.e upstream, top of dam and downstream (see figure 6). Slope stability has been simulated with three different cohesion (Fig.7-9). Slope stability also has been simulated with earthquake loading (Fig.10). From the results, factor of safety value is parallel with cohesion value. Comparison between cohesion after grouting and factor of safety is shown in fig. 10. Cohesion after grouting should be ≥24 kN/m^2 for getting factor of safety to reach SNI standard namely ≥1,3 (Fig. 10).

Table 7 Slope stability analytical results comparing with SNI Standard

No.	Dam condition	FK	Requirement of safety value (SNI)	Information
1	Empty dam without earthquake loading at upstream	1.766	1.4	satisfy
2	Empty dam with earthquake loading at upstream	1.570	1.2	satisfy
3	Empty dam without earthquake loading at downstream	1.512	1.4	satisfy
4	Empty dam without earthquake loading at upstream	1.367	1.2	satisfy
5	Maximum water level dam without earthquake loading at upstream	2.538	1.3	satisfy
6	Maximum water level dam with earthquake loading at upstream	2.041	1.1	satisfy
7	Maximum water level dam without earthquake loading at downstream	1.206	1.3	unsatisfy
8	Maximum water level dam with earthquake loading at downstream	1.081	1.1	unsatisfy
9	Sudden water level drop without earthquake at upstream	1.300	1.2	satisfy
10	Sudden water level drop without earthquake at downstream	1.402	1.2	satisfy

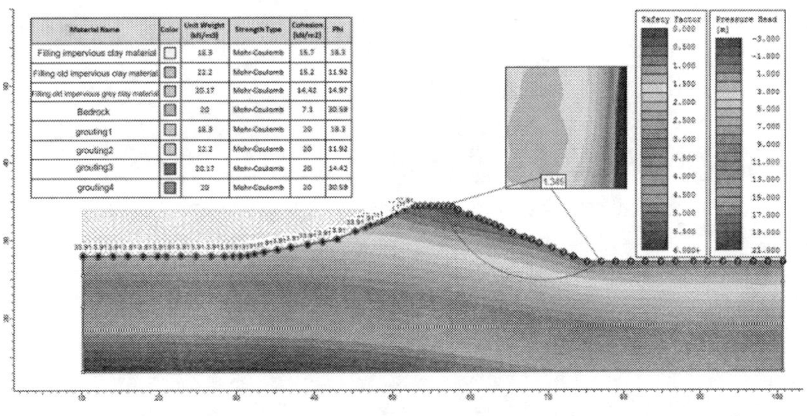

Fig. 5 Grouting simulation result of embankment which is encounter sliding. Cohesian value is 17 kN/m^2.

https://doi.org/10.1051/matecconf/201814707001

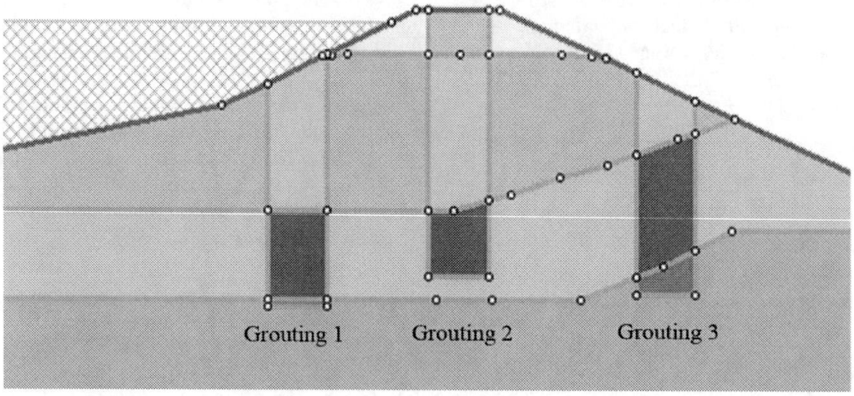

Fig.6 Grouting location points.

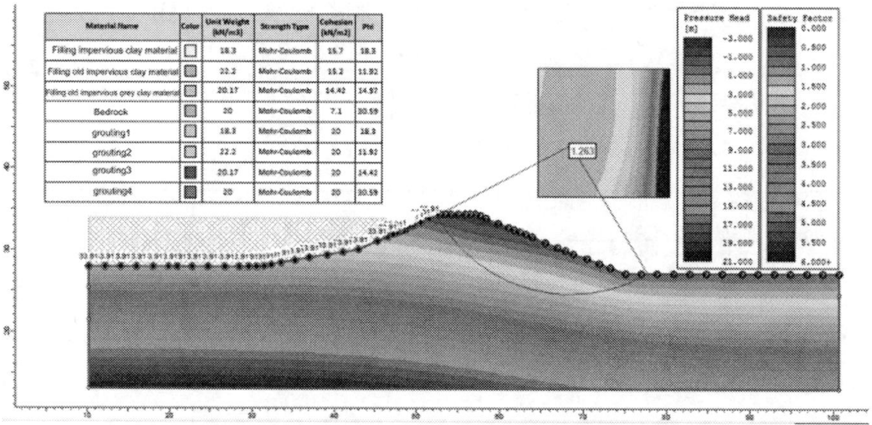

Fig. 7 Slope stability analysis after grouting is being assembled. Cohesion is 20 kN/m^2 after grouting.

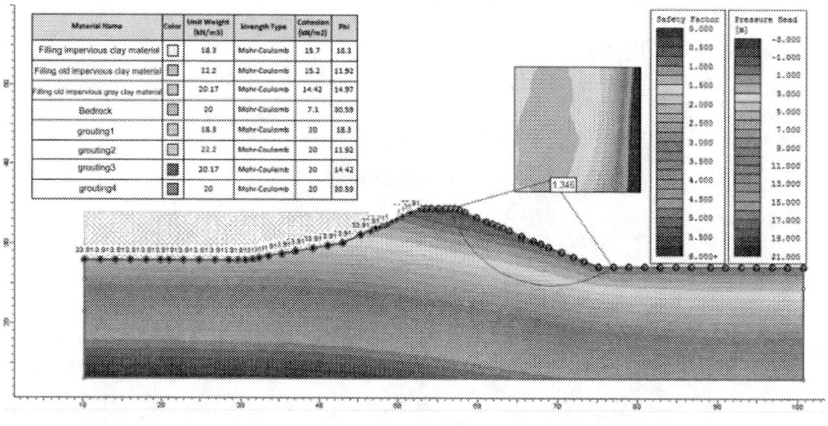

Fig. 8 Slope stability analysis after grouting is being assembled. Cohesion is 30 kN/m^2 after grouting.

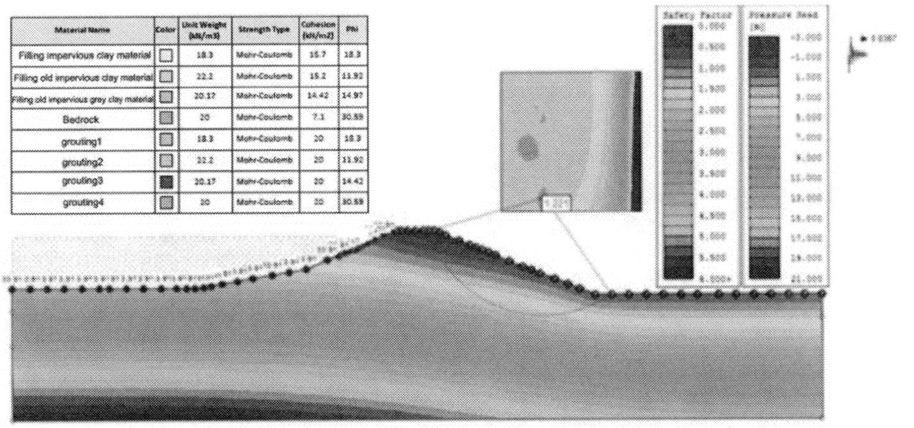

Fig. 9 Slope stability analysis after grouting is being assembled. Cohesion value is 20 kN/m^2 with earthquake loading.

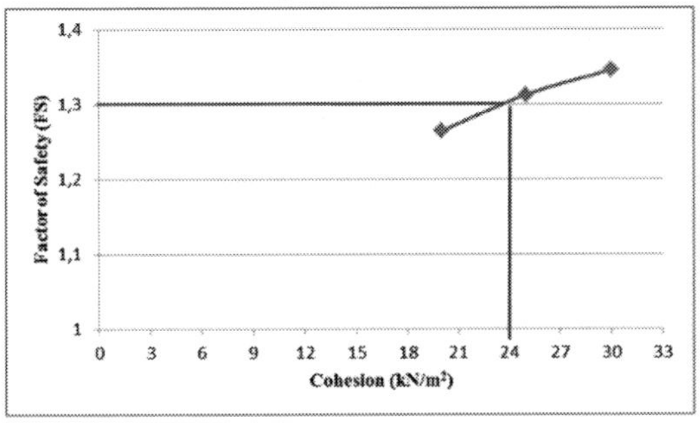

Fig.10. Safety Factor versus Cohesion.

4. Conclussions

We conclude that Factor of safety (FS) of dam in condition Unstable dam maximum water level without and with earthquake downstream part is not satisfy with FS of SNI standard i.e 1.206 and 1.081, respectively. After grouting with cohesian value ≥24 kN/m^2 , FS has fulfill SNI standard (≥1,3).

We thankfully appreciate for any supports from PT. Selimut Bumi Adhi Cipta and BWWS Solo, helped us in data and discussion. Research has supported by PNBP DIPA Diponegoro University No: SP DIPA-042.01.2.400898/2016, 7 Desember 2015, Fiscal year 2016.

References

1. R. Fell, P., MacGregor, D., Stapledon, G., Bell. *Geotechnical Engineering of Dams* (A.A. Balkema Publishers, Singapore, 2005)
2. J.,Warner, *Practical Handbook of Grouting Soil, Rock and Structures* (John Wiley & Sons, Mariposa, California, 2004)
3. P.G., Nicholson, *Soil Improvement and Ground Modification Methods* (Elsevier Inc., New York, 2015).
4. S.,Turkmen, *Eng. Geo.* **68**,159–169 (2003).
5. C.P.,Yang, *Eng. Geo.***75**, 1 –14 (2004).
6. M.Y, Fattah, M.M, Al-Ani, M.T.A., Al-Lamy, *Soils and Foundations J.* **54**,3, 396–404 (2014).
7. Keputusan Menteri Permukiman dan Prasarana Wilayah Nomor : 360/KPTS/M/ 2004 tentang *Analisis stabilitas bendungan tipe urugan akibat beban gempa*, Pd T-14-2004-A (2004).

https://doi.org/10.1051/matecconf/201814707002

Modeling Slope Topography Using Unmanned Aerial Vehicle Image Technique

Fu-Hsuan Yeh[1,*], *Chun-Jia* Huang[1], *Jen-Yu* Han[1], and *Louis* Ge[1]

[1]Department of Civil Engineering, National Taiwan University, 10617 Taipei, Taiwan

Abstract. Nowadays, a wide range of site planning, field investigation and slope analysis need to be carried out for slope protection and landslide-related disaster reduction. To enhance the efficiency of topography modeling, unmanned aerial vehicle (UAV) has become a new surveying technique to obtain spatial information. This study aims to determine the topography of a slope by using a digital camera mounted on UAV to photograph with a high degree of overlap. The 3D point clouds data were generated through image feature point extraction integrated with accurate GPS ground control points. It is found in this study that the obtained Digital Surface Model (DSM) data, compared with the traditional field surveying techniques, has much superior performance. The resolution of the DSM has reached 1.58 cm. Also, the error of elevation and distance between DSM and actual 3D coordinates obtained by traditional total station survey is acceptance. It is clear that such a UAV surveying technique can replace conventional surveying methods and provide complete and accurate 3D topography information in automatic and highly efficient manner for most geotechnical applications.

1 Introduction

Recently, unmanned aerial vehicle (UAV) has become a new surveying technique to obtain spatial information. Many researches have been focusing on applying UAVs to cope with disasters and acquire three-dimensional (3D) mapping and modeling. For example, Turner et al. [1] have compared their UAV mapping with those obtained by traditional aerial or satellite platforms. It was concluded that UAV-acquired datasets have better resolutions in both temporal and spatial aspects. For typical mapping tasks, their geometry and associated attributes should be identified and corrected first. Laliberte and Rango [2] mapped a land cover of an experimental range in New Mexico, USA. Wallace et al. [3] conducted forest inventory in the University of Tasmania in Australia. Lucieer et al. [4] and Turner et al. [5] mapped and monitored moss beds in the Antarctic area. Remondino et al. [6] concluded that 3D results like DSM/DTM, contour lines, textures 3D models, vector data, can be produced in a reasonable automated way by using UAV platforms. Abdelkader et al. [7] conducted real-time flash-flood monitoring in the Jeddah hydrological basin in Saudi Arabia. Siebert and Teizer [8] built a UAV system to rapidly and autonomously acquire mobile 3D mapping data. Goncalves and Henriques [9] used a UAV to map and monitor sand dunes and beaches in sensitive coastal areas. Liu et al. [10] collected information, including the potential applications of UAV to seismic risk assessment, transportation, disaster response, construction management, surveying and mapping, and flood monitoring and assessment. It can be seen that the above-mentioned environmental applications have been used in broad areas. More applications of various kinds can be expected in the near future.

In Taiwan, rock fall is a common geo-disaster that threatens people's properties and lives. The survey area is a slope land near the Qingshui cliff, located in Xiulin Township, Hualien County, in eastern Taiwan, where a railway track is passing by (see Fig.1). The survey area is about 0.29 ha. For the sake of slope protection and train safety, a rock-fall protection fence, named ultra-light fence, is planned to build along the slope. Before planning and design, topographic surveys need to be carried out to evaluate the site condition and suitability of the survey area. Therefore, it was accomplished by gathering spatial information including elevations and surface features of the surrounding terrain. To enhance the efficiency of topography modeling, unmanned aerial vehicle (UAV) has become a popular surveying technique to acquire spatial information. Data for orthophotos, 3D models, and point clouds in high density with great accuracy can be obtained. By using the information, any slope's cross section can be plotted for further design.

In addition, the conventional method, where the spatial information is obtained from a total station, is carried out in this case. The goal of this paper is to evaluate the UAV technique in comparison with the conventional method, and assess the difference between the both methods.

* Corresponding author: stillya81@gmail.com

 https://doi.org/10.1051/matecconf/201814707002

Fig.1. The condition of the survey area

2 Methodology

Fig.2 shows the procedure of the method to obtain the slope topographic information by using UAV images. Firstly, flight route plan should be developed and ground control points should be preselected for preparation. Hardware and flight controller of the UAV system, including aircraft structure, battery, sensors and signal, need to be checked.

Afterwards, the flight route and surrounding environment are examined. The quadrocoper can be turned on, where the flight planning is ready to start. After the quadrocoper takes off, it will follow the operator's instruction and take photos. When the expected flight planning is finished, the UAV returns automatically to its lift-off location by switching "coming home" mode or being manually operated by the operator. If the survey cannot be performed in one trip, additional flights can be carried out. It is noted that the flying height should remain the same, so as to ensure the images that can be merged seamlessly.

Additionally, GPS control survey should be conducted before or after the UAV work finished. After that, these images can be post-processed using photo-grammetry technique, where the spaitial information can be generated. Pix4Dmapper Pro is software uses the photo-grammetry technique. It is customizable, timely, and can complement a wide range of applications. It can turn images into geo-referenced 2D maps and 3D models with high precisions. The UAV images, therefore, can be processed by using this analysis software.

2.1 Definition of UAV system

UAV is an aircraft without a human pilot aboard. The DJI Phantom 4 was used in this work. Shown in Fig.3, the UAV is approximately 1380-g in weight, consisting of four light flying wings, frame structure, multiple sensors (including GPS, gyroscope, barometer and accelerometer), high performance camera, an advanced 3-axis gimbal with flight controller and telemetry control system. The UAV camera has a 12-Megapixel (4000×3000 pixels) sensor. Table 1 shows the detailed specification for DJI Phantom 4.

DJI Phantom 4 is capable of staying up in the air for about 28 minutes and can fly around 5 km of range, with complete control and a live 720p HD view of everything the camera sees. In this case, the surveying assignments of the operational altitude was about 30 m. In a flight with an overlap more than 80%, 198 photos were obtained for further analysis. The spatial resolution was about 1.58 cm/pixel (see Table 2).

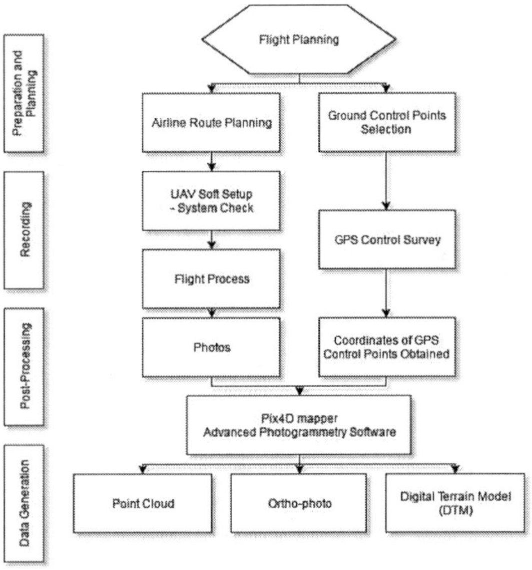

Fig.2. UAV flight methodology: preparation, recording, post-flight data processing, and data generation

Fig.3. DJI Phantom 4

Table 1. Specification of DJI Phantom 4

Item	Content	Item	Content
Effective pixels	12 Megapixels	Image size	4000 × 3000
Len	FOV94° 20 mm	Flight weight	1380 g
Sensor	1/2.3' CMOS	Diagonal Size (Propellers Excluded)	350 mm

Table 2. The information of operation

Item	Content
Flight height (m)	30
Number of shots	198
Spatial resolution (cm/ pixel)	1.58

2.2 Total Station

Pentax-R202NE was set up for the sake of comparing with the UAV technique. It is a surveying equipment combining Electromagnetic Distance Measuring Instrument and electronic theodolite. The instrument can be used to measure horizontal and vertical angles as well as sloping distance of object to the instrument, and collect all the necessary spatial information of points without using a prism or a pole. The accuracy of distance measurement is 3 mm + 2 ppm. The accuracy of angle measurement varies from 2 to 6 seconds.

2.3 Aerial Triangulation

This paper uses the UAV image to obtain the object space coordinates of tirrena. The method incoporates the aerial triangulation theory. The aerial Triangulation in Photogrammetry is a method determining and calculating 3D object coordinates by using photographs exposed from different positions, covering the same object. With aerial triangulation in aerial photogrammetry, it might be able to calculate 3D coordinates for object elements on almost any object (see Fig.4). It needs at least three ground control points with known positions that are visible in at least some of the photos. The ground control points have to be a part of the aerial triangulation (e.g. Hamid [11]).

The calculation method uses rigorous collinearity equations in photogrammetry to establish the object-UAV image relationship. In addition, the lens distortion needs to be considered. The UAV camera must be rigorously calibrated. Equation (1) uses to calculate a space intersection, with known interior orientation parameters of a camera. In equation (1), x, y are the image points, and f is the focal length. Δx , Δy denote the camera calibration parameters. The external orientation parameters, X_L, Y_L, Z_L, are the perspective center coordinates, while m_{11} to m_{13} are rotation matrix coefficients. The unknowns are the object space coordinates (X_L, Y_L, Z_L) to which image points correspond (Mikhail [12]).

Therefore, the 3D object coordinates are calculated by using the spatial intersections. Finally, the image dense matching is used to generate 3D point clouds and Digital Surface Model (DSM).

$$
\begin{aligned}
x-\Delta x &= -f\frac{m_{11}(X-X_L)+m_{12}(Y-Y_L)+m_{13}(Z-Z_L)}{m_{31}(X-X_L)+m_{32}(Y-Y_L)+m_{33}(Z-Z_L)} \\
y-\Delta y &= -f\frac{m_{21}(X-X_L)+m_{22}(Y-Y_L)+m_{23}(Z-Z_L)}{m_{31}(X-X_L)+m_{32}(Y-Y_L)+m_{33}(Z-Z_L)}
\end{aligned} \quad (1)
$$

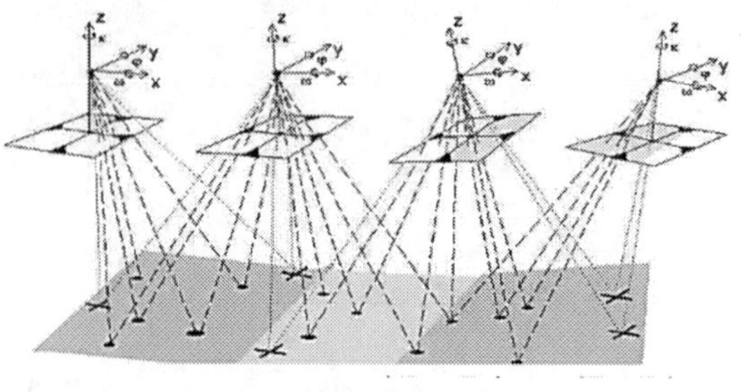

Fig.4. Aerial triangulation (wolf [13])

2.3.1 Ground control survey

If UAV system is used for surveying work, global positioning system (GPS) control points need to be set up for acquiring data with high precision. These control points are set manually before the UAV flies and are created for generating a Geographic coordinate system (Taiwan Datum 1997). They are used to correct the image orientation. In the survey area covered by natural features, it is not appropriate to use trees and rocks as control points; therefore, artificial points needed to be placed on the ground surface. The man-made features should be sufficiently clear and fixed carefully to avoid being damaged before GPS operation and UAV photographing. A minimum of three ground control points are required, so in this study, four ground control points were set up with a uniform distribution (see Fig.5).

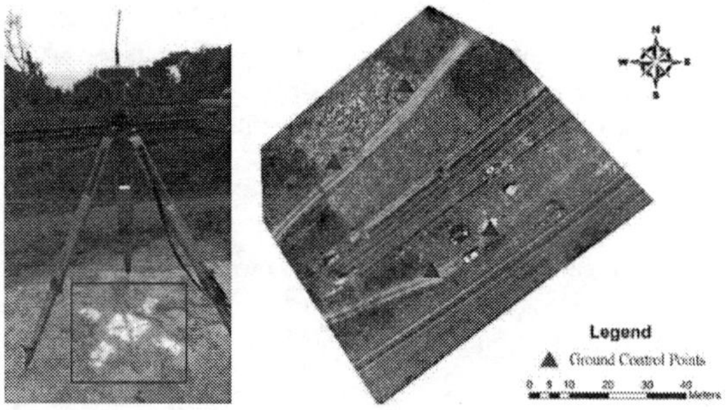

Fig.5. Distribution of GPS control points

3 Result and Analysis

3.1 Photogrammetric data processing

The data processing is simple. All aerial images from UAV need to be imported from camera to computer. It is very important that the images have adequate overlaps between each other. Also, the 3D coordinates of GPS should be applied in the program. In this study, the

Average Ground Sampling Distance (GSD), which is also called spatial resolution, the distance between pixel centers as measured on the ground, was approximately 1.58 cm. The analysis result shows that these images are sufficient for good quality matching.

It is essential to export the results as 3D colored point clouds or orthophotos, as shown in Fig.6 and Fig.7, respectively. Fig.8 shows the DSM, which is generated by 3D point clouds, represents the MSL elevations of the reflective surfaces. The terrain profiles can be obtained from the DSM data. DSM data can be used for producing relief maps and be useful for surface terrain modeling, dip direction and angle inclination of slope, and any visualization applications. The topographic map contour lines can be automatically drawn by using DSM data as well.

Fig.6. Point cloud generation in Pix4D

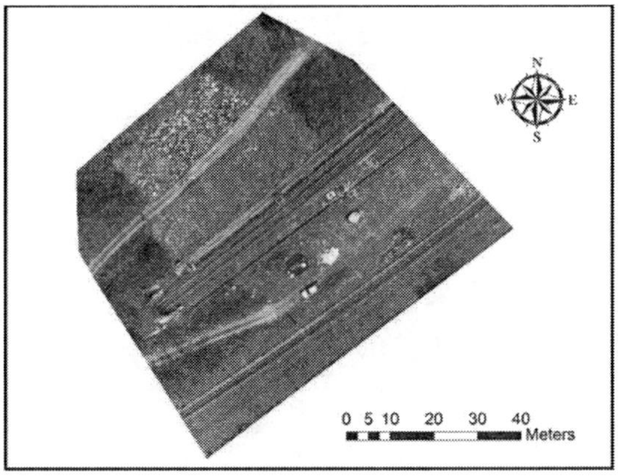

Fig.7. Orthophoto generation from Pix4D

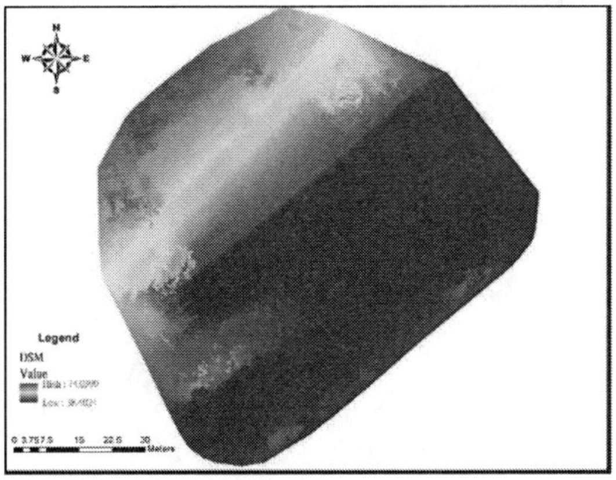

Fig.8. Digital surface model (DSM)

3.2 Comparison with UAV Technique and Total Station

In order to compare the pros and cons between UAV technique and total station, the results from both methods are presented in the following section. Fig.9 shows the three terrain profiles for comparison.

Fig.9. Terrain profile of the slope

3.2.1 Terrain profile survey with Total Station

The terrain profiles can be surveyed by using total station. A measuring line and sight along the line were selected. When a target is sighted, horizontal angles, zenith angle as well as horizontal distances are measured and recorded by surveyors accordingly. After obtaining the field data by the total station, the data, such as geographic coordinate system and elevation, can be processed in office.

3.2.2 Terrain profile survey with UAV Data

The terrain profiles from UAV data can be generated by the DSM information, as refer to 3.1 However, surface vegetation is filtered out by manual editing. The Digital Elevation Model can be generated after this procedure. The result is shown in Fig.10.

https://doi.org/10.1051/matecconf/201814707002

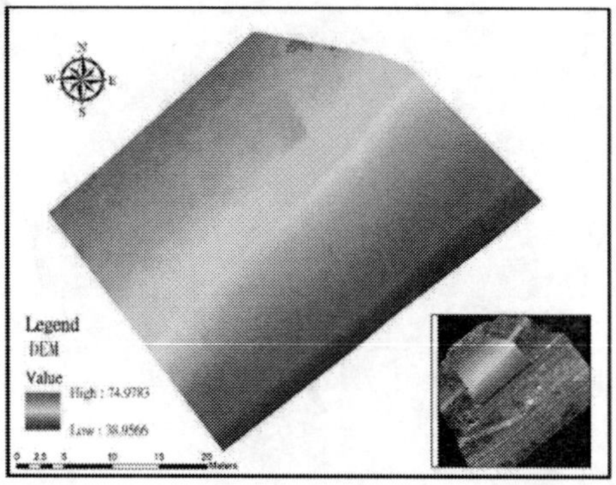

Fig.10. Digital Elevation model (DEM)

3.2.3 Comparison of Terrain Results

Table 3 shows the results of the terrain from the total station and UAV. The obtained spatial information from UAV is more complete than total station. In the field work, time spent and manpower for UAV technique are less than that for total station method. In general, the advantages of using total stations are "good accuracy of measurement" and "manual errors involved in reading and recording are eliminated". The disadvantages of using total stations are getting high cost and taking longer time and vice versa. Taking the total station as a reference, the maximum error of UAV result is 0.151 m and minimum error is 0.008 m. There is 0.061 m root-mean-square error between the elevation results of UAV and total station. In this study, centimeter error variance is sufficient for this slope survey application. Fig.11 shows the terrain results from the UAV technique and total station method. It can be seen that the results of the two method are very close. In addition, UAV technique can capture more information as detailed as possible.

Table 3. Comparison terrain results between total station and UAV

Method	Quantities (points)	Time period of field work	Manpower
Total Station	50	3 hours	2~3
UAV	3000	20 min	1

Method	maximum error	minimum error	Root-mean-square error
	(m)	(m)	(m)
Total Station	-	-	-
UAV	0.151	0.008	0.061

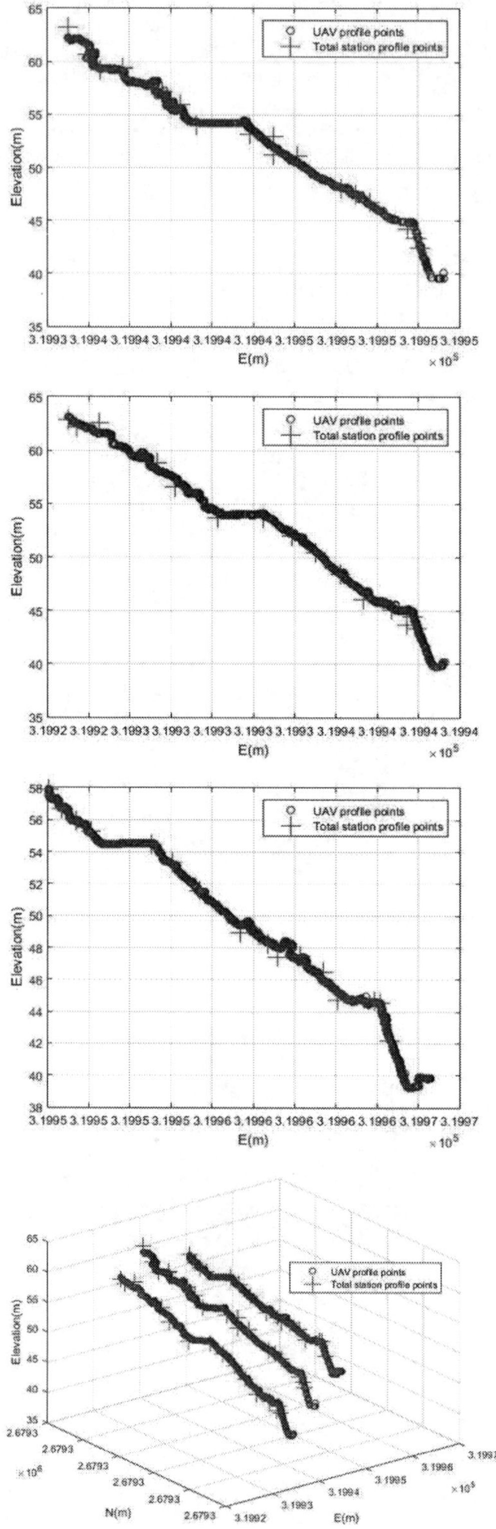

Fig.11. Comparison terrain results between total station and UAV

4 Conclusion

Generally, the conventional surveying technique, the total station method, can collect information more accurately but needs much more time, money and operating skill. In this study, the conventional method with the total station and UAV methods were compared and evaluated. The result shows that there is 0.061 m root-mean-square error between the elevation results of UAV and total station. In such slope survey for site condition evaluation and design, centimeter error variance is acceptable.

Using the UAV technology not only obtains high-spatial-resolution observations and contribute a new version for constructing high-fidelity 3D models, but also gets low operating costs. It is more efficient to generate the topographic maps completely. It provides many opportunities in civil engineering applications.

References

1. Turner, D., Lucieer, A. and Wallace, L., IEEE Trans. Geosci. Remote Sens., **52(5)**, 2738-2745 (2013).
2. Laliberte, A.S and Rango, A., IEEE T. Geosci. Remote Sens., **47(3)**, 761-770 (2009).
3. Wallace, L., Lucieer, A., Watson, C. and Turner, D., Remote Sensing, **4(6)**, 1519-1543 (2012).
4. Lucieer, A, Robinson, S. and Turner, D.J., *Proceedings of the 34th International Symposium on Remote Sensing of Environment* (ISRSE34, 2011).
5. Turner, D., Lucieer, A. and Watson, C., Remote Sensing, **4(5)**, 1392-1410 (2012).
6. Remondino, F., Barazzetti, L., Nex, F., Scaioni, M. and Sarazzi, D., Remote Sensing and Spatial Information Sciences, **38(1/C22)** (ISPRS ICWG I/V, 2011).
7. Abdelkader, M., Shaqura, M., Claudel, C.G. and Gueaieb, W., *Proceedings of the 2013 International Conference on Unmanned Aircraft Systems* (2013).
8. Siebert, S. and Teizer, J., Automation in Construction, **41**, 1-14 (2014).
9. Gonçalves, J.A. and Henriques, R., ISPRS P&RS, **104**, 101-111 (2015).
10. Liu, P., Chen, A.Y., Huang, Y.N., Han, J.Y., Lai, J.S., Kang, S.C., Wu, T.H., Wen, M.C. and Tsai, M.H., Smart Struct. Syst., **13(6)**, 1065-1094 (2014).
11. Hamid, E., *Advanced Analytical Aerial Triangulation, 1nd ed.* (K.N.Toosi, Unversity of Technology, 2006)
12. Mikhail, E.M., James, S. B. and McGlone, J. C., *Introduction to modern photogrammetry, 1nd ed.* (Wiley, 2001)
13. Wolf, P.R. and Dewitt, B.A., *Elements of Photogrammetry with Applications in GIS, 3nd ed.* (McGraw-Hill, 2000).

Bearing Capacity of very Expansive Soils at Jatinangor Area, West Java, Indonesia

David Simangunsong[1], *Satrio* Wibowo[1], and *Zufialdi* Zakaria[2]

[1]Student of Bachelor Degree, Faculty of Geological Engineering, Padjadjaran University
[2]Engineering Geology and Geotechnics Laboratory, Faculty of Geological Engineering, Padjadjaran University

Abstract. Expansive soil is a kind of soil that has ability to shrinkage and swelling. According to Ronny (2014) Jatinangor area has expansive soil that is so very influential in the planning of infrastructure construction. This research aimed to measure the bearing capacity of the very expansive soils in Jatinangor area and to determine the correlation between activity number of soil and its bearing capacity. The method used is to collect the soil physics and mechanics data. Based on the soil mechanics data, the research location is divided into three zones of allowable bearing capacity, those are zone with allowable bearing capacity < 4 T/m2, zone with allowable bearing capacity 4-7 T/m2, and zone with allowable bearing capacity > 7 T/m2. The correlation between activity number and bearing capacity of soil follows the equation qa = -1.9505(A) + 6.957 with correlation coefficient is -0.7911.

1 Introduction

Expansive soil is a kind of soil that has ability to shrinkage and swelling. in relation to changes in water content. Soils that exhibit swelling and shrinking behavior contain expansive clay minerals, such as smectite, that absorb water, the more of this clay a soil contains the higher its swell potential and the more water it can absorb (Jones and Jefferson, 2012). The behavior of swelling on the soil can result in the differential settlement of the foundation resulting in damage to the infrastructure.

Jatinangor is an area that has rapidly developed into regional education, industry, government and commerce. It can be seen with a lot of infrastructure construction in this area. Jatinangor region is geologically composed of young volcanic rocks, so the soil produced from weathering is loose. This thing causes landslide around Jatinangor area. In addition, swelling of clay minerals on soil from weathering also influences the frequency of soil movement.

This research is located around Padjadjaran University, Jatinangor District, Sumedang, West Java. Ronny (2014) has classified soils at Jatinangor to four classes based on its expansiveness, which are low expansive soil, medium expansive soil, high expansive soil and very high expansive soil. This research aimed to measure the bearing capacity of the very expansive soils in Jatinangor area and to determine the relation between activity number of soil and its bearing capacity.

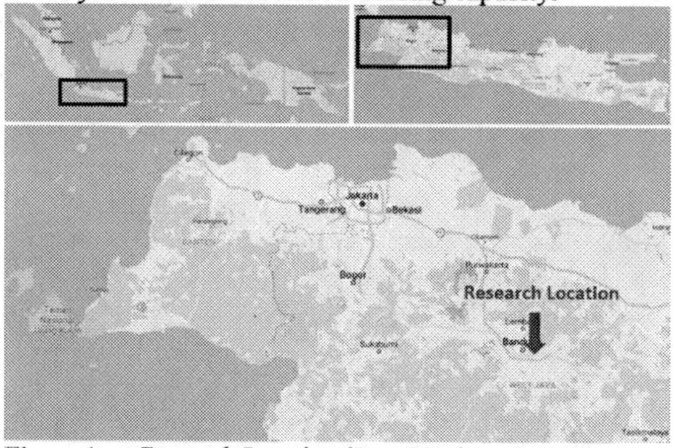

Figure 1 Research Location (June,2017)

* Corresponding author: david14001@mail.unpad.ac.id

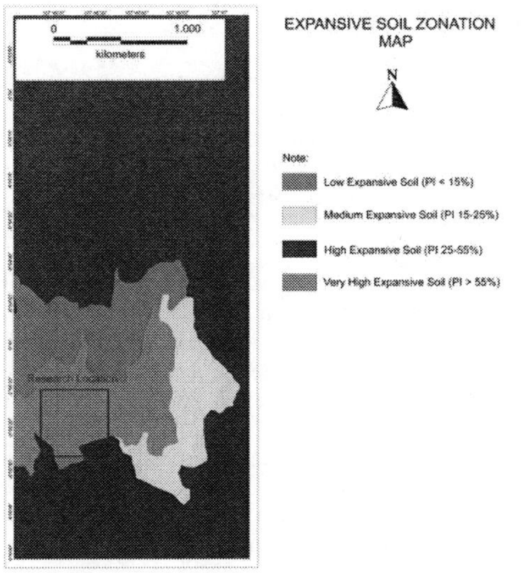

Figure 2 Expansive Soil Zonation Map at Research Location (Ronny, 2014)

1.1 Regional Geology

Based on the regional geological map of Bandung (Silitonga, 1973), the research location is composed of young volcanic rocks, tuffaceous sand, Lapilli, Lava, and agglomerat originating from Mount Tangkuban Parahu and Mount Tampomas. These rocks belong to the Quarternary volcano rocks.

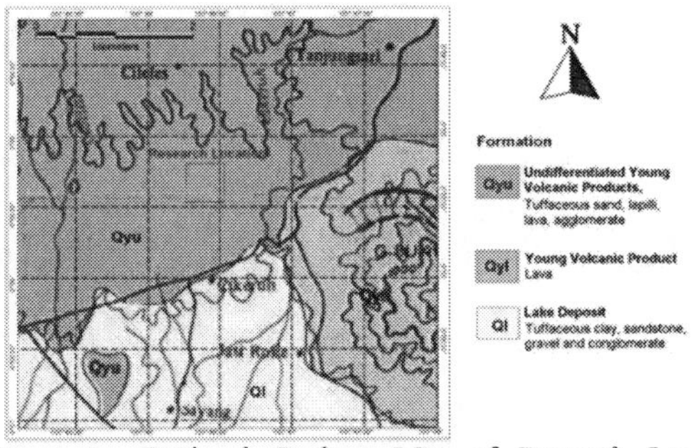

Figure 3 Regional Geology Map of Research Location (Silitonga,1973)

2 Methodology

The method used in this paper is to collect the secondary data consists of the soil physics data such as weight of soil content (γ), plasticity index, clay content in percent and soil mechanics data such as cohesion (c) and friction angle (ϕ) of very expansive soils at research location. The expansiveness of the soil can be known through the calculation of activity number (A). The activity number (Skempton, 1958 in Das, 1988) was formulated as a comparison of plasticity index (PI) and clay grain percentage (<0.002 mm) as follows:

$$A = \frac{PI}{\%clay} \quad (1)$$

Unit weight (γ), cohesion (c) and friction angle (ϕ) then used to calculate the allowable bearing capacity of very expansive soils on a continuous foundation with the following equation based on Terzaghi (1943, in Zakaria, 2014):

$$qult = c.Nc + \gamma.D.Nq + 0{,}5\gamma.B.N\gamma \quad (2)$$

$$qa = \frac{qult}{F} \quad (3)$$

From these data, the zonation map of bearing capacity of very expansive soils is made using Surfer 11 software, where column A is X coordinate, column B is Y coordinate, and column C is the bearing capacity of soils. The graph of the relationship between the activity number and the bearing capacity of soils is made using Microsoft Office Excel 2016 software.

3 Result and Discussion

The research area is a very expansive soil zone based on an expansive soil zonation map by Ronny (2014). This very expansive soil zone has a plasticity index value ranged from 24.63% to 69.67%, whereas the more plastic a soil sample, the higher the expansive of the soil. Activity number (A) from soil samples ranged from 0.44 to 1.37 where the soil is thought to contain the minerals of illite clay, kaolinitic and montmorillonite.

Table 1 Table of Activity Number and Allowable Bearing Capacity

Sample	USCS Classification	Coordinate X	Coordinate Y	IP (%)	%clay	A	Allowable bearing capacity (T/m²)
Z1aBT1	CH	107.772265	-6.923446	33.3	58.39	0.57	5.57
Z1aBT2	CH	107.771761	-6.923107	50.16	46.72	1.07	4.46
Z2aBT1	CH	107.773038	-6.922692	58.6	48.58	1.21	4.97
Z2aBT3	CH	107.772639	-6.922679	69.67	50.67	1.37	4.33
Z2bBT1	CH	107.772628	-6.92175	60.41	56.08	1.08	4.08
Z2bBT2	CH	107.772276	-6.921537	57.64	55.4	1.04	5.33
Z3BT2	MH	107.775193	-6.924046	24.63	55.77	0.44	6.31
Z3BT3	MH	107.775415	-6.923594	45.65	55.9	0.82	5.16
Z3BT4	MH	107.775005	-6.923444	49.54	54.7	0.91	5.81

From the various samples contained in the research area, the bearing capacity of very expansive soils is calculated with the type of continuous foundation, the depth of 1 meter and width of 1 meter as follows:

- A bearing capacity zonation map of very expansive soils is constructed with the following range of values:
- Zone with allowable bearing capacity < 4/m2. This zone is scattered in the northwest, north and southeast of the research area.

- Zone with allowable bearing capacity 4-7 T/m2. This zone is scattered almost throughout the research area.
- Zone with allowable bearing capacity > 7 T/m2. This zone is scattered in the southern part of the research area.

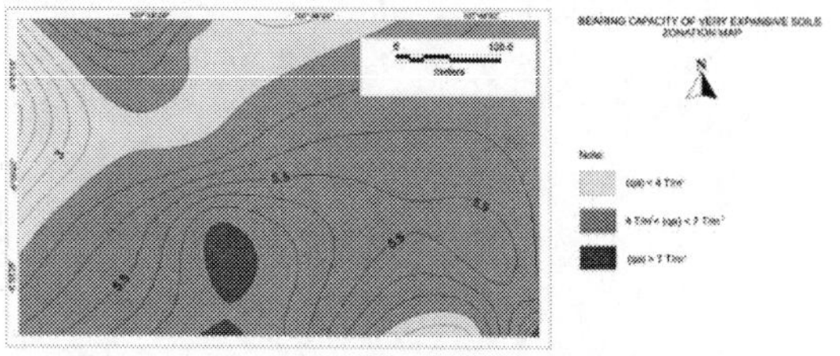

Figure 4 Bearing capacity zonation map of very expansive soils

The correlation between activity number and bearing capacity of very expansive soils in research area is shown in graph with equation qa = -1.9505 (A) + 6.957 with R= -0.7911. Based on the value of negative correlation coefficient, it can be interpreted that the bearing capacity of the soil will decrease as the activity rate increases. This is due to the value of the activity number is affected by the value of the plasticity index. The higher the plasticity index, the higher the activity number.

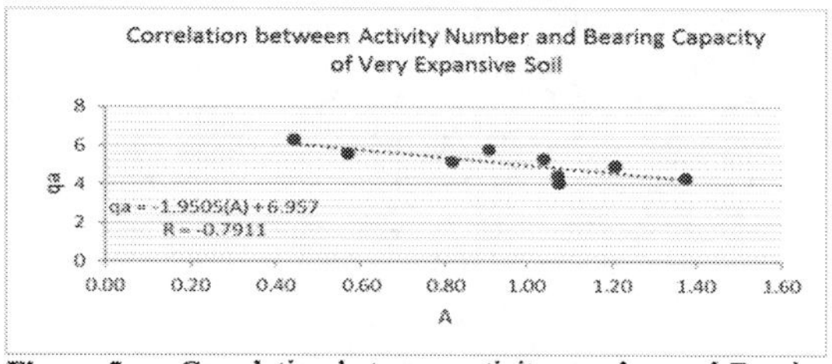

Figure 5 Correlation between activity number and Bearing capacity

The correlation between plasticity index and soil bearing capacity at research area shows correlation with equation qa = -0.041 (PI) + 7.162 with R= -0.783. It also can be interpreted that the soil bearing capacity will decrease when the plasticity index increases.

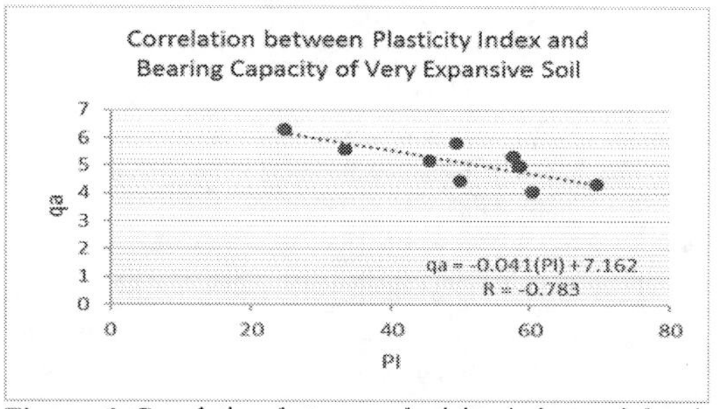

Figure 6 Correlation between plasticity index and bearing capacity

4 Conclusion

A bearing capacity zonation map of very expansive soils consists of three zones based on the value of allowable bearing capacity, which are zone with allowable bearing capacity < 4 T/m2, Zone with allowable bearing capacity 4-7 T/m2, and zone with allowable bearing capacity > 7 T/m2.

The correlation between activity number and bearing capacity of very expansive in research area is shown in graph with equation qa = -1.9505 (A) + 6.957 with R= -0.7911. Based on the value of negative correlation coefficient, it can be interpreted that the bearing capacity of the soil will decrease as the activity number increases. To build infrastructure on expansive soils, stabilization is required. Stabilization on expansive soils can be divided into 3 types: mechanical stabilization, physical stabilization, and chemical stabilization (Ingles and Metcalf, 1972).

- Mechanical stabilization is done by increasing the strength and carrying capacity of the existing soil by increasing its density.
- Chemical stabilization by adding certain chemicals resulting in chemical reactions.
- Physical stabilization is by adding geomembrane over expansive soil.

All authors want to say thank you to God because of His bless, this paper can be done. Authors also want to thank all people who involved of this paper, Padjadjaran University and also to Bandung Institute of Technology to give us chance to submit this paper.

5 Nomenclature

A	= Activity Number
B	= width of foundation (m)
c	= cohesion (T/m2)
D	= depth of foundation (m)
F	= Factor of Safety =3
Nc, Nq, Nγ	= factor of bearing capacity that depends on friction angle
PI	= plasticity index
qult	= ultimate bearing capacity (T/m2)
qa	= allowable bearing capacity (T/m2)
R	= correlation coefficient
γ	= Unit weight (T/m3)
ϕ	= friction angle (o)
% clay	= percentage of clay content

References

1. P.H Silitonga, *Peta Geologi Lembar Bandung*, (1973)
2. L. D. Jones, I. F. Jefferson, *Expansive Soil. Institution of Civil Engineers Manuals series*. (2012)
3. O. G. Ingles, J. B Metcalf, *Soil stabilization: principles and practice,* (1972)

4. F. A. Ronny, *Potensi Tanah Mengembang Wilayah Jatinangor*, (2014)
5. B. M. Das, *Mekanika Tanah (Prinsip-Prinsip Rekayasa Geoteknis)*.**1**. (1988)
6. Z. Zakaria, *c dan ϕ untuk Daya Dukung Fondasi Dangkal*, (2010)
7. Y.P. Putra, *Zonasi Daya dukung Tanah Fondasi Dangkal Kawasan Universitas Padjadjaran Jatinangor Sumedang Jabar*. (2015)

Probabilistic Analysis of Cut-Slope Stability for Tropical Red Clay of Depok, West Java as an Effect of Rainfall Duration and Intensity

Fathiyah Hakim Sagitaningrum[1*], *Widjojo* Adi Prakoso[2], and *Erly* Bahsan[2]

[1]Graduate Student, Department of Civil Engineering, Universitas Indonesia, Depok 16424, Indonesia
[2]Department of Civil Engineering, Universitas Indonesia, Depok 16424, Indonesia

Abstract. Landslide in Indonesia, specifically in Java island, occurs during rainy seasons. In Java island, it is known that the tropical red clay has the ability to stand at steep angles, while in stability analysis due to rainfall, practitioners only consider the rise of groundwater table. Previous studies states that one of the factor affecting factor of safety (FS) for tropical red clay slopes is the formation of saturated zones due to matric suction. This research studies the effect of rainfall intensity and duration to FS of cut-slopes as parametric study with probabilistic analysis for different height of 10m, 20m, and 30m also slope angles of 27°, 45°, 55°, and 70°. Rainfall parameter are taken from FTUI rainfall station for advanced pattern and three-days duration of rain. Analysis of seepage uses SEEP/W and slope stability uses SLOPE/W. It is known that the significant increase of probability of failure due to the three-days rainfall is achieved at the 10m height and 70°-angled slope. Increase of the probability of failure is mainly due to rainfall infiltration which saturates the surface and pore water pressure increase until certain time where infiltration stops and turn into surface run-off.

1 Introduction

In Indonesia, according to National Board of Disaster Management (BNPB) landslide data throughout 2013-2016, there are at least 184 landslides happened in West Java, for example, whereas 150 landslides out of the 184 landslides occurred due to the effect of rainfall and unstable structure of the soil.

Rainfall intensity is one of the primary factor of landslides causes [1]. This is closely related to the condition of Indonesia as one of the tropical nations which have high intensity rainfall. In the concept of unsaturated soil mechanics, sudden high intensity rainfall would infiltrate the soil and reduces the matric suction.

Cut-slopes in Indonesia, under the guidelines for excavated slope or cut-slopes in volcanic and residual soil by the Ministry of Public Works, accommodates only the change of groundwater table for rainfall effect [2]. This method is not entirely suitable to all cases whereas one of the causes of landslides caused by rainfall is due to the formation of saturated zones near the ground surface [3, 4].

The aim of the research is to examine the effect of rainfall intensity and duration as well as the effect of variation of cut-slope geometry to the factor of safety (FS) of the tropical red clay as one of the typical residual soil found in Java.

2 Literature Review

2.1. Unsaturated Soil

The concept of unsaturated soil stated that soil is not in a fully saturated condition which explains that there is negative pore water pressure above the groundwater table level. Negative value of the pore water pressure is due to the existence of matric suction, which is a reaction of cohesion-adhesion that fills in between the soil particle due to the existence of both air and water at the same time (e.g [5]).

The void in unsaturated soil consists of two components, matric suction and osmotic suction, which changes the formulation of shear strength, seepage, and volume change of soil above the ground water level. Matric suction also gives changes in the formulation of volumetric water content and hydraulic conductivity function which is shown in Soil-Water Characteristic Curve (e.g.: [5]).

2.2 Tropical Red Clay

Tropical red clay is one of the residual soil group type which is usually found in Sumatera and Java island as one of the unique soil which can stand still at an angle more than 45°. Wesley categorized the residual soil with

* Corresponding author: fathiyah.hakim@gmail.com

Plasticity Index and Liquid Limit from the Atterberg Limit tests as below [4]:

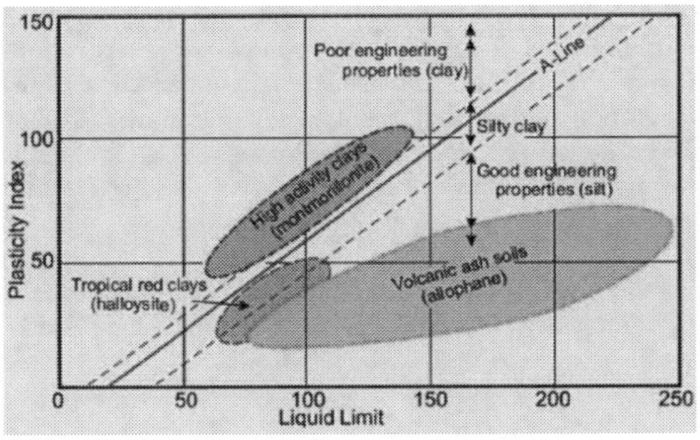

Fig. 1. Plasticity graph for residual soil classification [4]

It is known that in Wesley's previous studies, the mineral composition and soil formation of residual soil allows the groundwater table elevation to rise quickly during rainfall which reduces the FS of slopes quite fast [6]. Research of Tohari and Sarah in Syahbana et al otherwise stated that the main reason of slope stability failure is due to the saturated zones forming inside the body of the slope [3]. Other notes taken for the tropical red clay soil is the slip surface failure which is usually translational and near to zero probability for deep-seated [4].

2.3 Infiltration and Infiltration Rate Function

Infiltration is tightly related to rainfall, which is usually defined as rainfall intensity (mm/hr). Some factors affecting the infiltration rate are hydraulic properties, storage capacity of soil, soil property, rainfall intensity, and infiltration rate of antecedent rainfall [8].

At certain conditions, when rainfall intensity exceeds the soil's infiltration capacity, rainfall will no longer infiltrates into the soil, but turned into surface run-off instead. At this transient condition, the pore water pressure of the surface is defined as zero [9].

2.4 Probabilistic Assessment

In slope stability analysis, the parameters used to formulate the safety factor are full of uncertainties. This leads to illogical results for applying the formulated safety factor. The probabilistic assessment and reliability analysis approach are methods which develops rationally to accommodate the uncertainties of various sources and estimate the probability of calculating performance in a systematic order [8].

One of the parameters which affects the probabilistic and reliability analysis method are geotechnical parameters in certain distribution functions, such as normal and lognormal distribution. The calculated safety factor is then distributed according to the function, generating not only the number of the safety factor, but also the probability of failure of the slope in forms of percentage and reliability index [9].

The software in this research, SLOPE/W accommodates the Monte-Carlo simulation method where the uncertain parameters taken into account are unit weight and shear strength parameters (cohesion and friction angle). In SLOPE/W, the number of probability of failure of the slope is defined as the percentage of the Probability Density Function (PDF) area of the calculated safety factor under the value of 1.0.

2.5 Related Studies

For comparison with previous studies of slope stability due to rainfall in different geometry conditions, Huat et al study of unsaturated slope stability chart is taken into account [7].

Huat et al study concerns the Malaysia residual soil which is simplified into a coupling analysis of SEEP/W and SLOPE/W. Huat et al calculates the FS of a slope after a 24-hr rainfall with factors affecting the formulation of the FS are slope angle variation tanβ, slope height, effective friction angle of the slope (°), ratio of suction friction angle and effective friction angle (ϕ^b/ϕ'), dimensionless unit of $c'/\gamma H$, and groundwater table condition according to rainfall analysis from SEEP/W. The formulation of the FS of the slope is then calculated as [7]:

$$F = f - s \tan \phi^b \tag{1}$$

As F is the FS of the slope after 24-hr rainfall, f and s as the stability coefficient, and ϕ^b as the suction friction angle of the soil (°). In the formulation, it is known that after 24-hr rainfall, the higher the slope height, the bigger the value of f and s which indicates that the FS formulated will become higher.

3 Research Method

3.1. Soil Parameters and Slope Geometry

The soil material in this study is taken to be a homogenous silty clay material type of tropical red clay which are classified according to Fig.1 in an unsaturated condition of 60% water content and saturated hydraulic conductivity of 1 x 10^{-9} m/s. The index parameter properties are taken from Wesley [10] while the shear strength parameter is taken from Triaxial CU data around Depok, West Java and are distributed into a normal distribution function truncated at 2σ from its mean value. Particularly, for the means of comparing the slope in different geometry, a dimensionless unit of $c'/\gamma H = 0.05$ is used for determining the effective cohesion of each slope height. The overall property of the tropical red clay of Depok, West Java, is taken as Table 1 below:

Table 1. Summary of soil parameters

		Slope Height		
		10 m	20 m	30 m
Effective Cohesion	**c' (kPa)**	9	18	27
	COV	44.97%	44.97%	44.97%
	Standard Deviation	4.05	8.09	12.14
Effective Friction Angle	**ϕ' (°)**	23°	23°	23°
	COV	23.49%	23.49%	23.49%
	Standard Deviation	5.4	5.4	5.4
Soil Type		Silty Clay		
γ_{wet} (kN/m³)		18 kN/m³		
ϕ^b (°)		10°		

The unsaturated condition, is assumed to be in the function of silty clay soil for Fredlund-Xing estimation and without any residual water content. The matric suction function of volumetric water content and hydraulic conductivity of SEEP/W is set as a range from 0.01 kPa to 100 kPa.

As for the slope geometry, the height of 10 m, 20 m, and 30 m are taken with various angles such as 27°, 45°, 55°, and 70°. The geometries taken are considered as typical geometry of cut-slope for road constructions around West Java according to [3, 11]. The geometry of the model taken in consideration is as shown in Fig.2.

From Fig.2, the definition of H is the height of the slope, which are differentiated into 10 m, 20m, and 30m, ½ H is the elevation for groundwater table which is considered at a quite deep elevation as a distinct characteristic of slope for tropical red clay, 2H is the elevation of hard soil, and L is the width of slope according to the angle of the slope which is noted as α.

3.2 Boundary Condition

The analysis uses a coupling program of SEEP/W and SLOPE/W. The boundary conditions for SEEP/W is differentiated in steady-state conditions where all the boundaries in Fig.2 is considered, assumed as the initial condition of the slope, and transient conditions where only the rainfall intensity is considered with no ponding formed. The pore water pressure is reviewed at two points, as in Fig.2, at point 2-8 and 3-9. For the SLOPE/W, slope stability is reviewed with the entry-exit method which spans from point 1 to the middle of line 2-3 as the entry range and point 4 to the middle of line 2-3 as the exit range with Bishop analysis.

3.3 Rainfall Parameters

The rainfall parameter is taken form Faculty of Engineering, Universitas Indonesia rain station for a three-days rainfall period. To acquire the rainfall pattern, the rainfall data from July 2014 – March 2015 are divided into a three-days rainfall for each month. The three-days rainfall are taken into consideration where in Java island, a consecutive rainfall from three days are very rare but still have a possibility of happening [4].

The three-days rainfall then divided into 36 intervals which accommodates a two-hourly average rainfall. As for the intensity value, the rainfall intensity is taken from the 24-hour rainfall of 10 years return period from the rain station. The value is then taken as the highest intensity from the three-days rainfall and the rainfall distribution, named as advanced pattern because the highest intensity is achieved at the first day of the three-days rainfall, is as shown in Fig.3.

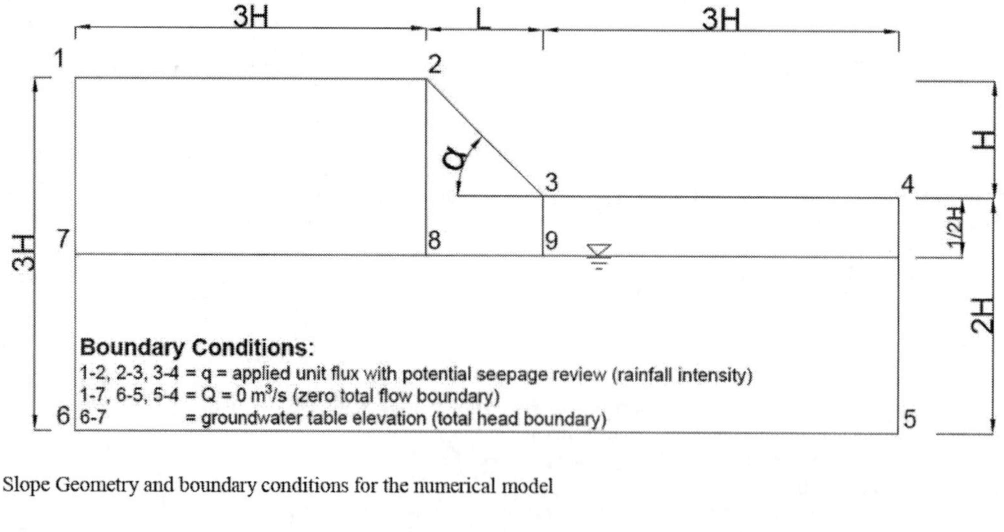

Fig.2. Slope Geometry and boundary conditions for the numerical model

 https://doi.org/10.1051/matecconf/201814707004

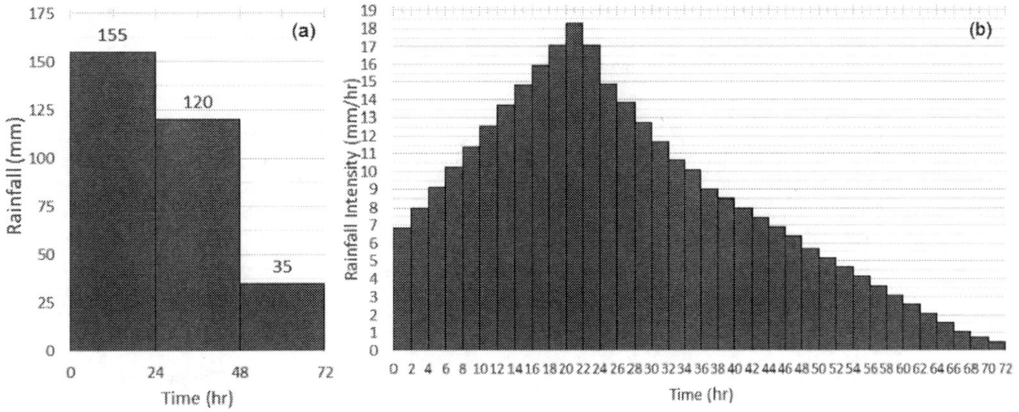

Fig.3. Advanced rainfall pattern for (a) daily and (b) hourly distribution

4 Results and Analysis

4.1 Comparison between change of slope angle and probability of failure

For each height of slope, taken 10 m, 20 m, and 30 m, the change of the slope's probability of failure every 12 hour of the rainfall duration for each slope angle is reviewed. The results are shown below:

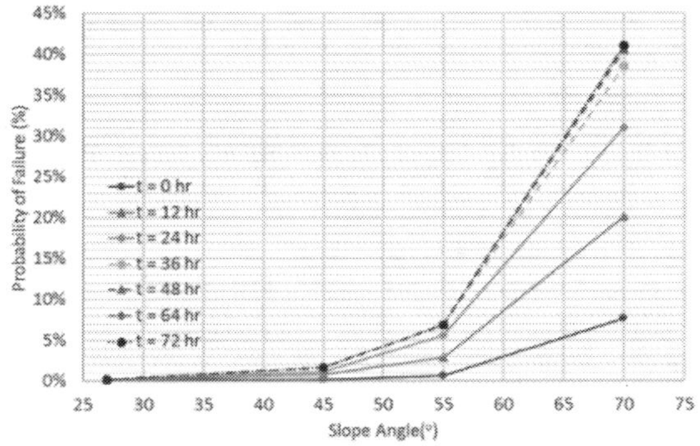

Fig. 4. Probability of failure through time for different angle at 10 m height

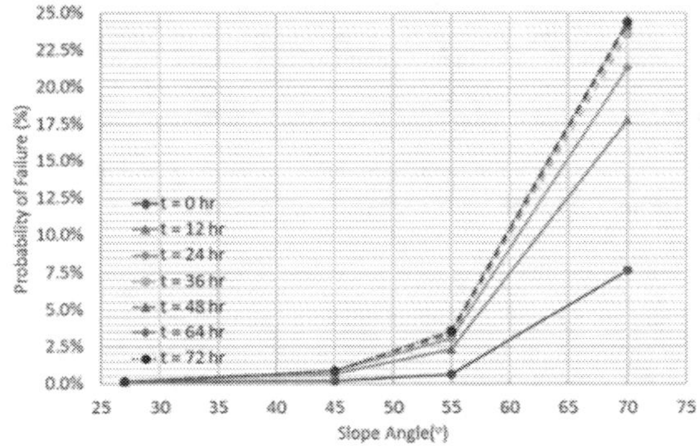

Fig. 5. Probability of failure through time for different angle at 20 m height

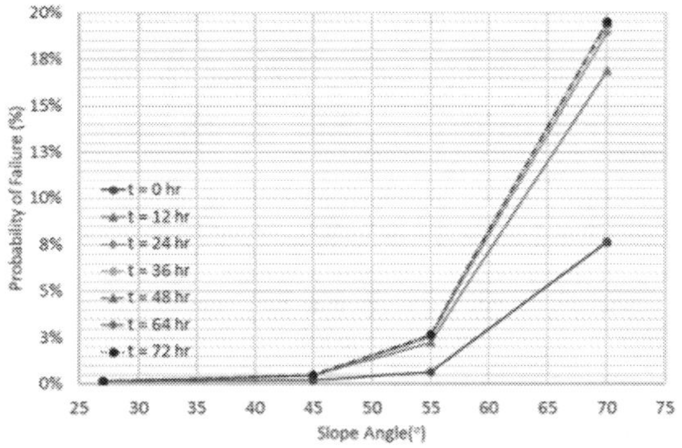

Fig. 6. Probability of failure through time for different angle at 30 m height

From Fig.4, Fig.5, and Fig.6, it is known that the behaviour for each of the height of the slope is the same. For each height, the increase of initial probability of failure of the slope is directly proportional with the increase of the angle of the slope which indicates that the steeper the slope, the greater the probability of the slope to fail.

It is also known that the most significant change of the probability of failure during rainfall is achieved at the 70° slope angle. The probability of failure significantly increases at the first 24 hours, different for 30 m height that increases significantly at the first 12 hours, and become stagnant at the end of the rainfall. This observed behaviour is proportional with the behaviour of the rainfall where the highest intensity is at the first 24 hours and no longer receive high intensity rainfall after it.

Lastly, it is known that the value of $c'/\gamma H$ takes effect on the value of the probability of failure of the slope. In order to achieve the $c'/\gamma H$ value for each height, the cohesion value of higher slopes is increased. This condition resulted into a lower probability of failure which means that the rise of cohesion value and height of slope is not uniform.

4.2 Comparison between change of slope height and probability of failure

For each angle of slope, taken 27°, 45°, 55°, and 70°, the change of the slope's probability of failure every 12 hour of the rainfall duration for each slope height is reviewed. The results are shown as below:

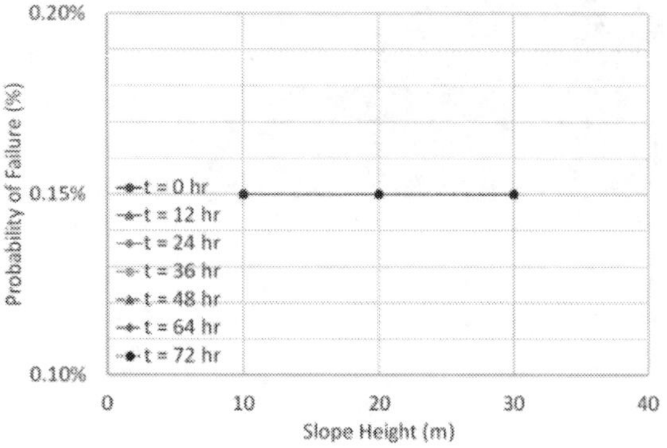

Fig. 7. Probability of failure through time for different height at slope angle 27°

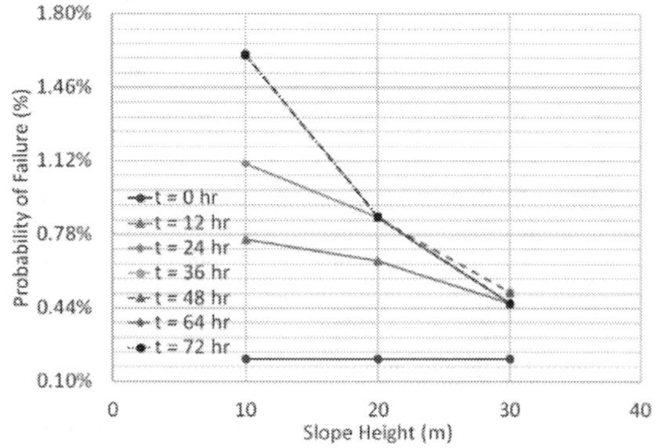

Fig. 8. Probability of failure through time for different height at slope angle 45°

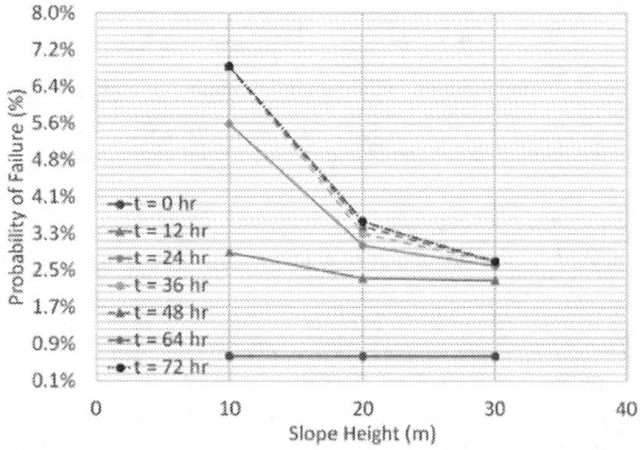

Fig. 9. Probability of failure through time for different height at slope angle 55°

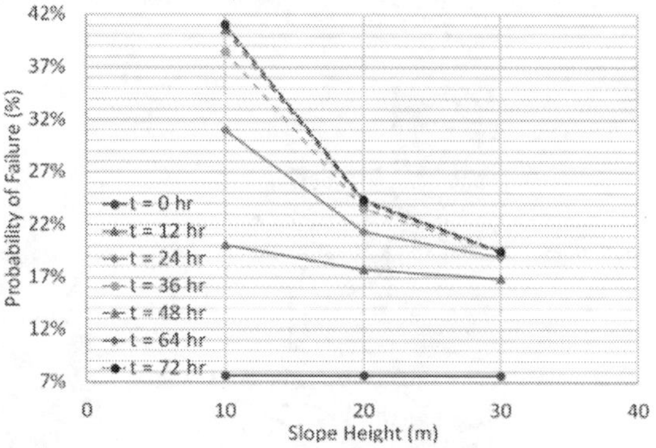

Fig. 10. Probability of failure through time for different height at slope angle 70°

From Fig.7, Fig.8, Fig.9, and Fig.10, it is known that the behaviour for each of the height of the slope is the same. For each height, the increase of initial probability of failure of the slope is directly proportional with the increase of the angle of the slope which indicates that the steeper the slope, the greater the probability of the slope to fail.

It is also known that the initial value of probability of failure of each slope height at each angle is the same. This is due to the given geometry of the slope in the stated $c'/\gamma H$ and geometry boundary of this research which gives proportional ratio of vertical and horizontal of the slope resulting in different value of FS, but the same value of probability of failure.

Lastly, it is known that the most significant change of the probability of failure during rainfall is achieved at the 70° slope angle. The probability of failure significantly increases at the first 24 hours, different for 30 m height that increases significantly at the first 12 hours, and become stagnant at the end of the rainfall. This is proportional with the behaviour of the rainfall where the highest intensity is at the first 24 hours and no longer receive high intensity rainfall after it.

4.3 Comparison between change of pore water pressure and slope height

From the two sections above, it is known that the most significant change of probability of failure is achieved at the 70° slope angle and the height of 10 m. In examining the highest probability of failure significant change of 10 m height, the pore water pressure profile during rainfall is considered.

The pore water pressure profile to be reviewed is taken at the 70° slope angle for each height of 10 m, 20 m, and 30 m every 12 hour. The results are shown below:

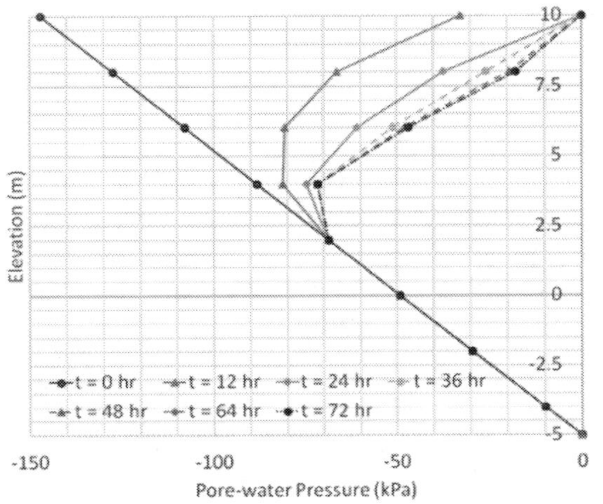

Fig. 11. Change of pore-water pressure at Line 2-8 (Top) for 70° angle of 10-meter-high slope

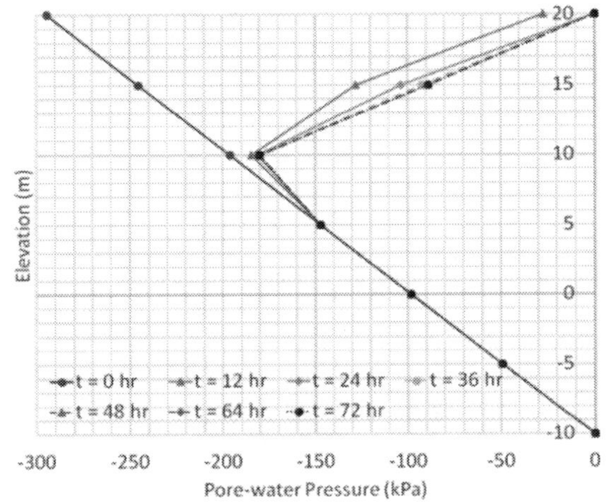

Fig. 12. Change of pore-water pressure at Line 2-8 (Top) for 70° angle of 20-meter-high slope

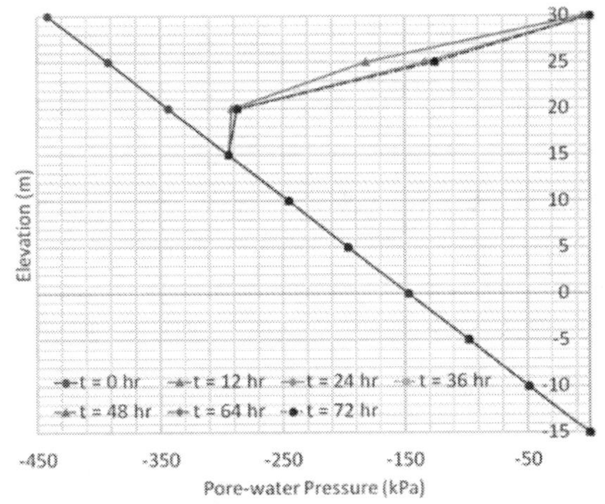

Fig. 13. Change of pore-water pressure at Line 2-8 (Top) for 70° angle of 30-meter-high slope

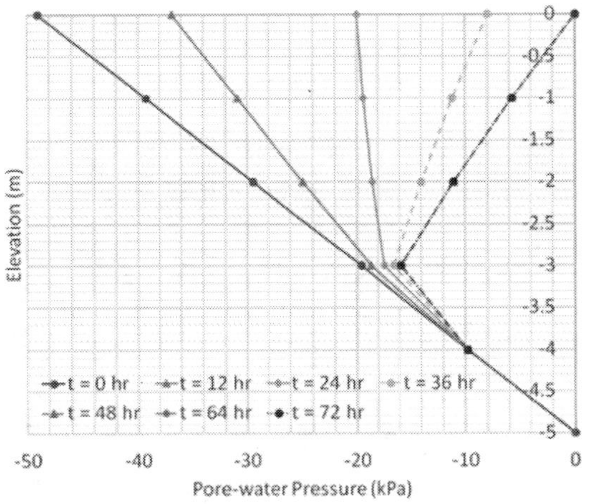

Fig. 14. Change of pore-water pressure at Line 3-9 (Bottom) for 70° angle of 10-meter-high slope

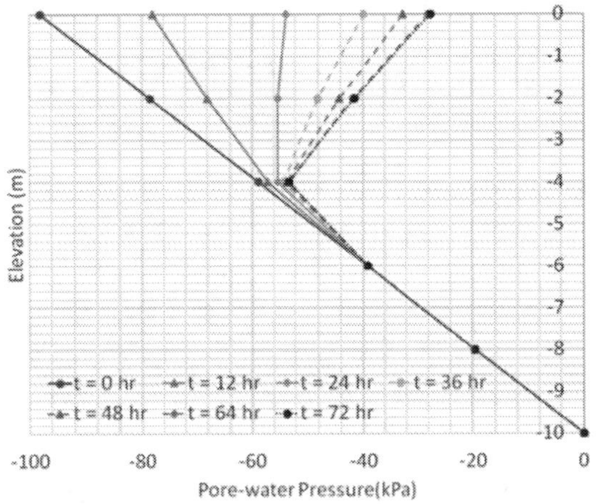

Fig. 15. Change of pore-water pressure at Line 3-9 (Bottom) for 70° angle of 20-meter-high slope

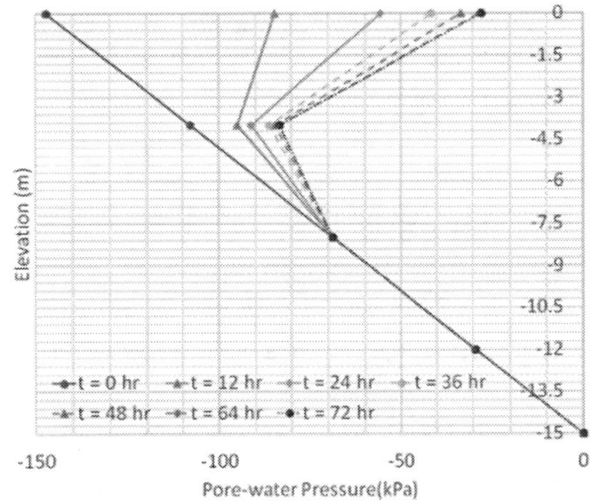

Fig. 16. Change of pore-water pressure at Line 3-9 (Bottom) for 70° angle of 30-meter-high slope

From Fig.11, Fig.12, Fig.13, Fig.14, Fig.15, and Fig. 16, it is known that the maximum negative pore water pressure for different height is different whereas the greater the height of the slope, the greater the maximum negative pore water pressure. During rainfall, infiltration occurs until the depth of ± 5 m which reduces the negative pore water pressure near the surface. The depth of infiltration of ± 5 m reaches the body of the slope for 10 m, but in ratio will only change the pore water pressure at the surface for the height of 20 m and 30 m. These findings are consistent with Section 4.2 where the probability of failure increases more significantly for the 10 m height.

It is also known that at the surface, each of the slope height reaches the 0 kPa pore water pressure as compared to [9] is natural for transient condition where the rainfall intensity is bigger than the hydraulic conductivity of the soil. The value of 0 kPa also indicates that the infiltration at the surface can no longer occur and rainfall will then be complete surface run-off. This change of behaviour explains the stagnant value of probability of failure after the first 24-hour rainfall and gives precautions on handling cut-slopes with such geometries to be careful of landslide cause change from saturated landslide due to rainfall to erosion.

5 Conclusions

From the previous sections, there are some conclusions that can be noted.

First, the initial probability of failure until the angle of 55° for each slope height is quite small but becomes quite big at the angle of 70° which indicates that the recommended cut-slope angle to be constructed is under 70°.

Second, the most significant change of FS during rainfall is achieved at the height of 10 m due to the depth of rainfall infiltration that saturates into the body of the slope. The taller the slope, the slower the infiltration reaches the body of the slope which indicates faster failure for shallow slope.

Lastly, due to high rainfall intensity and small hydraulic conductivity, the surface of the slope saturates faster which makes rainfall infiltration occurs quite fast and turned into surface runoff after the high rainfall intensity rain occurs. This notes that for cut-slope of tropical red clay soil, it is important to know that landslides may occur due to erosion after a consecutive high rainfall intensity occurrence.

The authors would like to thank Civil Engineering Department, Universitas Indonesia, especially Soil Mechanic Laboratory staffs and lecturers for the spirit and input of this research.

References

1. H. Rahardjo, T.H. Ong, R.B. Rezaur, E.C. Leong, J Geotech Geoenviron, **133(12)**, 1532-1543 (2007)
2. Ministry of Public Works (PU), *Perencanaan teknis lereng galian pada material vulkanik dan tanah residual* (2015)
3. A.J. Syahbana, A. Tohari, E. Soebowo, D. Sarah, K. Sugianti, JLBG, V **4(1)**, 33-47 (2013)
4. L.D. Wesley, *Geotechnical Engineering in Residual Soils* (John Wiley & Sons Inc., New Jersey, 2010)
5. D.G. Fredlund, *The emergence of unsaturated soil mechanics* (Fourth Spencer J. Buchanan Lecture, Texas A&M, 1996)
6. L.D. Wesley, Geotechnique, **27**, 125-126 (1997)
7. B.B.K. Huat, F.H. Ali, R.S.K. Rajoo, Am J Environ Sci, **2(4)**, 154-160 (2006)
8. G.-r. Zhang, Y.-j. Qian, Z.-c. Wang, B. Zhao, Sci. World J., **2014**, 7 pg (2014)
9. L. Zhang, J. Li, X. Li, J. Zhang, H. Zhu, *Rainfall-induced soil slope failure: stability analysis and probabilistic assessment* (CRC Press: Taylor & Francis Group, Boca Raton, 2016)
10. L.D. Wesley, Geotechnique, **23**, 471-494 (1973)
11. M.E. Woodbridge, U.R. Cook, B.S. Saroso, *PIARC World Conference Kuala Lumpur* (1999)

Assessment of Bridge Substructure in Java Island

Pringga Satria Panji[1,*], *Tommy* Ilyas[2], and *Erly* Bahsan[2]

[1]Graduate Student, Civil and Environmental Engineering, University of Indonesia
[2]Lecturer, Civil and Environmental Eengineering Department, University of Indonesia

Abstract. Bridges play important roles in transportation system. Hazard to substructure of bridges become amplified. Based on previous year data, at least one-third of total bridges in Indonesia are damaged in certain level of defect. The performance of existing bridges shall be in-check everytime. Substructure of bridge itself is prone to hazards such as ground displacement, slope instability, seismic-related hazard and scour. Traffic data and soil investigation data will be used to analyse ground displacement and slope instability where the bridge located. Local geological and seismic data will be used to assess seismic-related hazard. Quantitative data is technical information and analyses from geotechnical aspects. Qualitative data is an expert system collected from bridge expert and local goverment. The expert system will have a rating system for each aspect. Fuzzy-based method is an effective tool for modeling some vague datas and this ease the decision-making process. Fuzzy Analytical Hierarchy Process (FAHP) will be used to analyses geotechnical aspect and expert system. Hazard identification, risk rating, risk analysis, and risk assessment are steps conducted in FAHP method. Classification and rating of risk can be done with proposed method. This assessment can be a tool or recommendation for local government where the bridge located.

1 Introduction

In a road network system, bridges play an important role as a complementary road infrastructure. Bridges can be the backbone of infrastructure that links one region to another (Andric et al, 2015). Bridges can connect road that is separated by rivers, lakes, ravines, valleys, straits, highways and railways. In its development, the bridge undergoes the evolution from wooden and simple stone bridges to bridges with more complex structural systems and advanced materials. The advanced materials used in more complex bridge structure (such as concrete, steel and cables) that continue to develop will encourage the construction of bridges with more complex technology than ever before.

In Indonesia, Directorate General of Highways of Ministry of Public Works and Housing (*Dirjen Bina Marga Kementerian PU-PERA*) is responsible for bridge management through archipelago. The Directorate General of Highways uses the Bridge Management System (BMS) for more systematic monitoring and planning of the bridge. The BMS developed and owned by the Directorate General of Highways serves as a tool for the process of storing bridge-related data; such as design work, construction, rehabilitation and monitoring of bridge condition. For the purposes of the bridge survey, Indonesia's Directorate General of Highways has a Working Unit of Planning and Supervision of National Roads and located in each province.

According to data collected from the Ministry of Public Works and Housing Statistics Information Book (*Buku Informasi Statistik Kementerian PUPERA*) year 2015, the total number of bridges in Indonesia recorded by the Ministry per year 2014 is 14710 bridges with various conditions. Here are the details of the bridge conditions recorded:

- Good condition: 6609 bridges (45%)
- Medium condition: 3137 bridges (21.3%)
- Lighty-damaged condition: 3253 bridges (22.1%)
- Heavily-damaged condition: 1360 bridges (9.2%)
- Critical condition: 314 bridges (2.1%)
- Collapse / break-up: 37 bridges (0.3%)
- Total bridges recorded: 14710 bridges

From the data above, the condition of the bridge in Indonesia is composed of various conditions, ranging from good conditions until the condition of collapsed or broke. The bridges listed above are bridges that have been designed and built by the Directorate General of Highways. Those bridges have various shaping materials with varying length and spans.

When reviewed from the number of bridges with damaged conditions (light, heavy and critical damage), then the percentage of bridges with damaged condition

* Corresponding author: pringgas@gmail.com, pringga.satria61@ui.ac.id

https://doi.org/10.1051/matecconf/201814707005

in Indonesia is about 33.4% of the total bridges. Approximately one-third of bridges in Indonesia are in damaged condition. From this condition, the bridge maintenance and monitoring system have to be evaluated. Good systems, wise decision makers and sufficient funds will be able to reduce the percentage of damaged bridges in Indonesia. Decision makers (central and local government) play an important role in the care and monitoring of bridge conditions.

Assessment process to determine the current state of the bridge can be done by collecting various related data and then those data can be analyzed and generate a value that will be able to assist in the decision-making process. Related research on the structure of the bridge (such as deck, frame and bridge pier or pier) has relatively much research on it. Research on the assessment of the sub-structure of bridge is not as much as upper-structure of bridges. Damage to upper and lower structures of jembaatan will result in disruption of the service life of the bridge.

Use a two-column format, and set the spacing between the columns at 8 mm. Do not add any page numbers.

2 Literature Study

To evaluate sub-structure condition of a bridges, several methods can be conducted. Then, after evaluation process, assessment process shall be done. In this paper, assessment process is done by Fuzzy Analytical Hierarchy Process (Fuzzy AHP). Fuzy AHP is based on fuzzy set that is developed by Lotfi Zadeh in 1965. The result of Fuzzy AHP is rating of each criterion. Rating that is respected to value of each criteria is the result of assessment process. Based on the rating, decision making can be done with those rating.

In this paper, evaluation criteria will be described based on text book. Slope stability, ground movement and Fuzzy AHP will be described below.

2.1. Basic Concept of Slope Stability Evaluation

According to Duncan and Wright (2005), the process of slope stability evaluation needs to be done to determine the safety factor of a slope. Clear and comprehensive evalution results should also be done for the following reasons:

- Evaluation results should be checked by a few engineers and experts from relevant institutions. Multiple examinations are intended to minimize errors that may occur during the evalution process and to gain a different perspective on a problem.
- Evalution results must be clear and understandable by the client
- Responsibility for engineering evalution results is usually given to engineers at an institution or company. The engineer must understand the results of the evaluation and understand the basis of the decisions taken in the analysis and evaluation.
- Evaluation results should be well documented. Good documentation will make it easier if the data at any time required in the future.

It is inevitable that each slope location evaluated has different characteristics from one another. In Java, soil conditions in each province will be different. Geological detail plays an important role in slope stability, and for that, geological information of a region is very important (Duncan & Wright, 2005). Then, the next step in the evaluation of the stability of the slope is the evalution of the soil property. Evaluation of soil properties will be closely related to geotechnical investigations. Geotechnical investigation work will include field and laboratory work. The properties obtained are quite diverse, among others are: property of soil shear strength, soil stiffness, soil physical characteristics and others. In slope stability, shear strength parameters and soil density are prioritized.

Pore water pressure also plays an important and critical role in the evalution of slope stability. Normally, pore water pressure (PWP) analysis is based on the groundwater level obtained through the information in the bor-log or piezometer measurement test. The boundary conditions also need to be specified in the PWP analysis. If necessary, seepage analysis is required to determine the behavior of pore water movement or flow. Flow charts or PWP contours and total heads need to be displayed and clearly documented during the PWP analysis process.

Occasionally, some slopes have special features that affect slope stability, such as the addition of slope reinforcement. Slope reinforcement can be soil nails, geogrid, geotextile, anchor, pile and other types of reinforcement. Basically, the addition of slope reinforcement is intended to increase the value of the safety factor (FS). Details on slope reinforcement should be well documented in an effort to evaluate slope stability. The final result of slope stabilization is slope geometry, land property and FS final value.

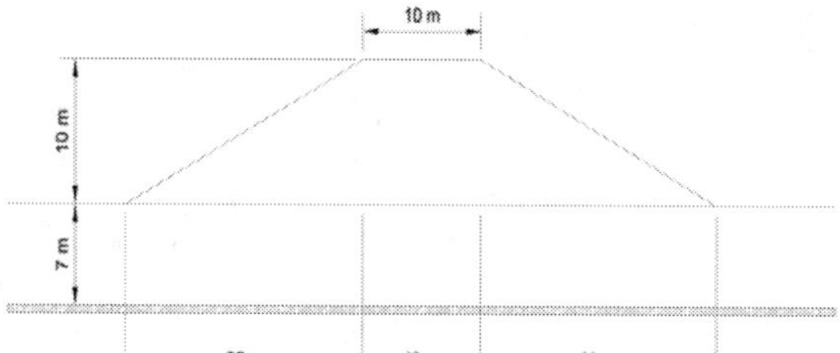

Figure 1 Example of geometry and soil properties in case of embankment stability

Basically, calculation or analysis of slope stability can be done manually or using geotechnical software. Manual computations are calculated using several methods such as Bishop, Taylor, Spencer, Fellenius, Morganstern and other methods. The whole method of

https://doi.org/10.1051/mateconf/201814707005

manual calculation is based on the concept of equilibrium limit. Then, if using the software, the geotechnical software widely used and will be used in this research is Plaxis and Geoslope. The concept of calculation on Plaxis is based on the finite element method (FEM), while the concept of calculation on Geoslope is based on the concept of equilibrium limit.

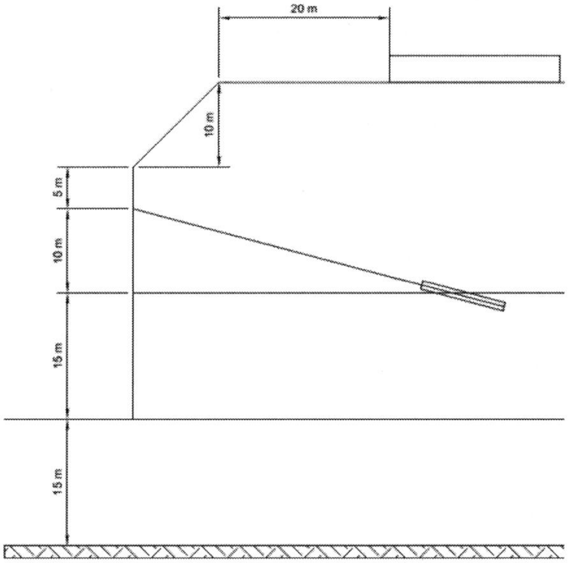

Figure 2 Sample case evaluation of slope stability with ground anchor

2.2 Basic Concept of Ground Movement Evaluation

Ground movement is an effect that can actually be expected in a bridge project. Ground movement on a bridge can be caused by several things, among others:

- Road traffic load
- Earthquake
- Land consolidation
- Abutment instability
- Open-cut work around the location of the bridge
- Unstable ground conditions

Traffic load on a bridge is already planned for the maximum load that occurs, but sometimes in a time, the maximum load continues to occur and combined with other causal factors such as unstable ground conditions, then soil displacement in the bridge can occur. For the case of bridges built on soft soils, the displacement due to primary consolidation becomes a major issue. Primary consolidation results from the dissipation of pore water on the soil. In case of pore water dissipation, the volume of soil will shrink and there will be soil displacement.

Another cause of of ground movement is earthquake. Basically, the earthquake load has been included into the calculation process, but sometimes the earthquake that occurs exceeds the prediction in the calculation. If that happens, then the displacement of the soil beyond predicted value can not be avoided. Then, instability of the slope at the abutment of bridge causes displacement in that location. Open excavations or open-cut conducted around the location of the bridge will also cause soil displacement around the bridge. Open excavations are often done during the construction period. Open excavations during the actual operation of the bridge should be avoided as much as possible.

The most common characteristic that can be observed when sub-structure of bridges displaced beyond the predicted value is by observing the condition of bridge structural elements (such as pier condition, beam and bridge deck). Another feature is to observe unnatural movements in the ground around the location of the bridge. Another case is to observe the overall geological conditions that allow the movement of the soil around the bridge, such as the case of bridges located in the basin or valley whose slopes are steep and unstable; If the slopes around the basin or valley move significantly, then the soil around the bridge base can move anyway.

2.3 Basic Concept of Fuzzy AHP Method for Assessing Bridge Substructure Condition

Fuzzy AHP method is based on Fuzzy logic developed by Lotfi Zadeh in 1965. Fuzzy logic itself is a component in the formation of soft computing. Fuzzy logic consists of sets which are then called Fuzzy sets. In Fuzzy sets, each member in a Fuzzy set has its own value known as the membership degree or membership function. In everyday life applications, Fuzzy logic is used to map problems from input data to expected output data (Kusumadewi, 2010).

Figure 3 Simple mapping of input-output process

In Figure 3, there is a "blackbox" box between the input box and the output box. The blackbox can be designated as Fuzzy logic. Clearly, the blackbox turns the input box into an output box. The blackbox contains the methods or set of rules used to process the input data into output data. Here are some reasons according to Cox (1994) regarding the use of Fuzzy logic:

- Fuzzy logic concept is relatively easy to understand. This is because Fuzzy logic uses the basic set theory, so the mathematical concepts underlying fuzzy reasoning will be easy to understand.
- Flexibility of Fuzzy logic. Fuzzy logic is able to adapt to the changes and uncertainties that accompany the problem.

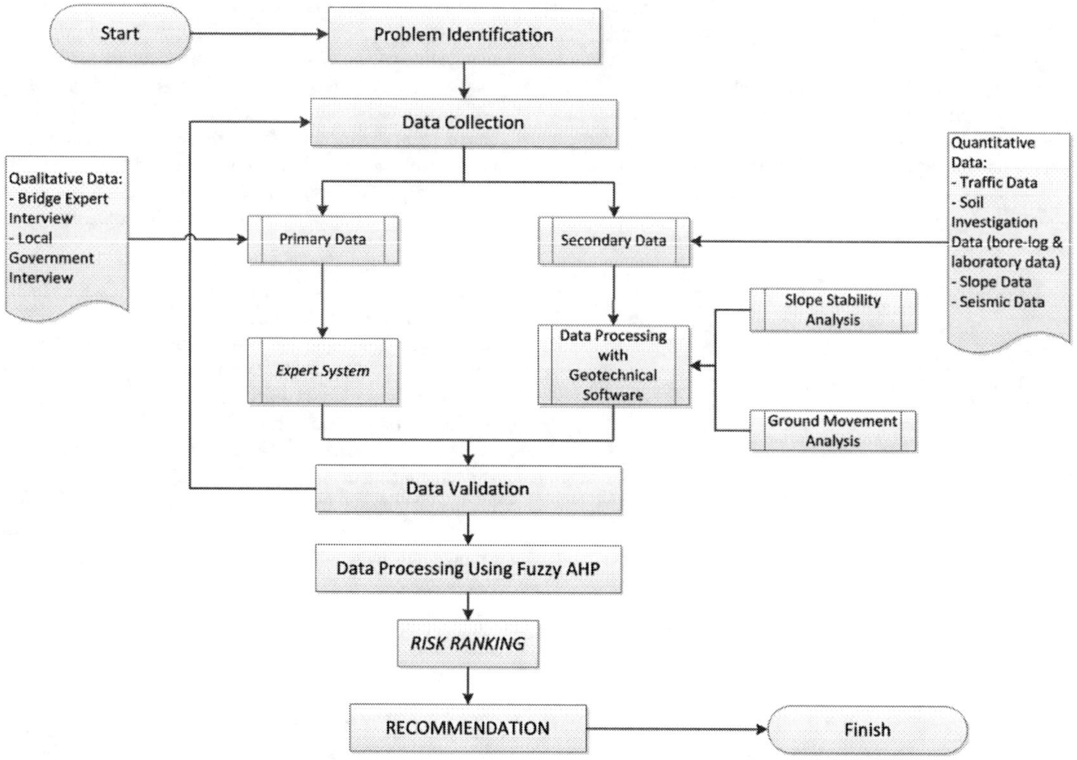

Figure 4 Research Flowchart

- High level of tolerance to exclusive data. If in a homogeneous set of data, then there is one exclusive data (different data from the other data itself), then Fuzzy logic is able to handle that exclusive data.
- Non-linear modeling. Fuzzy logic is able to model non-linear functions
- Inclusion expert experience. In Fuzzy logic, expert experience will be included and become an important component in Fuzzy logic
- Relatively wide application in engineering scope
- The use of natural language which is relatively easy to understand

In contrast to bivalent logic which is an absolute value between 0 or 1. Choice of value is only absolute value 0 (zero) or 1 (one). If it is not a value of 0, then its value should be 1 and vice versa. The absolute value will cause inaccuracy for a certain range of data. In Fuzzy logic, there is no absolute value, there is an absolute limit. The absolute limit in Fuzzy logic is 0 and 1. The value belonging to the data into membership function is the value that is in the range 0 to 1. The value can be 0.1, 0.25, 0.931 and so on.

3 Overview in Analyzing and Designing Fuzzy System Using *Fuzzy Toolbox* in MatLab Enviromnment

MATLAB software is a very powerful tool with various toolbox that accompany it. MATLAB is widely used to solve mathematical problems with graphic illustrations. Basically, MATLAB is an executable programming language-based software. MATLAB has both built-in and add-on toolboxes. Especially for data processing with Fuzzy AHP method, the use of toolbox in the form of Fuzzy Toolbox becomes obligation. Fuzzy Toolbox is an add-on that must be purchased separately on the Mathwoks website (MATLAB publisher); This is because Fuzzy Toolbox does not become the built-in toolbox in MATLAB.

In Fuzzy Toolbox, there is a system called Fuzzy Inference System (FIS). This FIS is the "blackbox" as described in the previous section. FIS is in charge of inference of input data provided. FIS has several derived functions, namely:

- FIS Editor (as sub-function input inference)
- Rule Editor (as sub-function input inference)
- Membership Function Editor (as input sub-function input)
- Rule Viewer (as output sub-function and read-only)
- Surface Viewer (as output sub-function and read-only)

https://doi.org/10.1051/matecconf/201814707005

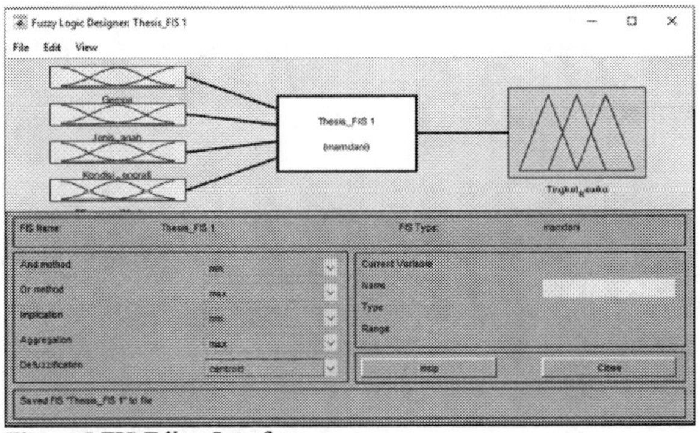

Figure 5 FIS Editor Interface

4 Research Instrument

This study uses a variety of research instruments. This diverse research instrument is based on differences in the two types of data to be used in the study. Two types of data used are qualitative data (which later will form expert system) and quantitative data (in the form of analysis of slope stability and ground movement).

For qualitative data, this type of data will be primary data. The instrument used in order to obtain this primary data is by interview. Respondents who were used as resource persons in the interviews were pre-determined bridge experts (purposive sampling). The interview will use a questionnaire and will be done at the site where the respondent reside or work The bridge experts who will be the respondents are experts who served in various institution, such as:

- Ministry of Public Works
- Teachers / lecturers at the University or Institute
- Professional or practitioner

There are several questions that will be asked as a list of questions. Here's an outline of the list of questions to be made:

- The general condition of the bridge (especially the sub-structural component) in the region x
- Frequency of maintenance / repair / rehabilitation of bridges (including sub-structural components) within one-budget year
- Funds that should be provided for the maintenance and / or rehabilitation of bridges (especially sub-structural components)
- Damage that often occurs in structural components under the bridge
- The response time between the damage with the repair effort

For qauntitative data, this type of data will be secondary data. The instrument used to obtain this secondary data is by working with third parties who have the data. This data is an archive owned by a third party. The researcher will use the archive data in accordance with the given permission. The secondary data plan to be used is:

- Bor-log data (Standard Penetration Test, SPT). This data was obtained from a third party conducting a geotechnical investigation on the bridge project under study. Typical geotechnical investigation data for bridge projects will include SPT, CPT and laboratory data. Data sources can be individual, consultant planners, contractors, ground mechanics laboratories, project owners and others
- Cone Penetration Test (CPT). This data will be obtained simultaneously with SPT data
- Laboratory data. This data will be obtained simultaneously with SPT data
- Geological data. Geological data obtained from the government agency (Geospatial Information Agency) that organizes geological mapping for the area around the bridge that became the object of research.
- Seismic data. Seismic data is obtained from the Center for Settlement Development (Puskim). The seismic data will be the spectral design at the location of the bridge studied. The data contained in Puskim is detailed up to the location coordinates, so that the spectral data of the earthquake can be more specific and precession.

5 Sampling Design

Sampling design to be conducted by researchers is as follows:

- Population. The population here is in the form of existing bridges in Java Island (with administrative area of Banten Province, Jakarta, West Java, Central Java, Yogyakarta and East Java).
- Samples. Researchers will conduct samples with targets of 5 to 10 bridges with medium span to long span bridges. Sampling of the bridge will be done randomly. The type of bridge under study will adjust to the availability of secondary data, so there is no reference to the number of bridges per province. It could be one province taken two or three samples whereas in other provinces not taken sampling due to the absence of data. Sampling will be done randomly and following the availability of third party data.
- Variables. For the variables in the sample, the constituent material of the bridge sub-structure component will be differentiated into concrete and steel with varying degrees of damage.

https://doi.org/10.1051/matecconf/201814707005

6 Data Processing

Data processing to be conducted by researchers is as follows:

Figure 6 Qualitative Data Processing

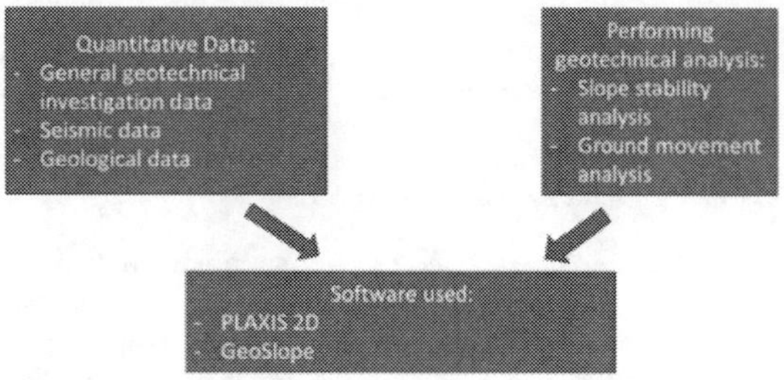

Figure 7 Quantitative Data Processing

8 Conclusion

In this research, Fuzzy AHP will be used to assess all possible hazard in the sub-structure of bridges. General geotechnical aspects will be analyzed. Authors propose Fuzzy AHP method to classify, conduct risk rating and assess the sub-structure of bridge. The end product will be risk or hazard classification of each bridge. This assessment can be a tool or recommendation for local government where the bridge located. Local goverment may able to make decision making process more precise and faster. More precise and faster the decision made the better bridge will be saved. The research is conducted for several bridges in Java Island, Indonesia.

References

1. An, Min et al, *A Fuzzy Reasoning anf Fuzzy-Analytical Hierarchy Process Based Approach to the Process of Railway Risk Information: A Railway Risk Management System*, Information Sciences **181**, pp. 3946-3966 (2011)
2. Attaway, Stormy, *MATLAB: A Practical Introduction to Programming and Problem Solving*, Oxford - UK: Bitterworth – Heinermann - Elsevier Inc. (2009)
3. Andric, Jelena M. et al, *Risk Assessment of Bridges Under Multiple Hazards in Operation Period,* Journal of Safety Science – Elsevier **83**, pp. 80-92 (2016)
4. Carbonari, Sandro et al, *Soil-Structure Interaction Effect in Single Bridge Piers Founded on Inclined Pile Groups*, Soil Dynamics and Earquake Engineering **92**, pp. 52-67 (2017)
5. Cook, Wesley et al, *Bridge Failure Rate*, Journal of Performance of Constructed Facility **29**, pp. 1-8 (2015)
6. Corotis, Ross B., *An Overview of Uncertainty Concepts Related to Mechanical and Civil Engineering,* Journal of Risk and Uncertainty in Engineering System Vol. **1**, pp. 1-12 (2015)
7. Craig, Robert F., *Soil Mechanics*, 7th edition, New York – USA: Spon Press (2004)
8. Duncan, J. M. and Stephen G. Wright, *Soil Strength and Slope Stability*, New Jersey – USA: John Wiley & Sons Inc. (2005)
9. Fagundes, Diego F. et al, *Load Transfer Mechanism and Deformation of Reinforced Piled Embankment*, Geotextiles and Geomembranes **45**, pp. 1-10 (2017)
10. Kusumadewi, Sri et al, *Fuzzy Multi-Attribute Decision Making (Fuzzy MADM)*, Yogyakarta – Indonesia: Penerbit Graha Ilmu (2006)
11. Kusumadewi, Sri, *Analisis & Desain Sistem Fuzzy Menggunakan Toolbox Matlab*, Yogyakarta – Indonesia: Penerbit Graha Ilmu (2002)
12. ________, *Aplikasi Logika Fuzzy Untuk Pendukung Keputusan*, 2nd edition, Yogyakarta – Indonesia: Penerbit Graha Ilmu (2010)
13. Lee, Sangwook, *Determination of Priority Weight Under Multi-Attribute Decision-Making Situation: AHP versus Fuzzy AHP*, Journal of Construction Engineering and Management **141**, pp. 1-9 (2015)
14. Marwoto, Sofa, *Pemodelan Sistem Pendukung Keputusan Untuk Pemeliharaan Jembatan Beton*, Jurnal Teknik Sipil, Volume **II** No. 1 (2014)
15. Maula, B. H. and Zhang, L., *Assessment of Embankment Factor of Safety Using Two Commercially Available Programs in Slope Stability Analysis,* Procedia Engineering **14**, pp. 559-566 (2011)
16. Nezarat, Hamidreza et al, *Ranking of Geological Risks in Mechanized Tunneling by Using Fuzzy Analytical Hierarchy Process (FAHP),* Tunneling and Underground Space Technology **50**, pp. 358-364 (2015)
17. Patel, D. A. et al, *Hazard Assessment Using Consistent Fuzzy Preference Relation Approach*, Journal of Construction Engineering and Management **142**, pp. 1-10 (2016)
18. Senouci, Ahmed et al, *Fuzzy-Based Model for Predicting Failure of Oil Pipelines,* Journal of Infrastructure System **20**, pp. 1-11 (2014)
19. Zadeh, Lotfi A., *Is There a Need for Fuzzy Logic,* Journal of Information Sciences No. **178**, pp. 2751-2779 (2008)
20. Zatar, W. A. et al, *Bridge Embankment. I: Seismic Risk Assessment and Rating*, Journal of Performance of Constructed Facility **22**, pp. 171-180 (2008)

Development of dynamic impact cone for assessing induced partial saturation (IPS) as liquefaction mitigation technique

Fritz R. P. Nababan [1,*]

[1]Soil Mechanics Laboratory – ITB, Bandung, Indonesia

Abstract. Liquefaction of saturated loose sands during an earthquake has been a great concern for practicing earthquake engineer because of its damaging effects. Current liquefaction mitigation techniques used in practice are expensive. A new innovative mitigation technique which is cost effective was proposed by Yegian et al. [1] called "Induced Partial Saturation" (IPS). This method involves in generating gas bubbles within the pores of fully-saturated sand and transformed it to partially-saturated sand. This paper described the development of a dynamic impact cone to assess the effect of this IPS treatment on saturated loose sand specimens. This dynamic impact cone consists of a cone, rod, ruggedized pressure transducer, string pot, and DAQ. The cone which has pressure transducer incorporated at its tip was driven into fully and partially (treated) saturated sand specimens prepared in the laboratory. The excess pore pressures and penetrations depth per impact of the two specimens were compared. The tests results such as the excess pore pressure and the penetration indicate that the impact penetration on the fully-saturated specimens liquefied sand, but could not liquefy the partially-saturated sand. Therefore the concept and development of this dynamic impact cone has promising future applications.

1 Introduction

Liquefaction of saturated loose sands during earthquake event poses a great concern for practicing geotechnical earthquake engineer because tremendous damage it can cause. Researches on finding effective and cost effective liquefaction mitigation techniques are still undergoing in many part of the world. One promising new mitigation technique which is innovative, practical, and cost effective was proposed by Yegian et al. [1] called "Induced Partial Saturation" (IPS). This technique involves generating small amount of minute gas bubbles within the pores of initially fully-saturated sand, thus reducing the saturation degree and at the same time increasing the compressibility of the soil pores. Reduction of the pore's compressibility reduces the tendency of the pore fluid to generate excess pore water pressures. These generated gas bubbles in the pores are made by chemical reaction using environmentally friendly agent.

Small scale laboratory testing has shown the benefit of IPS as described in Yegian et al. [1] and Eseller-Bayat et al. [2],[3]. Most recently the benefit of IPS treatment was also observed and proven in large scale laminar box tests in NEES University of Buffalo facilities, these results were analyzed and presented by Nababan [4].

This paper describes idea and development of IPS verification technique to be used both in the laboratory and in the field. A verification method of IPS treatment which utilizes a dynamic impact cone was explored, which can verify the presence of gas bubbles in the pores by measuring the behavior of the specimens before and after the treatment. The cone is considered simple and practical to implement in the laboratory and in the field for specimens before and after IPS treatment. The detail of the developed cone is described herein as well as the typical results of the test.

2 Dynamic Impact Cone

The basic concept for IPS verification is comparing the excess pore pressure generation and the soil resistance if dynamic loading introduced to a specimen. A drop weight was chosen to generate dynamic impact load on the cone. The postulate is drop weight will impart large enough shear strains that it will liquefy the specimen and pore pressure transducer incorporated at the cone tip will measure a high excess pore pressures, enough to reduce the vertical stresses close to zero for the untreated specimen (i.e. liquefaction). It is expected that if the same cone is driven into the same specimen after being treated with IPS, the presence of gas bubbles in the pores will considerably reduce the generation of excess pore pressure. Comparing the difference of excess pore pressure records between the fully saturated and IPS treated specimens can then be considered as evidence of IPS treatment. Another hypothesis is that cone penetration due to impact of weight drop would be larger in the fully saturated (before IPS treatment) compare to that of partially saturated specimens (after IPS treatment).

2.1 Prototype Cone

A prototype of dynamic impact cone had been developed as shown in Figure 1 following. This cone

* Corresponding author: fritzuu@gmail.com

consists of: an aluminum rod, aluminum cone, pressure transducer and gravity driven hammer system. The cone is attached to a 63.5cm long hollow rod with diameter of Ø 3.2cm. The cone is pointed at 60 degree, while the backside of the cone is a male thread to connect to the rod. The pressure transducer sits inside the cone facing towards the point and is fixed in placed by silicone. The tip of the transducer connected to a chamber with four holes on its side.

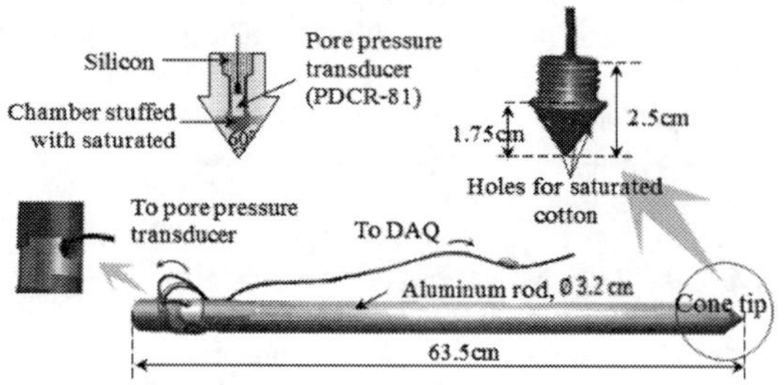

Fig. 1. Prototype cone.

In order to keep sand particles entering the chamber in which the transducer is fixed, saturated cotton is filled inside the chamber. The pore pressure transducer (PPT) used in this prototype is PDCR-81 (input 5V and output 0-100mV, range pressure 0-5 psig), which contains a diaphragm with 4 strain gauges attached in full bridge configuration. This diaphragm is protected by a porous stone.

The specimens in general were prepared using Ottawa sand and de-aired water into a plastic barrel with known dimensions. Firstly, a measured quantity of de-aired water was placed in the barrel, and measured quantity dry Ottawa sand was rained into the barrel until no more water was left on top of the specimen. Knowing the weights and volumes of the sand, the water, and the final specimen height, the void ratio and the degree of saturation were computed using soil phase relations. The partial saturation specimen was prepared by mixing the chemical with de-aired water prior to sand pouring; the degree of saturation achieved for this specimen was about 70%.

The summary of the tests done with this prototype cone in fully saturated as well as partially saturated specimens just proven the initial postulates. In the fully saturated specimen the impact caused the specimen to liquefy at the cone tip level which was verified by measured excess pore pressures at the tip. The measured excess pore pressure reached the effective overburden pressure at the same level. The measured excess pore pressure for IPS treated specimen were approximately 30% of the effective overburden pressure.

The limitations encountered of this cone were: 1) pressure transducer used for the cone is found not sturdy enough taking impact load as observed several of the transducers quitted during testing, 2) soils enter the chamber during testing as the stuffed saturated cotton is not enough to filter the soil.

2.2 Developed Dynamic Impact Cone

The dynamic impact cone should work for both laboratory and field applications and has to overcome the limitations of the prototype cone mentioned above. This cone is design to be used with drilling rod, E-rod, made of harden steel with diameter of 1.3 inch, small enough to be used in the laboratory and sturdy enough to be used in the field. The following figure shows detail of the cone developed.

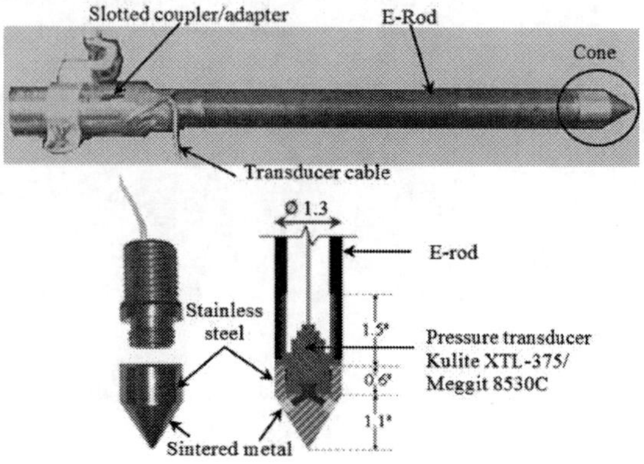

Fig. 2. Dynamic impact cone design used with drilling E-rods (diam. 1.3 inch).

The developed dynamic impact cone consists of heavy duty ruggedized pressure transducer, saturation and calibration chamber, data acquisition system, and data processing software. Details of this cone are described in the following.

2.2.1 *Heavy Duty Pressure Transducer*

The pressure transducer needed for the system has to be able to withstand heavy beating and very high acceleration caused by the impact of the drop weight. The research on the suitable pressure transducer candidate came up with at least two types of proper commercial pressure transducer as shown in following figure 3. These are not pore pressure transducer but regular pressure transducers which have no porous stones. Instead of porous stones on their faces, they have stainless steel screens. This type of transducer can work well to measure liquid as well as gas pressures. The two miniature pressure transducers used for the cone were: Kulite XTL 375 with pressure range 0-25psia (absolute), and Meggit 8530C-100 with pressure range of 0-100psia (absolute), reference pressure relative to the perfect vacuum. Both transducers were very sturdy and rugged, capable of working in extreme environments, and can withstand up to 10,000g impact load and up to 1,000g sinusoidal load. The output of both pressure transducers were in mV (millivolt), which required the use of the signal conditioner for data acquisition.

https://doi.org/10.1051/matecconf/201814707006

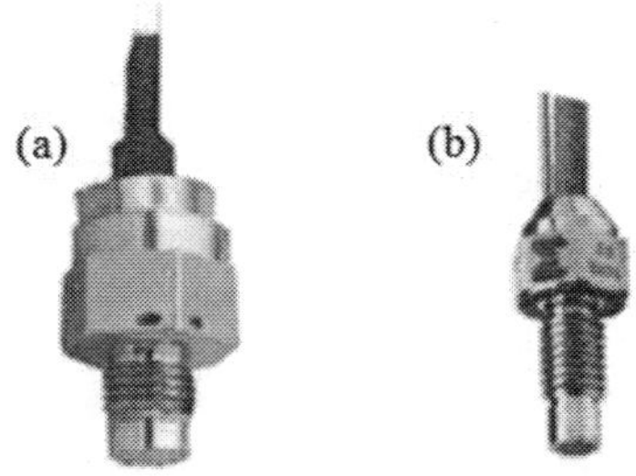

Fig. 3. Heavy duty ruggedized types of pressure transducers were explored (a) Kulite XTL 375 (b) Meggit 8530C-100.

2.2.2 *Saturation and Calibration Chamber*

The filter used to separate the cone chamber with surrounding soil is sintered metal, four sintered metal were installed at the tip of the cone as shown in figure 2 earlier. The chamber is filled with mixture of glycerin and water, this liquid act as medium in which the pressure from surrounding soil is transferred to transducer. It is essential to keep the sintered metal fully saturated with glycerin and water prior to each testing to avoid measurement delay and error. A chamber was made for saturation and calibration of the cone as illustrated in figure 4 the following. Air pump and vacuum were used for the chamber, air pressures were used for calibration and vacuum were used for cone saturation.

Fig. 4. Saturation and calibration chamber, air pump and vacuum, and glycerin.

The steps describe preparation and assemblies of the dynamic impact cone are as follow:

1- Saturation of the cone (sintered metal/ filter). The apparatus shown in figure 4 is utilized for this purpose. The chamber is filled halfway with 50% de-aired water and 50% glycerin and then the cone is placed inside and 20 psi vacuum is applied for 6-8 hours.
2- The pressure transducer is then installed in its housing (see figure 2). The cable of the transducer is extended before placing and shrink wrap is used to make the connections watertight.
3- Assembling of transducer housing and the cone tip is done in a small container filled with saturation fluid, to ensure that the system remains saturated.
4- After the cone is assembled then it has to be recalibrated using the same chamber as shown in figure 4. Then the cone is inserted through the hole on the lid and air tight seal is created. The transducer cable is than connected to a portable data acquisition (PDAQ) and a laptop with installed data processing software, LabVIEW is used. Calibration is made by applying pump pressure and comparing with the transducer reading. Adjustments are made in the conditioner of the PDAQ system to ensure reliable transducer readings.
5- After calibration the AW rod segments are prepared to a length equal to an anticipated depth of a test. At this stage, the cone is ready for testing.

2.2.3 *Portable data acquisition (PDAQ) system for field application*

Portable data acquisition (PDAQ) system was developed alongside the dynamic impact cone as shown in figure 5 below. A case was used to house the PDAQ system, which is watertight, airtight, dustproof, crushproof.

The outside dimensions of Length 19-1/8", Width 15-7/16", Depth 9/16", and empty weight is about 4 kg. Overall the components of the developed PDAQ system entail:

- Portable NI USB-6216 M
  Powered by USB cable from the field laptop, this USB-NI 6216 has 16 analog input channels, sampling rate of 400kS/s, and powered by USB cable out from the laptop.
- Signal Conditioners
  Signal conditioner used for the PDAQ is Ultra Slim Pak G448

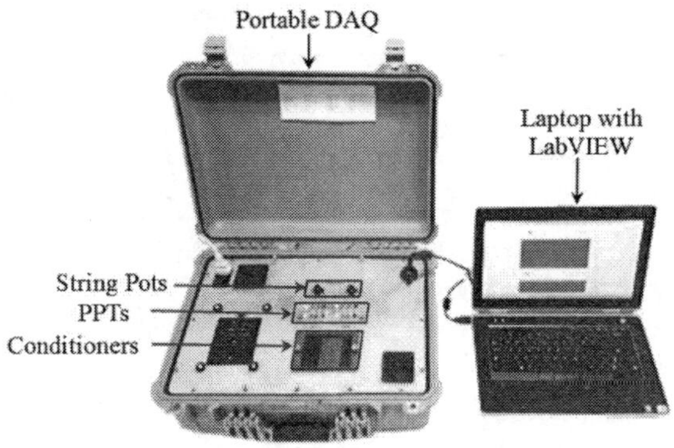

Fig. 5. Portable data acquisition (PDAQ) developed for dynamic cone testing.

- Auto recharge battery and Converter
  A small and compact battery 24VDC was used to power the conditioners and the fan. AC to DC converter was also used to convert the voltage from AC wall to DC battery.

- Connectors
  Two types of connectors were made available in the front panel: first type is four units of 2-pin connector for the cone, and the second type is 2 units of shell-type connector for anticipated utilization of displacement transducer.

2.2.4 *Laptop and Data Processing Software*

Laptop is used with National Instrument (NI) data acquisition software, LabVIEW, for data recording and processing.

3 Test and Results

The developed dynamic impact cone then tested first in small scale test in the laboratory and followed by field testing. Test results of small scale specimen in the laboratory only presented in this paper.

3.1 Specimen Preparation and Test Set Up

The specimens were prepared using Ottawa sand and de-aired water in quite transparent barrel with a known dimensions. In brief, the preparation processes are as follow: 1) Measurement of quantity of de-aired water and Ottawa sand, 2) Placement of water inside the barrel, 3) Raining dry Ottawa sand into the barrel, 4) Reading the initial condition of the specimen. Given the weights and volumes of Ottawa sand, de-aired water, final specimen height, and free water height on top of the specimen, then the properties of specimen prepared such as void ratio, porosity, unit weight, and degree of saturation can be computed by mean of soil phase relations. Partially saturated specimen is prepared using a mixture of de-aired water and certain quantity of the environmental friendly chemical powder was used.

As stated earlier, proving the usefulness of dynamic impact cone in assessing IPS treated specimen, ones need to performed similar tests for both fully saturated and partially saturated specimens with the same initial relative density. Therefore, a standard set-up was made in the laboratory to test both specimens and to get comparable results. Figure 6 following illustrated the set up used for the test. In general, the set up entails of fixed frame, drop hammer with guide rod, height indicator, and the specimen barrel. This set up ensure that the impact energy delivered by drop hammer for every blows is identical and thus different responses of the specimens can be reasonably compared. The impact energy was maintained constant during the test by keeping the falling height equal for each blow. In laboratory test the drop weight used was 2.3kg with falling height of 30cm.

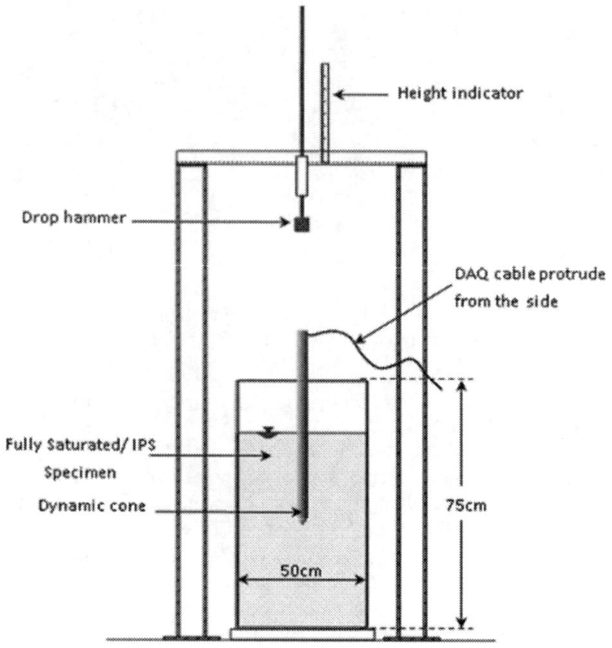

Fig. 6. Standard set up used in laboratory testing.

3.2 Test Results

Two tests of dynamic impact cone on fully and partially specimens were conducted and the results were compared. These specimens were prepared by method described previously with the same amount of sand and water. The resulting degree of saturation of the partially saturated sand was about 65%. Both specimens had similar relative densities (D_r) of about 50%. The numbers on the photographs shown in figure 7 indicates top view of five test locations for each of the specimen where dynamic impact cone tests took place.

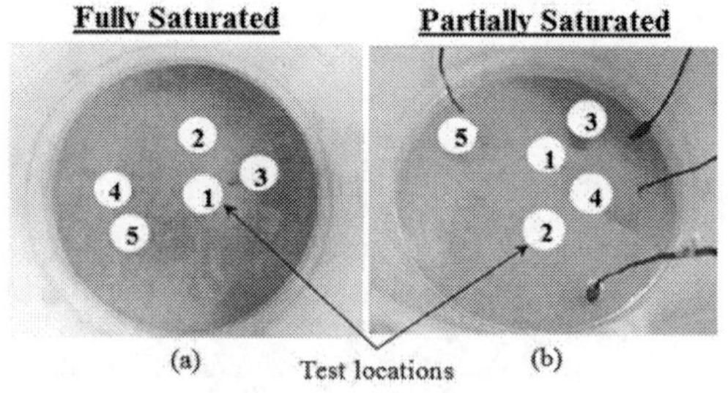

Fig. 7. Locations of cone tests in fully and partially saturated specimens.

Figure 8 shows the test results from the fully saturated sand specimen, when the cone was driven at locations of Tests 1 and 5. Initial relative density before Test 1 and 5 approximately are 50% and 65% consecutively. The spikes in the pressure transducer data show excess pore pressures developed during driving of the cone. The ratios of excess pore pressure and initial vertical effective stress were about 90%, indicating that the cone almost liquefied the fully saturated sand specimen. While figure 9 shows results from testing the partially saturated sand specimen at location Test 1 and 2, where with 1% weight of environmentally friendly chemical. The initial relative density of the partially saturated specimen Test 1 and 2 were approximately the same with the fully saturated specimen test stated above, 50% and 65%. It is noted that the maximum excess pore pressure generated by driving of the dynamic impact cone in the partially saturated specimen was about 35%, this was significantly smaller than 90% recorded in the fully saturated specimen.

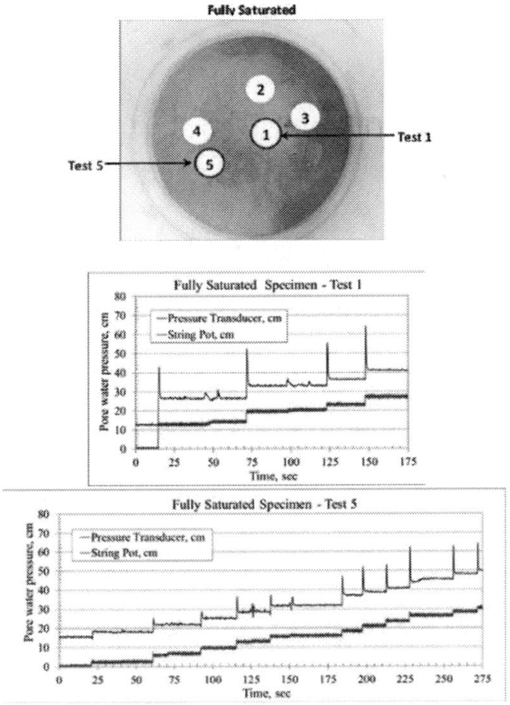

Fig. 8. Typical results of fully saturated specimen, Tests 1 and 5.

Again these results demonstrated that dynamic impact cone can picked up the behavior differences of fully saturated and partially saturated specimens and therefore verify the effectiveness of IPS technique.

Another measured parameter from the tests that were useful in confirming the presence of gas bubbles and the effectiveness of IPS treatment is the penetration per blow of the cone. The initial postulate that the amount of penetration per blow of IPS treated specimen should be lower than fully saturated specimen was again confirmed as illustrated by figure 10. This figure shows that in order to penetrate approximately 25cm of the prepared specimens, it took 4 blows for fully saturated specimen and 17 blows for partially saturated specimen. This behavior indicated that the presence of gas bubbles in pores of IPS treated specimen reduced the tendency of excess pore pressure generation during impact thus increase its bearing capacity if compare to fully saturated specimen, this explains why IPS treated specimen can take more beating to achieved the same penetration as the fully saturated specimen.

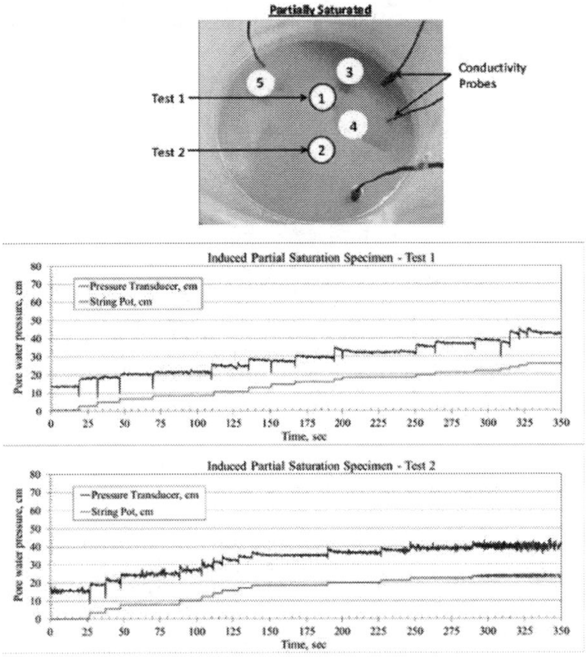

Fig. 9. Typical results of partially saturated specimen, Tests 1 and 2.

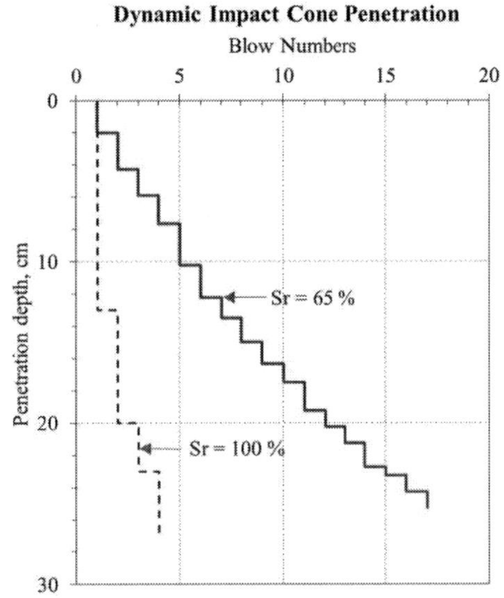

Fig. 10. Comparison of cone penetration of fully saturated and IPS treated specimen.

 https://doi.org/10.1051/matecconf/201814707006

4 Conclusions and Recommendations

The concept of using a dynamic impact cone to verify the presence and effect of partial saturation on a liquefaction potential sand specimen has been tested and had been proven effective in assessing the presence of gas bubbles in the pores of the specimen. There are two measured parameters that were used to verify IPS treatment: 1) excess pore pressure generated due to impact, and 2) amount of penetration per impact. The test results indicated that the presence of gas bubbles in the soil pores effectively reduced the tendency of the specimen to generate excess pore pressure and increase bearing capacity of the specimen relatively compare to fully saturated specimen.

It appears that laboratory test results shows promising future application of dynamic impact cone to be used for field application, mainly but not limited to assessing IPS treatment. Future research on this topic would be challenging at the same time promising.

Great appreciation goes to peoples that involved in this research to name a few: Hadi Kazemi, Seda Gokyer, Michael McNeil, Kurt Braun, and Mishac K. Yegian.

References

1. M. K. Yegian, E. Eseller-Bayat, A. Alshawabkeh, S. Ali, J. Geotech. Geoenviron. Eng., **133 (4)**, 372 (2007).
2. E. Esseler-Bayat, M. K. Yegian, A. Alshawabkeh, S. Gokyer, J. Geotech. Geoenviron. Eng., **139 (6)**, 863 (2013)
3. E. Esseler-Bayat, M. K. Yegian, A. Alshawabkeh, S. Gokyer, J. Geotech. Geoenviron. Eng., **139 (6)**, 872 (2013)
4. F. R. P. Nababan, Ph.D Diss., (2015)

MATEC Web of Conferences **147**, 07007 (2018) https://doi.org/10.1051/matecconf/201814707007
SIBE 2017

Observational Method for More Reliable Settlement Prediction for Reclamation on the Holocene Marine Clay Deposit in Jakarta Bay

Budiwan Adi Tirta[1,*], *M. Adecar* Nugroho[1], *Bonifacius* Parahita[1], and *Dirk-Jan* Jaspers Focks[2]

[1]PT. Witteveen+Bos Indonesia, Park View Plaza, 6th Floor, Jl. Taman Kemang No. 27, Jakarta Selatan 12730, Indonesia
[2]Witeeveen+Bos, Leeuwenbrug 27, 7411 AE, Deventer, the Netherlands

Abstract. Construction of the 276 and 312 ha reclamation on the Holocene marine clay deposit in Jakarta Bay is on-going. Settlement has been measured during the construction in order to satisfy the requirement of residual settlement. This paper briefly discusses limitation of accuracy of settlement prediction based on purely parameters derived from laboratory and in situ testing, in particular the Holocene marine clay. A reliable 'observational' method which makes full use of the monitoring data is described and estimates reliable residual settlement. More accurate settlement method is then also briefly described taken into account complex loading history. Finally, the factors, which are source of inaccurate settlement predictions and its implications, are included in the discussion.

1 Introduction

1.1 Geological setting

Jakarta bay is defined by the capes of Tanjung Krawang to the east and T. Jawa to the west. The central sector of the bay is the place for discharge area of several fluvial systems (Sungai Angke, S. Sunter, S. Ciliwung, S. Bekasi and S. Cikarang). The other two larger systems (S. Cisadane and S. Citarum) discharge peripherally from T. Krawang and T. Jawa [1]. The coastal plain of the bay is widely represented by river channel [2] and beach deposits [3]. This delta flanking Jakarta Bay, like other places in north Java, consists largely of clays - their sand content is small [4]. These deposits were formed during the Holocene. The Holocene is the most recent geological age corresponding to the current interglacial period extending from about 12,000 years ago up to now. This upper part of Holocene marine clay (HMC) is underline with the lower part of marine clay, which was formed about 120,000 years ago. These clay formations are likely found in low-laying coastal area in Indonesia [5].

The properties of this upper HMC are varying from place to place. However, it has high water content and void ratio. It is generally classified as very soft to soft, is close to normally consolidated and has high compressibility and well know lateral extent. Typical range of cone resistance for UMC is 0.1 to 0.2 MPA with mean effective stress of 350 kPa. Normally, when performing Standard Penetration Test (SPT), it will result in a normalized SPT blow count (N-SPT) of 0 to 2. As can be seen from Figure 1, variation of water content and its saturated weight of soil are shown in four different places in North Java.

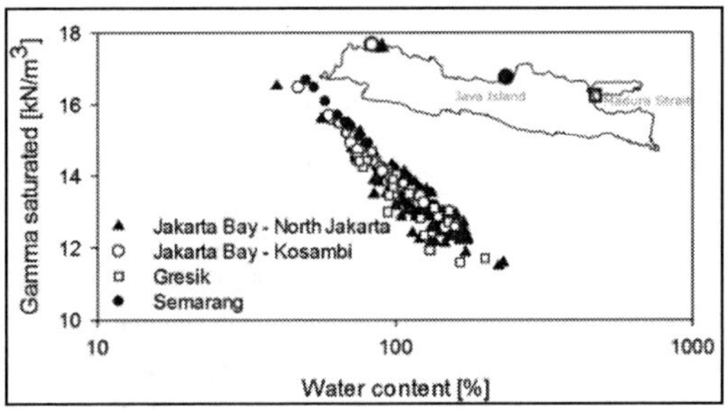

Fig. 1. Saturated weight and water content of Holocene soft clay in different locations in north part of Java (N-SPT=0-2).

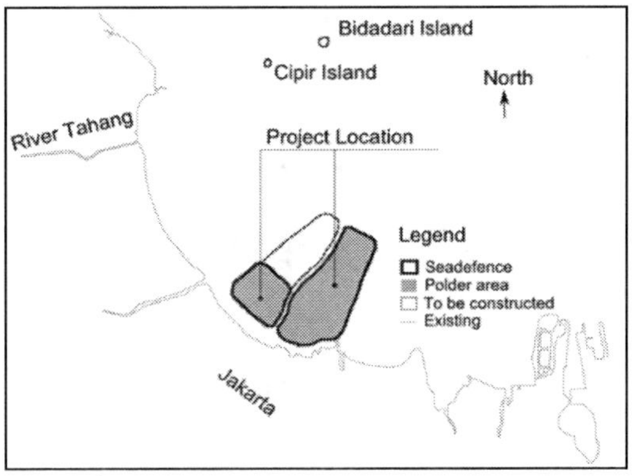

Fig. 2. Location of the two reclamation islands (Island 2A and 2B with part to be constructed in the future) for a polder system on the Jakarta Bay.

1.2 Reclamation islands

Construction of the two artificial islands on Jakarta Bay was started in 2012 and is still on-going. Their final

* Corresponding author: budiwan.tirta@witteveenbos.com

https://doi.org/10.1051/matecconf/201814707007

shape for a polder system has almost reached as shown in Figure 2. The system consists of outer sea defence (a dike for protection) and inner polder area for canal, drainage systems, pumping house and other infrastructure and buildings.

The reclamation islands on the low laying coastal area are underlay by the more than 10-metres-thick-layer of very soft to soft clays as shown in Figure 3. The soft clays are underlay clayey sand from the river mouth, and followed by tuffaceous sediments consisted of silty clay and silt ranging from soft to very stiff and hard. Tuffaceous sands with consistency of dense to very dense (cemented and none cemented) are also found in between the sediments. Table 1 provides an overview of the soil profile and properties respectively. This was derived based on results of several in situ tests consisting of boring and Standard Penetration Test (SPT) and some of the classification parameters.

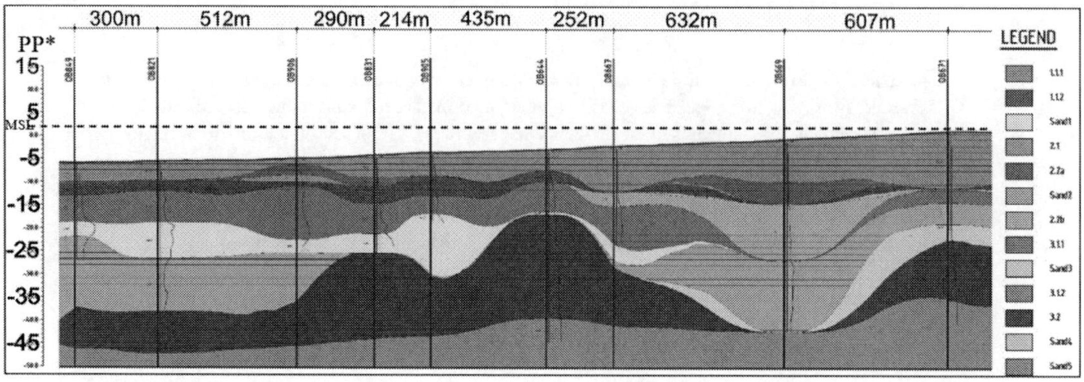

Fig. 3. A cross section showing soil layers along the north (left) and south (right) part between the two islands where the mean sea level (MSL) at +1.2 m from the project reference level, PP*.

Table 1. Soil profile and parameters with a lower characteristic value.

Unit [-]	Soil Description [-]	Unit weight [kN/m³]	Water content [%]	N-SPT blow counts
1.1.1	Silty CLAY, very soft	13.6	122	0-2
1.1.2	Silty CLAY, soft	14.0	116	0-2
Sand-1 from river mouth	Clayey SAND	17.0	45	10-30
2.1	Tuffaceous Silty CLAY, soft to medium stiff	15.5	73	2-6
2.2	Tuffaceous Silty CLAY, medium stiff to stiff	15.6	68	6-15
Sand-2 and -3 lenses between 2.2. to 3.2	Tuffaceous SAND, dense to very dense, cemented and non cemented	15.7 to 17.2	43-47	30-40
3.1.1	Tuffaceous SILT, medium stiff to stiff	15.2	61	15-30
3.1.2	Tuffaceous Silty CLAY, stiff to very stiff	15.6	59	>30
3.2	Tuffaceous Silty CLAY, hard	15.8	49	>30
Sand lense 4-5	TUFFACEOUS SAND, very dense, cemented	15.7-17.2	34-42	50-80

1.3 Fill stages and requirement

This paragraph described in general how the islands are constructed. However, the description was limited to aspects related to settlement prediction for polder area. Some definitions for this paragraph are given based on Figure 4 [6] and will be used more often for the coming chapters.

In the first stage of fill period, sand was sprayed and submerged to the bottom of the sea. In this earlier stage no soil improvement was carried out yet. Therefore, a required thickness, shape and certain waiting period were foreseen in order to avoid instability. Once, the sand was already above the water level, monitoring instrument was installed. Improvement works for the clay layers were also carried out to accelerate consolidation and gain strength in the soft clay layers during waiting period. In some area, a temporary fill excavation was carried out during the waiting period. This was due to a levelling area for accessibility road. A Prefabricated Vertical Drains (PVD) with a triangular grid and a centre-to-centre distance of 1.2 m was penetrated reaching layer with a N-SPT blow count of

15. Deep penetration of the PVD was varying from 15-20 m over the entire two islands. A settlement during construction time (termed here as surcharge) occurred, adding required total fill material during the construction time or so called bruto fill – so netto fill and the surcharge. In most of the fill area, it was also required to have additional surcharge to reach requirements with shorter waiting period. The requirements are in terms of a residual settlement after a handover time and design level. The residual settlement of 0.30 m for the polder area has been chosen for period of 50 years of service time. This is started after the handover time. Since most of the building on the island will be constructed on pile, therefore, the handover time is started in the end of construction fill time. It is expected that after removal of the applied additional surcharge, limited residual settlement might reach the requirements. After service life of 50 years, the polder is expected to be at a design level, which was determined not only regarding end settlement but also adding a predicted land subsidence in the vicinity of the islands that was assumed to be constant with a rate of 2 cm/year (so 1.0 m in 50 years). Other aspects related in determining the design level are water management and coastal engineering. However, they are out of scope of this paper, focusing only on geotechnical aspect.

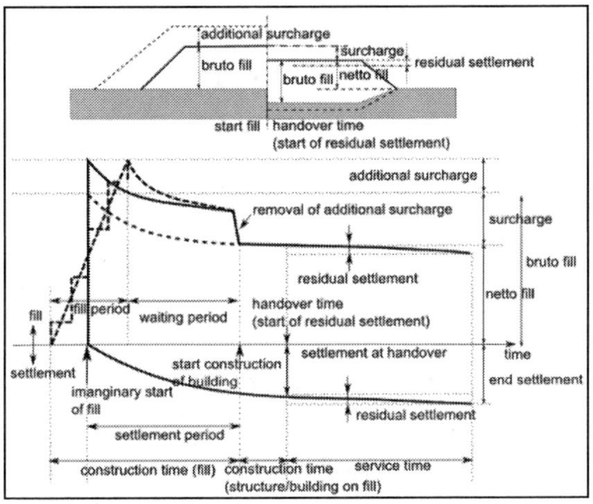

Fig. 4. Scheme (top) and terminology (bottom) fill-settlement-time diagram according to CROW 204 definition [6].

1.2 Problem definition and structure of the paper

Significant consolidation of the upper HMC with a depth of 20-50 m below the surface will occur during the placed fill material, giving the greatest portion of the total settlement. The soft clays with very low permeability imply two big consequences: very low consolidation and high creep rate. The increasing water pressure due to the fill has to be released in order to avoid instability until the end of the construction time. Very low consolidation process in the clays will be an obstacle to apply another fill material – larger interval within fill or longer waiting period per fill. This waiting period, however, is influenced with application of vertical drainage. This method in accelerates consolidation period to gain sufficient strength of the sub soil and also to reach the required residual settlement with shorter waiting period. The second geotechnical consequence is related to creep, which takes place also in the upper young deposit layers. Consolidation and creep occur also in a deeper layer (deeper than 20-50 m) due to stress increasing (deep ground water extraction). These last two components were treated as the land subsidence and were not covered within the regular settlement prediction. However, these were already considered when determining design levels of the islands, as described previously. This paper discusses all of these aspects and elaborates them with any geotechnical consequences, which might occur when dealing with the reclamation project underlain the very soft soil. Importance of an observational method through a monitoring program to reduce/avoid risk is also part of the discussion as proven to give accurate settlement prediction to approve the construction when the required residual settlement is achieved.

2 Observational method

In this project, two observational methods were carried out in terms of monitoring stations including instrumentations and also additional soil investigation and its supervision.

2.1 Monitoring stations

Monitoring station consisted of several monitoring instruments. Coupling of monitoring data gave essential insight on the behaviour of the subsoil towards the construction activity in the fill period. Instruments like, settlement plate, piezometer, stand pipe and, if necessary, inclinometer, were included in monitoring station. A settlement plate covers a grid which equals to an area of 100-by-100 square meters. In total there are almost 400 settlement plates have been placed. In most of the sea defence, piezometer and inclinometer were also installed. Whereas, some stand pipes were installed to cover evenly for both the islands. In general all the instrumentation installed and measured as soon as sand fills were above the water. The amount of settlement before the plate in placed was estimated by having CPTu. By using of this information and also the previous soil investigation (prior construction), the bottom level of sand can be determined. Moreover, the settlement can be determined by subtracting the different of the level. This was used to re-construct the data gap before the measurements become available. Earlier stage of settlement measurements was considered as valuable data to verify the parameter sets used in the design phase.

2.2 Additional soil investigation and supervision

During the design work it posed a lot of uncertainties. The soft marine clays have water content that is higher

https://doi.org/10.1051/matecconf/201814707007

than the liquid limit, indicating that the undisturbed soil samplings are difficult to guarantee. Therefore, in the earlier stage of construction, additional soil investigations with a supervision program have been planned in order to verify the soil parameters based on the previous soil investigation during the design process. The CPTu has been also used to determine the actual soil profile when sand fill was already above the water. Changing the soil profile due to fill process can effect also the settlement prediction.

3 Ground uncertainty and risk

Risk ground engineering that influences accuracy of the settlement prediction related the following major aspects: the subsoil parameters. The most critical parameters for this aspect are the compression indices and the pre-consolidation pressure. Another important aspect is consolidation which its parameter was the coefficient of consolidation. This is, actually, not related to the soil parameter but a 'process' parameter since permeability and stiffness response during consolidation [6]. Chosen consolidation and settlement model will also influence the accuracy. Other aspects were related to fill sequence, groundwater fluctuation during construction, quality of fill material (that leads to efficiency of PVD) and possible changing of the actual soil profile during filling period. All of these aspects are given in Figure 5.

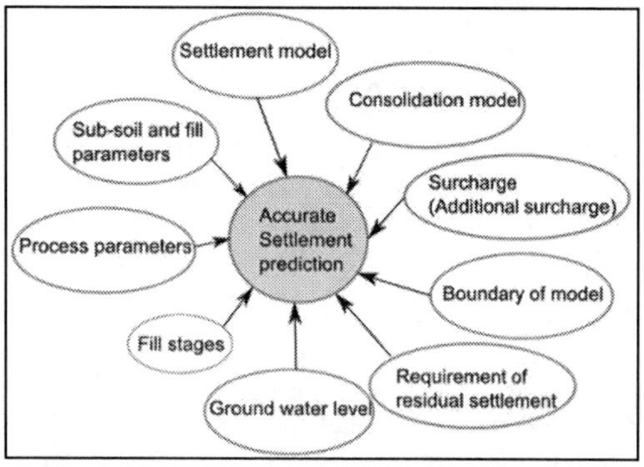

Fig. 5. Aspects related to input parameters for accurate settlement prediction.

Deviation in any settlement prediction (inaccurate) might occur due to the fact that the risk ground engineering does always exist. The risk is associated to uncertainty as an absence of information about parts of a system under consideration, which is always present [7]. This uncertainty was embedded in input parameters for settlement prediction. For each aspect of settlement prediction, risk was defined. Consequence of risk associated to settlement prediction is then briefly described in the following sub-chapters. Advantage of having observational approach is also elaborated here to reduce or even avoid the risks.

3.1 Calculation models

3.1.1 Settlement model

The NEN-Bjerrum isotache model supports the common linear strain parameters C_r, C_c and C_α. The reloading/swelling index C_r describes the elastic stiffness during unloading and reloading (below preconsolidation pressure). The primary compression index C_c and the coefficient of secondary compression C_α describe respectively the idealized elasto-plastic deformation and the viscous creep rate during virgin loading. These parameters can be determined from the oedometer tests which was previously described in Dutch Norms NEN 5118 [8] and 6740 [9]. The interrelation ship with Bjerrum (1967) is the reason behind the name of the model. Den Haan [10] has developed the full mathematical formulation by application of the small strain limits derived from the a,b,c isotache natural strain model. The parameters C_r and C_c are in fact related to changes in void ratio: becoming

$$RR = C_r / (1+e_0) \quad (1)$$

$$CR = C_c / (1+e_0) \quad (2)$$

respectively. Cα is however directly related to changes in linear strain. Please note that this definition of the Cα complies with common practice, but differs from the original definition by Mesri [11].The shared isotache formulation implies that all inelastic compression results from visco-plastic creep. The NEN-Bjerrum model, therefore, assumes that creep rate will reduce with increasing overconsolidation and that overconsolidation will grow by unloading and by ageing. Bjerrum's name is attached to this model, because he was the first to notice that creep rate depends on both overconsolidation ratio and age. The NEN-Bjerrum model is suited for cases with un- and reloading, by using a rate-type visco-plastic isotache formulation (all plastic compression results from creep). This model solves the problem of any simplified interpolation model, such as Koppejan [12], that tends to underestimate the creep time considering a complex fill/load stage that involving also re-loading [13].

The creep effect is taken into account in the settlement prediction when the stress is higher than the pre-consolidation stress [13]. In Figure 6 a typical result of settlement prediction with ($C_\alpha \neq 0$) and without the effect of the creep ($C_\alpha = 0$) in a 5 m thick of layer 1.1.1 with the NEN-Bjerrum settlement model. For the sake of simplicity no PVD effect has been taken into account in the models. As can be seen from both graphs in Figure 6, the creep already occurs at low stress level, and significantly starts at date around 100 days.

https://doi.org/10.1051/matecconf/201814707007

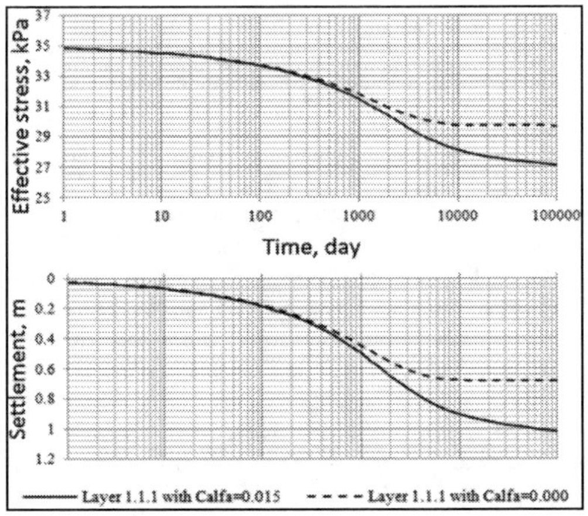

Fig. 6. Creep effect in NEN-Bjerrum settlement model, a simplified case with a 5 m thickness of layer 1.1.1 underlay none-compressible sand layer.

3.1.2 Consolidation model

Due to a surcharge, excess water in cohesive soil will occur. Consolidation is estimated by how much water flows out from the soil. Without any PVD the excess water will only flows out vertically. By introducing PVD, the water will also flows out horizontally through strips. In the latter case, consolidation will accelerate and faster than the former case. In the classic Terzaghi [14] method, effect of the PVD is estimated by adjusting coefficient of vertical consolidation. Barron [15] solved the former solution to estimate the effect in radial direction. Carillo [16] combined both solution by Terzaghi and Barron and. The method is so called Terzaghi-Barron-Carillo method and has been widely applied in the practice. Another alternative is to consider Darcy method. The main different between both methods is that Terzaghi method does not take into account effect of creep in the consolidation process, in the other hand, Darcy method does take into account creep effect in the consolidation. However, both methods will be equal considering a case with a single soil layer with a constant stiffness and ground water table where no creep is applied.

The used approach is combined use of NEN-Bjerrum for settlement model with a linear rek and Darcy solution for consolidation model for the case with PVD and additional (temporary) surcharge. These models are available in the commercial D-series program version 14.1 developed by Deltares. The settlement prediction is estimated by means of the software.

3.2 Soil and process parameters

3.2.1 Compressibility indices and process parameter

Ground sampling in very soft soil always is difficult in a good way and always results in sample disturbance. Sampling breaks the contact between soil particles causing stress reduction [10]. Therefore, 'undisturbed samples' often mentioned simply do not exist. These give significant contribution of risk when these samples are tested, then to derive parameters.

The soil parameters in the design phase were based on soil data obtained in several soil surveys consisting of boreholes, SPT's and laboratory tests. During monitoring settlement, it was observed that the compressibility derived from the laboratory test was found to be higher than it was expected. Based on the investigation it can be concluded that the classification tests, description of soil profile and units (Table 1) were consistent based on such quite simple tests such as SPT, bore-logging, Atterberg limits. However, for more complex parameters such as compressibility, it was not the case. As shown in the Figure 7, a widely spread of the results of *CR* in relation to the water content was observed giving a very big spreading with a margin of up to 30%, which was unexpected.

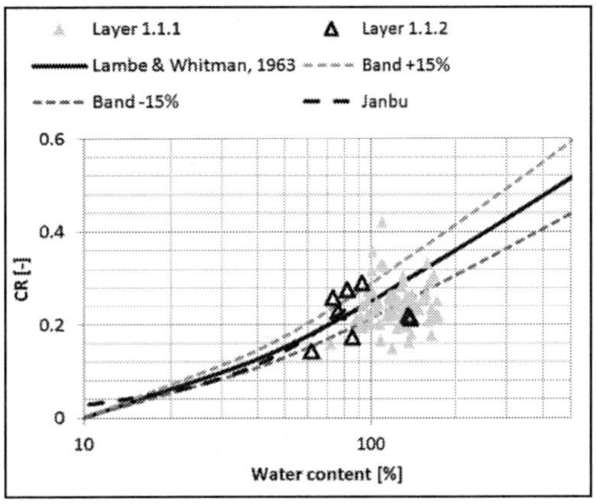

Fig. 7. Correlation of *CR* based on the result of several consolidation tests of layer 1.1.1 and 1.1.2 and its corresponding water content.

As consequence, the soil behaviour based on the soil parameter from the consolidation tests differed significantly from the prediction, as can be seen in Figure 8. During further investigation, this was found out due to in proper handling of the soft soil samples for such a long period causing significant disturbance in the sampling before consolidation tests.

3.2.2 Earlier stage of fitting

During the earlier stage of fitting, the parameter were updated accordingly based on correlation such as Lambe & Whitman [17] and Janbu [18] to determine the compressibility parameters *CR*, *RR* and C_α with index properties (water content and Atterberg limits) including Liquid Limit (*LL*), Plastic Limit (*PL*) and Plasticity Index (*PI*). The ratio of *CR*/*RR*=6.0 was considered for all layers. For the update of parameters, the typical ratio for soft clay and stiff clay is *CR*/*RR*=5.0 and 8.0 respectively. Similar procedure in determining C_α, it was used the typical ratio of C_α/CR from 0.06 for soft clay until 0.02 for stiff clay. The update of the parameters

 https://doi.org/10.1051/matecconf/201814707007

was including also the process parameters with a constant C_h/C_v=1.5. The C_v value for the first two layers was found to be varied and, therefore, divided into three zones covering the north, centre and south part of Island 2A. The pre-overbourden pressure (*POP*) of the soils was reduced for layer 1.1.1 and 1.1.2 with respect to the result of the laboratory test, while the other layers are not. Table 2 provides the complete parameters after adjustment based on the earlier stage of fitting.

Table 2 Adjusted NEN-Bjerrum compressibility parameters after an earlier stage of settlement-fits.

Unit [-]	*CR* [-]	*RR* [-]	C_α [-]	*POP* [kPa]	C_v [m²/s]	C_h/C_v [-]
1.1.1	0.26	0.050	0.015	5.0	2.0-6.0 10^{-8}	1.5
1.1.2	0.25	0.050	0.014	5.0	4.0-8.0 10^{-8}	1.5
2.1	0.20	0.035	0.008	41.5	5.0 10^{-7}	1.5
2.2	0.19	0.030	0.007	43.5	5.0 10^{-7}	1.5
3.1.1	0.18	0.025	0.006	63.3	2.34 10^{-6}	1.5
3.1.2	0.18	0.025	0.006	68.3	1.89 10^{-6}	1.5
3.2	0.16	0.020	0.005	90.4	2.82 10^{-6}	1.5

These updated parameters provided a good match with the field results as shown in Figure 8 and were used also for the fitting process of Island 2B as well.

3.2.3 Next stage of fitting

More data has been available in Island 2A after the earlier stage of the fitting process that was used for validating the soil and process parameters. At least two CPTu data were also available to verify the amount of settlement before the monitoring data available. The last CPTu data was to check the recent soil profile as this more relevant for long term prediction. Based on the fitting, predicted settlement were in-line with measured settlement, and therefore, were still consistent as shown in Figure 9. After the removal of the additional surcharge, the residual settlement meets the requirement in 50 years.

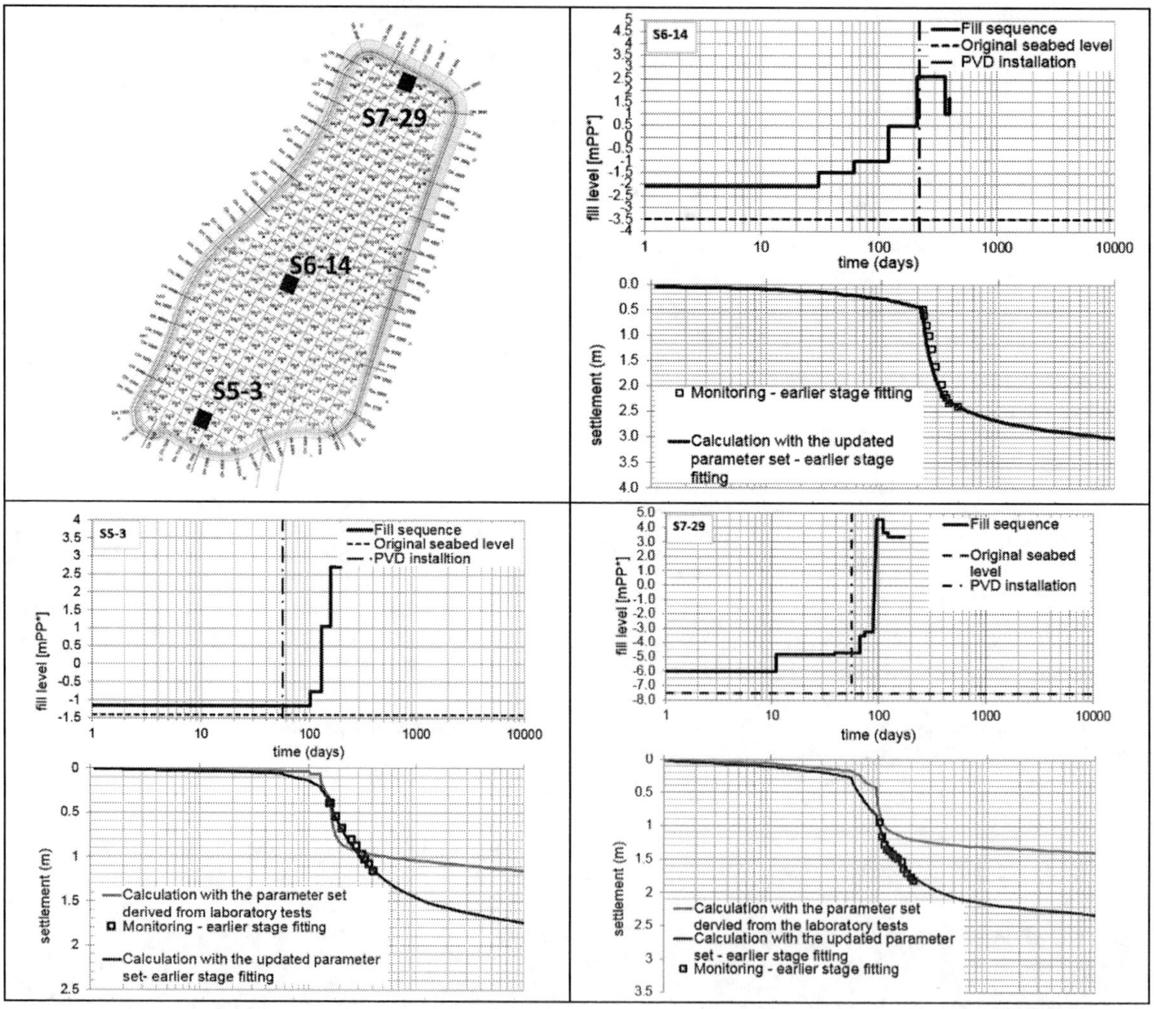

Fig. 8. Typical earlier stage of settlement fits in the north, middle and south part of Island 2A with the design and adjusted soil parameters.

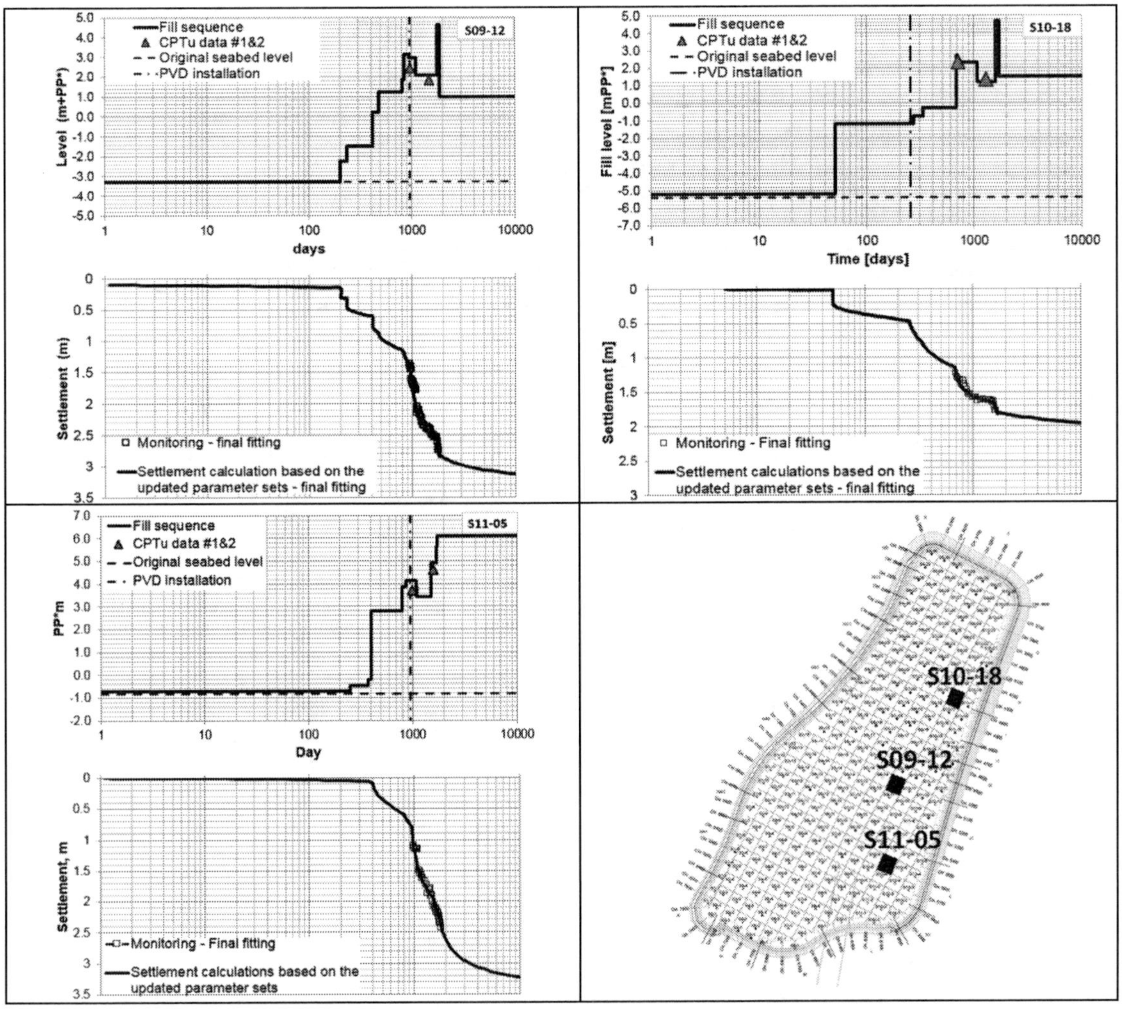

Fig. 9. Typical settlement fit Island 2A in the final phase of the fill stage - with the adjusted soil parameters and location of the measurement.

3.3 Other aspects

During the next stage of fitting in Island 2B, it was found that there were deviation between the prediction and measurement, although the parameter sets were already adjusted. There were two important factors that caused the deviation. One of the aspects was the quality of the fill material that might affect the consolidation process. This related to quite high fine content embedded in the sand fill, which was found to be higher than 15% giving low friction with high friction ratio in comparison to those of typically clean sand. The second aspect was due to the temporary water pumping during compaction works in a quite long period during the earlier stage of PVD installation. Based on the example of the fittings shown in Figure 10, an adjustment of process parameter C_v is necessary to have good fit. In addition, fluctuation of ground water level during the period of temporary pumping was also taken into account to provide better results. Having pore pressure measurement into the fitting made this possible. Based on the results of the fitting, the consolidation occurs slowly than expected and leads to quite high residual settlement, with an adjustment of total C_v with factor of 1.2-3.0. The settlement period of Island 2B is still on-going before the end of the construction time or starting of the hand over. However, additional surcharge can still be planned with certain waiting period in order to achieve the required residual settlement; otherwise longer waiting period can also be an option to achieve depending on the construction planning of the island.

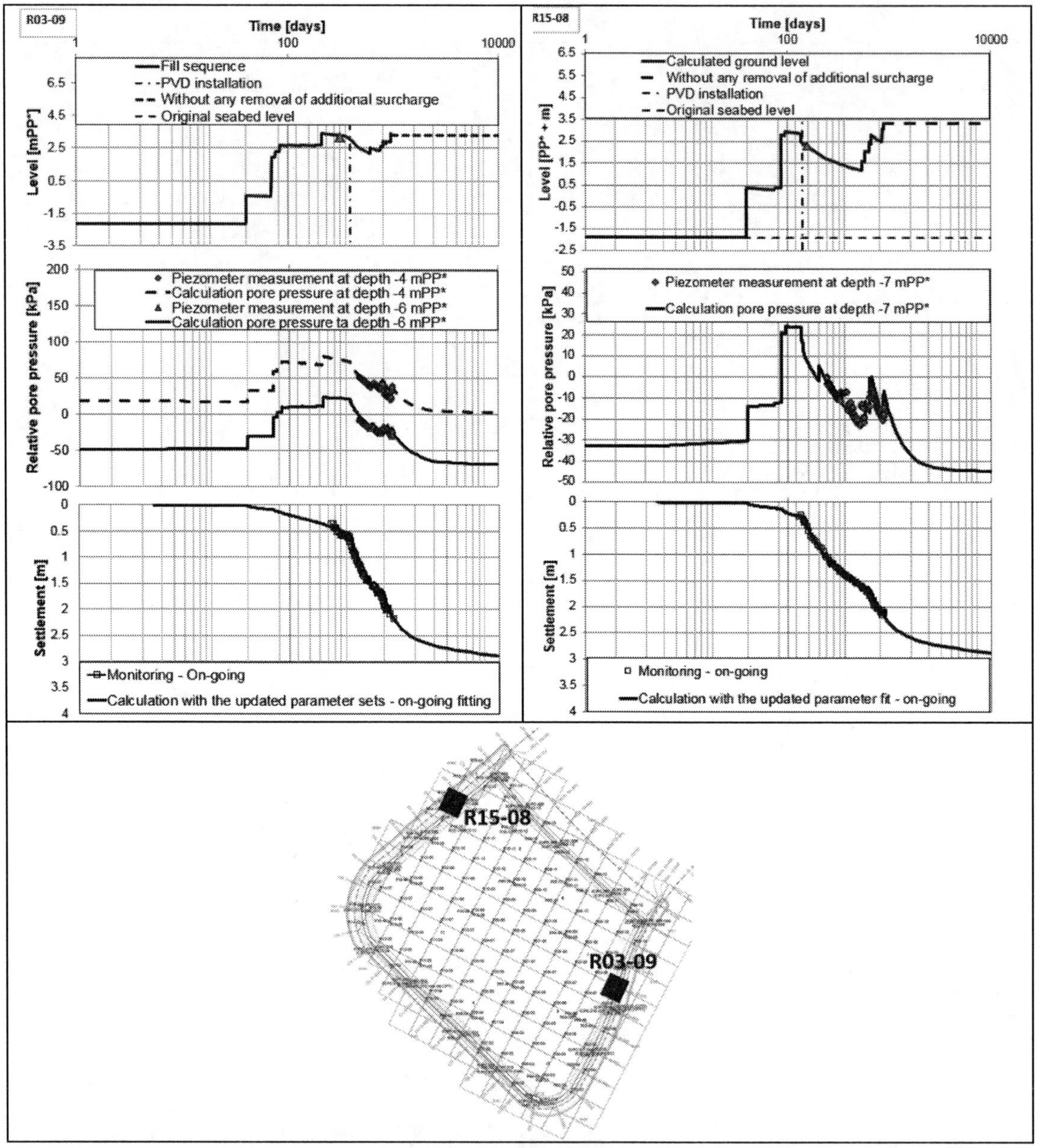

Fig. 10. Typical settlement and pore pressure fits Island 2B during the phase of the fill - with the adjusted soil parameters and further adjustment on C_v and water level fluctuation and location of the measurement.

4 Conclusion and recommendation

Risk associated to ground engineering, in this particular case related to settlement prediction was pre-defined prior to the start-up of the reclamation islands on Jakarta Bay. Each aspects of the risk were registered and were anticipated within the project phases: from the earlier stage towards the completion of the projects through the monitoring programs. The observational program, which was part of the construction phase, has been proved to be a powerful approach to calibrate the sub-soil soils parameter and the process parameters related to the PVD application. These two parameters were the main source of uncertainty in the settlement prediction. With proper settlement and consolidation models dealing with such complex loading history, the approach resulted in more reliable parameters that lead to better settlement prediction. The settlement results were not only successively predicted the immediate settlement due to fill with very complex loading history but also the effect of creep as the used method consider it. The application of the additional surcharge along with the PVD with certain waiting period was able to limit the residual settlement. This was also the good measure to successively achieve the strict residual settlement requirement of 30 cm within 50 years design lift time for the polder islands. Other important aspects were related to the quality of fill material and ground water fluctuation that were affecting the consolidation process. As the settlement period is still on-going measure can still be proposed based on the actual condition based on the fitting.

Although, the full set of parameters were already validated with the settlement measurements including the creep effect, however, a long-term measurement has been placed in Island 2A (will be also followed in Island 2B) to closely monitor the creep. In this case, settlement plates have been placed in the area where the surcharge already removed to the desired hand over level. Natural rek models such as *a-b-c* isotache or finite element models, such as Soft Soil Creep model, could be also an approach for settlement model considering higher rek cases. These two subjects are not covered yet within this paper, and will be kept for the future research.

Nomenclature

a, b	=	Stiffness parameters of soil
c	=	Creep parameter of soil
C_c	=	Primary compression index
C_h	=	Horizontal consolidation coefficient
C_r	=	Elastic stiffness during unloading and reloading (below preconsolidation pressure)
CR	=	Compression ratio
CPTu	=	Cone Penetration Test with pore pressure (*u*) measurement
C_v	=	Vertical consolidation coefficient
C_a	=	Secondary compression index. Similar to the common Ca.
e_0	=	Initial void ratio
HMC	=	Holocene Marine Clay
LL	=	Liquid Limit
NSPT	=	Normalized Standard Penetration Test
SPT	=	Standard Penetration Test
PI	=	Plasticity Index
POP	=	Pre Overburden Pressure
PL	=	Plastic Limit
PVD	=	Pre-fabricated Drains
RR	=	Reloading/swelling ratio
SPT	=	Standard Penetration Test

References

1. T.M., Williams, J. Rees, & D. Setiapermana, *Land-derived contaminant influx to Jakarta Bay, Indonesia. Volume 1: Geochemestry of marine water and sediment*, Technical Report WC/97/19 Overseas Geology Series, Natural Environment Research Council (1997)
2. R.w. van Bemmelen, *The geology of Indonesia - General Geology of Indonesia and Adjacent Archipelagoes*, ***IA***, Government Printing Office (1949)
3. C.R. Meinardi, *Hydrogeological considerations with regard to drinking water supply and land improvement in Desa Totoran (Kab. Indramayu)*, OTA-33 **9** (year?)
4. J.H.J. Terwijndt , P.G.E.F. Augustinus, J. R. Boermsa, and P. Hoekstra, *Mud-discharge, dispersion, and deposition in a mansoon-dominated coastal environment*, Proceedings International Conference Coastal Sediment, Coastal Sediments **87** (1987)
5. M. Dobbie, *Site investigation of the Holocene Marine Clay of Southeast Asia using the CPT*, Proceeding of Soft Soils (2014)
6. CROW **204**, *"Betrouwbaarheid van zettingsprognoses"*, in Dutch, (2004)
7. Smallman, *Crisis and Risk Management*, NIMBAS University (2000)
8. NEN **5118**, *"Bepaling een-dimensionel samendrukkingseigenschappen van de grond"*, in Dutch (1991)
9. NEN **6740**, *"Basis eisen en belastingen"* ,in Dutch (2006)
10. E.J. den Haan, J.B. Sellmeijer, *Calculation of Soft Ground Settlement with an Isotache Model*, Proceedings Soft Ground Technology Conference, ASCE Geotechnical nr. **112**, 94-104 (2000)
11. M.A. Mesri, *Coefficient of Secondary Compression*, Journal of Soil Mechanics and Foundations Division, **277**, pages 123–137 (1973)
12. W.A. Koppejan, *A formula combining the Terzaghi load compression relationship and the Buisman secular time effect*, Proc. 2nd Int. Conf. Soil Mech. and Fnd. Eng., **272**, 32–37 (1948)
13. E.J. den Haan, *"Het a,b,c-isotachenmodel: hoeksteen van een nieuwe aanpak voor zettingsberekeningen"*, Geotechniek (**4**), p 28-35, in Dutch (2003)
14. K. Terzaghi, & R.B. Peck, *Soil Mechanics in Engineering Practice* (1967)
15. R.A. Barron, *Consolidation of fine-grained soils by drainwells*, Trans ASCE **113**, 718-742 (1948)
16. N. Carillo, *Simple two and three-dimensional cases in the theory of consolidation of soils*, Journal of Math. Phys. **21**, 1-5 (1942)
17. T. Lambe and R.V. Whitman, *Soil Mechanics*, John Willey and Sons, 553 p (1969)
18. N. Janbu, *Soil Compressibility as Determined by Oedometer & Triaxial Tests*, Proceedings, 3rd European Conference on Soil Mech. And Found. Eng., Vol. **1**, p 19-25 (1963)

Insight look the subsidence impact to infrastructures in Jakarta and Semarang area; Key for adaptation and mitigation

Heri Andreas[1,*], *Hasanuddin* Zainal Abidin[1], *Dhota* Pradipta[1], *Dina* Anggreni Sarsito[1] and *Irwan* Gumilar[1]

[1]Geodesy Research Division, Faculty of Earth Science and Technology, Institute of Technology Bandung, Indonesia

Abstract. Land subsidence is not a new phenomenon for Jakarta and Semarang areas. According to some publications, the value of Jakarta's subsidence may reach 2-4 meter in certain place. Meanwhile, more than 2 meter of subsidence is taking place in northern part of Semarang. Some techniques are being used to derive land subsidence information in these areas such as repeated leveling measurements, GPS surveys, and InSAR measurements. The impact of land subsidence in Jakarta and Semarang could be seen in several ways, such as sea inundation ("Rob" in Javanese), problem on infrastructures, the wider expansion of flooding areas, etc. Cracking on housing, street, "sinking" on the bridges and dyke, problems on drainage are examples of infrastructures problems due to the land subsidence. Huge costs have already been spent to fixing those infrastructures problems both in Jakarta and Semarang. Since mostly liniear pattern of land subsidence are recognized in both area today and probably in the next years to come, in this case the impact especially on infrastructures are probably will getting worse. Insight look the subsidence impact to the infrastuctures in Jakarta and Semarang area would raisen the concern, and become key for adaptation and mitigation.

1 Introduction

Jakarta is the capital city of Indonesia, inhabiting an area of about 662 km^2 on Java Island with a population of about 12 million people (fig.1). Geographically Jakarta located at 6^0 10' South Latitude and 106^0 49' East Longitude. Geologically Jakarta is a low land flood basin area with ticked sediment to about 400 meter depth. In the last four decades, urban development of Jakarta has grown very rapidly in the sectors of industry, trade, transportation, housing, and many others. In accordingly land subsidence is quite well known phenomenon for Jakarta. The occurrence of land subsidence was recognized at least in the early the development of the city. The evidence for subsidence was based on repeated leveling measurements, GPS (Global Positioning System) surveys, InSAR (Interferometric Synthetic Aperture Radar) measurements, extensometer, etc. According to some publications [1-4] the yearly value of Jakarta's subsidence generally ranging from 1 to10 centimeter per-year and may reach 20-26 centimeter in certain place, especially in northern part of Jakarta for the recent years.

Semarang is the capital city of Central Java province, located in the northern coast of Java island, Indonesia (fig.1). It is centered at the coordinates of about 6° 58' South Latitude and 110° 25' East Longitude, and covers an area of about 37,367 hectares or 374 km2, with the population of about 1.55 million people in 2010. Topographically, Semarang consists of two major landscapes, namely lowlands and coastal areas in the north and hilly regions in the south. Geologically, Semarang has three main lithologies, namely, volcanic rock, sedimentary rock, and quite ticked alluvial deposits. Land subsidence is also quite well known phenomenon for Semarang. Increases in the population and urban development in the area have accelerated land subsidence through excessive groundwater extraction, and load from buildings and structures. Based on the leveling surveys conducted by the Centre of Environmental Geology from 1999 to 2003 it was found that relatively large subsidence was detected around Semarang Harbor, Pondok Hasanuddin, Bandar Harjo and around Semarang Tawang Railway station, with the rates ranging from 1 to 17 cm/year. Results derived from GPS show that land subsidence in Semarang has spatial and temporal variations. In general, subsidence rates in Semarang have an average rate of about 6 to 7 cm/year, with maximum rates that can go up to 14-19 cm/year at certain locations. The estimation based on the PS (Permanent Scattered) InSAR technique also revealed that the areas close to the shoreline have subsidence rates of more than 8 cm/year [5]. The result is based on the PS InSAR based velocity data derived from 28 ERS-2 and ENVISAT-ASAR radar scenes recorded between 27 November 2002 and 23 August 2006. In another study [6], after processing 22 ascending ALOS-SAR images during January 2007 to January 2009 plus 2 descending SAR images acquired on 6 June 2006 and 17 June 2007, it found maximum subsidence rates of about 8 cm/year in the northern region of Semarang. They are high magnitude of subsidence rate. Surprisingly for the recent

* Corresponding author: heri@gd.itb.ac.id

years it reported that the trend of subsidence is acceleration of instead of linier trend in the nortern eastern area of Semarang City. The area is coinsize with factory areas.

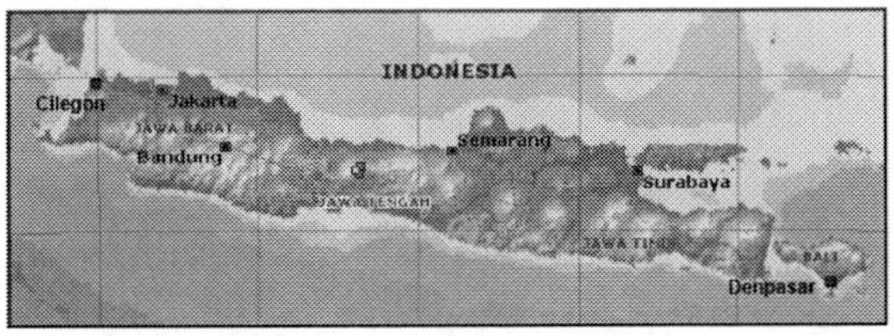

Fig. 1. Geographically location of Jakarta and Semarang city in the Island of Java Indonesia

The impact of land subsidence in Jakarta and Semarang can be seen in several forms, such as sea inundation ("Rob" in Javanese), problem on infrastructures, the wider expansion of flooding areas, etc. Cracking on the housing, building, street, "sinking" on the bridges and dyke, problem on the drainage are examples of infrastructures problem due to the land subsidence. The subsidence it also badly influences the quality and amenity of the living environment and life (e.g. health and sanitation condition) in the affected areas. The economic losses caused by land subsidence in Jakarta and Semarang are enormous; since many buildings and infrastructure severely affected by land subsidence and its collateral coastal flooding disasters are located in the dense housing, industrial zone, and vital infrastructures.

Since mostly liniear pattern of land subsidence are recognized with quite significant rate per year in both area today and probably in the next years to come, in this case the impact especially on infrastructures are probably will getting worse a head too. Certainly, it would lead to more disaster and costs in the near future and absolutely need to be concern. Insight look the subsidence impact to the infrastuctures in Jakarta and Semarang area would raisen the concern, and it is also become key for adaptation and mitigation.

2 Methods

The places with the largest subsidence value, and the boundary condition between the largest and the less (location of differential subsidence), the depth variation on pile foundation, either in Jakarta or Semarang city is prone to the impacts. Cracking on the housing, building, street, "sinking" on the bridges and dyke, problem on the drainage, etc. likely can be seen in these areas. In this case, as a method, investigating the impact to infrastructures in both areas would be focused within these areas. Figure 2 shows map of land subsidence in Jakarta where we can see the largest magnitude to the smallest subsidence in the area over periode 1925 to 1915, while figure 3 shows map of land subsidence for Semarang area for around 8 years measurement (2008-2016). Table 1 and 2 notes the rate and the magnitude of land subsidence each in Jakarta and Semarang area, to give the illustration how large and how fast the subsidence exists. The value of Jakarta's subsidence reached 2-4 meter in certain place from 1925 to 2015, while around 1 meter is taking place in northern part of Semarang from 2008 to 2016. Both of these subsidence will probaly continuing in the quite some times.

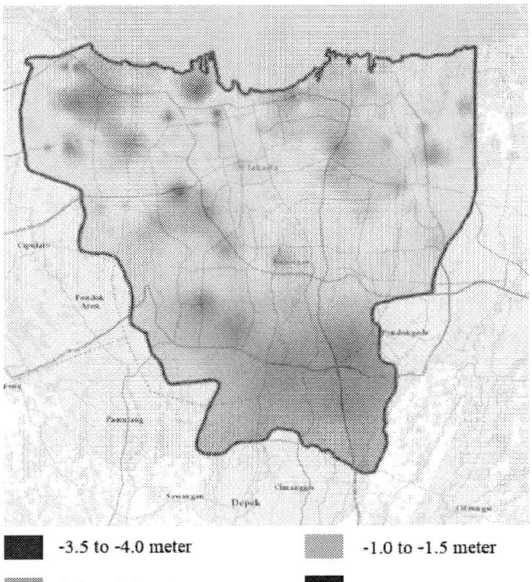

Fig. 2. Map of land subsidence in Jakarta from 1925 to 2015 base on integration of leveling measurements data, GPS surveys, and InSAR.

Table 1. Summary of rate and magnitude of land subsidence in Jakarta area.

No	Segment	Avg. Rate from 1925-1975 (m/years)	Avg. Rate from 1975-2015 (m/years)	Max. Magnitude 1925-2015 (meter)
1	Northern Part	-0.003	-0.10	-4.75
2	Western Part	-0.003	-0.08	-4.36
3	Eastern Part	-0.003	-0.07	-3.12
4	Central Part	-0.002	-0.02	-0.98
5	Southern part	-0.001	-0.01	-0.58

Table 2. Summary of rate and magnitude of land subsidence in Semarang area

No	Segment	Avg. Rate from 2008-2016 (m/years)	Max. Magnitude 2008-2016 (meter)
1	Northern Part	-0.10	-1.00
2	Western Part	-0.05	-0.45

No	Segment	Avg. Rate from 2008-2016 (m/years)	Max. Magnitude 2008-2016 (meter)
3	Eastern Part	-0.06	-0.48
4	Southern Part	-0.01	-0.10

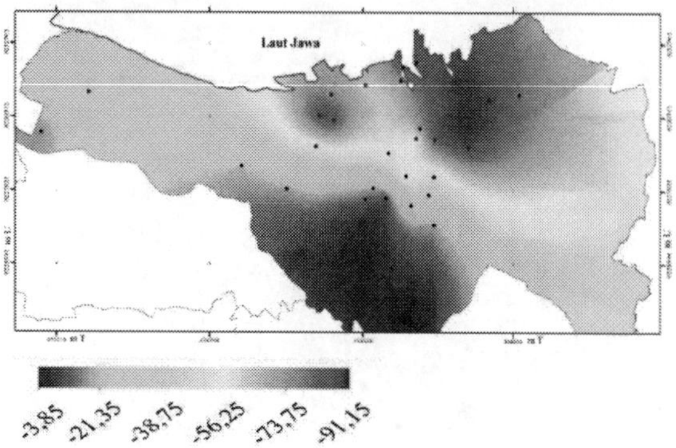

Fig. 3. GPS derived subsidence in Semarang in the periods of 8 year measurement (2008-2016) in centimeter

Since both Jakarta and Semarang area located in the coastal area, in this case chance of inundated infrastructures is big. So, as another method, investigating the impact to infrastructures in Jakarta and Semarang would be focused also within these areas. The inundation comes from the impact of land sinking due subsidence and the sea level rise. Result from satellite altimetry and the tide gauge observations shows the sea level rise at the Java Sea (including north of Jakarta bay and Semarang bay) is around 6 to 10 millimeter per years [7]

3 Results

Below we can see the result from insight look the subsidence impact to infrastructures in Jakarta and Semarang area. We divided the result into several sub chapter each explain about insight look of inundated infrastructures, cracks on building and houses, insight look on the bridges and drainage, on the roads, on dykes, and insight look on the others affected infrastructures. These all results can become important keys for adaptation and mitigation [8]

3.1 Insight looks the inundated infrastructures

Areas at northern part of Jakarta such as Kamal Muara, Pantai Indah Kapuk, Muara Angke, Muara Karang, Muara Baru, Ancol, Tanjung Priok and Marunda are prone to sea inundation or "Rob". On the high tide some area can be flooded up to one meter. Some places even now have permanently inundated. In northern of Semarang, place likes Pantai Marina, Tanjung Mas, Tambak Loro, Kaligawe, etc., the "Rob" situation are even worse than Jakarta. The inundation is literally destroying the infrastructures. Slowly but sure the inundation it's like a silent killer to the infrastructures.

Figure 4 shows pictures of inundated infrastructure in Jakarta, while figure 5 shows pictures in Semarang. Figure 6 shows abandon houses due to the most effect that already existed.

Fig. 4. Some pictures of inundated infrastructures in Jakarta (taken in year 2010). The inundation is slowly damaging the infrastructures if no action of adaptation or mitigation

Fig. 5. Some pictures of inundated infrastructures in Semarang (taken in year 2016). The inundation is slowly damaging the infrastructures if no action of adaptation or mitigation

Fig. 6. The silent killer has already given most of the effect. Some houses and buildings in Semarang have already been abandoned (pictures taken in year 2016)

3.2 Insight looks the cracks on buildings and houses

In the northern part of Jakarta and Semarang, especially around boundary area of the larger and the less subsidence value, and the shallow and the deep infrastructures where they meet, the cracks on building and housing are appearing. Figure 7 shows pictures of cracks on the building and houses in Jakarta, while figure 8 shows pictures in Semarang. Many have suffering and fixing the buildings and the houses due to the impact.

Fig. 7. Cracks on the buildings and houses around northern part of Jakarta (pictures taken in 2009). Many have suffering and fixing from the subsidence impact

Fig. 8. Cracks on the buildings and houses around northern part of Semarang (pictures taken in 2012). Many have suffering and fixing from the subsidence impact

3.3 Insight looks on bridges and drainage

In the first initiation, the bridge level is design around 1 to 1.5 meter or even more above the water level so the water would easily pass through and flowing. Nevertheless, the subsidence with rates for about 10-20 centimeter per year has made bridge is slowly "touching" the water. In the end, the flow will be disturbed, and even the water will pass the bridge's floor. Figure 9 shows the "sinking" bridge in Jakarta, while figure 10 shows similar situation in Semarang. Many have suffering and fixing the bridges due to the subsidence impact and the costs are quite expensive.

In Jakarta area at least there are ten bridges which in the near future might be slowly destroying by the water underneath. Even some of them will pass by the water regularly on the high tide. Two bridges are located in Kamal Muara, one in Muara Baru, several bridges around Ancol and Martadinata, several bridges located around Gunung Sahari. We have seen also bridges that newly elevated which are one located in Muara Angke, and one in Muara Karang. Base on the news, cost for elevated the bridges is around 1 million US dollar each. We can imagine if in the future more bridges need to be elevated, in this case many dollar should spent.

In Semarang area at least there are five bridges which in the near future might be slowly destroying by the water underneath. Even some of them are passed by the water permanently now (fig. 10). This situation is very danger for the people whom use the bridge, especially during the rainy season when the water level even higher. Two bridges are located in Raden Patah, one in Tanjung Mas, several bridges around Kaligawe. We expected around five years from now the government should spend at least 5 million US dollar for fixing the bridges.

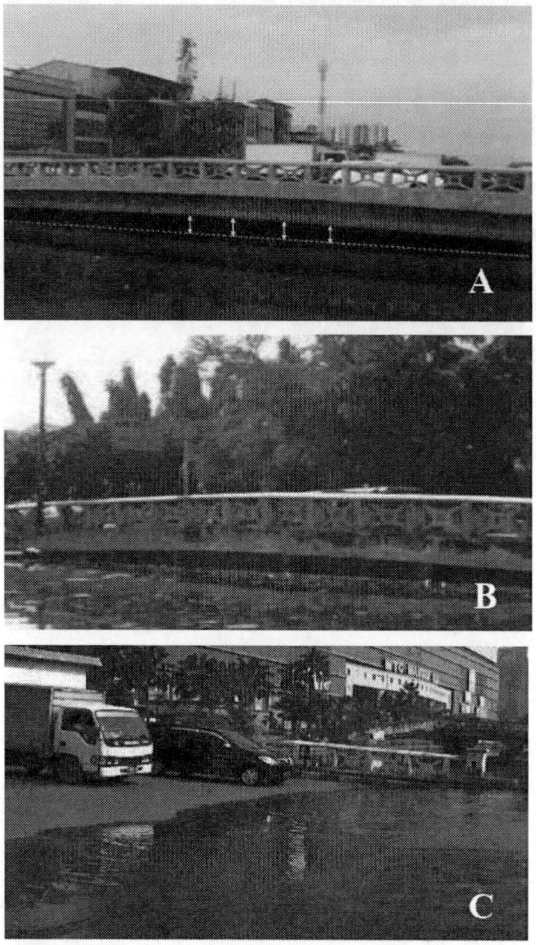

Fig. 9. Illustration of "sinking" bridge in Jakarta. Figure A, B, C shows the bridge is slowly touching the water, and finally the water can pass the bridge

Fig. 10. Illustration of "sinking" bridge in Semarang area. We can see the bridge is passed by the water. The bridge can be flooded even permanently.

Land subsidence can be disturbed the drainage system as well. Due to differential subsidence, water can stop flowing or reverse or even has to pump out to the higher level of others drainage or canal and river. Figure 11 shows pictures of dysfunction of drainage system in Jakarta and Semarang. The situation is problem for water management, it can be increase the flood potential, etc.

Fig. 11. Dysfunction of drainage system in Jakarta and Semarang. Water is stop flowing or reverse or even has to pump out to higher level of other drainage or canal and river

3.4 Insight looks on roads

Due to the impact from subsidence, many roads have been damage both in Jakarta and Semarang. The differential subsidence might wrinkle the flatter of the road. Figure 12 shows picture of wrinkle road in Jakarta and Semarang. The roads maybe also destroy by heavy weight from the load, and frequently inundated by the water. Many have suffering and fixing the road in Jakarta and Semarang due to the subsidence impact. Some even need to elevate for even a meters thickness to avoid flooding. The costs are remarkably expensive. These situations will continue if the subsidence is continuing through times.

In Jakarta area at least we identify more than 20 kilometer of road are suffering the impact from subsidence, while in Semarang area also more than 20 kilometer as well. Both mostly located near the coastal area. Base on the news, cost for fixing the road is around 1 million US dollar for ten kilometer. We can calculate the length of road need to be repair and how many dollars should spend.

 https://doi.org/10.1051/matecconf/201814708001

Fig. 12. Impact of subsidence to the road in Jakarta and Semarang. The road is wrinkling and damaging. Many have suffering and fixing. The costs are remarkably expensive.

3.5 Insight looks on dykes

Dykes are supposed to be protecting the land from "Rob", but they are sinking. Figure 13 shows picture of 1 meter of dyke which was failed to protect Jakarta from "Rob" in 2007, and after raised 1 meter more in 2008, it is slowly sinking and about to failed again to protect Jakarta after 2013. For the recent years indeed it has been elevated again for more than 1 meter. Figure 14 shows situation in Semarang where dyke is failed to protect the land from inundation. So, there are interesting fact that dyke is considered not a final solution for subsidence mitigation.

Fig. 13. Illustration of dyke "sinking" in northern part of Jakarta. In 2007, around 1 meter of dyke failed to protect the land, while 2008 another 1 meter elevated dyke would soon after 2013 would probably failed again.

Fig. 14. Situations where "sinking" dykes failed to protect northern area of Semarang from inundation or "Rob"

In table 3 below, we try to calculate the sinking dyke around Jakarta and Semarang. We can see the rate and the future magnitude of sinking within 10 to 20 years. In this case, we can see the longevity of dyke protection in both areas. For the information, Jakarta now has a "giant sea wall" program on protecting the city from the subsidence and inundation. But, if the sea wall is subsiding and sinking, then how effective the program is an important question to be answered.

Table 3. Rate and future magnitude of dykes sinking in Jakarta and Semarang

No	Segment	Rate Sinking (m/years)	Magnitude after 10 year (meter)	Magnitude after 20 year (meter)
1	Pantai Mutiara Jakarta	0.10	1.0	2.0
2	Muara Baru Jakarta	0.12	1.2	2.4
3	Ancol Jakarta	0.05	0.5	1.0
4	Tanjung Mas Semarang	0.08	0.8	1.6
5	Tambak Loro Semarang	0.10	1.0	2.0
6	Kaligawe Semarang	0.12	1.2	2.4

3.6 Insight looks on others affected infrastructures

There are still many infrastructures type that can be impacted or affected by the subsidence instead of some that have already explained above. The piping for instance can be impacted by the subsidence, and even the breakwater that is supposed to protect beach from abrasion can be also impacted. Figure 15 shows Google and pictures of sinking breakwaters in Pantai Ancol north of Jakarta. In 2004 around 1 meter height of breakwaters are manage to protect the beach from abrasion, while in 2010 and recent years the breakwaters are drowning to the sea.

Fig. 15. The sinking of breakwater in Pantai Ancol Jakarta. In 2004 around 1 meter height of breakwaters manage to protect the beach from abrasion, while in 2010 and the recent years they are suffering drowning to the sea.

3.7 Map of the impacts

We have plotted on map many of the impacts of the land subsidence to the infrastructures in Jakarta and Semarang (see figure 16 and 17). We can see that areas at northern part of Jakarta such as Kamal Muara, Pantai Indah Kapuk, Muara Angke, Muara Baru, Ancol, etc. are prone to the impacts, while area in northern of Semarang, place likes Tanjung Mas, Kaligawe, etc., are experiencing the same situation.

As mentioned earlier in Jakarta at least there are ten bridges which are affected by the subsidence, more than 20 kilometer of roads are suffering the impact from subsidence. We surveyed all around northern part of Jakarta each in 2009 and in 2012, and we found more than five hundreds houses are suffering cracks due to subsidence impacts. We found more than 200 hectare areas are suffering tidal inundation even there are already dykes. If dyke is not established, it surely more hectares would be suffering.

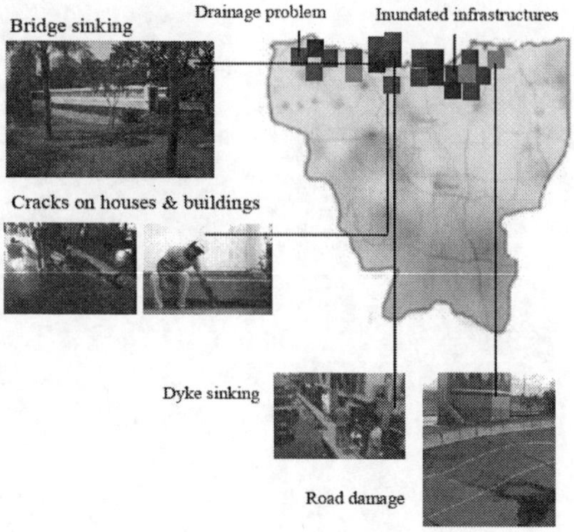

Fig. 16. Map plotted the impacts of land subsidence to the infrastructures in Jakarta area

Also mentioned earlier in Semarang at least there are five bridges which are affected by the subsidence, more than 20 kilometer of roads are suffering the impact from subsidence. We surveyed all around northern part of Semarang area each in 2009 and in 2016, and we found more than five hundreds houses are suffering cracks due to subsidence impacts. We also found many houses are being elevated in every five year for about 1 meter to avoid inundation. We found more than 800 hectare areas are suffering tidal inundation so far.

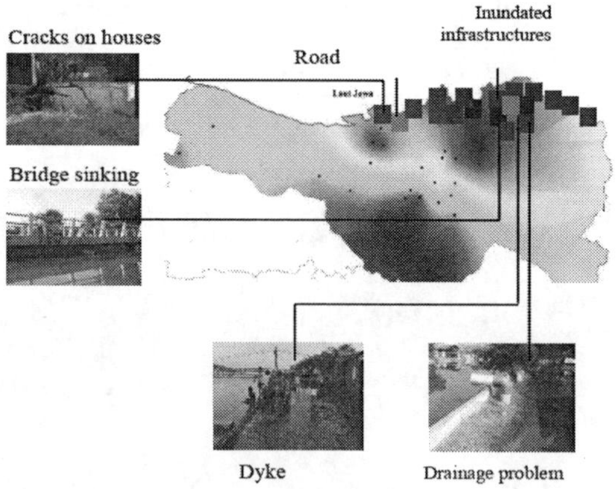

Fig. 17. Map plotted the impacts of land subsidence to the infrastructures in Semarang area

4 Discussions

From insight look the subsidence impact to infrastructures in Jakarta and Semarang area, we considered the impact is already a disaster. Some adaptation and mitigation have been carried out (e.g. build the dyke, fixing the infrastructures, etc), but it seem still not quite sufficient. We can see clear example from the dyke sinking and failed to protect land from inundation. This situation is because land subsidence in still continuing in to the future. Indeed on figure 18 we can see the linier trend of land subsidence in Jakarta and Semarang for recent years and most probably for sometimes in the future. Considered strongly we need to stop the land subsidence in both areas for best mitigation.

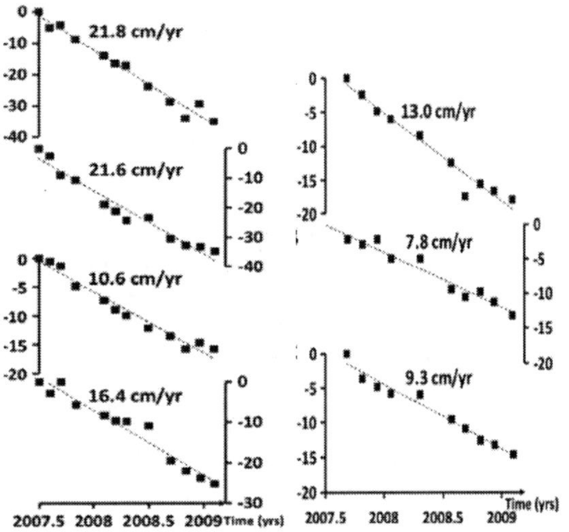

Fig. 18. Linier trend of land subsidence in Jakarta and Semarang areas which are probably still continuing for the future

If only we built the dykes and fixing the infrastructures, with the fact of linier trend of subsidence, in this case we always wait for silent disaster to slowly returned given a disaster. There are several example on others city around the world who manage to stop the land subsidence (e.g. Tokyo and Osaka Japan, Venice Italy, San Joaquin USA, etc.). By stopping the groundwater abstraction, mostly the land subsidence is stopped. Nevertheless stopping groundwater abstraction in Jakarta and Semarang is stillJ., Arnaud a big homework.

5 Conclusions

Cracking on housing, building, street, "sinking" on the bridges and dyke, problems on drainage, etc. are infrastructures problems due to the land subsidence in Jakarta and Semarang area. Huge costs have already been spent to fixing those infrastructures problems in both areas. Base on the news, cost for elevated the bridges is around 1 million US dollar each. We can imagine if ten or even more bridges need to be elevated, in this case many dollars absolutely should be spend. Base on the news, cost for fixing the road is around 1 million US dollar for ten kilometer. We can calculate the length of road need to be repair and how many dollars should spend. For Jakarta and Semarang, in each area at least there is already more than 20 kilometer of road being affected by the land subsidence impact.A.

Since mostly liniear pattern of land subsidence are recognized continously in both area, in this case the impact especially on infrastructures are probably will getting worse. Certainly, it would lead to more costs in the near future and absolutely need to be concern. Insight look the subsidence impact to the infrastuctures in these areas would raisen the concern, and it is also become key for adaptation and mitigation. We cannot wait for any five years from now for silent disaster of land subsidence given finally the significant disaster to Jakarta and Semarang. We need better adaptation and mitigation immediately. There are plan for giant sea wall in Jakarta, and sea wall in Semarang. Nevertheless until now we found it still in visibilities studies and master plan. Considered strongly we need to stop the land subsidence in both areas for best mitigation.

References

1. D. Murdohardono, H. Tirtomihardjo. Penurunan tananh di Jakarta dan rencana pemantauannya. In: Proceedings of the 22nd annual convention of the Indonesian association of geologists, Bandung, 6–9 December, **pp 346–354** (1993)
2. H.Z. Abidin. Suitability of levelling, GPS and INSAR for monitoring land subsidence in urban areas of Indonesia. GIM Int 19(7):12–15 (2005)
3. H.Z. Abidin, H. Andreas, I. Gumilar, Y. Fukuda, Y.E. Pohan, T. Deguchi. Land subsidence of Jakarta (Indonesia) and its relation with urban development. Nat. Hazards **59(3)**, 1753–1771(2011)
4. F.N. Koudogbo, J. Duro, A. Arnaud, P. Bally, H.Z. Abidin, and H. Andreas. Combined X- and L band PSI analyses for assessment of land subsidence in Jakarta, Proc. SPIE, Remote Sensing for Agriculture, Ecosystems, and Hydrology **XIV, 8531, 853107**, doi:10.1117/12.974821 (2012)
5. F. Kuehn, D. Albiol, G. Cooksley, J. Duro, J. Granda, S. Haas, A. Hoffmann-Rothe, and D. Murdohardono. Detection of land subsidence in Semarang, Indonesia, using stable points network (SPN) technique. Environmental Earth Sciences, DOI 10.1007/s12665-009-0227-x (2009)
6. A.M. Lubis, T. Sato, N. Tomiyama, N. Isezaki, and T. Yamanokuchi. Ground subsidence in Semarang-Indonesia investigated by ALOS–PALSAR satellite SAR interferometry. Journal of Asian Earth Sciences, 40, 5, 30 March, **pp. 1079-1088** (2011)
7. S.L. Nurmaulia, L. Fenoglio-Marc, M. Becker. Long term sea level change from satellite altimetry and tide gauges in the indonesian region, paper presented at the EGU General Assembly 2010, 2–7 May. Vienna, Austria (2010)
8. H. Andreas, H.Z. Abidin, I. Gumilar, T.P. Sidiq, B. Yuwono. Adaptation and mitigation of land subsidence in Semarang. AIP Conference Proceedings (2017)

Utilization of alternative fuels and materials in cement kiln towards emissions of benzene, toluene, ethyl-benzene and xylenes (BTEX)

Ulfi Muliane* and *Puji* Lestari

Department of Environmental Engineering, Institut Teknologi Bandung (ITB), 40132 Bandung, Indonesia

Abstract. Co-processing in cement industry has benefits for energy conservation and waste recycling. Nevertheless, emissions of benzene, toluene, ethyl-benzene, and xylenes (BTEX) tend to increase compared to a non co-processing kiln. A study was conducted in kiln feeding solid AFR (similar to municipal solid waste, MSW) having production capacity 4600-ton clinker/day (max. 5000 ton/day) and kiln feeding biomass having production capacity 7800-ton clinker/day (max. 8000 ton/day). The concentration of VOCs emissions tends to be higher at the raw mill on rather than the raw mill off. At the raw mill on, concentration of total volatile organic carbon (VOCs) emission from cement kiln stack feeding Solid AFR 1, biomass, Solid AFR 2, and mixture of Solid AFR and biomass is 16.18 mg/Nm^3, 16.15 mg/Nm^3, 9.02 mg/Nm^3, and 14.11 mg/Nm^3 respectively. The utilization of biomass resulted in the lower fraction of benzene and the higher fraction of xylenes in the total VOCs emission. Operating conditions such as thermal substitution rate, preheater temperature, and kiln speed are also likely to affect BTEX emissions.

1 Introduction

Cement is an energy-intensive industry that takes into account at around 35% for energy cost [1]. It requires a high need of funding for fossil fuel consumption. There is a technique called co-processing that utilized alternative fuels in the manufacturing process in order to decrease the cost of fossil fuel consumption [2]. The similar characteristics between cement kiln and incinerator make the cement kiln operated as disposal facilities as well as utilized waste as secondary fuel [3]. A ton of coal can be substituted by 1.6-ton rice husk as AFR in the cement plant [4]. In addition to resources and energy conservation, the CO_2 emissions from cement kiln stack through co-processing activities can reduce up to 0.118 ton CO_2/ton clinker in 40% thermal substitution rate (TSR) [5].

In spite of its benefit, BTEX emissions from co-processing kiln are likely higher compared to a non co-processing kiln. BTEX are considered as main sources of secondary organic aerosols [6] that contribute to the attenuation of regional visibility, climate change, and potential health hazards [7]. Based on a research in Canada, benzene is associated with an increased risk of cancer (leukemia and lungs) and cancer-related mortality [8].

BTEX emissions from non-coprocessing kiln are in the range 2.346 – 3.458 mg/Nm^3 [9] while from co-processing kiln utilized various of waste are in the range 1.859 – 46.804 mg/Nm^3 [9]. Based on Fatimah [9] findings, the average of BTEX concentrations from kiln using AFR varies from 0.5 – 20 times higher than kiln utilized coal only. BTEX emissions from cement industry are formed by decomposition of the organic constituent in a preheater [10, 11]. Emission factor of benzene from cement kiln using ESP as air pollution control is 0.0016 kg/ton cement while for toluene, ethyl-benzene, and xylene is 0.00010 kg/ton cement, 9.5 x 10^{-6} kg/ton cement, and 6,5 x 10^{-5} kg/ton cement respectively [12]. For overall VOCs, in European countries, the emission factor of VOCs is 52.4 g/ton clinker with exhaust volume of kiln gas 2300 Nm^3/ton clinker [13].

Based on *EU Directive 2000/76/EC*, the daily average limit for BTEX emissions along with other volatile organic compounds (VOCs) from incineration or co-incineration process is 10 mg/m^3 (dry, 273 K, 1013 hPa, 10% O_2) as total organic carbon (TOC) parameter. The high temperature (1200 to 2000°C) and residence time over two seconds at that temperature result in complete destruction of toxic compound in secondary fuel up to 99.99% inside the kiln [3]. Nonetheless, VOCs emission from cement kiln stack still exceeds the limit of the threshold. The mechanism of BTEX formation from cement kiln with AFR utilization is not clear. Therefore, it is essential to conduct a study of waste utilization as AFR in cement manufacturing and its relationship with BTEX emission as a preliminary study for further application towards the sustainable industry.

* Corresponding author: ulfimuliane@tl.itb.ac.id

https://doi.org/10.1051/matecconf/201814708002

2 Materials and Method

The research began with collecting gas samples from kiln stack and AFR from silo prior to preheater/pre-calciner systems followed by laboratory analysis. Gas samples from cement kiln stack were collected through active method sampling using sampling train. BTEX is a group of organic volatile compounds (VOCs) that became the target of sampling. As VOCs, BTEX exists in the gas phase from cement kiln stack thus sampling was conducted in a non-isokinetic condition in the flow rate of 1 liter per minute for 30 minutes. Activated Charcoal Type G (Dräger Safety AG & Co.KgaA) along with DESAGA Gas Samplers GS 212 are used to collect the samples. Then, samples were analyzed using GC-MS (gas chromatography – mass spectrometry) referred to **EN 13649** (*Determination of the mass concentration of individual gaseous organic compounds: activated carbon and solvent desorption method*).

In regard to find out the relationship between BTEX emission and utilization of AFR, calculation of TSR in which the amount of energy replaced by AFR is important. Eq. 1 shows the calculation of TSR where q_{AF} is the quantity of AFR [ton/hour], q_C is the quantity of coal [ton/hour], CV_{AFR} is the calorific value of AFR [MJ/kg], and CV_C is the calorific value of coal [MJ/kg].

$$TSR = \frac{\sum_1^n (q_{AF} \times CV_{AF})}{\sum_1^n (q_{AF} \times CV_{AF}) + \sum_1^n (q_C \times CV_C)} \tag{1}$$

In addition to flue gas and AFR analysis, operating conditions during cement manufacturing process were also recorded including raw mill condition, preheater temperatures, kiln temperatures, and kiln speed. For kiln residence time, it is calculated as a function of kiln speed as shown in Eq. 2, where t is material residence time [min], L is kiln length [m], Θ is angle of repose of material [°], v is slope of kiln [°], D is kiln shell diameter [m], Lining is refractory thickness [m], rpm is kiln rotation per minute [rpm], and F is factor of kiln diameter constant.

$$t = \frac{1{,}77 \times L \times \sqrt{\Theta}}{v \times (D - 2 \times Lining) \times rpm} \times F \tag{2}$$

This research was conducted in an actual scale of cement plant having production capacity 5.6-million-ton cement/year. There is two kiln involved in the study, first kiln feeding solid AFR (similar to MSW) having production capacity 4600-ton clinker/day (max. 5000 ton/day) and second kiln feeding biomass having production capacity 7800-ton clinker/day (max. 8000 ton/day).

Each kiln stack is equipped with ESP as air pollution control. Kiln 1 has a 4.5-meter-high stack with 15.904 m^2 of diameter and kiln 2 has a 5.6-meter-high stack with 25.518 m^2 of diameter. There is two sampling condition involving ESP namely raw mill on and raw mill off. In raw mill on condition, the exhaust gas from the burner and preheater pass through the raw mill then the ESP. Dust from ESP will be returned to the preheater while the clean gas will go out from the stack. When sampling in the raw mill off condition, the exhaust gas from burner and preheater will directly go to the conditioning tower followed by ESP.

3 Results and Discussion

The source of the air pollutant from cement kiln stack can be classified into three types i.e. raw materials, fuels and process inside kiln system [14]. In addition to AFR characteristics, there is also an influence of operating condition towards emission from cement kiln stacl. TSR, preheater temperature, kiln speed, and kiln residence time show a relationship with each compound in the flue gas.

From the flue gas analysis, it was found that not only BTEX presented in the cement kiln stack emission. Other VOCs such as nonane, octane, and decane were also existed and it will also be discussed as non-BTEX emission. The BTEX and non-BTEX emission henceforward will be discussed together in term of VOCs emission.

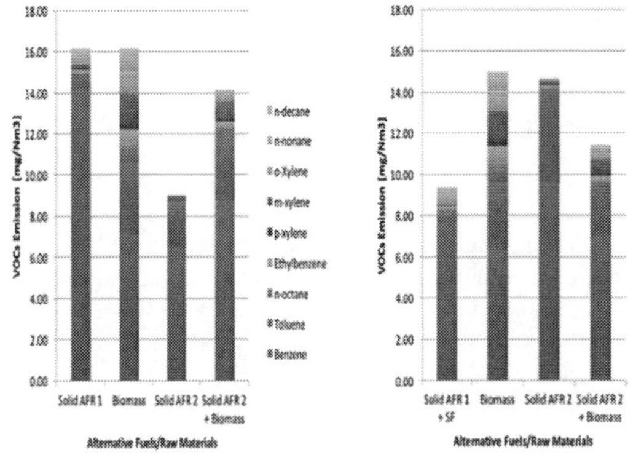

Fig. 1. VOCs emission at different sampling condition: (a) raw mill on and (b) raw mill off

Fig. 1 presents the average concentration of VOCs element in two different sampling condition i.e. raw mill on and raw mill off. The concentration of total VOCs at the raw mill on condition is relatively higher than at the raw mill off, except for Solid AFR 2 fuels. Total VOCs from kiln fed Solid AFR 1 at the raw mill on is 16.18 mg/Nm^3 and 9.41 mg/Nm^3 at raw mill off. For kiln substituted with biomass, the VOCs emission is 16.15 mg/Nm^3 with the raw mill on while raw mill of is slightly lower at 14.98 mg/Nm^3. Then for a kiln with a mixed fuel of Solid AFR 2 and biomass, the total concentration of VOCs is also higher at raw mill compared to raw mill off at 14.11 mg/Nm^3 and 11.18 mg/Nm^3 respectively. On the other hand, VOCs emission from kiln fed with Solid AFR 2 only is higher at raw mill off rather than raw mill on at 14.66 mg/Nm^3 and 9.02 mg/Nm^3 respectively.

In addition to VOCs concentration, the fraction of each VOCs element could be expected from Fig. 2. It can be seen that the largest fraction amongst VOCs is benzene followed by toluene at both conditions. However, the fraction of benzene and toluene are different depending on the types of AFR.

The fraction of benzene in Solid AFR 1 kiln is around 80% of total VOCs emission both for the raw mill on and raw mill off condition. In the Solid AFR 2 kiln, the portion of benzene is around 65% – 70% of the total emission. For kiln using mixed fuel of Solid AFR 2 and biomass, benzene only takes part of around 60% of total VOCs. Then for the kiln fed with biomass only, the fraction of benzene is the smallest amongst other types of fuel at around 40% of total VOCs. It is expected that the utilization of biomass resulted in the lower fraction of benzene in the total VOCs emission.

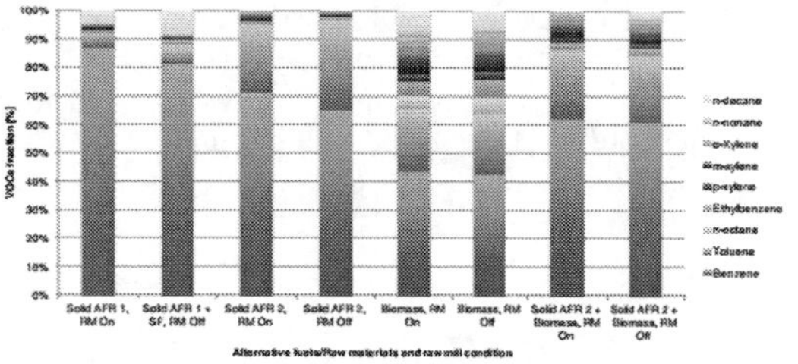

Fig. 2. Comparison of VOCs element fraction based on types of AFR

Another interesting finding from the VOCs element fraction is xylenes. The proportion of xylenes (p-xylene, m-xylene, and o-xylene) when biomass is involved tend to increase. Kiln substituted with solid AFR only shows the fraction of xylenes at around 2% - 3%. A significant increment is shown by the kiln fed with mixed fuel contained biomass that raises up to around 10% of xylenes from total VOCs. The highest fraction of xylenes is presented by the kiln utilized biomass only as AFR at around 15% - 16% of its emission.

The total VOCs emission shows that there is a change between raw mill on and raw mill off condition. Nevertheless, the fraction of each VOCs element does not present any prominent difference in both conditions. The variation of VOCs element fraction more likely depends on the types of the fuel. Therefore, it can be stated that raw mill condition shows an impact towards VOCs concentration, yet do not affect the fraction of VOCs element.

Table 1 shows the mass balance of cement production that utilized Solid AFR or Biomass as alternative fuels and/or raw materials. It can be seen that in each condition the input materials including coals, AFR, and raw meal generate clinker as the product at around half of the amount of the input. Kiln 2 has larger production capacity so that the input of raw meal is almost one-half of kiln 1.

The substitution rate of AFR could not be calculated through mass balance. From the amount of AFR put on the burner, the calculation of AFR substitution rate must involve the heating value. By using Eq. 1, the maximum substitution rate in term of energy is 15.76%, 21.23% for kiln 1 when raw mill on and off respectively, 16.43%, and 19.38% for kiln 2 when raw mill on and off respectively.

Table 1. Balance of cement production

Parameter	Unit	Kiln 1		Kiln 2	
		Raw Mill On	**Raw Mill Off**	**Raw Mill On**	**Raw Mill Off**
Fuels:					
Coal (kiln)	ton/hr	18.20	15.33	13.16	12.86
Coal (calciner)	ton/hr	14.69	13.65	-	-
Coal (inline cyclone preheater)	ton/hr	-	-	9.78	8.47
Coal (side-line cyclone preheater)	ton/hr	-	-	28.85	24.08
Solid AFR	ton/hr	4.66	4.68	-	-
SF	ton/hr	0	1.28	-	-
Biomass	ton/hr	-	-	7.59	7.13
Heating value:					
Coal	kcal/kg	4525	4525	4492	4492
Solid AFR	kcal/kg	4439	4439	-	-
SF	kcal/kg	-	4492	-	-
Biomass	kcal/kg	-	-	5362	5362
Materials:					
Raw meal	ton/hr	329.05	329.93	522.69	500.54
Production:					
Clinker	ton/hr	190.64	191.16	314.85	301.46

Operating condition of cement kiln plan was recorded to show its influence towards BTEX emission. Record of operating condition was changing hourly. In Table 2, the record of the operating parameter is presented for maximum TSR condition.

During the record of operating condition, it can be inferred that kiln temperature was stable at 1366.7°C for kiln 1 and 1362.65°C for kiln 2. Nonetheless, preheater temperature tends to fluctuate every hour from 822.92°C to 860.03°C. The kiln speed changed by the time but only under 10%. Residence time is calculated as the function of kiln speed so that it has a similar trend with the kiln speed.

Table 2. Kiln operating condition at maximum TSR

Parameter	Unit	Kiln 1		Kiln 2	
		Raw Mill On	**Raw Mill Off**	**Raw Mill On**	**Raw Mill Off**
TSR	%energy	15.76	21.23	16.43	19.38
Operating Condition:					
Kiln temperature	°C	1366.7	1366.7	1362.65	1362.5
Preheater temperature	°C	839.74	839.86	848.22	844.39
Kiln speed	rpm	2.28	2.38	2.66	2.67
Residence time	min	30.72	29.43	26.33	26.23

The utilization of synthetic fuel at raw mill off condition is less than one-third of the whole amount of AFR (Table 1). The trend of VOCs emission between raw mill on and raw mill off condition is not affected by feeding SF into the kiln. In line with it, the trend of VOCs element fraction in the emission is not changed by the presence of SF. Thus, it is assumed that the utilization of SF as an additive at the raw mill off did not give significant impact to the VOCs emission.

Fatimah [9] have found that the utilization of AFR as substituted fuel has increased BTEX emissions up to 20 times compared to the kiln fed with coal only. The increment of VOCs emission from cement kiln stack is suspected due to the presence of AFR. To study the relationship between AFR and VOCs emission, physical and chemical characteristics of AFR as well as the operating condition of manufacturing process were analyzed. Physical and chemical characteristics of AFR along with kiln operating condition show an influence towards VOCs emission.

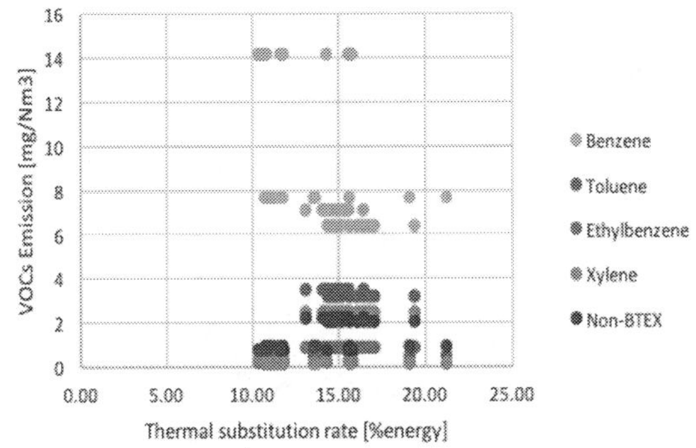

Fig. 3. Influence of thermal substitution rate towards VOCs emission

TSR represents the quantity of AFR involved in the combustion process inside the kiln. The higher the TSR imply the higher amount of carbon originated from AFR that suspected the source of VOCs emission. Fig. 3 shows how TSR affect VOCs emission.

It can be seen that increase of AFR substitution rate result in the increase of toluene, ethyl-benzene, xylene, octane, nonane, and decane emission, except benzene. Benzene is the only organic compound tends to decrease in emission by the increase in substitution rate of alternative fuels. The explanation of this condition may relate to the relationship between VOCs emissions and organic carbon content in the alternative fuels. Emission of benzene is expected not only comes from thermal decomposition of organic carbon content in the alternative fuels but also from other organic materials inside the preheater.

Fig. 4 shows the relationship between VOCs emission and preheater temperature. Each VOCs element i.e. toluene, ethylbenzene, xylene, octane, nonane, and decane tend to decrease with the increase of preheater temperature, except benzene. As previous findings, benzene shows different result compared to other compounds. Emission of benzene tends to increase when preheater temperature is increased.

Preheater temperatures are varied from 823 to 860°C inside the preheater during the sampling period. The release of organic compounds in the kiln system occurred in temperature between 400 – 600°C [15]. At the temperatures over 600°C, the organic compounds inside burner tend to be destructed. Thus, the higher the temperature the lower the VOCs emission.

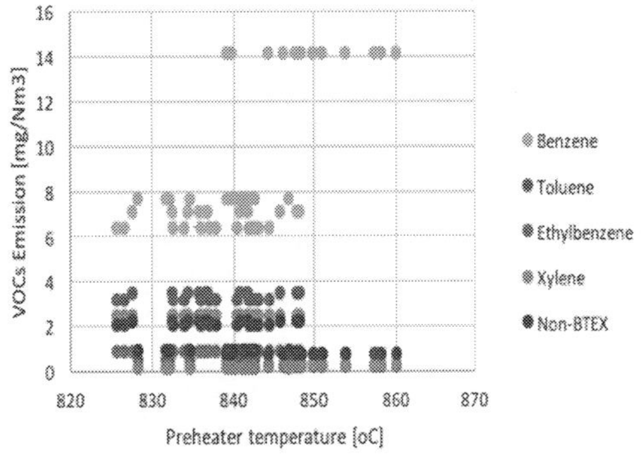

Fig. 4. Influence of preheater temperature towards VOCs emission

The special case about benzene is that at the temperature range of 823°C and 860°C inside preheater, its concentration tends to increase. When VOCs are exposed to high temperatures, it will crack into lighter hydrocarbon species (HCs), but some of the HCs will not be oxidized and the remaining HC could be considered as a whole and called total hydrocarbons (THC) (Salvador et al., 2006). Referred to Kara [18], oxidation of VOCs may occur if flame temperatures in the burner are kept around 1000°C - 2000°C. Thus, there is a possibility that the destruction temperature of benzene is not achieved at the temperature range of preheater.

The unachieved destruction temperature for benzene is expected due to formation rate of is higher than its oxidation rate. Benzene is the most stable compound compared to other VOCs detected in the flue gas due to its ring structure. The more stable the compound, the higher the required energy to oxidize it. Therefore, benzene is the most difficult compound to be oxidized compared to other VOCs element. It is described in the standard (25°C) combustion enthalpy of benzene in the gas phase ($-\Delta_c H^{\circ}_{gas}$) that has smallest value 789,08±0,10 kcal/mol (toluene = 943,58±0,12 kcal/mol; ethylbenzene = 1101,13±0,17 kcal/mol; p-xylene = 1098,29±0,22 kcal/mol; m-xylene = 1098,12±0,15 kcal/mol; o-xylene = 1098,54±0,24 kcal/mol; n-octane = 1317,45±0,16 kcal/mol; n-nonane = 1474,90±0,18 kcal/mol; and n-decane = 1632,34±0,21 kcal/mol) [17]. Based on the comparison of standard combustion enthalpy value, benzene has the smallest amount of released energy due to high energy required to oxidize it. It could explain why the rate of benzene oxidation is the least compared to other compounds so that emission of benzene tends to increase when the temperature is increased.

The influence of kiln speed towards VOCs emission is shown in Fig. 5. Emissions of toluene, ethylbenzene, xylene, and other VOCs non-BTEX tend to increase when kiln speed is increased. However, benzene is the only compound that tends to decrease when kiln speed is increased.

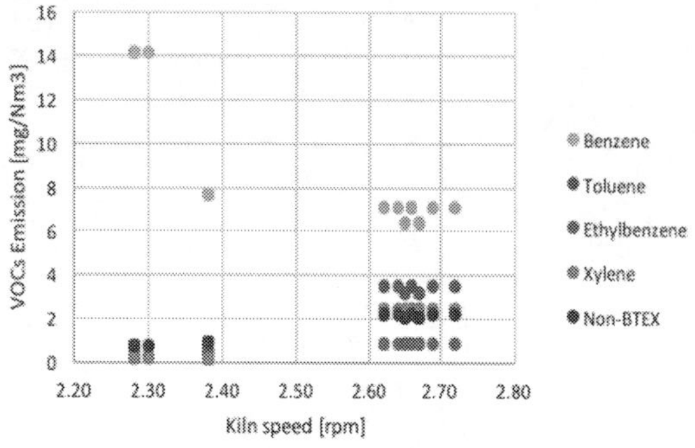

Fig. 5. Influence of kiln speed towards VOCs emission

Kiln speed is related to turbulence, which is an indicator showing the effectiveness of contact between material and oxygen inside rotary kiln [18]. It indicates that the higher the kiln speed, the higher the airflow rate inside burner so that turbulence becomes higher. As the turbulence increase, the effectiveness of oxygen contact with materials inside kiln is also increased. However, increase in airflow rate due to an increase of kiln speed, induce lower global fuel richness inside the flame of the burner, whose temperature will decrease [19]. As implied in the relationship between preheater temperature and VOCs emission, the decrement of temperature could lead to increase of VOCs emission.

For the benzene case, it is expected that increase of kiln speed result in the increase of benzene oxidation rate. As the kiln speed higher, the turbulence, as well as airflow rate inside the burner, become higher. The airflow rate supports materials oxidation as well as the gas inside the burner. Hence, an increase of kiln speed affects to the decreased of benzene concentration.

Fig. 6 represents the relationship between material residence time and VOCs emission. It is showed that increase of material residence time results in the decrease of VOCs emission except for benzene. Similar to the previous findings, benzene always shows different results with other VOCs.

The material residence time is obtained from a calculation involving the data of kiln speed. From the Eq. 2 it can be seen that material residence is inversely proportional to the kiln speed. The longer the time of material in the kiln, the lower the kiln speed. It is expected that increase of material residence time will increase the degree of effectiveness of the oxidation process both for materials and its gas. Therefore, the longer the material residence time results in the lower VOCs emission.

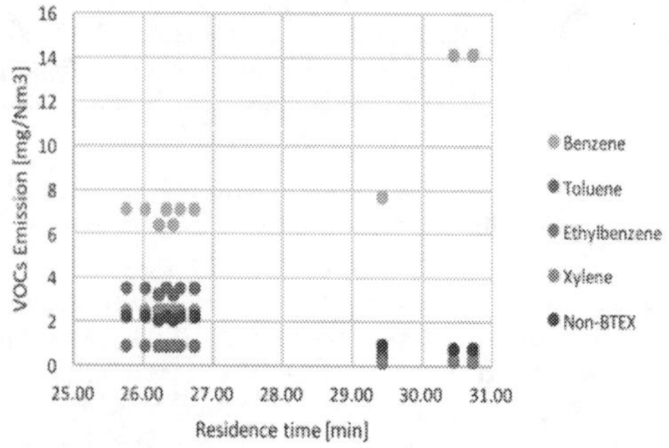

Fig. 6. Influence of residence time towards VOCs emission

The relationship between material residence time and VOCs emissions i.e. toluene, ethyl-benzene, xylene, and the non-BTEX, is not applicable for benzene. The concentration of benzene tends to increase by the increase of material residence time. The material residence time is related to the degree of effective oxidation. Nonetheless, the oxidation of materials and the exhaust gas not only depends on the residence time but also the temperature. Benzene is the compound that requires the highest energy for oxidation compared to other VOCs. Thus, the increase of residence time does not result in the decrease of benzene concentration due to high oxidation energy demand of benzene.

4 Conclusion

VOCs emissions from cement kiln stack consisted of benzene, toluene, ethyl-benzene, p-xylene, m-xylene, o-xylene, n-octane, n-nonane, and n-decane are vary based on alternative fuels. VOCs emissions from kiln feeding biomass, biomass mixed solid AFR, solid AFR (2014) and solid AFR (2015) are 15.68±2.23 mg/Nm^3; 13.1±4.42 mg/Nm^3; 12.40±3.25 mg/Nm^3; and 11.96±4.32 mg/Nm^3 consecutively. Overall, benzene and toluene are compounds that exist in the highest concentration from all alternative fuels. Benzene concentration from kiln feeding solid AFR, biomass, or biomass mixed with solid AFR are in range 6.46 – 14.14 mg/Nm^3 (43.29 – 85.31%) and toluene are in range 0.71 – 3.51 mg/Nm^3 (4.73 – 29.46%).

The cement kiln stack feeding alternative fuels tends to generate higher VOCs emission at the condition of the raw mill on rather than the raw mill off. The condition of raw mill affects the total VOCs concentration but does not change the fraction of each VOCs element. The fraction of VOCs element more depends on the types of alternative fuels. The utilization of biomass shows a decrement of benzene proportion and an increment of xylenes proportion of all VOCs.

A relationship between each VOCs emission and operating condition, for instance, thermal substitution rate, preheater temperature, kiln speed, and material residence time exist. Nonetheless, benzene always shows different trend amongst other VOCs. The high oxidation energy demand of benzene compared to the other compounds is suspected to be the main reason of benzene shows a distinct behavior.

5 Nomenclature

AFR	= alternative fuels and/or raw materials
BTEX	= benzene, toluene, ethyl-benzene, and xylenes
ESP	= electrostatic precipitator
MSW	= municipal solid waste
NOx	= nitrogen oxides
RM	= raw mill
SF	= synthetic fuel
TSR	= thermal substitution rate
VOCs	= volatile organic compounds

References

1. BPS, *Statistik Industri Besar dan Sedang* (2012)
2. W. Queiroz Lamas et al, *Ren. and Sust. Energy Reviews*, **19**, 200-207 (2013)
3. A. Ottoboni, I. De Souza, G.J. Menon, R.J. Da Silva, *Energy Convers. Mgmt*, **39**, 1899-1909 (1998)
4. G. Mardiana, R. Mahardika, *Seminar Rekayasa Kimia dan Proses* (2010)
5. F.D. Qonitan, *Scientific Reports Dept. of Env. Eng.* (2013)
6. L. Zi-feng, H. Ji-ming, D. Jing-chun, L. Jun-hua, **30**, 969-975 (2009)
7. L. Kun, W. Weigang, G. Maofa, L. Jiangjun, W. Dong, *Scientific Reports* (2014)
8. P. Villeneuve, M. Jerrett, J. Su, R. Burnett, H. Chen. J. Brook, *Env. Pol.*, **183**, 30-39 (2013)
9. R. Fatimah, *Scientific Reports Dept. of Env. Eng.* (2008)
10. A. Chatterjee, *Cem. and Concr. Rsrch*, **41**, 624-641 (2011)
11. VDZ, *Scientific Reports (2008)*
12. EPA, *Scientific Reports (1994)*
13. WBCSD, World Business Council for Sustainable Development (2012)
14. W. Greer, *Portland Cement Association Research* (2003)
15. EU, IPPC Bureau (2013)
16. Y. Kara, The`se de Doctorat (Ph.D.) (2003)
17. E. Prosen, W. Johnson, F. Rossini, *J. Res. NBS*, **36**, 455-461 (1946)
18. D.G. Wilson, Handbook of Solid Waste Management, (1977)
19. S. Salvador, Y. Kara, J. Commandre, *Applied Thermal Engineering*, **26**, 1640-1651 (2006)

https://doi.org/10.1051/matecconf/201814708003

Stream Sediment Geochemical Survey of Selected Element In Catchment Area Of Saguling Lake

Eka Wardhani[1,*], *Suprihanto* Notodarmojo[2], and *Dwina* Roosmini[2]

[1]Study Program of Environmental Engineering, Faculty of Civil and Environmental Engineering, Post-Graduate Program, Institut Teknologi Bandung 40132, Indonesia,
[2]Departement of Environmental Engineering, Faculty of Civil and Environmental Engineering, Institut Teknologi Bandung 40132, Indonesia

Abstract. Saguling Lake is one of the largest lakes in West Java Province that accommodates domestic and non-domestic wastes via the Citarum River as its main water source. This study aims to determine the geochemical background concentration (Cbg) in water catchment area of Saguling Lake. The knowledge of the Cbg of heavy metals is essential for defining pollution, identifying the source of contamination, and for establishing reliable environmental quality criteria for sediments. The value of Cbg will be used for assessment of the sediment quality in Saguling Lake. Assessment of sediment quality is very important to determine the actual condition of water in the lake and as the basis for management of waters environment in the future. The search was taken at 22 sampling points in the unpolluted water catchment area. Samples were collected and analyzed for Cd, Cr, Cu, and Pb. Each sample was digested in agua regia and analyzed by ICP-EOS. Results showed Cbg which are: Cd 0.34 ± 0.10 mg/kg, Cr 110.57 ± 28.61 mg/kg, Cu 49.93 ± 9.28 mg/kg, and Pb 18.62 ± 9.83 mg/kg. Based on the assessment result, it is concluded that the sediment quality in Saguling Lake is categorized as polluted by Cd, Cr, Cu, and Pb metals.

1 Introduction

The Sediments are the layers of relatively finely divided matter covering the bottoms of river, streams, lake, and oceans. Sediment typically consists of mixture of fine, medium, and coarse grained minerals, including clay, silt, and sand mixed with organic matter. They may vary in composition from pure mineral matter to predominantly organic matter. Sediments are lake of a variety of biological, chemical, and pollutant detritus in bodies of water and are repositories of pollutants such as heavy metals and organic compounds. Of particular concern is the transfer of chemical species from sediment into aquatic food chains via organisms that spend significant parts of their life cycles in contact with or living in sediment [10].

Sediment constitutes an integral and dynamic part of a stream basin, originating from weathering of minerals and soils up stream. Natural concentrations of heavy metal as a result of the weathering process of mineral deposits can be quite high in stream sediments close to the deposit, but decrease with increasing distance downstream, due to dissipation energy and dilution of sediments from other not polluting source [12].

The term "geochemical background" was first introduced by geochemistry explorers in the mid-twentieth century to distinguish between the un-mineralized element abundance and the mineralized one from rock formations. In the last few decades "Geochemical background concentration (Cbg)" has become one of the most important terms in environmental science. Cbg is used to distinguish pollution occurring in environment whether the source of pollutant comes from human activities or from natural phenomena (geogenic/biogenic) [13]. The adaptation process of Cbg in environmental science generates wider meanings and applications. Currently Cbg is applied not only to rocks, minerals, and sediments but also to water and air [13,7].

Cbg is the concentration of metal derived from the weathering of original rocks of the earth's crust without any addition of anthropogenic activity [9]. Cbg is very specific depending on the structure of rocks forming it. There are many Cbg terms in geology such as ambient background, area background, natural background, naturally occurring background, Pedogeochemical background, and pre-industrial background [13]. Evaluation of Cbg is essential since it generates impact for assessment of environmental risk and the determination of regulations related to the environmental baseline of elemental pollution [13].

The purpose of this research is to identify the Cbg of 4 types of heavy metals i.e. Cd, Cr, Cu, and Pb from water catchment area of Saguling Lake. The value of Cbg will be used as a basis to determine the sediment quality in the lake. Monitoring on sediment quality

* Corresponding author: ekw_wardhani@yahoo.com

should become an integral part of water quality management and water pollution control program considering that sediment is the site for accumulation of pollutants including heavy metals.

2 Material and methods

2.1 Sample collection and chemical analysis

Stream sediment samples were collected using hand trowel from 22 different points in Catchment area of Saguling lake, in November 2015 sampling location are presented in Table 1. The sampling locations were recorded (latitudinal and longitudinal position) using hand-held Global Positioning System (GPS). The location of sampling were selected based on criteria that these locations are in areas that have not been contaminated by human activity. Sampling sites are scattered in the water catchment area of Saguling Lake and concentrated in Southern part which is relatively less contaminated than the Northern or Eastern part.

Table 1. Sampling location catchment area of Saguling lake

No	South	East	Location
1	6°47'34.66"	107°34'39.07"	Kertawangi
2	6°48'52.80"	107°42'30.66"	Pasir Buleud
3	6°52'24.84"	107°42'52.00"	Melatiwangi
4	6°56'32.30"	107°32'09.70"	Nanjung
5	6°58'30.00"	107°36'30.00"	Cisirung
6	6°59'28.50"	107°37'35.30"	Dayeuhkolot
7	7°01'58.20"	107°40'36.28"	Ciheulang
8	7°04'29.00"	107°44'53.00"	Wangisagara
9	7°08'43.05"	107°41'57.00"	Babakan
10	7°09'13.14"	107°42'34.17"	Pacet
11	7°10'21.76"	107°39'41.60"	Halimun Mt
12	7°11'36.62"	107°38'46.49"	Margamukti
13	7°12'06.02"	107°39'55.00"	Cibeureum
14	7°12'30.14"	107°39'20.16"	Cisanti Lake 1
15	7°12'35.87"	107°39'24.10"	Cisanti Lake 2
16	7°12'40.12"	107°39'31.71"	Cisanti Lake 3
17	7°13'03.60"	107°39'22.50"	Cisanti Lake 4
18	7°10'09.12"	107°42'36.69"	Cihawuk
19	7°13'22.78"	107°41'26.65"	Guha Mt 1
20	7°13'22.78"	107°41'26.65"	Guha Mt 2
21	7°12'40.44"	107°39'09.01"	Wayang Mt 1
22	7°12'40.44"	107°39'09.01"	Wayang Mt 2

Samples were then placed into polyethylene bags, and refrigerated at 4°C. They were analyzed in the laboratory. Each sediment sample was dried at 50°C, the samples were then grounded using pestle and mortar until all particles passed a 200-mesh nylon sieve. This fraction is mainly contained silt and clay; Thus, this fraction may adsorb a higher heavy metal concentration. The extraction of heavy metals were made using conventional digestion procedure of sediment samples, which consists of digesting aliquots of 1 g of dry sediment was weighted into Teflon beakers, to which a mixture of concentrated HNO_3 : HCl (3:1) was added and digested in a water bath during 3 h at 120°C. The concentration of Cd, Cr, Cu, and Pb, were determined for each sample using an Inductively Coupled Plasma Optical Emission Spectrometry or *ICP-OES*.

Currently there are various geochemical indices to be used in assessment of the quality of sediments in waters related to heavy metal pollution. The indices include Contamination Factor (CF). Both indexes assess the quality of sediment at each sampling point. Data of Cbg shall be used to determine the quality of sediments at 12 points in Saguling lake. Assessment of sediment quality in Saguling lake uses the method of CF. The 12 points of sediment sampling locations around Saguling lake are presented in Table 2 and Figure 1.

Table 2. Sampling locations

No	South	East	Location
1A	06°56'29,8"	107°32'10.7"	Citarum River Nanjung section
1B	06°54'58,9"	107°28'32.3"	Sungai Citarum *Trash Boom* Batujajar
2	06°53'13,5"	107°28'32.3"	Cihaur Kampung Cipeundeuy
3	06°53'13,4"	107°27'09.0"	Cimerang
4	06°53'13,0"	107°25'54.4"	Muara Cihaur Kampung Maroko
5	06°56'07,6"	107°27'25.5"	Muara Cipantik
6	06°57'14,6"	107°26'03.8"	Muara Ciminyak-lokasi perikanan jaring terapung
7	06°56'14,9"	107°24'50.8"	Muara Cijere
8	06°56'00,4"	107°22'22.4"	Muara Cijambu
9	06°54'54,4"	107°22'26.3"	Dekat *Intake Structure*
10A	06°51'49,8"	107°20'57.0"	*Tailrace*
10B	06°51'10,8"	107°20'58.0"	Bantar Caringin

Fig. 1. Study area and sampling stations in Saguling Lake West Java

The level of sediment contamination by heavy metals is often expressed in CF in the following equation:

$$CF = \frac{Cme}{Cbg} \quad (1)$$

Where C_{Me} is the concentration of metals in the sediment samples (mg/kg) [17]. This method compares the concentration of one type of heavy metal to the background heavy metal which is the concentration of the original heavy metals derived from the earth's crust. The results of the calculation by this CF method are compared with some grade of pollution levels as presented in Table 3.

Table 3. Catagory class of sediment quality with CF [17]

CF value	Level of Pollution
CF <1	Low
1 ≤ CF ≥ 3	Moderated
3 ≤ CF ≥ 6	Considerable
CF> 6	Very high

3 Results and disscussion

3.1 General and local geological setting

The catchment area of Saguling lake is a basin area of Bandung. The greater Bandung area is a large intra-montane basin surrounded by volcanic highlands. The central part of basin has an altitude of about 665 m and surrounded by up to 2.400 m. Tertiary and quaternary volcanic terrain. The basin measures about 45 x 45 km and has a total area of 1.830 km^2. The basin is surrounded by volcanic mountains, which are the lowest at the north-west side. On the side, the Saguling lakes is located. The main river, the Citarum springs at Situ Cisanti 78 km upstream of the lake. In the basin about 200 major tributaries with corresponding sub basins can be distinguished. Citarum river with its tributaries forms the main drainage systems of the basin catchment. Deposits in the basin comprise of course volcanic lastics, fluvial sediments and notably a thick series of lacustrine deposits. The climate of the basin of the basin is relatively cool compare to the rest of Indonesia, with temperature mostly between 20 and 30°C. The annual rainfall varies from 1,200 to 3,000 mm, with an average of 2,215 mm. The monthly rainfall during the wet season (November-April) is about 250 mm, varying from 100-500 mm. During the dry season (July-September) monthly rainfall usually is less than 50 mm [16].

The greater Bandung basin area encompasses five administrative units which are Bandung municipality, Cimahi municipality, Bandung regency, West Bandung regency plus part of Sumedang regency (3 village). Large population growth and urbanization is taking place in the greater Bandung Basin. In 1995, the population was about 2.5 million and in 2010 the population grew to 7.8 million. It is commonly believed that this urbanization will continue, but it is very uncertain to what extent. In addition with expansion of manufacturing and textile industries in Bandung basin urbanization was increased. The population growth need for an enormous increase of settlements in the basin. Various types of industries scattered in regencies and cities in the Saguling Lake water catchment area comprises 556 industries, of which 442 industries are textile that are the largest. In addition to textile industry, there are 8 units of paper industry, 11 units of leather tanning, 16 units of plastic and rubber industry, 14 units of chemical industries, 40 units of metal and electroplating, 25 units of food and beverages. They are spread in various regions, mostly in Bandung regency of 325 industries followed by Cimahi City 101 industries and 90 industries in Bandung City. Source of pollution. The textile industry, paper industry, leather industry and food-beverage industry are potential to become the sources of high organic substance pollution load, while metal and metal coating industries are potential to dispose of heavy metals [16].

3.2 Background concentration

Concentrations of Cd, Cr, Cu, and Pb at 22 Stream sediments located in the Saguling lake water catchment are presented in Figure 2.

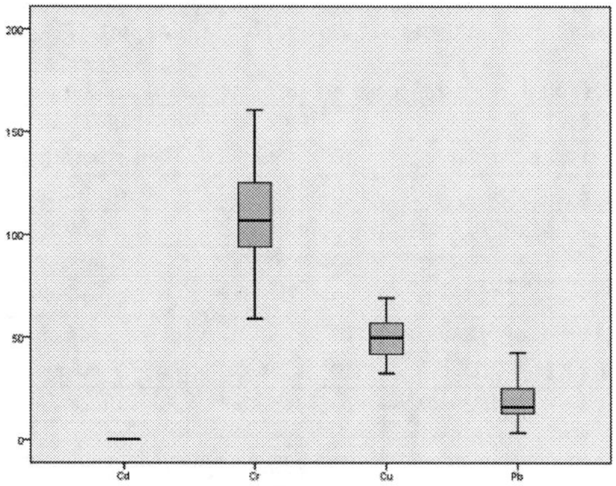

Fig. 2. Concentrations of Cd, Cr, Cu, and Pb in Catchment Area of Saguling lake (mg/kg)

Concentration of Cd in catchment area of Saguling lake ranges from 0.17-0.53 mg/kg, the mean value 0.34 mg/kg. The lowest concentration in sampling point 7 (Ciheulang) and the highest in sampling point 11 (Halimun Mountain). Concentration of Cd in the earth's crust is 0.1 mg/kg while in average the concentration of Cd in the world's soil is 0.3 mg/kg [11,19]. Cd concentration in the Catchment area of Saguling lake is relatively higher than the concentration value in the earth's crust, however when compared to the average Cd concentration contained in the soil based on Mc Bride, 1994, the sampling points of 4 (Nanjung), 5 (Cisirung), 6 (Dayeuhkolot), 7 (Ciheulang), 8 (Wangisagara), and 15 (Cisanti Spring) are lower than the average Cd concentration contained in soil around the world

Cr concentration in the catchment area of Saguling lake range from 59.13-160.76 mg/kg, the mean value of 110.57 mg/kg. The lowest concentration is at sampling point 6 (Dayeuhkolot) and the highest at sampling point 11 (Halimun Mountain). Cr concentration in the earth's crust is 126 mg/kg while the average Cr concentration in the earth's soil is 90 mg/kg [11,19]. Cr concentrations in the Catchment area of Saguling lake when compared to the concentration values contained in the Earth's crust are mostly lower, there are only the five sampling points above 90 mg/kg i.e. the sampling point 10 (Pacet), 11 (Halimun Mountain), 12 (Margamukti), 16 (Cisanti Springs 3), and 19 (Guha Mountain). Cr concentration contained in Catchment area of Saguling lake is higher than that of in world's soil

Cu concentrations in catchment area of Saguling lake range from 32.60-69.16 mg/kg, the mean value 49.93 mg/kg. The lowest concentration is at sampling point 16 (Cisanti Springs 3) and the highest at sampling point 6 (Dayeuh Kolot). Cu concentration in the earth's crust is 25 mg/kg while the average Cu concentration in the earth's soil is 45 mg/kg [11,19]. Pb concentrations in Catchment area of Saguling lake range from 3.80-43.63 mg/kg, the mean value 18.62 mg/kg. The lowest concentration is at sampling point 21 (Wayang Mountain 1) and the highest at sampling point 16 (Cisanti Springs 3). The concentration of Pb in the earth's crust is 14.8 mg/kg while the mean concentration of Pb in the world's soil is 20 mg/kg [11,19]. Table 4 describes the statistical parameters of heavy metals stream sediment samples from the Catchment area of Saguling lake

Table 4. Statistical parameters of heavy metals streams sediment samples from the catchment area of Saguling lake expressed in mg/kg

Element	Min	Max	Mean	Median	St Dev
Cd	0.17	0.53	0.34	0.32	0.10
Cr	59.13	160.76	110.57	107.06	28.61
Cu	32.60	69.16	49.93	49.81	9.28
Pb	3.80	43.63	18.62	16.18	9.83

Table 5 describes the Cbg of some heavy metals from various sources.

Table 5. Cbg of element in sediment expressed in mg/kg

Area	Element			
	Cd	Cr	Cu	Pb
Crust [(a)]	0.1	126	25	14.8
World Average[(b)]	0,3	90	45	20
River Sediment[(c)]	1.26	62.1	11.5	57.8
Range in soil [(d)]	0.06-1.1	7-221	6-80	10-84
River Sediment[(e)]	0.66 ± 0.69	nd	49.3 ± 39.3	7.3 ± 9.6
River Sediment[(f)]	0.075	145	11.9	10.8
River Sediment[(g)]	0.42	94.90	90.10	38.10
River Sediment[(h)]	0.65	11.75	9.58	17.78
This study	0.34 ± 0.10	110.57 ± 28.61	49.93 ± 9.28	18.62 ± 9.83

[a][19], [b][11], [c][15], [d][6], [e][4], [f][5], [g][3], [h][14]

Based on literature of similar research, the value of Cbg at the site of research was determined by using the average value from the research location [18], and using the minimum concentration value [8]. In this research, the value of Cbg is determined using the average metal concentration values from 22 sampling points referring to the study of [18] because the value is not much different to the value of metal concentration contained in the earth's crust and in the world's soil. Table 6 describes summary of research on quality assessment related to heavy metal contamination in several countries.

Table 6. Summary of research on quality assessment related to heavy metal contamination in several countries

Parameters	Reference		
	[18]	[8]	This study
Research site	Turkey's Upper Tigris River	Lake Veeranam, India	Lake Saguling West Java Province
Types of metal	As, Cd, Co, Cr, Cu, Ni, Pb, Zn	Cd, Cr, Cu, Ni, Pb, and Zn	Cd, Cr, Cu, Pb
Number of sampling points	6	28	22
The method of determinin g Cbg	Cbg using the average value of heavy metal concentration in uncontaminat ed sediments	Cbg based on the minimum concentrati on at the site of the research.	Cbg using the average value of heavy metal concentration in uncontaminat ed sediments.

Based on the comparison of Cbg referring to [8,12,18,19], the Cbg in Catchment area of Saguling lake takes the average value of results since it is not much different to the Cbg in the earth's crust and the average value in the world. The results of the analysis showed that the Cbg specifically: Cd 0.34 ± 0.10 mg/kg, Cr 110.57 ± 28.61 mg/kg, Cu 49.93 ± 9.28 mg/kg, Pb 18.62 ± 9.83 mg/kg.

3.3 Application of Cbg

Heavy metals in waters usually exist in low concentrations however the anthropogenic activity will increase their concentration, and cause problems to the waters. The sediment layer is an important part for tracking various pollutants such as heavy metals. The sediment layer can act as a source of pollutants in water body and potential to release heavy metals and other pollutants from their sedimentary bonds into the water column [20]. The water pollution level by heavy metals can be estimated by analyzing the concentrations of heavy metals present in water, sediments and aquatic organisms [2].

Concentrations of heavy metals contained in Saguling Lake sediment are presented in Table 7. The results of analysis showed that Cd concentration ranges from 13.22 to 23.2 mg/kg, Cr ranges from 95.85-197.98 mg/kg, Cu ranges from 37.94-152.53 mg/kg, and Pb Ranges from 20.31-32.74 mg/kg.

Table 7. Concentrations of heavy metal in Saguling lake (mg/kg)

Location Sampling	Cd	Cr	Cu	Pb
1A	23.25	197.98	63.32	23.79
1B	22.92	184.39	152.53	25.79
2	20.92	160.10	135.35	29.11
3	20.62	169.78	64.96	20.31
4	22.90	191.02	87.25	32.74
5	22.00	174.25	82.92	28.92
6	19.15	157.97	132.80	27.9
7	20.61	189.56	83.04	25.18
8	18.64	152.16	73.36	21.18
9	19.81	163.75	126.71	26.27
10A	13.22	95.85	37.94	31.31
10B	14.58	107.82	42.89	28.29
Average	19.89	162.05	90.26	26.73
Maximum	23.25	197.98	152.53	32.74
Minimum	13.22	95.85	37.94	20.31
Standard *	1.50	80.00	65.00	50.00

* ANZECC ISQG-Low, 1997

The calculation of sediment quality applied the (CF) using Cbg based on the result of this research, specifically: Cd: 0.34±0.10 mg/kg, Cr: 110.57 ± 28.61 mg/kg, Cu: 49.93 ± 9.28 mg/kg, Pb: 18.62 ± 9.83 mg/kg. The results of calculation on sediment quality in Saguling lake are presented in Table 8. The table shows that the sediment has been contaminated by Cd at very high category, contaminated by Cr at moderate category, contaminated by Cu at moderate-considerable category, lastly contaminated by Pb at moderate category.

Table 8. Contamination Factor Category

Sampling Location	Cd	Cr	Cu	Pb
1A	Very High	Moderate	Moderated	Moderate
1B	Very High	Moderate	Considerable	Moderate
2	Very High	Moderate	Considerable	Moderate
3	Very High	Moderate	Moderate	Moderate
4	Very High	Moderate	Moderate	Moderate
5	Very High	Moderate	Moderate	Moderate
6	Very High	Moderate	Moderate	Moderate
7	Very High	Moderate	Moderate	Moderate
8	Very High	Moderate	Moderate	Moderate
9	Very High	Moderate	Moderate	Moderate
10A	Very High	Moderate	Moderate	Moderate
10B	Very High	Moderate	Moderate	Moderate

4 Conclusion

Based on the results of the analysis, the Cbg of this study are: Cd 0.34±0.10 mg/kg, Cr 110.57±28.61 mg/kg, Cu 49.93±9.28 mg/kg, Pb 18.62±9.83 mg/kg. The values of Cbg are determined to assess the sediment quality of Saguling Basin by applying the CF method. Based on the CF category, the sediment has been contaminated by Cd at very high category, contaminated by Cr at moderate category, contaminated by Cu at moderate-considerable category, lastly contaminated by Pb at moderate category.

This research has been financially supported by State Minister for Research and Technology and Higher Education. Authors thank the anonymous reviewer for their critical and constructive comments of the manuscript.Equations should be centred and should be numbered with the number on the right-hand side.

References

1. ANZECC (Australian and New Zealand Environment and Conservation Council) ANZECC interim *sediment quality guidelines* (1997).
2. N. Belzile, Y.W. Chen, J.M. Gunn, S.S. Dixit. Environ Pollut; **130** (2004).
3. F. Botson, A.P. Karageorgis, E.M. Dassenakis, and M. Scoullos. Marine Pollutant Bulletin, **62** (2011)
4. A. Cobelo-Gracia and R. Prego. Marine Pollution Bulletin, **46** (2003)
5. V. Dauvelter and S. Rognerrud. Chemosphare. **42** (2011)
6. A.S. Ekwere, and A.A. Eluez. Research Journal of Applied Sciences Engineering and Technology, **4** (2012).
7. A. Galuszka. Poland Environmental Geology, **52** (2007).
8. G. Suresh, P. Sutharsan, V. Ramasamy, R. Venkatachalapathy. Ecotoxicology and Environmental Safety **84** (2012).
9. H.E. Hawkes and Webb, J.E. *Geochemistry in mineral exploration* New York: Harper (1962).
10. S. Manahan. E *Environmental Chemistry 8 Edition* USA, (2004).
11. J. Matschullat, R. Ottenstein and C. Reimann Environmental Geology **39**, (2000)
12. M.B. Mc Bride. *Environmental Chemistry of Soils* New York: Oxford University Press. 406. (1994)
13. G.S. Plumlee and M.J. Logsdon. Reviews in Economic Geology, **6A** (1995).
14. C. Reimann, P. Filzmoser and R.G. Garret. Science of the total environment, **346** (2005)
15. L. Ren-Ying, Y. Hao, Z. Zhi-Bao, L. Jun-Jie, S. Xiao-Hua, and J. Feng. Pedosphere, **17** (2007).
16. Report PT Indonesia Power UBP Saguling 1. http://www.indonesiapower.co.id
17. Sanjay Sakan, Gordan Devis, Environ Geochem Health **37** (2015).
18. D.C. Tomilson, J.G. Wilson, C.R. Haris and D.W. Jefferey, Helgol. Wiss. Meeresunters, **33**, (1980)
19. M. Varol, B. Şen.Catena **92**, (2012).
20. K.H. Wedepohl. Cosmochimical. Acta **59** (1995).
21. Z Wang, R. Sun, H. Zhang, L. Chen. Environ. Sci. Eng. 2015, **9** (2) (2015).

Assessment of Infectious Waste Management Practices at Hospital with Excellent Accreditation Level in Bandung, Cimahi and East Jakarta, Indonesia

Novi Fitria[1, 2,*], *Enri* Damanhuri[1], and *Indah R.S* Salami[1]

[1]Faculty of Civil and Environment Engineering, Bandung Institute of Technology, Jl. Ganesha 10, Bandung 40132, Indonesia
[2]Bakti Asih Higher education School of Analyst Bakti Asih Bandung, Jl. Padasuka Atas 233, Bandung 40192, Indonesia

Abstract. This study includes the procedures available and methods of handling and disposing of infectious waste at Military hospital with Excellent Accreditation level in Bandung, Cimahi and East Jakarta, Indonesia. A total three (3) military hospitals with equal type of hospital and level accreditation were surveyed during the course of this research. The methods consisted of survey and interview with the authorities of the hospital and the personal involved in the management of the generated waste. The information was collected using forms specially developed for this purpose. Site visits were conducted to support and supplement information gathered in the survey. Assessment of infectious waste handling divided into six parameters: Hospital policy at organizational structure, status of cleaning services worker, classification/segregation process, collect and transport the infectious waste, condition of temporary storage of infectious waste and disposal phase of infectious waste. The result showed that the hospital with highest level of accreditation have less appropriate practices when it comes to segregation, collecting, storage and disposal of waste generated in comparison to developed country. It appears that hospital authorities should pay better attention to educational planning, organizational resources and supervision at infectious waste management.

1 Introduction

Infectious waste from hospital has proven ability to transmit disease among healthcare workers and other individuals. Oroei [1] studied that infectious waste accounts for approximately 15-25% of all medical waste; however it has the potential to negatively affect human safety and health. A definition of infectious based on Stankovic [2], US EPA [3], Komilis [4], Nwachukwu [5] and government regulation PP 101/2014 [6] separate different types of waste.

Infectious waste includes anything potentially infectious, such as body fluids or secretion (e.g., blood, pleural fluid, semen, vaginal secretion, vomit, *feces* or urine), contaminated sharp object (e.g., contaminated needles, syringe and surgical blades), biological laboratory waste (e.g., cultures, stocks and growth media), pathological waste (e.g., human tissue, organs or body fluids), and single use disposable equipment, utensils and instrument soiled with potentially infectious agents. Hospital's facility or area that become source of infectious waste (e.g., isolation wards, emergency room (IGD), operating theatres, intensive care units (ICU), pathological and biological laboratories (Clinic pathology)

The World Health Organization (WHO) [7] has advocated that hospital waste should be treated as special wastes from Bdour [8] research. because it's volume and the complexity of hospital infectious waste increase, the risk of transmitting disease through unsafe handling and disposal practices also increases. The rise in incidence of such disease as AIDS, hepatitis B and C makes the possibility of infection of personal handling these waste and risk to public health arising from the transport of infectious waste. Hospitals are the biggest source of infectious waste.

Based on Fitria [9] studied that in one governance hospital in Bandung, in past two years (2015 and 2016) The Increased of medical waste generation varies greatly with range 35-556%. Infectious waste extremely increased in body tissue, syringe and intravenous bag category (556%, 533% and 444%). The Infectious waste now growing globally to impose stricter controls on handling and disposal of waste generated by hospital. This is an extension of the common concern for hospital to improve a monitoring and evaluating on infectious waste management should be an integral part of hospital management.

In Indonesia, to achieved a better practice in services to patients but still keep the integrity to environment concern, in 2012 based on regulation of the ministry of health (Permenkes no 12 /2012) [10], hospital must have accreditation level. The purpose from hospital accreditation program is to increase quality assurance to patients and environment. Hospital accreditations have three (3) level of assessment. Level one (1) includes 5 steps of assessment, level two (2) includes 12 steps of assessment and level three (3) of accreditation

* Corresponding author: novie.fitria@gmail.com

assessment includes 16 steps. The highest level of hospital accreditation assessment is level 3 with 16 steps of assessment excellent level. This condition describe the hospital with highest level of accreditation already settle the obligation to implement the appropriate rules and procedures in hospital management include the rules and procedures of infectious waste handling and disposal.

However in developing country such as Indonesia, the management hospital was in consist to implement the commitment, especially in infectious disease control. Sometimes budget issue is the reason for misconduct of infectious waste handling and disposal. Patwari [11] has studied, in many countries hazardous and medical waste are still handled and disposed of together with domestic, thus creating a health risk to healthcare worker, the public and environment.

Very few studies have been conducted on infectious waste in Indonesia that related to relationship between hospital accreditation level with its implementation especially on the handling and disposal of infectious hospital waste management. Furthermore, most our knowledge of hospital infectious waste management is limited. Thus, research required to establish database and information on hospital infectious waste handling and disposal. The objective of this study are directed towards: (1) conducting a survey of available procedures, techniques and methods of handling and disposal of hospital infectious waste that related to hospital accreditation level; and (2) Identifying the important characteristics and their possible influence of misconduct the handling and disposal implementation on hospital infectious waste.

2 Materials and Methods

This study was conducted at three (3) hospitals that have same class, type and level of hospital accreditation. The three hospitals were coded to maintain data confidentiality. The characteristic of the hospitals are military hospital with type of hospital "B" and excellent level accreditation. Hospital that located in Bandung, Cimahi and East Jakarta, Indonesia were coded with hospital "A", "B" and "C". Hospital facilities in this study assumed practice a good management of infectious waste and concern to environmental issue. Furthermore we have to explore how far the level of accreditation could make any progress in infectious waste management. Data collection lasted for three months – January until March, 2017 after receiving approval from Ethics Research Community of Dustira Hospital, Cimahi. The methods consisted of survey and interview with the authorities of the hospital and the personal involved in the management of the generated waste. The information was collected using forms specially developed for this purpose. Site visits were conducted to support and supplement information gathered in the survey. Interviews and site visits were helpful in obtaining information about common practices in the management of the infectious waste (e.g., handling, segregation, transport, storage, treatment, and final disposal). Qualitative analysis used as research methods.

Table 1 shows the hospital code, types and other information about study area.

Table.1. Summary of studied hospital

Hospital Code	Owner	Class	Accreditation level	Number of Beds
A	Military-Governmental Bandung	B	Excellent	192
B	Military-Governmental Cimahi	B	Excellent	496
C	Military-Governmental East Jakarta	B	Excellent	138

3 Results and Discussion

3.1 Organizaional Structure

Organizational structure is an arrangement or relationship between components if various parts and positions within an organization, the components contained in an organization have a dependency. So if there is a good component it will affect the other components and of course will affect the organization. In hospital A and C, the handling of infectious waste carried out under the supervision of environmental health department and evaluated by control of infectious diseases department (CID's) as showed at Fig. 1. But the opposite, in hospital B, infectious waste handling and control infectious diseases under in same authority which is health and safety department as showed at Fig. 2.

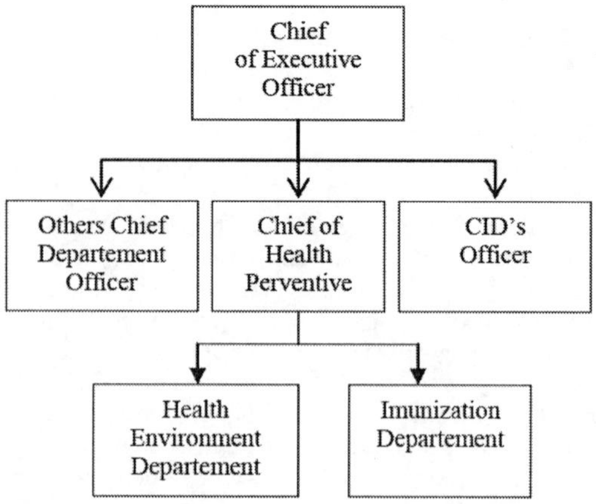

Fig.1. Organizational structure and infectious waste at Hospital A and C.

* Corresponding author: novie.fitria@gmail.com

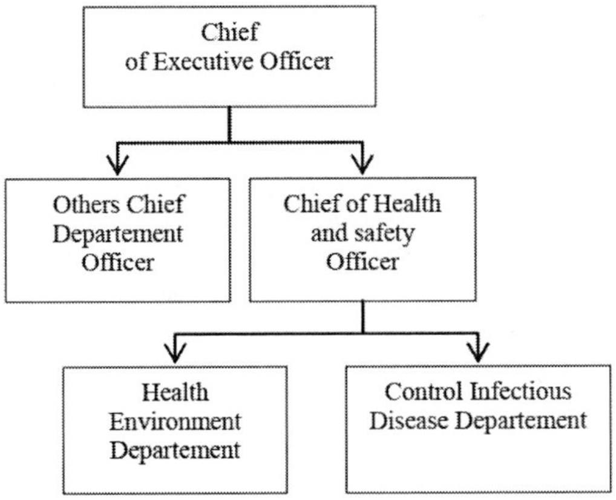

Fig.2. Organizational structure and infectious waste at Hospital B.

3.2 Hospital Cleaning Service Status

Hospital cleaning service personnel have the main duty to always maintain the cleanliness of the hospital. One of their tasks is to transport infectious waste from outpatient and inpatient facilities. Transportation of infectious waste is scheduled twice a day. In the morning, transport of infectious waste scheduled at 7 o'clock am and on the afternoon, transport is scheduled at 5 o'clock pm. The status of cleaning service personnel in the hospital varies. Hospital A and C have their cleaning service personnel status is as contract labor officer.

The management of worker status of the cleaning service personnel is still under the authority of the hospital's human resources department. The opposite, in hospital B the status of hospital cleaning service personnel uses outsourcing services, their status not under the authority of the hospital's human resources department. Differences in hospital cleaning service personnel status can have an impact on the implementation of infectious hospital waste management.

3.3 Segregation, Transport, Storage and Disposal of Hospital Infectious Waste

In general, the separation of infectious waste in the three hospitals has the same procedure. The three hospitals simply divide infectious waste into two categories, infectious and non-infectious waste. In infectious waste there is no further separation, although it has a different container between sharp and non-sharp infectious wastes, eventually disposing of both are placed in the same container that is in a yellow plastic bag. While non-infectious waste placed on a black plastic bag.

Based on International regulation, Kumari [12] studied there are six (6) segregation of hospital infectious waste (Table 2). In this study the category of infectious waste color coding divided into yellow and black plastic bag (Table 3).

Table.2. Different categories of waste along with international regulation color coding.

Category	Items	Color Coding
General waste	Paper, wrappers, peels of fruit and vegetables, remains of food & edibles etc.	Black
Bio - Medical waste		
Non-plastic infectious waste	Human and animal anatomical waste, waste contaminated with blood, and body fluids including cotton, dressings, soiled plaster casts, lines beddings, other material contaminated with blood.	Yellow
Plastics infectious waste	Disposable items such as glucose bottle, hub remove syringe tube, catheters, intravenous sets, gloves etc.	Red
Sharp waste	Needles, scalpels, blade, etc. that may cause puncture and cuts, this includes both used and unused sharp etc.	Puncture proof red container
Glass waste	Glass materials like bottle, ampoules, slide, tubes, etc.	White
Liquid waste	Waste generated from laboratory and washing, cleaning, housekeeping and disinfecting activities, Chemical used in production of biological, chemicals used in disinfection, as insecticides etc.	Blue

Based on the results of interviews shows that the hospital management difficulty getting a supplier of red and purple plastic bags. In addition, the price of purple and red plastic bags is more expensive than the black or yellow plastic bags. The higher price of plastic bag in other color becomes the influencing factor. Sometimes it doesn't fit with the purchase budget.

The practice of separation of infectious and non-infectious waste hospital generally has the same problem, which infectious waste and non-infectious waste has been mixed starting from the source. This is due to the depletion of yellow plastic stock as a hospital infectious waste container. The delays caused the nurse used the black plastic bag to collect the infectious waste.

Hospitals that do not use outsourcing services in their hygiene worker status (hospitals A and C), have run out of plastic bags a few times due to hospital purchase delays. But hospital B that used outsourcing services, almost never run out of yellow plastic bags. Because based on on-going contracts with outsourcers, all infectious and non-infectious containers have been provided automatically. In this case prove that using outsourcing services due to infectious waste handling more safe and effective.

* Corresponding author: novie.fitria@gmail.com

https://doi.org/10.1051/matecconf/201814708004

Table. 3. Segregation practice based on color coding at Hospital A,B and C.

Category	Items	Color Coding
General waste	Paper, wrappers, peels of fruit and vegetables, remains of food & edibles etc.	Black
Bio - Medical waste		
Non-plastic infectious waste	Human and animal anatomical waste, waste contaminated with blood, and body fluids including cotton, dressings, soiled plaster casts, lines beddings, other material contaminated with blood.	Yellow
Plastics infectious waste	Disposable items such as glucose bottle, hub remove syringe tube, catheters, intravenous sets, gloves etc.	Yellow
Sharp waste	Needles, scalpels, blade, etc., that may cause puncture and cuts, this includes both used and unused sharp etc.	Puncture proof Yellow container
Glass waste	Glass materials like bottle, ampoules, slide, tubes, etc.	Yellow
Liquid waste	Waste generated from laboratory and washing, cleaning, housekeeping and disinfecting activities, Chemical used in production of biological, chemicals used in disinfection, as insecticides etc.	White and yellow plastic container

However the hospital management should start to concern about quality of yellow and black plastic bag. Every year the price of plastic bags is increasing and it is not accompanied with the budget to purchase the plastic bags, it makes the hospital management selected the lower quality of plastic bags. The impact of this decision, yellow plastic bags easily tear and spill out the content.

At transportation phase, the transport of infectious and non-infectious waste from the treatment room to the waste bin is temporarily practiced in different ways although the procedure for transporting infectious waste is already referring to the correct standards based on Chen [13]. Fig. 3 shows the standard procedure for transporting infectious waste. In the procedure there is a phase that infectious waste should through the sterilization process before weighing process done.

Fig. 4 shows that in hospital A have mismanagement of the transport of infectious and non-infectious waste from treatment room to temporary storage is practiced on the same garbage trolley. Yellow and black plastic bags were brought in the same trolley to temporary storage.

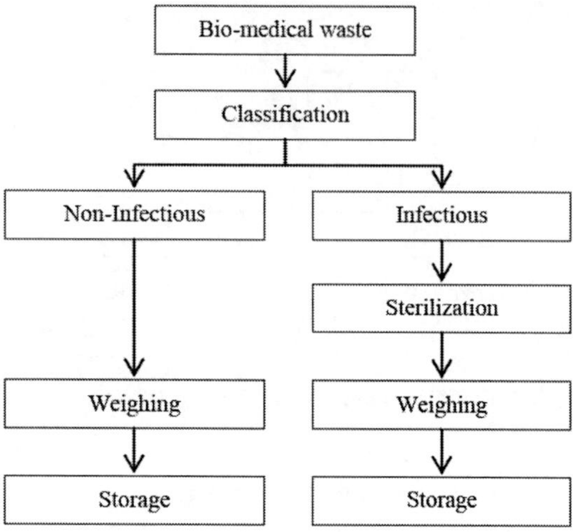

Fig.3. Standard treatment processes for infectious waste and noninfectious waste

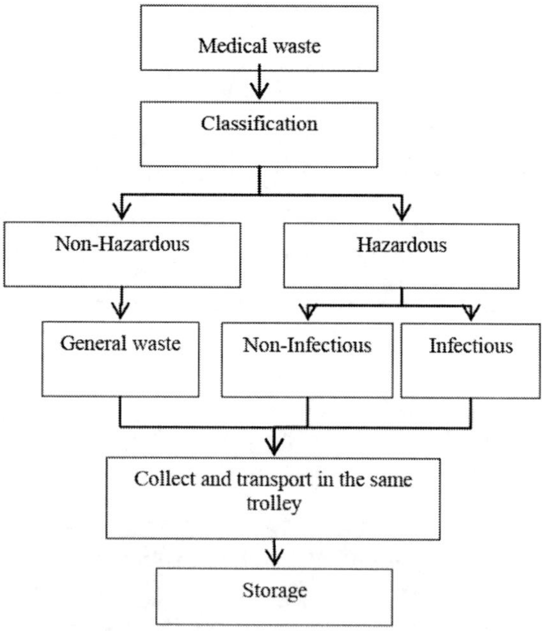

Fig.4. Treatment processes for infectious waste and non-infectious waste at Hospital A

Hospital B using outsourcing services to transport infectious and non-infectious waste, both are transported by different trolleys. Fig.5 shows the path of infectious waste handling in real activities at hospital B. After the infectious waste arrived at temporary waste, the cleaning services sorting between plastic and non-plastic infectious waste to reduce the volume of infectious waste, after that, the non-plastic infectious waste will be incinerated. At hospital C, the transport of infectious and

* Corresponding author: novie.fitria@gmail.com

non-infectious waste practiced separately, but before the infectious waste placed in infectious temporary storage area, infectious waste is transported to a place where non-infectious waste is located. This happens because there is a practice of sorting on infectious waste before being placed in temporary storage.

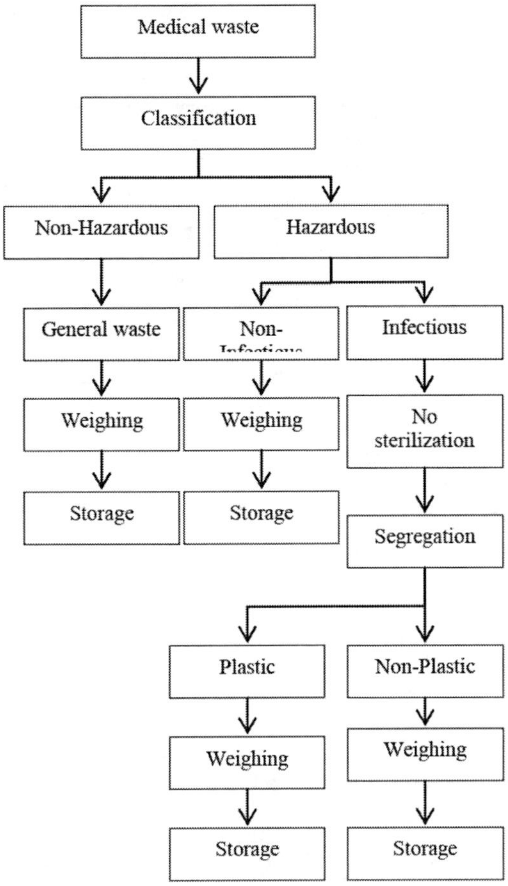

Fig.5. Treatment processes for infectious waste and non infectious waste at Hospital B

Based on the regulation of infectious waste transportation, the infectious waste must through the sterilization process before weighing and storage phase. However none the hospitals practiced the sterilization process, because the price of sterilization equipment for infectious waste is very expensive. Disadvantages of infectious waste transport practices at hospital C as illustrated in Fig.6 shows that the cleaning services workers carrying infectious and non-infectious waste at the same time. Procedurally, infectious waste carried separately with non-infectious waste. The impact of this infectious waste transport practice will lead to general waste and non-infectious waste being automatically became infectious waste. At the end the general waste and non-infectious waste that already change the status became infectious waste will be transport from temporary storage to open dumping landfill area.

If the nurse cannot find yellow plastic bag for infectious waste, they will use black plastic bag to collect the infectious waste from treatment room. Cleaning services worker will not recheck what inside of black plastic bag. They will be direct put the black plastic bag that contain infectious waste in general waste temporary storage. Infectious waste will arrived along the municipal waste to open dumping area.The reason they conduct this practices because the temporary storage of infectious waste and non-infectious waste are in same location and there is no warning from their supervisor. Interview result with cleaning services worker show, that actually they know and notice about the infectious wate standard procedures, but they ignore the procedures with efficiency and effectiveness times reason and they can finish their job faster.

Infectious waste transport practices in hospital B as illustrated in Figure 9 show that the cleaning services worker conduct the infectious waste trasport as the standard procedures. But there is activity that the cleaning services worker sorted infectious waste in temporary storage 3. The cleaning services worker seperated between plastic infectious and non-plastic infectious waste before disposal process done. This activity will risk the cleaning services worker health exposure by infectious diseases. They conduct the seperation activity because instruction from the hospital management in purpose to reduce over budget for infectious waste disposal. Hospital B has an incinerator which can use to incinerated non-plastic infectious waste. At the end the hospital will only pay to third party to dispose their infectious waste in plastic category only.

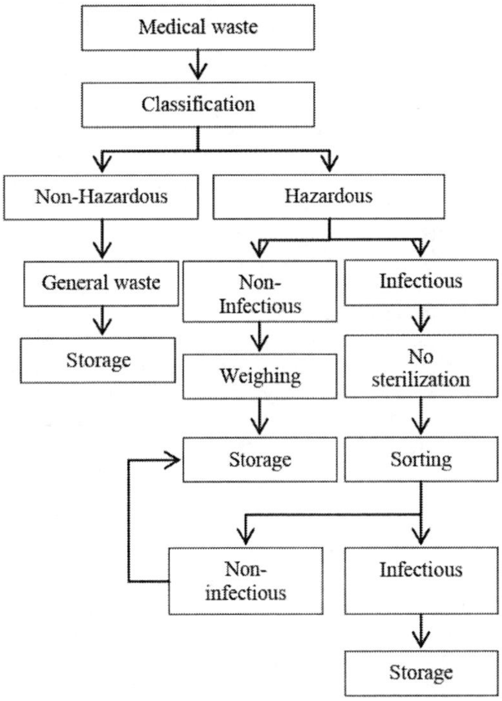

Fig.6. Treatment processes for infectious waste and non infectious waste at Hospital C

* Corresponding author: novie.fitria@gmail.com

As illustrated in Fig.6, hospital C the cleaning services worker sorted infectious waste in temporary storage . If the cleaning services worker found non-infectious and general waste in yellow plastic bag, they will took it from the yellow plastic bag and then collect it in black plastic bag. Procedurally, infectious waste that already collects in yellow plastic bag from treatment room cannot be open again, and keep it that way until arrived at infectious waste temporary storage. The impact of this infectious waste sorting practice will lead to general and non-infectious waste being automatically became infectious waste. At the end the general waste and non-infectious waste that already change the status became infectious waste will be transport from temporary storage to open dumping landfill area. Otherwise the cleaning services worker will risk exposure infectious diseases.

Temporary storage phase, condition of infectious temporary storage also varies between these three hospitals. At hospital A, the temporary storage between general wastes, non-infectious waste and infectious waste are in the same location. There is no door at temporary storage, this condition possible anyone and animal go inside to temporary storage. The temporary storage floor at hospital A is not in ceramic coated. This condition makes the cleaning services worker difficulty to clean the storage floor.

At hospital B and C, the temporary storage between general wastes, non-infectious waste and infectious waste are in different location. The conditions for temporary storage of infectious waste are in accordance with the regulations. They lock the temporary storage of infectious, only in charge worker who can enter the temporary storage. The temporary storage also already ceramic coated, this condition makes the worker easy to clean up the temporary storage. Infectious and non – infectious located in different storage, this condition makes the lower risk that infectious waste will contaminated the non-infectious waste and general waste. At hospital A, B and C there is no cooler room facility, to storage infectious waste. The average temperature at temporary storage was 28-31°C.

Disposal Phase, Hospital A, B and C have the same registered waste disposal companies who collect and transport the infectious waste. The disposal companies (outsourcing company) collect and transport at hospital A, B and C twice a week. In this study show the infectious waste will stay more than 48 hours in temporary storage. Based on government regulation, if the infectious waste stay in temporary storage more than 48 hours, the infectious waste must located in the storage room with cooler facility to prevent the bacteria or other pathogen microorganism growth.

At hospital A, if the disposal companies who collect infectious waste from temporary storage being late from the schedule, the temporary storage will over capacity and the cleaning services worker will put the infectious container outside from temporary storage. The collection of infectious waste schedules are twice a week with a total of 14 container of infectious waste.

Standard regulation mention that the contents of the container should not exceed 2/3 of the total volume of the, but in this study show that containers are overloaded, due this condition the outsourcing worker will step on the infectious waste before they put the container inside the infectious waste vehicle.

3.4 Assessment of infectious waste management practices at Hospital A, B and C

Evaluating the handling practices of infectious waste in Hospital is important to improve infectious waste management. Hospital with excellent accreditation level should conduct the infectious waste handling according to standard procedures. Comparison the result between the hospitals shows in Table 4.

Table.4. Assessment of Infectious waste management practices at Hospital A, B and C

Parameter of Assessment	Hospital A		Hospital B		Hospital C	
	NC	C	NC	C	NC	C
Organizational Structure		√		√		√
Status of Cleaning services worker		√		√		√
Classification/segregation Phase	√			√		√
Collect and transport phase	√			√	√	
Storage phase	√			√		√
Disposal phase		√		√		√
% Comply		50		100		83

NC = Not comply according to standard procedures
C = Comply according to standard procedures

4 Conclusions

The infectious waste generated by hospital, which is heavily contaminated with pathogen microorganism, seldom receives the attention it deserves from researchers. Therefore, this study evaluated the practices of infectious waste at hospital with the highest accreditation level using observation and formal interview method. The infectious waste management practices assessment divided into six parameters: Organizational structure, status of cleaning services worker, classification/segregation, collect and transport phase, storage phase and disposal phase.

Failure of infectious waste treatment process in hospital is generally caused by human error. The status of outsourced cleaning services workers has a higher level of awareness and cautious of infectious waste danger compared to non-outsourcing cleaning service worker.

This study provides useful information for workers, decision-makers and public regarding the proper management of Infectious waste management and the sustainable development of waste management. From three hospital that have been observed, the hospital B achieved the highest score (100%) for the infectious waste management procedures against six assessment parameters. Other side, hospital A and C scored 50% and

* Corresponding author: novie.fitria@gmail.com

83%. This research proves that the level of excellent accreditation does not guarantee the implementation of infectious waste handling in accordance with the procedure. This study also increases the understanding of the potential problems associated with infectious waste, thereby increasing awareness of how to improve management of hospital infectious waste to better protect worker, public and environment.

The authors would like to thank to Department of health environment the hospital in Bandung, Cimahi and East Jakarta for the warm reception and valuable contribution given during the study. The works are financially supported by Minister of Research and Education of Indonesia and Department of Environmental engineering, Bandung Institute of Technology.

References

1. M. Oroei, M. Momeni, C.J. Palenik, M. Danaei, M. Askarian. Journal of infectious and Public Health, J. **7**, pp 192-198, (2014)
2. A. Stankovic, D. Niki, M. Nikolic, Waste Management and Reasearh, **26,** pp 309 – 313, (2008)
3. US Environmental Protection Agency (US EPA)., Medical waste. Available at : https://www.epa.gov/rcra/medical-waste (1992)
4. D. Komilis, A. Fouki, D. Papadopoulus, Waste Management, J. **32**, pp 1434-1441, (2012)
5. N.C. Nwachukwu, F.A. Orji, O.C. Ugbogu, Epub.DOI : 10.5772/53196 (2013)
6. Peraturan pemerintah Indonesia (PP) 101/2014., Available at : https://www.kemenkopmk.go.id/sites/default/files/produkhukum/PP%20Nomor%20101%20Tahun%202014_0.pdf (URL link)
7. World Health Organization (WHO)., Medical waste, Available at : http://www.who.int/topics/medical_waste/en/, (URL link).
8. A. Bdour, B. Altrabsheh, N. Hadadin, M. Al-shareif, Waste Management, J. **27**, pp 746-759, (2007)
9. N. Fitria, E. Damanhuri, 3rd Symposium of the Asian Regional Branch of International Waste Working Group, Departement of Civil and Environmental Engineering (ICEE), pp.59, (2017).
10. Ministry of health Indonesia (Permenkes 12/2012) Available at : http://perpustakaan.bappenas.go.id/lontar/file?file=digital/122029-[_Konten_]-Permenkes%20No%2012%2020120001.pdf, (URL link)
11. M.A. Patwary, W.T. O'Hare, G. Street, K.M. Elahi, S.S. Hossain, M.H. Sarker, Waste Management, J. **29,** pp 2392-2397, (2009)
12. R Kumari, K. Srivastava, A. Wakhlu, A. Singh, Clinical epidemiology and global health, J, **1**, pp 131-136, (2013)
13. Y.C. Chen, P.Y. Tsai, Waste Management & Research, J. **35** pp 1-9, (2017)

* Corresponding author: novie.fitria@gmail.com

 https://doi.org/10.1051/matecconf/201814708005

Study of whole effluent acute toxicity test (*Daphnia magna*) as an evaluation of Ministry of Environment and Forestry Decree No. 3 In 2014 concerning industrial performance rank in environmental management

Neng Rohmah[1], *Dwina* Roosmini[2], and *Mochamad Adi* Septiono[2]

[1]Program Studi Teknik Lingkungan, Fakultas Teknik Sipil dan Lingkungan, Institut Teknologi Bandung, Jl Ganesha 10 Bandung 40132
[2]Program Studi Teknik Lingkungan, Fakultas Teknik Sipil dan Lingkungan, Institut Teknologi Bandung, Jl Ganesha 10 Bandung 40132

Abstract. Only 15% of the industries in Citarum Watershed, specifically in Bandung Regency, West Bandung Regency, Sumedang Regency, Bandung City and Cimahi City, are registered as PROPER industries. They must comply to indicators as set in the Minister of Environment and Forestry Decree No. 3 In 2014 concerning Industrial Performance Rank in Environmental Management, as a requirement to apply for PROPER. Wastewater treatment and management, referencing to Minister of Environment and Forestry Decree No. 5 In 2014 concerning Wastewater Effluent Standards, must be performed to be registered as PROPER industries. Conducting only physical-chemical parameter monitoring of wastewater is insufficient to determine the safety of wastewater discharged into the river, therefore additional toxicity tests involving bioindicator are required to determine acute toxicity characteristic of wastewater. The acute toxicity test quantifies LC_{50} value based on death response of bioindicators from certain dosage. *Daphnia magna* was used as bioindicator in the toxicity test and probit software for analysis. In 2015-2016, the number of industries that discharged wastewater exceeding the standard was found greater in non-PROPER industries than in PROPER industries. Based on the toxicity level, both PROPER and non-PROPER industries have toxic properties, however PROPER industries of 2015-2016 is more toxic with $LC50_{96}$ value reaching 2.79%.

1 Introduction

PROPER (*Program Penilaian Peringkat Kinerja Perusahaan/Assessment of Industrial Performance Rank Program*) as industrial rank incentive of environmental aspect is one of efforts implemented by Indonesian Ministry of Environment and Forestry (MoEF) to pursue compliance of environmental management regulations. Registered PROPER industries in Upper Citarum watershed, specifically Bandung Regency; West Bandung; Sumedang Regency; Bandung City and Cimahi City, amount to only 15% of all industries around those locations (BPLHD West Java, 2015). PROPER industry participants must comply to certain parameters in Minister of Environment and Forestry Decree No. 3 In 2014 (*PermenLKH No. 3/2014*) to be registered. In regard to wastewater, there are further tests to be conducted and effluent standards to be complied, according to Minister of Environmental and Forestry Decree No. 5 In 2014 (*PermenLKH No. 5/2014*) concerning wastewater effluent standard, including physical, chemical and biological parameters. Wastewater quality measurements in physical parameters have burden of solid dimensions, while chemical parameters require another chemical addition. Therefore, further biological tests exposing certain dosage on bioindicators is needed to determine toxicity level of discharged wastewater.

Physical and chemical parameter assessment alone would be insufficient in pointing out pollution impacts on aquatic organisms (Ningrum, 2012). Government regulations obligates to figure wastewater acute toxicity through death responses measurement of test animals from different dosage of wastewater. Concentration that causes 50% of bioindicator death response is used to estimate LC_{50} value in a period of observation time, according to Government Regulation No. 18 In 1999 (PP No. 18/1999). Acute and chronic toxicity is determined through toxicity tests (William and Burson, 1995). In this research, *Daphnia magna* is used as bioindicator to analyze acute toxicity, then using probit software in analysis.

* Corresponding author: [1]ftslnengrohmah@gmail.com, [2]dwinaroosmini@yahoo.com, [3]tioseptiono@gmail.com

2 Methodology

Whole Effluent Toxicity (WET) tests was conducted with grab samples from registered PROPER and non-PROPER industries of 2015-2016 in Upper Citarum watershed, in Laboratory of Industrial Hygiene of Environmental Engineering ITB. Measurements of environmental parameters, namely pH, dissolved oxygen, conductivity and temperature, were conducted. Names and classification/type of industry samples is shown in Figure 1. Two liters of wastewater samples was preserved at 4°C without additional preservatives, to prevent chemical transformation and microorganism degradation.

In these toxicity tests, *Daphnia magna* was used as bioindicator due to its simple method of cultivation in the lab and its important role in the food chain and sensitivity to various pollutants (Mark, 1997). WET Tests include 7 days of cultivation phase using bottled water instead of tap water to avoid chlorine contamination, which poisons *Daphnia magna* (US.EPA, 2002). During this process bioindicators were fed with yeast powder and sanitized properly.

WET test was conducted with variation of wastewater concentration, with each combination tripled and in static non-renewal, in which test organisms are exposed to a fresh solution of the same concentration of sample, either by transferring the test organisms from one test chamber to another, or by replacing all or a portion of solution in the test chambers (USEPA, 2002).

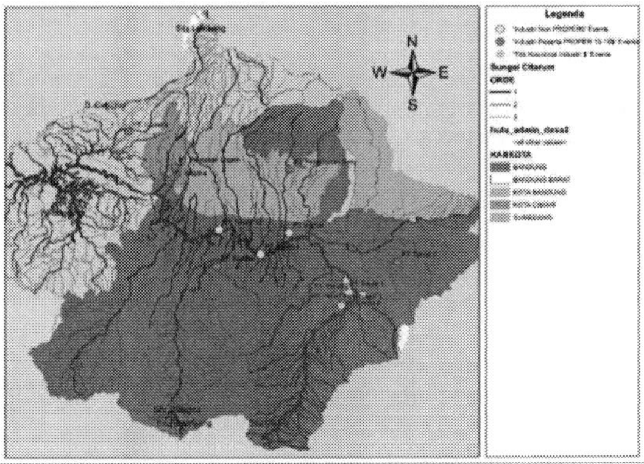

Figure 1 Location study

There are two stages in this method, namely range finding test and definitive test. Range finding test exposes test animal for 48 hours to the samples. Definitive test requires 96 hours in total, with number of death responses and environmental parameters need to be observed at each concentration every 24 hours. Standard measurement method of physical and chemical properties of the waste is shown in Table 1. In addition to those two stages of the test, samples underwent references toxicant test, which aims to determine level of sensitivity and health of test organisms over time. Mount, et al (1997) suggested that LC_{50} reference value of a toxicity test is 4.7 g/L, which means the WET tests are adequate to be conducted if the actual LC_{50} value of this test are close to the reference value.

Table 1 Characteristics parameter measurement method of physical-chemical

Parameter	Measurement Method	Standard of Measurement
pH	pH meter	SNI 06-6989.2-2009
Phosphate	Spectrophotometer	SNI 6989.76-2009
COD	Closed Reflux	SNI 6989.2-2009
Hardness	Complexometry - EDTA	SNI 06-6989.12-2004
Alkalinity	Acid-Base Titration	SNI 06-2420-1991
Ammonium	Spectrophotometer	SNI 19-6603-2001
TSS	Gravimetry	SNI 06-6989.3-2004
Conductivity	Conductivity meter	
Temperature	Thermometer	
DO	DO meter	
Phenol	Spectrophotometer	SNI 06-6989.21-2004
TDS	Gravimetry	SNI 06-6989.3-2004

In the range finding test, wastewater concentrations are shown in Table 2, which involved 40 mL of wastewater sample and six (6) *Daphnia magna* in each test chamber. Range finding test resulted in minimum and maximum concentration, which later were used in the definitive test. New set of concentration in the definitive test was measured using Equation 1. By multiplying R value to the smallest range, new concentration was able to be determined for definitive test.

Table 2 Variation of concentration in range finding test

No.	Waste Concentration (%)	Wastewater (mL)	Dilution Water (mL)
1	6,25	2,5	37,5
2	12,5	5	35
3	25	10	30
4	50	20	20
5	100	40	0

$$r = \left(\frac{a}{b}\right)^{\frac{1}{n-1}} \quad (1)$$

Description: r = constant
a = the highest concentration with mortality of 50%
b = the lowest concentrations with mortality below 50%
n = number of concentrations

3 Result and Discussion

3.1 Participant of PROPER 2014-2015

PermenLHK No. 3/2014 states PROPER is an evaluation of compliance and performance beyond compliance of industries or any activity in regards of pollution control, environmental impact assessments, and/or hazardous waste management. PROPER rates these performances into color-rated ranks namely black, as the least

https://doi.org/10.1051/matecconf/201814708005

preferred, red, blue, green, and gold, as the most preferred.

PROPER industry ranks in this study area are shown in Figure 2. There had been 51 industries participated in PROPER in 2014 and 50 industries in the following year. There was 34 blue-ranked PROPER industries, then in 2015 the number increased to 37. However, there had been declining numbers in green and gold ranks PROPER industries.

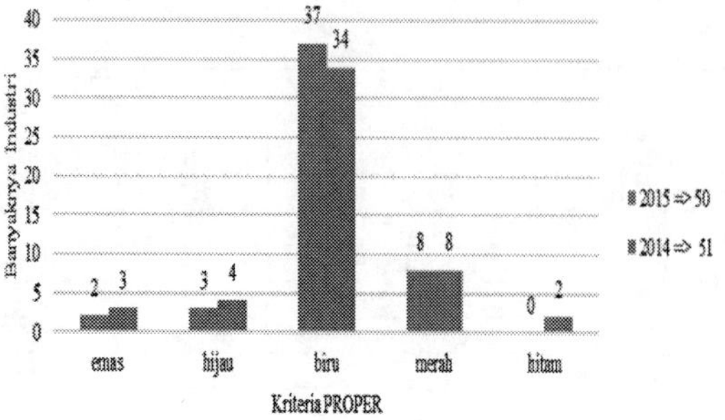

Figure 2 The Value PROPER In Citarum Hulu

Figure 3 shows PROPER industries based on their ranks around Upper Citarum area. Textile industries have the most blue-ranked industries, reaching to 10, while red-ranked were hotels with 4 hotels.

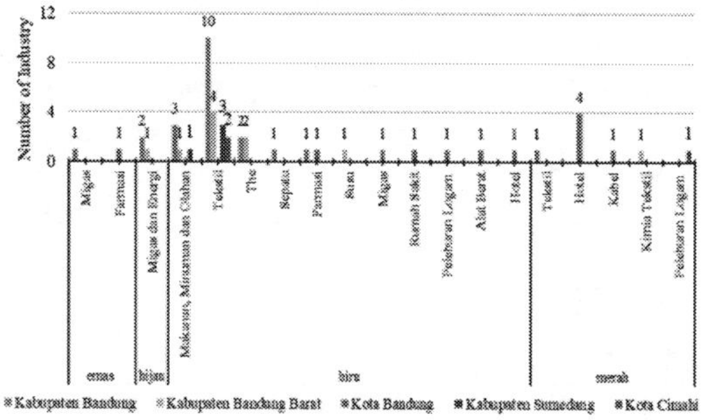

Figure 3 Type Industry Based on Value PROPER in Citarum Hulu

Blue-ranked PROPER shows that industries have conducted environmental management assessment and carry out programs as required in the policy and regulations, including wastewater discharge permit and its monitoring plans based on PermenLHK No. 3/2014. However, these control efforts had not yet reached 100%. Several of these industries had not yet been in compliance to effluent standards on PermenLKH No.5/2014 as they should.

3.2 Wastewater physical-chemical properties of PROPER registered industries in 2015-2016 and Non-PROPER industries

3.2.1 pH

pH value measures hydrogen ions, indicating the acidity or alkalinity scale of a solution. Figure 4 shows to meet the standard, pH must be within the range of 6-9, according to PermenLKH No.5/2014. Alkaline or acid water may be harmful to living organism due to its interference to metabolism and respiration system. Ideal pH for aquatic organisms varies between 7 to 8.5.

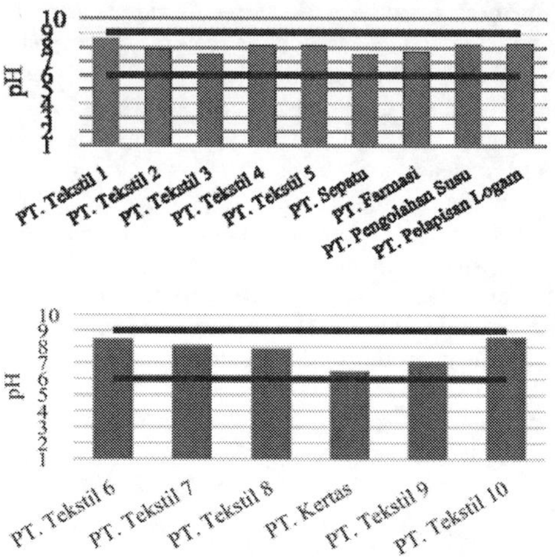

Figure 4 Concentration of pH wastewater effluent. Above: industry participants PROPER 2015 - 2016 and below: industrial non PROPER

3.2.2 Conductivity/Electric conductivity (EC)

Conductivity measures strength of a solution to pass electrical flow through conductive ion contents. PT. *Pelapisan Logam* (PT. Metal Coating) had the highest conductivity reaching 3950 µs/cm, while PT. Tekstil 4 had the smallest conductivity, which is389 µs/cm for PROPER industries. Non-PROPER industries have higher value with the highest reached 4030 µs/cm (PT. Tekstil 8) and the lowest value of 645 µs/cm (PT. Tekstil 10). Full result is shown in Figure 5. High value of conductivity in wastewater effluent means ther is rich mineral content. Conductivity value depends on content of inorganic ions (Tchobanoglous, 2003).

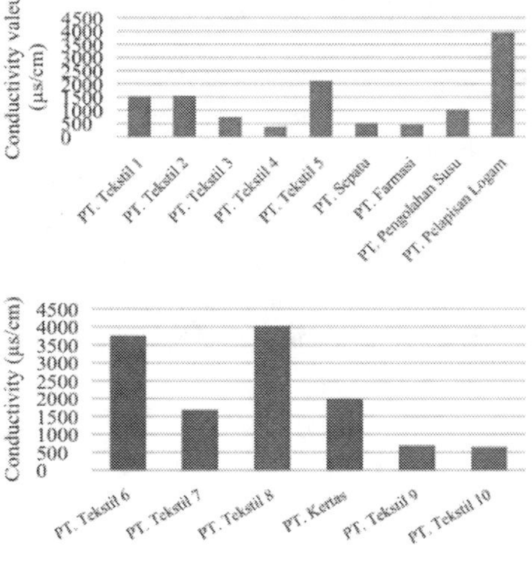

Figure 5 Concentration of Conductivity wastewater effluent. ABove: industry participants PROPER 2015 - 2016 and below: industrial non PROPER

3.2.3 Turbidity

Turbidity shows inorganic or organic material content in a solution or water. This inorganic or organic material can affect the ecosystem of aquatic organism. Figure 6 shows PROPER industries have the greatest value of turbidity of 135 NTU (PT. Tekstil 1), while the lowest is 19.133 NTU (PT. Tekstil 3). Non-PROPER industries have higher range on turbidity value, reaching 331 NTU (PT. Paper) as the highest and 52.2 NTU (PT. Tekstil 8) for the least turbid. Turbidity indicates fine particle contents suspended in a solution or wastewater and causing light not able to propagate straight into solutions. High value of turbidity indicates greater level of pollution in bodies of water.

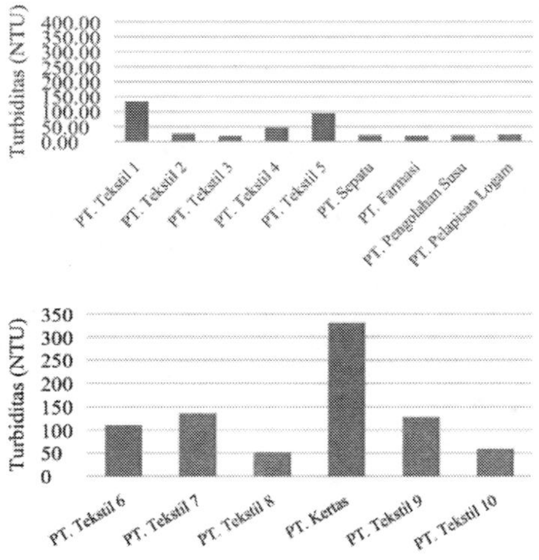

Figure 6 Concentration of Turbidity in wastewater effluent. Above: PROPER industries 2015 - 2016 and below: non-PROPER industries

3.2.4 Total Suspended Solids/TSS

TSS is residual total solids with maximum particle size of 2 µm or larger than the size of the colloidal particles. Figure 7 shows TSS in wastewater from PROPER and non-PROPER industries in 2015-2016. Lower range was found in PROPER industries, as 456 mg /L (PT. Tekstil 5) for the most TSS content and 86 mg/L (PT. Tekstil 2) for the least, and higher range was from non-PROPER industries with 592 mg/L (PT. Kertas) as the highest and 224 mg/L (PT. Tekstil 10) as the lowest. TSS content will reduce sun penetration into the water, affecting oxygen regeneration from photosynthesis and may cause disruption to aquatic organism growth.

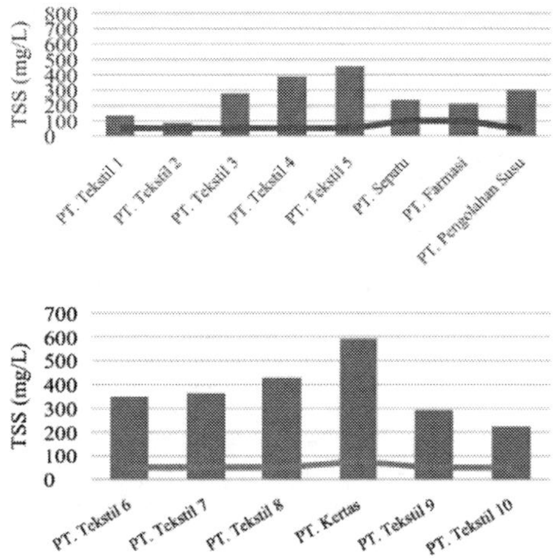

Figure 7 Concentration of TSS wastewater effluent. Above: PROPER industries 2015 - 2016 and below: non-PROPER industries

3.2.5 Total Dissolved Solis/TDS

TDS indicates dissolved content, whether organic or inorganic substances in liquid solution. Figure 8 shows TDS value in PROPER industries in 2015-2016and non-PROPER industries. For PROPER industries in 2015 - 2016, the highest value was 3480 mg/L (PT. Metal Coating) and the smallest value was 144 mg/L (PT. Pharmacy). As for non-PROPER industry, the highest value was 3004 mg / L (PT. Textile 6) and the lowest was 640 mg/L (PT. Textile 10). TDS value measures inorganic materials in ion. Conductivity correlates with TDS due to the concentration of ionized dissolved solids in water.

 https://doi.org/10.1051/matecconf/201814708005

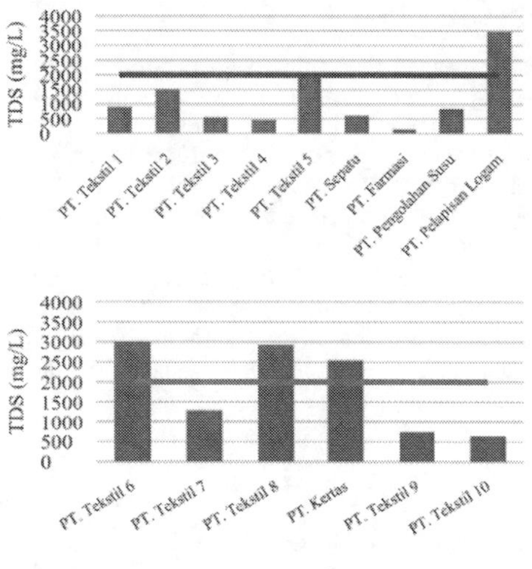

Figure 8 Concentration of TDS wastewater effluent. Above: PROPER industries 2015 - 2016 and below: non-PROPER industries

3.2.6 Chemical Oxygen Demand (COD)

COD represents the amount of oxygen required to oxidize organic substances contained in water (Boyd, 1990). Figure 9 shows COD value of PROPER industries in 2015-2016 and non-PROPER industries. All PROPER industries in 2015-2016 discharged wastewater with COD content meeting the standards in PermenLKH No. 5/2014, with the highest value of 198.4 (PT. Tekstil 1) and 121.6 mg/L (PT. Tekstil 2) as the lowest. Non-PROPER industries show relatively higher COD with wider range from 320 mg/L (PT. Tekstil 10) to 3840 mg/L (PT. Tekstil 9). Greater amount of COD in wastewater causes DO to decrease, leading to death for test animals (Effendi, 2003).

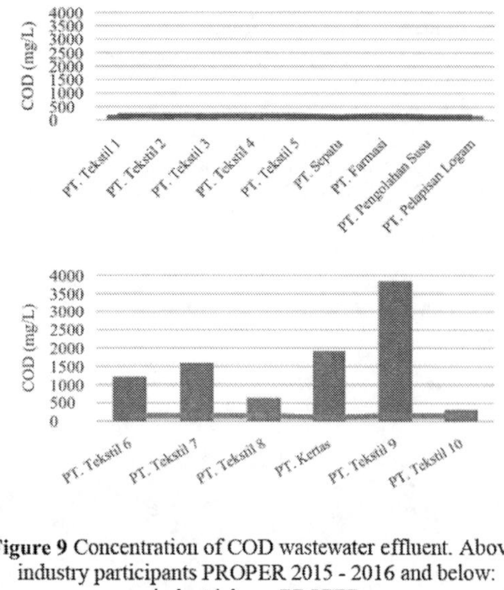

Figure 9 Concentration of COD wastewater effluent. Above: industry participants PROPER 2015 - 2016 and below: industrial non PROPER

3.2.7 Total Hardness

Total hardness of PROPER industries of 2015-2016 and non-PROPER industries (Figure 10), which is the amount of Ca^{2+} and Mg^{2+} ions, was determined by EDTA titration and used cation-sensitive indicators. PROPER industries had the range of 9.34 mg/L $CaCO_3$ (PT. Pharmacy) to 214.21 mg/L $CaCO_3$ (PT. Textiles 1), as for non-PROPER industries from 20.31 mg/L $CaCO_3$ (PT. Tekstil 10) to 253.81 mg/L CaCO3 (PT. Textile 6). Based on the toxicity level, greater hardness of water may limit fish growth while low hardness can increase fish sensitivity to toxic metals.

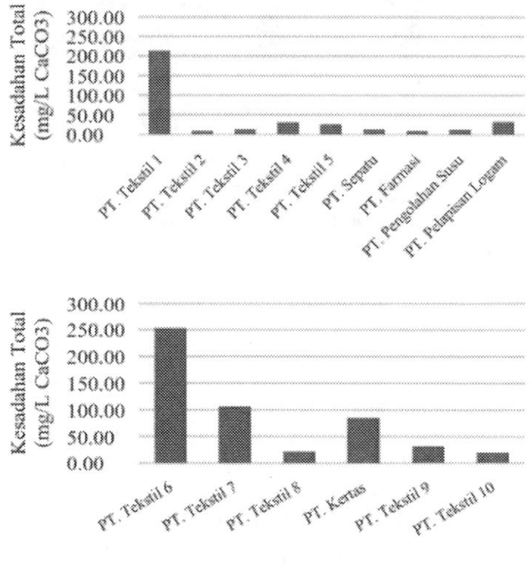

Figure 10 Concentration of Total Hardness wastewater effluent. Above: PROPER industries 2015 - 2016 and below: non-PROPER industries

3.2.8 Total Alkalinity

The alkalinity is the ability of wastewater to neutralize acid or anion quantity, which neutralizes hydrogen cations. Figure 11 shows that PROPER industries have relatively lower content of total alkalinity compared to non-PROPER industries, ranging from 35 mg/L $CaCO_3$ (PT. Metal Coating) to 970 mg/L $CaCO_3$ (PT. Textile 5). Non-PROPER industries had relatively higher value rangeing from 1200 mg/L $CaCO_3$ to 3,700 mg/L $CaCO_3$ (PT. Tekstil 6) and the lowest value of 1,200 mg/L $CaCO_3$. Optimum alkalinity ranges from 90 to150 mg/L $CaCO_3$, while lower alkalinity is limited to the value of 5 mg/L $CaCO_3$. Aquatic organisms tend to avoid high water with alkalinity, which is also proportional to hardness.

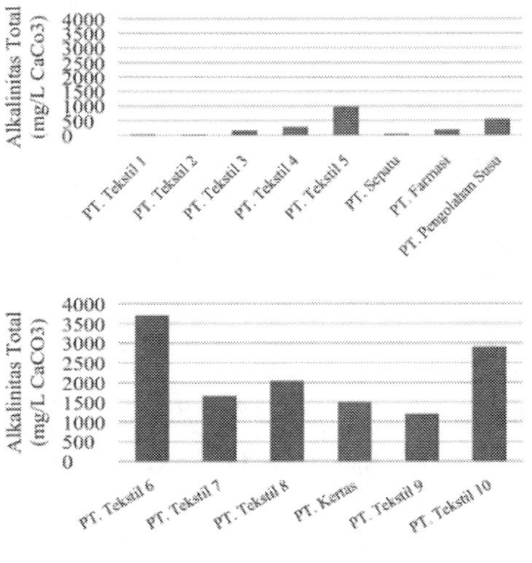

Figure 11 Concentration of Total Alkalinity wastewater effluent. Above: PROPER industries 2015 - 2016 and below: non-PROPER industries

3.2.9Ammonium

Figure 12 shows total ammonium in PROPER industries of 2015-2016 and non-PROPER industries, with the highest content reaching 4.06 mg/L (PT. Textiles 1) for PROPER industries of 2015-2016 and 13.64 mg/L (PT. Textile 10) for non-PROPER industries. Lowest value was reaching 1.73 mg/L (PT. Textile 6) for PROPER industries and 4.35 mg/L for non-PROPER. At low concentration, high ammonium content may lead to lower rate of physical development and, accumulatively, death in fish and shrimp (Da Cruz, 2007).

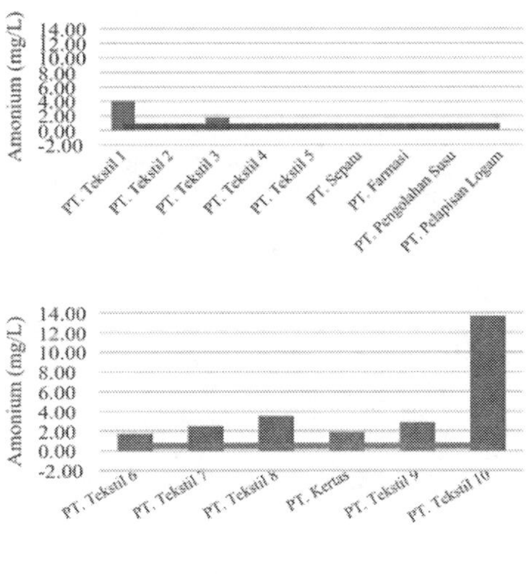

Figure 12 Concentration of Ammonium wastewater effluent. Above: PROPER industries 2015 - 2016 and below: non-PROPER industries

3.2.10 Total Phosphate

Phosphate is major source of potassium and nitrogen that are not soluble in water. Figure 13 shows that PROPER industries of 2015-2016 with the most total phosphate content was reaching 1.25 mg/L (PT. Tekstil 4), while non-PROPER ranged from 2.03 mg/L (PT. Textiles 9 & PT. Paper) to 0.21 mg/L (PT. Textile 8).

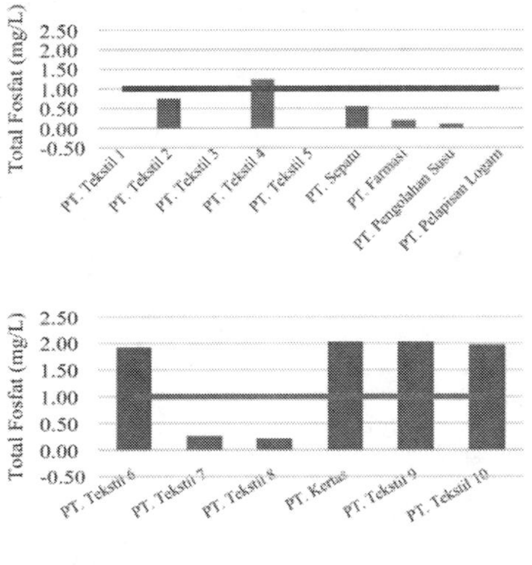

Figure 13 Concentration of Total Phosphate wastewater effluent. Left: PROPER industries 2015 - 2016 and right: non-PROPER industries

3.3 Whole Effluent Toxicity Test: Reference test

During reference test, *Daphnia magna* was exposed with NaCl for 24 hours as control reference. Concentration varied between 5 g/L, 4.8 g/L, 4.6 g/L, 4.4 g/L and 4.2 g/L. Reference test results showed 50% death of *Daphnia magna*, shown in Table 3.

Table 3 LC50 Value of The Reference Test

No.	Date of Reference Test	LC50 Value	Average	Deviation Standard
1	12 June 2016	4,903	4,685	0,584
2	16 July 2016	5,128		
3	16 August 2016	4,023		

Results shows that average LC50 from three reference tests is 4.685, which is close to the value mentioned in literatures, therefore it is adequate to do the WET test.

3.4 Range Finding Test

Results of range finding test (Table 4 and Table 5) shows the average LC_{50} value from nine industry is 37.50% with the lowest score of 8.652% and the highest of 100%. Values closer to 100% indicates more toxic wastewater. This narrows down new combination of concentration in the definitive test.

Table 4 LC50-48 Hours Test Range Finding for Industry Participants PROPER 2015 - 2016

No.	Name of Industry	LC50 96 hour (%)	Range (%)
1	PT. Textile 1	12,023	6,25 – 12,5
2	PT. Textile 2	22,054	12,5 - 25
3	PT. Metal Coating	46,72	25 - 50
4	PT. Pharmacy	19,347	12,5 - 25
5	PT. Textile 3	10,854	6,25 – 12,5
6	PT. Textile 4	100,916	50 - 110
7	PT. Milk Processing	56,691	50 - 100
8	PT. Shoe	60,222	50 - 100
9	PT. Textile 5	8,652	6,25 – 12,5

Table 5 LC50-48 Hours Test Range Finding. For Industrial Non PROPER

No.	Name of Industry	LC50 96 hour (%)	Range (%)
1	PT. Textile 6	24,152	12,5 - 25
2	PT. Textile 7	34,351	25 - 50
3	PT. Textile 8	20,987	12,5 - 25
4	PT. Paper*	35 x 10^9	
5	PT. Textile 9	43,908	25 - 50
6	PT. Textile 10*	0	

3.5 Definitive Test

Definitive test results were analyzed using probit software. $LC_{50\text{-}96}$ was obtained from software analysis, then reviewed to find whether there was any link between the value LC50-96 hours with PROPER value criteria. Variations of concentration in this test were obtained from result range on the range finding test, obtained by multiplying R value in Equation 1 with initial value range. Table 6 and Table 7 shows $LC_{50\text{-}96}$ value in definitive test.

Table 6 Value LC50-96 Hour Definitive test results. For Industry Participants PROPER 2015 - 2016

Name of Industry	LC50 96 hour Definitif Test (%)	TUa	PROPER Value
PT. Tekstil 1	6,357	15,73	Biru
PT. Tekstil 2	13,555	7,377	Biru
PT. Pelapisan Logam	20,567	4,862	Merah
PT. Farmasi	12,967	7,712	Biru*
PT. Tekstil 3	2,719	36,78	Biru
PT. Tekstil 4	88,399	1,131	Biru
PT. Pengolahan Susu	45,097	2,217	Merah*
PT. Sepatu	49,959	2,002	Biru
PT. Tekstil 5	7,585	13,18	Biru

Table 7 Value LC50-96 Hour Definitive test results for Industry Non PROPER

Name of Industry	LC50 96 hour Definitif Test (%)	TUa
PT. Tekstil 6	17,25	5,80
PT. Tekstil 7	9,96	10,04
PT. Tekstil 8	9,05	11,05
PT. Tekstil 9	9,31	10,74

According to Maryland *Department of the Environment* (2012), WET data interprets toxic in samples when generating LC_{50} or EC_{50} value less than or equal to 100% effluent concentration. Table 5 shows that $LC_{50\text{-}96}$ value of all samples from industry wastewater effluent was less than 100%, indicating they are not fit to be discharged.

3.6 Environment Parameter

USEPA (2002) recommends to conduct environmental parameter measurement for pH, temperature, DO and conductivity. These assessment is intended to monitor abrupt environmental changes due to toxicity. The toxicity in this test is measured by using Daphnia Magna.

3.6.1 Dissolved oxygen (DO)

Daphnia sp. lives in condition with DO content above 3 mg/l (Ebert, 2005). DO is required for metabolism. Figure 14 shows average value of DO in samples from both PROPER industries of 2015-2016 (ranges between 5.0–7.0) and non-PROPER industries (between 6.0–8.0) are in the optimum range for *Daphnia* to survive.

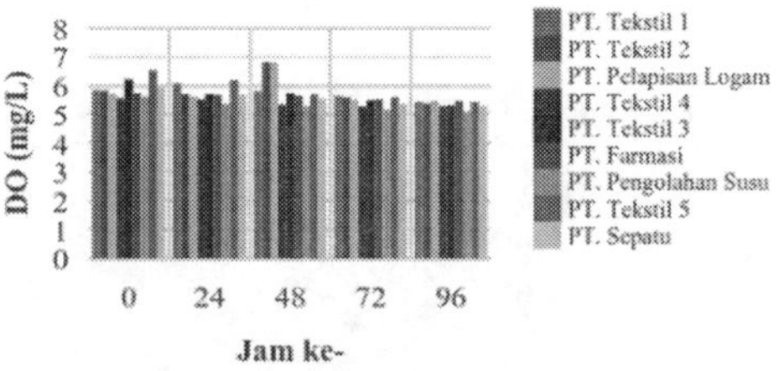

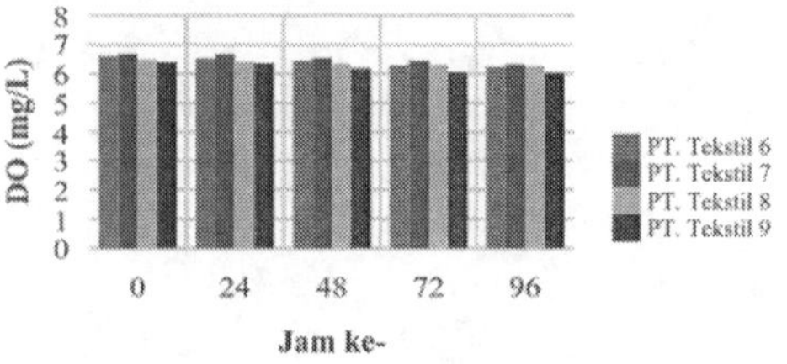

Figure 14 Measurement Parameters DO during test Definitive. Above: PROPER industries 2015 - 2016 and below: non-PROPER industries

3.6.2Temperature

Optimum temperature ranges from 20-30°C for phytoplankton growth in water environment (Effendi, 2003). Temperature affects organism survival ability, for example algae grows well in the temperature range of 30-35°C and 20-30°C. Room temperature is a right condition for organism to metabolize and reproduce.

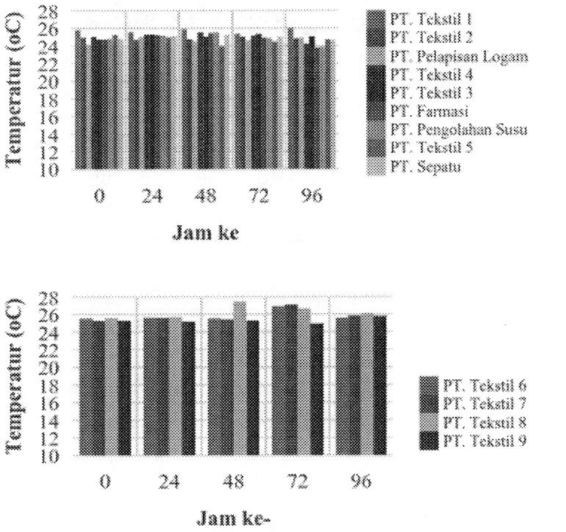

Figure 15 Measurement Parameters Temperature during Test Definitively. Above: PROPER industries 2015 - 2016 and below: non-PROPER industries

Figure 15 shows different range among PROPER industries of 2015-2016 (24-26 °C) and non-PROPER industries (24-27 °C), however, both are still among optimum temperature range for *Daphnia sp.* to live. Condition with temperature out of the optimum range will disrupt its growth and development.

3.6.3Conductivity/Electrical Conductivity (EC)

Conductivity, which is influenced by free ion content such as heavy metal ions released by pollutants in wastewater (Fonridest Environmental, 2014), shown in Figure 16 indicates PT. Metal Coating had the highest EC value amongst all PROPER industries measurements for 48 hour (1000 - 2200 µs/cm). PT. Tekstil 8 excelled amongst all non-PROPER industries (700-900 µs/cm).

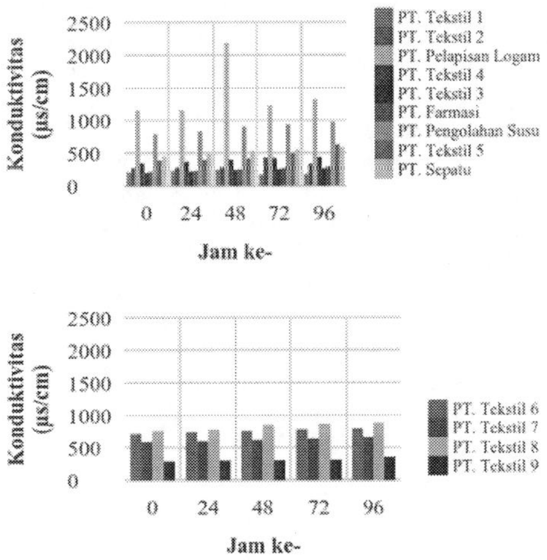

Figure 16 Measurement parameters Conductivity during Definitive Test. Above: PROPER industries 2015 - 2016 and below: non-PROPER industries

*3.6.4*pH

According to Pennak (1989), optimum pH growth for of *Daphnia sp.* ranges from 6.5 to 8.5. Figure 17 shows PT. Milk Processing as the highest pH (in the range of 8.90–9.20) amongst other PROPER industries, which were in the range of 8.0 to 8.7. For non-PROPER industries pH were in the range be 7.5 - 8.5. Both PROPER and non-PROPER industries had pH value in the optimum growth range for *Daphnia sp.*, except for PT. Milk Processing and PT. Tekstil. pH value above or below the optimum range growth may cause death (USEPA, 2002).

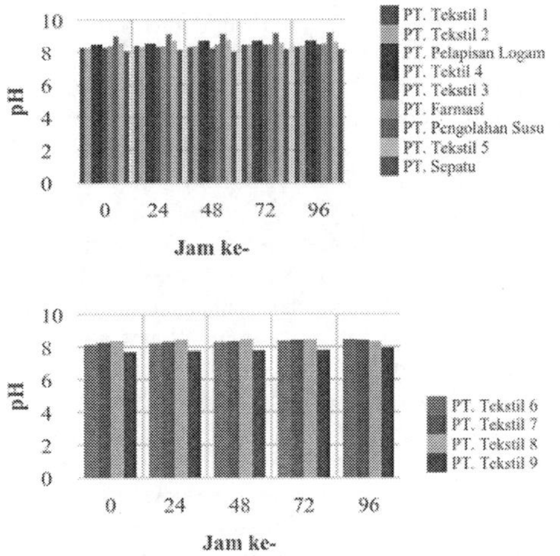

Figure 17 Measurement parameters pH during Definitive Test, Above: PROPER industries 2015 - 2016 and below: non-PROPER industries

3.7 Evaluation of PROPER 2015 – 2016 Criteria Based Value Acute Toxicity and Industrial Wastewater Physical-Chemical Characteristics

The smaller LC_{50} value points out more toxic properties in the wastewater. Figure 18 shows acute toxicity proportion for blue-category PROPER, indicating $LC_{50\text{-}96}$ value below 100%. PT. Tekstil 3 had the least LC_{50} value reaching 2.719%, which had not met reference standard of $LC_{50\text{-}96}$ value based on Maryland *Department of the Environment* (2012). Based on this result, blue-category PROPER received by industries would not indicate low toxicity level, because toxicity test standards have not been included into indicator of environmental assessments.

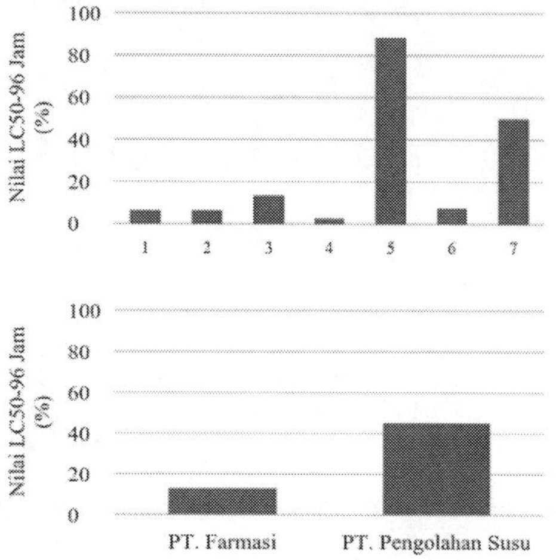

Figure 18 Value Acute Toxicity Criteria LC50-96 hours against PROPER. Left: Industry Participants PROPER and Right: New Participants PROPER Industry 2015 - 2016

Both graphics on Figure 18 contains information on all 9 PROPER industries of 2015-2016 with 2 new industries registered as PROPER industries in 2016 based on their current report, namely PT. Farmasi (blue-catogory PROPER) and PT. Pengolahan Susu (red-category PROPER). Figure 18 shows lower $LC_{50\text{-}96}$ value for PT. Farmasi than PT. Pengolahan Susu. Better ranking, such as blue-category PROPER, did not indicate lower toxicity in the wastewater. Different raw material from both industries may be a contributing factor to this condition, considering chemicals is mostly used for PT. Farmasi.

Figure 18 also shows both blue and red category of PROPER have $LC_{50\text{-}96}$ value less than 100%, pointing out toxicity level of the discharged wastewater. This result draws one example of blue-category PROPER, that have implemented environmental assessment and management as regulated in PermenLHK No. 3/2014, contrarily discharged toxic wastewater. The figure is also drawing red-category PROPER example that also discharged toxic wastewater, despite the undergoing efforts in environmental management. Physical and chemical parameters of environmental monitoring standards still did not met the standard in PermenLKH No. 5/2014 even as the toxicity tests are ruled out.

4 CONCLUSION

PROPER and non-PROPER industries of 2015-2016 discharged wastewater with some of its parameters exceeding the standard regulated in PermenLKH No. 5/2014. These following physical-chemical parameters in PROPER industries of 2015-2016 wastewater were exceeding the standards regulated in PermenLKH No. 5/2014, namely COD (67% of the industries), TSS (100%), TDS (22%), phenol (78%), ammonium (22%) and phosphate (11%). Several non-PROPER industries of 2015-2016 had parameters exceeding standard as well, namely COD (100% of the industries), TSS (100%), ammonium (100%), phosphate (67%) and TDS (50%).

$LC_{50\text{-}96}$ value results from Whole Effluent Toxicity (WET) tests shows 100% wastewater from all industries are toxic. PT. Tekstil 3 and PT. Tekstil 8 were the most toxic wastewater with TUa value of 36.78 and 11.05.

Blue-category PROPER industries, which are 78% of all PROPER industries of 2015-2016, still discharged toxic wastewater. Higher rank on PROPER categories did not indicate low toxicity of wastewater.

Several PROPER-registered industries discharged wastewater with parameters exceeding the standard, as regulated in PermenLKH No. 3/2014.

5 REFERENCES

1. Davraz, Metin., Koru, Murat., and Akdag, Ali Ekrem. *The Effect of Physical Properties on Thermal Conductivity of Lightweight Aggregate*. Procedia Earth and Planetary Science, Suleyman Demirel University, Turkey **15**, 85 – 92 (2015)
2. Ebert, D. Ecology. *Epidemiology and Evaluation of Parasitism in Daphnia*. National Liblary of Medicine (US)-National Center for Biotechnology Information, Bethesda. (2005)
3. Effendi, H. *Telaah Kualitas Air Bagi Pengelolaan Sumberdaya dan Lingkungan Perairan*. Kanisius. Yogyakarta. 258 p (2003)
4. Habibi, B.J. *Peraturan Pemerintah Republik Indonesia Nomor 18 Tahun 1999 Tentang Pengelolaa Limbah Berbahaya dan Beracun*. Jakarta (1999)
5. Kambuaya, Balthasar. Peraturan Menteri Lingkungan Hidup Republik Indonesia Nomor 3 Tahun 2014 Tentang Program Penilaian Peringkat Kinerja Perusahaan dalam Pengelolaan Lingkungan Hidup. Jakarta (2014)
6. KLHK. *Hasil Evaluasi Kinerja Pengelolaan Lingkungan Proper 2015 – 2016 Rapor Sementara PT. Sanbe Farma*. Jakarta (2016)
7. Mark, U dan John Solbe. *Analysis of the Ecotoc Aquatic Toxicity (EAT) Database V – the Relevance of Daphnia magna as a Representative Test Species*. Chemosphere, **36**, 1, pp. 155 – 156 (1997)
8. Mokoginta, I. *Budidaya Pakan Alami Air Tawar*.

Direktorat Pendidikan Menengah Kejuruan Direktorat Jendral Pendidikan Dasar dan Menengah Departemen Pendidikan Nasional. Jakarta. 2003.
9. Mount, D.R., Gulley, D.D., Hocket, J.R., Garrison, T.D., Evans, J.M. *Statistical Models to Predict The Toxicity of Major Ions to CerioDaphnia dubia, Daphnia magna and Pomephales promelas (Fatted Minnows).* Environmental Toxicology and Chemistry, **16** (1997).
10. Ningrum, Sri Rahayu Widiya. *Validasi Uji Toksisitas Akut Metode Organization for Economic Cooperation and Development (OECD) 425 pada Mencit Betina Menggunakan Tembaga (III) Sulfat Pentahidrat*. Universitas Indonesia. Depok (2012)
11. Pennak, R.W. Freshwater Invertebrate of The United States (3rd ed). John Wiley & Sons. New York (1989)
12. Pratiwi. A. D. Syah., Hardjito., L. Goreti. *Fatty Acid Syntetys by Indonesian Marine Diatom. Hayati Journal of Biosciences. **16(4)**, 151-156 (*2009)
13. Salmiyatun. Panduan Pembuangan Limbah Perbekalan Farmasi. Buku Kedoketaran EGC. Jakarta (2003)
14. Tongur, Suheyla. and Yildirim, Rifat. *Acute Toxicity Assessment of Antibiotics in Water by Luminiscene Bacteria and Lepidium Sativum*. World Multidisciplinary Earth Sciences Symposium, WMESS 2015. Procedia Earth and Planetary Science. Selcuk University, Department of Enviromental Engineering, Konya. **15**, 468 – 473 (2015)
15. Tyagi., Chopra., Durgapal., and Kumar. *Evaluation of Daphnia magna as an Indicator of Toxicity and Treatment Efficacy of Municipal Sewage Treatment Plant.* Delhi, India. JASEM ISSN 1119-8362, **11**, 61 – 67 (2007)
16. U.S. Environmental Protection Agency. *Methods for Measuring the Acute Toxicity of Effluents and Receiving Waters to Freshwater and Marine Organisms*. Final Report. Office of Water, Washington DC. 5th ed. EPA/821/R-02/012 (2002)
17. William PL, Burson JL. *The Basic Science of Toxicity*. Macmillan Publ. Co, New York, **54** (1995)
18. Wolfe, D., Schorr, M., Hanson, M., Nelson, C.H. and Richards, S.M. *Hazard Assessment for a harmaceutical Mixture Detected in The Upper Tennessee River Using Daphnia magna*. Global J. Environ. Sci. Manage., Department of Biological and Environmental Sciences, University of Tennessee at Chattanooga, USA. **1** (1), 1 – 14, (2015)
19. Tchobanoglous, G., Burton, Frankin L., dan Stensel, H. David. *Wastewater Engineering : Treatment and Reuse (Fourth Edition)*. RRC : Metcalft and Eddy, Inc (2003)
20. Boyd, C.E.. *Water Quality in Ponds For Aquaculture, Alabama Agricultural Experiment Station.* Auburn University. Alabama. 482 p, (1990)
21. Costa Da Cruz, R. *Nitrification and Oxidation of Carbon Sources and Nitrogen Removal in Ammonia Rich Effluents*. Portugal. Technical University of Lisbon, Instituto Superior Tecnico. (2007)
22. Maryland Department of the Environment (MDE). *Guidance: Effluent Biotoxicity Testing Protocol for Industrial and Municipal Effluents*. Water Management Administration, Washington Blvd, Baltimore (2012)

The progressive realization of human rights to water: the legal basis, policy implications, and monitoring challenge

Anindrya Nastiti[1,*] and *Teddy* Prasetiawan[2]

[1] Environmental Management Technology Research Group, Faculty of Civil and Environmental Engineering, Institut Teknologi Bandung, Jalan Ganeca 10 Bandung
[2] Research Center of Expertise Agency, The House of Representatives of the Republic of Indonesia, Central Jakarta, Indonesia

Abstract. Since 2010, the United Nations General Assembly had explicitly recognized the human right to water and sanitation and obliged States to provide for its progressive realization and entitles everyone to sufficient, safe, acceptable, physically accessible and affordable water for essential personal and domestic uses. This paper scrutinizes the legal basis and the policy implication for human right to water in Indonesia, before and following the annulment of the Water Resource Law 7/2004. This paper considers that one of the greatest challenges is to find an appropriate and internationally-comparable methodology in measuring the progressive realization of human rights to water and sanitation. We also highlight the importance of composite indicators and concludes that single variable indicators are insufficient to capture the range of issues involved in the realization of the human rights to water.

1 1 Human Rights to Water

A long journey towards the international recognition of water as a basic human rights started in decades ago. The International Covenants on Economic, Social and Cultural Rights (ICESCR) did not specifically mention water as human rights, but water is indispensable part the right to an adequate standard of living. In 1993, the Vienna World Conference on Human Rights avowed that extreme poverty constitutes a violation of human dignity. Although water was not also specifically recognized as basic human rights, the association of poverty with water shortage is well established—poverty is prevalent mostly in water-short area. The victory of a global level recognition of water as human rights was finally celebrated with the issuance of UN Resolution 64/292 by the UN General Assembly, 28th of July 2010. The right to water is now a legal right, rather than a merely charitable service. Internationally-binding convention will drive States to focus more on their responsibility in fulfilling this right to their citizens (Nurcahyono et al., 2015).

But what is a right to water? Scanion et al. (2014) explained the two contents of the human right to water: substantive right and procedural right (Fig. 1).

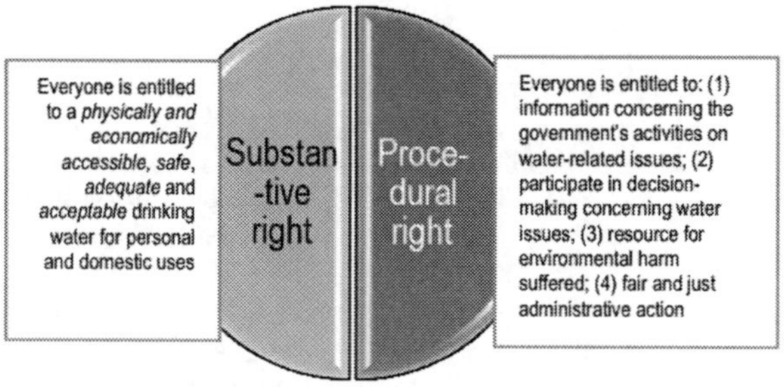

Fig. 1. The contents of the human right to water

Substantive right constitutes how a human right should be defined. The definition of the human right to water was firstly declared during the 1977 Mar del Plata Conference: "all people, whatever their stage of development and their social and economic conditions, have the rights to drinking water in quantities and of a quality equal to their basic need." The UN Resolution 64/292 further mentions that everyone is entitled to a physically and economically accessible, safe, adequate and acceptable drinking water for personal and domestic uses. Physically accessible means that water should be within safe physical reach. The National Statistical Agency (2013) requires water collection time is no more than 30 minutes a roundtrip. Economically accessible means that costs and charges associated with water must be affordable and must not prevent a person from accessing safe water and should not compromise his or her ability to enjoy other basic rights. Safe means water

* Corresponding author: anindrya@tl.itb.ac.id

should be free from contaminations than endanger health. Water should also be adequate and continuous for drinking, cooking, and maintaining basic hygiene (United Nations, 2007). Aside from those requirements, the human right to water also holds a principle of nondiscrimination in which water should be accessible for all: the most vulnerable and marginalized groups of population, regardless of their gender, race, and socioeconomic status, should have access to water services. Here, accessibility includes both physical and economic accessibility.

Even so, the recognition of a right to water does not immediately eliminate the barriers in extending this right to citizens (Wutich et al., 2016). Water supply in many countries often does not meet the requirements of physical and economic access, quality, quantity, and continuity (see Bain et al., 2014; Kumpel & Nelson, 2014; Banerjee et al., 2008; Kayaga & Franceys, 2007; Vollaard et al., 2005). In Indonesia, access to piped water as the best form of provision only accounts for 10.2% of the total population by 2015 (National Statistical Agency, 2015). In addition, the existing house connections often fail to deliver contaminant-free and reliable water on daily basis. Non-piped water sources, such as deep groundwater and commercial water (including water delivered from vendors and all forms of bottled water) are expensive to obtain, thus limiting economic accessibility of the poor citizen to such sources.

Meanwhile, procedural right deals with the enforcement of the substantive rights; a well-established of the former will serve as a protective shield above the latter. According to Scanion et al. (2014), the core procedural right to water includes the rights of individuals to: (1) information concerning the government's activities on water-related issues; (2) participate in decision-making concerning water issues; (3) resource for environmental harm suffered; (4) fair and just administrative action. All of these rights are inter-related. Scanion et al. (2014) mention that access to information enables citizens to participate meaningfully in an accountable and transparent decision-making and effective enforcement in the water sector. In other words, the procedural right includes the institutional aspects in realizing a physically and economically accessible, safe, adequate, and non-discriminative water service for all.

In the discussion about institutional factors of the water sector, privatizations in the operational aspects of water supply services had been in the spotlight. Albuquerque and Roaf (2012) suggests that privatization of water and sanitation service utilities can prove detrimental to human rights, where there is not a simultaneous implementation of specific measures to regulate service provision and to ensure that there are sufficient incentives to expand and improve access in low-income areas.

This paper aims to see whether the national regulations have incorporated the substantive right and procedural rights to water and link it with the issue of privatization. The remainder of the paper is as follow. Section 2 follows the legal basis of the human right to water in Indonesia, before and following the annulment of the Water Resource Law 7/2004. Section 3 elaborates on the human right to water in the context of SDGs.

1 2 Legal Dimensions of the Human Right to Water in Indonesia

The Constitution of Indonesia resolutely expresses that water is a prime resource for the sustenance of life, with Clause 33 verse (3) firmly claims the state's ownership of water. This clause is a parental ruling for all the forthcoming water resource-related regulations. However, it is inevitable that the development of international conventions, mainly exploited as a legal landing during many water resources advocacies, largely influence the tidal courses of Indonesian water resource regulations.

In this section, we discuss the evolution of the national legal regimes for the protection and promotion of human rights to water. We specifically discuss three legal regimes: (1) The 1974 Watering Act; (2) The 2014 Water Resource Act (WRA); and (3) The Post-annulment of the 2014 Water Resource Act (WRA). The debates over the legal framework of the human right to water centers around the notion of the ownership of water.

The Watering Act Number 11 Year 1974 is a very first lex specialis, or law governing a specific subject matter, comprehensively regulated all water resources affairs in Indonesia. The previous colonial water regulation, the 1936 Algemeen Waterreglement was focused mainly to irrigation and only applied to some parts of the country. In the context of human right, the Watering Act highlighted the social function of water and that water should be used predominantly for the prosperity of the society. Water is owned by the state in which central and local governments serves as executing bodies in implementing water resource management.

In 2004, the Water Resource Act Number 7, was enacted. Clause 5 mentions that: "The State guarantees individual rights to obtain water for basic needs towards a healthy, hygienic, and productive life"; a statement of accountability that was never mentioned in the previous Watering Act. Clause 26 verse (7) states that all water resource utilization shall rank the social function of water first; hence, this Act prioritizes individual basic needs and irrigated people agriculture in water allocations. However, aside from the social function, the economic and environmental functions of water were also recognized in this 2004 Act (Fig. 2). The economic function of water, often narrowly understood as water privatization, had since become the main focal point in the debate stage of human rights to water.

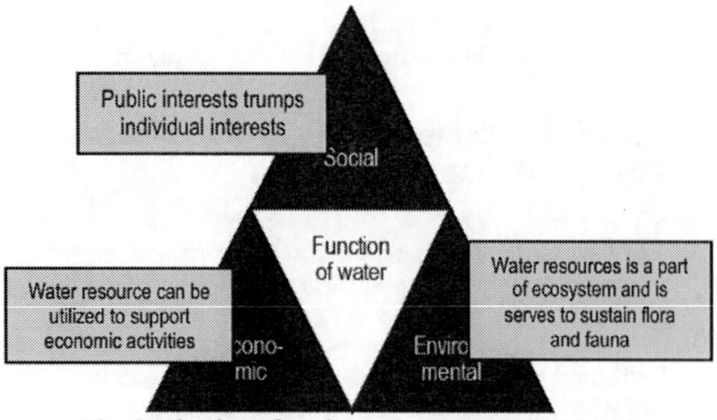

Fig. 2. The three functions of water mentioned in the 2004 Water Resource Act: social, economic, and environmental functions.

Both the 1974 and 2004 legal regimes, and all their derivative regulations, do not firmly regulate the 'what's-allowed and what's-not' in the context of private sector privatization. For example, although the 1974 Watering Act put forth the social function of water, it also allows private sector participation by granting the executive role in water resource management to individuals or groups. In fact, the first water privatization in Indonesia (North Serang, Batam, and Jakarta) took place during this regime. The sketchy ruling of private sector participations is suspected to sustain obscured privatizations (Al' Afghani, 2012).

Water sector privatization is what brought the 2004 Act into the spotlight by human right activists. Indeed, the evidence against water privatizations had shown that such an institutional approach in water resource management fail to achieve intended development goals, particularly water service for the poor (Bakker, 2008). McIntosh (in KruHa, 2016) suggests that private water companies were reluctant to invest on expanding piped water network towards poor residential areas. Swyngedow even went further by suggesting that private companies cherry-picked those who were willing to pay high for mediocre service quality (KruHa, 2016). These debauched reputations of water sector privatization serve as a basis of community pressure to erase all forms of water resource commercialization.

The strong opposition towards water privatization finally led to the annulment of the 2014 Water Resource Act and the re-enactment of the 1974 Watering Law through the Decree of Constitutional Court Number MK No. 85/PUU-XI/2013. This Decree further declares six strict principles in water resource management:

1. Control over water resource shall never disturb, forsake, or negate the human rights to water;
2. The State is responsible in realizing the human rights to water for the people, in which access to drinking water is a standalone basic rights;
3. Environmental conservation remains one of basic human rights as regulated in the 1945 Constitutional Act.
4. Absolute command and control over water belongs to the State;
5. The priority of commercial operational activities is given to the central state- or local state-owned companies; and
6. Tightly regulated private sector privatization is permissible.

In regard to the context of drinking water service, the newly enacted Government Regulation Number 122 year 2015 re-emphasizes that the drinking water provision is conducted in a manner to fulfil the citizens' right to drinking water. The substantive right is included in the Clause (1) of the regulation, where: "The development of the drinking water supply system … in order to fulfil the requirements of quantity, quality, and continuity of drinking water supply," and Clause 2, where: "..to achieve a high quality drinking water supply management and service in an affordable manner.

Meanwhile, the sketchy ruling of water privatization was undertaken by the enactment of the Government Regulation Number 121 year 2015 regarding Water Resource Commercial Operations. This regulation practically adopts the Constitutional Court's six principles of water resource management. Private sector participations becomes 'the last resort' when public sectors are unable to provide service albeit with 'certain and tight conditions'. But these conditions are not specified anywhere in this 121/2015 regulation, thus undesired multi-interpretations of this regulation is plausible. The previously mentioned regulation, 122 year 2015, at least provides conditions that allow private sector involvement in drinking water supply management services. Private sector can participate under certain principles: firstly, the state agencies partnering with the private agencies are the ones that hold the Water Extraction Permit, rather than the private agencies. Secondly, the partnerships prioritize low-income communities. Private sector also can only participate in the form: (1) of the expansion and management of the raw water supply and production units of drinking water supply system; and (2) the investment of operation and maintenance technologies and systems that subsequently operated and managed by the state agencies; and (3) the investment of operation and maintenance technologies and systems based on the performance-based contract mechanisms. However, the procedural rights to ensure that any private sector involvement will not violate the principle of the human right to water is unclear.

At present, the House of Representatives of Indonesia (DPR RI) is attempting to pass the draft of the new Water Resource Act towards the 2017 Priority Legal Program. This new draft was prepared in the spirit of returning legal and policy framework to gravitate into the ideologies of human rights to water. The new Water Resource Act calls for incorporating the Constitutional Court's six principles of water resource management in explicit manners in order to leave multi-interpretations behind.

1 3 Human Rights to Water and the SDGs: A Progressive Realization

The Millennium Development Goals had been valuable in raising awareness among States of the situation that too many people suffer because of the lack of access to safe water (and sanitation). However, the MDGs did not reflect the principles of the human right to water as it only called for a proportion of those without access to drinking water rather than universal access (Albuquerque & Roaf, 2012). The nondiscrimination principle of access to water becomes a foundation for the Sustainable Development Goals (SDGs) that calls for a universal access to water.

Meanwhile, following the recognition of water as a human right, the post-2015 development framework also suggests that States are responsible and legally accountable to use the maximum resources available to ensure that universal access to "sufficient, safe, acceptable, physically accessible and affordable water". The SDGs calls for equitable access to water. 'Equitable' indicates a progressive reduction and elimination of inequity between population subgroups, not only in term of physical accessibility, but also economic accessibility, quality, continuity, and quantity. Therefore, the SDGs also concerns about the distributive outcomes of water supply provision is without discrimination.

In the midst of the anti-privatization that successfully led to the annulment of the 2014 Water Resource Act, Indonesia has streamlined the goal in achieving universal access to water, enacted through the Presidential Regulation Number 2 year 2015 regarding the National Mid-term Development Plan or RPJMN 2015-2019. The RPJMN emphasizes the non-discriminatory principle in drinking water provision system by stating that the fulfilment of the basic rights, including drinking water and sanitation, for the poor households needs to be the center of attention. According to Luh et al. (2013), a State's compliance with the right to water is not determined by the level of achievement of water-related goals, but rather through the principle of progressive realization. Since States may be constrained by prior conditions and resources available in realizing the human right to water, a compliance is attained when a State shows that it has "taken steps", or made progress, towards realizing universal access (Luh et al., 2013). Thus, the rate of change of certain human rights criteria is similarly important in the level of achievement of such a criteria.

According to the RPJMN, the developed drinking water system should consider the four principles: quality, quantity, continuity, and affordability. The National Statistical Agency (2016) assesses potential indicators for the SDGs' water target. To measure physical accessibility, the Government of Indonesia still relies on the concept of "minimum basic service" by defining populations with "access" as those who use water from improved sources that are accessible within 30 minutes per trip. This single variable indicator is insufficient to capture the range of issues involved in the realization of the human rights to water. However, a comprehensive monitoring system with composite indicators to ensure the substantive right to water, including the one that also take into account water adequacy, quality, and affordability, has yet to be developed. In the context of affordability principle, the RPJMN also calls for the full-cost recover-based tariff system. More studies need to be conducted in order to assess the effect of this tariff system towards access to water for the poor and marginalized community.

Meanwhile, a path towards enabling the procedural right to water subsists in the RPJMN document. This document mentions that knowledge management is done through an integrated data and information management through NAWASIS that involve active participations from the related stakeholders. But the specific regulatory framework, let alone the monitoring, of such procedural rights, has yet to be seen. Future researches should also focus in analyzing whether Indonesia has enough regulatory instruments that specify not only substantive right, but also procedural right, and how the available rules through which distributive outcomes ensue accommodate the principle of the human right to water.

1 4 Conclusion

We conclude that explicit codifications of human rights to water in the national (and local) regulations are crucial in bridging the access gap and service quality of water service, particularly for the poor. The commencement of the SDGs that demands universal access to water for all put forward the issue of human rights to water in the center of the global stage of water. The substantive right to water requires water should not only be physically accessible, but also economically accessible, safe, and adequate for all without discrimination. For this to be realized, the current challenge is to ensure that the distributive outcomes of water service provision, such as physical access, costs, reliability, and water quality, is fair. Finding appropriate and internationally-comparable methodology in measuring the distributive outcomes of the progressive realization of human rights to water is in itself challenging.

The efforts in doing so are upscaled when not only those distributive outcomes needs to be constantly monitored, but also the rules through which these outcomes ensue needs to be assessed. At the moment, it is unclear whether the Indonesian regulatory framework has accommodated the procedural right to water sufficiently. We suggest future studies of the progressive realization of human rights to water in Indonesia to also focus on developing the way the institutional and regulatory aspects can be best evaluated.

References

1. MM. Al 'Afghani. "Anti-Privatisation Debates, paque Rules and 'Privatised' Water Services

Provision: Some Lessons from Indonesia". IDS Bulletin **43**, 2 (2012).
2. K. Bakker. "The Ambiguity of Community: Debating Alternatives to Private- Sector Provision of Urban Water Supply". Water Alternatives **1**,2 (2008).
3. A. Wutich, M. Beresford, C. Carvajal. Can informal water vendors deliver on the promise of a human right to water? Results from Cochabamba, Bolivia. World Development **79** (2016).
4. UN General Assembly, International Covenant on Economic, Social and Cultural Rights of 16 December (1966). http://www.ohchr.org/Documents/ProfessionalInterest/cescr.pdf
5. C. de Albuquerque, V. Roaf. On the right track: good practices in realising the rights to water and sanitation (2012).
6. Scanion, J., Cassar, A., & Nemes, N. Water as a Human Right? IUCN, Gland Switzerland and Cambridge, UK (2004).
7. A. Nurcahyono, H. Syam, Y. Sundaya. Hak Atas Air dan Kewajiban Negara dalam Pemenuhan Akses terhadap Air. MIMBAR, Jurnal Sosial dan Pembangunan (2015).
8. United Nations, A/HRC/6/3 Annual Report of the United Nations High Commissioner for Human Rights and Reports of the Office of the United Nations High Commissioner for Human Rights and the Secretary-General, Human Rights Council, General Assembly of the United Nations, 32 pages. (2007). Accessed 12 June 2015 at http://www2.ohchr.org/english/issues/water/iexpert/docs/A-CHR-6-3_August07.pdf.
9. National Statistical Agency, Welfare Statistics, Jakarta:Sub-Directorate of Households Statistics, National Statistical Agency (2015).
10. R. Bain, J. Wright, H. Yang, S. Gundry, S. Pedley, J. Bartram , "Improved but not necessarily safe: water access and the Millennium Development Goals", in P W R Quentin Grafton (editor), *Global Water: Issues and Insights*, Australian National University Press, Canberra, (2014).
11. E. Kumpel, KL. Nelson. Mechanisms affecting water quality in an intermittent piped water supply. Environmental science & technology, **48,** 5(2014).
12. S. Banerjee, Q. Wodon, A. Diallo, T. Pushak, E. Uddin, C. Tsimpo, V. Foster. "Access, affordability, and alternatives: Modern infrastructure services in Africa", Background Paper 2 (Phase I), Africa Infrastructure Country Diagnostic, (2008).
13. S. Kayaga, R. Franceys. Costs of urban utility water connections: Excessive burden to the poor. Utilities Policy, **15**,4, (2007). http://doi.org/10.1016/j.jup.2007.06.002
14. AM. Vollaard, S. Ali, J. Smet, H. van Asten, S. Widjaja, LG. Visser, C. Surjadi, JT. van Dissel, "A Survey of the Supply and Bacteriologic Quality of Drinking Water and Sanitation in Jakarta, Indonesia", Southeast Asian Journal of Tropical Medicine and Public Health, **36**, 6(2005).
15. National Statistical Agency. The National Socio-economic Survey 2013. Jakarta: National Statistical Agency, (2013).
16. Kruha, *Kajian Hak Atas Air*. Jakarta: Indie Book Corner (2016).
17. The Government of Indonesia. The Watering Act Number 11 Year 1974.
18. The Government of Indonesia. The Water Resource Act Number 7 Year 2005.
19. Constitutional Court (2013). The Decree of Constitutional Court Number MK No. 85/PUU-XI/2013.
20. The Government of Indonesia. The Government Regulation Number 121 Year 2015 regarding the Commercialization of Water Resources.
21. The Government of Indonesia. The Government Regulation Number 122 Year 2015 regarding the Drinking Water Provision System.
22. The Government of Indonesia. The Presidential Regulation Number 2 Year 2015 regarding the National Mid-term Development Plan or RPJMN 2015-2019.
23. J. Luh, et al., Equity in water and sanitation: Developing an index to measure progressive realization of the human right. Int. J. Hyg. Environ. Health (2013), http://dx.doi.org/10.1016/j.ijheh.2012.12.007
24. National Statistical Agency. (2016). *Indonesia - Survei Sosial Ekonomi Nasional 2016 Maret (KOR).* http://microdata.bps.go.id/mikrodata/index.php/catalog/769/datafile/F2 (Accessed 31 March 2017).

Acidic Wet Deposition in Bandung City Indonesia

Nia Yuniarti Hasan[1], *Driejana*[2] , *Aminudin* Sulaeman[3] and *Herto Dwi* Ariesyady[4]

[1]Environmental Engineering Doctoral Program, Institut Teknologi Bandung, Jl. Ganesa 10, Bandung 40132.
[2]Air and Waste Management Research Group, Faculty of Civil and Environmental Engineering, Institut Teknologi Bandung, Jl. Ganesa 10, Bandung 40132.
[3]Analytical Chemistry Research Group, Faculty of Mathematics and Natural Science, Institut Teknologi Bandung, Jl. Ganesa 10, Bandung 40132.
[4]Environmental Technology and Management Research Group, Faculty of Civil and Environmental Engineering, Institut Teknologi Bandung, Jl. Ganesa 10, Bandung 40132.

Abstract. Acidic deposition (acid rain) has been known to cause negative impact to environment. Measurement of rain chemical composition was done in March to August 2016 to investigate temporal variation of atmospheric deposition in Bandung City. Rainwater samples are collected at four sampling sites. The chemical properties of the rain determined were pH and ionic concentrations of SO_4^{2}-, NO_3-, Cl^-, Ca^{2+}, Mg^{2+}, Na^+, K^+ and NH_4^+. The rainwater was typically acidic with the range of mean pH between 5.29 – 5.56 and mean electric conductivity value range between 16.37 – 19.96 µS/cm. Temporal variation showed higher SO_4^{2+}, NO_3^+ and NH_4^+ concentration in August when rainfalls were the lowest (214.88 mm). Similar temporal variation for the mean pH (5.98) and electric conductivity (25.36 µS/cm) in August. Ca^{2+} highest concentration was observed in June. Generally, the the major component of precipitation chemistry were SO_4^{2-} and NO_3^-, Ca^+ and NH_4^+, for anion and cations, respectively. The largest proportions of the total anions in all sampling sites was SO_4^{2-} ranged from 32% (Lembang) to 37% (Buah Batu) followed by NO_3^- (13 – 15%) and and Cl^- (4 – 7%). For cations, NH_4^+ dominated with proportion ranged between 21 – 27% and subsequently Ca^{2+} (13 – 19%) and followed by Na^+, Mg^+, K^+ and H^+. The Neutralizing Factor (NF) results reveal that Ca^{2+} and NH_4^+ are the dominant neutralization substances in the rainwater. However it should be noted that reduced nitrogen (NH_4^+) act as mobile nitrogen that further will acidify the environment, particularly to soil ecosystem downstreams.

1 Introduction

There are two meanings of wet deposition terms determined by The World Meteorological Organization (WMO). Firstly, wet deposition is a process of scavenging of any gases and/or particles from the atmosphere by liquid (i.e., water droplets) and solid (i.e., ice crystals) phases. The process involves removal of any species by droplets/ice crystals within clouds (i.e., in-cloud scavenging), and by falling drops/snowflakes (i.e., below-cloud scavenging). Second, wet deposition is the mass of material deposited from the atmosphere to the underlying surface in precipitation (over unit of area during unit of time) [1].

Acid deposition in the general is precipitation with a pH level of less than 5.6 and thus is considered to be acidic, and when it falls on the Earth it is considered to be "acid deposition." Acid deposition from atmospheric precipitation can makes soils, lakes and ponds become acidic. Sulfur dioxide and nitrogen oxides, typical air pollutants are transformed into sulfuric and nitric acid compounds in the atmosphere, with resultant changes of their properties. These acid compounds are carried by the wind as fine particles (aerosols) on sunny days. They dissolve in rain drops on rainy days or within clouds, then fall to the Earth's surface, making the environment acidic. Research has detected this transport on a continental scale, with effects extending beyond national borders [2].

Acid rain is a mild combination of mainly sulfuric and nitric acid. Sulfurous acid and nitrous acid are less stable and are present only in very low amounts. Sulfuric acid and nitric acid are the main acids present in acid rain. The problem occurs when rainwater combines with gaseous oxides of sulfur, nitrogen, and phosphoric and hydrochloric acid mists. The latter two and sulfur are released into the atmosphere from automobile exhausts, industries and electric power plants [3].

Nitrogen deposition has had a detrimental impact on temperate ecosystems resulting in soil acidification and a reduction in plant biodiversity. Soils release base cations during acidification process, such as calcium and magnesium, neutralizing the increase in acidity, if base cations have been depleted, aluminium was released from the soils, often reaching toxic levels was superseded by iron release into soil water [4].

Preliminary study evaluation of air quality in Bandung City showed factor characteristic of airborne

* Corresponding author: driejana@tl.itb.ac.id

https://doi.org/10.1051/matecconf/201814708007

particulate matter samples were attributed to sea salt, soil dust, vehicular emissions and biomass burning and industrial emissions. Many scale factories are located in and around the city with several pocket of slum areas, lacking waste disposal system, rapid growth of industry, residential area, population and motor vehicle has caused air pollutant problem in Bandung City [5].

Wet deposition is the most effective scavenging process to removing particulate and gaseous air pollutant from atmosphere in Bandung City. We present the precipitation chemistry and wet deposition that occurred at four sampling sites in Bandung City from February to August 2016. Weekly bulk precipitation chemistry samples were analyzed for the anion of SO_4^{2-}, NO_3^-, Cl^- and for the cations of Ca^{2+}, Mg^{2+}, Na^+, K^+ and NH_4^+ in additional to pH and conductivity measurement.

2 Material and Method

Rain water samples were collected on a weekly basis in four sampling stations from March 2016 to August 2016. These stations were distributed in Bandung City. The locations of the sampling sites were chosen in order to represent various geographical conditions and expected pollution levels and are presented in Figure 1.

Three of the sampling sites are placed in urban areas. The first installed in an altitude of 1271 m asl at Lembang farming area. The second was installed at an altitude of 803 m asl in Coblong area and the third was installed at an altitude of 742 m asl in Sumur Bandung . The fouth was installed at an altitude of 642 m asl at Buah Batu area.

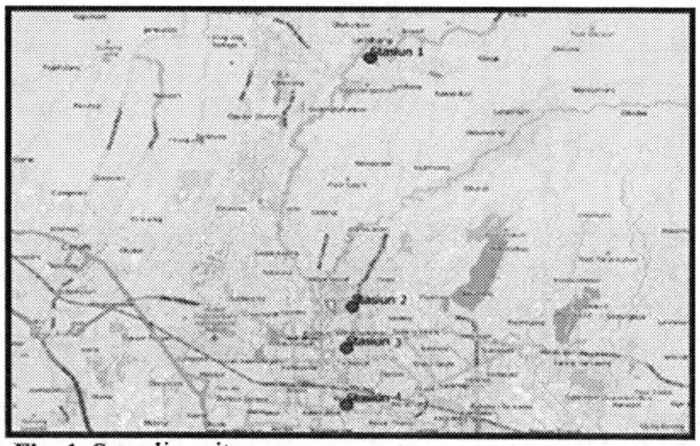

Fig. 1. Sampling site

2.1. Sampling Equipment and Methods

The rainwater samples was collected manually using a polyethylene bottle (1000 mL) with a polyethylene funnel (12 mm in diameter) and put on a stainless steel cylinder. The funnel equipped with a filter plug (1 mm mesh) to protect large debris. The inside wall of canister painted with black color to diminish light levels around the collecting bottle and to inhibit biological growth [6].

Rainwater sampling was collected on a weekly basis from March to August 2016. The bulk collector bottle was replaced with a clean one and funnels were rinsed with deionized water. Samples bottle were placed in a cool box with blue ice and were transported from the sampling sites to the laboratory.

2.2 Laboratory Analysis Method

Upon arrival at the laboratory, samples were immediately weighed for volume determination, followed by pH measurement (Sartorius Professional Meter PP-20) and electric conductivity (Agilent Technologies 3200 C). The pH meter and conductivity meter were calibrated before measurement. Samples were sealed from the atmosphere and stored in the dark at 4°C in the refrigerator until analysis. Major ions (SO_4^{2}, NO_3^-, Cl^-, Ca^{2+}, Mg^{2+}, Na^+, K^+ and NH_4^+) were analysis using ion chromatograph DIONEX ICS 500 DP.

2.3 Quality Assurance

2.3.1 Ionic Balance (IB)

The accuracy of laboratory analysis was determined using ion balance analysis. Ionic Balance is the sum of major anions in a water sample that is theoreticaly should equal to sum of major cations. In reality, factors such as the presence of unknown ions that have not been analyzed, analytical error or measurement precision could cause differences in ion balance (IB) was calculate by Equation 1 [6]:

$$IB = (\sum_{anions} - \sum_{cations})/(\sum_{anion} + \sum_{cations})\ x\ 100\ \%_s \qquad (1)$$

It is suggested that the difference between ions were IB ± 8% if sum ions > 100 μeq/L, IB ± 15% if sum ions 50 – 100 μeq/L and IB ± 30% if sum of ions < 50 μeq/L [2].

2.3.2 Precipitation Weight Mean (PWM) Concentration

In order to determine the influence of air pollution to rainwater chemistry, precipitation weighted mean (PWM) concentration was calculated to eliminate the effect of rainwater amount for each ion analyzed. PWM is carried out by summing the product of ion concentration and the precipitation amount, obtained from rainwater sampler, for each data pair and dividing the result by the sum of precipitation amount for a certain period [6].

$$C_{w,j} = \Sigma C_{j,k} \,.\, P_k / \Sigma P_{k\,s} \qquad (2)$$

where :

$C_{w,j}$: precipitation weighted concentration for ion j ($meqL^{-1}$)

Cj,k: ion j concentration in precipitation of time k ($meqL^{-1}$)

Pk = precipitation amount of time k (mm $time^{-1}$)

3 Result and Discussion

3.1. Precipitation Chemistry

Generally SO_4^{2-} and NO_3^- for anion and Ca^{2+} and NH_4^+ for cations, were found as major component in precipitation chemistry, at all sampling sites (Table 1). Table 1 present mean, minimum and maximum of pH, conductivity and concentration of major ions in rainwater in Bandung City. Mean concentration of SO_4^{2-} ranged from 39.90 µeq/L (at Lembang) to 78.76 µeq/L (at Sumur Bandung) and mean concentration of NO_3^- ranged from 19.47 (at Lembang) to 27.41 µeq/L (at Sumur Bandung). Mean concentration of Ca^{2+} ranged from 31.43 µeq/L (at Coblong) to 51.61 µeq/L (at Lembang) and NH_4^+ ranged from 29.07 µeq/L (at Lembang) to 49.70 µeq/L (at Sumur Bandung).

Gaseous pollutant such as SO_2 and NO_x are released into the atmosphere when fuels such as oil and coal are burnt. SO_2 and NO_x are the major causative agents (precursors) of acid deposition. The high concentrations of alkaline ions (mainly Ca^{2+}, NH_4^+) in the atmosphere have played an important role to neutralize the acidity of rainwater, but it is worth noting that the emission of NO_x from the automobile exhaust is increasing and becoming the important precursor of acid rain [7].

We observed that the lowest mean rainwater pH value in Buah Batu was 5.28 which was less than pH value of 5.6 of unpolluted rainwater at equilibrium with atmospheric CO_2 [8]. Carbonic acid is a weak acid, it brings down the pH of the rainwater to 6.0 – 5.2. With pH levels ranging between 6.0 – 5.2, rainwater is acidic, but still not dangerous. This is a reversible reaction. Sometimes, the pH level can even become as low as 2. This phenomenon of acidic rainwater precipitation, is called acid rain.

The mean electric conductivity in rainwater samples ranged from 15.67 µS/cm at Buah Batu – 19.37 µS/cm at Sumur Bandung. The common range of electric conductivity for precipitation sample is from < 0.5 to 100 mS/cm [2]. Precipitation conductivity is mainly contributed by water soluble ions, the value being related to the total sum of anions and cations in the rainwater. Low precipitation conductivity is an indicator of good atmospheric environmental quality [9].

In term of ionic balance, more than 65% of the data is characterized by positive ionic balance value, which means the sum of anions in the samples exceeds that of cations. The chemical determination of heavy metals, aluminum, mercury, and organic compounds will aid in the characterization of precipitation if the ion balance not meet criteria [2].

3.2 Temporal Variation of Precipitation Chemistry

Monthly variation of precipitation chemistry is shown in Figure 2. Higher SO_4^{2-} and NO_3^- anions concentration were observed in August (89.41 µeq/L and 35.69 µeq/L) and cation NH_4^+ (77.76 µeq/L) when the lowest rainfall was observed (214.88 mm). But cation Ca^{2+} higest concentration (94.27 µeq/L) was observed in June (521.67 mm).

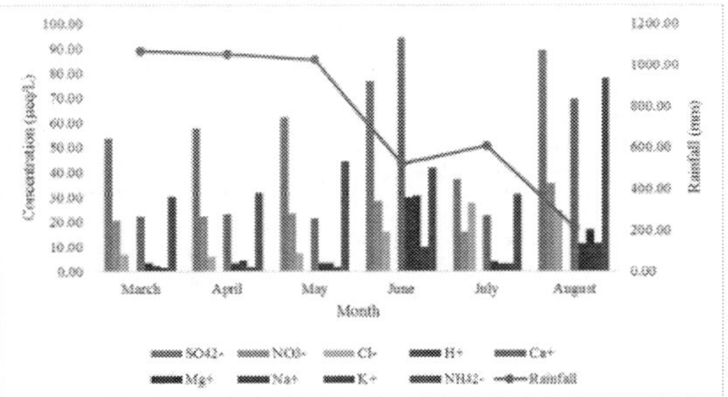

Fig. 2. Temporal variation of precipitation chemisty in Bandung City.

3.3 Temporal Variation of pH and Electric Conductivity

Temporal variation of pH and electric conductivity presented in Figure 3. Data show mean pH (5.98) and mean electric conductivity (25.36 µS/cm) were slightly higher in less rainy periods in August. This can be explained by the fact that, in drought periods, the atmosphere is considerably more polluted, so the wet scavenging process carries a higher load of compounds which increases both measured parameters.

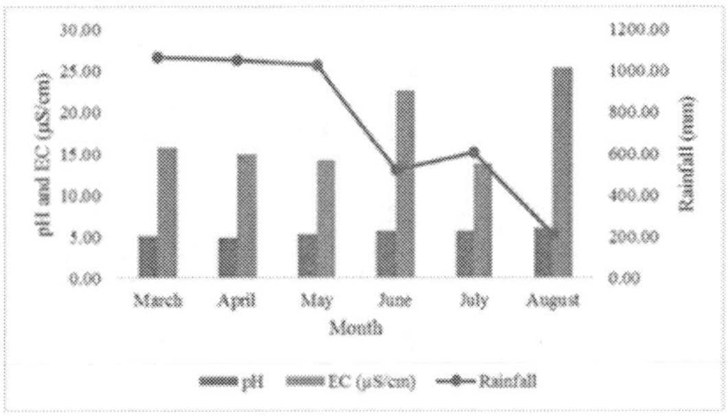

Fig. 3. Temporal variation of pH and electric conductivity in Bandung City emporal variation of precipitation chemisty in Bandung City

https://doi.org/10.1051/matecconf/201814708007

Table 1. Mean, minimum and maximum of pH, conductivity (µS/cm) and concentration (µeq/L) of major ions in rainwater collected four site from March to August 2016
(n = the number of samples)

	Lembang n=24		Coblong n = 24		Sumur Bandung n = 24		Buah Batu n = 24	
	Mean	Min-Max	Mean	Min-Max	Mean	Min-Max	Mean	Min-Max
SO_4^{2-}	39.90	9.22-81.84	64.70	18.00-135.38	78.76	32.35-192.33	65.36	20.82-125.97
NO_3^-	19.47	0.00-67.70	27.22	11.12-67.61	27.41	14.16-55.29	22.62	9.01-41.68
Cl^-	13.59	2.23-71.62	13.29	3.39-77.26	11.69	3.95-39.26	9.80	3.33-30.42
Ca^+	51.61	4.41-827.8	31.43	7.65-91.81	45.67	15.69-204.97	44.34	16.64-132.60
Mg^+	21.77	0.49-440.44	5.26	0.58-18.91	6.72	2.30-28.53	6.34	2.43-16.20
Na^+	23.46	0.07-398.37	5.98	0.13-26.19	6.84	0.87-29.25	6.14	0.74-25.03
K^+	9.09	0.61-76.33	3.93	0.51-29.04	2.89	0.97-48.36	3.50	0.87-15.79
NH_4^+	29.07	0.00-70.66	49.22	16.30-206.16	49.70	23.06-86.45	37.75	0.00-88.45
H^+	0.01	0.00-0.04	0.01	0.00-0.06	0.01	0.00-0.04	0.01	0.00-0.02
pH	5.63	4.36-7.71	5.41	4.24-6.79	5.38	4.10-6.29	5.28	4..68-6.36
Cond.	17.88	4.94-134.90	17.71	5.34-50.45	19.37	12.84-40.41	15.67	5.61-27.7

2.4 Precipitation-weight Mean Concentrations (PWM)

The proportion of Precipitation Weight Mean (PWM) concentrations in the rainwater sample shown the most anion components of rainwater sample in all sampling sites was SO_4^{2-}, which contributed for 32% (at Lembang) – 37% (at Buah Batu) to the total anions, followed by NO_3^- (13 – 15%) and Cl^- (4 – 7%). Both NH_4^+ (21 – 27%) and Ca^+ (13 – 19%) were the major cations components, followed by Na^+, Mg^+, K^+ and H^+.

Precipitation Weight Mean (PWM) concentrations in the rainwater sample for each sampling site location as shown in Figure 4, Figure 5, Figure 6, and Figure 7.

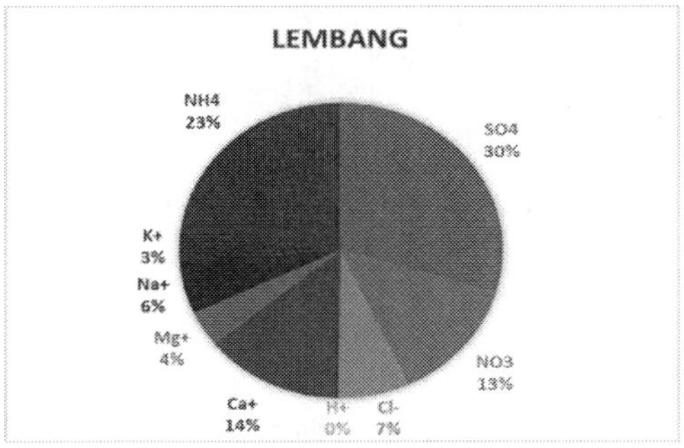

Fig. 4. Precipitation Weight Mean (PWM) concentration in Lembang

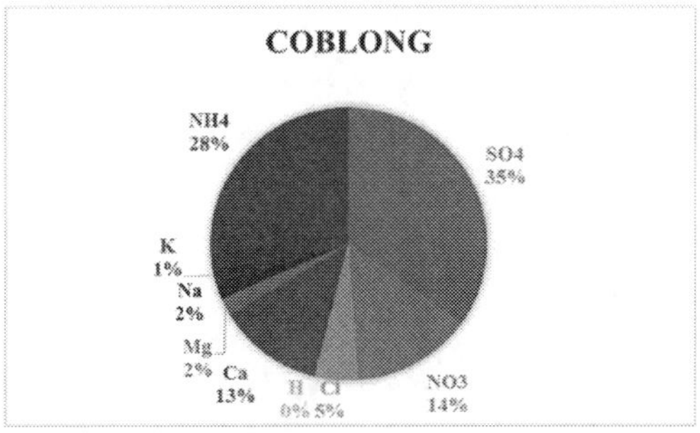

Fig. 5. Precipitation Weight Mean (PWM) concentration in Coblong

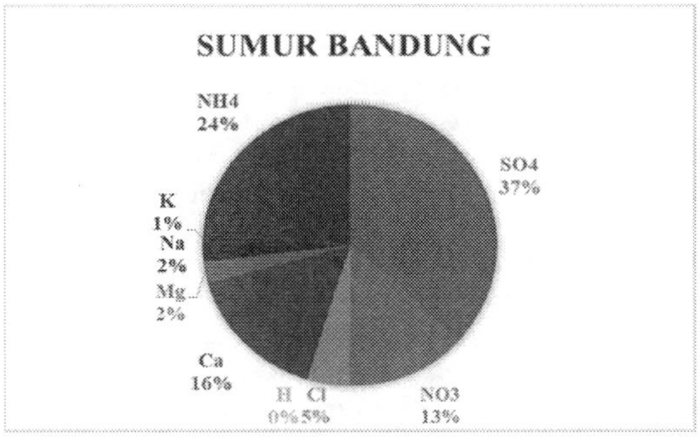

Fig. 6. Precipitation Weight Mean (PWM) concentration in Sumur Bandung

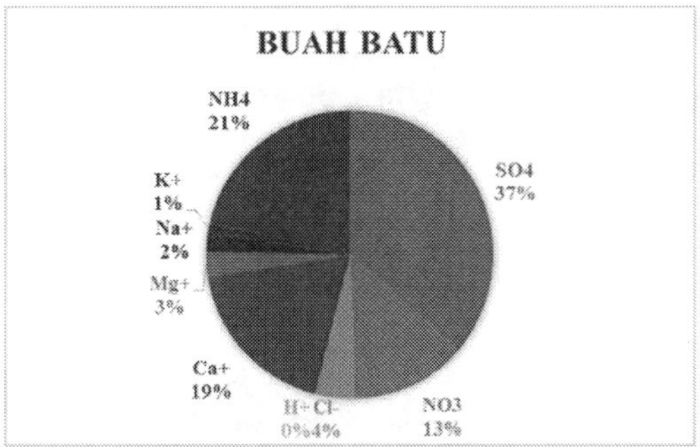

Fig. 7. Precipitation Weight Mean (PWM) concentration in Buah Batu

2.5 Neutralizing Factor (NF)

Neutralization factors (NF) can be used to evaluate the neutralization of precipitation by Ca^{2+}, Mg^{2+} and NH_4^+, which are calculated by the following Equation 3 [10,11,12,13,6] :

$$NF_{Xi} = [X_i^{\ 2+}] / [SO_4^{2-} + NO_3^-] \qquad (3)$$

where Xi is the chemical component of interest, with all the ions expressed in μeq/L.

The NF values for Ca^{2+}, Mg^{2+} and NH_4^+ in rainwater of the study area are 0.87, 0.37 and 0.49 (at Lembang); 0.34, 0.06 and 0.54 (at Coblong); 0.43, 0.06 and 0.47 (at sumur Bandung) and 0.50, 0.07 and 0.43 (at Buah Batu). The results reveal that Ca^{2+} and NH_4^+ are the dominant neutralization substances in the rainwater.

2.6 Sources of Emission

To identify the sources of emission which affects the quality of atmospheric precipitation was used to quantifies the sources emissions [14]. The result found ratio SO_4^{2-}/NO_3^- were 2.05 (at Lembang), 2.38 (at Coblong), 2.87 (at Sumur Bandung) and 2.89 (at Buah Batu). The results showed that sulfuric emissions still dominated, however the ratio is not as high as the one found in cities affected by industry. Ratio SO_4^{2-}/NO_3^- at Dhanbad City, known as coal city in India was 16.94 due the fact that the site are located beside or in industrialized urban areas [14]. SO_2 and NO_x come from electric power generation that relies on burning fossil fuels, like coal [3].

Nitrate and sulfate are the major component of acid rain and their concentrations are increasing rapidly in the atmosphere growing to anthropogenic activities such as the burning of coal and petroleum fuels. Nitrogen in precipitation is present mainly as ammonium (NH_4^+) and nitrate (NO_3^-). NH_3 emissions sources from soils, oceans, and animal excreta. Sulfate in precipitation in different areas study were sea-salt sulfate, emissions of SO_2 from industrialized areas and local sources as coal combustion and biogenic sulfur [15].

2.7 Wet Deposition Rate

Wet deposition rate for certain study periods were present in **Table 2**. The highest wet deposition rate of anion due to SO_4^{2-} with mean ranged between 15.90 kg/ha in Lembang to 45.45 kg/ha in Sumur Bandung. For cations, wet deposition rate of Ca^{2+} ranged between 8.58 kg/ha in Lembang to 11.60 kg/ha in Buah Batu. NH_4^+ has significant rate for cation in all sampling sites with range values between 4.60 kg/ha in Lembang to 10.77 kg/ha in Sumur Bandung.

Calcium is a major netralising agent that may play an important role in the precipitation chemistry in Bandung City. Calcium in rainwater may be contributed from soil dust, coal combustion and cement factories [16]. The neutralization capacity of soil dust from desert and semi-arid areas northern China was largely related to the calcite ($CaCO_3$) content of the soil, so precipitation pH in this part of China has remained high despite the high sulfur emissions and large quantities of sulfuric acid in the precipitation. The chemical composition of the anthropogenic emitted calcium dust are not known, but is likely a combination of $CaCO_3$, CaO and $CaSO_4$ [17].

Table 2. Wet deposition rate (kg/ha)

Sampling Site	Rainfall (mm)	Anion (kg/ha)			Cations (kg/ha)				
		Cl	NO_3	SO_4	Ca	Mg	Na	K	NH_4
Lembang	829.88	4.00	10.02	15.90	8.58	2.20	4.48	2.95	4.35
Coblong	1155.61	5.44	19.51	35.91	7.28	0.74	1.59	1.78	10.26
Sumur Bandung	1201.52	4.98	20.42	45.45	11.00	0.98	1.89	3.36	10.77
Buah Batu	1304.94	4.53	18.30	40.97	11.60	1.01	1.84	1.79	8.89

4 Conclusions

The total amount of wet precipitation measured during the sampling period (March to August 2016) for Bandung City were ranged between 214.99 mm – 1067.32 mm. The mean pH showed that samples had their majority acidic properties. This acidity was directly associated with SO_4^{2-} and NO_3^-, where SO_4^{2-} that were found to be the main contributor to this characteristic.

On the other hand, the neutralization process mainly originated from the presence of Ca^{2+} and NH_4^+. Cation wet deposition rate of Ca has highest. Calcium is a major netralising agent that may play an important role in the precipitation chemistry in Bandung City. Ammonium (NH_4^+) came the second and showed significant rate for cation in all sampling site. There are not many agricultural areas in the city of Bandung. It is suggested that ammonia is originated from decomposition process of waste and wastewater mix in open channel in Bandung City are. However, this is still need further investigation.

This paper is part of the research "Prediksi Dampak Kesehatan Lingkungan Deposisi Basah Logam Berat di Wilayah Perkotaan Bandung" funded by Riset KK dan Inovasi ITB 2017.

References

1. A World Meteorogy Organization, Manual For The GAW Precipitation Chemistry Programme Guilines, Data Quality Objectives and Standard Operating Procedures, GAW Precipitation Chemistry Science Advisory Group, 2004 Amanded (2015)
2. M Acid Deposition Monitoring Network (EANET), Techical Manual for Wet Deposition Monitoring in East Asia, Network Center for EANET, (2015)
3. S. Bhargava, and S. Bhargava, Journal of Applied Chemistry **5**, pp. 19 – 24, (2013)
4. W.D. Bowman, C.C. Cleveland, L. Halada, J. Hres, and J.S. Baro, Nature Geoscience **1**, pp. 767 – 770, (2008)
5. D.D. Lestiani, M. Santoso, S. Kurniawati, and A. Markwitz, Indo. J. Chem., **13** (3), pp. 271 – 277, (2013)
6. D.W. Driejana, I.L. Raper, Gee, *Air Pollution VIII*, (WIT Press, Salt Lake City, UT, 2000)
7. H. Wang and G. Han, Atmospheric Research **99**, pp. 190 – 196, (2011)
8. Seinfeld, H. John, *Atmospheric Chemistry and Physics of Air Pollution*, Jhon Wiley and Son, Inc., New York, (1986)
9. Y. Cheng, Y. Liu, M. Huo, Q. Sun, H. Wang, Z. Chen, Bai., Journal of Environmental Sciences, **23** (8), pp. 1350 – 1358, (2011)
10. K. Ali, G. Momin, S. Tiwari, P. Safai, D. Chate, P. Rao, Atmospheric Environment **38** (25), pp. 4215-4222, (2004)

11. M. Possanzini, P. Buttini, V. Di Palo, Science of Total Environment **74**, pp. 111 – 120, (1988)
12. T. Özsoy, P. Türker, S. Örnektekin, Water, Air, & Soil Pollution **189**, pp. 69–83, (2008)
13. P.R. Salve, A.A. Maurya, S.R. Wate, S. Devotta, Bulletin Environmental Contaminant Toxicology **80**, pp. 242 – 246, (2008)
14. P. Rai, International Journal of Current Engineering and Technology, **4** (2), pp. 673 – 679, (2014)
15. C.Q. Liu, and H.Y. Xiao, Atmospheric Environment **36**, pp. 5121–5130, (2002)
16. Z. Xu, Y. Wu, W.J. Liu, C.H. Liang, J. Ji, T. Zhao, and X. Zhang, Atmospheric Research, **164 – 165**, pp. 278 – 285, (2015)
17. T. Larsen and G.R. Carmichael, Environmental Pollution **110**, pp. 89 – 102, (2000)
18. R.M. Ceron, J.G. Ceron, C.G. Carballo, C.A. Aguilar, C. Montalvo, J.A. Benitez, Y.J. Villareal, and M.M. Gomez, Journal of Environmental Protection **4**, pp. 50 – 56, (2013)
19. F.R.F. Cerqqueira, M.F. Pinto, I.N. Derossi, W.T. Esteves, M.D.R.S. Santos, R.C. Matos, Atmospheric Pollution Research **5**, pp. 253-261, (2014)

AUTHOR INDEX

AUTHOR INDEX

AUTHOR INDEX